人工智能之數學

玄心居士丙午書

Copyright and Credits

Mathematics for AI and Machine Learning

Author ⊙ **Fuheng Wu**

Producer ⊙ Siqi Lee

Techinical Editor ⊙ Simon Lin

Graphic Designer ⊙ 2Peanuts Studio

Publisher ⊙ Math4AI Press

Website ⊙ https://math4ai.org

Built with Peanutbook ⊙ https://peanutbook.com

First published ⊙ 2026 March (Produce ID: `cfc4d73579953ae9219e6c9a35812a8e648f51bd`)

Friends & Supporters:

To my parents, Chuntao He and Zongyuan Wu.

- *Fuheng Wu*

I hope everyone can find the courage

希望大家鼓起勇气,

to pursue what they truly enjoy doing.

勇敢地去做自己喜欢做的事情。

Don't become fixated on whether you are the very best.

不要去执着于自己是不是最优秀的。

Keep a balanced mindset,

放平心态,

and you may find yourself going even farther.

你反而能够走得更远。

Wang Hong

王虹

Mathematician

2026 Fields Medalist

2026.7.26

Preface

This book provides a rigorous mathematical foundation for modern artificial intelligence and machine learning. The mathematics underlying AI is not exotic—it draws from linear algebra, calculus, probability, and optimization—but these classical fields are rarely taught together in a way that reveals their deep connections. This book gathers the essential mathematical tools into one coherent narrative, bridging theory with practical applications in deep learning, reinforcement learning, large language models, and generative AI.

The premise is simple: a surprisingly small set of mathematical concepts reappears across seemingly different AI methods. Linear algebra provides the language of representation. Calculus and optimization describe how learning happens. Probability theory captures uncertainty and enables reasoning about distributions. And stochastic dynamics—the mathematics of randomness evolving in time—connects sampling, generation, and the continuous limits of iterative algorithms. Once you see the common mathematical structure, methods that seemed unrelated begin to rhyme.

Who This Book Is For

This book is designed for undergraduate and graduate students, AI/ML researchers, and practitioners seeking a rigorous mathematical foundation. It also supports Olympiad-style mathematical training and can serve as a supplementary reference for advanced AI competition preparation at the national and international level. The primary audience includes those who want to understand AI and machine learning at a mathematical level—not just apply algorithms, but derive them, modify them, and reason about why they work.

The book assumes familiarity with undergraduate calculus; linear algebra is developed from first principles starting with the definition of vectors. You do not need to read the book linearly. If you are comfortable with linear algebra and want to jump to optimization or probability, you can. If your interest is specifically in generative models and diffusion, Part IV can serve as a self-contained introduction once you have the prerequisites from earlier parts.

Goals of This Book

This book grew from a simple observation: while classical and modern AI both rest on deep mathematics, there was no single text that systematically covered that mathematics as one progression. The aim is to fill that gap. More specifically, this book has three goals:

1. **Build a working toolkit.** The mathematics here—linear algebra, matrix calculus, probability, and optimization—should be sufficient to read, understand, and derive results in modern ML papers. The emphasis is on forms that are directly useful, not on maximal generality.

2. **Unify perspectives.** Many concepts that appear distinct are deeply connected: likelihood and KL divergence, score functions and Langevin dynamics, SDEs and their ODE limits, sampling and generation. A major aim of this book is to reveal these connections and present them as variations on common themes—from classical foundations through contemporary methods.

3. **Stay implementation-aware.** Even when presenting mathematics abstractly, we emphasize forms that are numerically stable, dimensionally consistent, and directly translatable to code. The gap between mathematical notation and working implementation should be small.

Structure of the Book

The book spans 21 chapters organized into four parts, each building on the previous. The following overview summarizes each part and chapter; use it as a map when jumping to a topic or planning a course.

Part I: Linear Spaces and Representations (Chapters 1–10) develops the linear algebra foundation. We begin with vector spaces and inner products, progress through matrix operations and subspaces, and culminate in the major decompositions: QR, LU, eigendecomposition, and SVD. These tools establish the mathematical foundation for representation in AI—describing how systems represent data, parameters, and transformations.

Chapter 1, Vector Space and Inner Product, defines vectors, inner products, norms, and the geometric language of representation in $\mathbb{R}^d$.

Chapter 2, Matrix, introduces matrix algebra, multiplication views, and the operations that appear throughout ML.

Chapter 3, Subspaces and Orthogonality, develops subspaces, projections, and orthonormal bases—the geometry behind least squares and PCA.

Chapter 4, QR Decomposition and Numerical Rank, covers QR factorization, orthonormalization, and rank-revealing decompositions.

Chapter 5, Square Matrix and LU Decomposition, treats determinants, invertibility, and LU/PLU factorization for linear systems.

Chapter 6, Geometric Transformation and Eigen Decomposition, connects linear maps to eigenvalues, eigenvectors, and diagonalization.

Chapter 7, Symmetric Matrix, focuses on real symmetric matrices, quadratic forms, and spectral theorems.

Chapter 8, Singular Value Decomposition (SVD), presents the SVD, low-rank approximation, and pseudoinverses.

Chapter 9, Matrix Decompositions and Beyond, surveys additional factorizations and when to use each.

Chapter 10, Matrix Block Partitioning, introduces block matrices and partitioned operations used in large-scale and structured models.

Part II: Differentiation and Optimization (Chapters 11–12) covers matrix calculus and optimization methods. Gradients, Jacobians, and Hessians provide the language for describing how functions change; optimization algorithms translate this local information into learning. This part formalizes learning as structured search in parameter space.

Chapter 11, Matrix Calculus, develops gradients, Jacobians, Hessians, and the chain rule in matrix form.

Chapter 12, Optimization Methods, covers first- and second-order methods, convexity, and the algorithms behind training.

Part III: Probability and Information for Learning (Chapters 13–16) introduces probabilistic thinking. We develop probability theory, information-theoretic quantities like entropy and KL divergence, the Evidence Lower Bound (ELBO), variational inference and latent variable models, and the Bellman equations underlying reinforcement learning. The perspective shifts from fitting functions to modeling distributions.

Chapter 13, Probability and Random Variables, builds distributions, conditioning, transformations, and limit theorems.

Chapter 14, Information Theory, introduces entropy, KL divergence, mutual information, and the ELBO.

Chapter 15, Bellman Equations and Operators, formalizes dynamic programming and value functions for sequential decision making.

Chapter 16, Variational Inference and Latent Variables, covers mean-field VI, the reparameterization trick, and VAE-style models.

Part IV: Score, Dynamics, and Diffusion (Chapters 17–21) explores the mathematics of modern generative models. Score functions, Langevin dynamics, stochastic differential equations with Itô calculus, ODE/SDE continuous limits of algorithms, and Fokker–Planck equations describe how probability distributions evolve. This part frames generative modeling as the study of distributional dynamics.

Chapter 17, Score Function and Energy-Based Models, defines scores, energy functions, and their role in generative modeling.

Chapter 18, Langevin Dynamics and Sampling, connects scores to MCMC, Langevin updates, and practical samplers.

Chapter 19, Stochastic Differential Equations, develops Itô calculus, Brownian motion, and SDE models.

Chapter 20, ODE, SDE, and Continuous Limits of Algorithms, links discrete updates to continuous-time flows (e.g., gradient flow, diffusion limits).

Chapter 21, Fokker–Planck and Distribution Dynamics, describes how densities evolve under drift and diffusion—the PDE view of generative dynamics.

The progression is largely linear: later chapters depend on earlier ones. However, each part is relatively self-contained, so you can enter at the part most relevant to your current needs.

How to Use This Book

For a **two-semester course**, the natural division is Parts I–II in the first semester (linear algebra, calculus, optimization) and Parts III–IV in the second (probability, information theory, stochastic dynamics).

For **self-study**, we recommend working through the chapters in order, at least on first reading. The exercises are essential—mathematics is learned by doing, not by reading. If you find yourself comfortable with a topic, skim the chapter and focus on the exercises.

For **reference**, the detailed table of contents and index should help you locate specific results. The notation section below establishes conventions used throughout.

To Get the Most Out of This Book

You will get the most from this book if you **work the exercises** and **sketch proofs or derivations** before reading every detail in the text. Mathematics is not a spectator sport: the goal is to make notation feel inevitable, not memorized.

Prerequisites. Comfort with undergraduate calculus is assumed. Linear algebra is developed from definitions, but prior exposure helps you move faster through Part I. Basic Python and NumPy are useful for the optional computational exercises and for exploring ideas on the companion site, but the core narrative is mathematical—you can follow the book with pencil and paper alone.

Companion website. Interactive visualizations and supplementary material are available at https://math4ai.org/. The companion site is designed to build intuition for key ideas (decompositions, distributions, flows, SDEs) without replacing the proofs in the book. When a chapter has a demo, try the sliders and plots after you have read the definitions once.

Reading paths. For a **two-semester course**, Parts I–II fit naturally in the first term and Parts III–IV in the second. For **generative modeling**, read Part I for notation, skim Part II as needed, then Part III (especially Chapters 14 and 16) before Part IV. For **reinforcement learning**, Part I plus Chapters 13 and 15 provide the probability and Bellman background.

Code and figures. Example code and color figures are maintained in the book's repository. If you are reading a print edition, high-resolution color versions of selected diagrams may be available from the support site.

Conventions Used in the Text

Several typographic conventions appear throughout the book:

Bold marks important terms when they are first introduced or when referring to UI elements in software discussions.

Italic is used for emphasis within a sentence and for symbolic names such as file names or repository paths.

`Monospace` or code blocks set commands, function names, and short code snippets, for example:

```
import numpy as np
A = np.linalg.qr(X)[0]
```

Displayed equations use numbered labels where cross-references matter; inline math uses standard LaTeX notation. Chapter and section references use `\ref{...}` labels in the PDF (e.g., Chapter~11).

Support and Feedback

We welcome feedback from readers.

Companion site and updates: https://math4ai.org/

Source and examples: https://github.com/henrywoo/math4ai

Errata and suggestions: If you find an error or have a suggestion, please open an issue at https://github.com/henrywoo/math4ai/issues. Include the chapter and section when you can—it helps us fix mistakes quickly.

General questions: For questions about notation, exercises, or how a chapter connects to ML practice, the issue tracker is the best place to reach us; others may have the same question.

We are grateful for readers who report typos, unclear definitions, or missing steps in derivations. Every correction makes the book better for the next reader.

Special Callouts

Throughout the book you will see four kinds of boxed callouts. **Notes** contain mathematical remarks, caveats, or optional context that can deepen understanding without interrupting the main argument. **Important** callouts highlight key definitions, formulas, or cautions that are worth remembering. **Warnings** flag common mistakes or pitfalls. **AI/ML** callouts mark applications and connections to machine learning practice—algorithms, models, training tricks, and related tooling. You can read notes and AI/ML asides at your leisure; important callouts and warnings are worth a pause. Examples:

This is what a note looks like.

This is what an important callout looks like.

This is what a warning looks like.

This is what an AI/ML callout looks like.

Exercises

Each chapter concludes with exercises designed to reinforce key ideas and extend them in meaningful directions. The exercises range from straightforward computations to more substantial problems that require combining multiple concepts. Full-color figures throughout the book illuminate complex ideas, while extensive problem sets support both self-study and classroom instruction.

We strongly encourage you to attempt the exercises. Active problem-solving is the fastest way to transform notation into intuition. Worked solutions will be provided separately.

Legend

The following legend illustrates the visual conventions used throughout this book to represent different types of matrices. Each matrix type is distinguished by its shape, color, and border style. The list below (and the figure that follows) spells out these conventions:

- **Rectangular Matrix (M×N)**: Pink fill — general rectangular matrices
- **Square Matrix (A)**: Pink fill with light blue border — used for general square matrices
- **Symmetric Matrix (A = A^T)**: Dark red background — symmetric matrices where $A = A^\top$
- **Lower Triangular (L)**: Pink fill — lower triangular matrices
- **Lower Triangular with Unit Diagonal (L_1)**: Pink fill with dark blue dashed diagonal line and "1" label — rectangular lower triangular matrices with unit diagonal (diagonal elements equal to 1)
- **Upper Triangular (R)**: Light green fill — standard upper triangular matrices
- **Generalized Upper Triangular (R)**: Light green fill with dark yellow dashed border — rectangular upper triangular matrices (used in QR decomposition for non-square matrices)
- **Diagonal Matrix (Λ)**: Light green fill — diagonal matrices (often containing eigenvalues)
- **Orthogonal Matrix (Q)**: Purple fill — represents matrices with orthonormal columns/rows
- **Permutation Matrix (P)**: Orange fill with light blue border — permutation matrices used for row/column reordering (e.g., in PLU decomposition)
- **Eigenvector Matrix (V)**: Dark blue fill with white text — eigenvector matrices used in eigendecomposition $A = V\Lambda V^{-1}$
- **Non-Negative Matrix (A ⩾ 0)**: Red fill with white text — matrices with all non-negative entries

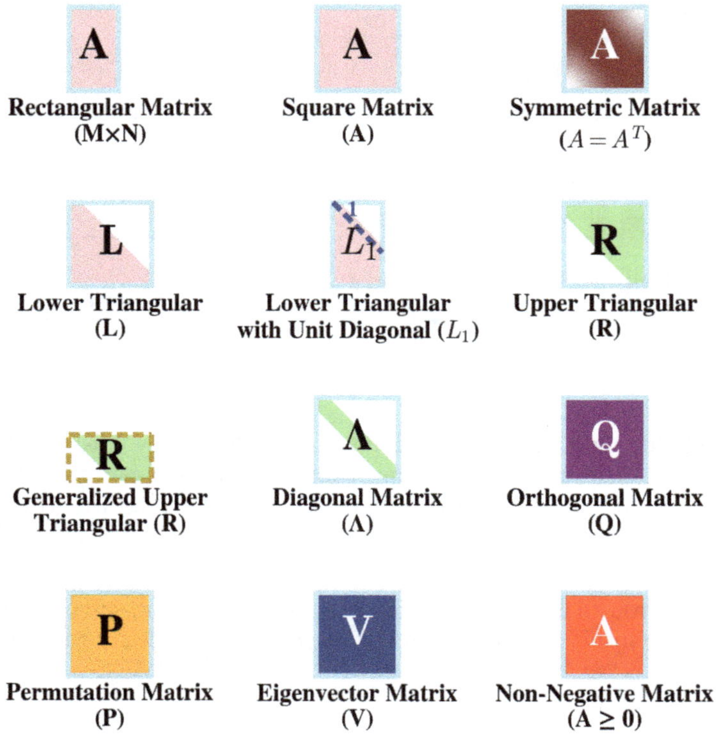

The figure and list above summarize these conventions so you can quickly identify matrix types in decomposition diagrams and throughout the text.

Notation and Conventions

The following summarizes our notation for scalars, vectors, matrices, and derivatives. We follow standard practice in machine learning and applied mathematics and avoid ambiguous notation where we can.

Scalars, Vectors, and Matrices

- **Scalars** are denoted by lowercase letters: $x, y, \lambda \in \mathbb{R}$.
- **Vectors** are denoted by boldface lowercase letters: $\mathbf{v}, \mathbf{x}, \mathbf{y} \in \mathbb{R}^d$. We use boldface notation $\mathbf{v} \in \mathbb{R}^{d\times 1}$ to explicitly denote column vectors. For simplicity, we may also write $\mathbf{v} \in \mathbb{R}^d$ with the understanding that vectors are column vectors by default. Both $\mathbf{v}$ (boldface) and $\vec{v}$ (arrow notation) are common ways to denote vectors in mathematical literature; we primarily use boldface notation $\mathbf{v}$ in this book, but may occasionally use $\vec{v}$ for clarity or emphasis.
- **Matrices** are denoted by uppercase letters: $A \in \mathbb{R}^{M\times N}$.
- **Transpose** of a vector or matrix is denoted by $(\cdot)^T$ or $(\cdot)^\top$.
- **Norms**: The Euclidean norm (also called ℓ_2 norm) is denoted by $\|\mathbf{v}\|_2$. We consistently use double vertical bars $\|\cdot\|$ for norms (using \| in LaTeX), and single vertical bars $|\cdot|$ only for absolute values.

- **Inner Product**: The standard inner product in $\mathbb{R}^d$ is defined as $\langle \mathbf{u}, \mathbf{v} \rangle = \mathbf{u}^\top \mathbf{v}$.
- **Vector space**: When we say "$\mathbb{R}^d$ is a vector space," we mean it satisfies the usual axioms (closure under addition and scalar multiplication, existence of zero and additive inverses, etc.); we omit the full list here.

Subscripts and Superscripts

To avoid confusion between vector components, matrix rows, and data samples, we distinguish them as follows:

- **Subscripts** $(X_0, X_1, \dots, X_{N-1})$ denote **components of a single vector**. For example, if $\mathbf{X} \in \mathbb{R}^{N\times 1}$ is a random vector, then X_i refers to the i-th component of that vector.
- **Superscripts** serve two purposes:
  - **Matrix rows**: For a matrix $A \in \mathbb{R}^{M\times N}$, we use $A^{(i)}$ or $\mathbf{a}^{(i)}$ to denote the i-th **row** of the matrix as a row vector. This distinguishes rows from columns, which are typically denoted using subscripts (e.g., A_j for the j-th column) or bracket notation (e.g., $A(:, j)$).
  - **Samples/observations**: $(X^{(0)}, X^{(1)}, \dots, X^{(K-1)})$ denote **different samples or observations** from the same random variable or vector. These are independent realizations or data points.

This convention avoids ambiguity: if we used $X_0, X_1, \dots, X_{K-1}$ for samples, it would conflict with the notation for vector components. When working with vector samples, we can combine both notations: $X_i^{(k)}$ denotes the i-th component of the k-th sample vector.

Examples:

- For empirical mean calculations, we write $\hat{\mu} = \frac{1}{K}\sum_{k=0}^{K-1} X^{(k)}$, where $X^{(k)}$ are K independent observations. If each $X^{(k)}$ is itself a vector, we can refer to its components as $X_i^{(k)}$.
- For a data matrix $A \in \mathbb{R}^{M\times N}$ where each row is a data point, $A^{(i)}$ denotes the i-th row (data point), and $A_j^{(i)}$ or $[A]_{i,j}$ denotes the j-th component of the i-th row.

Indexing Conventions

- **0-based** (ML/code): vector/matrix indices and loops (e.g. $v_0, \dots, v_{d-1}, \sum_{i=0}^{N-1}$), as in NumPy and deep-learning code.
- **1-based** (classical math): abstract enumerations (e.g. $A_1, A_2, \dots, Z_1, \dots, Z_d$, Precision@$k$).

Special Matrices

- The **identity matrix** is denoted by I (or I_N when the dimension matters).
- **All-ones**: column vector $\mathbf{1}_N$ (or $\mathbf{1}$ when N is clear); matrix $\mathbf{1}_{M\times N}$ with dimensions always shown. J denotes Jacobian matrices.

Calculus

Full definitions and derivations are given in Chapter~11.

We use standard notation for derivatives. The **gradient** $\nabla_{\mathbf{x}} f$ of a scalar function $f : \mathbb{R}^N \to \mathbb{R}$ is always a **column vector** in $\mathbb{R}^{N\times 1}$. For vector-valued functions we use the **Jacobian** $\frac{\partial \mathbf{f}}{\partial \mathbf{x}}$ (shape $M \times N$), and for second derivatives the **Hessian** $\mathbf{H}_f = \nabla_{\mathbf{x}}^2 f \in \mathbb{R}^{N\times N}$. For composite functions we write the chain rule as $\frac{\partial \mathbf{f}}{\partial \mathbf{g}}\frac{\partial \mathbf{g}}{\partial \mathbf{x}}$, so dimensions align under matrix multiplication. We adopt the usual ML convention: gradients in denominator layout (column vectors), Jacobians in numerator layout.

For matrix-valued functions $f : \mathbb{R}^{M\times N} \to \mathbb{R}^{P\times Q}$, the **Fréchet derivative** at X is the linear operator $f'(X)$ satisfying $f(X+\Delta) = f(X)+f'(X)[\Delta]+o(\|\Delta\|)$, where Δ denotes a perturbation matrix. We use Δ (or Δ_X, Δ_A when needed) for such perturbations; full definitions and their relation to gradients and Jacobians are developed in that chapter.

Summary of Key Shapes (Linear Algebra and Calculus)

The following table summarizes the shapes of key objects in linear algebra and calculus:

Object	Notation	Shape
Gradient	$\nabla_{\mathbf{x}} f$	$N \times 1$
Jacobian	$\mathbf{J_f}(\mathbf{x})$	$M \times N$
Hessian	$\mathbf{H}_f = \nabla^2_{\mathbf{x}} f$	$N \times N$
Fréchet derivative	$f'(X)[\Delta]$ (linear operator)	Maps $\mathbb{R}^{M\times N} \to \mathbb{R}^{P\times Q}$
Perturbation matrix	Δ	$M \times N$
Identity	I_N	$N \times N$
All-ones vector	$\mathbf{1}_N$	$N \times 1$
All-ones matrix	$\mathbf{1}_{M\times N}$	$M \times N$

Probability and Random Variables

Conventions below are developed in Chapter~13 and Chapter~14.

We use the following notation for probability and information theory:

- **Random variables**: Uppercase letters X, Y, Z for scalar random variables; $\mathbf{X}, \mathbf{Y}$ for random vectors. Lowercase letters x, y, z denote samples (observed values) from these random variables. When we write $X \sim \mathcal{N}(\mu, \sigma^2)$, this means the random variable X is distributed according to the specified distribution. A sample x from X is a specific observed value.
- **Probability**: Uppercase P denotes probability values, e.g., $P(X = x)$, $P(X \leq x)$, $P(X = x \mid Y = y)$.
- **Probability Mass Function (PMF)**: Lowercase p for discrete random variables, e.g., $p_X(x) = P(X = x)$, $p_{X,Y}(x, y)$, $p_{X|Y}(x \mid y)$.
- **Probability Density Function (PDF)**: Lowercase f for continuous random variables, e.g., $f_X(x)$, $f(x, y)$, $f_{X|Y}(x \mid y)$.
- **Expectation, Variance, Covariance**: $\mathbb{E}[X]$, $\mathrm{Var}[X] = \mathbb{E}[(X - \mathbb{E}[X])^2]$, $\mathrm{Cov}[X, Y] = \mathbb{E}[(X - \mathbb{E}[X])(Y - \mathbb{E}[Y])]$.
- **Information Theory Notation**:
  - **Entropy**: $H(X)$ for entropy of X, $H(X \mid Y)$ for conditional entropy, $H(X, Y)$ for joint entropy (comma denotes joint distribution).
  - **Mutual Information**: $I(X; Y)$ uses a semicolon to distinguish it as a measure of relationship (not a joint distribution). The semicolon emphasizes that mutual information quantifies the relationship between X and Y, while $H(X, Y)$ with a comma represents the entropy of their joint distribution.

We use uppercase P for probability values and lowercase p (PMF) or f (PDF) for the corresponding functions; $p(x)$ or $f(x)$ may be used as shorthand when the random variable is clear. In information-theoretic notation, a semicolon in $I(X; Y)$ indicates mutual information, and a comma in $H(X, Y)$ indicates the joint distribution.

Reinforcement Learning Notation

Used in Chapter~15.

For reinforcement learning and sequential decision making we use the following notation:

- **Random variables**: Uppercase letters denote random variables: S_t (state at time t), A_t (action at time t), R_{t+1} (reward received after transitioning from S_t to S_{t+1}), G_t (return at time t). Following the Sutton & Barto convention, we use R_{t+1} to denote the reward received after taking action A_t in state S_t and transitioning to S_{t+1}.
- **Realizations**: Lowercase letters denote specific values or realizations of random variables: s (a specific state), a (a specific action), r (a specific reward), s_t (the state value at time t in a trajectory).
- **Conditional expectations**: In expressions like $\mathbb{E}[G_t|S_t = s]$, S_t is the random variable and s is its realization. For brevity and when the context is clear, we may also write $\mathbb{E}[G_t|s_t = s]$ where s_t represents the state at time t in the trajectory.
- **Sets and spaces**: Calligraphic letters denote sets: $\mathcal{S}$ (state space), $\mathcal{A}$ (action space).
- **Value functions**: Value functions are denoted as $V_\pi(s)$ and $Q_\pi(s, a)$, where s and a are specific state and action values. We use subscripts (e.g., π, θ, $*$) to indicate dependencies on policies, parameters, or optimality, rather than superscripts. Note that in other textbooks and literature, you may encounter the superscript notation (e.g., V^π, Q^π, V^*) which is also common and equivalent.
- **Double subscripts**: When we need to specify both a policy/parameter subscript and an index subscript, we use parentheses to clarify the grouping. For example, $(\mathbf{v}_\pi)_i$ denotes the i-th element of the vector $\mathbf{v}_\pi$, and $(P_\pi)_{ij}$ denotes the (i, j)-th element of the matrix P_π.

As in probability theory generally, uppercase letters mark random variables and lowercase letters their realizations. We follow the Sutton & Barto convention of writing R_{t+1} for the reward received after the transition, so that the reward is clearly tied to the outcome of the action.

DNN Input Format Conventions

Relevant to chapters that discuss deep learning, optimization, and neural network applications (e.g., Chapter~11, Chapter~12, and later).

Tensor shapes in deep learning follow domain-specific conventions; we use these:

- **Computer Vision (CV)**: We use the **NCHW** format:
  - N: Batch size
  - C: Number of channels
  - H: Height
  - W: Width
  - Example: An image batch has shape (N, C, H, W), e.g., $(32, 3, 224, 224)$ for 32 RGB images of size 224×224.
- **Natural Language Processing (NLP)**: We use **BLD** or **BLHD** format:
  - B: Batch size
  - L: Sequence length
  - D: Feature/embedding dimension
  - H: Number of attention heads (when considering multi-head attention)
  - Examples:
    - Single-head attention: Input tensors have shape (B, L, D)

  * Multi-head attention: Query, key, value tensors have shape (B, L, H, d_k) where $d_k = D/H$ is the per-head dimension, or equivalently (B, H, L, d_k) depending on the implementation. The total dimension satisfies $D = d_k \times H$.

In NLP we use B for batch size and L for sequence length so they are not confused with the matrix dimension N used elsewhere; in CV we use N for batch size to match common practice in the vision literature.

The conventions above are used consistently throughout the book and are intended to balance rigor, clarity, and compatibility with modern machine learning practice.

Acknowledgments

This book benefited from conversations with colleagues and feedback from early readers. Special thanks to Juncong Jack Wu for contributing to Chapter~11 and for reviewing Part~II; to Skyler Xue (Yale, CS and Biomedical Engineering) for reviewing Part~I; to Yijie Jin (Math PhD, GIT) for reviewing Part~III; and to Dingshen Zhang (Math PhD, OSU) for reviewing Part~IV. We thank Haimei Shao (Math PhD, CFU) for an early review of the book. We also thank Xuanxin Jushi from the North American Society for Calligraphy and Aesthetics for the calligraphy that graces this book. We are grateful to everyone who pointed out errors, suggested improvements, or simply asked good questions that forced clearer thinking. Any remaining mistakes are, of course, our own.

Foreword

Modern artificial intelligence rests on a relatively compact set of mathematical ideas, including linear algebra, calculus, optimization, probability and statistics, and the dynamics of probability distributions. As artificial intelligence becomes part of everyday life, understanding its mathematical foundations matters more than ever. When representation, learning, uncertainty, interpretability, and numerical behavior become practical concerns in real systems, mathematical reasoning remains one of the main tools we have for judging whether a method is sound.

I first met Fuheng Wu at Uber ATG, where we were developing artificial intelligence technologies for autonomous driving. In that setting, mathematical reasoning and engineering practice were inseparable. We had to think carefully about how observations were represented and transformed, how model architectures could be designed and optimized for better predictive performance, how uncertainty should be quantified and carried through to the final output, and how algorithms would behave numerically under real-world constraints.

That environment demanded a particular habit of mind, one that is also evident throughout this book: the ability to reduce a difficult problem to its underlying structure and to treat geometry, algebra, probability, and computation as parts of a common language. Writing a book of this scope requires the author to confront several difficult questions. What truly belongs in the mathematical foundation of artificial intelligence? In what order should the ideas be introduced? And how can mathematical rigor be maintained without losing the intuition that gives the subject meaning?

Most books on machine learning are organized around artificial intelligence methods, introducing the necessary mathematics as it is needed. This book takes the opposite approach. Mathematics provides its underlying structure, while machine learning and artificial intelligence supply the applications and motivation. The classical foundations are developed gradually and carried from traditional machine learning into modern AI, from matrix methods and optimization to neural networks, statistical models, and generative models. Recurring ideas such as projection, decomposition, gradients, and evolving probability distributions return throughout the book, taking on richer meanings as the discussion progresses.

The mathematics also remains closely connected to practical questions in AI training, evaluation, and inference. The first chapter, for example, relates Cauchy sequences to the convergence of model parameters during training. Chapter 7 uses style transfer in computer vision to explain Gram matrices. The book also discusses block tensor structures that arise in distributed inference. These connections help readers understand abstract concepts while revealing the common structure behind mathematical tools that may initially appear unrelated.

Artificial intelligence will continue to evolve, but the mathematical foundations covered in this book will have a much longer life. They will help readers understand current methods, evaluate new ones as they emerge, and approach the future research and practice of artificial intelligence with greater care and clarity.

I am pleased to introduce *Mathematics for AI and Machine Learning.* It is an ambitious and substantial book, and a valuable contribution to education in artificial intelligence and mathematics.

Yu Guo
Ph.D., Harvard University

July 2026

Contents

I

Linear Spaces and Representations

Chapters 1–10

Vector spaces, linear structures, spectra, and compression - the complete mathematical foundation for representation in AI

The geometry of intelligence: where data becomes meaning

Modern AI systems operate in high-dimensional spaces where data, parameters, and representations are inherently vectorial. Linear algebra provides the language to describe these spaces, the transformations between them, and the geometric structures that govern stability and expressiveness.

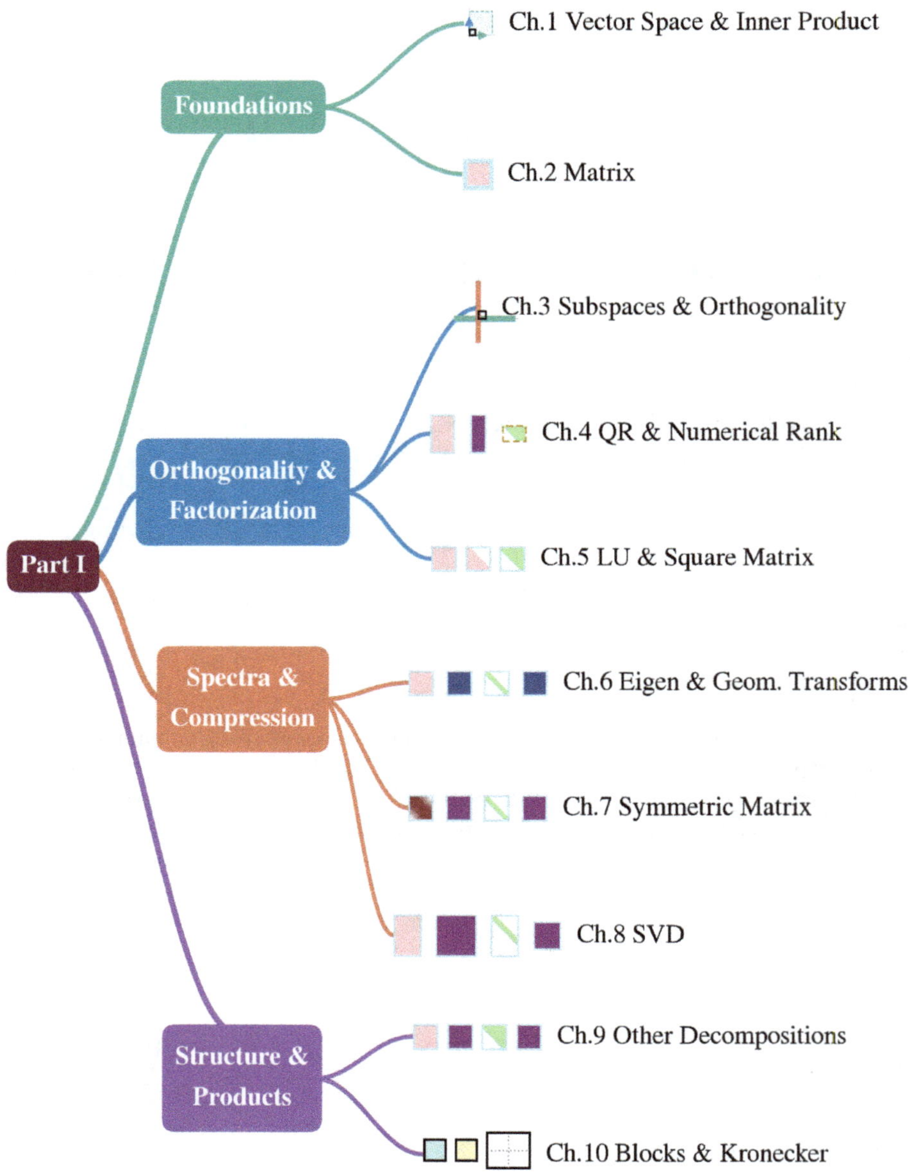
Part I
Foundations
Ch.1 Vector Space & Inner Product
Ch.2 Matrix
Orthogonality & Factorization
Ch.3 Subspaces & Orthogonality
Ch.4 QR & Numerical Rank
Ch.5 LU & Square Matrix
Spectra & Compression
Ch.6 Eigen & Geom. Transforms
Ch.7 Symmetric Matrix
Ch.8 SVD
Structure & Products
Ch.9 Other Decompositions
Ch.10 Blocks & Kronecker

Chapter 1

Vector Space and Inner Product

Foundations of linear algebra: vector spaces, linear combinations, and basis

"Mathematics is the queen of the sciences."

— Carl Friedrich Gauss

- ▷ `Inner product Algebra` $\langle \mathbf{u}, \mathbf{v} \rangle = \langle \mathbf{v}, \mathbf{u} \rangle = \mathbf{u}^\top \mathbf{v} = \sum_{i=0}^{d-1} u_i v_i$
- ▷ `Inner product Geometry` $\langle \mathbf{u}, \mathbf{v} \rangle = \|\mathbf{u}\|_2 \cdot \|\mathbf{v}\|_2 \cos\theta$
- ▷ `Outer product` $\mathbf{A} = \mathbf{u}\mathbf{v}^\top$ where $(\mathbf{A})_{ij} = u_i v_j$ (rank-1 matrix)
- ▷ `Lp norm` $\|\mathbf{v}\|_p = \left(\sum_{i=0}^{d-1} |v_i|^p\right)^{1/p}$ for $p \geq 1$
- ▷ `L1 norm` $\|\mathbf{v}\|_1 = \sum_{i=0}^{d-1} |v_i|$ (Manhattan norm)
- ▷ `L2 norm` $\|\mathbf{v}\|_2 = \sqrt{\sum_{i=0}^{d-1} v_i^2} = \sqrt{\langle \mathbf{v}, \mathbf{v} \rangle}$ (Euclidean norm)
- ▷ `Energy` $\|\mathbf{v}\|_2^2 = \langle \mathbf{v}, \mathbf{v} \rangle = \sum_{i=0}^{d-1} v_i^2$
- ▷ `L∞ norm` $\|\mathbf{v}\|_\infty = \max_i |v_i|$ (Chebyshev norm)
- ▷ `Frobenius norm` $\|\mathbf{v}\|_F = \|\mathbf{v}\|_2$ (for vectors)
- ▷ `Orthogonality` $\mathbf{u} \perp \mathbf{v}$ if $\langle \mathbf{u}, \mathbf{v} \rangle = 0$
- ▷ `Unit vector` $\hat{\mathbf{v}} = \frac{\mathbf{v}}{\|\mathbf{v}\|_2}$
- ▷ `Projection and rejection` $\mathrm{Proj}_{\hat{\mathbf{v}}}(\mathbf{u}) = \langle \mathbf{u}, \hat{\mathbf{v}} \rangle \hat{\mathbf{v}} = \frac{\langle \mathbf{u}, \mathbf{v} \rangle}{\|\mathbf{v}\|_2^2} \mathbf{v}$; $\mathbf{d}_\perp = \mathbf{u} - \mathrm{Proj}_{\hat{\mathbf{v}}}(\mathbf{u})$
- ▷ `Cauchy inequality` $|\langle \mathbf{u}, \mathbf{v} \rangle| \leq \|\mathbf{u}\|_2 \cdot \|\mathbf{v}\|_2$
- ▷ `Triangle inequality` $\|\mathbf{u} + \mathbf{v}\|_2 \leq \|\mathbf{u}\|_2 + \|\mathbf{v}\|_2$
- ▷ `Energy additivity (general)` $\|\mathbf{u} + \mathbf{v}\|_2^2 = \|\mathbf{u}\|_2^2 + \|\mathbf{v}\|_2^2 + 2\langle \mathbf{u}, \mathbf{v} \rangle$
- ▷ `Gram-Schmidt` $\mathbf{e}_0 = \frac{\mathbf{v}_0}{\|\mathbf{v}_0\|_2}$, $\mathbf{R}_i = \mathbf{v}_i - \sum_{j=0}^{i-1} \langle \mathbf{v}_i, \mathbf{e}_j \rangle \mathbf{e}_j$, $\mathbf{e}_i = \frac{\mathbf{R}_i}{\|\mathbf{R}_i\|_2}$
- ▷ `Hyperplane distance` $d = \frac{\theta^\top \mathbf{x} - b}{\|\theta\|_2}$
- ▷ `Banach space` Complete normed vector space + Cauchy sequence converges
- ▷ `Hilbert space` Banach + inner product (norm induced by inner product)

1.1 Vector Basics

A **vector** $\mathbf{v} \in \mathbb{R}^d$ can be understood from two equivalent perspectives:

1. **Algebraic perspective**: A vector is an ordered collection of d real numbers (called **components** or **coordinates**), represented as a column vector:

$$\mathbf{v} = \begin{bmatrix} v_0 \\ v_1 \\ \vdots \\ v_{d-1} \end{bmatrix} \in \mathbb{R}^{d\times 1},$$

where $v_0, v_1, \dots, v_{d-1} \in \mathbb{R}$ are scalars.

2. **Geometric perspective**: A vector is a quantity with both **magnitude** (length) and **direction** (the concept of unit vector, which precisely captures direction, will be introduced later). Importantly, a vector exists as an abstract geometric object independent of any coordinate system; its coordinates are merely a representation. In $\mathbb{R}^d$, a vector can be visualized as an arrow from the origin to the point $(v_0, v_1, \dots, v_{d-1})$, where the magnitude is $\|\mathbf{v}\|_2 = \sqrt{\sum_{i=0}^{d-1} v_i^2}$ and the direction is determined by the relative proportions of its components (e.g., vectors $(1, 2)$ and $(2, 4)$ point in the same direction but have different magnitudes).

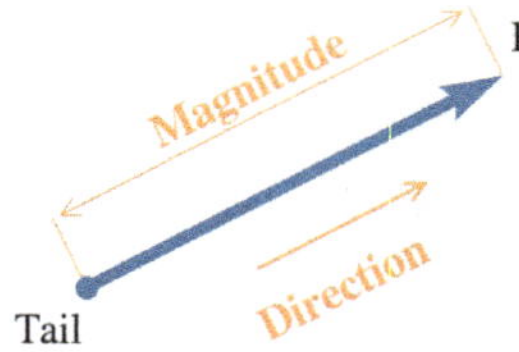

These two perspectives are equivalent: the algebraic representation (components) uniquely determines the geometric properties (magnitude and direction), and vice versa. For simplicity, we may write $\mathbf{v} \in \mathbb{R}^d$ with the understanding that vectors are column vectors by default.

By convention, **column vector** is represented using a bold lowercase letter (e.g., `$\mathbf{v}$` for $\mathbf{v}$ in LaTeX), and a bold lowercase letter with a superscript *(e.g., `$\mathbf{v}^\ast$` for $\mathbf{v}^*$) denotes a **row vector**.

1.1.1 Transpose

The transpose of a column vector $\mathbf{v} \in \mathbb{R}^{d\times 1}$ is a row vector:

$$\mathbf{v}^\top = \begin{bmatrix} v_0 & v_1 & \cdots & v_{d-1} \end{bmatrix} \in \mathbb{R}^{1\times d}.$$

We will sometimes use the notation $\mathbf{v} = [v_0, v_1, \dots, v_{d-1}]^\top$ to represent a column vector, where the transpose symbol indicates that the vector $\mathbf{v}$ should be interpreted as a column (vertical) vector rather than a row (horizontal) vector. Note that transposing twice returns the original vector: $(\mathbf{v}^\top)^\top = \mathbf{v}$.

In languages like Numpy/PyTorch, vectors are represented as **arrays/tensors**. A 1D array can be interpreted as either a row or column vector depending on context. To explicitly represent a column vector (shape `(d, 1)`), use `v = v.reshape(-1, 1)` or create a 2D array directly.

1.1.2 Inner Product

Inner products introduce fundamental geometric concepts: the length of a vector, the angle between two vectors, and orthogonality(when $\theta = \pi/2$).

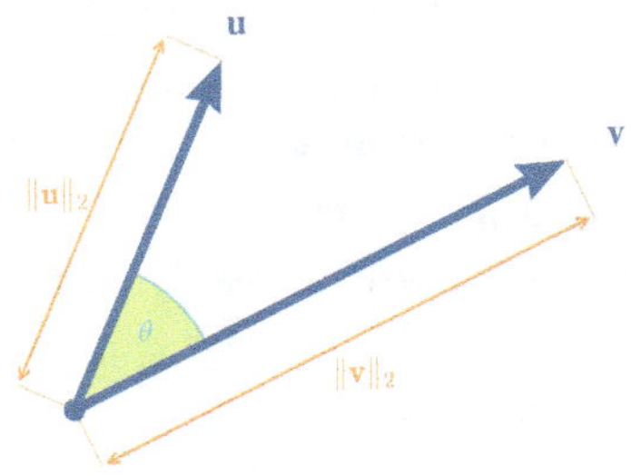

1.1.2.1 Dot Product and Similarity

The most common inner product is the **dot product** (also called **scalar product**) in $\mathbb{R}^d$:

$$\langle \mathbf{u}, \mathbf{v} \rangle = \mathbf{u}^\top \mathbf{v} = \sum_{i=0}^{d-1} u_i v_i. \tag{1.1}$$

The dot product is a specific instance of a more general concept. We now define inner products formally.

Geometric interpretation:

In $\mathbb{R}^d$ with the standard dot product, we have the following geometric relationship:

$$\langle \mathbf{u}, \mathbf{v} \rangle = \|\mathbf{u}\|_2 \cdot \|\mathbf{v}\|_2 \cos\theta \tag{1.2}$$

where θ is the angle between the two vectors, and $\|\cdot\|_2$ denotes the ℓ_2 (Euclidean) norm, which will be defined in Section~1.1.3. This relationship shows that the dot product measures both the magnitudes and the alignment (angle) between vectors. When vectors are normalized (unit length), the dot product reduces to $\cos\theta$, which is known as the **cosine similarity**:

$$\cos\theta = \frac{\langle \mathbf{u}, \mathbf{v} \rangle}{\|\mathbf{u}\|_2 \cdot \|\mathbf{v}\|_2} = \frac{\mathbf{u}^\top \mathbf{v}}{\|\mathbf{u}\|_2 \cdot \|\mathbf{v}\|_2}$$

Cosine similarity is a fundamental measure of similarity in machine learning, information retrieval, and natural language processing. Both the dot product and cosine similarity are used for measuring similarity in different contexts: the dot product is sensitive to both direction and magnitude, making it suitable when vector magnitudes carry meaningful information (e.g., Transformer attention mechanisms use dot products by default), while cosine similarity focuses purely on directional alignment and is preferred when magnitudes should be normalized away (e.g., in text similarity tasks where document length should not affect similarity scores).

Coordinate independence: The dot product is a geometric property that does not depend on the choice of coordinate system. While the component-wise formula $\langle \mathbf{u}, \mathbf{v} \rangle = \sum_{i=0}^{d-1} u_i v_i$ appears to depend on coordinates, the geometric interpretation $\langle \mathbf{u}, \mathbf{v} \rangle = \|\mathbf{u}\|_2 \cdot \|\mathbf{v}\|_2 \cos\theta$ shows that it depends only on the magnitudes and the angle between vectors—quantities that are invariant under coordinate transformations. The same two vectors will have the same dot product regardless of which orthonormal basis is used to represent them (orthonormal bases are formally defined in *1.12*), as long as the dot product is computed with respect to that basis.

1.1.2.2 Dot Product vs. Inner Product

While the dot product is the most common inner product, there are other inner products that satisfy the following **inner product requirements**. A mapping $\langle \cdot, \cdot \rangle$ that takes two vectors and returns a real number is an **inner product** if it satisfies (where V denotes a set of vectors):

1. **Bilinearity**: $\langle \lambda\mathbf{u} + \psi\mathbf{v}, \mathbf{w} \rangle = \lambda\langle \mathbf{u}, \mathbf{w} \rangle + \psi\langle \mathbf{v}, \mathbf{w} \rangle$ and $\langle \mathbf{u}, \lambda\mathbf{v} + \psi\mathbf{w} \rangle = \lambda\langle \mathbf{u}, \mathbf{v} \rangle + \psi\langle \mathbf{u}, \mathbf{w} \rangle$ for all $\mathbf{u}, \mathbf{v}, \mathbf{w} \in V$ and $\lambda, \psi \in \mathbb{R}$.
2. **Symmetry**: $\langle \mathbf{u}, \mathbf{v} \rangle = \langle \mathbf{v}, \mathbf{u} \rangle$ for all $\mathbf{u}, \mathbf{v} \in V$. Since the result is a scalar, and a scalar's transpose equals itself, $\mathbf{u}^\top \mathbf{v} = (\mathbf{u}^\top \mathbf{v})^\top = \mathbf{v}^\top \mathbf{u}$.

3. **Positive definiteness:** $\langle \mathbf{u}, \mathbf{u} \rangle > 0$ for all $\mathbf{u} \in V \setminus \{\mathbf{0}\}$, and $\langle \mathbf{0}, \mathbf{0} \rangle = 0$.

For example, for vectors in $\mathbb{R}^2$, the mapping $\langle \mathbf{u}, \mathbf{v} \rangle := 3u_0v_0 + u_1v_1$ defines an inner product that is different from the dot product, as it satisfies all three requirements above.

Convention: Throughout this book, unless otherwise stated, "inner product" and "dot product" are used interchangeably, referring to the dot product in $\mathbb{R}^d$.

Inner products are the foundation of **Hilbert spaces** (see Section~1.4), which are complete inner product spaces and provide the mathematical framework for attention mechanisms, kernel methods, and function approximation.

1.1.3 Norms

1.1.3.1 Property: Norm Axioms

For $\mathbf{v}, \mathbf{u} \in \mathbb{R}^d$ and $\alpha \in \mathbb{R}$, a function $\|\cdot\|$ is called a **norm** if it satisfies the following axioms:

1. **Non-negativity:** $\|\mathbf{v}\| \geq 0$
2. **Definiteness:** $\|\mathbf{v}\| = 0 \iff \mathbf{v} = \mathbf{0}$
3. **Homogeneity:** $\|\alpha\mathbf{v}\| = |\alpha|\|\mathbf{v}\|$
4. **Triangle inequality:** $\|\mathbf{u} + \mathbf{v}\| \leq \|\mathbf{u}\| + \|\mathbf{v}\|$

These axioms hold for all ℓ_p norms with $p \geq 1$ (including $p = 1, 2, \infty$). The norm is the foundation of Banach spaces (see Section~1.3), which are complete normed vector spaces. Banach spaces provide the mathematical framework for analyzing convergence, optimization, and function spaces in machine learning and numerical analysis.

1.1.3.2 Definition: ℓ_p Norm (Minkowski Norm)

For $p \geq 1$, the **ℓ_p norm** (also called the **Minkowski norm**) of a vector $\mathbf{v} \in \mathbb{R}^d$ is defined as

$$\|\mathbf{v}\|_p = \left(\sum_{i=0}^{d-1} |v_i|^p \right)^{1/p} \tag{1.3}$$

Special cases:

$$\|\mathbf{v}\|_1 = \sum_{i=0}^{d-1} |v_i|, \quad \|\mathbf{v}\|_2 = \sqrt{\sum_{i=0}^{d-1} v_i^2} = \sqrt{\langle \mathbf{v}, \mathbf{v} \rangle}, \quad \|\mathbf{v}\|_\infty = \max_i |v_i|.$$

The ℓ_2 norm (also called the **Euclidean norm**) is the most commonly used norm in this book. It is related to the inner product via $\|\mathbf{v}\|_2 = \sqrt{\langle \mathbf{v}, \mathbf{v} \rangle}$. The ℓ_1 norm (also called the **Manhattan norm** or **taxicab norm**) is defined as $\|\mathbf{v}\|_1 = \sum_{i=0}^{d-1} |v_i|$ and is particularly useful for sparse solutions in machine learning (see LASSO regularization). When $p \to \infty$, the ℓ_p norm converges to the **Chebyshev norm** (also called the **maximum norm** or **ℓ_∞ norm**): $\|\mathbf{v}\|_\infty = \max_i |v_i|$, which measures the largest component magnitude.

1.1.3.3 Definition: Mahalanobis Norm (M-norm)

For a vector $\mathbf{v} \in \mathbb{R}^d$ and a positive definite covariance matrix $\Sigma \in \mathbb{R}^{d \times d}$, the **Mahalanobis norm** (denoted as **M-norm** or **Mahalanobis distance**) is defined as

$$\|\mathbf{v}\|_M = \sqrt{\mathbf{v}^\top \Sigma^{-1} \mathbf{v}} \tag{1.4}$$

The M-norm accounts for correlations between dimensions through the covariance matrix Σ. When $\Sigma = I$ (identity matrix), it reduces to the standard Euclidean norm. The unit ball of the M-norm is a rotated and scaled ellipse (or ellipsoid) whose orientation and shape are determined by the eigenvectors and eigenvalues of Σ. In machine learning, the M-norm is used in anomaly detection, clustering (e.g., Gaussian mixture models), and when data has correlated features.

1.1.3.4 Definition: Standardized Euclidean Norm (Std-L2)

For a vector $\mathbf{v} \in \mathbb{R}^d$ and positive scaling factors $\sigma_1, \ldots, \sigma_d > 0$, the **standardized Euclidean norm** (denoted as **Std-L2**) is defined as

$$\|\mathbf{v}\| = \sqrt{\sum_{i=0}^{d-1} \left(\frac{v_i}{\sigma_i}\right)^2} \tag{1.5}$$

This norm accounts for different scales along different dimensions. When $\sigma_i = 1$ for all i, it reduces to the standard Euclidean norm. The unit ball of Std-L2 is an axis-aligned ellipse in $\mathbb{R}^2$ (or an axis-aligned ellipsoid in higher dimensions).

1.1.3.5 Visualization: Unit Balls of Different Norms

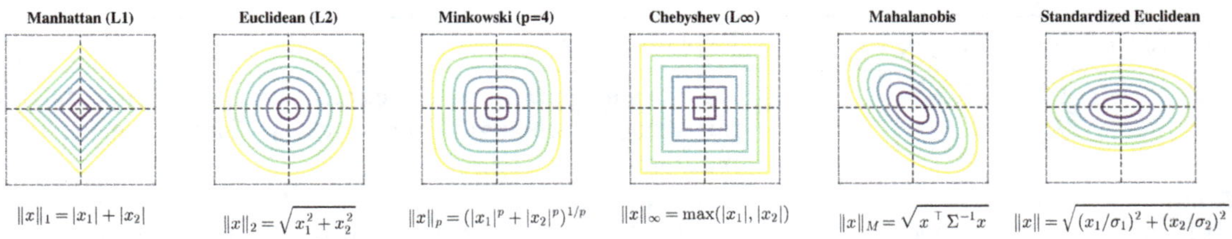

Figure 1.1: 2D Visualization of Different Norms

Figure 1.1 visualizes the unit balls (contours of constant norm value) for different norms in $\mathbb{R}^2$, shown in the following order:

#	Norm Name	Formula	Geometric Shape
1	**Manhattan (L1)**	$\|\mathbf{v}\|_1 = \|v_1\| + \|v_2\|$	Diamond-shaped unit ball
2	**Euclidean (L2)**	$\|\mathbf{v}\|_2 = \sqrt{v_1^2 + v_2^2}$	Circular unit ball
3	**Minkowski (L4)**	$\|\mathbf{v}\|_4 = (\|v_1\|^4 + \|v_2\|^4)^{1/4}$	"Squircle" unit ball
4	**Chebyshev (L∞)**	$\|\mathbf{v}\|_\infty = \max(\|v_1\|, \|v_2\|)$	Square-shaped unit ball
5	**Mahalanobis**	$\|\mathbf{v}\|_M = \sqrt{\mathbf{v}^\top \Sigma^{-1} \mathbf{v}}$	Rotated elliptical unit ball
6	**Std-L2**	$\|\mathbf{v}\| = \sqrt{(v_1/\sigma_1)^2 + (v_2/\sigma_2)^2}$	Axis-aligned elliptical unit ball

1.1.3.6 Insights: Std-L2 and M-norm in AI/ML*

This section provides advanced insights connecting norms to AI/ML. It is mathematically and conceptually dense, covering optimization algorithms, normalization techniques, and geometric perspectives that build on concepts from later chapters (e.g., matrix operations in Chapter 2, polar decomposition in Chapter 8, matrix calculus in Chapter 11, Fisher Information Matrix in Chapter 14). If you find this section challenging, feel free to skip it now and revisit it after completing the later chapters—the core definitions of norms remain independent of these applications though.

The **Std-L2 norm** and **M-norm** are fundamental to modern AI/ML, though they often appear under different names or in implicit forms. Understanding their geometric properties provides deep insights into how machine learning algorithms work.

Classical Machine Learning Applications: In algorithms like K-nearest neighbors (KNN), K-means clustering, and support vector machines (SVM), **Z-score normalization**[1] (also called **standard scaling**) is essential. After dividing each feature by its standard deviation, computing the Euclidean distance between standardized features is mathematically equivalent to computing the **Std-L2 distance** in the original feature space. This prevents features with larger scales from dominating the distance calculation, ensuring that all features contribute meaningfully to the algorithm's decisions.

Deep Learning Architecture: Batch Normalization (BN) and **Layer Normalization (LN)** implicitly create a Std-L2-like working space. While they do not explicitly define a distance or norm, they transform activations to have zero mean and unit variance, effectively inducing a coordinate system that is geometrically equivalent to Std-L2 geometry. This stabilizes gradient flow and improves training stability by ensuring that subsequent weight matrices operate in an approximately isotropic space, preventing the vanishing or exploding gradient problems that plague deep networks.

Optimizer Design (Geometric Perspective): Perhaps the most elegant connection is how different optimizers can be understood through their underlying geometric structures. The progression from **SGD** $\rightarrow$ **Adam** $\rightarrow$ **Muon/Shampoo**[2] corresponds to increasingly sophisticated geometric structures:

Optimizer	Norm(Shape)	Explanation
SGD	L2 (Circle)	Assumes all parameter directions are equally important, but struggles when gradient scales vary significantly across parameters.
Adam RMSProp	Std-L2 (Axis-aligned ellipse)	Normalizes by the second moment (variance) of gradients, effectively performing standardization in parameter space. This addresses scale differences but ignores correlations between parameters.
Muon Shampoo	M-norm (Rotated ellipse)	Uses polar decomposition (Muon) or Kronecker-factored covariance (Shampoo) to handle parameter correlations, achieving a form of whitening that accounts for non-axis-aligned dependencies.

Adam's advantage over SGD comes from its use of **Std-L2 geometry**, adapting learning rates per parameter based on historical variance. Muon's advantage over Adam (particularly in Transformer linear layers) comes from upgrading to **M-norm-like geometry**, handling non-axis-aligned correlations through the orthogonal factor of polar decomposition (see Chapter 8 for details on polar decomposition and the Muon optimizer).

M-norm Applications: The **M-norm** plays a crucial role in AI/ML, both explicitly and

[1]**Z-score normalization** (also called **standard scaling** or **standardization**): A data preprocessing technique that transforms features to have zero mean and unit variance. For a feature vector $\mathbf{x} \in \mathbb{R}^d$ with mean μ and standard deviation σ, the Z-score normalized vector is $\mathbf{z} = (\mathbf{x} - \mu)/\sigma$, where division is element-wise. This ensures all features are on the same scale, preventing features with larger magnitudes from dominating distance calculations or gradient updates. After Z-score normalization, computing Euclidean distance between normalized features is mathematically equivalent to computing the Std-L2 distance in the original feature space. Z-score normalization is essential for algorithms like KNN, K-means, and SVM that rely on distance metrics, and it forms the basis for normalization techniques in deep learning such as Batch Normalization and Layer Normalization.

[2]Second-order optimizer with Kronecker-factored preconditioners; see Anil et al., 2020, https://arxiv.org/abs/2002.09018.

implicitly. In **anomaly detection** and **out-of-distribution (OOD) detection**, the Mahalanobis distance is used directly to identify samples that deviate significantly from the training data distribution. When features are highly correlated (e.g., height and weight), Euclidean distance fails, but Mahalanobis distance automatically accounts for these correlations by measuring along the data's principal directions. In **metric learning** (e.g., face recognition, vector search), neural networks are trained to learn embeddings where M-norm-like geometry is "baked in," allowing efficient inference using simple Euclidean distance or cosine similarity. The M-norm also forms the mathematical foundation of **natural gradient descent** and second-order optimization methods, where the Fisher Information Matrix (or its approximations) acts as a preconditioner, effectively implementing an M-norm metric on the parameter space.

Unifying Perspective: This geometric perspective reveals that **optimization can be viewed as performing steepest descent under different Riemannian metrics**: Euclidean (SGD), diagonal Riemannian (Adam), and block/low-rank Riemannian (Muon/Shampoo). Each optimizer implicitly defines a geometry on the parameter space, and understanding these geometries provides insights into why certain optimizers work better for specific problems.

1.1.3.7 Definition: Unit Vector

A **unit vector** (also called a normalized vector or direction vector) is a vector with unit norm. For any non-zero vector $\mathbf{v} \in \mathbb{R}^d$, the unit vector in the direction of $\mathbf{v}$ is

$$\hat{\mathbf{v}} = \frac{\mathbf{v}}{\|\mathbf{v}\|_2} \tag{1.6}$$

In component form:

$$\hat{\mathbf{v}} = \begin{bmatrix} \frac{v_0}{\sqrt{\sum_{i=0}^{d-1} v_i^2}} & \frac{v_1}{\sqrt{\sum_{i=0}^{d-1} v_i^2}} & \cdots & \frac{v_{d-1}}{\sqrt{\sum_{i=0}^{d-1} v_i^2}} \end{bmatrix}^\top .$$

Several useful properties follow from this definition. First, a unit vector has unit norm:

$$\|\hat{\mathbf{v}}\|_2 = 1, \quad \text{or equivalently,} \quad \hat{\mathbf{v}}^\top \hat{\mathbf{v}} = 1.$$

Second, the inner product of a vector with its unit vector gives the vector's norm:

$$\mathbf{v}^\top \hat{\mathbf{v}} = \|\mathbf{v}\|_2.$$

This follows from $\mathbf{v}^\top \hat{\mathbf{v}} = \mathbf{v}^\top \frac{\mathbf{v}}{\|\mathbf{v}\|_2} = \frac{\|\mathbf{v}\|_2^2}{\|\mathbf{v}\|_2} = \|\mathbf{v}\|_2$.

A fundamental geometric decomposition is that any vector equals its **direction times its length**:

$$\mathbf{v} = \|\mathbf{v}\|_2 \cdot \hat{\mathbf{v}},$$

where $\hat{\mathbf{v}} = \mathbf{v}/\|\mathbf{v}\|_2$ is the unit vector (direction) and $\|\mathbf{v}\|_2$ is the magnitude (length). The unit vector $\hat{\mathbf{v}}$ points in the same direction as $\mathbf{v}$ but has length 1, lying on the unit circle (in 2D) or unit sphere (in 3D). This decomposition separates directional information (which way the vector points) from magnitude information (how long the vector is), making unit vectors particularly useful for representing directions, orientations, and computing angles between vectors via the inner product: $\langle \hat{\mathbf{u}}, \hat{\mathbf{v}} \rangle = \cos\theta$ where θ is the angle between the directions of $\mathbf{u}$ and $\mathbf{v}$.

1.1.4 Definition: Energy and Frobenius Norm (F-Norm)

For a vector $\mathbf{v} \in \mathbb{R}^d$, the **Frobenius norm** (also called the **F-norm**) is defined as

$$\|\mathbf{v}\|_F = \sqrt{\sum_{i=0}^{d-1} v_i^2} = \|\mathbf{v}\|_2.$$

The Frobenius norm will be extended to matrices in Chapter 2, where it generalizes the ℓ_2 norm to two-dimensional arrays.

The **energy** of a vector $\mathbf{v} \in \mathbb{R}^d$ is defined as the squared ℓ_2 norm:

$$\|\mathbf{v}\|_2^2 = \|\mathbf{v}\|_F^2 = \langle \mathbf{v}, \mathbf{v} \rangle = \sum_{i=0}^{d-1} v_i^2 \tag{1.7}$$

The term "energy" comes from signal processing and physics, where the energy of a signal or system is proportional to the sum of squared values. In machine learning, the energy of a vector often represents the total magnitude or "strength" of features, which is used in regularization (e.g., ℓ_2 regularization penalizes large parameter values) and in normalization techniques to control the scale of activations in neural networks. The Frobenius norm is crucial for measuring the magnitude of weight matrices, computing regularization terms, and analyzing the energy content of feature representations across layers.

1.1.5 Theorem: Energy Additivity

For any **orthogonal** vectors $\mathbf{u}, \mathbf{v} \in \mathbb{R}^d$,

$$\|\mathbf{u} + \mathbf{v}\|_2^2 = \|\mathbf{u}\|_2^2 + \|\mathbf{v}\|_2^2,$$

or equivalently in component form:

$$\sum_{i=0}^{d-1} (u_i + v_i)^2 = \sum_{i=0}^{d-1} u_i^2 + \sum_{i=0}^{d-1} v_i^2.$$

For **non-orthogonal** vectors, the energy formula includes a cross term:

$$\|\mathbf{u} + \mathbf{v}\|_2^2 = \|\mathbf{u}\|_2^2 + \|\mathbf{v}\|_2^2 + 2\langle \mathbf{u}, \mathbf{v} \rangle,$$

which follows from expanding $\|\mathbf{u} + \mathbf{v}\|_2^2 = \langle \mathbf{u} + \mathbf{v}, \mathbf{u} + \mathbf{v} \rangle = \langle \mathbf{u}, \mathbf{u} \rangle + 2\langle \mathbf{u}, \mathbf{v} \rangle + \langle \mathbf{v}, \mathbf{v} \rangle$. When $\mathbf{u} \perp \mathbf{v}$, the inner product term $\langle \mathbf{u}, \mathbf{v} \rangle = 0$, recovering the orthogonal case.

1.1.6 Definition: Orthogonality

Two vectors $\mathbf{u}, \mathbf{v} \in \mathbb{R}^d$ are **orthogonal** (denoted $\mathbf{u} \perp \mathbf{v}$) if

$$\langle \mathbf{u}, \mathbf{v} \rangle = \sum_{i=0}^{d-1} u_i v_i = 0.$$

1.1.7 Definition: Vector Addition

Let $\mathbf{u}, \mathbf{v} \in \mathbb{R}^{d \times 1}$. Then

$$\mathbf{u} + \mathbf{v} = \begin{bmatrix} u_0 + v_0 \\ u_1 + v_1 \\ \vdots \\ u_{d-1} + v_{d-1} \end{bmatrix}.$$

As illustrated in Figure 1.2, vector addition follows the **parallelogram rule**: place the vectors $\mathbf{u}$ and $\mathbf{v}$ tail-to-tail, complete the parallelogram, and the diagonal from the common tail to the opposite vertex represents $\mathbf{u} + \mathbf{v}$ (shown in red). Alternatively, using the **tail-to-head method**, place the tail of $\mathbf{v}$ at the head of $\mathbf{u}$, and the sum $\mathbf{u} + \mathbf{v}$ is the vector from the tail of $\mathbf{u}$ to the head of $\mathbf{v}$.

Since vector addition is defined, vector subtraction is naturally defined as the inverse operation: if $\mathbf{c} = \mathbf{u} + \mathbf{v}$, then $\mathbf{u} = \mathbf{c} - \mathbf{v}$ and $\mathbf{v} = \mathbf{c} - \mathbf{u}$. As illustrated in Figure 1.2, subtracting $\mathbf{v}$ from $\mathbf{c}$ gives $\mathbf{u}$: geometrically, place $\mathbf{c}$ and $\mathbf{v}$ tail-to-tail, and the difference $\mathbf{c} - \mathbf{v}$ is the vector from the head of $\mathbf{v}$ to the head of $\mathbf{c}$ (which equals $\mathbf{u}$). More generally, for any vectors $\mathbf{u}$ and $\mathbf{v}$, the difference $\mathbf{u} - \mathbf{v}$ can be computed as $\mathbf{u} + (-\mathbf{v})$, where $-\mathbf{v}$ is the vector with the same magnitude as $\mathbf{v}$ but opposite direction.

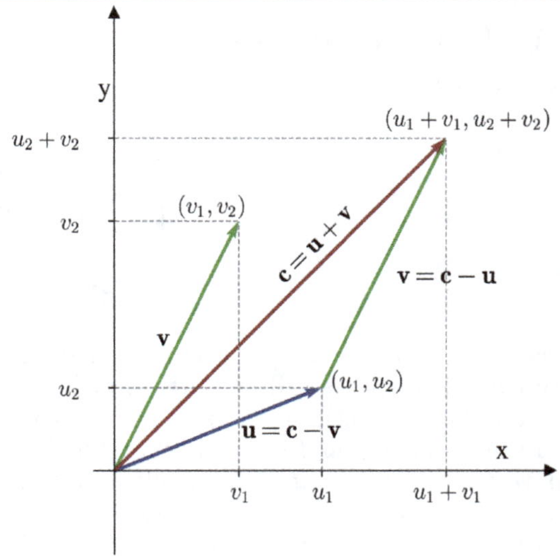

Figure 1.2: Vector Addition

1.1.8 Definition: Scalar Multiplication

Let $\mathbf{v} \in \mathbb{R}^{d\times 1}$ and $c \in \mathbb{R}$. Then

$$c\mathbf{v} = [cv_0, cv_1, \dots, cv_{d-1}]^\top.$$

1.1.9 Theorem: Triangle Inequality

For any $\mathbf{u}, \mathbf{v} \in \mathbb{R}^d$,

$$\|\mathbf{u} + \mathbf{v}\|_2 \leq \|\mathbf{u}\|_2 + \|\mathbf{v}\|_2.,$$

or equivalently in component form:

$$\sqrt{\sum_{i=0}^{d-1}(u_i + v_i)^2} \leq \sqrt{\sum_{i=0}^{d-1} u_i^2} + \sqrt{\sum_{i=0}^{d-1} v_i^2}.$$

1.1.10 Theorem: Cauchy–Schwarz Inequality

For any $\mathbf{u}, \mathbf{v} \in \mathbb{R}^d$, we have

$$|\langle \mathbf{u}, \mathbf{v} \rangle| \leq \|\mathbf{u}\|_2 \cdot \|\mathbf{v}\|_2 \tag{1.8}$$

or equivalently in component form:

$$\left|\sum_{i=0}^{d-1} u_i v_i\right| \leq \sqrt{\sum_{i=0}^{d-1} u_i^2}\sqrt{\sum_{i=0}^{d-1} v_i^2}.$$

Taking the square of both sides, we obtain the squared form:

$$\langle \mathbf{u}, \mathbf{v} \rangle^2 \leq \|\mathbf{u}\|_2^2 \cdot \|\mathbf{v}\|_2^2,$$

or equivalently in component form:

$$\left(\sum_{i=0}^{d-1} u_i v_i\right)^2 \leq \left(\sum_{i=0}^{d-1} u_i^2\right)\left(\sum_{i=0}^{d-1} v_i^2\right).$$

1.1.11 Outer Product

For $\mathbf{u} \in \mathbb{R}^{M\times 1}$ and $\mathbf{v} \in \mathbb{R}^{N\times 1}$,

$$\mathbf{u}\mathbf{v}^\top = \begin{bmatrix} u_0v_0 & u_0v_1 & \cdots & u_0v_{N-1} \\ u_1v_0 & u_1v_1 & \cdots & u_1v_{N-1} \\ \vdots & \vdots & \ddots & \vdots \\ u_{M-1}v_0 & u_{M-1}v_1 & \cdots & u_{M-1}v_{N-1} \end{bmatrix}$$

The outer product $\mathbf{u}\mathbf{v}^\top$ is a **rank-1 linear operator**, not necessarily an orthogonal projection. The "projection" interpretation above is a way to understand how this operator acts on vectors, but the operator itself is a general linear transformation. Orthogonal projection is a special case when certain conditions are met. (The concept of **rank** is formally defined in Chapter 2, where we show that $\text{rank}(\mathbf{u}\mathbf{v}^\top) = 1$.)

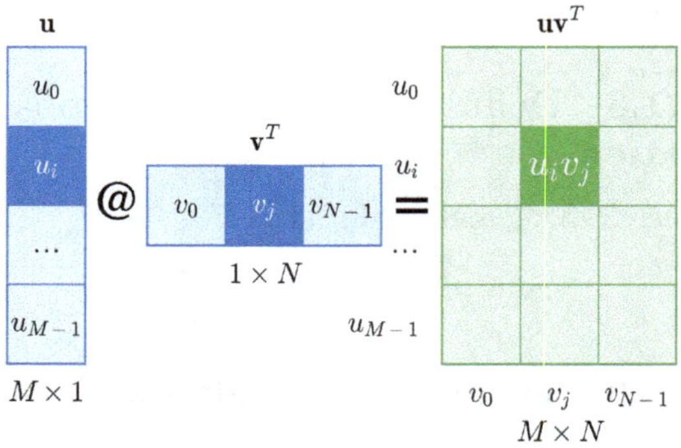

Figure 1.3: Outer Product

1.1.11.1 Example: Outer Product in $\mathbb{R}^2$

Given $\mathbf{u} = \begin{bmatrix} 1 \\ 2 \end{bmatrix}$ and $\mathbf{v} = \begin{bmatrix} 3 \\ 4 \end{bmatrix}$, compute the outer product $\mathbf{u}\mathbf{v}^\top$ and explain why its rank is 1.

Solution

The outer product is

$$\mathbf{u}\mathbf{v}^\top = \begin{bmatrix} 1 \\ 2 \end{bmatrix} \begin{bmatrix} 3 & 4 \end{bmatrix} = \begin{bmatrix} 3 & 4 \\ 6 & 8 \end{bmatrix}.$$

The columns are $\begin{bmatrix} 3 \\ 6 \end{bmatrix} = 3\mathbf{u}$ and $\begin{bmatrix} 4 \\ 8 \end{bmatrix} = 4\mathbf{u}$, both multiples of $\mathbf{u}$. The column space is $\text{span}\{\mathbf{u}\}$, which is one-dimensional, so the rank is 1.

Alternatively, for any vector $\mathbf{x} = \begin{bmatrix} x_1 \\ x_2 \end{bmatrix}$,

$$(\mathbf{u}\mathbf{v}^\top)\mathbf{x} = \begin{bmatrix} 3x_1 + 4x_2 \\ 6x_1 + 8x_2 \end{bmatrix} = (3x_1 + 4x_2)\mathbf{u}.$$

The output always lies along $\mathbf{u}$, so all outputs are on one line. A linear map with one-dimensional output has rank 1.

The outer product is a special case of matrix multiplication, but we give it a special name because it produces the simplest nontrivial matrix (rank-1), isolates a single output direction, and serves as the building block for matrix decompositions, SVD (Chapter 8), attention mechanisms, and low-rank models.

1.1.11.2 Property: Projection Operation's Perspective

Consider $\mathbf{u} \in \mathbb{R}^{M\times 1}$ and $\mathbf{v} \in \mathbb{R}^{N\times 1}$. Then $\mathbf{u}\mathbf{v}^\top$ can be interpreted as the following sequence of operations:

1. Consider any vector $\mathbf{x}$ in the first space $\mathbb{R}^N$, compute its projection to the unit vector $\hat{\mathbf{v}}$: $\hat{\mathbf{v}}^\top\mathbf{x}$.
2. Use the above quantity to scale the unit vector $\hat{\mathbf{u}}$ in the second space $\mathbb{R}^M$: $\hat{\mathbf{u}}(\hat{\mathbf{v}}^\top\mathbf{x})$.
3. Scale the above vector by $\|\mathbf{u}\|_2 \cdot \|\mathbf{v}\|_2$.

Example: Rotation Find a pair of 2-dim vectors $\mathbf{u}, \mathbf{v} \in \mathbb{R}^2$, such that for any 2-dim vector $\mathbf{x}$,

1. $\mathbf{u}\mathbf{v}^\top$ rotates the x-component of $\mathbf{x}$ counterclockwise by θ and ignores the y-component.
2. $\mathbf{u}\mathbf{v}^\top$ rotates the y-component of $\mathbf{x}$ counterclockwise by θ and ignores the x-component.

Solution

1. $\mathbf{u} = \begin{bmatrix}\cos\theta\\ \sin\theta\end{bmatrix}, \mathbf{v} = \begin{bmatrix}1\\0\end{bmatrix}.$
2. $\mathbf{u} = \begin{bmatrix}-\sin\theta\\ \cos\theta\end{bmatrix}, \mathbf{v} = \begin{bmatrix}0\\1\end{bmatrix}.$

Example: Projection and Scaling Find a pair of 2-dim vectors $\mathbf{u}, \mathbf{v} \in \mathbb{R}^2$, such that $\mathbf{u}\mathbf{v}^\top$ projects any 2-dim vector to the direction $\widehat{\theta}$ and scales its length by λ.

Solution

$$\mathbf{u} = \begin{bmatrix}\cos\theta\\ \sin\theta\end{bmatrix}, \quad \mathbf{v} = \lambda\begin{bmatrix}\cos\theta\\ \sin\theta\end{bmatrix}.$$

Example: Component Extraction and Placement Find a pair of d-dim vectors $\mathbf{u}, \mathbf{v} \in \mathbb{R}^d$, such that for any d-dim vector $\mathbf{x} \in \mathbb{R}^d$, the operator $\mathbf{u}\mathbf{v}^\top$ extracts the i-th component x_i from the input and places it at the j-th position in the output, with all other components being zero.

In other words, the output vector has the form $[0, \dots, 0, x_i, 0, \dots, 0]^\top$ where x_i appears at position j.

Solution

We set $\mathbf{v}$ to be a unit vector along axis i (the i-th component is 1 and all other components are 0), i.e., $\mathbf{v} = \mathbf{e}_i$. We set $\mathbf{u}$ to be a unit vector along axis j (the j-th component is 1 and all other components are 0), i.e., $\mathbf{u} = \mathbf{e}_j$.

Then $(\mathbf{u}\mathbf{v}^\top)\mathbf{x} = \mathbf{e}_j\mathbf{e}_i^\top\mathbf{x} = \mathbf{e}_j x_i$, which gives a vector with x_i at position j and zeros elsewhere.

The outer product is fundamental in linear algebra. Every matrix can be expressed as a sum of rank-1 matrices (outer products), which is the foundation of the **Singular Value Decomposition (SVD)**. SVD will be introduced in Chapter 8, and matrix norms (including the Frobenius norm, which measures matrix energy) are discussed in Chapter 2. The outer product $\mathbf{u}\mathbf{v}^\top$ is a rank-1 matrix, and understanding it helps build intuition for matrix decompositions and low-rank approximations used extensively in machine learning, dimensionality reduction, and data compression.

1.1.12 Applications and Examples

1.1.12.1 Vector-Vector Projection and Distance

Define the unit vector $\hat{v} = \frac{\mathbf{v}}{\|\mathbf{v}\|_2}$. The projection of $\mathbf{u}$ onto $\hat{v}$ is

$$\mathbf{p} = \langle \mathbf{u}, \hat{v} \rangle \hat{v} = \frac{\langle \mathbf{u}, \mathbf{v} \rangle}{\|\mathbf{v}\|_2^2} \mathbf{v} \tag{1.9}$$

In component form, for each component $k = 0, \dots, d-1$: $[\mathbf{p}]_k = \frac{v_k}{\sum_{i=0}^{d-1} v_i^2} \sum_{j=0}^{d-1} v_j u_j$.

The length of the projection (with sign) is

$$\|\mathbf{p}\|_2 = \langle \mathbf{u}, \hat{v} \rangle = \frac{\langle \mathbf{u}, \mathbf{v} \rangle}{\|\mathbf{v}\|_2} \tag{1.10}$$

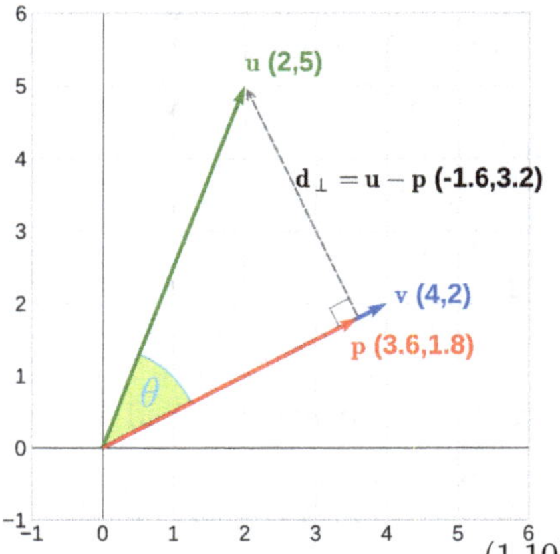

Figure 1.4: Vector Projection and Distance

The **orthogonal component** from the projection is

$$\mathbf{d}_\perp = \mathbf{u} - \mathbf{p} \tag{1.11}$$

This vector $\mathbf{d}_\perp$ represents the component of $\mathbf{u}$ that is orthogonal to $\mathbf{v}$ (and thus orthogonal to the projection $\mathbf{p}$).

In component form: $[\mathbf{d}_\perp]_k = u_k - \frac{v_k}{\sum_{i=0}^{d-1} v_i^2} \sum_{j=0}^{d-1} v_j u_j$, where $[\mathbf{d}_\perp]_k$ is the k-th component of the orthogonal component $\mathbf{d}_\perp$.

Geometric Proof via Cosine:

Let θ be the angle between $\mathbf{u}$ and $\mathbf{v}$. By the geometric interpretation of the inner product,

$$\langle \mathbf{u}, \mathbf{v} \rangle = \|\mathbf{u}\|_2 \cdot \|\mathbf{v}\|_2 \cos\theta.$$

The projection of $\mathbf{u}$ onto $\mathbf{v}$ has length $\|\mathbf{u}\|_2 \cos\theta = \frac{\langle \mathbf{u}, \mathbf{v} \rangle}{\|\mathbf{v}\|_2}$ and direction $\hat{v} = \frac{\mathbf{v}}{\|\mathbf{v}\|_2}$. Therefore,

$$\mathbf{p} = \frac{\langle \mathbf{u}, \mathbf{v} \rangle}{\|\mathbf{v}\|_2} \hat{v} = \langle \mathbf{u}, \hat{v} \rangle \hat{v},$$

where the last equality follows from $\langle \mathbf{u}, \hat{v} \rangle = \frac{\langle \mathbf{u}, \mathbf{v} \rangle}{\|\mathbf{v}\|_2}$.

Inner product as projection: The inner product $\theta^\top \mathbf{x}$ computes the scalar projection of $\mathbf{x}$ onto the direction of θ. When θ is a unit vector, the projection equals the inner product; otherwise, it is scaled by $\|\theta\|_2$. This connection—**inner product is projection**—is fundamental to understanding geometric operations in machine learning.

1.1.12.2 Example: Orthogonal Distance from Point to Hyperplane

Find the **orthogonal distance** from a point $\mathbf{u} \in \mathbb{R}^d$ to the hyperplane defined by $\theta^\top \mathbf{x} = b$, where $\theta \in \mathbb{R}^d$ is the normal vector, $b \in \mathbb{R}$ is the **offset** (a constant that shifts the hyperplane away from the origin; when $b = 0$, the hyperplane passes through the origin), and $\mathbf{x} \in \mathbb{R}^d$ denotes an arbitrary point on the hyperplane.

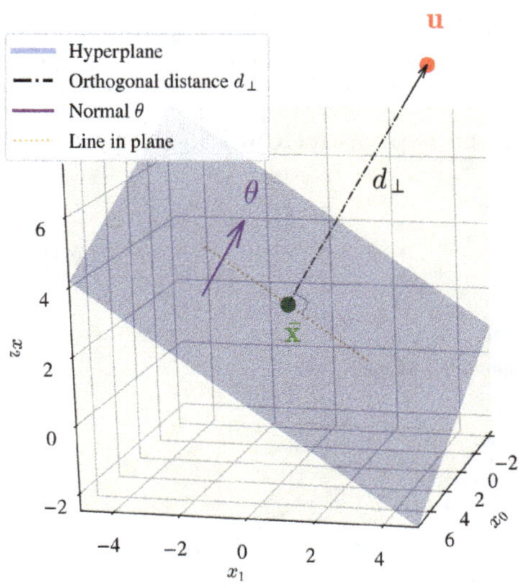

Figure 1.5: Orthogonal distance to hyperplane

Solution

The vector θ is called the **normal vector** because it is perpendicular to the hyperplane. To see this, consider any two points $\mathbf{x}_1, \mathbf{x}_2$ on the hyperplane (both satisfying $\theta^\top \mathbf{x}_1 = b$ and $\theta^\top \mathbf{x}_2 = b$). Their difference $\mathbf{x}_1 - \mathbf{x}_2$ lies in the hyperplane, and we have:

$$\theta^\top(\mathbf{x}_1 - \mathbf{x}_2) = \theta^\top \mathbf{x}_1 - \theta^\top \mathbf{x}_2 = b - b = 0,$$

which means θ is orthogonal to any vector lying in the hyperplane.

Let $\bar{\mathbf{x}}$ be the **foot of the perpendicular** dropped from $\mathbf{u}$ to the hyperplane, and let $d_\perp$ be the **orthogonal distance** from $\mathbf{u}$ to the hyperplane (positive if $\mathbf{u}$ is on the same side as θ, negative otherwise). Since $\bar{\mathbf{x}}$ lies on the hyperplane, we have $\theta^\top \bar{\mathbf{x}} = b$.

The **orthogonal component** $\mathbf{u} - \bar{\mathbf{x}}$ is parallel to the normal vector, so:

$$\mathbf{u} - \bar{\mathbf{x}} = \hat{\theta} d_\perp,$$

where $\hat{\theta} = \frac{\theta}{\|\theta\|_2}$ is the unit vector in the direction of θ. Taking the dot product with θ:

$$\theta^\top(\mathbf{u} - \bar{\mathbf{x}}) = \theta^\top \hat{\theta} d_\perp = \|\theta\|_2 d_\perp.$$

Since $\theta^\top \bar{\mathbf{x}} = b$, we have:

$$\theta^\top \mathbf{u} - b = \|\theta\|_2 d_\perp,$$

which gives the general formula:

$$d_\perp = \frac{\theta^\top \mathbf{u} - b}{\|\theta\|_2}$$

Connection to Linear Layers in Deep Neural Networks: A linear regression model and a DNN linear layer share the same functional form (where $\mathbf{x} \in \mathbb{R}^d$, $\theta \in \mathbb{R}^d$, $b \in \mathbb{R}$, and $\hat{y} \in \mathbb{R}$):

$$\hat{y} = \theta^\top \mathbf{x} + b.$$

This defines an **affine mapping** from input space to output space. The **prediction residual** (or **regression residual**) $r = y - \hat{y}$ measures the prediction error along the output axis.

> In Section~1.1.12.1, we defined the **orthogonal component** $\mathbf{d}_\perp = \mathbf{u} - \mathbf{p}$ as the component of a vector orthogonal to its projection. Here, the **prediction residual** (or **regression residual**) $r = y - \hat{y}$ is a scalar that measures the difference between the observed value y and the predicted value $\hat{y}$ in regression. The term "residual" is reserved for prediction errors in regression, following statistical convention, while "orthogonal component" refers to the geometric component in vector projection.

Geometrically, the equation

$$\theta^\top \mathbf{x} - y + b = 0$$

defines a hyperplane in the augmented space $(\mathbf{x}, y) \in \mathbb{R}^{d+1}$. If we write the hyperplane in the form $\tilde{\theta}^\top \mathbf{u} = \tilde{b}$ where $\tilde{\theta} = \begin{bmatrix} \theta \\ -1 \end{bmatrix}$ and $\mathbf{u} = \begin{bmatrix} \mathbf{x} \\ y \end{bmatrix}$, then the orthogonal distance from a point to this hyperplane is **proportional** to the prediction residual:

$$d_\perp = \frac{r}{\sqrt{\|\theta\|_2^2 + 1}} = \frac{y - \hat{y}}{\sqrt{\|\theta\|_2^2 + 1}}.$$

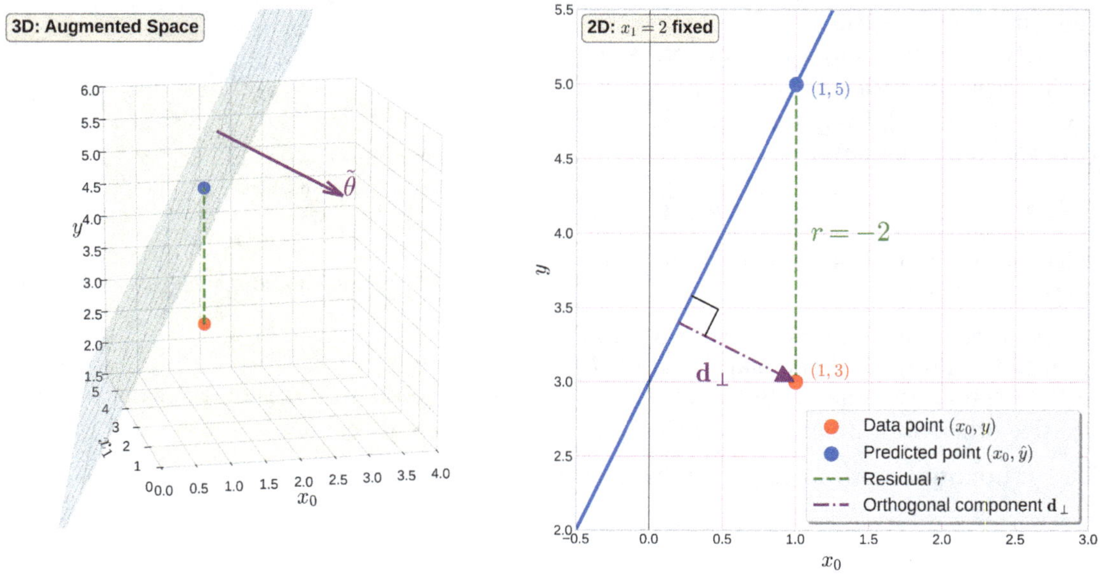

Figure 1.6: Linear Regression in Augmented Space

Figure 1.6 illustrates this geometric relationship in both 3D augmented space and a 2D cross-section. The left panel shows the regression hyperplane in the augmented space (x_0, x_1, y), with the data point, predicted point, and prediction residual r (green dashed line) clearly visible. The normal vector $\tilde{\theta}$ to the hyperplane is also shown. The right panel shows a 2D view with $x_1 = 2$ fixed, displaying both the prediction residual (vertical green dashed line) and the orthogonal component $\mathbf{d}_\perp$ (purple dash-dot line), which is the perpendicular distance from the data point to the regression line.

For example, consider a linear regression model with parameters $\theta = \begin{bmatrix} 2 \\ -1 \end{bmatrix}$ and intercept $b = 5$. The prediction function is $\hat{y} = 2x_0 - x_1 + 5$. For a data point with $\mathbf{x} = \begin{bmatrix} 1 \\ 2 \end{bmatrix}$ and $y = 3$, we have $\hat{y} = 2 \cdot 1 - 1 \cdot 2 + 5 = 5$, so the prediction residual is $r = 3 - 5 = -2$.

The hyperplane in augmented space is defined by $\theta^\top \mathbf{x} - y + b = 0$, or equivalently $\begin{bmatrix} 2 \\ -1 \\ -1 \end{bmatrix}^\top \begin{bmatrix} x_0 \\ x_1 \\ y \end{bmatrix} = -5$. The orthogonal distance from the point $\mathbf{u} = \begin{bmatrix} 1 \\ 2 \\ 3 \end{bmatrix}$ to this hyperplane (using the normal vector

$\tilde{\theta} = \begin{bmatrix} 2 \\ -1 \\ -1 \end{bmatrix}$) is $d_\perp = \frac{\begin{bmatrix} 2 \\ -1 \\ -1 \end{bmatrix}^\top \begin{bmatrix} 1 \\ 2 \\ 3 \end{bmatrix} - (-5)}{\sqrt{2^2+(-1)^2+(-1)^2}} = \frac{2}{\sqrt{6}}$. We can verify that $|d_\perp| = |r|/\sqrt{\|\theta\|_2^2+1} = 2/\sqrt{6}$, confirming that the orthogonal distance is **proportional** to the prediction residual, with scaling factor $1/\sqrt{\|\theta\|_2^2+1}$. The sign convention depends on the orientation chosen for the normal vector in the augmented space.

The least squares objective minimizes the sum of squared prediction residuals, which corresponds to minimizing the sum of squared distances from data points to the regression hyperplane up to a constant scaling factor.

Residual vs. Orthogonal Distance and Regression vs. Classification: An important distinction arises between what linear regression minimizes and what classification methods like support vector classification (SVC) minimize. This distinction is clearly visible in Figure 1.6.

Linear regression minimizes prediction residuals along the output axis y. For a fixed input $\mathbf{x}$, the error is measured as $r = y - \hat{y}$, which is the vertical distance shown as the green dashed line in Figure 1.6. The least squares objective is:

$$\min_{\theta,b} \sum_i (y_i - \theta^\top \mathbf{x}_i - b)^2 = \min_{\theta,b} \sum_i r_i^2.$$

This measures **output prediction error**, not the shortest geometric distance. Linear regression is suitable for numerical prediction and continuous targets, where we care about how far off our predictions are in the output space.

In contrast, support vector classification (SVC) optimizes **orthogonal distances** from data points to the decision boundary. For an SVC hyperplane $w^\top \mathbf{x} + b = 0$, the orthogonal distance from a point $\mathbf{x}$ to the hyperplane is:

$$d_\perp(\mathbf{x}) = \frac{|w^\top \mathbf{x} + b|}{\|w\|_2}.$$

This is measured along the normal vector direction (the purple dash-dot line in Figure 1.6), making it a purely geometric, rotation-invariant quantity. SVC maximizes the minimum orthogonal distance (margin):

$$\max_{w,b} \min_i \frac{y_i(w^\top \mathbf{x}_i + b)}{\|w\|_2}.$$

Terminology: SVM (Support Vector Machines) is the general term for the family of methods that includes both **SVC** (Support Vector Classification) for classification tasks and **SVR** (Support Vector Regression) for regression tasks. They share the same geometric optimization framework (minimizing $\|w\|_2^2$ subject to constraints), but differ in their constraint formulations: SVC uses margin-based constraints for classification, while SVR uses ε-insensitive tube constraints for regression.

The above two examples illustrate vector-to-vector projection and vector-to-hyperplane projection. We will cover matrix and subspace projections in Section~2.10 of Chapter~2 and Section~6.1 of Chapter~6.

1.1.12.3 Example: Scaled Dot-Product Attention in Transformers

In the Transformer architecture, **scaled dot-product attention** computes attention scores between query and key vectors using inner products. Given a query vector $\mathbf{q} \in \mathbb{R}^d$ and a key vector $\mathbf{k} \in \mathbb{R}^d$, the attention score is computed as:

$$\text{score}(\mathbf{q}, \mathbf{k}) = \frac{\mathbf{q}^\top \mathbf{k}}{\sqrt{d}}.$$

The inner product $\mathbf{q}^\top \mathbf{k}$ measures how well the query aligns with the key. A large positive value indicates strong alignment (the query is highly relevant to this key), a large negative value indicates strong anti-alignment, and values near zero indicate weak correlation. The scaling factor $1/\sqrt{d}$ prevents the dot products from growing too large in magnitude, which helps stabilize gradients during training.

Transformers use dot products rather than cosine similarity $\frac{\mathbf{q}^\top \mathbf{k}}{\|\mathbf{q}\|_2 \|\mathbf{k}\|_2}$ for several reasons. First, **magnitude matters**: the magnitude of query and key vectors carries semantic information about signal strength, which is lost when normalizing to unit vectors. Second, **computational efficiency**: dot products can be computed in parallel for all query-key pairs via matrix multiplication $QK^\top$. Third, **effective scaling**: the fixed factor $1/\sqrt{d}$ scales the dot products to a range where gradients flow effectively, facilitating the learning of the weight matrices W_Q and W_K that project inputs into query and key spaces. Finally, **gradient flow**: dot products provide better gradient signals during backpropagation compared to normalized cosine similarity, which can have vanishing gradients when vectors are nearly orthogonal.

In the full attention mechanism, this score is computed for all query-key pairs and used to weight the corresponding value vectors, allowing the model to focus on the most relevant information.

1.1.12.4 Example: Kernel - Similarity Measure

In machine learning, we often need to measure the similarity between two data points $\mathbf{x}^{(m)}$ and $\mathbf{x}^{(n)}$. However, sometimes the raw data space is not suitable for measuring similarity directly. How can we define a similarity measure that works in a transformed feature space?

Solution: We use a feature map $\phi(\cdot)$ that transforms data points into a (possibly higher-dimensional) feature space, and define the kernel function as the inner product in that space:

$$\kappa(\mathbf{x}^{(m)}, \mathbf{x}^{(n)}) \triangleq \phi(\mathbf{x}^{(m)})^\top \phi(\mathbf{x}^{(n)}),$$

where $\phi(\cdot)$ maps each data point to a column vector in the feature space.

Interpretation: The kernel function $\kappa(\mathbf{x}^{(m)}, \mathbf{x}^{(n)})$ measures the similarity between $\mathbf{x}^{(m)}$ and $\mathbf{x}^{(n)}$ in the feature space. Since it's an inner product, it inherits the geometric interpretation: large positive values indicate similar directions (high similarity), values near zero indicate orthogonality (low similarity), and large negative values indicate opposite directions (high dissimilarity). This is the foundation of kernel methods in machine learning, such as support vector machines (SVMs) and kernel PCA.

1.2 Vector Space

1.2.1 Definition: Linear Combination

Let $\mathbf{v}_0, \dots, \mathbf{v}_{N-1} \in \mathbb{R}^d$.

Then the following expression is a **linear combination** of these vectors:

$$\sum_{n=0}^{N-1} c_n \mathbf{v}_n,$$

where $c_0, \dots, c_{N-1} \in \mathbb{R}$ are called **weights**. Sometimes we call this expression a **weighted sum**.

1.2.2 Definition: Linear Independence

Vectors $\mathbf{v}_0, \dots, \mathbf{v}_{N-1} \in \mathbb{R}^d$ are **linearly independent** if the only solution $(c_0, \dots, c_{N-1})$ to the equation $\sum_{n=0}^{N-1} c_n \mathbf{v}_n = \mathbf{0}$ is $c_0 = \dots = c_{N-1} = 0$.

1.2.3 Definition: Vector Space

A **vector space** (also called a **linear space**) is a set V equipped with two operations:

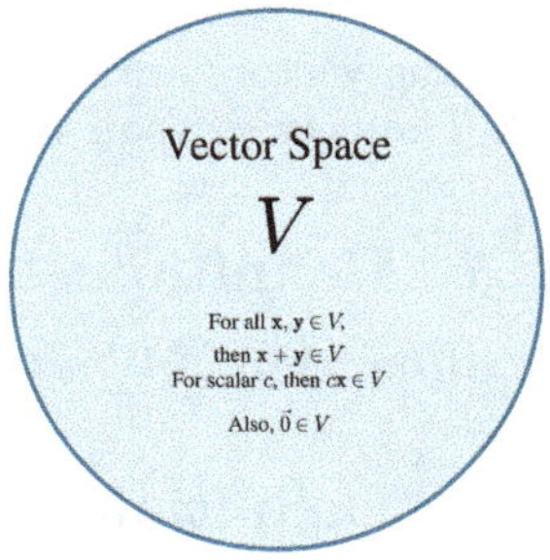

1. **Vector addition**: A binary operation $+ : V \times V \to V$ that maps any pair $(\mathbf{u}, \mathbf{v})$ to $\mathbf{u} + \mathbf{v} \in V$.
2. **Scalar multiplication**: An operation $\cdot : \mathbb{R} \times V \to V$ that maps any pair $(c, \mathbf{v})$ to $c\mathbf{v} \in V$.

These operations must satisfy the following **vector space axioms**:

Addition axioms:
1. **Commutativity**: $\mathbf{u} + \mathbf{v} = \mathbf{v} + \mathbf{u}$ for all $\mathbf{u}, \mathbf{v} \in V$.
2. **Associativity**: $(\mathbf{u} + \mathbf{v}) + \mathbf{w} = \mathbf{u} + (\mathbf{v} + \mathbf{w})$ for all $\mathbf{u}, \mathbf{v}, \mathbf{w} \in V$.
3. **Existence of zero vector**: There exists $\mathbf{0} \in V$ such that $\mathbf{v} + \mathbf{0} = \mathbf{v}$ for all $\mathbf{v} \in V$.
4. **Existence of additive inverses**: For each $\mathbf{v} \in V$, there exists $-\mathbf{v} \in V$ such that $\mathbf{v} + (-\mathbf{v}) = \mathbf{0}$.

Scalar multiplication axioms:
5. **Distributivity over vector addition**: $c(\mathbf{u} + \mathbf{v}) = c\mathbf{u} + c\mathbf{v}$ for all $c \in \mathbb{R}$ and $\mathbf{u}, \mathbf{v} \in V$.
6. **Distributivity over scalar addition**: $(c + d)\mathbf{v} = c\mathbf{v} + d\mathbf{v}$ for all $c, d \in \mathbb{R}$ and $\mathbf{v} \in V$.
7. **Associativity of scalar multiplication**: $(cd)\mathbf{v} = c(d\mathbf{v})$ for all $c, d \in \mathbb{R}$ and $\mathbf{v} \in V$.
8. **Identity element**: $1 \cdot \mathbf{v} = \mathbf{v}$ for all $\mathbf{v} \in V$.

The set $\mathbb{R}^d$ of all d-dimensional real vectors forms a vector space under component-wise addition and scalar multiplication, as defined above.

1.2.4 Definition: Linear Subspace

A set W is a **linear subspace** of $\mathbb{R}^d$ if:

1. The zero vector $\mathbf{0} \in W$.
2. For any $\mathbf{u}, \mathbf{v} \in W$, $\mathbf{u} + \mathbf{v} \in W$.
3. For any $\mathbf{v} \in W$ and $c \in \mathbb{R}$, $c\mathbf{v} \in W$.

While conditions 2 and 3 imply that $\mathbf{0} \in W$ (by taking $c = 0$), it is standard practice to explicitly state condition 1 for clarity. This ensures that the definition is self-contained and avoids any potential ambiguity in mathematical proofs and algorithmic implementations.

Connection to Hyperplanes: A hyperplane defined by $\theta^\top \mathbf{x} = b$ is a **linear subspace** (vector space) if and only if $b = 0$ (the hyperplane passes through the origin). When $b \neq 0$, the hyperplane is not a vector space because it does not contain the origin (violating condition 1), and it is not closed under scalar multiplication (multiplying a point on the hyperplane by 0 gives the origin, which is not on the hyperplane). In this case, the hyperplane is an **affine space**—a translated linear subspace. The set of all vectors $\mathbf{x}_1 - \mathbf{x}_2$ where $\mathbf{x}_1, \mathbf{x}_2$ lie on the hyperplane forms a linear subspace (the direction space of the hyperplane), but the hyperplane itself is not a vector space when $b \neq 0$.

1.2.5 Definition: Linear Span

The **linear span** of vectors $\mathbf{v}_0, \dots, \mathbf{v}_{N-1} \subset \mathbb{R}^d$ is defined as

$$\left\{ \sum_{n=0}^{N-1} c_n \mathbf{v}_n \middle| c_0, \dots, c_{N-1} \in \mathbb{R} \right\}$$

1.2.6 Definition: Basis of a Space W

A set of vectors is called a **basis** of W if:

1. The vectors are linearly independent.
2. The vectors span W.

1.2.7 Definition: Dimension of a Space W

The **dimension** of W is the number of vectors in a basis of W.

1.2.8 Definition: Orthonormal Basis of Space W

Let W be a space of dimension r. A set of vectors $\mathbf{v}_0, \dots, \mathbf{v}_{r-1}$ is an **orthonormal basis** of space W if

$$\langle \mathbf{v}_i, \mathbf{v}_j \rangle = \delta_{ij}, \quad \forall i, j \in 0, \dots, r-1 \tag{1.12}$$

where δ_{ij} is the **Kronecker delta** (a scalar indicator function), defined as $\delta_{ij} = 1$ if $i = j$ and $\delta_{ij} = 0$ if $i \neq j$. This condition means that each vector has unit length ($\|\mathbf{v}_i\|_2 = 1$ for all i) and different vectors are orthogonal ($\langle \mathbf{v}_i, \mathbf{v}_j \rangle = 0$ for $i \neq j$).

1.2.9 Gram–Schmidt Algorithm for Orthonormalization

Let W be a space of dimension r, and let $\mathbf{v}_0, \dots, \mathbf{v}_{r-1}$ be a basis of W. We construct an **orthonormal basis** $\mathbf{e}_0, \dots, \mathbf{e}_{r-1}$ as follows.

1.2.10 Algorithm: Gram–Schmidt

Algorithm 1 Gram–Schmidt Algorithm

Input: A set of linearly independent vectors $\{\mathbf{v}_0, \dots, \mathbf{v}_{r-1}\}$ that form a basis for a subspace W
Output: An orthonormal basis $\{\mathbf{e}_0, \dots, \mathbf{e}_{r-1}\}$ of W

```
e_0 ← v_0 / ||v_0||_2                          ▷ Normalize the first vector
for i = 1, ..., r − 1 do
    R_i ← v_i − Σ_{j=0}^{i−1} ⟨v_i, e_j⟩ e_j    ▷ Compute residual
    e_i ← R_i / ||R_i||_2                      ▷ Normalize residual
end for
```

The vectors $\mathbf{v}$ must be linearly independent. Otherwise, if some vector is a linear combination of previous ones, the residual $\mathbf{R}_i$ will be the zero vector (i.e., $\|\mathbf{R}_i\|_2 = 0$), and the normalization step will fail. This linear independence requirement is fundamental in Gram–Schmidt orthogonalization and QR decomposition, which are used extensively in numerical linear algebra, least squares problems, and in neural network initialization schemes that require orthonormal weight matrices.

1.3 Banach Space

1.3.1 Definition: Cauchy Sequence

A sequence $\{\mathbf{v}_n\}_{n=0}^{\infty}$ in a normed vector space is called a **Cauchy sequence** if for every $\epsilon > 0$, there exists an integer N such that for all $m, n \geq N$, we have $\|\mathbf{v}_m - \mathbf{v}_n\| < \epsilon$. Intuitively, a Cauchy sequence is one where the terms become arbitrarily close to each other as the sequence progresses.

Figure 1.7 visualizes a Cauchy sequence in $\mathbb{R}^2$ formed by a geometric series of vectors. Starting from an initial vector $\mathbf{v}_0$, each subsequent vector is obtained by rotating and scaling the previous one: $\mathbf{v}_{n+1} = \alpha R(\theta)\mathbf{v}_n$ where $\alpha = 0.88$ (decay factor) and $\theta = -45°$ (rotation angle). The sequence of partial sums $S_n = \sum_{i=0}^{n} \mathbf{v}_i$ forms a spiral that converges to a fixed point as $n \to \infty$, demonstrating the Cauchy property: consecutive terms become arbitrarily close, and the sequence converges in the complete space $\mathbb{R}^2$.

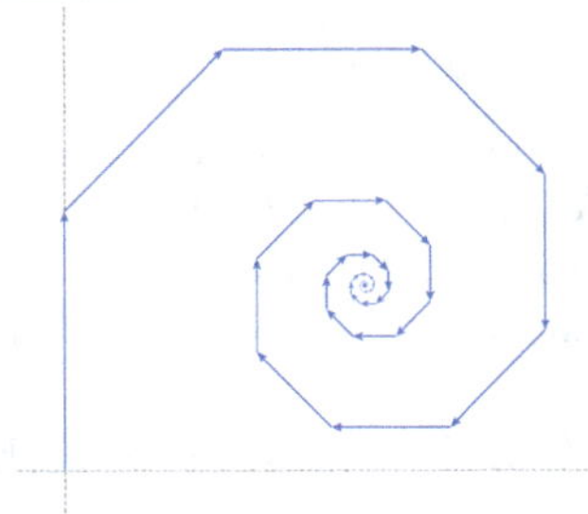

Figure 1.7: 2D Vector Cauchy Sequence

In machine learning, ideally the sequence of weight matrices $\{\mathbf{W}_k\}_{k=0}^{\infty}$ during training is itself a Cauchy sequence. This means that in later training steps, the parameter changes become progressively smaller.

If $\{\mathbf{W}_k\}$ is Cauchy, then necessarily $\|\mathbf{W}_{k+1} - \mathbf{W}_k\|_F \to 0$ as $k \to \infty$. The converse is **not** true: adjacent differences tending to zero is a **necessary but not sufficient** condition for Cauchy convergence (e.g., the harmonic partial sums $a_n = \sum_{i=1}^{n} \frac{1}{i}$ satisfy $|a_{n+1} - a_n| = \frac{1}{n+1} \to 0$ yet diverge). The gradient descent example below is different: its updates form a **contractive** mapping, which additionally guarantees that $\{\mathbf{W}_k\}$ converges and is therefore Cauchy.

1.3.1.1 Definition: Complete

A space with a distance function (metric) is **complete** if every Cauchy sequence converges to a limit within the space. For a normed vector space V with norm $\|\cdot\|$, the distance is given by $d(\mathbf{u}, \mathbf{v}) = \|\mathbf{u} - \mathbf{v}\|$. A normed vector space is complete if for every Cauchy sequence $\{\mathbf{v}_n\}_{n=0}^{\infty}$ in V, there exists $\mathbf{v} \in V$ such that $\|\mathbf{v}_n - \mathbf{v}\| \to 0$ as $n \to \infty$.

In finite-dimensional spaces like $\mathbb{R}^d$, every Cauchy sequence converges, so completeness is automatic. However, in infinite-dimensional spaces, there may exist Cauchy sequences that do not converge within the space, making completeness a non-trivial requirement.

1.3.2 Definition: Banach Space

A **Banach space** is a complete normed vector space. More precisely, a Banach space B is:

1. A vector space equipped with a norm $\|\cdot\| : B \to \mathbb{R}$ that satisfies the norm axioms (positive definiteness, absolute homogeneity, and triangle inequality).
2. **Complete** with respect to this norm: every Cauchy sequence in B converges to a limit in B.

In finite dimensions, every normed vector space is automatically complete (since all norms on finite-dimensional spaces are equivalent and finite-dimensional normed spaces are complete). Therefore, $\mathbb{R}^d$ with any norm (e.g., ℓ_1, ℓ_2, ℓ_∞) is a Banach space.

The completeness requirement becomes important in infinite-dimensional spaces. In finite-dimensional settings like $\mathbb{R}^d$, this distinction is less critical, but the terminology is still used. Banach spaces are fundamental in functional analysis, optimization theory, and numerical analysis. Every Hilbert space is a Banach space (since the norm induced by an inner product satisfies the norm axioms), but not every Banach space is a Hilbert space (a Banach space may not have an inner product that induces its norm).

1.3.2.1 Example: Cauchy Sequences in $\mathbb{R}$ and $\mathbb{R}^d$

The sequence $\{v_n\}_{n=1}^{\infty}$ in $\mathbb{R}$ with $v_n = 1 - \frac{1}{2^n}$ is Cauchy and converges to 1, so $\mathbb{R}$ is complete. The sequence $\{\mathbf{v}_n\}_{n=1}^{\infty}$ in $\mathbb{R}^d$ with $\mathbf{v}_n = \frac{1}{n}\mathbf{1}_d$ (where $\mathbf{1}_d$ is the all-ones vector in $\mathbb{R}^d$) is Cauchy and converges to $\mathbf{0}$, so $\mathbb{R}^d$ is complete.

In contrast, the rational numbers $\mathbb{Q}$ (with the absolute value norm) are **not** complete: there exist Cauchy sequences of rationals that converge to irrational numbers. For example, the sequence of decimal approximations $v_n = 1.4, 1.41, 1.414, 1.4142, ...$ (approximating $\sqrt{2}$) is Cauchy in $\mathbb{Q}$ but converges to $\sqrt{2} \notin \mathbb{Q}$. This is why $\mathbb{R}$ (the completion of $\mathbb{Q}$) is needed for analysis.

1.3.2.2 Example: Weight Matrix Space as a Banach Space

In machine learning, weight matrices are updated iteratively during optimization. Consider a sequence of 2×2 weight matrices $\{\mathbf{W}_k\}_{k=0}^{\infty}$ that represents the training trajectory. For example, let the optimal weight matrix be $\mathbf{W}^* = \begin{bmatrix} 1.0 & -2.0 \\ 0.5 & 3.0 \end{bmatrix}$ and the initial weight matrix be $\mathbf{W}_0 = \begin{bmatrix} 5.0 & 5.0 \\ -4.0 & -3.0 \end{bmatrix}$.

During gradient descent on the quadratic loss $\mathcal{L}(\mathbf{W}) = \|\mathbf{W} - \mathbf{W}^*\|_F^2$, we update $\mathbf{W}_{k+1} = \mathbf{W}_k - \eta \nabla \mathcal{L}(\mathbf{W}_k) = \mathbf{W}_k - 2\eta(\mathbf{W}_k - \mathbf{W}^*)$, where η is the learning rate and $\mathbf{W}^*$ is the optimal weight matrix.

As training progresses, the distance to the optimum decreases: $\|\mathbf{W}_k - \mathbf{W}^*\|_F \to 0$ as $k \to \infty$. This implies that the parameter updates get smaller: $\|\mathbf{W}_{k+1} - \mathbf{W}_k\|_F = 2\eta\|\mathbf{W}_k - \mathbf{W}^*\|_F \to 0$, which means the sequence $\{\mathbf{W}_k\}$ is Cauchy with respect to the Frobenius norm $\|\cdot\|_F$ (the square root of the sum of squared matrix elements).

The training sequence converges to the optimal weight matrix $\mathbf{W}^*$, and the space of 2×2 matrices equipped with the Frobenius norm is a Banach space. Figure 1.8 shows how the distance to the optimum $\|\mathbf{W}_k - \mathbf{W}^*\|_F$ decreases as training progresses, visually demonstrating convergence. Since $\|\mathbf{W}_k - \mathbf{W}^*\|_F \to 0$, the sequence converges to $\mathbf{W}^*$ and is therefore Cauchy; consequently $\|\mathbf{W}_{k+1} - \mathbf{W}_k\|_F \to 0$.

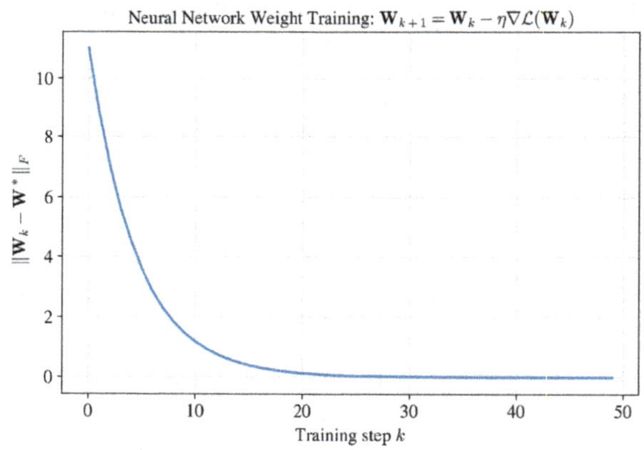

Figure 1.8: Matrix Cauchy Sequence

Why completeness matters in optimization: When training neural networks, we want the weight sequence $\{\mathbf{W}_k\}$ to be Cauchy so that it converges to a limit matrix $\mathbf{W}^*$ within the space. In a Banach space, completeness guarantees that every Cauchy sequence has such a limit. This does **not** follow merely from shrinking updates ($\|\mathbf{W}_{k+1} - \mathbf{W}_k\|_F \to 0$); additional structure—such as the contractive gradient descent map in this example—is needed. Without completeness, we could have a Cauchy sequence with no limit within the space—a pathological case that can occur in non-standard function spaces or ill-conditioned parameterizations. The Banach space structure ensures that convergent Cauchy sequences of parameters remain within the space and have well-defined limits.

Weight decay and geometry: Adding L2 weight decay[3] ($\lambda > 0$) changes the update to

[3]**Weight decay** is a regularization technique that penalizes large parameter values by adding a term proportional to the squared norm of parameters to the loss function. The update rule becomes $\mathbf{W}_{k+1} = \mathbf{W}_k - \eta\nabla\mathcal{L}(\mathbf{W}_k) - \eta\lambda\mathbf{W}_k$, where $\lambda > 0$ is the **weight decay coefficient**. Weight decay helps prevent overfitting and can improve generalization by encouraging smaller parameter values. Geometrically, it introduces a radial contraction force that changes

$\mathbf{W}_{k+1} = \mathbf{W}_k - \eta\nabla\mathcal{L}(\mathbf{W}_k) - \eta\lambda\mathbf{W}_k$. This is not merely adding a regularization penalty; it fundamentally changes the geometry of the optimization space. Without weight decay, optimization occurs in a flat Euclidean space where all directions are equivalent. With weight decay, the space becomes a Euclidean space with a radial potential field centered at the origin: $V(\mathbf{W}) = \frac{\lambda}{2}\|\mathbf{W}\|_F^2$. This creates a "quadratic bowl" geometry where points far from the origin experience a stronger pull back toward the origin. This geometric change makes the optimization problem strongly convex, confines parameter trajectories to a compact set, and makes the sequence more likely to be Cauchy—parameters cannot wander indefinitely and are pulled toward a bounded region.

Figure 1.9 visualizes this geometric transformation. The left plot shows the 3D surface of the quadratic bowl potential $V(\mathbf{W}) = \frac{\lambda}{2}\|\mathbf{W}\|_F^2$, demonstrating how the potential increases quadratically with distance from the origin. The training trajectory (red line) follows the surface of this bowl, showing how parameters are pulled toward the origin while simultaneously converging toward the optimum.

Technical nuance: With weight decay, the total loss becomes $\mathcal{L}(\mathbf{W}) = \|\mathbf{W} - \mathbf{W}^*\|_F^2 + \frac{\lambda}{2}\|\mathbf{W}\|_F^2$. The true minimum shifts from $\mathbf{W}^*$ (green star) to $\mathbf{W}_{opt} = \frac{2}{2+\lambda}\mathbf{W}^*$ (orange diamond), which is slightly "inward" toward the origin. This reflects the trade-off between fitting the data (being near $\mathbf{W}^*$) and regularization (being near the origin). The trajectory correctly converges to $\mathbf{W}_{opt}$, not $\mathbf{W}^*$.

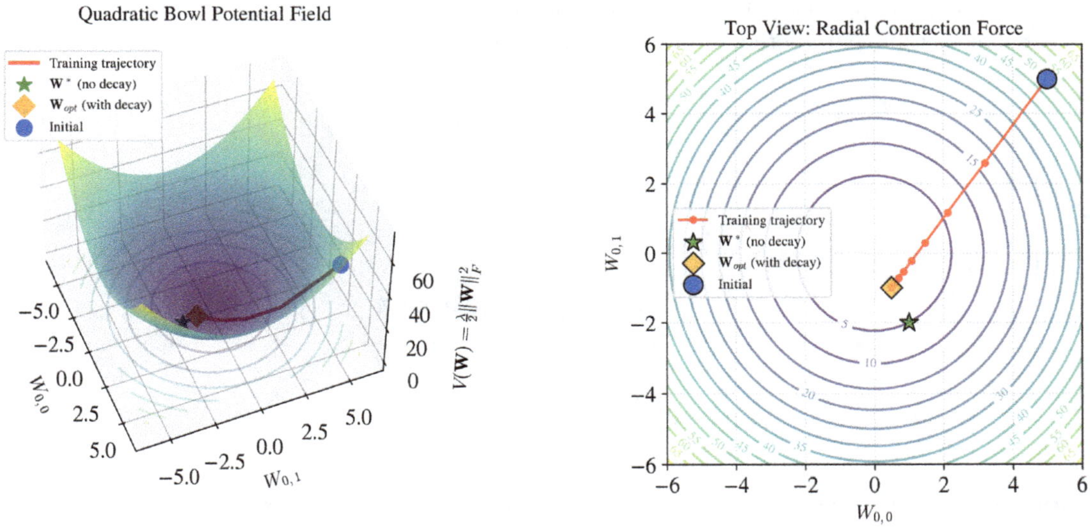

Figure 1.9: Weight Decay Quadratic Bowl Geometry

The right plot shows the top-down contour view, revealing the radial symmetry of the potential field. The concentric circles represent equipotential surfaces, and the trajectory demonstrates the radial contraction force: parameters are pulled toward the origin while simultaneously moving toward the shifted optimum. This visualization makes concrete the abstract notion that weight decay changes the optimization geometry from flat Euclidean space to a space with a radial potential field—the "quadratic bowl" that confines parameter trajectories and promotes Cauchy convergence.

---

the optimization space from flat Euclidean to a space with a radial potential field centered at the origin. **L2 weight decay** is equivalent to L2 regularization (Ridge regularization). **L1 weight decay** is equivalent to L1 regularization (LASSO).

1.3.2.3 Counterexample: Loss Decreases but Sequence is NOT Cauchy

In practice, loss can decrease while the parameter sequence is **not** Cauchy. This occurs when using SGD or Adam[4] with a **constant learning rate** and **stochastic gradient noise**. Consider the same setup as before, but with SGD updates:

$$\mathbf{W}_{k+1} = \mathbf{W}_k - \eta\left((\mathbf{W}_k - \mathbf{W}^*) + \epsilon_k\right),$$

where ϵ_k represents stochastic gradient noise from minibatch variance. Even near the optimum, the gradient estimate has non-zero variance.

Figure 1.10 shows three key observations:

1. **Plot (A)**: Loss decreases initially and then plateaus at a noise floor, but does not approach zero. This is common in training: loss stabilizes but does not continue decreasing.
2. **Plot (B)**: The consecutive differences $\|\mathbf{W}_{k+1} - \mathbf{W}_k\|_F$ do **not** approach zero. They oscillate around a non-zero mean (approximately 0.05 to 0.2 in this example), meaning parameters keep moving and never "tighten together." This violates the Cauchy property: if the sequence were Cauchy, we would see $\|\mathbf{W}_{k+1} - \mathbf{W}_k\|_F \to 0$.
3. **Plot (C)**: The distance to optimum $\|\mathbf{W}_k - \mathbf{W}^*\|_F$ decreases initially but then wanders in a small ball around the optimum, never converging to a fixed matrix. This is the typical "loss decreased but parameters didn't converge" training state.

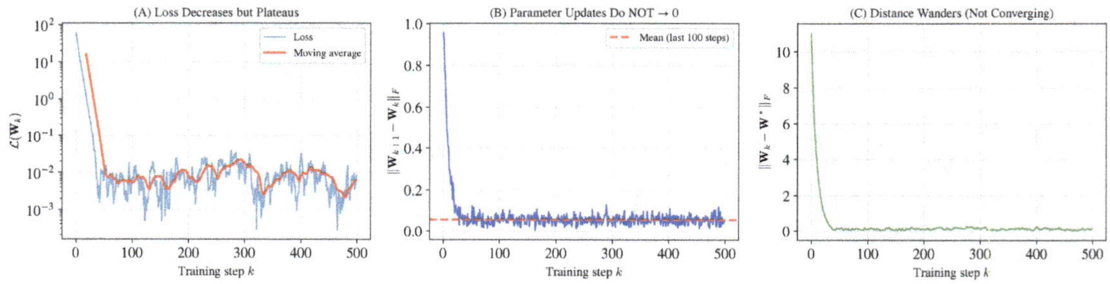

Figure 1.10: Non-Cauchy Training Sequence

Root cause: With constant learning rate η and non-zero noise variance, we have:

$$\|\mathbf{W}_{k+1} - \mathbf{W}_k\|_F = \eta\|(\mathbf{W}_k - \mathbf{W}^*) + \epsilon_k\|_F \approx \eta\|\epsilon_k\|_F,$$

which does not approach zero even when near the optimum. To make the sequence Cauchy (and ensure convergence), we need **learning rate decay** ($\eta_k \to 0$) or reduced noise variance (e.g., increasing batch size).

[4] **SGD (Stochastic Gradient Descent)** and **Adam (Adaptive Moment Estimation)** are optimization algorithms used in machine learning. **SGD** updates parameters using gradients computed on random minibatches of data: $\mathbf{W}_{k+1} = \mathbf{W}_k - \eta_k \nabla\mathcal{L}(\mathbf{W}_k)$, where η_k is the learning rate and $\nabla\mathcal{L}(\mathbf{W}_k)$ is the gradient estimate from a minibatch. The stochasticity comes from using a subset of data, which introduces variance in the gradient estimates. **Adam** adaptively adjusts learning rates for each parameter using estimates of first and second moments of gradients: $\mathbf{W}_{k+1} = \mathbf{W}_k - \eta\frac{\mathbf{m}_k}{\sqrt{\mathbf{v}_k}+\epsilon}$, where $\mathbf{m}_k$ and $\mathbf{v}_k$ are exponentially moving averages of gradients and squared gradients, respectively. Adam often provides faster convergence and better stability than SGD, especially in the early stages of training.

Connection to optimizers: The fundamental role of optimizers and learning rate schedules is to control whether the parameter update sequence $\|\mathbf{W}_{k+1} - \mathbf{W}_k\|_F$ forms a Cauchy sequence in an appropriate norm. Learning rate decay ensures $\eta_k \to 0$, making $\|\mathbf{W}_{k+1} - \mathbf{W}_k\|_F \to 0$ even with noise. Adam implicitly controls an effective norm by normalizing gradients, making the sequence more likely to be Cauchy, though potentially in a different norm than the Frobenius norm. This is why Adam often appears more stable initially, while SGD with decay can achieve better generalization by ensuring Cauchy convergence in the standard Frobenius norm.

Weight decay as a geometric intervention: Weight decay can help make sequences Cauchy by introducing a radial contraction force. However, there is a subtle geometric inconsistency when combining Adam with naive L2 weight decay: Adam's gradient updates operate in an adaptive (coordinate-dependent) geometry, while weight decay acts directly on parameters in the standard Euclidean geometry. This geometric mismatch can distort optimization paths. This is why AdamW was developed: it decouples weight decay from the gradient update, applying it as a parameter-space contraction rather than mixing it into the adaptive gradient geometry. The key insight is that weight decay doesn't just "shrink parameters"—it changes which directions are geometrically preferred, making the origin a special point and redefining what "far" and "near" mean in the optimization landscape.

1.4 Hilbert Space

A **Hilbert space** is a complete inner product space. More precisely, a Hilbert space H is:

1. A vector space equipped with an inner product $\langle \cdot, \cdot \rangle : H \times H \to \mathbb{R}$ (or $\mathbb{C}$ for complex Hilbert spaces).
2. **Complete** with respect to the norm induced by the inner product: $\|\mathbf{v}\| = \sqrt{\langle \mathbf{v}, \mathbf{v} \rangle}$.

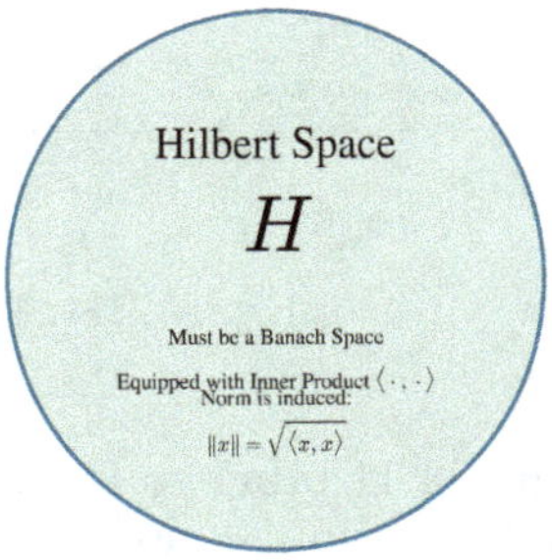

In finite dimensions, every inner product space is automatically complete (since all norms on finite-dimensional spaces are equivalent and finite-dimensional normed spaces are complete). Therefore, $\mathbb{R}^d$ with the standard inner product $\langle \mathbf{u}, \mathbf{v} \rangle = \mathbf{u}^\top \mathbf{v}$ is a Hilbert space.

The completeness requirement becomes important in infinite-dimensional spaces (e.g., function spaces like L^2). In finite-dimensional settings like $\mathbb{R}^d$, this distinction is less critical, but the terminology is still used. Hilbert spaces are fundamental in functional analysis, quantum mechanics, and signal processing.

1.4.1 Example: $\mathbb{R}^d$ as a Hilbert Space

The space $\mathbb{R}^d$ with the standard inner product $\langle \mathbf{u}, \mathbf{v} \rangle = \mathbf{u}^\top \mathbf{v}$ is a Hilbert space. It is a vector space equipped with an inner product, and in finite dimensions it is automatically complete. In infinite dimensions, examples include the space $L^2([0,1])$ of square-integrable functions on $[0,1]$ with inner product $\langle f, g \rangle = \int_0^1 f(x)g(x)dx$.

1.4.1.1 Counterexample: ℓ_1 Norm and LASSO Regularization

Not all Banach spaces are Hilbert spaces. Consider $\mathbb{R}^d$ equipped with the ℓ_1 norm $\|\mathbf{v}\|_1 = \sum_{i=0}^{d-1} |v_i|$. This space is a Banach space (it is complete with respect to the ℓ_1 norm), but it is not a Hilbert space because the ℓ_1 norm does not satisfy the parallelogram law: $\|\mathbf{u}+\mathbf{v}\|_1^2 + \|\mathbf{u}-\mathbf{v}\|_1^2 = 2\|\mathbf{u}\|_1^2 + 2\|\mathbf{v}\|_1^2$ does not hold in general. This means there is no inner product that induces the ℓ_1 norm (induced p-norms will be covered in Chapter 2).

In machine learning, the ℓ_1 norm appears in LASSO[5] (Least Absolute Shrinkage and Selection Operator) regularization, where the objective function includes a penalty term $\lambda\|\theta\|_1$. Unlike ℓ_2 regularization (which preserves the Hilbert space structure), ℓ_1 regularization operates in a Banach space setting. The geometric structure of the ℓ_1 unit ball (a diamond in two dimensions, a cross-polytope in higher dimensions) has sharp corners at the coordinate axes, which leads to sparse solutions[6]—parameters are driven exactly to zero. This sparsity property is a direct consequence of the non-Hilbert geometry: the ℓ_1 norm's non-smoothness at zero creates a natural preference for solutions on the coordinate axes, where many parameters vanish. The following figure illustrates this geometric difference and its consequences.

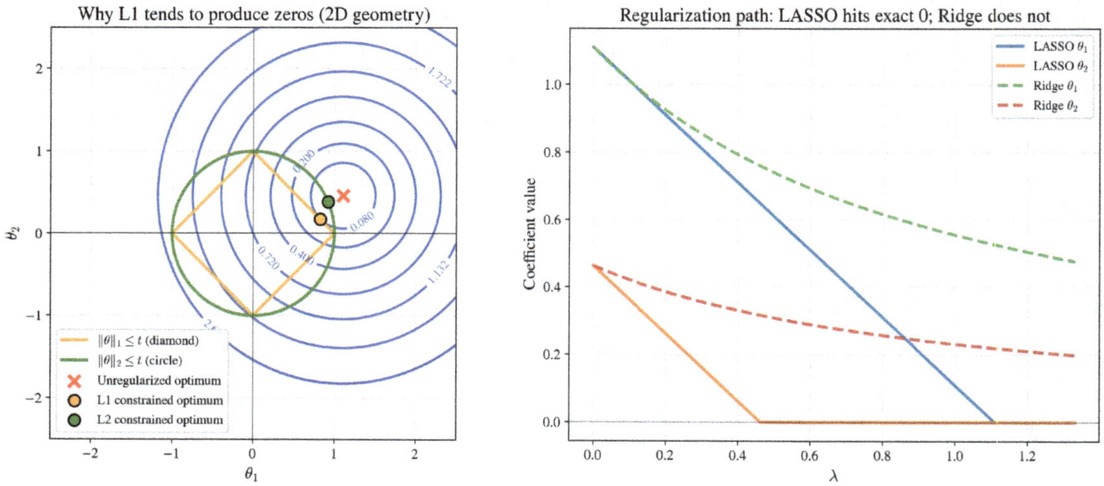

Figure 1.11: L1 vs L2 Regularization: Geometry and Regularization Paths

Figure 1.11 provides two complementary visualizations of why ℓ_1 regularization produces sparse solutions while ℓ_2 does not. The left panel illustrates the geometric difference: the blue contours represent the loss function (a quadratic bowl centered away from the origin), and the constraint sets are shown as a diamond (ℓ_1 norm) and a circle (ℓ_2 norm). As the loss contours expand, they first contact the diamond at its sharp corners (located on the coordinate axes), where one coordinate is exactly zero—this is the geometric source of sparsity. In contrast, the circle has no corners, so the optimal solution lies on a smooth arc where both coordinates are typically non-zero.

---

[5]**LASSO (Least Absolute Shrinkage and Selection Operator)**: A regularization technique that uses the ℓ_1 norm of parameters as a penalty term. The regularized objective is $\mathcal{L}(\theta) + \lambda\|\theta\|_1$, where $\theta \in \mathbb{R}^d$ is the parameter vector (e.g., weights in linear regression or neural networks), and $\lambda > 0$ is the regularization strength. Unlike ℓ_2 regularization (Ridge[^ridge]), LASSO produces sparse solutions by driving many parameters exactly to zero. This sparsity property makes LASSO particularly useful for feature selection, as it automatically identifies and eliminates irrelevant features. The geometric reason for sparsity is that the ℓ_1 unit ball has sharp corners at the coordinate axes, causing the optimal solution to lie on these corners where some coordinates are exactly zero.

[6]**Sparse solution (sparsity)**: A solution where most parameters are exactly zero, with only a few non-zero values. For example, if a model has 100 features, a sparse solution might have only 5 non-zero weights, with the remaining 95 being exactly zero. In contrast, a dense solution (typical of ℓ_2 regularization) has most or all parameters as non-zero, though some may be very small. Sparse solutions offer three key advantages: (1) automatic feature selection by identifying which features are truly relevant, (2) improved interpretability since only a few factors contribute to predictions, and (3) computational efficiency as operations involving zero parameters can be skipped or compressed. The geometric origin of sparsity in ℓ_1 regularization comes from the sharp corners of the ℓ_1 unit ball at coordinate axes, forcing solutions to lie on these corners where many coordinates are exactly zero.

The right panel shows the regularization paths: how the coefficients θ_1 and θ_2 change as the regularization parameter λ increases. For LASSO (solid lines), the coefficients follow piecewise linear paths that hit exact zero at finite λ values and remain zero thereafter—this is the signature behavior of sparsity. For Ridge regression[7] (dashed lines), the coefficients decay smoothly but never reach exactly zero, only approaching zero asymptotically as $\lambda \to \infty$. This visualization makes concrete the abstract geometric insight: the non-smooth geometry of the ℓ_1 ball (its sharp corners) translates directly into the piecewise-linear regularization paths that produce exact zeros, while the smooth geometry of the ℓ_2 ball produces smooth, non-sparse paths.

This geometric insight explains why ℓ_1 regularization (LASSO) is preferred when feature selection is needed—to identify which features are completely irrelevant—while ℓ_2 regularization (Ridge) is typically more robust when the primary goal is preventing overfitting without requiring sparsity.

1.4.1.2 Example: Transformer Attention in Hilbert Space

In machine learning, particularly in Transformer architectures[8], the attention mechanism operates naturally in a Hilbert space framework. Consider a sequence of token embeddings $\mathbf{x}_1, \dots, \mathbf{x}_n \in \mathbb{R}^d$, where each $\mathbf{x}_i$ is a vector in the embedding Hilbert space $\mathcal{H} \cong \mathbb{R}^d$ equipped with the standard inner product $\langle \mathbf{u}, \mathbf{v} \rangle = \mathbf{u}^\top \mathbf{v}$.

Notation in this example: Elsewhere in this book, vectors are **column vectors** $\mathbf{v} \in \mathbb{R}^{d\times 1}$ and we write $\mathbf{y} = A\mathbf{x}$. In Transformer papers and code, each token is often treated as a **row** $\mathbf{x}_i \in \mathbb{R}^{1\times d}$, with linear maps as right multiplication $\mathbf{q}_i = \mathbf{x}_i W_Q$ and batched matrices $Q \in \mathbb{R}^{n\times d_k}$ storing one token per row. Subscripts i, j here index **tokens**, not vector components. The column-vector form $\mathbf{q}_i = W_Q^\top \mathbf{x}_i$ (with $\mathbf{x}_i \in \mathbb{R}^{d\times 1}$) is equivalent.

Through linear transformations, we obtain query, key, and value vectors:

$$\mathbf{q}_i = \mathbf{x}_i W_Q, \quad \mathbf{k}_i = \mathbf{x}_i W_K, \quad \mathbf{v}_i = \mathbf{x}_i W_V,$$

where $W_Q, W_K \in \mathbb{R}^{d\times d_k}$ and $W_V \in \mathbb{R}^{d\times d_v}$ are learned weight matrices. All vectors $\mathbf{q}_i, \mathbf{k}_i, \mathbf{v}_i$ remain in finite-dimensional Hilbert spaces (subspaces of $\mathbb{R}^d$; see Definition: Linear Subspace).

Attention as inner product and projection: The attention mechanism computes similarity scores using inner products:

$$s_{ij} = \langle \mathbf{q}_i, \mathbf{k}_j \rangle = \mathbf{q}_i^\top \mathbf{k}_j.$$

The matrix $S = QK^\top \in \mathbb{R}^{n\times n}$ contains all pairwise inner products between query and key vectors: $S_{ij} = \langle \mathbf{q}_i, \mathbf{k}_j \rangle$. After scaling by $\frac{1}{\sqrt{d_k}}$ and applying softmax normalization, we obtain attention weights:

$$\alpha_{ij} = \frac{\exp(\mathbf{q}_i^\top \mathbf{k}_j / \sqrt{d_k})}{\sum_{t=1}^n \exp(\mathbf{q}_i^\top \mathbf{k}_t / \sqrt{d_k})}.$$

The attention output for the i-th token is then:

$$\mathbf{y}_i = \sum_{j=1}^n \alpha_{ij} \mathbf{v}_j.$$

[7]**Ridge regression**: A regularization technique that uses the ℓ_2 (Euclidean) norm of parameters as a penalty term. The regularized objective is $\mathcal{L}(\theta) + \lambda\|\theta\|_2^2$, where $\theta \in \mathbb{R}^d$ is the parameter vector (e.g., weights in linear regression or neural networks), and $\lambda > 0$ is the regularization strength. Ridge regression is equivalent to L2 regularization and L2 weight decay. Unlike LASSO, Ridge regression produces dense (non-sparse) solutions where parameters shrink smoothly toward zero but rarely reach exactly zero. This makes Ridge regression particularly useful when the primary goal is preventing overfitting without requiring feature selection.

[8]Notation matches Vaswani et al., *Attention Is All You Need* (NeurIPS 2017): token embeddings are rows of $X \in \mathbb{R}^{n\times d}$, with $Q = XW^Q$, $K = XW^K$, $V = XW^V$ (their Eq. (1)). Elsewhere in this book we use column vectors and $\mathbf{y} = A\mathbf{x}$.

From a geometric perspective, the attention mechanism can be understood as a projection-like operation in Hilbert space. The output $\mathbf{y}_i$ is always in the linear span of the value vectors: $\mathbf{y}_i \in \text{span}\{\mathbf{v}_1, \dots, \mathbf{v}_n\}$. The inner products $\langle \mathbf{q}_i, \mathbf{k}_j \rangle$ measure geometric similarity through direction alignment, and the softmax-normalized weights create a convex combination of values. This is analogous to orthogonal projection (see Equation (1.9)) $P_S(\mathbf{x}) = \sum_i \langle \mathbf{x}, \mathbf{e}_i \rangle \mathbf{e}_i$, but uses a non-orthogonal basis (the keys) and softmax normalization instead of strict orthogonality.

From a linear operator perspective, for fixed query and key matrices Q and K, the attention mechanism defines a linear operator $\mathcal{T}_{Q,K} : \mathbb{R}^{n \times d_v} \to \mathbb{R}^{n \times d_v}$ that maps value matrices to output matrices: $\mathcal{T}_{Q,K}(V) = \text{softmax}(QK^\top / \sqrt{d_k})V$. The attention matrix $A = \text{softmax}(QK^\top / \sqrt{d_k})$ is row-stochastic (each row sums to 1; see Chapter 14, Definition: Stochastic Matrix), which ensures that each output $\mathbf{y}_i$ is a convex combination of values and provides geometric stability.

Probabilistically, the softmax normalization creates a probability distribution over keys for each query. For fixed i, the attention weights $(\alpha_{i1}, \dots, \alpha_{in})$ form a probability distribution, and the output $\mathbf{y}_i = \sum_{j=1}^{n} \alpha_{ij} \mathbf{v}_j$ is the conditional expectation (see Chapter 13) of values under this distribution: $\mathbf{y}_i = \mathbb{E}[\mathbf{v}_J \mid \text{query } i]$, where J is a random index with distribution $\mathbb{P}(J = j \mid i) = \alpha_{ij}$. This probabilistic view connects attention to kernel methods and Gaussian processes, where similar weighted averaging operations appear.

This Hilbert space perspective unifies the geometric, algebraic, and probabilistic interpretations of attention: inner products measure similarity, the similarity matrix $S = QK^\top$ encodes geometric relationships, and the output is a geometrically-informed weighted projection of values onto a data-dependent subspace. The attention mechanism demonstrates how fundamental Hilbert space concepts—inner products, projections, and linear operators—naturally appear in deep learning architectures.

1.5 Mindmap

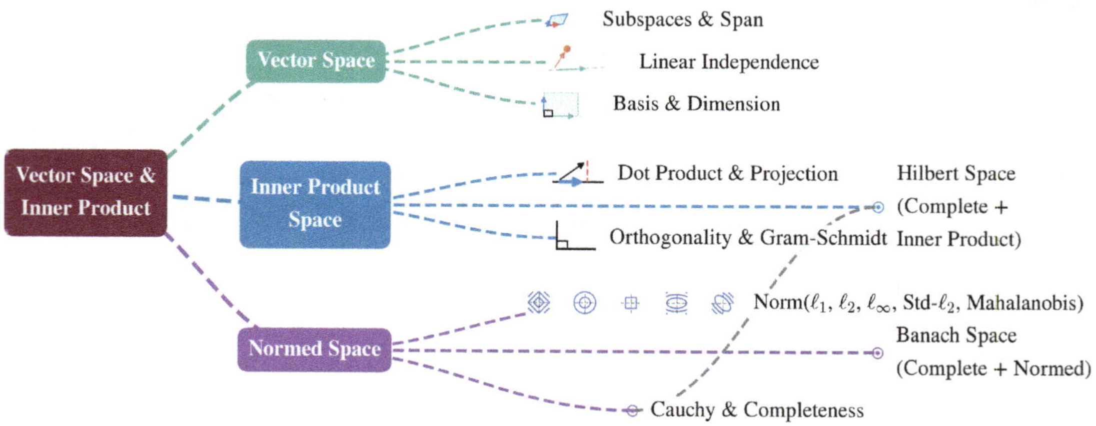

Figure 1.12: Mind map of vector space and inner product topics

1.6 Exercises

1.6.1 Testing Linear Independence of Vector Sets

For each of the following sets of vectors, determine whether they are linearly independent. Show your work.

1. $\mathbf{v}_0 = \begin{bmatrix} 1 \\ 2 \\ 3 \end{bmatrix}, \mathbf{v}_1 = \begin{bmatrix} 4 \\ 5 \\ 6 \end{bmatrix}, \mathbf{v}_2 = \begin{bmatrix} 7 \\ 8 \\ 9 \end{bmatrix}$
2. $\mathbf{v}_0 = \begin{bmatrix} 1 \\ 0 \\ 0 \end{bmatrix}, \mathbf{v}_1 = \begin{bmatrix} 0 \\ 1 \\ 0 \end{bmatrix}, \mathbf{v}_2 = \begin{bmatrix} 1 \\ 1 \\ 0 \end{bmatrix}$
3. $\mathbf{v}_0 = \begin{bmatrix} 1 \\ 1 \\ 1 \end{bmatrix}, \mathbf{v}_1 = \begin{bmatrix} 1 \\ -1 \\ 0 \end{bmatrix}, \mathbf{v}_2 = \begin{bmatrix} 0 \\ 1 \\ -1 \end{bmatrix}$

1.6.2 Verifying Linear Subspaces

For each of the following sets, determine whether it is a linear subspace of $\mathbb{R}^3$. Justify your answer by checking the three subspace conditions.

1. $W_1 = \left\{ \begin{bmatrix} x \\ y \\ z \end{bmatrix} \in \mathbb{R}^3 \middle| x + y + z = 0 \right\}$
2. $W_2 = \left\{ \begin{bmatrix} x \\ y \\ z \end{bmatrix} \in \mathbb{R}^3 \middle| x^2 + y^2 + z^2 = 1 \right\}$
3. $W_3 = \left\{ \begin{bmatrix} x \\ y \\ z \end{bmatrix} \in \mathbb{R}^3 \middle| x = 2y \text{ and } z = 0 \right\}$

1.6.3 Finding Basis and Dimension

Given the vectors $\mathbf{v}_0 = \begin{bmatrix} 1 \\ 2 \\ 1 \end{bmatrix}, \mathbf{v}_1 = \begin{bmatrix} 2 \\ 1 \\ 3 \end{bmatrix}, \mathbf{v}_2 = \begin{bmatrix} 3 \\ 3 \\ 4 \end{bmatrix}$:

1. Find a basis for $\text{span}\{\mathbf{v}_0, \mathbf{v}_1, \mathbf{v}_2\}$.
2. What is the dimension of this span?
3. Express $\mathbf{v}_2$ as a linear combination of the basis vectors from part 1.

1.6.4 Inner Product and Norm

Given $\mathbf{u} = \begin{bmatrix} 3 \\ 4 \end{bmatrix}$ and $\mathbf{v} = \begin{bmatrix} 1 \\ -2 \end{bmatrix}$:

1. Compute the inner product $\langle \mathbf{u}, \mathbf{v} \rangle$.
2. Compute the norms $\|\mathbf{u}\|_2$ and $\|\mathbf{v}\|_2$.
3. Verify the Cauchy–Schwarz inequality: $|\langle \mathbf{u}, \mathbf{v} \rangle| \leq \|\mathbf{u}\|_2 \cdot \|\mathbf{v}\|_2$.

1.6.5 Orthogonality and Projection

Given $\mathbf{u} = \begin{bmatrix} 1 \\ 2 \\ 3 \end{bmatrix}$ and $\mathbf{v} = \begin{bmatrix} 1 \\ 1 \\ -1 \end{bmatrix}$:

1. Determine whether $\mathbf{u}$ and $\mathbf{v}$ are orthogonal.
2. Compute the unit vector $\hat{\mathbf{v}} = \frac{\mathbf{v}}{\|\mathbf{v}\|_2}$.
3. Compute the projection $\text{Proj}_{\hat{\mathbf{v}}}(\mathbf{u})$.
4. Compute the residual $\mathbf{R} = \mathbf{u} - \text{Proj}_{\hat{\mathbf{v}}}(\mathbf{u})$.

1.6.6 Gram–Schmidt Process

Apply the Gram–Schmidt algorithm to construct an orthonormal basis $\{\mathbf{e}_0, \mathbf{e}_1, \mathbf{e}_2\}$ from the following linearly independent vectors:

$$\mathbf{v}_0 = \begin{bmatrix} 1 \\ 1 \\ 0 \end{bmatrix}, \quad \mathbf{v}_1 = \begin{bmatrix} 1 \\ 0 \\ 1 \end{bmatrix}, \quad \mathbf{v}_2 = \begin{bmatrix} 0 \\ 1 \\ 1 \end{bmatrix}.$$

Show all intermediate steps: compute each residual $\mathbf{R}_i$ and the corresponding orthonormal vector $\mathbf{e}_i$ for $i = 0, 1, 2$.

1.6.7 Vector Operations and Norms

Let $\mathbf{u} = \begin{bmatrix} 2 \\ -1 \\ 3 \end{bmatrix}$ and $\mathbf{v} = \begin{bmatrix} 1 \\ 4 \\ -2 \end{bmatrix}$.

1. Compute $\mathbf{u} + \mathbf{v}$ and $3\mathbf{u} - 2\mathbf{v}$.
2. Compute $\|\mathbf{u}\|_1$, $\|\mathbf{u}\|_2$, and $\|\mathbf{u}\|_\infty$.
3. Compute $\|\mathbf{v}\|_1$, $\|\mathbf{v}\|_2$, and $\|\mathbf{v}\|_\infty$.
4. Verify the triangle inequality: $\|\mathbf{u} + \mathbf{v}\|_2 \leq \|\mathbf{u}\|_2 + \|\mathbf{v}\|_2$.

1.6.8 Computing Different Norms

For the vector $\mathbf{w} = \begin{bmatrix} 3 \\ -4 \\ 0 \\ 5 \end{bmatrix}$:

1. Compute $\|\mathbf{w}\|_1$ (Manhattan norm).
2. Compute $\|\mathbf{w}\|_2$ (Euclidean norm).
3. Compute $\|\mathbf{w}\|_\infty$ (Chebyshev norm).
4. Compute $\|\mathbf{w}\|_F$ (Frobenius norm) and verify it equals $\|\mathbf{w}\|_2$.

1.6.9 3D Norm Visualization and Formulas

Figure 1.13 shows the 3D unit balls (surfaces of constant norm value) for different norms in $\mathbb{R}^3$:

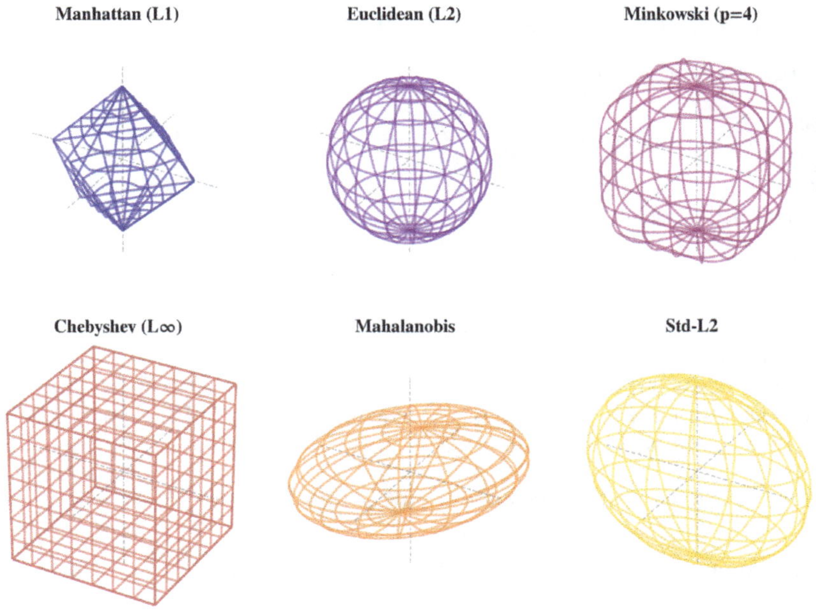

Figure 1.13: 3D Visualization of Different Norms

Compare with the 2D case, 3D unit balls are the natural extensions: they maintain the same geometric properties (sharp edges, smoothness, symmetry) but in one higher dimension.

For a vector $\mathbf{v} = \begin{bmatrix} v_1 \\ v_2 \\ v_3 \end{bmatrix} \in \mathbb{R}^3$, write the explicit formula for each norm:

1. **Manhattan (L1) norm**: The unit ball is an octahedron (8 triangular faces). Write $\|\mathbf{v}\|_1$.
2. **Euclidean (L2) norm**: The unit ball is a sphere. Write $\|\mathbf{v}\|_2$.
3. **Minkowski (Lp) norm**: The unit ball is a rounded cube (sqube) when $p = 4$. Write $\|\mathbf{v}\|_p$ for $p = 4$.
4. **Chebyshev (L∞) norm**: The unit ball is a cube (6 square faces). Write $\|\mathbf{v}\|_\infty$.
5. **Mahalanobis (M-norm)**: The unit ball is a rotated ellipsoid. Write $\|\mathbf{v}\|_M$ using a 3×3 positive definite covariance matrix Σ.
6. **Standardized Euclidean (Std-L2) norm**: The unit ball is an axis-aligned ellipsoid. Write $\|\mathbf{v}\|$ using scaling factors $\sigma_1, \sigma_2, \sigma_3 > 0$.

1.6.10 Outer Product Applications

Let $\mathbf{a} = \begin{bmatrix} 1 \\ 2 \\ 3 \end{bmatrix}$ and $\mathbf{b} = \begin{bmatrix} 4 \\ 5 \end{bmatrix}$.

1. Compute the outer product $\mathbf{a}\mathbf{b}^\top$.
2. Verify that the rank of $\mathbf{a}\mathbf{b}^\top$ is 1.
3. Compute the energy of the matrix $\mathbf{a}\mathbf{b}^\top$ (sum of squares of all elements).
4. Express $\mathbf{a}\mathbf{b}^\top$ as a sum of rank-1 matrices using the SVD perspective.

1.6.11 Dot Product vs Cosine Similarity

Given three word embedding vectors:

- $\mathbf{w}_1 = \begin{bmatrix} 3 \\ 4 \end{bmatrix}$ (word: "machine")
- $\mathbf{w}_2 = \begin{bmatrix} 6 \\ 8 \end{bmatrix}$ (word: "learning")
- $\mathbf{w}_3 = \begin{bmatrix} 4 \\ 3 \end{bmatrix}$ (word: "algorithm")

1. Compute the dot products $\langle \mathbf{w}_1, \mathbf{w}_2 \rangle$ and $\langle \mathbf{w}_1, \mathbf{w}_3 \rangle$.
2. Compute the cosine similarities between $\mathbf{w}_1$ and $\mathbf{w}_2$, and between $\mathbf{w}_1$ and $\mathbf{w}_3$.
3. Compare the results: which pair has a larger dot product? Which pair has a larger cosine similarity? Explain why they differ.
4. In a Transformer attention mechanism, query and key vectors are $\mathbf{q} = \begin{bmatrix} 2 \\ 1 \end{bmatrix}$ and $\mathbf{k}_1 = \begin{bmatrix} 4 \\ 2 \end{bmatrix}$, $\mathbf{k}_2 = \begin{bmatrix} 1 \\ 2 \end{bmatrix}$. Compute the attention scores $s_1 = \mathbf{q}^\top \mathbf{k}_1$ and $s_2 = \mathbf{q}^\top \mathbf{k}_2$. Why does Transformer use dot product instead of cosine similarity for attention?

1.6.12 Norms and Inner Products as Foundations of Banach and Hilbert Spaces

1. **Verifying norm axioms for ℓ_2 norm**: Show that the ℓ_2 norm $\|\mathbf{v}\|_2 = \sqrt{\sum_i v_i^2}$ satisfies all four norm axioms (non-negativity, definiteness, homogeneity, and triangle inequality) for vectors in $\mathbb{R}^d$. Explain why this makes $\mathbb{R}^d$ with the ℓ_2 norm a Banach space.

2. **Inner product induces norm**: Given the standard inner product $\langle \mathbf{u}, \mathbf{v} \rangle = \mathbf{u}^\top \mathbf{v}$ on $\mathbb{R}^d$, show that the induced norm $\|\mathbf{v}\| = \sqrt{\langle \mathbf{v}, \mathbf{v} \rangle}$ equals the ℓ_2 norm. Explain why this makes $\mathbb{R}^d$ with the standard inner product a Hilbert space.

3. **Cauchy sequences and completeness**: Consider the sequence $\{\mathbf{v}_n\}$ in $\mathbb{R}^2$ where $\mathbf{v}_n = \begin{bmatrix} \frac{1}{n} \\ \frac{1}{n^2} \end{bmatrix}$.

   - Show that this sequence is Cauchy with respect to the ℓ_2 norm (i.e., for any $\epsilon > 0$, there exists N such that $\|\mathbf{v}_m - \mathbf{v}_n\|_2 < \epsilon$ for all $m, n > N$).
   - Find the limit $\lim_{n\to\infty} \mathbf{v}_n$ and verify it is in $\mathbb{R}^2$.
   - Explain why this demonstrates that $\mathbb{R}^2$ is complete, making it both a Banach space (with the ℓ_2 norm) and a Hilbert space (with the standard inner product).

4. **Machine learning connection**: In optimization, we often work in function spaces. Explain why the concept of a norm (and thus Banach space) is essential for:

   - Defining convergence of optimization algorithms
   - Measuring distances between parameter matrices
   - Understanding regularization (e.g., L1 and L2 weight decay)

Chapter 2

Matrix

Matrix operations and linear transformations: from vectors to matrices

"The study of mathematics, like the Nile, begins in minuteness but ends in magnificence."

— Charles Caleb Colton

- ▷ `Prod.` $[C]_{mn} = \sum_{i=0}^{d-1} a_{mi} b_{in}$ where $C = AB \in \mathbb{R}^{M\times N}$, $A \in \mathbb{R}^{M\times d}$, $B \in \mathbb{R}^{d\times N}$
- ▷ `Hadamard prod.` $[A \odot B]_{ij} = a_{ij} b_{ij}$ for $A, B \in \mathbb{R}^{M\times N}$
- ▷ `Trans.` $(A^\top)^\top = A$; $[A^\top]_{ij} = a_{ji}$; $(A^\top)^{-1} = (A^{-1})^\top$ for invertible A
- ▷ `Inverse` $AA^{-1} = A^{-1}A = I$ for invertible square $A \in \mathbb{R}^{N\times N}$
- ▷ `Prod. trans./inv.` $(AB)^\top = B^\top A^\top$; $(AB)^{-1} = B^{-1}A^{-1}$ for invertible A, B
- ▷ `Diag. matrix` $A = \mathrm{diag}(d_0, \ldots, d_{N-1})$ where $a_{ij} = 0$ for $i \neq j$; $a_{ii} = d_i$
- ▷ `Diag. power` $D^k = \mathrm{diag}(d_0^k, \ldots, d_{N-1}^k)$ for diagonal D and integer k ($d_i \neq 0$ if $k < 0$)
- ▷ `Diag. det./inv./trace` $\det(D) = \prod_{i=0}^{N-1} d_i$; $D^{-1} = \mathrm{diag}(1/d_0, \ldots)$; $\mathrm{tr}(D) = \sum_{i=0}^{N-1} d_i$
- ▷ `Singular vals` $\sigma_i(A) = \sqrt{\lambda_i}$ for $i = 0, \ldots$ where λ_i are eigenvals. of $A^\top A$
- ▷ `Spec. norm/radius` $||A||_2 = \sigma_0(A)$; $\rho(A) = \max\{|\lambda_i|\}$ for $A \in \mathbb{R}^{N\times N}$
- ▷ `Proj. matrix` $P = X(X^\top X)^{-1}X^\top \in \mathbb{R}^{M\times M}$ where $X \in \mathbb{R}^{M\times N}$ has full column rank
- ▷ `Rejection` $R = A - PA$ where P projects A onto $\mathrm{span}(X)$; $X^\top R = \mathbf{0}$
- ▷ `Basis change` $\mathbf{x} = V\mathbf{z}$ where $\mathbf{x}/\mathbf{z}$ are in old/new basis, V is transition matrix.
- ▷ `Orthonorm. basis` $V^\top V = I$; $V^{-1} = V^\top$ when V is square
- ▷ `Induced p-norm` $||A||_p = \sup_{\mathbf{x}\neq 0} \frac{||A\mathbf{x}||_p}{||\mathbf{x}||_p}$ for $A \in \mathbb{R}^{M\times N}$
- ▷ `Cond. num.` $\kappa_p(A) = ||A||_p \cdot ||A^{-1}||_p$ for invertible A; $\kappa(A) = \kappa_2(A) = ||A||_2 \cdot ||A^{-1}||_2$
- ▷ `Trace` $\mathrm{tr}(A) = \sum_{n=0}^{N-1} a_{nn}$ for square $A \in \mathbb{R}^{N\times N}$
- ▷ `F-norm` $\|A\|_F^2 = \sum_{m=0}^{M-1}\sum_{n=0}^{N-1} a_{mn}^2 = \mathrm{tr}(A^\top A) = \sum_{i=0}^{N-1} \lambda_i = \sum_{i=0}^{\min(M,N)-1} \sigma_i(A)^2$
- ▷ `Rank of outer prod.` $\mathrm{rank}(\mathbf{u}\mathbf{v}^\top) = 1$ for nonzero $\mathbf{u} \in \mathbb{R}^{M\times 1}$, $\mathbf{v} \in \mathbb{R}^{N\times 1}$
- ▷ `Lin. reg` $\hat{\mathbf{y}} = X\hat{\theta}$; $\hat{\theta} = (X^\top X)^{-1}X^\top \mathbf{y}$; $\mathcal{L} = \frac{1}{2M}||X\theta - \mathbf{y}||^2$; $\nabla\mathcal{L} = \frac{1}{M}X^\top(X\theta - \mathbf{y})$

2.1 Matrix Representation

A matrix $A \in \mathbb{R}^{M \times N}$ is represented as

$$A = \begin{bmatrix} a_{00} & a_{01} & \cdots & a_{0(N-1)} \\ a_{10} & a_{11} & \cdots & a_{1(N-1)} \\ \vdots & \vdots & \ddots & \vdots \\ a_{(M-1)0} & a_{(M-1)1} & \cdots & a_{(M-1)(N-1)} \end{bmatrix}.$$

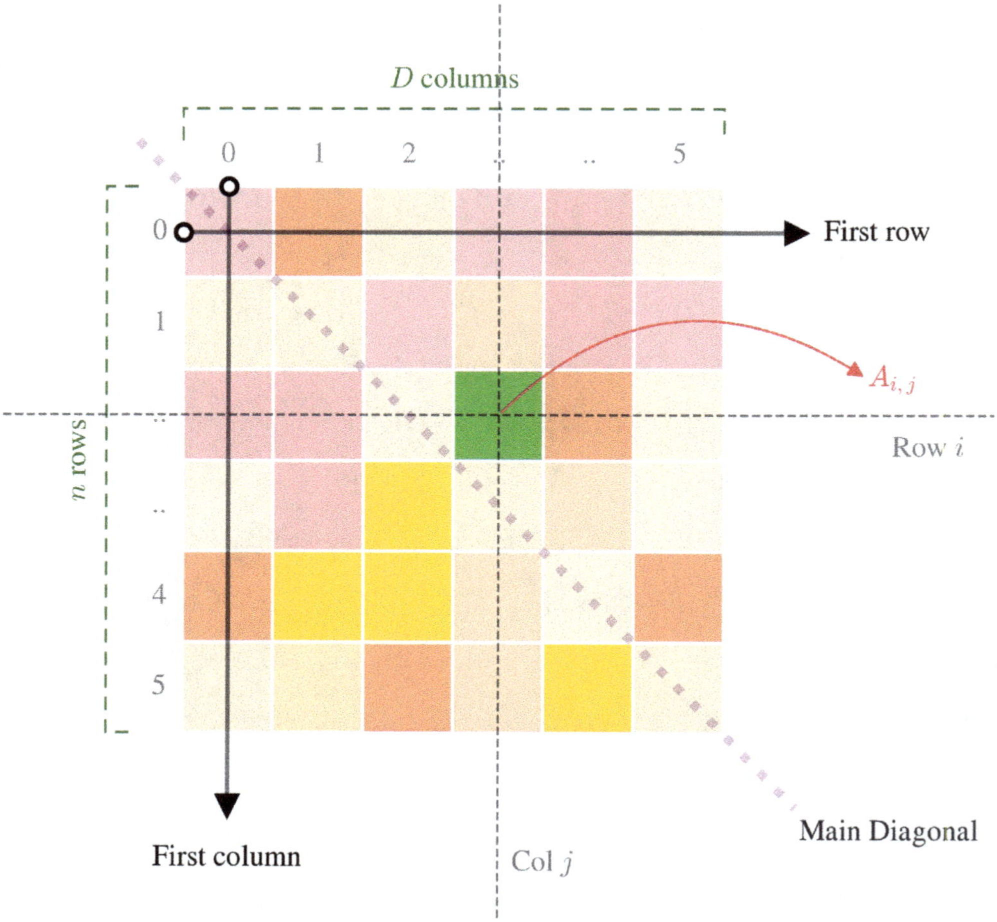

Figure 2.1: Matrix Structure

Figure 2.1 illustrates the fundamental structure of a matrix A with n rows and D columns. Each cell represents an element $A_{i,j}$ located at the intersection of row i and column j. The matrix indexing convention begins with row 0 and column 0, as indicated by the arrows marking the first row and first column. The highlighted green cell shows a specific entry at position (i, j), demonstrating how individual elements are referenced.

The purple dotted diagonal line identifies the **main diagonal** (or **principal diagonal**), which consists of elements a_{ii} where the row index equals the column index. The **anti-diagonal** (or **secondary diagonal**) consists of elements where $i + j = N - 1$.

2.1.1 Definition: Matrix as Horizontal Stack of Column Vectors

Given column vectors $\mathbf{a}_0, \dots, \mathbf{a}_{N-1} \in \mathbb{R}^{M\times 1}$ the matrix

$$A = \begin{bmatrix}\mathbf{a}_0 & \cdots & \mathbf{a}_{N-1}\end{bmatrix} \in \mathbb{R}^{M\times N}$$

is formed by horizontally stacking these vectors.

2.2 Transpose

The transpose operation flips a matrix along its main diagonal, converting rows into columns and columns into rows. For a matrix $A \in \mathbb{R}^{M\times N}$, the transpose $A^\top$ is an $N \times M$ matrix where the element at position (i, j) in A becomes the element at position (j, i) in $A^\top$. This operation is fundamental in linear algebra, appearing frequently in matrix operations, inner products, and the formulation of symmetric matrices.

The transpose of $A \in \mathbb{R}^{M\times N}$ is

$$A^\top = \begin{bmatrix} a_{00} & a_{10} & \cdots & a_{(M-1)0} \\ a_{01} & a_{11} & \cdots & a_{(M-1)1} \\ \vdots & \vdots & \ddots & \vdots \\ a_{0(N-1)} & a_{1(N-1)} & \cdots & a_{(M-1)(N-1)} \end{bmatrix} \in \mathbb{R}^{N\times M}.$$

2.2.1 Theorem: Double Transpose

$$(A^\top)^\top = A \tag{2.1}$$

2.3 Summation and Scalar Multiplication

2.3.1 Definition: Matrix Addition

Let $U, V \in \mathbb{R}^{M\times N}$. Then

$$U + V = \begin{bmatrix} u_{00} + v_{00} & \cdots & u_{0(N-1)} + v_{0(N-1)} \\ \vdots & \ddots & \vdots \\ u_{(M-1)0} + v_{(M-1)0} & \cdots & u_{(M-1)(N-1)} + v_{(M-1)(N-1)} \end{bmatrix}.$$

2.3.2 Definition: Scalar Multiplication

Let $U \in \mathbb{R}^{M\times N}$ and $c \in \mathbb{R}$. Then

$$cU = \begin{bmatrix} cu_{00} & \cdots & cu_{0(N-1)} \\ \vdots & \ddots & \vdots \\ cu_{(M-1)0} & \cdots & cu_{(M-1)(N-1)} \end{bmatrix}.$$

2.4 Matrix-Vector Multiplication - Two Views

2.4.1 Definition: Matrix-Vector Multiplication

For a matrix $A \in \mathbb{R}^{M\times d}$ and a vector $\mathbf{x} \in \mathbb{R}^{d\times 1}$, the **matrix-vector product** (or **matrix-vector multiplication**) $A\mathbf{x}$ is defined as the vector $\mathbf{y} \in \mathbb{R}^{M\times 1}$ where each component y_i is computed as the dot product of the i-th row of A and the vector $\mathbf{x}$.

The matrix-vector multiplication is only defined when the number of columns of A equals the dimension of $\mathbf{x}$ (both equal to d in this case). The resulting vector $A\mathbf{x}$ has dimension M, where M is the number of rows of A.

A matrix A represents a **linear mapping** (or linear transformation) from $\mathbb{R}^{d\times 1}$ to $\mathbb{R}^{M\times 1}$. The matrix-vector product $A\mathbf{x}$ maps a vector $\mathbf{x} \in \mathbb{R}^{d\times 1}$ to a vector $A\mathbf{x} \in \mathbb{R}^{M\times 1}$. This perspective is fundamental in understanding linear layers in neural networks, attention mechanisms, and other linear transformations in machine learning. For the geometric meaning of linear transformations (including rotations, scaling, reflections, and their effects on areas/volumes), see Chapter 6, Section 6.1.

2.4.2 View 1 (Row-wise)

$$A\mathbf{x} = \begin{bmatrix} \mathbf{a}_0^* \\ \mathbf{a}_1^* \\ \vdots \\ \mathbf{a}_{M-1}^* \end{bmatrix} \mathbf{x} = \begin{bmatrix} \mathbf{a}_0^*\mathbf{x} \\ \mathbf{a}_1^*\mathbf{x} \\ \vdots \\ \mathbf{a}_{M-1}^*\mathbf{x} \end{bmatrix} \in \mathbb{R}^{M\times 1}.$$

Each entry $\mathbf{a}_i^*\mathbf{x}$ is a **dot product** (inner product) between the i-th row of A (as a row vector $\mathbf{a}_i^*$) and the column vector $\mathbf{x}$. The result is a scalar for each row, giving the i-th component of the output vector.

Examples

1. Feature extraction in AI
2. Resource consumption in economics

2.4.3 View 2 (Linear combination of columns)

$$A\mathbf{x} = \begin{bmatrix} \mathbf{a}_0 & \mathbf{a}_1 & \cdots & \mathbf{a}_{d-1} \end{bmatrix} \begin{bmatrix} x_0 \\ x_1 \\ \vdots \\ x_{d-1} \end{bmatrix} = \sum_{i=0}^{d-1} \mathbf{a}_i x_i.$$

Examples

1. **Classification head**: In a neural network's final layer, each column $\mathbf{a}_i$ represents the learned weights/pattern for class i, and x is the feature vector from previous layers. The output $A\mathbf{x}$ gives class scores/logits, where each component is the dot product between the class pattern and the input features.

2. **Attention mechanism**: In transformers, each column $\mathbf{a}_i$ represents a key vector, and x_i is the attention weight. The matrix-vector product computes the weighted sum of key vectors based on attention scores.

3. **Word embeddings**: In natural language processing, columns represent word embedding vectors, and x_i are the coefficients from a context vector. The result is a weighted combination of embeddings representing semantic meaning.

2.5 Matrix Multiplication - Four Views

2.5.1 Definition: Matrix Multiplication

For matrices $A \in \mathbb{R}^{M\times d}$ and $B \in \mathbb{R}^{d\times N}$, the **matrix product** (or **matrix multiplication**) AB is defined as the matrix $C \in \mathbb{R}^{M\times N}$ where each entry $[C]_{mn}$ is computed as the dot product of the m-th row of A and the n-th column of B.

The matrix multiplication is only defined when the number of columns of A equals the number of rows of B (both equal to d in this case). The resulting matrix $C = AB$ has dimensions $M \times N$, where M is the number of rows of A and N is the number of columns of B.

2.5.2 Theorem: Matrix Multiplication Properties

1. $A(BC) = (AB)C$ (Associativity)
2. $A(B + C) = AB + AC$ (Distributivity)
3. $(AB)^\top = B^\top A^\top$ (Transpose of Product)

> In general, $AB \neq BA$ (matrix multiplication is non-commutative). This has important implications in machine learning: in **neural networks**, layer order matters ($W_2W_1\mathbf{x} \neq W_1W_2\mathbf{x}$); in **attention mechanisms**, $QK^\top$ has different dimensions than $K^\top Q$; in **optimization**, operation order affects gradient flow. Special cases that commute: two diagonal matrices commuting with each other; identity matrix, and scalar multiples commuting with any general matrix A.

2.5.3 Computation Rules

The matrix product $C = AB \in \mathbb{R}^{M\times N}$ has entries

$$[C]_{mn} = \sum_{i=0}^{d-1} a_{mi}b_{in}, \quad m = 0, \dots, M-1, \quad n = 0, \dots, N-1 \tag{2.2}$$

In matrix form:

$$C = \left[\sum_{i=0}^{d-1} a_{mi}b_{in}\right]_{M\times N}.$$

2.5.4 View 1 (Row × column)

Let $C = AB$. Then

$$C = \begin{bmatrix} \mathbf{a}_0^* \\ \vdots \\ \mathbf{a}_{M-1}^* \end{bmatrix} \begin{bmatrix} \mathbf{b}_0 & \cdots & \mathbf{b}_{N-1} \end{bmatrix} = \left[\mathbf{a}_m^*\mathbf{b}_n\right]_{M\times N}.$$

Each entry (m, n) in the output matrix C is computed as a **dot product** between row m of A (as $\mathbf{a}_m^*$) and column n of B (as $\mathbf{b}_n$). That is, $[C]_{m,n} = \mathbf{a}_m^*\mathbf{b}_n$. This means each entry is computed independently, one at a time.

Geometric interpretation: Matrix multiplication AB represents the **composition of transformations**: AB applies transformation B (from $\mathbb{R}^{N\times 1}$ to $\mathbb{R}^{d\times 1}$) followed by transformation A (from $\mathbb{R}^{d\times 1}$ to $\mathbb{R}^{M\times 1}$), mapping vectors from $\mathbb{R}^{N\times 1}$ to $\mathbb{R}^{M\times 1}$ in a single step. This composition view is fundamental in deep learning, where multiple linear layers are composed to form complex transformations. For the geometric interpretation of matrix multiplication (including how it affects areas, volumes, and orientations), see Chapter 6, Section 6.1.

2.5.5 View 2 (Column-wise)

$$C = \begin{bmatrix} A\mathbf{b}_0 & A\mathbf{b}_1 & \cdots & A\mathbf{b}_{N-1} \end{bmatrix}.$$

Geometric interpretation: Each column of AB is the result of applying transformation A to the corresponding column of B (i.e., $[AB]_{:,n} = A\mathbf{b}_n$). This column-wise transformation view shows that matrix multiplication transforms each column vector of B independently through the linear transformation represented by A.

2.5.6 View 3 (Row-wise)

$$C = \begin{bmatrix} \mathbf{a}_0^* B \\ \mathbf{a}_1^* B \\ \vdots \\ \mathbf{a}_{M-1}^* B \end{bmatrix}.$$

Geometric interpretation: Each row of AB is the result of applying transformation B to the corresponding row of A (i.e., $[AB]_{m,:} = \mathbf{a}_m^* B$). This row-wise transformation view shows that matrix multiplication transforms each row vector of A independently through the linear transformation represented by B.

2.5.7 View 4 (Outer-product view)

$$C = \sum_{i=0}^{d-1} \mathbf{a}_i \mathbf{b}_i^* \tag{2.3}$$

The summation is over the inner dimension d (the number of columns of A and rows of B), not the outer dimension N (the number of columns of B). This is crucial for understanding matrix multiplication: the outer product view $AB = \sum_{i=0}^{d-1} \mathbf{a}_i \mathbf{b}_i^\top$ shows that each term $\mathbf{a}_i \mathbf{b}_i^\top$ is a rank-1 matrix, and the sum combines these rank-1 components.

> In deep learning, this perspective helps understand how linear layers decompose transformations into rank-1 components, which is fundamental for low-rank approximations and matrix factorization techniques.

2.5.7.1 Example: Linear Regression

Let $X = \begin{bmatrix} \mathbf{x}^{(0)\top} \\ \mathbf{x}^{(1)\top} \\ \vdots \\ \mathbf{x}^{(N-1)\top} \end{bmatrix} \in \mathbb{R}^{N \times d}$ and $\mathbf{y} = \begin{bmatrix} y^{(0)} \\ y^{(1)} \\ \vdots \\ y^{(N-1)} \end{bmatrix} \in \mathbb{R}^N$.

The matrix X is called the **design matrix** (or **data matrix**). It contains N data samples (rows) and d features (columns). Each row $\mathbf{x}^{(n)\top}$ represents one data point, and each column represents one feature across all samples.

The linear regression model assumes

$$\mathbf{y} = X\theta + \epsilon \tag{2.4}$$

where $\theta \in \mathbb{R}^d$ is the parameter vector and $\epsilon \in \mathbb{R}^N$ is the error term. Equivalently, for each sample $n = 0, \dots, N-1$:

$$y^{(n)} = \mathbf{x}^{(n)\top}\theta + \epsilon^{(n)}.$$

The least squares objective is to minimize the **squared error** (SE):

$$\mathcal{L}(\theta) = ||X\theta - \mathbf{y}||_2^2 = (X\theta - \mathbf{y})^\top (X\theta - \mathbf{y}) \tag{2.5}$$

This is the sum of squared errors, not the mean squared error (MSE). MSE would be $\frac{1}{N}||X\theta - \mathbf{y}||_2^2$. However, minimizing SE is equivalent to minimizing MSE (since N is constant), so both yield the same optimal solution.

Expanding the objective:

$$\mathcal{L}(\theta) = \theta^\top X^\top X\theta - 2\theta^\top X^\top \mathbf{y} + \mathbf{y}^\top \mathbf{y}.$$

Taking the gradient with respect to θ:

$$\nabla_\theta \mathcal{L}(\theta) = 2X^\top X\theta - 2X^\top \mathbf{y}.$$

The computation of this gradient uses matrix calculus rules (e.g., $\nabla_\theta(\theta^\top X^\top X\theta) = 2X^\top X\theta$), which are formally covered in Chapter 11.

Setting the gradient to zero gives the **normal equations**:

$$X^\top X\theta = X^\top \mathbf{y} \tag{2.6}$$

The normal equations are a system of linear equations derived from minimizing the least squares objective. They provide the necessary conditions for the optimal parameter vector θ that minimizes $\mathcal{L}(\theta)$.

Assuming $X^\top X$ is invertible (the definition of matrix inverse is introduced later in the "Matrix Inverse" section), the optimal estimator for linear regression is

$$\hat{\theta} = (X^\top X)^{-1}X^\top \mathbf{y} = \left(\sum_{n=0}^{N-1} \mathbf{x}^{(n)}\mathbf{x}^{(n)\top}\right)^{-1} \sum_{n=0}^{N-1} \mathbf{x}^{(n)}y^{(n)} \tag{2.7}$$

Common Pitfalls 1(Scalar Algebra vs. Linear Algebra)

Square of matrices: If square matrices A and B satisfy $A^2 = B^2$, we **cannot** conclude that $A = \pm B$. For non-square matrices A, $A^\top A$ or $AA^\top$ can be considered as the "square" of A. If non-square matrices A and B satisfy $A^\top A = B^\top B$ or $AA^\top = BB^\top$, we still **cannot** conclude that $A = \pm B$.

Square of sum: In scalar algebra, $(a+b)^2 = a^2 + 2ab + b^2$. For square matrices A and B, the square of their sum expands as $(A+B)^2 = (A+B)(A+B) = A^2 + AB + BA + B^2$, where AB and BA **cannot** be combined in general. For non-square matrices, $(A+B)^\top(A+B) = A^\top A + A^\top B + B^\top A + B^\top B$, which is different from $(A+B)(A+B)^\top = AA^\top + AB^\top + BA^\top + BB^\top$.

Matrix equality: In scalar algebra, if $a \neq 0$ and $ab = ac$, we can derive $a(b-c) = 0$ and conclude $b = c$. However, in matrix multiplication, if $A \neq \mathbf{0}$ and $AB = AC$, we can derive $A(B-C) = \mathbf{0}$, but we **cannot** directly conclude that $B = C$.

Zero matrix: In scalar algebra, if $ab = 0$, we know that $a = 0$ or $b = 0$. However, if $AB = \mathbf{0}$, we **cannot** conclude that $A = \mathbf{0}$ or $B = \mathbf{0}$.

Product may be a scalar: Matrix products can result in a scalar (a 1×1 matrix). Examples include $\mathbf{x}^\top \mathbf{y}$, $\mathbf{x}^\top \mathbf{x} = \|\mathbf{x}\|_2^2$, $\mathbf{1}^\top \mathbf{x}$, $\mathbf{x}^\top Q\mathbf{x}$, and $\mathbf{v}^\top \Sigma \mathbf{v}$. A 1×1 matrix is equal to its transpose. When encountering scalar matrix products, consider whether they represent some form of "distance" or norm.

2.6 Hadamard Product (Element-wise Multiplication)

2.6.1 Definition: Hadamard Product

Let $A, B \in \mathbb{R}^{M\times N}$ be matrices of the same dimensions. The **Hadamard product** (also called **element-wise product** or **entry-wise product**) of A and B, denoted $A \odot B$, is defined as:

$$A \odot B = \begin{bmatrix} a_{00}b_{00} & a_{01}b_{01} & \cdots & a_{0(N-1)}b_{0(N-1)} \\ a_{10}b_{10} & a_{11}b_{11} & \cdots & a_{1(N-1)}b_{1(N-1)} \\ \vdots & \vdots & \ddots & \vdots \\ a_{(M-1)0}b_{(M-1)0} & a_{(M-1)1}b_{(M-1)1} & \cdots & a_{(M-1)(N-1)}b_{(M-1)(N-1)} \end{bmatrix} \in \mathbb{R}^{M\times N}.$$

Entry-wise, the Hadamard product is:

$$[A \odot B]_{ij} = a_{ij}b_{ij}, \quad i = 0, \dots, M-1, \quad j = 0, \dots, N-1.$$

2.6.2 Properties of Hadamard Product

1. **Commutativity**: $A \odot B = B \odot A$.
2. **Associativity**: $(A \odot B) \odot C = A \odot (B \odot C)$.
3. **Distributivity over addition**: $A \odot (B + C) = A \odot B + A \odot C$.
4. **Identity element**: $A \odot \mathbf{1}_{M \times N} = A$, where $\mathbf{1}_{M \times N}$ is the matrix of all ones (same dimensions as A).
5. **Zero element**: $A \odot \mathbf{0} = \mathbf{0}$, where $\mathbf{0}$ is the zero matrix (same dimensions as A).
6. **Scalar multiplication compatibility**: $c(A \odot B) = (cA) \odot B = A \odot (cB)$ for any scalar c.

The Hadamard product (denoted by `\odot` in LaTeX) is fundamentally different from standard matrix multiplication: it requires matrices of identical dimensions and multiplies corresponding entries, whereas standard matrix multiplication AB requires compatible dimensions (number of columns of A equals number of rows of B) and computes dot products. The Hadamard product is commutative ($A \odot B = B \odot A$) and associative ($(A \odot B) \odot C = A \odot (B \odot C)$), unlike standard matrix multiplication which is non-commutative.

In programming languages like NumPy, the Hadamard product is performed using the `*` operator (for arrays) or the `numpy.multiply()` function. In machine learning, the Hadamard product appears in attention mechanisms (element-wise masking), activation functions (element-wise application), gating mechanisms, and various element-wise operations in neural networks.

2.7 Diagonal Matrices

2.7.1 Definition: Diagonal Matrix

$A \in \mathbb{R}^{N \times N}$ is **diagonal** if

$$a_{ij} = 0 \quad \text{for } i \neq j.$$

That is, all off-diagonal entries are zero. A diagonal matrix has the form:

$$A = \begin{bmatrix} a_{00} & 0 & \cdots & 0 \\ 0 & a_{11} & \cdots & 0 \\ \vdots & \vdots & \ddots & \vdots \\ 0 & 0 & \cdots & a_{(N-1)(N-1)} \end{bmatrix} = \operatorname{diag}(a_{00}, a_{11}, \dots, a_{(N-1)(N-1)}).$$

Notation: We often write $\operatorname{diag}(d_0, d_1, \dots, d_{N-1})$ to denote a diagonal matrix with diagonal entries $d_0, d_1, \dots, d_{N-1}$.

2.7.2 Properties of Diagonal Matrices

1. **Matrix-vector multiplication**: For a diagonal matrix $D = \text{diag}(d_0, d_1, \dots, d_{N-1})$ and vector $\mathbf{x} \in \mathbb{R}^{N\times1}$:

$$D\mathbf{x} = \begin{bmatrix} d_0 x_0 \\ d_1 x_1 \\ \vdots \\ d_{N-1}x_{N-1} \end{bmatrix}.$$

Each component is scaled independently by the corresponding diagonal entry.

2. **Matrix multiplication**: The product of two diagonal matrices $D_1 = \text{diag}(d_1^{(0)}, \dots, d_1^{(N-1)})$ and $D_2 = \text{diag}(d_2^{(0)}, \dots, d_2^{(N-1)})$ is also diagonal:

$$D_1 D_2 = \text{diag}(d_1^{(0)} d_2^{(0)}, \dots, d_1^{(N-1)} d_2^{(N-1)}).$$

Diagonal matrices commute: $D_1 D_2 = D_2 D_1$.

3. **Matrix power**: For a diagonal matrix $D = \text{diag}(d_0, d_1, \dots, d_{N-1})$ and integer $k \geq 0$:

$$D^k = \text{diag}(d_0^k, d_1^k, \dots, d_{N-1}^k).$$

Each diagonal entry is raised to the k-th power. For negative powers, if all $d_i \neq 0$, then $D^{-k} = (D^{-1})^k = \text{diag}(d_0^{-k}, d_1^{-k}, \dots, d_{N-1}^{-k})$.

4. **Inverse**: If all diagonal entries are non-zero, then $D^{-1} = \text{diag}(1/d_0, 1/d_1, \dots, 1/d_{N-1})$.

5. **Determinant**: $\det(D) = \prod_{i=0}^{N-1} d_i$ (product of diagonal entries).

6. **Trace**: $\text{tr}(D) = \sum_{i=0}^{N-1} d_i$ (sum of diagonal entries).

2.8 Definition: Singular Values and Vectors

For a matrix $A \in \mathbb{R}^{M\times N}$ (which can be rectangular, i.e., $M \neq N$, or square), the **singular values** of A, denoted $\sigma_0(A), \sigma_1(A), \dots, \sigma_{\min(M,N)-1}(A)$, are the square roots of the eigenvalues of the Gram matrix $A^\top A$ (or equivalently, of $AA^\top$). (Gram matrices are formally defined in Chapter 7.)

Specifically, if $\lambda_0 \geq \lambda_1 \geq \dots \geq \lambda_{r-1} > 0$ are the positive eigenvalues of $A^\top A$, where $r = \text{rank}(A)$, then the singular values are defined as

$$\sigma_i(A) = \sqrt{\lambda_i(A^\top A)}, \quad i = 0, \dots, \min(M, N) - 1,$$

with $\sigma_i(A) = 0$ for $i \geq r$. The singular values are ordered such that $\sigma_0(A) \geq \sigma_1(A) \geq \dots \geq \sigma_{\min(M,N)-1}(A) \geq 0$.

The largest singular value $\sigma_0(A)$ is also called the **spectral norm** of A: $||A||_2 = \sigma_0(A)$. We will revisit the spectral norm in Chapter 8 (SVD), where we explore its connection to singular value decomposition and its applications in optimization.

> The term "singular values" is distinct from "singular matrix" (a non-invertible square matrix, defined earlier). However, they are related: for a square matrix A, if A is singular (non-invertible), then A has at least one zero singular value. Conversely, if all singular values of A are positive, then A is nonsingular (invertible).

2.8.1 Left and Right Singular Vectors

Associated with each singular value $\sigma_i(A)$ are two vectors:

- **Left singular vectors** $\mathbf{u}_i \in \mathbb{R}^M$: These are the orthonormal eigenvectors of $AA^\top$ (the **row Gram matrix**), corresponding to eigenvalues $\lambda_i = \sigma_i^2$. This follows from the fact that if $\mathbf{v}_i$ is an eigenvector of $A^\top A$ with eigenvalue σ_i^2, then $AA^\top(A\mathbf{v}_i) = \sigma_i^2(A\mathbf{v}_i)$, so $A\mathbf{v}_i$ (and hence $\mathbf{u}_i = \frac{1}{\sigma_i}A\mathbf{v}_i$) is an eigenvector of $AA^\top$ with the same eigenvalue σ_i^2. They form an orthonormal basis for the output space $\mathbb{R}^M$ (or can be extended to one).
- **Right singular vectors** $\mathbf{v}_i \in \mathbb{R}^N$: These are the orthonormal eigenvectors of $A^\top A$ (the **column Gram matrix**), also corresponding to eigenvalues $\lambda_i = \sigma_i^2$. Equivalently, for each non-zero singular value $\sigma_i > 0$, the right singular vector can be used to define the left singular vector as $\mathbf{u}_i = \frac{1}{\sigma_i}A\mathbf{v}_i$. They form an orthonormal basis for the input space $\mathbb{R}^N$.

Both Gram matrices $A^\top A$ and $AA^\top$ share the same non-zero eigenvalues $\lambda_i = \sigma_i^2$, which is why both yield the same singular values $\sigma_i = \sqrt{\lambda_i}$. The relationship $A\mathbf{v}_i = \sigma_i\mathbf{u}_i$ shows that right singular vectors $\mathbf{v}_i$ are mapped by A to the corresponding left singular vectors $\mathbf{u}_i$, scaled by the singular value σ_i.

> Unlike **eigenvalues** (covered in Chapter 6), which are only defined for square matrices, **singular values** are defined for any matrix (rectangular or square). This makes singular values more general and widely applicable. Singular values provide a fundamental characterization of a matrix's geometric properties. They measure the "stretching" factors of the linear transformation represented by A: if A maps a unit vector to a vector of length σ, then σ is a singular value. The singular value decomposition (SVD), which we will cover in Chapter 8, expresses any matrix as $A = U\Sigma V^\top$, where Σ is a diagonal matrix containing the singular values, U contains the left singular vectors, and V contains the right singular vectors. Singular values are crucial in many applications: they determine the rank of a matrix (the number of non-zero singular values), enable low-rank approximations, and are used in principal component analysis (PCA) and dimensionality reduction techniques.

2.9 Matrix Inverse

2.9.1 Definition: Inverse Matrix

Let $A \in \mathbb{R}^{N\times N}$ be a square matrix. If there exists a matrix $B \in \mathbb{R}^{N\times N}$ such that

$$AB = I_{N\times N},$$

then B is called the **inverse** of A, denoted by A^{-1}.

A square matrix A is **invertible** (or **nonsingular**) if A^{-1} exists. Otherwise, A is **non-invertible** (or **singular**).

The inverse matrix is fundamental in solving linear systems $A\mathbf{x} = \mathbf{b}$: if A is invertible, then $\mathbf{x} = A^{-1}\mathbf{b}$ is the unique solution. In machine learning, matrix inverses appear in least squares problems (normal equations), ridge regression, and various optimization algorithms. However, computing the full inverse is often avoided in practice due to numerical instability and computational cost; instead, we solve linear systems directly or use factorizations like QR or Cholesky decomposition. The concept of invertibility is closely related to rank: a square matrix is invertible if and only if it is full rank (i.e., $\text{rank}(A) = N$). Inverse matrices are further explored in Chapter 3, where we show that for orthonormal matrices, the inverse equals the transpose ($A^{-1} = A^\top$).

2.9.2 Theorem: Left Inverse Equals Right Inverse for Square Matrices

Let $A \in \mathbb{R}^{N\times N}$. If there exists a matrix $B \in \mathbb{R}^{N\times N}$ such that $AB = I_{N\times N}$, then A is invertible, $B = A^{-1}$, and consequently $BA = I_{N\times N}$.

In particular, for any invertible matrix A, $AA^{-1} = A^{-1}A = I_{N\times N}$.

2.9.3 Theorem: Uniqueness of Inverse

If A is invertible, then its inverse A^{-1} is unique.

Proof

Suppose B and C are both inverses of A, so $AB = I$ and $AC = I$. By the left inverse equals right inverse theorem, we have $BA = I$ and $CA = I$. Then $B = BI = B(AC) = (BA)C = IC = C$.

2.9.4 Theorem: Inverse of Transpose

If A is invertible, then $A^\top$ is also invertible, and

$$(A^\top)^{-1} = (A^{-1})^\top \tag{2.8}$$

Proof

We need to show that $(A^{-1})^\top A^\top = I$ and $A^\top (A^{-1})^\top = I$.

$$\begin{aligned}(A^{-1})^\top A^\top &= (AA^{-1})^\top = I^\top = I,\\ A^\top (A^{-1})^\top &= (A^{-1}A)^\top = I^\top = I.\end{aligned}$$

2.9.5 Theorem: Inverse Properties

For invertible square matrices A and B:

1. **Inverse of product:** $(AB)^{-1} = B^{-1}A^{-1}$.

   Proof: We need to show that $(B^{-1}A^{-1})(AB) = I$ and $(AB)(B^{-1}A^{-1}) = I$.

   $$(AB)^{-1} = B^{-1}A^{-1} \tag{2.9}$$

   Verification:

   $$\begin{aligned}(B^{-1}A^{-1})(AB) &= B^{-1}(A^{-1}A)B = B^{-1}IB = B^{-1}B = I,\\ (AB)(B^{-1}A^{-1}) &= A(BB^{-1})A^{-1} = AIA^{-1} = AA^{-1} = I.\end{aligned}$$

2. **Inverse of transpose:** $(A^\top)^{-1} = (A^{-1})^\top$.

   (This is already proven in the "Inverse of Transpose" theorem above.)

3. **Inverse of inverse:** $(A^{-1})^{-1} = A$.

   Proof: Since $A^{-1}A = AA^{-1} = I$, by definition A is the inverse of A^{-1}, so $(A^{-1})^{-1} = A$.

These properties are fundamental for working with matrix inverses. They allow us to manipulate expressions involving inverses algebraically, which is essential in solving linear systems, computing projections, and various optimization problems in machine learning.

Unlike scalar arithmetic, **matrix division is not directly defined**. Instead, division-like operations are performed using the **matrix inverse**. For invertible square matrices $A, B \in \mathbb{R}^{N\times N}$, the expression A/B (or $\frac{A}{B}$) is not a standard notation, but if it were to be interpreted, it would typically mean AB^{-1} (right division) or $B^{-1}A$ (left division), depending on context. The matrix inverse (defined above) serves as the analog of division in matrix algebra, allowing us to "undo" matrix multiplication or solve linear systems.

Related matrix operations: Other matrix operations that extend scalar functions to matrices are covered in Chapter 6, Section~6.4: **matrix power** A^k (for diagonalizable matrices, see equation *6.10*), **matrix exponential** $\exp(A) = e^A$, **matrix logarithm** $\log(A)$, and **matrix square root** $A^{1/2}$. These operations are defined through eigendecomposition and provide powerful tools for analyzing linear transformations and solving differential equations.

Common Pitfalls 2(Scalar Algebra vs. Linear Algeba)
Scalar multiplication: When dealing with scalar multiplication, note that $(kA)^{-1} = k^{-1}A^{-1}$ and $(kA)^\top = kA^\top$ for non-zero scalar k. Additionally, A^{-1} represents the matrix inverse, not a "reciprocal" in the scalar sense. Therefore, A^{-1} **cannot** be written as $\frac{1}{A}$ or $1/A$, as these are not defined in linear algebra.
No cancellation: When a matrix product is a scalar and appears in both numerator and denominator (e.g., $\frac{\mathbf{x}^\top A\mathbf{x}}{\mathbf{x}^\top \mathbf{x}}$ or $\frac{\mathbf{x}^\top A\mathbf{x}}{\mathbf{x}^\top B\mathbf{x}}$), the common factors **cannot** be canceled. In expressions like $\frac{\mathbf{x}^\top AQA\mathbf{x}}{\mathbf{x}^\top A^\top A\mathbf{x}}$, neither $\mathbf{x}^\top$ and $\mathbf{x}$ nor $A^\top$ and A can be canceled.

2.10 Projection Matrix

In Chapter 1, we covered vector projection onto a vector and projection onto a hyperplane. Here, we introduce the **projection matrix**, which is an expansion of the vector projection concept from Chapter 1. The projection matrix allows us to project both vectors and matrices onto a subspace in a unified, matrix-based framework. This is fundamental in many applications, including least squares problems, dimensionality reduction, and feature extraction in machine learning.

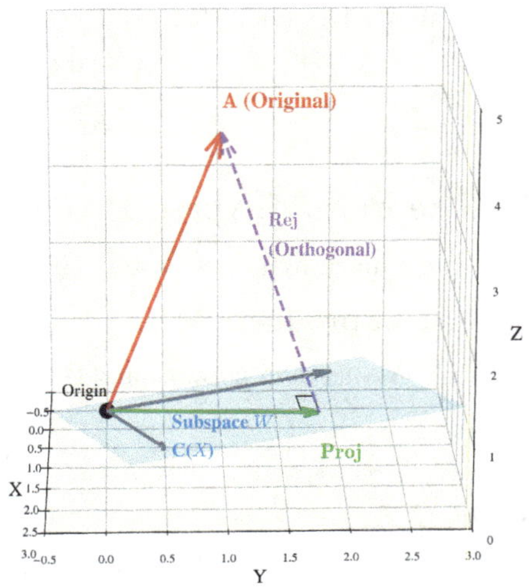

Figure 2.2: Projection Matrix

2.10.1 Definition: Projection Matrix

Let $X \in \mathbb{R}^{M\times N}$ be either a single vector ($N = 1$) or a matrix with multiple columns ($N \geq 1$). If X is a matrix, let $X = \begin{bmatrix}\mathbf{x}_0 & \cdots & \mathbf{x}_{N-1}\end{bmatrix}$. If X is a single vector, we require $X \neq \mathbf{0}$.

Assumption: The columns of X are linearly independent (so that $X^\top X$ is invertible). For a single vector, this means $X \neq \mathbf{0}$.

Let $W = \mathrm{C}(X)$ denote the **column space** of X (the subspace spanned by the columns of X, or the span of the single vector if $N = 1$). The column space is formally defined in Chapter 3.

The **projection matrix** onto W is:

$$\boxed{P = X(X^\top X)^{-1}X^\top \in \mathbb{R}^{M\times M}} \tag{2.10}$$

The projection matrix P projects any vector or matrix onto W.

As shown in Figure 2.2, the projection $P\mathbf{A}$ is the "shadow" of vector $\mathbf{A}$ on the subspace W, and the residual $\mathbf{R} = \mathbf{A} - P\mathbf{A}$ is orthogonal to W.

2.10.2 Projection of Matrices

For a matrix $A \in \mathbb{R}^{M\times K}$, the projection onto W (applied column-wise) is:

$$PA = X(X^\top X)^{-1}X^\top A.$$

Residual:

$$R = A - PA.$$

Orthogonality:

$$X^\top R = \mathbf{0}.$$

Each column of the residual matrix R is orthogonal to all columns of X, meaning each column lies in the orthogonal complement of W.

When A has a single column ($K = 1$), A is a vector and the formula still applies. The projection matrix formula $P = X(X^\top X)^{-1}X^\top$ provides a unified framework that works for both cases: projecting a vector or matrix onto the span of a single vector (when X is a vector, i.e., $N = 1$) or onto the column subspace of a matrix (when X has multiple columns, i.e., $N > 1$).

2.10.3 Properties of Projection Matrix

1. **Idempotency**: $P^2 = P$.

   Proof: $P^2 = [X(X^\top X)^{-1}X^\top][X(X^\top X)^{-1}X^\top] = X(X^\top X)^{-1}(X^\top X)(X^\top X)^{-1}X^\top = X(X^\top X)^{-1}X^\top = P$.

2. **Symmetry**: $P^\top = P$.

   Proof: $P^\top = [X(X^\top X)^{-1}X^\top]^\top = X(X^\top X)^{-\top}X^\top = X(X^\top X)^{-1}X^\top = P$, where we used $(X^\top X)^{-\top} = (X^\top X)^{-1}$ since $X^\top X$ is symmetric.

3. **Orthogonality of residual (vector case)**: For $\mathbf{u} \in \mathbb{R}^{M\times 1}$ and residual $\mathbf{R} = \mathbf{u} - P\mathbf{u}$,

   $$X^\top \mathbf{R} = \mathbf{0}.$$

   Proof: $X^\top \mathbf{R} = X^\top(\mathbf{u} - P\mathbf{u}) = X^\top \mathbf{u} - X^\top X(X^\top X)^{-1}X^\top \mathbf{u} = X^\top \mathbf{u} - X^\top \mathbf{u} = \mathbf{0}$.

4. **Orthogonality of residual (matrix case)**: For $A \in \mathbb{R}^{M\times K}$ and residual $R = A - PA$,

   $$X^\top R = \mathbf{0}.$$

   Proof: $X^\top R = X^\top(A - PA) = X^\top A - X^\top X(X^\top X)^{-1}X^\top A = X^\top A - X^\top A = \mathbf{0}$.

The projection is **unique**: the minimizer is well-defined as a single vector or matrix, not a set. This follows from the fact that the objective function $f(\mathbf{y}) = ||\mathbf{y} - \mathbf{x}||_2^2$ is strictly convex. To see this, note that $f(\mathbf{y}) = (\mathbf{y} - \mathbf{x})^\top(\mathbf{y} - \mathbf{x})$, and its Hessian is $\nabla^2 f(\mathbf{y}) = 2I$, which is positive definite (all eigenvalues equal to $2 > 0$). By the Hessian test, a function is strictly convex if its Hessian is positive definite everywhere.

> The uniqueness of the projection is fundamental in optimization and machine learning: it ensures that projecting onto a subspace is a well-defined operation. In neural networks, projection operations appear in attention mechanisms, where queries are projected onto key-value subspaces, and in dimensionality reduction techniques like PCA, where data is projected onto principal component subspaces.

2.11 Change of Basis

Let the original basis be $E = \begin{bmatrix}\mathbf{e}_0 & \cdots & \mathbf{e}_{d-1}\end{bmatrix}$, and the new basis (also called the **target basis** or **latent basis**)

$$E' = \begin{bmatrix}\mathbf{e}'_0 & \cdots & \mathbf{e}'_{l-1}\end{bmatrix}.$$

The **basis transformation matrix** V expresses each new basis vector as a linear combination of the original basis vectors. Specifically, each column of V contains the coordinates of a new basis vector $\mathbf{e}'_i$ expressed in the original basis E:

$$V = \begin{bmatrix} \mathbf{v}_0 & \cdots & \mathbf{v}_{l-1} \end{bmatrix},$$

where $\mathbf{v}_i \in \mathbb{R}^{d\times 1}$ is the representation of $\mathbf{e}'_i$ in the original basis, i.e., $\mathbf{e}'_i = \sum_{j=0}^{d-1} v_{ji}\mathbf{e}_j$.

2.11.1 Definition: Change of Basis Representation

If $\mathbf{x} \in \mathbb{R}^{d\times 1}$ and $\mathbf{z} \in \mathbb{R}^{l\times 1}$ represent the same vector in different bases, where:

- $\mathbf{x}$ is the vector expressed in the original basis E (coordinates in $\mathbb{R}^d$)
- $\mathbf{z}$ is the same vector expressed in the new basis E' (coordinates in $\mathbb{R}^l$)

then the relationship between the coordinate representations is:

$$\mathbf{x} = V\mathbf{z}.$$

This means: to convert from the new basis coordinates $\mathbf{z}$ to the original basis coordinates $\mathbf{x}$, multiply by the transformation matrix V.

2.11.2 Theorem: Orthonormal Bases

When both the original basis E and the new basis E' are **orthonormal** (i.e., all basis vectors are unit vectors and mutually orthogonal), the transformation matrix V has a special property: its columns are orthonormal. This is expressed as:

$$V^\top V = I_{l\times l}.$$

This condition means that the columns of V form an orthonormal set: each column has unit length ($||\mathbf{v}_i||_2 = 1$) and different columns are orthogonal ($\mathbf{v}_i^\top \mathbf{v}_j = 0$ for $i \neq j$). When $l = d$ (i.e., V is square), this also implies that V is an **orthonormal matrix**, and we have $V^{-1} = V^\top$, making the inverse transformation particularly simple: $\mathbf{z} = V^\top \mathbf{x}$.

2.12 Matrix Norms

Matrix norms provide a way to measure the "size" of a matrix, analogous to how vector norms measure the length of vectors. While vector norms measure the distance of a point from the origin, matrix norms (specifically induced norms) measure the worst-case amplification factor of a linear transformation.

2.12.1 Definition: Matrix Norm

A **matrix norm** is a function $\|\cdot\| : \mathbb{R}^{M\times N} \to \mathbb{R}$ that assigns a non-negative real number to each matrix. For matrices $A, B \in \mathbb{R}^{M\times N}$, a matrix norm $\|\cdot\|$ satisfies the following 4 axioms:

1. **Non-negativity**: $\|A\| \geq 0$
2. **Definiteness**: $\|A\| = 0 \iff A = 0$
3. **Homogeneity**: For any $\alpha \in \mathbb{R}$, $\|\alpha A\| = |\alpha| \cdot \|A\|$
4. **Triangle inequality**: $\|A + B\| \leq \|A\| + \|B\|$

2.12.1.1 Sub-multiplicativity

Many commonly used matrix norms (including induced norms and the Frobenius norm, both defined below) also satisfy **sub-multiplicativity**:

$$\|AB\| \leq \|A\| \cdot \|B\|$$

This is an additional property, not a general axiom for all matrix norms.

For example, the **max norm** $\|A\|_{\max} = \max_{i,j} |a_{ij}|$ satisfies the four basic axioms but not sub-multiplicativity: for $A = \begin{bmatrix} 1 & 1 \\ 1 & 1 \end{bmatrix}$, we have $\|A\|_{\max} = 1$ but $\|A^2\|_{\max} = 2 > 1 = \|A\|_{\max}^2$.

Matrix sub-multiplicativity is analogous to the vector **Cauchy-Schwarz inequality** $|\langle \mathbf{u}, \mathbf{v} \rangle| \leq \|\mathbf{u}\| \cdot \|\mathbf{v}\|$ in Chapter 1: both bound a "product" (matrix product or inner product) by the product of norms. The proof of sub-multiplicativity for induced norms relies on the Cauchy-Schwarz inequality. Sub-multiplicativity is crucial in analyzing the stability of iterative algorithms, bounding the growth of errors in numerical computations, and establishing convergence rates in optimization.

In deep learning, sub-multiplicativity helps analyze gradient flow through neural networks and provides bounds on the **Lipschitz constant** of composite functions. A function $f : \mathbb{R}^d \to \mathbb{R}^m$ is L-Lipschitz if $||f(\mathbf{x}) - f(\mathbf{y})|| \leq L||\mathbf{x} - \mathbf{y}||$ for all $\mathbf{x}, \mathbf{y}$. For a linear transformation $f(\mathbf{x}) = A\mathbf{x}$, the Lipschitz constant equals the spectral norm $||A||_2$, which bounds how much the output can change relative to input perturbations. This is essential for adversarial robustness, gradient clipping, and spectral normalization in GANs.

2.12.2 Definition: Induced p-Norm

For a matrix $A \in \mathbb{R}^{M \times N}$ and any non-zero vector $\mathbf{x} \in \mathbb{R}^N$, the induced p-norm is defined as

$$\|A\|_p = \sup_{\mathbf{x} \neq 0} \frac{||A\mathbf{x}||_p}{||\mathbf{x}||_p}.$$

The supremum (denoted sup) is the least upper bound of a set. For a continuous function over a compact set, the supremum equals the maximum. Here, $\sup_{\mathbf{x} \neq 0}$ means we take the supremum over all non-zero vectors $\mathbf{x} \in \mathbb{R}^N$. The induced norm finds the vector $\mathbf{x}$ that maximizes the ratio $\frac{||A\mathbf{x}||_p}{||\mathbf{x}||_p}$, which represents how much the transformation A can "stretch" or "amplify" vectors. The induced p-norm $||A||_p$ answers the question: "What is the maximum factor by which A can amplify the length of a vector?" If $||A||_p = 2$, then A can at most double the length of any vector (measured in the p-norm). This is crucial for understanding the stability and conditioning of linear transformations.

For specific values of p, the induced p-norm has simpler forms:

- $p = 1$

  $$||A||_1 = \max_{0 \leq n \leq N-1} \sum_{m=0}^{M-1} |a_{mn}|.$$

  This is the maximum column sum (sum of absolute values in each column).

- $p = 2$ **(Spectral norm)**

  $$||A||_2 = \sqrt{\lambda_{\max}(A^\top A)}.$$

  This is the square root of the largest eigenvalue of $A^\top A$, which equals the largest singular value of A. The matrix $A^\top A$ is the **column Gram matrix** (inner products between columns of A), which we will cover in Chapter 8.

- $p = \infty$

$$||A||_\infty = \max_{0 \le m \le M-1} \sum_{n=0}^{N-1} |a_{mn}|.$$

This is the maximum row sum (sum of absolute values in each row).

2.12.2.1 Example

Given the matrix

$$A = \begin{bmatrix} 1 & 2 \\ 3 & 4 \end{bmatrix},$$

compute the induced 1-norm $||A||_1$ and induced ∞-norm $||A||_\infty$.

Solution:

For the induced 1-norm, we compute the maximum column sum:

$$||A||_1 = \max_j \sum_i |a_{ij}| = \max(|1| + |3|, |2| + |4|) = \max(4, 6) = 6.$$

For the induced ∞-norm, we compute the maximum row sum:

$$||A||_\infty = \max_i \sum_j |a_{ij}| = \max(|1| + |2|, |3| + |4|) = \max(3, 7) = 7.$$

2.12.3 Definition: Condition Number

For an invertible square matrix $A \in \mathbb{R}^{N \times N}$, the **condition number** (with respect to the induced p-norm) is defined as

$$\kappa_p(A) = ||A||_p \cdot ||A^{-1}||_p.$$

When $p = 2$ (the spectral norm), we often simply write $\kappa(A) = \kappa_2(A) = ||A||_2 \cdot ||A^{-1}||_2$.

The condition number measures how "ill-conditioned" a matrix is: a large condition number indicates that the matrix is close to singular (non-invertible), making the linear system $A\mathbf{x} = \mathbf{b}$ sensitive to small perturbations in $\mathbf{b}$ or A. A well-conditioned matrix has a condition number close to 1, while an ill-conditioned matrix has a very large condition number.

> Induced matrix norms are fundamental in AI/ML. First, they measure the **sensitivity** of linear transformations to input perturbations: if $||A||_p$ is large, small changes in input can cause large changes in output, leading to numerical instability. In **neural networks**, weight matrices with large norms can cause exploding gradients during backpropagation, making training unstable. The spectral norm $||A||_2$ (the largest singular value) is particularly important: it bounds the Lipschitz constant of the linear layer, which is essential for **adversarial robustness** (ensuring small input perturbations don't cause large output changes) and **gradient clipping** (preventing gradient explosion). In **attention mechanisms** (Transformers), the induced norm of attention matrices determines how much information can flow between tokens, affecting the model's ability to capture long-range dependencies. In **optimization**, induced norms help analyze convergence rates: for gradient descent on $f(\mathbf{x}) = \frac{1}{2}\mathbf{x}^\top A\mathbf{x}$, the convergence rate depends on the condition number $\kappa(A) = ||A||_2 \cdot ||A^{-1}||_2$, which measures how "ill-conditioned" the problem is. **Spectral normalization** in GANs explicitly constrains the spectral norm of weight matrices to stabilize training and improve generation quality. Understanding induced norms enables practitioners to design more stable architectures, diagnose training issues, and develop regularization techniques.

2.12.4 Definition: Frobenius Norm

For a matrix $A \in \mathbb{R}^{M \times N}$, the Frobenius norm is defined as

$$||A||_F = \sqrt{\sum_{m=0}^{M-1} \sum_{n=0}^{N-1} a_{mn}^2}.$$

The Frobenius norm treats the matrix as an MN-dimensional vector and applies the ℓ_2 norm to it. Unlike induced norms, which measure **geometric transformation strength** (how much the matrix can amplify vectors), the Frobenius norm measures **element-wise magnitude**: it considers each matrix entry a_{mn} individually, computes its squared magnitude a_{mn}^2, and sums these squared magnitudes across all entries. This gives a measure of the total "size" of the matrix based on the magnitudes of its individual elements, without considering how the matrix transforms vectors.

2.12.5 Definition: Trace

For a square matrix $A \in \mathbb{R}^{N \times N}$, the **trace** of A, denoted $\text{tr}(A)$, is the sum of its diagonal entries:

$$\text{tr}(A) = \sum_{n=0}^{N-1} a_{nn}.$$

The trace is a fundamental operation on square matrices. It will be used in expressing the Frobenius norm and appears in many important identities. The trace has the key property that $\text{tr}(AB) = \text{tr}(BA)$ for compatible matrices, and for square matrices, the trace equals the sum of eigenvalues. These and other properties are developed in Chapter 5.

2.12.6 Theorem: Equivalent Forms of the Frobenius Norm

$$||A||_F^2 = \text{tr}(A^\top A) = \sum_{i=0}^{N-1} \lambda_i(A^\top A) = \sum_{i=0}^{\min(M,N)-1} \sigma_i(A)^2 \tag{2.11}$$

where $\lambda_i(A^\top A)$ are the eigenvalues of $A^\top A$ and $\sigma_i(A)$ are the singular values of A.

Proof The first equality follows from the definition of trace:

$$\text{tr}(A^\top A) = \sum_{m=0}^{M-1} \sum_{n=0}^{N-1} a_{mn}^2 = ||A||_F^2.$$

The second equality uses the fact that the trace equals the sum of eigenvalues:

$$\text{tr}(A^\top A) = \sum_{i=0}^{N-1} \lambda_i(A^\top A).$$

The third equality follows from the relationship between eigenvalues of $A^\top A$ and singular values of A: $\lambda_i(A^\top A) = \sigma_i(A)^2$ for $i = 0, \ldots, \min(M, N) - 1$, and $\lambda_i(A^\top A) = 0$ for $i \geq \min(M, N)$.

Using the cyclic property of trace (which will be established in Chapter 5), we also have $\text{tr}(AA^\top) = \text{tr}(A^\top A) = ||A||_F^2$.

2.12.6.1 Example

Given the matrix

$$A = \begin{bmatrix} 1 & 2 \\ 3 & 4 \end{bmatrix},$$

verify the equivalent forms of the Frobenius norm $||A||_F^2$.

Solution:

1. **From definition** (sum of squares of all elements):

$$||A||_F^2 = 1^2 + 2^2 + 3^2 + 4^2 = 1 + 4 + 9 + 16 = 30.$$

2. **From trace of Gram** ($\text{tr}(A^\top A)$):

$$A^\top A = \begin{bmatrix} 1 & 3 \\ 2 & 4 \end{bmatrix} \begin{bmatrix} 1 & 2 \\ 3 & 4 \end{bmatrix} = \begin{bmatrix} 10 & 14 \\ 14 & 20 \end{bmatrix},$$

$$||A||_F^2 = \text{tr}(A^\top A) = 10 + 20 = 30.$$

3. **From eigenvalues of Gram** ($A^\top A$): The eigenvalues of $A^\top A = \begin{bmatrix} 10 & 14 \\ 14 & 20 \end{bmatrix}$ are $\lambda_0 = 15 + \sqrt{221}$ and $\lambda_1 = 15 - \sqrt{221}$ (solving $\det(A^\top A - \lambda I) = 0$, where det denotes the determinant, which will be formally defined in Chapter 5). However, we can verify:

$$\lambda_0 + \lambda_1 = \text{tr}(A^\top A) = 30,$$

which matches the Frobenius norm squared.

4. **From singular values of A**: The singular values of A are the square roots of the eigenvalues of $A^\top A$: $\sigma_0 = \sqrt{\lambda_0}$ and $\sigma_1 = \sqrt{\lambda_1}$. Therefore,

$$||A||_F^2 = \sigma_0^2 + \sigma_1^2 = \lambda_0 + \lambda_1 = 30.$$

All four methods yield the same result: $||A||_F^2 = 30$, hence $||A||_F = \sqrt{30}$.

2.12.7 Properties of Orthonormal Matrices wrt Matrix Norms

An **orthonormal matrix** (also called **orthogonal matrix**) is a square matrix $A \in \mathbb{R}^{N\times N}$ such that $A^\top A = I_{N\times N}$. This means the columns of A form an orthonormal set: each column has unit length and different columns are orthogonal. Equivalently, $A^{-1} = A^\top$ (as mentioned in the "Change of Basis" section). (Orthonormal matrices are formally defined in Chapter 3, where we also cover column-orthonormal matrices for rectangular matrices.)

Orthonormal matrices have special properties with respect to matrix norms, which we explore below.

Let $A \in \mathbb{R}^{N\times N}$ be an orthonormal matrix (i.e., $A^\top A = I_{N\times N}$).

1. **Vector norm preservation**

   For any vector $\mathbf{x} \in \mathbb{R}^{N\times 1}$,

$$||A\mathbf{x}||_2 = ||\mathbf{x}||_2.$$

   Proof

$$||A\mathbf{x}||_2^2 = (A\mathbf{x})^\top (A\mathbf{x}) = \mathbf{x}^\top A^\top A\mathbf{x} = \mathbf{x}^\top I_{N\times N}\mathbf{x} = \mathbf{x}^\top \mathbf{x} = ||\mathbf{x}||_2^2.$$

   Taking square roots gives $||A\mathbf{x}||_2 = ||\mathbf{x}||_2$.

2. **Spectral norm**

$$||A||_2 = 1.$$

Proof

Since $A^\top A = I_{N\times N}$, all eigenvalues of $A^\top A$ are 1. The spectral norm is the square root of the largest eigenvalue: $||A||_2 = \sqrt{\lambda_{\max}(A^\top A)} = \sqrt{1} = 1$.

3. **Frobenius norm**

$$||A||_F = \sqrt{N}.$$

Proof

$$||A||_F^2 = \mathrm{tr}(A^\top A) = \mathrm{tr}(I_{N\times N}) = N.$$

Taking square roots gives $||A||_F = \sqrt{N}$.

4. **Multiplication invariance (Frobenius norm)**

For any compatible matrix B:

- **Left-multiplication**: If $B \in \mathbb{R}^{N\times M}$, then $||AB||_F = ||B||_F$.
- **Right-multiplication**: If $B \in \mathbb{R}^{M\times N}$, then $||BA||_F = ||B||_F$.

Proof

Left-multiplication:

$$||AB||_F^2 = \mathrm{tr}((AB)^\top (AB)) = \mathrm{tr}(B^\top A^\top AB) = \mathrm{tr}(B^\top I_{N\times N} B) = \mathrm{tr}(B^\top B) = ||B||_F^2.$$

Right-multiplication:

$$||BA||_F^2 = \mathrm{tr}((BA)^\top (BA)) = \mathrm{tr}(B^\top BAA^\top) = \mathrm{tr}(B^\top B I_{N\times N}) = \mathrm{tr}(B^\top B) = ||B||_F^2,$$

where we used the cyclic property of trace ($\mathrm{tr}(XYZ) = \mathrm{tr}(YZX)$) and the fact that for square orthonormal matrices, $AA^\top = I_{N\times N}$.

Taking square roots gives the results.

2.12.7.1 Example

Consider the rotation matrix

$$A = \begin{bmatrix} \cos\theta & -\sin\theta \\ \sin\theta & \cos\theta \end{bmatrix} = \frac{1}{\sqrt{2}} \begin{bmatrix} 1 & -1 \\ 1 & 1 \end{bmatrix},$$

where $\theta = \frac{\pi}{4}$ (45 degrees). Verify that A is orthonormal and check the properties.

Solution:

1. **Orthonormality check**: Compute $A^\top A$:

$$A^\top A = \frac{1}{2} \begin{bmatrix} 1 & 1 \\ -1 & 1 \end{bmatrix} \begin{bmatrix} 1 & -1 \\ 1 & 1 \end{bmatrix} = \frac{1}{2} \begin{bmatrix} 2 & 0 \\ 0 & 2 \end{bmatrix} = \begin{bmatrix} 1 & 0 \\ 0 & 1 \end{bmatrix} = I_{2\times 2}.$$

Therefore, A is orthonormal.

2. **Vector norm preservation**: For $\mathbf{x} = \begin{bmatrix} 1 \\ 0 \end{bmatrix}$,

$$A\mathbf{x} = \frac{1}{\sqrt{2}} \begin{bmatrix} 1 & -1 \\ 1 & 1 \end{bmatrix} \begin{bmatrix} 1 \\ 0 \end{bmatrix} = \frac{1}{\sqrt{2}} \begin{bmatrix} 1 \\ 1 \end{bmatrix},$$

$$||A\mathbf{x}||_2 = \frac{1}{\sqrt{2}}\sqrt{1^2 + 1^2} = 1, \quad ||\mathbf{x}||_2 = 1.$$

   The norms are preserved: $||A\mathbf{x}||_2 = ||\mathbf{x}||_2 = 1$.

3. **Spectral norm**: Since $A^\top A = I_{2\times 2}$, all eigenvalues are 1, hence $||A||_2 = \sqrt{1} = 1$.

4. **Frobenius norm**: $||A||_F = \sqrt{2}$ (since $N = 2$). Direct computation:

$$||A||_F^2 = \frac{1}{2}(1^2 + (-1)^2 + 1^2 + 1^2) = \frac{1}{2} \cdot 4 = 2, \quad ||A||_F = \sqrt{2}.$$

5. **Multiplication invariance**: For $B = \begin{bmatrix} 2 & 3 \\ 4 & 5 \end{bmatrix}$,

   Left-multiplication:

$$AB = \frac{1}{\sqrt{2}} \begin{bmatrix} 1 & -1 \\ 1 & 1 \end{bmatrix} \begin{bmatrix} 2 & 3 \\ 4 & 5 \end{bmatrix} = \frac{1}{\sqrt{2}} \begin{bmatrix} -2 & -2 \\ 6 & 8 \end{bmatrix},$$

$$||AB||_F^2 = \frac{1}{2}((-2)^2 + (-2)^2 + 6^2 + 8^2) = \frac{1}{2} \cdot 108 = 54, \quad ||B||_F^2 = 54.$$

   Therefore, $||AB||_F = ||B||_F = \sqrt{54}$.

   Right-multiplication:

$$BA = \begin{bmatrix} 2 & 3 \\ 4 & 5 \end{bmatrix} \frac{1}{\sqrt{2}} \begin{bmatrix} 1 & -1 \\ 1 & 1 \end{bmatrix} = \frac{1}{\sqrt{2}} \begin{bmatrix} 5 & 1 \\ 9 & 1 \end{bmatrix},$$

$$||BA||_F^2 = \frac{1}{2}(5^2 + 1^2 + 9^2 + 1^2) = \frac{1}{2} \cdot 108 = 54, \quad ||B||_F^2 = 54.$$

   Therefore, $||BA||_F = ||B||_F = \sqrt{54}$.

The Frobenius norm is the most commonly used matrix norm in AI/ML. In **deep learning**, it is used for **weight regularization** (L2 regularization): $\lambda||W||_F^2$ penalizes large weights, preventing overfitting. The relationship $||A||_F^2 = \sum_i \sigma_i(A)^2$ connects it to matrix **rank**, making it useful for **low-rank approximation** and **model compression**. In **matrix factorization** (recommendation systems), minimizing $||X - WH||_F^2$ is the standard loss function. It also appears in **proximal gradient methods** and **gradient clipping**. Unlike the spectral norm, the Frobenius norm can be computed efficiently in $O(MN)$ time by summing the squares of all matrix elements, making it practical for large-scale deep learning where weight matrices have millions of parameters. In contrast, computing the spectral norm requires eigenvalue decomposition ($O(\min(M, N)^3)$) or iterative methods, and is often approximated or computed periodically rather than at every training step.
For **orthonormal matrices**, the norm preservation property $||A\mathbf{x}||_2 = ||\mathbf{x}||_2$ is crucial for **orthogonal weight initialization**, helping maintain gradient magnitudes and preventing vanishing/exploding gradients. The spectral norm $||A||_2 = 1$ ensures orthonormal transformations are 1-Lipschitz, used in **spectral normalization** for training stability in GANs. The Frobenius norm invariance $||AB||_F = ||B||_F$ when A is orthonormal preserves feature energy in **feature extraction** and **layer analysis**. In **PCA**, orthonormal transformation matrices preserve total variance. Orthonormal matrices also represent **rotations and reflections**, used in **data augmentation** and **equivariant**

neural networks.

2.13 Linear Independence and Rank

2.13.1 Definition: Linear Independence

A set of vectors $\{\mathbf{v}_0, \mathbf{v}_1, \dots, \mathbf{v}_{k-1}\} \subset \mathbb{R}^M$ is **linearly independent** if the only solution to

$$c_0\mathbf{v}_0 + c_1\mathbf{v}_1 + \cdots + c_{k-1}\mathbf{v}_{k-1} = \mathbf{0}$$

is $c_0 = c_1 = \cdots = c_{k-1} = 0$.

If there exist scalars $c_0, c_1, \dots, c_{k-1}$ (not all zero) such that the above equation holds, then the vectors are **linearly dependent**.

Linear independence is fundamental in understanding the structure of vector spaces and matrices.

In machine learning, linearly independent features provide unique information, while linearly dependent features are redundant. When columns of a data matrix are linearly dependent, the model has fewer effective parameters than apparent, leading to issues in optimization (e.g., non-unique solutions in least squares). Understanding linear independence helps in feature selection, dimensionality reduction, and diagnosing overfitting.

2.13.2 Definition: Rank

For a matrix $A \in \mathbb{R}^{M\times N}$:

- The **column rank** is the number of linearly independent column vectors.
- The **row rank** is the number of linearly independent row vectors.

A fundamental theorem states that **column rank equals row rank**. Therefore, we can simply refer to the **rank** of a matrix, denoted $\text{rank}(A)$, without ambiguity.

The rank of a matrix measures the "dimensionality" of the information contained in the matrix. It determines how many independent directions the matrix can span. In machine learning, rank determines the effective capacity of linear models, the number of independent parameters, and is crucial for understanding overfitting, model compression, and the expressiveness of linear transformations. A matrix with $\text{rank}(A) = \min(M, N)$ is called **full rank**, meaning all its rows (or columns) are linearly independent. **See also:** Rank is used extensively in Chapter~3 (Rank-Nullity Theorem and four fundamental subspaces) and Chapter~4 (numerical rank computation via QR decomposition).

2.13.3 Theorem: Upper Bound on Rank

For any matrix $A \in \mathbb{R}^{M\times N}$,

$$\text{rank}(A) \le \min\{M, N\}.$$

Proof

Each column lies in $\mathbb{R}^M$, so the column span has dimension at most M. There are N columns, so the span dimension is at most N. Hence,

$$\text{rank}(A) \le \min\{M, N\}.$$

2.13.4 Theorem: Rank of Transpose

For any matrix A,

$$\text{rank}(A^\top) = \text{rank}(A).$$

This follows directly from the fact that column rank equals row rank.

2.13.5 Theorem: Rank of Matrix Product

For $A \in \mathbb{R}^{M\times d}$ and $B \in \mathbb{R}^{d\times N}$,

$$\text{rank}(AB) \le \min\{\text{rank}(A), \text{rank}(B)\}.$$

Proof

The columns of AB are linear combinations of the columns of A: $(AB)_{:,j} = AB_{:,j} = \sum_{i=0}^{d-1} b_{ij}A_{:,i}$. Therefore, the column space of AB is contained in the column space of A, so $\text{rank}(AB) \le \text{rank}(A)$.

Similarly, the rows of AB are linear combinations of the rows of B: $(AB)_{i,:} = A_{i,:}B = \sum_{j=0}^{d-1} a_{ij}B_{j,:}$. Therefore, the row space of AB is contained in the row space of B, so $\text{rank}(AB) \le \text{rank}(B)$.

Combining both bounds, we have $\text{rank}(AB) \le \min\{\text{rank}(A), \text{rank}(B)\}$.

This theorem is crucial for understanding the information capacity of matrix compositions.

In neural networks, when multiple linear layers are composed (e.g., $W_2(W_1\mathbf{x})$), the rank of the composition is bounded by the minimum rank of the individual layers, limiting the network's effective capacity. This motivates low-rank factorizations and explains why deep networks can have bottlenecks even when individual layers are full rank.
Resolving rank bottlenecks: While this rank bound applies to purely linear compositions, several mechanisms in modern neural networks help overcome this limitation:

1. **Nonlinear activations**: Activation functions (ReLU, GELU, etc.) can increase the effective rank of representations. For example, softmax's exponential function can increase matrix rank (see the example on rank increase via exponential functions).
2. **Residual connections**: Skip connections (e.g., $\mathbf{y} = W_2(W_1\mathbf{x}) + \mathbf{x}$) use addition instead of pure composition, allowing information to bypass rank-constrained paths and preserving the full input dimensionality.
3. **Increased hidden dimensions**: While the rank of W_2W_1 is bounded, using wider hidden layers (larger d) provides more degrees of freedom and can capture richer representations within the rank constraint.
4. **Depth over width**: Deep networks with many layers can learn complex functions even when each composition has limited rank, as the depth allows for hierarchical feature extraction.
5. **Low-rank as a feature**: Rather than avoiding rank constraints, modern architectures (e.g., LoRA, low-rank attention) explicitly use low-rank factorizations to reduce parameters while maintaining performance, turning the constraint into a design principle.

2.13.6 Theorem: Rank of an Outer Product

Let $\mathbf{u} \in \mathbb{R}^{M\times 1}$ and $\mathbf{v} \in \mathbb{R}^{N\times 1}$ be nonzero vectors. Then

$$\text{rank}(\mathbf{u}\mathbf{v}^\top) = 1.$$

Proof

The matrix $\mathbf{u}\mathbf{v}^\top$ has columns that are scalar multiples of $\mathbf{u}$:

$$\mathbf{u}\mathbf{v}^\top = \begin{bmatrix} v_0\mathbf{u} & v_1\mathbf{u} & \cdots & v_{N-1}\mathbf{u} \end{bmatrix}.$$

Since all columns are multiples of the single vector $\mathbf{u}$, the column space is one-dimensional (spanned by $\mathbf{u}$), hence $\text{rank}(\mathbf{u}\mathbf{v}^\top) = 1$.

> This theorem connects to the outer product view of matrix multiplication (View 4): $AB = \sum_{i=0}^{d-1} \mathbf{a}_i \mathbf{b}_i^\top$ shows that any matrix product can be decomposed into a sum of rank-1 matrices. This perspective is fundamental for low-rank approximations, matrix factorization, and understanding how linear layers in neural networks decompose transformations into rank-1 components.

2.13.6.1 Example: Matrix Rank Increase via the Exponential Function in Softmax

The softmax operation (element-wise exponential followed by normalization) can increase the rank of a matrix. This is a key property that enables standard attention mechanisms to have stronger expressive power than linear attention.

Simple example: Consider the zero matrix:

$$M = \begin{pmatrix} 0 & 0 \\ 0 & 0 \end{pmatrix}$$

which has $\text{rank}(M) = 0$. After applying the element-wise exponential function e^M (this is the first step of softmax, before normalization), we get:

$$e^M = \begin{pmatrix} e^0 & e^0 \\ e^0 & e^0 \end{pmatrix} = \begin{pmatrix} 1 & 1 \\ 1 & 1 \end{pmatrix}$$

which has $\text{rank}(e^M) = 1$. The rank increased from 0 to 1. Note that the normalization step in softmax (dividing each row by its sum) does not change the rank, as it only scales rows by non-zero constants.

More realistic example: Consider a rank-1 matrix formed by an outer product:

$$M = \mathbf{x}\mathbf{x}^\top \in \mathbb{R}^{n \times n}$$

where $\mathbf{x} \in \mathbb{R}^n$ is a random vector. This matrix has $\text{rank}(M) = 1$ (since it's an outer product). However, after applying the element-wise exponential function e^M:

$$e^M = \begin{pmatrix} e^{x_0^2} & e^{x_0 x_1} & \cdots & e^{x_0 x_{n-1}} \\ e^{x_1 x_0} & e^{x_1^2} & \cdots & e^{x_1 x_{n-1}} \\ \vdots & \vdots & \ddots & \vdots \\ e^{x_{n-1} x_0} & e^{x_{n-1} x_1} & \cdots & e^{x_{n-1}^2} \end{pmatrix}$$

the rank typically increases significantly. For example, with $n = 100$ and random $\mathbf{x}$, we typically have $\text{rank}(e^M) \approx 17$–$18$, much higher than the original rank of 1.

Why this matters: In attention mechanisms, $QK^\top$ has rank at most $d_k \ll L$. The exponential function (the key component of softmax) can increase the rank of the attention matrix. Since normalization in softmax does not change rank, the full softmax operation preserves this rank increase, providing more expressive power than linear attention, which is constrained by the rank of $QK^\top$ without the exponential transformation.

2.13.6.2 Example: Rank Properties in Transformer Encoder vs Decoder

In Transformers[1], **encoder** and **decoder-only** models exhibit different rank properties in their attention mechanisms due to different masking strategies:

---

[1] Vaswani et al., *Attention Is All You Need*, NeurIPS 2017.

1. **Encoder (used in MLM - Masked Language Modeling)**:
   - Uses **bidirectional attention**: each position can attend to all positions in the sequence (both past and future)
   - **No causal mask needed**: Unlike decoder-only models, standard encoder-only models (like BERT) do not require a causal/FIM(Fill-in-the-Middle) mask for attention, as they are designed to process the entire sequence bidirectionally. (*Padding masks may be used to handle variable-length sequences, but these do not affect the structural rank properties discussed here. FIM masks may be used in specialized training setups to enable infilling capabilities, but the standard bidirectional attention without FIM mask is what we analyze here for rank properties.*)
   - In multi-head attention, each head computes an attention matrix $A \in \mathbb{R}^{L \times L}$ as $A = \text{softmax}(QK^\top/\sqrt{d_k})$ where $Q, K \in \mathbb{R}^{L \times d_k}$ are the query and key matrices for a single head, with $L \gg d_k$ (sequence length much larger than per-head dimension)
   - The attention score matrix $QK^\top$ has shape $L \times L$ (since $Q \in \mathbb{R}^{L \times d_k}$ and $K^\top \in \mathbb{R}^{d_k \times L}$). However, by the rank property of matrix products, $\text{rank}(QK^\top) \leq \min\{\text{rank}(Q), \text{rank}(K^\top)\}$. Since both Q and K have at most d_k linearly independent columns (or rows), we have $\text{rank}(Q) \leq d_k$ and $\text{rank}(K^\top) = \text{rank}(K) \leq d_k$, hence $\text{rank}(QK^\top) \leq d_k \ll L$. This means that despite being an $L \times L$ matrix, $QK^\top$ has rank at most d_k, making it a low-rank matrix. Here d_k is the per-head dimension, and the total attention dimension is $D = d_k \times H$ where H is the number of heads
   - **Softmax's role**: Although softmax (element-wise exponential followed by normalization) can technically increase the rank of $QK^\top$ to full rank, the **effective rank** or information capacity of the encoder's attention matrix is bottlenecked by d_k. The matrix A is generated from a low-rank matrix $QK^\top$, which constrains its degrees of freedom and leads to rapid spectral decay
   - This **effective low-rank constraint** is a fundamental limitation of bidirectional attention when $L \gg d_k$, limiting its expressive power despite the formal full-rank property
2. **Decoder-only (used in CLM - Causal Language Modeling)**:
   - Uses **causal attention**: each position can only attend to previous positions (including itself)
   - The attention is computed as $A = \text{softmax}((QK^\top/\sqrt{d_k}) + M)$ where M is the causal mask (typically $M_{ij} = -\infty$ for $j > i$ and $M_{ij} = 0$ otherwise). The mask is applied **after** computing $QK^\top$ but **before** softmax, ensuring that masked positions receive zero attention after normalization
   - The attention matrix has a **lower triangular structure** due to the causal mask: $A_{ij} = 0$ for $j > i$
   - **Key property**: For a lower triangular matrix, the determinant equals the product of diagonal elements. Since softmax ensures all diagonal elements $A_{ii} > 0$, we have $\det(A) > 0$, which implies $\text{rank}(A) = L$ (full rank)
   - This **guaranteed full-rank property** gives decoder-only architectures theoretically stronger expressive power than encoder architectures

The rank difference provides a **theoretical explanation** for why decoder-only models (like GPT) have become the dominant architecture for large language models[2]: they avoid the effective low-rank bottleneck that plagues bidirectional attention, while maintaining the ability to capture sequential dependencies through the causal structure. The guaranteed full-rank property of triangular attention matrices (regardless of d_k) provides a fundamentally different inductive bias that favors autoregressive generation.

[2]Zhang et al., *Encoder-Decoder or Decoder-Only? Revisiting Encoder-Decoder Large Language Model*, 2025; discusses why decoder-only became dominant and compares pretraining vs inference efficiency.

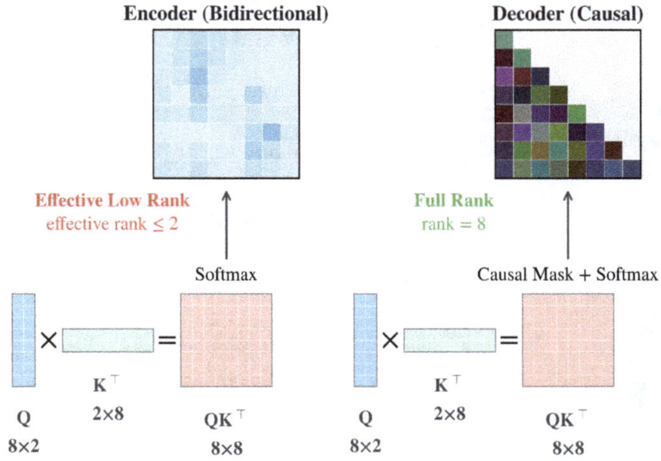

Figure 2.3: Encoder vs Decoder attention rank properties

Figure 2.3 shows a simple demo with $L = 8$ and $d_k = 2$ to illustrate the rank difference. In practice, $L \gg d_k$ makes the rank contrast even more pronounced. For example, Llama 3.1-128k[3] uses $L = 128,000$ and $d_k = 128$, while Llama 4 Scout-17B-16E[4] pushes this further with a context length of $L = 262,144$ (262k tokens) and $d_k = 128$ (Llama series typically maintains $d_k = 128$ for efficiency). If bidirectional attention (encoder) were used to process such lengths, it would face catastrophic "rank collapse": with $\text{rank}(QK^\top) \leq d_k = 128$ for a $262,144 \times 262,144$ attention matrix, the effective information capacity would be severely bottlenecked. In contrast, decoder-only architectures with causal attention maintain full rank ($\text{rank}(A) = L = 262,144$) regardless of d_k, demonstrating why decoder-only models are essential for ultra-long context processing.

While the full-rank property is a key theoretical advantage, other factors also contribute to decoder-only dominance, including training objective consistency (next-token prediction), KV cache efficiency for inference, and better in-context learning capabilities.

[3]Llama 3.1 with 128K context: Meta blog; Hugging Face model card (70B) for config (e.g. `max_position_embeddings = 131072`).

[4]Configuration from Llama-4-Scout-17B-16E model: `max_position_embeddings = 262144`, `head_dim = 128`.

2.14 Mindmap

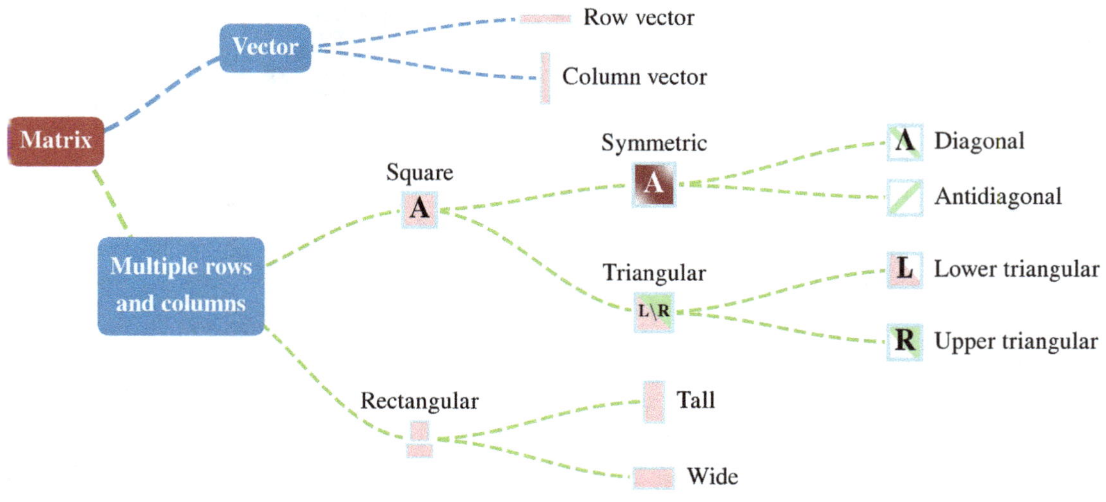

Figure 2.4: Mind map of matrix topics

2.15 Exercises

2.15.1 Matrix Operations in Neural Networks

Consider a simple neural network layer with weight matrix $W \in \mathbb{R}^{3\times 2}$ and bias vector $\mathbf{b} \in \mathbb{R}^3$:

$$W = \begin{bmatrix} 1 & 2 \\ 3 & 4 \\ 5 & 6 \end{bmatrix}, \quad \mathbf{b} = \begin{bmatrix} 1 \\ -1 \\ 0 \end{bmatrix}, \quad \mathbf{x} = \begin{bmatrix} 2 \\ -1 \end{bmatrix}.$$

1. Compute $\mathbf{h} = W\mathbf{x} + \mathbf{b}$ using the column-vector view (linear combination of columns).
2. Compute $\mathbf{h} = W\mathbf{x} + \mathbf{b}$ using the row-vector view (inner products with rows).
3. In backpropagation, we need to compute $W^\top \mathbf{g}$ where $\mathbf{g} \in \mathbb{R}^3$ is the gradient. If $\mathbf{g} = \begin{bmatrix} 1 \\ 2 \\ 3 \end{bmatrix}$, compute $W^\top \mathbf{g}$.
4. Prove that $(W\mathbf{x})^\top = \mathbf{x}^\top W^\top$ for any matrix $W \in \mathbb{R}^{m\times n}$ and vector $\mathbf{x} \in \mathbb{R}^n$.

2.15.2 Outer Product in Attention Mechanisms

In self-attention mechanisms (Transformers), the standard **Scaled Dot-Product Attention** computes attention as $\text{softmax}(QK^\top/\sqrt{d_k})V$, where $Q \in \mathbb{R}^{n\times d_k}$, $K \in \mathbb{R}^{n\times d_k}$, $V \in \mathbb{R}^{n\times d_v}$ are matrices whose **rows** are query, key, and value vectors respectively. The attention score matrix $QK^\top$ has elements $(QK^\top)_{ij} = \mathbf{q}_i^\top \mathbf{k}_j$ where $\mathbf{q}_i^\top$ and $\mathbf{k}_j^\top$ are the i-th and j-th rows of Q and K.

Here we use column vector notation for simplicity: let $\mathbf{q}, \mathbf{k}, \mathbf{v} \in \mathbb{R}^d$ be column vectors. The computation $\mathbf{q}^\top \mathbf{k}$ corresponds to one element of the attention score matrix $QK^\top$ (before softmax normalization). We focus on the mathematical structure of outer products and rank-1 decompositions.

Let:

$$\mathbf{q} = \begin{bmatrix} 1 \\ 2 \end{bmatrix}, \quad \mathbf{k} = \begin{bmatrix} 3 \\ 4 \end{bmatrix}, \quad \mathbf{v} = \begin{bmatrix} 5 \\ 6 \end{bmatrix}.$$

1. Compute the raw attention score $\alpha = \mathbf{q}^\top \mathbf{k}$ (scalar, before softmax normalization).
2. The weighted value is $\alpha \mathbf{v}$. Show that this can be written as $(\mathbf{v}\mathbf{q}^\top)\mathbf{k}$ using the outer product $\mathbf{v}\mathbf{q}^\top$.
3. Compute the rank of the matrix $\mathbf{v}\mathbf{q}^\top$. Prove that for any nonzero vectors $\mathbf{u} \in \mathbb{R}^m$ and $\mathbf{v} \in \mathbb{R}^n$, the rank of $\mathbf{u}\mathbf{v}^\top$ is 1.
4. For a batch of queries $Q = [\mathbf{q}_1, \mathbf{q}_2] \in \mathbb{R}^{d\times 2}$ and keys $K = [\mathbf{k}_1, \mathbf{k}_2] \in \mathbb{R}^{d\times 2}$, the raw attention matrix (before softmax) is $A = Q^\top K$. Show that A can be decomposed as a sum of rank-1 matrices.

2.15.3 L2 Regularization and Frobenius Norm

In deep learning, L2 regularization adds a penalty term $\lambda ||W||_F^2$ to the loss function to prevent overfitting. Consider a weight matrix:

$$W = \begin{bmatrix} 1 & 2 & 3 \\ 4 & 5 & 6 \end{bmatrix}.$$

1. Compute the Frobenius norm $||W||_F$ directly from the definition.
2. Compute $W^\top W$ and show that $||W||_F^2 = \text{tr}(W^\top W)$.
3. Compute the induced 1-norm $||W||_1 = \max_j \sum_i |w_{ij}|$ (maximum column sum).
4. Compute the induced ∞-norm $||W||_\infty = \max_i \sum_j |w_{ij}|$ (maximum row sum).
5. For a function $\mathcal{L}(W) + \lambda ||W||_F^2$ where $\mathcal{L}$ is differentiable, show that $\nabla_W(\mathcal{L}(W) + \lambda ||W||_F^2) = \nabla_W \mathcal{L} + 2\lambda W$.

2.15.4 Normal Equations and Ridge Regression

For linear regression with design matrix $X \in \mathbb{R}^{N\times d}$ and target vector $\mathbf{y} \in \mathbb{R}^N$, the **ridge regression** objective (L2-regularized least squares) is:

$$\mathcal{L}(\theta) = ||X\theta - \mathbf{y}||_2^2 + \lambda ||\theta||_2^2,$$

where $\lambda > 0$ is the regularization parameter.

1. Expand $\mathcal{L}(\theta)$ and show it can be written as:

$$\mathcal{L}(\theta) = \theta^\top (X^\top X + \lambda I)\theta - 2\theta^\top X^\top \mathbf{y} + \mathbf{y}^\top \mathbf{y}.$$

2. Take the gradient $\nabla_\theta \mathcal{L}(\theta)$ and show that:

$$\nabla_\theta \mathcal{L}(\theta) = 2(X^\top X + \lambda I)\theta - 2X^\top \mathbf{y}.$$

3. Prove that the optimal solution is:

$$\hat{\theta} = (X^\top X + \lambda I)^{-1} X^\top \mathbf{y}.$$

Why does adding λI guarantee invertibility even when $X^\top X$ is singular?

4. Prove that $X^\top X + \lambda I$ is invertible for any $\lambda > 0$, even when $X^\top X$ is singular.
5. Show that $\lim_{\lambda \to 0} \hat{\theta} = (X^\top X)^{-1} X^\top \mathbf{y}$ (when $X^\top X$ is invertible) and $\lim_{\lambda \to \infty} \hat{\theta} = \mathbf{0}$.

2.15.5 Spectral Norm and Singular Values

Given the weight matrix of a neural network layer:

$$W = \begin{bmatrix} 2 & 1 \\ 1 & 2 \end{bmatrix}.$$

1. Compute $W^\top W$ and find its eigenvalues λ_1, λ_2.
2. Compute the spectral norm $||W||_2 = \sqrt{\lambda_{\max}(W^\top W)}$.
3. Show that $||W||_2$ equals the largest singular value $\sigma_{\max}(W)$ of W. (Hint: The singular values of W are the square roots of the eigenvalues of $W^\top W$.)
4. Define $\tilde{W} = W/||W||_2$. Compute $\tilde{W}$ and show that $||\tilde{W}||_2 = 1$.
5. Prove that for any matrix $A \in \mathbb{R}^{m \times n}$ and vector $\mathbf{x} \in \mathbb{R}^n$, $||A\mathbf{x}||_2 \leq ||A||_2 \cdot ||\mathbf{x}||_2$.

2.15.6 Orthonormal Matrices and Initialization

Orthonormal weight initialization (e.g., in RNNs and Transformers) uses matrices W such that $W^\top W = I$.

1. Let $W \in \mathbb{R}^{N \times N}$ be orthonormal. Prove that $||W||_2 = 1$ and $||W||_F = \sqrt{N}$.
2. Prove that for any vector $\mathbf{x}$, $||W\mathbf{x}||_2 = ||\mathbf{x}||_2$ (orthonormal matrices preserve lengths).
3. For orthonormal matrices $W_1, W_2, \dots, W_L \in \mathbb{R}^{N \times N}$, prove that the composite transformation $f(\mathbf{x}) = W_L \cdots W_2 W_1 \mathbf{x}$ satisfies $||f(\mathbf{x})||_2 = ||\mathbf{x}||_2$ for any vector $\mathbf{x}$.

2.15.7 Lipschitz Constants and Adversarial Robustness

A function $f : \mathbb{R}^d \to \mathbb{R}^m$ is **L-Lipschitz** if $||f(\mathbf{x}) - f(\mathbf{y})||_2 \leq L||\mathbf{x} - \mathbf{y}||_2$ for all $\mathbf{x}, \mathbf{y}$.

1. For a linear layer $f(\mathbf{x}) = W\mathbf{x}$ where $W \in \mathbb{R}^{m \times d}$, prove that the smallest Lipschitz constant is $L = ||W||_2$ (the spectral norm). (Hint: Use the definition of induced norm and show both directions.)
2. Consider a two-layer network $g(\mathbf{x}) = W_2 \sigma(W_1 \mathbf{x})$ where σ is the ReLU activation (which is 1-Lipschitz). Using sub-multiplicativity, prove that the Lipschitz constant of g is at most $||W_2||_2 \cdot ||W_1||_2$.
3. For a matrix $W \in \mathbb{R}^{m \times d}$ with $||W||_2 \leq 1$, prove that $||W(\mathbf{x} + \delta) - W\mathbf{x}||_2 \leq ||\delta||_2$ for any vectors $\mathbf{x}, \delta \in \mathbb{R}^d$.
4. For matrices $W_1, W_2, \dots, W_L$ with $||W_i||_2 \leq 1$ for all i, and a 1-Lipschitz function σ (i.e., $||\sigma(\mathbf{x}) - \sigma(\mathbf{y})||_2 \leq ||\mathbf{x} - \mathbf{y}||_2$), prove that the composite function $f(\mathbf{x}) = W_L \sigma(W_{L-1} \sigma(\cdots \sigma(W_1 \mathbf{x})))$ is 1-Lipschitz.

2.15.8 Condition Number and Optimization Convergence

The **condition number** of a matrix A is $\kappa(A) = ||A||_2 \cdot ||A^{-1}||_2$ (for invertible A), measuring how "ill-conditioned" the problem is.

1. For the matrix $A = \begin{bmatrix} 2 & 1 \\ 1 & 2 \end{bmatrix}$, compute $\kappa(A)$.

2. Consider gradient descent on the quadratic objective $f(\mathbf{x}) = \frac{1}{2}\mathbf{x}^\top A\mathbf{x} - \mathbf{b}^\top \mathbf{x}$ where A is positive definite. The convergence rate depends on $\kappa(A)$. Prove that:
   - If $\kappa(A) = 1$ (i.e., A is a scalar multiple of identity), gradient descent converges in 1 step.
   - If $\kappa(A)$ is large, convergence is slow.

3. For a positive definite matrix A, if we use $P = A^{-1/2}$ as a preconditioner, show that the condition number of $P^\top AP$ is 1.

2.15.9 Matrix Norm Inequalities and Bounds

Prove the following inequalities:

1. **Frobenius-Spectral Inequality**: For any matrix $A \in \mathbb{R}^{M\times N}$,

$$||A||_2 \leq ||A||_F \leq \sqrt{\min(M,N)} \cdot ||A||_2.$$

   When does equality hold? (Hint: Use the relationship between Frobenius norm and singular values.)

2. **Sub-multiplicativity**: For matrices A, B with compatible dimensions, prove:

$$||AB||_F \leq ||A||_F \cdot ||B||_F, \quad ||AB||_2 \leq ||A||_2 \cdot ||B||_2.$$

3. **Triangle Inequality for Matrix Norms**: For any two matrices $A, B \in \mathbb{R}^{M\times N}$, prove:

$$||A+B||_F \leq ||A||_F + ||B||_F, \quad ||A+B||_2 \leq ||A||_2 + ||B||_2.$$

4. For matrices $W_1, W_2, \ldots, W_L$ with $||W_i||_2 = \alpha > 1$ for all i, prove that $||W_L W_{L-1} \cdots W_1||_2 \leq \alpha^L$.

Chapter 3

Subspaces and Orthogonality

Four fundamental subspaces and orthonormal matrices: rank-nullity theorem and orthogonality relationships

---

"The essence of mathematics lies in its freedom."

— Georg Cantor

- ▷ `Col span` $C(A) = \{A\mathbf{x} : \mathbf{x} \in \mathbb{R}^N\} = \mathrm{span}\{\mathbf{a}_0, \dots, \mathbf{a}_{N-1}\}$ for $A = [\mathbf{a}_0, \dots, \mathbf{a}_{N-1}] \in \mathbb{R}^{M\times N}$
- ▷ `Row span` $R(A) = C(A^\top) = \mathrm{span}\{\mathbf{a}_0^\top, \dots, \mathbf{a}_{M-1}^\top\}$ for $A = [\mathbf{a}_0^\top; \dots; \mathbf{a}_{M-1}^\top] \in \mathbb{R}^{M\times N}$
- ▷ `Null/L-null spaces` $N(A) = \{\mathbf{x} \in \mathbb{R}^N : A\mathbf{x} = \mathbf{0}\}$; $N(A^\top) = \{\mathbf{y} \in \mathbb{R}^M : A^\top\mathbf{y} = \mathbf{0}\}$
- ▷ `Rank-nullity` $\dim C(A) + \dim N(A) = N$ and $\dim C(A^\top) + \dim N(A^\top) = M$ for $A \in \mathbb{R}^{M\times N}$, $\mathrm{rank}(A) = r$
- ▷ `Dims.` $\dim C(A) = r$, $\dim C(A^\top) = r$, $\dim N(A^\top) = M - r$, $\dim N(A) = N - r$ for $A \in \mathbb{R}^{M\times N}$, $\mathrm{rank}(A) = r$
- ▷ `Ortho./direct sum` $C(A) \perp N(A^\top)$, $C(A) \oplus N(A^\top) = \mathbb{R}^M$; $C(A^\top) \perp N(A)$, $C(A^\top) \oplus N(A) = \mathbb{R}^N$
- ▷ `Proj./residual` $\mathbf{x} = \mathrm{proj}_A(\mathbf{x}) + \mathrm{res}_A(\mathbf{x})$ with $\mathrm{proj}_A(\mathbf{x}) \in C(A)$, $\mathrm{res}_A(\mathbf{x}) \in N(A^\top)$; $\mathrm{res}_A(\mathbf{x})^\top\mathbf{a}_n = 0$
- ▷ `Col-orthonormal` $A \in \mathbb{R}^{M\times N}$ $(M \ge N)$, $A^\top A = I_N$; if $M = N$: $AA^\top = I_N$ and $A^{-1} = A^\top$
- ▷ `Regularized inv.` $(AA^\top + \lambda I_M)^{-1} = \frac{1}{\lambda} I_M - \frac{1}{\lambda(1+\lambda)} AA^\top$ for col-orthonorm. A, $\lambda > 0$
- ▷ `Orthonorm. Q` $Q \in \mathbb{R}^{N\times N}$, $Q^\top Q = QQ^\top = I_N$; $Q^{-1} = Q^\top$ (rotation/reflection)
- ▷ `Σ outer prods` $Q = [\mathbf{q}_0, \dots, \mathbf{q}_{N-1}]$: $\sum_{n=0}^{N-1} \mathbf{q}_n\mathbf{q}_n^\top = I_N$
- ▷ `Inv. Σ wtd outers` $(\sum_{n=0}^{N-1} c_n\mathbf{q}_n\mathbf{q}_n^\top)^{-1} = \sum_{n=0}^{N-1} c_n^{-1}\mathbf{q}_n\mathbf{q}_n^\top$ for $c_n \neq 0$
- ▷ `IP/norm pres.` $\langle Q\mathbf{x}, Q\mathbf{y}\rangle = \langle \mathbf{x}, \mathbf{y}\rangle$; $||Q\mathbf{x}||_2 = ||\mathbf{x}||_2$
- ▷ `det(Q)` $\det(Q) = \pm 1$ ($+1$ rotation, -1 reflection)
- ▷ `eig(Q)` $|\lambda| = 1$ for eigenvalues $\lambda \in \mathbb{C}$; eigenvectors may be complex $\mathbf{x} \in \mathbb{C}^N$
- ▷ `Complex IP` $\langle \mathbf{x}, \mathbf{y}\rangle = \overline{\mathbf{x}}^\top\mathbf{y}$ for $\mathbf{x}, \mathbf{y} \in \mathbb{C}^N$

3.1 Definition: Four Fundamental Subspaces

In Chapter~2, we introduced matrix-vector multiplication and established that a matrix A represents a linear transformation mapping vectors from $\mathbb{R}^N$ to $\mathbb{R}^M$. For a matrix $A \in \mathbb{R}^{M\times N}$, we can now ask two fundamental questions about how A transforms vectors:

1. **What outputs can A produce?** (Which vectors $\mathbf{b} \in \mathbb{R}^M$ can be written as $A\mathbf{x}$ for some $\mathbf{x}$?)
2. **What inputs map to zero?** (Which vectors $\mathbf{x} \in \mathbb{R}^N$ satisfy $A\mathbf{x} = \mathbf{0}$?)

By considering both A and its transpose $A^\top$, we obtain four fundamental subspaces that completely characterize the structure of the linear transformation represented by A:

1. **Column space** $C(A)$ The span of all column vectors of A (see Section~1.2.5 for the definition of span). This is the set of all possible outputs: $C(A) = \{A\mathbf{x} : \mathbf{x} \in \mathbb{R}^N\}$.

2. **Row space** $R(A)$ (where $R(A) = C(A^\top)$) The span of all row vectors of A, which equals the column space of $A^\top$. This represents the "effective input space" that actually affects the output.

3. **Left null space** $N(A^\top)$ The set (span) of all vectors $\mathbf{y} \in \mathbb{R}^M$ such that

$$A^\top\mathbf{y} = 0. \tag{3.1}$$

   These are vectors in the output space that are orthogonal to all columns of A.

4. **Null space** $N(A)$ The set (span) of all vectors $\mathbf{x} \in \mathbb{R}^N$ such that

$$A\mathbf{x} = 0. \tag{3.2}$$

   These are vectors in the input space that map to zero under A.

3.2 Theorem: Rank-Nullity Theorem

For any matrix $A \in \mathbb{R}^{M\times N}$ with rank r:

1. **Column space and null space:**

$$\dim(C(A)) + \dim(N(A)) = N \tag{3.3}$$

Equivalently, $\operatorname{rank}(A) + \dim(N(A)) = N$.

2. **Row space and left null space:**

$$\dim(C(A^\top)) + \dim(N(A^\top)) = M \tag{3.4}$$

Equivalently, $\operatorname{rank}(A) + \dim(N(A^\top)) = M$.

The Rank-Nullity Theorem is one of the most fundamental results in linear algebra. It states that the dimension of the column space (the "range" of the transformation) plus the dimension of the null space (the "kernel") equals the number of columns. This theorem has profound implications: if a matrix has many independent columns (high rank), it has a small null space, meaning the transformation is "injective" on a large subspace. Conversely, if the rank is low, the null space is large, meaning many inputs map to zero. **See also:** The definition of rank is introduced in Chapter~2. Numerical methods for computing rank (via QR decomposition and SVD) are covered in Chapter~4.

> In machine learning, this theorem helps understand the capacity of linear models, the number of free parameters, and the geometry of solution spaces in optimization problems. The term "kernel" here refers to the **null space** (also called the kernel) of a linear transformation, which is the set of vectors that map to zero. This is **distinct from** the "kernel" in **kernel methods** (e.g., SVM with RBF kernel) or **convolutional kernels** (e.g., in CNNs), which refer to kernel functions $\kappa(\mathbf{x}, \mathbf{y})$ that compute similarity or to small filter matrices used in convolution operations. The null space kernel is a linear algebra concept (a subspace), while kernel functions and convolutional kernels are computational tools for feature mapping and image processing.

3.3 Theorem: Dimensions and Orthogonality of Four Fundamental Subspaces

Let $A \in \mathbb{R}^{M\times N}$ be a matrix of rank r. Then:

1. $\dim(C(A)) = r$.
2. $\dim(C(A^\top)) = r$.
3. $C(A) \perp N(A^\top)$, $C(A) \oplus N(A^\top) = \mathbb{R}^M$, hence

$$\dim(N(A^\top)) = M - r \tag{3.5}$$

4. $C(A^\top) \perp N(A)$, $C(A^\top) \oplus N(A) = \mathbb{R}^N$, hence

$$\dim(N(A)) = N - r \tag{3.6}$$

The notation $\oplus$ denotes the **direct sum** of subspaces: $U \oplus V$ means that every vector in the space can be written uniquely as the sum of a vector in U and a vector in V, and $U \cap V = \{\mathbf{0}\}$ (the subspaces intersect only at zero). When $U \perp V$ (orthogonal), we say U and V are **orthogonal complements**.

The column and row spaces both have dimension r (the rank). The column space and left null space are orthogonal complements in $\mathbb{R}^M$, and the row space and null space are orthogonal complements in $\mathbb{R}^N$. The direct sum relationships $C(A) \oplus N(A^\top) = \mathbb{R}^M$ and $C(A^\top) \oplus N(A) = \mathbb{R}^N$ imply the dimension formulas via the Rank-Nullity Theorem.

Intuition via projection: The orthogonal complement $N(A^\top)$ is not an abstract set—it consists of all vectors that remain after projecting onto $C(A)$. This extends the vector projection concept from Section~1.1.12.1: just as projecting a vector $\vec{u}$ onto $\hat{v}$ yields $\text{Proj}_{\hat{v}}(\vec{u})$ with rejection $\vec{R} = \vec{u} - \text{Proj}_{\hat{v}}(\vec{u})$ orthogonal to $\hat{v}$, projecting any vector $\mathbf{x} \in \mathbb{R}^M$ onto the subspace $C(A)$ yields $\text{proj}_A(\mathbf{x})$ with rejection $\mathbf{x} - \text{proj}_A(\mathbf{x})$ orthogonal to $C(A)$ and lying in $N(A^\top)$. The projection gives the "explainable" part, while the rejection represents the "unexplainable" part. This projection viewpoint makes the direct sum decomposition concrete: every vector decomposes uniquely into a component in $C(A)$ and a component in $N(A^\top)$.

3.3.1 Example: Four Fundamental Subspaces in 2D

Consider the matrix

$$A = \begin{bmatrix} 1 & 1 \\ 0 & 0 \end{bmatrix} \in \mathbb{R}^{2\times 2}.$$

This matrix has rank $r = 1$. Let us identify the four fundamental subspaces:

- **Column space** $C(A)$: The columns are $\begin{bmatrix}1\\0\end{bmatrix}$ and $\begin{bmatrix}1\\0\end{bmatrix}$ (they are the same). Thus, $C(A) = \text{span}\left\{\begin{bmatrix}1\\0\end{bmatrix}\right\}$, a 1-dimensional subspace (the x-axis).

- **Left null space** $N(A^\top)$: We solve $A^\top \mathbf{y} = \mathbf{0}$, i.e., $\begin{bmatrix}1 & 0\\1 & 0\end{bmatrix}\begin{bmatrix}y_1\\y_2\end{bmatrix} = \begin{bmatrix}0\\0\end{bmatrix}$, which gives $y_1 = 0$. Thus, $N(A^\top) = \text{span}\left\{\begin{bmatrix}0\\1\end{bmatrix}\right\}$, a 1-dimensional subspace (the y-axis).

- **Row space** $C(A^\top)$: The rows are $\begin{bmatrix}1 & 1\end{bmatrix}$ and $\begin{bmatrix}0 & 0\end{bmatrix}$. Thus, $C(A^\top) = \text{span}\left\{\begin{bmatrix}1\\1\end{bmatrix}\right\}$, a 1-dimensional subspace (the line $y = x$ in $\mathbb{R}^2$).

- **Null space** $N(A)$: We solve $A\mathbf{x} = \mathbf{0}$, i.e., $\begin{bmatrix}1 & 1\\0 & 0\end{bmatrix}\begin{bmatrix}x_1\\x_2\end{bmatrix} = \begin{bmatrix}0\\0\end{bmatrix}$, which gives $x_1 + x_2 = 0$, so $x_1 = -x_2$. Thus, $N(A) = \text{span}\left\{\begin{bmatrix}-1\\1\end{bmatrix}\right\}$, a 1-dimensional subspace (the line $y = -x$ in $\mathbb{R}^2$).

We verify the direct sum relationships:

- $C(A) \oplus N(A^\top) = \mathbb{R}^2$: Every vector $\begin{bmatrix}a\\b\end{bmatrix} \in \mathbb{R}^2$ can be written uniquely as $\begin{bmatrix}a\\b\end{bmatrix} = \begin{bmatrix}a\\0\end{bmatrix} + \begin{bmatrix}0\\b\end{bmatrix}$, where $\begin{bmatrix}a\\0\end{bmatrix} \in C(A)$ (the x-axis) and $\begin{bmatrix}0\\b\end{bmatrix} \in N(A^\top)$ (the y-axis). These subspaces are orthogonal and span $\mathbb{R}^2$.

- $C(A^\top) \oplus N(A) = \mathbb{R}^2$: Every vector $\begin{bmatrix}a\\b\end{bmatrix} \in \mathbb{R}^2$ can be written uniquely as $\begin{bmatrix}a\\b\end{bmatrix} = \frac{a+b}{2}\begin{bmatrix}1\\1\end{bmatrix} + \frac{b-a}{2}\begin{bmatrix}-1\\1\end{bmatrix}$, where the first term is in $C(A^\top)$ (the line $y = x$) and the second term is in $N(A)$ (the line $y = -x$). These subspaces are orthogonal and span $\mathbb{R}^2$.

This example illustrates that the column space and left null space are orthogonal complements in $\mathbb{R}^2$ (the x-axis and y-axis), and the row space and null space are orthogonal complements in $\mathbb{R}^2$ (the lines $y = x$ and $y = -x$).

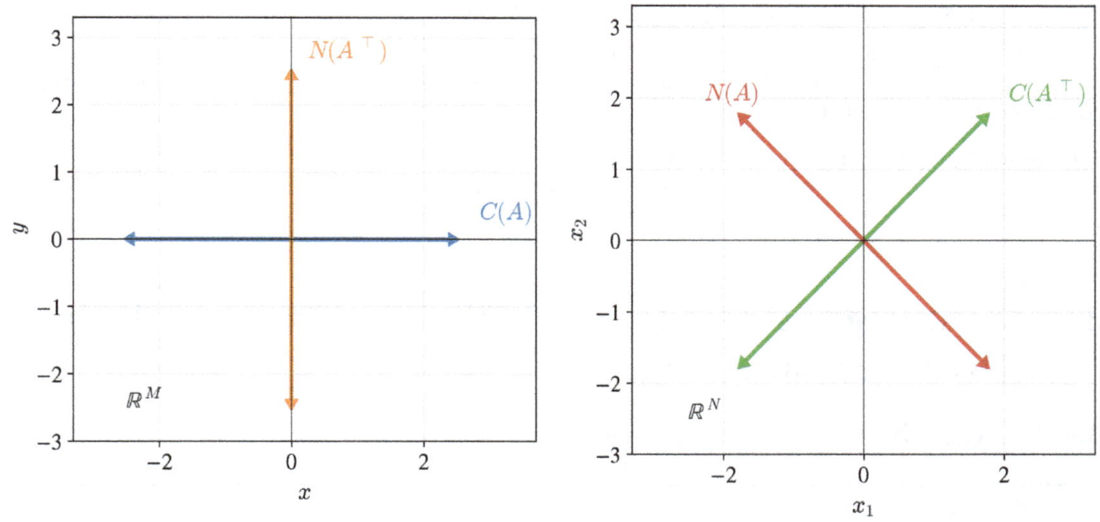

Figure 3.1: Four fundamental subspaces in 2D.

Figure 3.1 shows the four subspaces for the 2-by-2 example: the left panel is the output space $\mathbb{R}^M$ with column space $C(A)$ (horizontal axis) and left null space $N(A^\top)$ (vertical axis); the right panel is the input space $\mathbb{R}^N$ with row space $C(A^\top)$ (line $y = x$) and null space $N(A)$ (line $y = -x$).

3.3.2 Proof

We prove that

$$C(A) \perp N(A^\top) \quad \text{and} \quad C(A) \oplus N(A^\top) = \mathbb{R}^M. \tag{3.7}$$

Step 1: Orthogonality

We show that for any $\mathbf{x} \in C(A)$ and $\mathbf{y} \in N(A^\top)$,

$$\mathbf{y}^\top \mathbf{x} = 0.$$

Since $\mathbf{x} \in C(A)$, there exists $\mathbf{z} \in \mathbb{R}^N$ such that

$$\mathbf{x} = A\mathbf{z}.$$

Then, for any $\mathbf{y} \in N(A^\top)$,

$$\mathbf{y}^\top \mathbf{x} = \mathbf{y}^\top (A\mathbf{z}) = (\mathbf{y}^\top A)\mathbf{z} = 0^\top \mathbf{z} = 0,$$

where the third equality follows from $A^\top \mathbf{y} = 0$.

Step 2: Direct Sum

For any vector $\mathbf{x} \in \mathbb{R}^M$, let $\text{proj}_A(\mathbf{x})$ be its projection onto $C(A)$ and $\text{rej}_A(\mathbf{x}) = \mathbf{x} - \text{proj}_A(\mathbf{x})$ be the rejection.

We show that $\text{rej}_A(\mathbf{x}) \in N(A^\top)$. Since the rejection is orthogonal to every column of A, for each column $\mathbf{a}_n$,

$$\text{rej}_A(\mathbf{x})^\top \mathbf{a}_n = 0.$$

Equivalently, $\text{rej}_A(\mathbf{x})^\top A = 0$, which implies $A^\top \text{rej}_A(\mathbf{x}) = 0$. Therefore, $\text{rej}_A(\mathbf{x}) \in N(A^\top)$.

Since $\mathbf{x} = \text{proj}_A(\mathbf{x}) + \text{rej}_A(\mathbf{x})$ with $\text{proj}_A(\mathbf{x}) \in C(A)$ and $\text{rej}_A(\mathbf{x}) \in N(A^\top)$, and these subspaces are orthogonal (from Step 1), we have $C(A) \oplus N(A^\top) = \mathbb{R}^M$. This decomposition shows that $N(A^\top)$ is the set of all "leftover" directions after projection onto $C(A)$, making the orthogonal complement concrete rather than abstract.

The same argument applies to $C(A^\top)$ and $N(A)$: for any $\mathbf{x} \in \mathbb{R}^N$, we can decompose it uniquely as $\mathbf{x} = \text{proj}_{A^\top}(\mathbf{x}) + \text{rej}_{A^\top}(\mathbf{x})$ where $\text{proj}_{A^\top}(\mathbf{x}) \in C(A^\top)$ and $\text{rej}_{A^\top}(\mathbf{x}) \in N(A)$, giving $C(A^\top) \oplus N(A) = \mathbb{R}^N$.

3.4 Gram–Schmidt Algorithm to Compute a Null Space Orthonormal Basis

3.4.1 Algorithm: Null Space of $A^\top$

Algorithm 2 Null Space of $A^\top$

Input: Matrix $A \in \mathbb{R}^{M \times N}$
Output: An orthonormal basis of $N(A^\top)$
  Perform QR decomposition: $A = QR$
  $j \leftarrow \text{rank}(A)$
  for $m = 0, 1, \dots, M - 1$ **do**
    Let $\mathbf{e}_m \in \mathbb{R}^M$ be the m-th standard basis vector
    Project $\mathbf{e}_m$ onto existing orthonormal vectors:
    $r_{im} \leftarrow \mathbf{e}_m^\top \mathbf{q}_i$ for $i = 0, \dots, j - 1$
    Compute the rejection: $\text{rej}(\mathbf{e}_m) \leftarrow \mathbf{e}_m - \sum_{i=0}^{j-1} \mathbf{q}_i r_{im}$
    if $\text{rej}(\mathbf{e}_m) \neq 0$ **then**
      $\mathbf{q}_j \leftarrow \frac{\text{rej}(\mathbf{e}_m)}{\|\text{rej}(\mathbf{e}_m)\|_2}$
      $j \leftarrow j + 1$
    end if
    if $j = M$ **then**
      break
    end if
  end for

3.4.2 Output

An orthonormal basis of the left null space $N(A^\top)$ is

$$\begin{bmatrix} \mathbf{q}_{\text{rank}(A)} & \mathbf{q}_{\text{rank}(A)+1} & \cdots & \mathbf{q}_{M-1} \end{bmatrix}. \tag{3.8}$$

The algorithm presented above is mathematically correct and illustrates the geometric relationship between QR decomposition and null spaces. However, in practice, the null space is usually computed via **SVD** or **rank-revealing QR** for numerical stability, especially when dealing with nearly rank-deficient matrices or large-scale computations.

> Computing null space bases is essential for understanding solution spaces in overdetermined linear systems (least squares), analyzing the capacity of linear models (the null space represents directions where the model output doesn't change), and understanding why regularization works (it constrains solutions away from the null space). In neural networks, the null space of weight matrices relates to the network's effective capacity and redundancy. The left null space $N(A^\top)$ is particularly important in analyzing residuals and understanding when linear systems have no exact solution.

3.5 Orthonormal Matrices

Having established the structure of the four fundamental subspaces and their orthogonality relationships, we now turn to a special class of matrices that naturally arise in this context: **orthonormal matrices**. For square matrices, both columns and rows form an orthonormal basis, making them fundamental tools for representing rotations and reflections. For rectangular matrices, we use the term **column-orthonormal matrices**, where only the columns form an orthonormal basis (the rows do not). These matrices are essential for representing orthogonal projections and **distance-preserving embeddings** (mappings that preserve distances, norms, and angles between vectors).

3.6 General Properties of Orthonormal Matrix

3.6.1 Definition: Column Orthonormal Matrix

Let $A = \begin{bmatrix} \mathbf{a}_0 & \mathbf{a}_1 & \cdots & \mathbf{a}_{N-1} \end{bmatrix} \in \mathbb{R}^{M \times N}$, $M \geq N$. Matrix A is **column orthonormal** if the vectors $\{\mathbf{a}_0, \dots, \mathbf{a}_{N-1}\}$ form an orthonormal basis of an N-dimensional subspace of $\mathbb{R}^M$.

3.6.2 Theorem: Column Orthonormal Matrix Property

If $A \in \mathbb{R}^{M \times N}$ is column orthonormal, then

$$A^\top A = I_{N \times N} \tag{3.9}$$

3.6.3 Theorem: Inverse of Regularized Column Orthonormal Matrix

For any column orthonormal matrix $A \in \mathbb{R}^{M \times N}$ and any $\lambda > 0$,

$$(AA^\top + \lambda I_{M \times M})^{-1} = \frac{1}{\lambda} I_{M \times M} - \frac{1}{\lambda(1+\lambda)} AA^\top \tag{3.10}$$

This identity follows from the Woodbury formula using $A^\top A = I_{N \times N}$. For orthonormal matrices, this simplifies the matrix inversion identity, making it particularly useful in numerical linear algebra and optimization. In neural networks, this property is exploited in architectures that use orthonormal weight matrices to maintain gradient flow and prevent vanishing/exploding gradients.

See also: The basic definition of matrix inverse is introduced in Section~2.9. Complete theory of inverses, including detailed properties and proofs, is covered in Section~5.4.

3.7 Square Orthonormal Matrices

A square orthonormal matrix $A \in \mathbb{R}^{N \times N}$ satisfies both $A^\top A = I_{N \times N}$ and $AA^\top = I_{N \times N}$ (i.e., $A^{-1} = A^\top$). This is in contrast to column-orthonormal matrices, which only satisfy $A^\top A = I$ but $AA^\top \neq I$ (where $AA^\top$ is a projection operator).

3.7.1 Property: Geometric View

Let $A = \begin{bmatrix} \mathbf{a}_0 & \mathbf{a}_1 & \cdots & \mathbf{a}_{N-1} \end{bmatrix} \in \mathbb{R}^{N \times N}$ be orthonormal. For any $\mathbf{x} \in \mathbb{R}^{N \times 1}$,

$$A\mathbf{x} = \sum_{n=0}^{N-1} \mathbf{a}_n x_n \tag{3.11}$$

is a rotation or reflection of $\mathbf{x}$.

3.7.2 Property: Change of Basis

If $A \in \mathbb{R}^{N\times N}$ is orthonormal, then for any $\mathbf{x} \in \mathbb{R}^{N\times 1}$,

$$A\mathbf{x} = \sum_{n=0}^{N-1} \mathbf{a}_n x_n \tag{3.12}$$

represents the same geometric vector as $\mathbf{x}$, expressed in the orthonormal basis $\{\mathbf{a}_0, \ldots, \mathbf{a}_{N-1}\}$.

3.7.3 Theorem: Inverse of Square Orthonormal Matrix

If $A \in \mathbb{R}^{N\times N}$ is orthonormal, then

$$A^{-1} = A^\top \tag{3.13}$$

> **See also:** Matrix inverse is defined in Chapter~2. The complete theory of inverses is covered in Chapter~5.

3.7.4 Theorem: Sum of Outer Products for Orthonormal Matrix

If $A \in \mathbb{R}^{N\times N}$ is a real-valued orthonormal matrix, then

$$\sum_{n=0}^{N-1} \mathbf{a}_n \mathbf{a}_n^\top = I_{N\times N} \tag{3.14}$$

3.7.5 Theorem: Inverse of Weighted Sum of Outer Products

Let $A \in \mathbb{R}^{N\times N}$ be real-valued orthonormal and $c_0, \ldots, c_{N-1} \neq 0$. Then

$$\left(\sum_{n=0}^{N-1} c_n \mathbf{a}_n \mathbf{a}_n^\top\right)^{-1} = \sum_{n=0}^{N-1} c_n^{-1} \mathbf{a}_n \mathbf{a}_n^\top \tag{3.15}$$

See also: Matrix inverse is defined in Section~2.9. Complete theory of inverses is covered in Section~5.4.

3.7.6 Theorem: Inner-Product Preservation

If $A \in \mathbb{R}^{N\times N}$ is real-valued orthonormal, then for any $\mathbf{x}, \mathbf{y} \in \mathbb{R}^{N\times 1}$,

$$\langle A\mathbf{x}, A\mathbf{y}\rangle = \langle \mathbf{x}, \mathbf{y}\rangle \tag{3.16}$$

3.7.7 Theorem: Determinant of Orthonormal Matrix

If $A \in \mathbb{R}^{N\times N}$ is real-valued orthonormal, then

$$\det(A) = \pm 1 \tag{3.17}$$

This theorem relates the determinant (formally defined in Chapter~5) to the geometric properties of orthonormal matrices. The determinant ± 1 reflects the fact that orthonormal matrices preserve volumes (determinant magnitude) but may reverse orientation (determinant sign). A determinant of $+1$ corresponds to a rotation, while -1 corresponds to a reflection.

3.8 Complex Numbers and Rotations

When analyzing the eigenvalues of orthonormal matrices, we encounter complex eigenvalues and eigenvectors even though the matrices have real entries. This occurs because the characteristic polynomial $\det(A - \lambda I) = 0$ may have complex roots. While this book does not focus primarily on complex numbers, we introduce the essential complex number concepts needed to understand these eigenvalues and their connection to rotations. The key insight is that rotations in the plane correspond naturally to complex exponentials via Euler's formula.

3.8.1 Definition: Complex Numbers

A **complex number** $z \in \mathbb{C}$ is written as $z = a + bi$, where $a, b \in \mathbb{R}$ are the real and imaginary parts, and i is the imaginary unit satisfying $i^2 = -1$. The **complex conjugate** of $z = a + bi$ is $\overline{z} = a - bi$, and the **modulus** (or absolute value) is $|z| = \sqrt{a^2 + b^2} = \sqrt{z\overline{z}}$.

Complex numbers can be visualized in the **complex plane**, where the horizontal axis represents the real part and the vertical axis represents the imaginary part. A complex number $z = a + bi$ corresponds to the point (a, b) in $\mathbb{R}^2$.

3.8.2 Euler's Formula

Euler's formula establishes a fundamental connection between complex exponentials and trigonometric functions:

$$e^{i\theta} = \cos\theta + i\sin\theta, \tag{3.18}$$

where $\theta \in \mathbb{R}$ is a real angle. This formula implies that $|e^{i\theta}| = \sqrt{\cos^2\theta + \sin^2\theta} = 1$, so $e^{i\theta}$ lies on the unit circle in the complex plane. Multiplying a complex number by $e^{i\theta}$ rotates it counterclockwise by angle θ around the origin.

From Euler's formula, we can express trigonometric functions in terms of complex exponentials:

$$\cos\theta = \frac{e^{i\theta} + e^{-i\theta}}{2}, \quad \sin\theta = \frac{e^{i\theta} - e^{-i\theta}}{2i}. \tag{3.19}$$

3.8.3 Complex Numbers and 2D Rotation Matrices

There is a natural correspondence between complex number multiplication and 2D rotation matrices. Consider the rotation matrix that rotates a vector in $\mathbb{R}^2$ counterclockwise by angle θ:

$$R(\theta) = \begin{pmatrix} \cos\theta & -\sin\theta \\ \sin\theta & \cos\theta \end{pmatrix}. \tag{3.20}$$

Multiplying a 2D vector $\begin{pmatrix} x \\ y \end{pmatrix}$ by $R(\theta)$ is equivalent to treating (x, y) as the complex number $z = x + iy$ and multiplying by $e^{i\theta}$:

$$R(\theta)\begin{pmatrix} x \\ y \end{pmatrix} = \begin{pmatrix} x\cos\theta - y\sin\theta \\ x\sin\theta + y\cos\theta \end{pmatrix}, \tag{3.21}$$

which corresponds to $(x+iy)\cdot e^{i\theta} = (x+iy)(\cos\theta + i\sin\theta) = (x\cos\theta - y\sin\theta) + i(x\sin\theta + y\cos\theta)$.

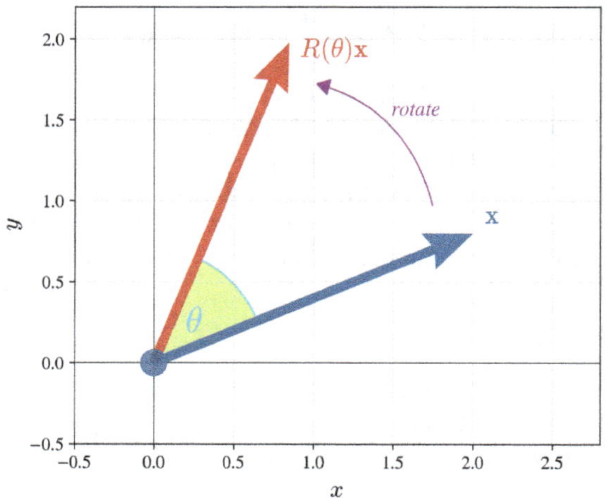

Figure 3.2: Rotation preserves norms

This correspondence extends to eigenvalues: the rotation matrix $R(\theta)$ has eigenvalues $e^{i\theta}$ and $e^{-i\theta}$, both with modulus 1, reflecting the fact that rotations preserve distances.

3.8.4 Eigenvalues of Orthonormal Matrices

When analyzing the eigenvalues of real orthonormal matrices, we may encounter **complex** eigenvalues and eigenvectors, even though the matrix itself has real entries. This occurs because the characteristic polynomial $\det(A - \lambda I) = 0$ may have complex roots. To properly handle complex eigenvectors, we need the complex inner product.

3.8.5 Definition: Inner Product of Complex Vectors

For $\mathbf{x}, \mathbf{y} \in \mathbb{C}^N$, define

$$\langle \mathbf{x}, \mathbf{y} \rangle = \overline{\mathbf{x}}^\top \mathbf{y}. \tag{3.22}$$

This extends the real inner product from Section~1.1.2 to complex vectors, where $\overline{\mathbf{x}}$ denotes the complex conjugate of $\mathbf{x}$.

3.8.6 Theorem: Eigenvalues of Orthonormal Matrices

Let $A \in \mathbb{R}^{N\times N}$ be real-valued orthonormal and consider

$$A\mathbf{x} = \lambda\mathbf{x}. \tag{3.23}$$

1. Each eigenvalue satisfies $\lambda \in \mathbb{C}$ and $|\lambda| = 1$.
2. The associated eigenvector may be complex, $\mathbf{x} \in \mathbb{C}^N$.

Geometric interpretation: This reflects the fact that orthonormal matrices represent rotations and reflections, which preserve distances and angles.

3.8.6.1 Example: Rotary Position Embedding (ROPE) in Transformers

Rotary Position Embedding (ROPE)[1] encodes positional information in Transformers using complex number rotations. For a query vector $\mathbf{q}_m \in \mathbb{R}^d$ at position m and a key vector $\mathbf{k}_n \in \mathbb{R}^d$ at position n, ROPE splits each d-dimensional vector into $d/2$ pairs: $(\mathbf{q}_m^{(0)}, \mathbf{q}_m^{(1)}), (\mathbf{q}_m^{(2)}, \mathbf{q}_m^{(3)}), \dots, (\mathbf{q}_m^{(d-2)}, \mathbf{q}_m^{(d-1)})$. Each pair $(q_m^{(2i)}, q_m^{(2i+1)})$ is treated as a complex number $q_m^{(2i)} + i \cdot q_m^{(2i+1)}$ and rotated by angle $m\theta_i$:

$$\begin{pmatrix} q_m^{(2i)} \\ q_m^{(2i+1)} \end{pmatrix} \leftarrow \begin{pmatrix} \cos(m\theta_i) & -\sin(m\theta_i) \\ \sin(m\theta_i) & \cos(m\theta_i) \end{pmatrix} \begin{pmatrix} q_m^{(2i)} \\ q_m^{(2i+1)} \end{pmatrix},$$

where $\theta_i = 10000^{-2i/d}$ is the base frequency for dimension pair i. The same rotation is applied to the corresponding key vector pairs with angle $n\theta_i$.

The rotation matrix $\begin{pmatrix} \cos\theta & -\sin\theta \\ \sin\theta & \cos\theta \end{pmatrix}$ corresponds to multiplication by $e^{i\theta} = \cos\theta + i\sin\theta$ in the complex plane, connecting to the eigenvalues of orthonormal matrices: rotations have eigenvalues $e^{\pm i\theta}$ with $|e^{\pm i\theta}| = 1$. When computing the attention score $(\mathbf{q}_m^{\text{rot}})^\top \mathbf{k}_n^{\text{rot}}$, the rotation angles combine such that the result depends on the relative position difference $m-n$ rather than absolute positions. This relative position encoding emerges naturally from the geometric properties of rotations, enabling the model to capture positional relationships without explicit position embeddings.

ROPE's reliance on complex rotations provides theoretical grounding: the rotation preserves vector norms (since $|e^{i\theta}| = 1$), and the relative position encoding generalizes to sequences longer than those seen during training, as the rotation angles scale continuously with position.

[1] Su et al., *RoFormer: Enhanced Transformer with Rotary Position Embedding*, 2021. https://arxiv.org/abs/2104.09864

3.9 Mindmap

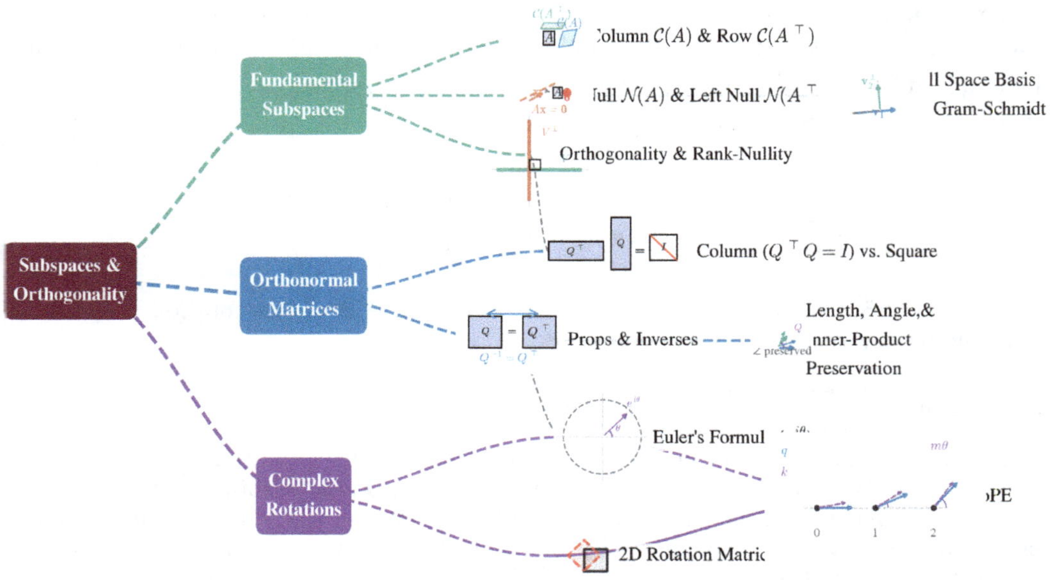

Figure 3.3: Mind map of subspace and orthogonality topics

3.10 Exercises

3.10.1 Feature Redundancy and Null Space in Linear Models

In a linear regression model, the design matrix $X \in \mathbb{R}^{N \times d}$ contains N samples and d features. When features are linearly dependent, the model has redundant parameters. Consider the design matrix:

$$X = \begin{bmatrix} 1 & 2 & 3 \\ 2 & 4 & 6 \end{bmatrix},$$

where each row is a sample and each column is a feature.

1. Compute the rank of X and identify which features are redundant.
2. Find a basis for the null space $N(X)$ of X. What does this tell you about the relationship between the features?
3. Find a basis for the column space $C(X)$ of X. What is the effective dimensionality of the feature space?
4. Find a basis for the left null space $N(X^\top)$ of X. What does this represent in the context of samples?
5. Verify that $\dim(C(X)) + \dim(N(X)) = 3$ (the number of features) and $\dim(C(X^\top)) + \dim(N(X^\top)) = 2$ (the number of samples).

3.10.2 Regularization and Solution Space in Ridge Regression

In ridge regression, we solve $(X^\top X + \lambda I)\theta = X^\top \mathbf{y}$ to find parameters θ. When $X^\top X$ is singular (not full rank), the unregularized problem has multiple solutions. The regularization term λI constrains the solution away from the null space.

Consider the design matrix:

$$X = \begin{bmatrix} 1 & 2 & 3 \\ 2 & 4 & 6 \end{bmatrix}.$$

1. Compute $X^\top X$ and show that it is singular (not invertible).
2. Find the null space $N(X)$ and verify that any vector $\mathbf{v} \in N(X)$ satisfies $X^\top X\mathbf{v} = \mathbf{0}$.
3. Show that for any $\lambda > 0$, the matrix $X^\top X + \lambda I$ is invertible. (Hint: Show that $X^\top X + \lambda I$ has no zero eigenvalues.)
4. Verify that $C(X^\top) \perp N(X)$ by showing that for any $\mathbf{u} \in C(X^\top)$ and $\mathbf{v} \in N(X)$, we have $\mathbf{u}^\top \mathbf{v} = 0$.
5. Explain why regularization "constrains solutions away from the null space" in terms of the orthogonality relationship.

3.10.3 Rank and Model Capacity in Neural Networks

The rank of a weight matrix determines the effective capacity of a linear layer. Consider a weight matrix in a neural network:

$$W = \begin{bmatrix} 1 & 2 & 3 \\ 2 & 4 & 6 \\ 0 & 0 & 0 \end{bmatrix}.$$

1. Compute $\text{rank}(W)$ and express the dimensions of the four fundamental subspaces in terms of $M = 3$ (number of rows) and $N = 3$ (number of columns).
2. Find a basis for $C(W)$ (the output space) and determine its dimension.
3. Find a basis for $N(W)$ (the input null space) and determine its dimension. What does this tell you about inputs that produce zero output?
4. Verify the Rank-Nullity Theorem: $\dim(C(W)) + \dim(N(W)) = N$.
5. If W represents a linear layer $f(\mathbf{x}) = W\mathbf{x}$, what is the maximum number of linearly independent outputs this layer can produce?

3.10.4 Orthonormal Weight Initialization in Transformers

Orthonormal weight initialization (e.g., in Transformer attention layers) uses matrices Q, K, V such that $Q^\top Q = I$, $K^\top K = I$, and $V^\top V = I$ (column orthonormal). This helps maintain gradient magnitudes and prevent vanishing/exploding gradients.

Determine whether the following matrices are column orthonormal by checking $A^\top A = I$:

1. $$Q_1 = \begin{bmatrix} \frac{1}{\sqrt{2}} & \frac{1}{\sqrt{2}} \\ \frac{1}{\sqrt{2}} & -\frac{1}{\sqrt{2}} \end{bmatrix}$$

2.

$$Q_2 = \begin{bmatrix} 1 & 0 \\ 0 & 1 \\ 1 & 0 \end{bmatrix}$$

3.

$$Q_3 = \begin{bmatrix} \frac{3}{5} & \frac{4}{5} \\ -\frac{4}{5} & \frac{3}{5} \end{bmatrix}$$

4. For each matrix that is column orthonormal, verify that $||Q\mathbf{x}||_2 = ||\mathbf{x}||_2$ for $\mathbf{x} = \begin{bmatrix} 1 \\ 1 \end{bmatrix}$ (or appropriate dimension).

3.10.5 Projection and Residual Analysis in Least Squares

In least squares problems, we project the target vector $\mathbf{y}$ onto the column space $C(X)$ of the design matrix X. The residual $\mathbf{r} = \mathbf{y} - X\hat{\theta}$ lies in the left null space $N(X^\top)$ and is orthogonal to all columns of X.

Let $X \in \mathbb{R}^{3\times 2}$ be a column orthonormal matrix (i.e., $X^\top X = I_{2\times 2}$):

$$X = \begin{bmatrix} \frac{1}{\sqrt{2}} & \frac{1}{\sqrt{2}} \\ \frac{1}{\sqrt{2}} & -\frac{1}{\sqrt{2}} \\ 0 & 0 \end{bmatrix}.$$

1. Verify that $X^\top X = I_{2\times 2}$.
2. Show that $XX^\top$ is a projection matrix by verifying $(XX^\top)^2 = XX^\top$.
3. For $\mathbf{y} = \begin{bmatrix} 1 \\ 1 \\ 1 \end{bmatrix}$, compute the projection $\hat{\mathbf{y}} = XX^\top\mathbf{y}$ and verify that $\hat{\mathbf{y}} \in C(X)$.
4. Compute the residual $\mathbf{r} = \mathbf{y} - \hat{\mathbf{y}}$ and verify that $\mathbf{r} \in N(X^\top)$ by showing $X^\top\mathbf{r} = \mathbf{0}$.
5. Verify that $\hat{\mathbf{y}} \perp \mathbf{r}$ (the projection and residual are orthogonal).

Chapter 4

QR Decomposition and Numerical Rank

QR decomposition and matrix rank: orthonormal bases and linear independence

> *"In mathematics you don't understand things. You just get used to them."*
>
> — John von Neumann

▷ `Rank of sum` $\text{rank}(A+B) \leq \text{rank}(A) + \text{rank}(B)$

▷ `Rank of product` $\text{rank}(AB) \leq \min\{\text{rank}(A), \text{rank}(B)\}$

▷ `Rank with orthonormal` If A orthonormal columns, $\text{rank}(AB) = \text{rank}(B)$; if B orthonormal rows, $\text{rank}(AB) = \text{rank}(A)$

▷ `Rank of Gram matrices` $\text{rank}(AA^\top) = \text{rank}(A^\top A) = \text{rank}(A)$

▷ `Numerical rank` $\text{rank}_\epsilon(A) = \min\{k : \sigma_k < \epsilon \cdot \sigma_0\}$ for singular values $\sigma_0 \geq \sigma_1 \geq \cdots$

▷ `Generalized upper triangular` $R \in \mathbb{R}^{r \times N}$, last nonzero row index $\bar{r}_n$ satisfies $\bar{r}_n \leq \bar{r}_{n+1} \leq \bar{r}_n + 1$; allows zero diagonals

▷ `QR decomposition` $A = QR$ where Q has orthonormal columns, R is generalized upper triangular; $\mathbf{a}_n = \sum_{i=0}^{\bar{r}_n} \mathbf{q}_i r_{in}$

▷ `Orthonormality of Q` $Q^\top Q = I$; if Q square, then $Q^\top = Q^{-1}$ (orthogonal)

▷ `Gram-Schmidt` $\mathbf{q}_0 = \frac{\mathbf{a}_0}{\|\mathbf{a}_0\|_2}$; for $n \geq 1$: $r_{in} = \mathbf{a}_n^\top \mathbf{q}_i$, $\text{res}(\mathbf{a}_n) = \mathbf{a}_n - \sum_i \mathbf{q}_i r_{in}$; if nonzero, normalize to $\mathbf{q}_r$

▷ `Householder matrix` $H = I - 2\hat{v}\hat{v}^\top$ for unit $\hat{v}$; reflects across $\hat{v}^\perp$: $H\mathbf{x} = \mathbf{x} - 2(\mathbf{x}^\top \hat{v})\hat{v}$

▷ `Householder properties` $H = H^\top$, $H^\top H = I$, $H^2 = I$; eigenvalues: -1 ($\hat{v}$), 1 ($\hat{v}^\perp$)

▷ `Householder QR` $H_{r-1} \cdots H_0 A = R$ where H_k zeros below diagonal in column k; $A = H_0 \cdots H_{r-1} R = QR$; numerically stable

▷ `Rank-1 perturbation` $T = I - \beta \mathbf{u}\mathbf{u}^\top$; eigenvalues: $1-\beta$ ($\mathbf{u}$), 1 ($\mathbf{u}^\perp$)

▷ `Rank-1 special cases` $\beta = 0$: I; $\beta = 1$: $P_{\mathbf{u}^\perp}$; $\beta = 2$: Householder H

▷ `Rank-1 properties` $T = T^\top$; invertible iff $\beta \neq 1$, $T^{-1} = I + \frac{\beta}{1-\beta}\mathbf{u}\mathbf{u}^\top$; $\det(T) = 1 - \beta$, $\text{tr}(T) = d - \beta$

4.1 Rank: Advanced Properties and Numerical Computation

The basic definition of rank and linear independence were introduced in Chapter~2. The Rank-Nullity Theorem and the relationship between rank and the dimensions of the four fundamental subspaces are covered in Chapter~3. Here we focus on advanced properties and numerical methods for computing rank.

4.1.1 Theorem: Rank of Sum

For $A, B \in \mathbb{R}^{M\times N}$,

$$\operatorname{rank}(A+B) \leq \operatorname{rank}(A) + \operatorname{rank}(B) \tag{4.1}$$

4.1.2 Theorem: Rank of Product

For $A \in \mathbb{R}^{M\times d}$ and $B \in \mathbb{R}^{d\times N}$,

$$\operatorname{rank}(AB) \leq \min\{\operatorname{rank}(A), \operatorname{rank}(B)\} \tag{4.2}$$

4.1.3 Theorem: Rank Preservation with Orthonormal Matrices

1. If A has orthonormal columns, then

$$\operatorname{rank}(AB) = \operatorname{rank}(B) \tag{4.3}$$

Proof:

Let $A \in \mathbb{R}^{M\times r}$ have orthonormal columns, so $\operatorname{rank}(A) = r$ and $A^\top A = I_{r\times r}$.

Upper bound: From equation 4.2, $\operatorname{rank}(AB) \leq \min\{\operatorname{rank}(A), \operatorname{rank}(B)\} = \min\{r, \operatorname{rank}(B)\} \leq \operatorname{rank}(B)$.

Lower bound: Since $A^\top A = I_{r\times r}$, we have $B = A^\top(AB)$. Applying the rank inequality:

$$\operatorname{rank}(B) = \operatorname{rank}(A^\top(AB)) \leq \min\{r, \operatorname{rank}(AB)\}.$$

From the upper bound, $\operatorname{rank}(AB) \leq r$, so $\min\{r, \operatorname{rank}(AB)\} = \operatorname{rank}(AB)$. Therefore, $\operatorname{rank}(B) \leq \operatorname{rank}(AB)$.

Combining both inequalities gives $\operatorname{rank}(AB) = \operatorname{rank}(B)$.

Alternative proof using column space:

Since $\operatorname{Col}(AB) = A \cdot \operatorname{Col}(B)$ and A has orthonormal columns, the linear transformation A preserves dimension. Therefore, $\dim(\operatorname{Col}(AB)) = \dim(\operatorname{Col}(B))$, which implies $\operatorname{rank}(AB) = \operatorname{rank}(B)$.

2. If B has orthonormal rows, then

$$\operatorname{rank}(AB) = \operatorname{rank}(A) \tag{4.4}$$

4.1.4 Theorem: Rank of Gram Matrices

If $A \in \mathbb{R}^{M \times N}$ has rank r, then

$$\text{rank}(AA^\top) = \text{rank}(A^\top A) = r \tag{4.5}$$

Proof:

Upper bound: From equation 4.2, we have:

$$\text{rank}(AA^\top) \leq \min\{\text{rank}(A), \text{rank}(A^\top)\} = r,$$

and similarly $\text{rank}(A^\top A) \leq r$.

Lower bound: The column space of $AA^\top$ is $\text{Col}(AA^\top) = A \cdot \text{Col}(A^\top) = A \cdot \text{Row}(A)$. Since $\text{Row}(A) \cap N(A) = \{\mathbf{0}\}$ (from Chapter~3, $\text{Row}(A) \perp N(A)$), the linear transformation A is injective when restricted to $\text{Row}(A)$. Therefore, $\dim(\text{Col}(AA^\top)) = \dim(\text{Row}(A)) = r$, so $\text{rank}(AA^\top) = r$. Similarly, $\text{rank}(A^\top A) = r$.

Combining both bounds gives $\text{rank}(AA^\top) = \text{rank}(A^\top A) = r$.

This theorem is crucial for understanding how rank is preserved under Gram matrix formation.

> In machine learning, Gram matrices $AA^\top$ and $A^\top A$ appear frequently (e.g., kernel matrices, covariance matrices). The fact that they preserve rank means that the dimensionality of the feature space is maintained, which is important for understanding the capacity of models and the geometry of data.

4.2 Numerical Rank Computation

In practice, computing the exact rank of a matrix is challenging due to floating-point arithmetic. Small numerical errors can make a rank-deficient matrix appear full rank, or vice versa.

4.2.1 Definition: Numerical Rank

For a matrix $A \in \mathbb{R}^{M \times N}$ with singular values $\sigma_0 \geq \sigma_1 \geq \cdots \geq \sigma_{\min(M,N)-1} \geq 0$ (see Chapter~8 for SVD), the **numerical rank** with tolerance $\epsilon > 0$ is defined as

$$\text{rank}_\epsilon(A) = \min\{k : \sigma_k < \epsilon \cdot \sigma_0\} \tag{4.6}$$

That is, we count only singular values that are at least ϵ times the largest singular value.

> Numerical rank is essential in machine learning where data matrices are often nearly rank-deficient due to correlated features or redundant information. Using exact rank computation would incorrectly identify such matrices as full rank. Numerical rank with an appropriate tolerance (e.g., $\epsilon = 10^{-12}$ or machine epsilon) correctly identifies the effective dimensionality. This is crucial for:
>
> - **Feature selection**: Identifying redundant features
> - **Dimensionality reduction**: Determining the intrinsic dimensionality of data
> - **Model compression**: Finding low-rank approximations of weight matrices
> - **Regularization**: Understanding when regularization is necessary

4.2.2 Computing Rank via QR Decomposition

QR decomposition provides a numerically stable way to compute rank. For a matrix $A = QR$ where R is a generalized upper triangular matrix (see below), the rank equals the number of nonzero diagonal entries of R (or more precisely, the number of columns in Q).

However, due to numerical errors, we use a tolerance: the rank is the number of diagonal entries r_{ii} such that $|r_{ii}| > \epsilon \cdot ||A||_F$ for some tolerance ϵ.

> QR decomposition is more stable than Gaussian elimination for rank computation, especially for ill-conditioned matrices. The rank-revealing QR decomposition (RRQR) explicitly identifies the rank by permuting columns to ensure the most important columns come first. This is covered in Chapter~9.

4.3 Generalized Upper Triangular Matrices

Motivation: This definition generalizes classical upper triangular matrices to the rectangular and rank-deficient case arising in QR decomposition. In classical QR decomposition, when A is square and full rank, R is a standard upper triangular matrix (nonzero entries only on or above the diagonal). However, for rectangular or rank-deficient matrices, we need a more general structure that captures the essential properties while allowing for zeros and rectangular shapes. This generalized structure naturally arises from the Gram-Schmidt process applied to linearly dependent column vectors.

4.3.1 Definition: Generalized Upper Triangular Matrix

Let $R \in \mathbb{R}^{r \times N}$. Define

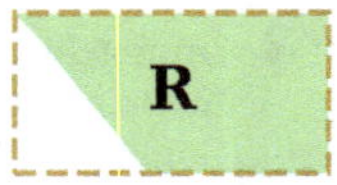

$$I_n \triangleq \max\{i : r_{in} \neq 0\}, \quad I_{-1} \triangleq -1,$$

and

$$\bar{I}_n = \max\{I_i : i = 0, \dots, n\}.$$

If

$$\bar{I}_n - \bar{I}_{n-1} \in \{0, 1\} \quad \forall n \in \{1, \dots, N-1\},$$

then R is a **generalized upper triangular matrix**.

> The generalized upper triangular matrix extends the standard upper triangular matrix to rectangular and rank-deficient cases. A **standard upper triangular matrix** is a square matrix $A \in \mathbb{R}^{N \times N}$ with $a_{ij} = 0$ for all $i > j$ (all entries below the main diagonal are zero). The **generalized version** allows rectangular shape ($r \times N$ with $r \leq N$), zero diagonal entries (rank-deficient case), and a more flexible "staircase" zero pattern where the last nonzero row index in each column increases by at most 1 from column to column. This structure naturally arises in QR decomposition of non-square or rank-deficient matrices, where the standard definition is too restrictive. (Standard upper triangular matrices are further studied in Chapter~5.)

4.3.1.1 Example: Generalized Upper Triangular Matrix

$$\begin{bmatrix} 3 & 1 & 2 & -5 & 0 & 4 & 1 \\ 0 & 5 & 0 & 1 & 8 & 7 & 0 \\ 0 & 0 & 0 & 1 & 2 & 3 & 2 \\ 0 & 0 & 0 & 0 & 1 & -2 & 0 \\ 0 & 0 & 0 & 0 & 0 & 0 & 5 \end{bmatrix}$$

4.4 QR Decomposition

4.4.1 Theorem: QR Decomposition

Let $A \in \mathbb{R}^{M \times N}$ with $\operatorname{rank}(A) = r$. Then

$$A = QR \tag{4.7}$$

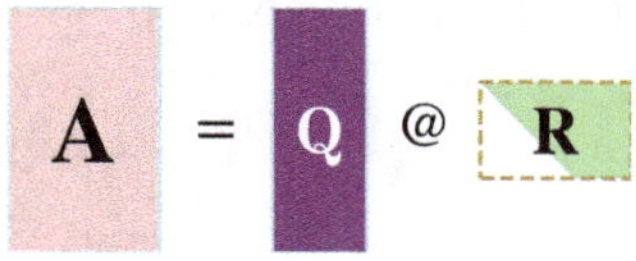

where:

1. $Q \in \mathbb{R}^{M \times r}$ consists of r **orthonormal** column vectors.
2. $R \in \mathbb{R}^{r \times N}$ is a generalized upper triangular matrix.

For each column n of A,

$$\mathbf{a}_n = \sum_{i=0}^{r-1} \mathbf{q}_i \cdot r_{in} = \sum_{i=0}^{\bar{r}_n} \mathbf{q}_i \cdot r_{in} \tag{4.8}$$

where the second equality holds since only the first $\bar{r}_n$ entries of column n in R are nonzero (by the definition of generalized upper triangular structure).

For a matrix $A = \begin{bmatrix} \mathbf{a}_0 & \cdots & \mathbf{a}_{N-1} \end{bmatrix} \in \mathbb{R}^{M \times N}$ (see Chapter~2 for the column vector representation), we ask two fundamental questions about the column span of A:

1. What is the dimension of $\operatorname{span}\{\mathbf{a}_0, \dots, \mathbf{a}_{N-1}\}$?
2. How do we find an orthonormal basis of this span?

QR decomposition provides direct answers to both questions:

1. The dimension of the column span is given by the rank r (the number of columns in Q).
2. The orthonormal basis of the column span is given by the columns of Q.

> This corresponds to a **reduced (thin) QR decomposition**, where Q contains only r orthonormal columns spanning the column space of A. This is the standard form used in practice, especially for rank-deficient or rectangular matrices.

4.4.1.1 Example: QR Decomposition

Do the QR decomposition of the following matrix:

$$A = \begin{bmatrix} 1 & -2 & 1 & -1 & 1 & -2 \\ 1 & -2 & 2 & 2 & 0 & 1 \\ -1 & 2 & 2 & 10 & -4 & 11 \\ 0 & 0 & 0 & 0 & 5 & 3 \\ 1 & -2 & 3 & 5 & -1 & 4 \end{bmatrix}.$$

Solution

We have

$$A = \begin{bmatrix} 1 & -2 & 1 & -1 & 1 & -2 \\ 1 & -2 & 2 & 2 & 0 & 1 \\ -1 & 2 & 2 & 10 & -4 & 11 \\ 0 & 0 & 0 & 0 & 5 & 3 \\ 1 & -2 & 3 & 5 & -1 & 4 \end{bmatrix} = \underbrace{\begin{bmatrix} \frac{1}{2} & 0 & 0 \\ \frac{1}{2} & \frac{1}{\sqrt{13}} & 0 \\ -\frac{1}{2} & \frac{3}{\sqrt{13}} & 0 \\ 0 & 0 & 1 \\ \frac{1}{2} & \frac{2}{\sqrt{13}} & 0 \end{bmatrix}}_{Q} \underbrace{\begin{bmatrix} 2 & -4 & 2 & -2 & 2 & -4 \\ 0 & 0 & \sqrt{13} & 3\sqrt{13} & -\sqrt{13} & 3\sqrt{13} \\ 0 & 0 & 0 & 0 & 5 & 3 \end{bmatrix}}_{R},$$

where $Q \in \mathbb{R}^{5 \times 3}$ has orthonormal columns and $R \in \mathbb{R}^{3 \times 6}$ is a generalized upper triangular matrix. The rank of A is $r = 3$, which equals the number of columns in Q and the number of rows in R.

4.4.2 Theorem: Orthonormality of Q in QR Decomposition

For QR decomposition $A = QR$ (defined above, *4.7*),

$$Q^\top Q = I_{r\times r} \tag{4.9}$$

This theorem states that the columns of Q are **orthonormal**, which is equivalent to $Q^\top Q = I_{r\times r}$. The (i,j)-th entry of $Q^\top Q$ is $\mathbf{q}_i^\top \mathbf{q}_j$ (the inner product of the i-th and j-th columns of Q). When $i = j$, we have $\mathbf{q}_i^\top \mathbf{q}_i = ||\mathbf{q}_i||_2^2 = 1$ (unit vectors), and when $i \neq j$, we have $\mathbf{q}_i^\top \mathbf{q}_j = 0$ (orthogonal vectors). Therefore, $Q^\top Q$ is the $r \times r$ identity matrix, confirming that the columns of Q form an orthonormal basis.

If Q is square ($M = r$), then $Q^\top = Q^{-1}$, meaning Q is an orthogonal matrix. For rectangular Q ($M > r$), we still have $Q^\top Q = I_{r\times r}$ (orthonormal columns), but $QQ^\top \neq I_{M\times M}$. This distinction is crucial: square orthogonal matrices act as global rotations or reflections, preserving geometry (lengths and angles) and mapping the entire space $\mathbb{R}^M$ onto itself, while rectangular matrices with orthonormal columns represent a **distance-preserving embedding** (also called an **isometric embedding**), strictly preserving **inner products** (and thus both norms and angles) when mapping vectors from $\mathbb{R}^r$ to a subspace of $\mathbb{R}^M$.

Specifically, if θ is the angle between vectors $\mathbf{x}$ and $\mathbf{y}$, the angle between $Q\mathbf{x}$ and $Q\mathbf{y}$ remains exactly θ. This follows from the **preservation of inner products** (see Chapter~3, Theorem: Inner-Product Preservation):

$$\langle Q\mathbf{x}, Q\mathbf{y}\rangle = \mathbf{x}^\top Q^\top Q\mathbf{y} = \mathbf{x}^\top I_{r\times r}\mathbf{y} = \mathbf{x}^\top \mathbf{y} = \langle \mathbf{x}, \mathbf{y}\rangle.$$

Since $\cos\theta = \frac{\langle \mathbf{x},\mathbf{y}\rangle}{\|\mathbf{x}\|_2\|\mathbf{y}\|_2} = \frac{\langle Q\mathbf{x},Q\mathbf{y}\rangle}{\|Q\mathbf{x}\|_2\|Q\mathbf{y}\|_2}$, the angle is invariant.

While **column-orthonormal matrices** (see Chapter~3) also preserve signal norms (acting as distance-preserving embeddings between layers of different sizes), **orthonormal matrices** (square, $M = r$) are uniquely important for Recurrent Neural Networks (RNNs). In iterative processes like RNNs, the input and output dimensions must match to allow repeated application of the same weight matrix W. The hidden state h_t represents the model's "memory" at the current time step: $h_t = \sigma(W \cdot h_{t-1})$, where W is an $M \times M$ orthonormal matrix. The hidden state vector rotates within the same M-dimensional space, maintaining its norm while exploring all directions—this is why **Orthogonal Initialization** is crucial for training deep recurrent networks.
In feed-forward and deep networks, column-orthonormal matrices also preserve signal norms. Consider layer-to-layer propagation: $h_l = W \cdot h_{l-1}$, where $h_{l-1} \in \mathbb{R}^r$ and $W \in \mathbb{R}^{M\times r}$ ($M > r$) is column-orthonormal ($W^\top W = I$). Although the dimension changes (from r to M), the signal's **energy (norm) and geometric structure (angles)** are perfectly preserved through this distance-preserving embedding. The signal does not "die out"—this is one reason why ResNet and other high-dimensional mapping networks work: as long as orthogonality is maintained, signals do not attenuate during forward propagation. In generative models (e.g., GANs or VAEs), a column-orthonormal matrix $Q \in \mathbb{R}^{M\times r}$ maps a low-dimensional latent code $\mathbf{z} \in \mathbb{R}^r$ to a high-dimensional image space $\mathbb{R}^M$. The generated images lie on a r-dimensional manifold embedded in $\mathbb{R}^M$, preserving the semantic structure of the latent space without geometric distortion.

4.5 Gram–Schmidt Algorithm for QR Decomposition

4.5.1 Algorithm: QR via Gram–Schmidt

The algorithm presented above is the classical Gram–Schmidt process. In practice, **modified Gram–Schmidt** or **Householder QR** is preferred for numerical stability, especially when dealing with nearly linearly dependent vectors or large-scale computations.

Algorithm 3 QR via Gram–Schmidt

Input: $A \in \mathbb{R}^{M \times N}$, $\mathbf{a}_0 \neq 0$
Output: $Q \in \mathbb{R}^{M \times r}$ (orthonormal columns) and $R \in \mathbb{R}^{r \times N}$ such that $A = QR$
  $\mathbf{q}_0 \leftarrow \frac{\mathbf{a}_0}{||\mathbf{a}_0||_2}$ ▷ Normalize first column
  $r_{00} \leftarrow ||\mathbf{a}_0||_2$
  $r \leftarrow 1$
  for $n = 1, \dots, N-1$ **do**
    Compute projections: $r_{in} \leftarrow \mathbf{a}_n^\top \mathbf{q}_i$ for $i = 0, \dots, r-1$
    Residual: $\text{res}(\mathbf{a}_n) \leftarrow \mathbf{a}_n - \sum_{i=0}^{r-1} \mathbf{q}_i r_{in}$
    if $\text{res}(\mathbf{a}_n) \neq 0$ **then**
      $\mathbf{q}_r \leftarrow \frac{\text{res}(\mathbf{a}_n)}{||\text{res}(\mathbf{a}_n)||_2}$ ▷ Normalize residual
      $r_{r,n} \leftarrow ||\text{res}(\mathbf{a}_n)||_2$
      $r \leftarrow r + 1$
    end if
  end for

4.6 Householder Matrix and Householder Transformation

The **Householder transformation** (also called **Householder reflection**) is a fundamental tool in numerical linear algebra for constructing orthonormal bases and performing QR decomposition. Unlike Gram–Schmidt, which builds orthonormal vectors incrementally, Householder transformations use geometric reflections to zero out matrix entries systematically.

4.6.1 Definition: Householder Transformation

Given a nonzero vector $\mathbf{v} \in \mathbb{R}^d$, define the unit vector $\hat{v} = \frac{\mathbf{v}}{||\mathbf{v}||_2}$. The **Householder transformation** that reflects any vector $\mathbf{x}$ across the hyperplane orthogonal to $\hat{v}$ is defined by the **Householder matrix**:

$$H = I - 2\hat{v}\hat{v}^\top \tag{4.10}$$

Equivalently, using the non-normalized vector $\mathbf{v}$, the Householder matrix can be written as:

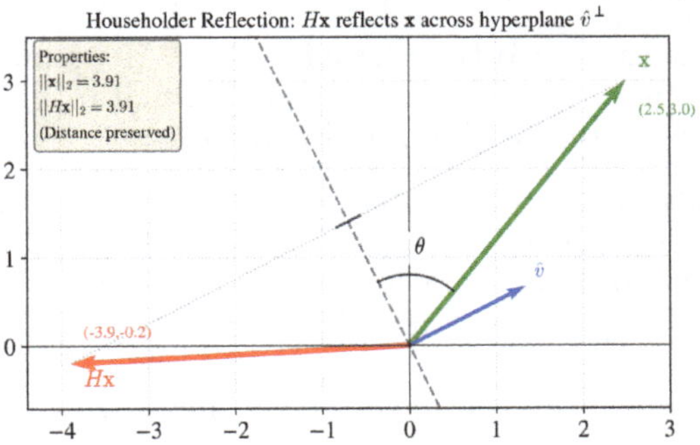

Figure 4.1: Householder Reflection

$$H = I - 2\frac{\mathbf{v}\mathbf{v}^\top}{\mathbf{v}^\top\mathbf{v}} \tag{4.11}$$

Proof of the Householder Reflection Formula:

We prove that $H = I - 2\hat{v}\hat{v}^\top$ reflects any vector $\mathbf{x}$ across the hyperplane $\hat{v}^\perp$.

Step 1: Decompose x into parallel and perpendicular components

Any vector $\mathbf{x}$ can be uniquely decomposed as:

$$\mathbf{x} = (\mathbf{x}^\top \hat{v})\hat{v} + (\mathbf{x} - (\mathbf{x}^\top \hat{v})\hat{v}),$$

where:

- $(\mathbf{x}^\top \hat{v})\hat{v}$ is the component parallel to $\hat{v}$ (the projection onto $\hat{v}$)
- $\mathbf{x} - (\mathbf{x}^\top \hat{v})\hat{v}$ is the component perpendicular to $\hat{v}$ (lying in the hyperplane $\hat{v}^\perp$)

Verification of the decomposition:

1. **The sum equals x:**

$$(\mathbf{x}^\top\hat{v})\hat{v} + (\mathbf{x} - (\mathbf{x}^\top\hat{v})\hat{v}) = (\mathbf{x}^\top\hat{v})\hat{v} + \mathbf{x} - (\mathbf{x}^\top\hat{v})\hat{v} = \mathbf{x}.$$

2. **The first term is parallel to $\hat{v}$:** Since $(\mathbf{x}^\top\hat{v})\hat{v}$ is a scalar multiple of $\hat{v}$, it is parallel to $\hat{v}$.
3. **The second term is perpendicular to $\hat{v}$:**

$$\hat{v}^\top(\mathbf{x} - (\mathbf{x}^\top\hat{v})\hat{v}) = \hat{v}^\top\mathbf{x} - (\mathbf{x}^\top\hat{v})(\hat{v}^\top\hat{v}) = \hat{v}^\top\mathbf{x} - (\mathbf{x}^\top\hat{v}) \cdot 1 = 0,$$

where we used $\hat{v}^\top\hat{v} = ||\hat{v}||_2^2 = 1$ (since $\hat{v}$ is a unit vector) and $\hat{v}^\top\mathbf{x} = \mathbf{x}^\top\hat{v}$ (both are scalars).

Step 2: Apply the Householder matrix to x

To reflect $\mathbf{x}$ across the hyperplane $\hat{v}^\perp$, we keep the perpendicular component unchanged but reverse the parallel component:

$$H\mathbf{x} = (I - 2\hat{v}\hat{v}^\top)\mathbf{x} = \mathbf{x} - 2\hat{v}(\hat{v}^\top\mathbf{x}) = \mathbf{x} - 2(\mathbf{x}^\top\hat{v})\hat{v}.$$

Substituting the decomposition:

$$H\mathbf{x} = [(\mathbf{x}^\top\hat{v})\hat{v} + (\mathbf{x} - (\mathbf{x}^\top\hat{v})\hat{v})] - 2(\mathbf{x}^\top\hat{v})\hat{v} = -(\mathbf{x}^\top\hat{v})\hat{v} + (\mathbf{x} - (\mathbf{x}^\top\hat{v})\hat{v}).$$

This shows that $H\mathbf{x}$ has:

- The perpendicular component unchanged: $\mathbf{x} - (\mathbf{x}^\top\hat{v})\hat{v}$ (same as in $\mathbf{x}$)
- The parallel component reversed: $-(\mathbf{x}^\top\hat{v})\hat{v}$ (opposite sign from $\mathbf{x}$)

Therefore, $H\mathbf{x}$ is the reflection of $\mathbf{x}$ across the hyperplane $\hat{v}^\perp$, which completes the proof.

Geometric interpretation: The Householder matrix H reflects any vector $\mathbf{x}$ across the hyperplane $\hat{v}^\perp$ (the subspace orthogonal to $\hat{v}$). Specifically, $H\mathbf{x}$ is the reflection of $\mathbf{x}$ across this hyperplane, as shown in Figure 4.1.

Application to matrices: Just like the projection matrix (see Chapter~2), the Householder matrix H can be applied to both vectors and matrices. When applied to a matrix $A \in \mathbb{R}^{d\times N}$, the Householder transformation HA reflects each column of A across the hyperplane $\hat{v}^\perp$. That is, if $A = \begin{bmatrix}\mathbf{a}_0 & \mathbf{a}_1 & \cdots & \mathbf{a}_{N-1}\end{bmatrix}$, then:

$$HA = \begin{bmatrix}H\mathbf{a}_0 & H\mathbf{a}_1 & \cdots & H\mathbf{a}_{N-1}\end{bmatrix},$$

where each column $H\mathbf{a}_n$ is the reflection of $\mathbf{a}_n$ across the hyperplane $\hat{v}^\perp$. This property is fundamental to Householder QR decomposition, where Householder matrices are applied to transform the columns of a matrix into upper triangular form.

4.6.2 Definition: Householder Matrix for Zeroing Entries

Given a vector $\mathbf{x} \in \mathbb{R}^d$ and a target vector $\mathbf{y} \in \mathbb{R}^d$ with $||\mathbf{y}||_2 = ||\mathbf{x}||_2$, we can construct a Householder matrix H such that $H\mathbf{x} = \mathbf{y}$.

A common choice is to reflect $\mathbf{x}$ to align with a coordinate axis. For example, to zero out all entries of $\mathbf{x}$ except the first, we set:

$$\mathbf{y} = \begin{bmatrix}||\mathbf{x}||_2 \\ 0 \\ \vdots \\ 0\end{bmatrix} \quad \text{or} \quad \mathbf{y} = \begin{bmatrix}-||\mathbf{x}||_2 \\ 0 \\ \vdots \\ 0\end{bmatrix}.$$

The Householder vector is then:

$$\mathbf{v} = \mathbf{x} - \mathbf{y},$$

and with $\hat{v} = \frac{\mathbf{v}}{\|\mathbf{v}\|_2}$, the Householder matrix is $H = I - 2\hat{v}\hat{v}^\top$, or equivalently $H = I - 2\frac{\mathbf{v}\mathbf{v}^\top}{\mathbf{v}^\top\mathbf{v}}$.

The choice of sign for $\mathbf{y}$ (positive or negative) affects numerical stability. Typically, we choose the sign opposite to the first component of $\mathbf{x}$ to avoid cancellation errors. This is known as the **signed Householder transformation**.

4.6.3 Properties of Householder Matrix

The Householder matrix is a cornerstone of numerical linear algebra and possesses several key properties:

1. **Symmetry**: $H = H^\top$.

   Proof: Since $H = I - 2\hat{v}\hat{v}^\top$ and $(I - 2\hat{v}\hat{v}^\top)^\top = I^\top - 2(\hat{v}\hat{v}^\top)^\top = I - 2\hat{v}\hat{v}^\top = H$.
2. **Orthogonality**: $H^\top H = I$ (equivalently, $H^{-1} = H^\top$).

   Proof:
   $$H^\top H = (I - 2\hat{v}\hat{v}^\top)(I - 2\hat{v}\hat{v}^\top) = I - 4\hat{v}\hat{v}^\top + 4\hat{v}\hat{v}^\top\hat{v}\hat{v}^\top.$$
   Since $\hat{v}$ is a unit vector, $\hat{v}^\top\hat{v} = ||\hat{v}||_2^2 = 1$, so:
   $$H^\top H = I - 4\hat{v}\hat{v}^\top + 4\hat{v}\hat{v}^\top = I.$$
3. **Involutory**: $H^2 = I$ (applying the reflection twice returns to the original).

   Proof: Since H is orthogonal and symmetric, $H^2 = HH = H^\top H = I$.
4. **Spectrum**: The eigenvalues of H are:
   - One eigenvalue of -1 with eigenvector $\hat{v}$ (the direction being reflected)
   - $d - 1$ eigenvalues of 1 with eigenspace $\hat{v}^\perp$ (the hyperplane of reflection)

   Proof:
   - For $\hat{v}$: $H\hat{v} = (I - 2\hat{v}\hat{v}^\top)\hat{v} = \hat{v} - 2\hat{v}(\hat{v}^\top\hat{v}) = \hat{v} - 2\hat{v} = -\hat{v}$, so $\hat{v}$ is an eigenvector with eigenvalue -1.
   - For any $\mathbf{w} \in \hat{v}^\perp$ (i.e., $\hat{v}^\top\mathbf{w} = 0$): $H\mathbf{w} = (I - 2\hat{v}\hat{v}^\top)\mathbf{w} = \mathbf{w} - 2\hat{v}(\hat{v}^\top\mathbf{w}) = \mathbf{w}$, so $\mathbf{w}$ is an eigenvector with eigenvalue 1.

The Householder matrix is a **reflection matrix**, which is a special type of orthogonal matrix. Unlike rotation matrices (which have determinant $+1$), reflection matrices have determinant -1. The spectrum property shows that Householder transformations preserve the norm of vectors in the hyperplane $\hat{v}^\perp$ while reversing the direction along $\hat{v}$. This geometric property makes Householder transformations ideal for zeroing out matrix entries while maintaining numerical stability.

4.6.4 Householder QR Decomposition

The **Householder QR algorithm** uses a sequence of Householder reflections to transform a matrix $A \in \mathbb{R}^{M\times N}$ into upper triangular form. The algorithm applies Householder matrices $H_0, H_1, \dots, H_{r-1}$ (where $r = \min(M, N)$) such that:

$$H_{r-1} \cdots H_1 H_0 A = R \tag{4.12}$$

where R is an upper triangular matrix. Since each H_k is orthogonal, we have:

$$A = H_0^\top H_1^\top \cdots H_{r-1}^\top R = QR \tag{4.13}$$

where $Q = H_0^\top H_1^\top \cdots H_{r-1}^\top$ is orthogonal (product of orthogonal matrices). Since each H_k is symmetric ($H_k = H_k^\top$), we can also write $Q = H_0 H_1 \cdots H_{r-1}$.

At each step k, the Householder matrix H_k is constructed to zero out entries below the diagonal in column k of the current matrix, without affecting previously zeroed entries.

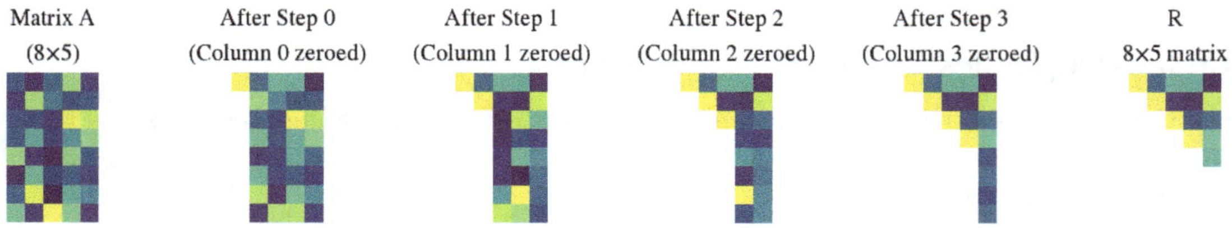

Figure 4.2: Householder QR Steps

Householder QR is preferred over Gram–Schmidt for numerical stability because:

1. **Backward stability**: Householder QR has better backward error bounds, meaning the computed Q and R satisfy $A + \Delta_A = QR$ where $||\Delta_A||$ is small relative to $||A||$.
2. **No accumulation of errors**: Unlike Gram–Schmidt, which accumulates rounding errors in the projection steps, Householder transformations are applied directly to the matrix.
3. **Efficiency**: For large matrices, Householder QR can be implemented efficiently using compact storage (storing only the Householder vectors, not the full matrices).

> In practice, libraries like LAPACK and NumPy use Householder QR (or its variant, the **compact WY representation**) for the `np.linalg.qr` function. This makes Householder QR the standard method for QR decomposition in production numerical linear algebra libraries.

4.7 Rank-1 Perturbation Theory

The Householder matrix $H = I - 2\mathbf{u}\mathbf{u}^\top$ is a special case of a more general class of transformations: **rank-1 perturbations of the identity matrix**. Understanding this generalization provides insights into how low-rank modifications affect matrix spectra and enables the design of learnable geometric transformations.

4.7.1 Definition: Rank-1 Perturbation of Identity

For a unit vector $\mathbf{u} \in \mathbb{R}^d$ (so that $||\mathbf{u}||_2 = 1$) and a scalar $\beta \in \mathbb{R}$, the **rank-1 perturbation of the identity** is defined as:

$$\boxed{T = I - \beta\mathbf{u}\mathbf{u}^\top \in \mathbb{R}^{d\times d}} \tag{4.14}$$

This operator represents a **rank-1 update** to the identity matrix, where the perturbation $\beta\mathbf{u}\mathbf{u}^\top$ is a rank-1 matrix (an outer product).

Rank-1 perturbations are fundamental in numerical linear algebra and machine learning. They enable efficient matrix updates (e. g., Sherman-Morrison formula for rank-1 updates to matrix inverses), low-rank approximations, and learnable transformations in neural networks. The parameter β controls the strength and nature of the perturbation, allowing interpolation between different geometric operations.

4.7.2 Theorem: Spectral Analysis of Rank-1 Perturbation

For the rank-1 perturbation $T = I - \beta \mathbf{u}\mathbf{u}^\top$ where $\mathbf{u}$ is a unit vector:

1. **Eigenvalues**:
   - One eigenvalue: $\lambda_0 = 1 - \beta$ with eigenvector $\mathbf{u}$
   - $d-1$ eigenvalues: $\lambda_i = 1$ for $i = 1, \dots, d-1$ with eigenspace $\mathbf{u}^\perp$
2. **Eigenvectors**:
   - The eigenvector corresponding to $\lambda_0 = 1 - \beta$ is $\mathbf{u}$ (the direction of perturbation)
   - The eigenvectors corresponding to $\lambda_i = 1$ span the hyperplane $\mathbf{u}^\perp$ (orthogonal to $\mathbf{u}$)

Proof:

1. **For u**:
$$T\mathbf{u} = (I - \beta \mathbf{u}\mathbf{u}^\top)\mathbf{u} = \mathbf{u} - \beta \mathbf{u}(\mathbf{u}^\top \mathbf{u}) = \mathbf{u} - \beta \mathbf{u} = (1-\beta)\mathbf{u}.$$
   Therefore, $\mathbf{u}$ is an eigenvector with eigenvalue $1 - \beta$.
2. **For any $\mathbf{w} \in \mathbf{u}^\perp$** (i.e., $\mathbf{u}^\top \mathbf{w} = 0$):
$$T\mathbf{w} = (I - \beta \mathbf{u}\mathbf{u}^\top)\mathbf{w} = \mathbf{w} - \beta \mathbf{u}(\mathbf{u}^\top \mathbf{w}) = \mathbf{w} - \beta \mathbf{u} \cdot 0 = \mathbf{w}.$$
   Therefore, any vector orthogonal to $\mathbf{u}$ is an eigenvector with eigenvalue 1.
3. **Dimension count**: Since $\mathbf{u}^\perp$ has dimension $d-1$ and $\mathbf{u}$ spans a 1-dimensional space, we have d linearly independent eigenvectors, confirming the complete eigensystem.

4.7.3 Special Cases: Geometric Interpretations

The parameter β controls the geometric nature of the transformation:

1. **$\beta = 0$: Identity**
$$T = I - 0 \cdot \mathbf{u}\mathbf{u}^\top = I.$$
   No transformation is applied; all eigenvalues equal 1.
2. **$\beta = 1$: Orthogonal Projection**
$$T = I - \mathbf{u}\mathbf{u}^\top = P_{\mathbf{u}^\perp}.$$
   This is the orthogonal projection matrix onto the hyperplane $\mathbf{u}^\perp$ (the orthogonal complement of $\mathbf{u}$). The spectrum is: one eigenvalue 0 ($\mathbf{u}$ is annihilated) and $d-1$ eigenvalues 1 ($\mathbf{u}^\perp$ is preserved). Subtracting the rank-one term $\mathbf{u}\mathbf{u}^\top$ to eliminate one direction while leaving the orthogonal subspace unchanged is the mechanism behind **deflation** in the power method (Chapter~6, Algorithm~7), where each computed eigencomponent is subtracted from A before the next one is sought.
3. **$\beta = 2$: Householder Reflection**
$$T = I - 2\mathbf{u}\mathbf{u}^\top = H.$$
   This is the Householder reflection matrix. The spectrum is: one eigenvalue -1 (direction $\mathbf{u}$ is reflected) and $d-1$ eigenvalues 1 (hyperplane is preserved).

4. **$\beta > 2$: Over-reflection**
   - For $\beta > 2$, the eigenvalue $1 - \beta < -1$, meaning the component along $\mathbf{u}$ is not just reflected but "over-reflected" (scaled by a factor greater than 1 in magnitude).
5. **$\beta < 0$: Amplification**
   - For $\beta < 0$, the eigenvalue $1 - \beta > 1$, meaning the component along $\mathbf{u}$ is amplified rather than reduced.

> The ability to interpolate between identity ($\beta = 0$), projection ($\beta = 1$), and reflection ($\beta = 2$) through a single learnable parameter β is powerful in machine learning. This enables networks to learn data-dependent geometric transformations that can smoothly transition between preserving information (identity), selectively forgetting information (projection), or inverting information (reflection). The spectral analysis shows that rank-1 perturbations are **selective**: they modify only one eigenvalue (along $\mathbf{u}$) while preserving all others, making them computationally efficient and interpretable.

4.7.4 Properties of Rank-1 Perturbation

For $T = I - \beta\mathbf{u}\mathbf{u}^\top$:

1. **Symmetry**: $T = T^\top$ (symmetric for all β).
2. **Invertibility**: T is invertible if and only if $\beta \neq 1$ (i.e., $1 - \beta \neq 0$). When invertible:

$$T^{-1} = I + \frac{\beta}{1-\beta}\mathbf{u}\mathbf{u}^\top \tag{4.15}$$

   This follows from the Sherman-Morrison formula for rank-1 updates.
3. **Determinant**:

$$\det(T) = 1 - \beta \tag{4.16}$$

   (since one eigenvalue is $1 - \beta$ and $d - 1$ eigenvalues are 1).
4. **Trace**:

$$\text{tr}(T) = d - \beta \tag{4.17}$$

   (sum of eigenvalues: $(1 - \beta) + (d - 1) \cdot 1 = d - \beta$).

> Rank-1 perturbations have several important applications in machine learning. In neural networks, rank-1 perturbations can replace standard identity shortcuts in residual connections, allowing the network to learn data-dependent geometric transformations with a single learnable parameter $\beta(\mathbf{x})$. Rank-1 updates also enable efficient incremental learning and online optimization, where model parameters are updated via rank-1 corrections rather than full matrix updates. Instead of strictly enforcing orthogonality (as in orthogonal neural networks), rank-1 perturbations allow soft, learnable deviations from orthogonality while maintaining interpretable geometric structure. Since rank-1 perturbations are parameterized by a single vector $\mathbf{u}$ and scalar β, they require only $O(d)$ parameters instead of $O(d^2)$ for a full matrix, enabling memory-efficient transformations in high-dimensional spaces. The theory of rank-1 perturbations bridges classical numerical linear algebra (Householder reflections, QR decomposition) with modern deep learning (learnable transformations, adaptive architectures). By understanding how a single parameter β controls the spectrum and geometry of the transformation, we can design more interpretable and efficient neural network architectures that leverage geometric structure while maintaining the flexibility to learn from data.

4.8 Mindmap

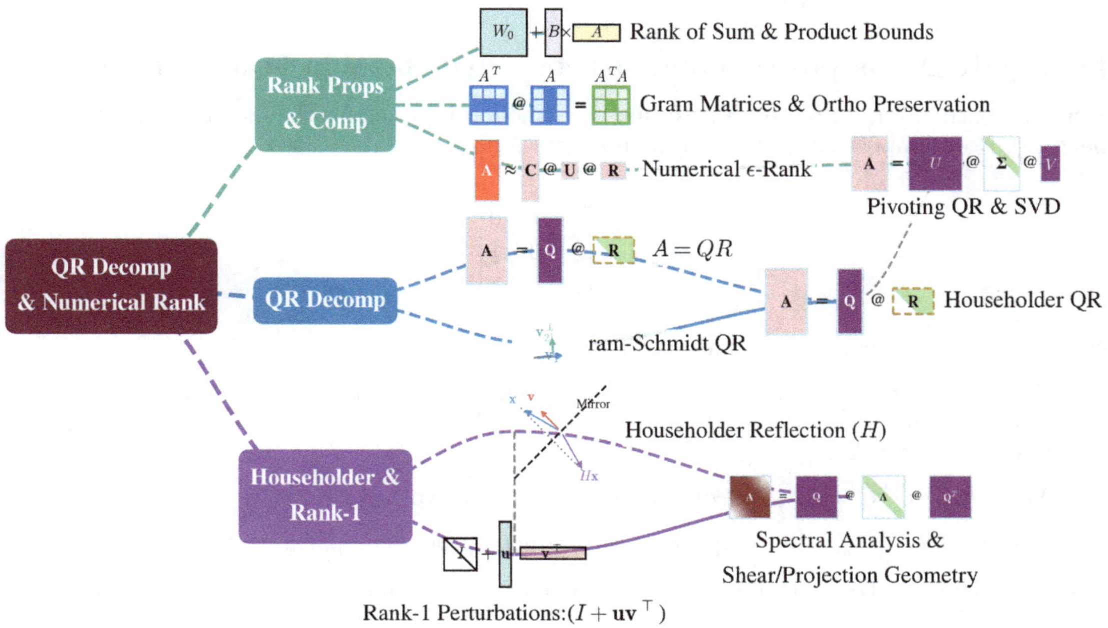

Figure 4.3: Mind map of QR decomposition and numerical rank topics

4.9 Exercises

4.9.1 Feature Redundancy and Rank in Linear Models

In a linear regression model, the design matrix $X \in \mathbb{R}^{N\times d}$ contains N samples and d features. When features are linearly dependent, the model has redundant parameters and $X^\top X$ becomes singular. Consider the design matrix:

$$X = \begin{bmatrix} 1 & 2 & 3 \\ 2 & 4 & 6 \\ 3 & 6 & 9 \end{bmatrix},$$

where each row is a sample and each column is a feature.

1. Compute $\text{rank}(X)$ and identify which features are redundant.
2. Show that $\text{rank}(X^\top X) = \text{rank}(X)$ by computing $X^\top X$ and its rank.
3. For the matrix $A = \begin{bmatrix} 1 & 0 & 1 \\ 0 & 1 & 1 \\ 1 & 1 & 0 \end{bmatrix}$, compute $\text{rank}(A)$ and determine whether $A^\top A$ is invertible.

4. For the matrix $B = \begin{bmatrix} 1 & 2 \\ 3 & 4 \\ 5 & 6 \end{bmatrix}$, compute rank($B$) and rank($B^\top B$).

4.9.2 QR Decomposition and Orthonormal Bases in Feature Extraction

In feature extraction, we often want to find an orthonormal basis for the column space of a data matrix. QR decomposition provides this directly. Consider a data matrix:

$$X = \begin{bmatrix} 1 & 1 & 0 \\ 1 & 0 & 1 \\ 0 & 1 & 1 \end{bmatrix},$$

where each column represents a feature vector.

1. Apply the Gram–Schmidt algorithm to compute the QR decomposition $X = QR$, where $Q \in \mathbb{R}^{3\times r}$ has orthonormal columns and $R \in \mathbb{R}^{r\times 3}$ is a generalized upper triangular matrix. (Here $r = \text{rank}(X)$.)
2. Verify that $Q^\top Q = I_{r\times r}$ by computing the product explicitly.
3. Verify that $X = QR$ by computing the product QR and comparing with X.
4. Show that the columns of Q form an orthonormal basis for the column space $C(X)$ of X.

4.9.3 Rank and Column Space in Overdetermined Systems

In least squares problems, we solve $X\theta = \mathbf{y}$ where $X \in \mathbb{R}^{N\times d}$ with $N > d$ (more samples than features). The solution exists and is unique if and only if X has full column rank (rank(X) = d). Consider the design matrix:

$$X = \begin{bmatrix} 1 & 2 & 3 \\ 2 & 4 & 6 \\ 1 & 1 & 1 \end{bmatrix}.$$

1. Compute rank(X) and determine whether X has full column rank.
2. Find a basis for the column space $C(X)$ using QR decomposition. Specifically, compute $X = QR$ and identify which columns of Q form the basis.
3. Express the third column $\mathbf{x}_2 = \begin{bmatrix} 3 \\ 6 \\ 1 \end{bmatrix}$ of X as a linear combination of the basis vectors from part 2.
4. Show that if X does not have full column rank, then $X^\top X$ is singular and the normal equations $X^\top X\theta = X^\top \mathbf{y}$ have infinitely many solutions.

4.9.4 Numerical Rank and Near-Rank Deficiency in Data Matrices

In practice, data matrices often have nearly linearly dependent columns due to correlated features. Numerical rank computation uses a tolerance to identify the effective rank. Consider a matrix with small numerical errors:

$$A = \begin{bmatrix} 1 & 2 & 2.0001 \\ 2 & 4 & 4.0002 \\ 3 & 6 & 6.0003 \end{bmatrix}.$$

1. Compute the exact rank of A (treating it as exact).
2. Show that if we use a tolerance $\epsilon = 10^{-3}$ and define numerical rank as the number of diagonal entries r_{ii} in the R matrix from QR decomposition such that $|r_{ii}| > \epsilon \cdot ||A||_F$, then the numerical rank may differ from the exact rank.
3. For the matrix $B = \begin{bmatrix} 1 & 1.001 \\ 1 & 1.002 \\ 1 & 1.003 \end{bmatrix}$, compute its QR decomposition and determine the numerical rank with tolerance $\epsilon = 10^{-4}$.
4. Explain why numerical rank is more useful than exact rank for feature selection in machine learning when features are highly correlated.

4.9.5 Rank Preservation and Orthonormal Transformations

In neural networks, orthonormal weight matrices preserve the rank of activations, which is important for maintaining information flow. Let $A \in \mathbb{R}^{M \times N}$ with $\text{rank}(A) = r$, and let $Q \in \mathbb{R}^{M \times M}$ be an orthonormal matrix (i.e., $Q^\top Q = I_{M \times M}$).

1. Prove that $\text{rank}(QA) = \text{rank}(A)$.
2. Prove that $\text{rank}(AQ^\top) = \text{rank}(A)$ when $Q \in \mathbb{R}^{N \times N}$ is orthonormal.
3. For $A = \begin{bmatrix} 1 & 2 & 3 \\ 2 & 4 & 6 \end{bmatrix}$ and $Q = \frac{1}{\sqrt{2}} \begin{bmatrix} 1 & 1 \\ 1 & -1 \end{bmatrix}$, verify that $\text{rank}(QA) = \text{rank}(A)$ by computing both ranks explicitly.
4. Show that if $A = QR$ is the QR decomposition of A, then $\text{rank}(A) = \text{rank}(R)$.

4.9.6 Gram Matrices and Rank in Kernel Methods

In kernel methods, we work with Gram matrices $K = XX^\top$ where $X \in \mathbb{R}^{N \times d}$ is a data matrix. The rank of K equals the rank of X, which determines the effective dimensionality of the feature space.

1. Prove that $\text{rank}(XX^\top) = \text{rank}(X)$ for any matrix $X \in \mathbb{R}^{N \times d}$.
2. For $X = \begin{bmatrix} 1 & 2 \\ 3 & 4 \\ 5 & 6 \end{bmatrix}$, compute $K = XX^\top$ and verify that $\text{rank}(K) = \text{rank}(X)$.
3. Show that if $X = QR$ is the QR decomposition of X, then $K = XX^\top = Q(RR^\top)Q^\top$ and $\text{rank}(K) = \text{rank}(RR^\top) = \text{rank}(R)$.
4. In kernel ridge regression, we solve $(K + \lambda I)\alpha = \mathbf{y}$ where $K = XX^\top$ is the Gram matrix. Prove that if X has full column rank d and $N > d$, then K has rank d (not full rank N), meaning K is singular. Explain why adding λI with $\lambda > 0$ makes $K + \lambda I$ invertible.

4.9.7 Householder Transformations and QR Decomposition

The Householder transformation is a fundamental tool for numerical QR decomposition. Consider a matrix $A \in \mathbb{R}^{4 \times 3}$:

$$A = \begin{bmatrix} 1 & 2 & 3 \\ 2 & 3 & 4 \\ 3 & 4 & 5 \\ 4 & 5 & 6 \end{bmatrix}.$$

1. Construct a Householder matrix H_0 that zeros out entries below the diagonal in the first column of A. Specifically, find the unit vector $\hat{v}$ such that $H_0 = I - 2\hat{v}\hat{v}^\top$ reflects the first column $\mathbf{a}_0 = \begin{bmatrix} 1 & 2 & 3 & 4 \end{bmatrix}^\top$ to align with $\begin{bmatrix} ||\mathbf{a}_0||_2 & 0 & 0 & 0 \end{bmatrix}^\top$.
2. Apply H_0 to A to get $A_1 = H_0 A$. Verify that the first column of A_1 has zeros below the first entry.
3. Show that H_0 is orthogonal (i.e., $H_0^\top H_0 = I$) and symmetric (i.e., $H_0 = H_0^\top$).
4. For a Householder matrix $H = I - 2\hat{v}\hat{v}^\top$ where $\hat{v} \in \mathbb{R}^d$ is a unit vector, prove that $H^2 = I$ (i.e., H is involutory). What does this mean geometrically?

4.9.8 Rank-1 Perturbations and Spectral Analysis

Rank-1 perturbations $T = I - \beta\mathbf{u}\mathbf{u}^\top$ generalize Householder transformations and have important applications in machine learning.

1. For a unit vector $\mathbf{u} = \frac{1}{\sqrt{3}} \begin{bmatrix} 1 & 1 & 1 \end{bmatrix}^\top \in \mathbb{R}^3$ and $\beta = 2$, compute the rank-1 perturbation $T = I - \beta\mathbf{u}\mathbf{u}^\top$ explicitly.
2. Verify that $\mathbf{u}$ is an eigenvector of T with eigenvalue $1-\beta = -1$. Find two linearly independent vectors orthogonal to $\mathbf{u}$ and verify that they are eigenvectors with eigenvalue 1.
3. Compute the determinant and trace of T from part 1, and verify that $\det(T) = 1 - \beta$ and $\operatorname{tr}(T) = d - \beta$ where $d = 3$.
4. For $\beta = 1$, show that $T = I - \mathbf{u}\mathbf{u}^\top$ is the orthogonal projection matrix onto the hyperplane $\mathbf{u}^\perp$. What are the eigenvalues and eigenvectors in this case? Explain why T is not invertible when $\beta = 1$.

Chapter 5

Square Matrix and LU Decomposition

Square matrices: triangular matrices, trace, determinants, and LU decomposition with variants and extensions

> *"Mathematics is not yet ready for such problems."*
>
> — Paul Erdős

- ▷ `Upper triangular` $U \in \mathbb{R}^{N \times N}$ with $u_{ij} = 0 \ \forall i > j$; $U \Leftrightarrow U^\top$ is l.t.
- ▷ `Triangular inverse` $(U^{-1})_{nn} = \frac{1}{u_{nn}}$ and U^{-1} is u.t.; U invertible $\Leftrightarrow u_{nn} \neq 0 \ \forall n$
- ▷ `Trace` $\mathrm{tr}(A+B) = \mathrm{tr}(A) + \mathrm{tr}(B)$, $\mathrm{tr}(\alpha A) = \alpha \mathrm{tr}(A)$, $\mathrm{tr}(AB) = \mathrm{tr}(BA)$
- ▷ `Cyclic trace` $\mathrm{tr}(ABC) = \mathrm{tr}(BCA) = \mathrm{tr}(CAB)$ for $A \in \mathbb{R}^{M \times N}$, $B \in \mathbb{R}^{N \times P}$, $C \in \mathbb{R}^{P \times M}$
- ▷ `Similarity invariance` $\mathrm{tr}(PAP^{-1}) = \mathrm{tr}(A)$ for invertible P
- ▷ `Trace derivative` $\frac{d}{dX}\mathrm{tr}(AX) = A^\top$ for $A \in \mathbb{R}^{M \times N}$, $X \in \mathbb{R}^{N \times M}$
- ▷ `Trace of AA.T` $\mathrm{tr}(AA^\top) = \sum_{m=0}^{M-1} \sum_{n=0}^{N-1} a_{mn}^2 = ||A||_F^2$
- ▷ `Determinant` $\det(A) = \det(A^\top)$, $\det(AB) = \det(A)\det(B)$, $\det(A^{-1}) = (\det A)^{-1}$; scaling row/column scales det; swapping flips sign; row addition unchanged
- ▷ `Laplace expansion` $\det(A) = \sum_{j=0}^{N-1} (-1)^{i+j} a_{ij} \det(A_{ij})$, A_{ij} removes row i, column j
- ▷ `Triangular determinant` If A triangular, $\det(A) = \prod_{n=0}^{N-1} a_{nn}$
- ▷ `Invertibility` For $A \in \mathbb{R}^{N \times N}$: $\det(A) \neq 0 \Leftrightarrow \mathrm{rank}(A) = N \Leftrightarrow A$ invertible
- ▷ `Leading principal minor` A_n is top-left $n \times n$ submatrix; $\det(A_n)$ is n-th LPM
- ▷ `LU decomp` If all LPMs non-zero, $A = LU$ where L with $l_{ii} = 1$
- ▷ `Determinant via LU` If $A = LU$, then $\det(A) = \prod_{n=0}^{N-1} u_{nn}$
- ▷ `Inverse via LU` If $A = LU$, then $A^{-1} = U^{-1}L^{-1}$
- ▷ `PLU decomp` $A \in \mathbb{R}^{M \times N}$, $A = PL_1U$ where P is permutation matrix
- ▷ `LDU decomp` If A admits LU, $A = L_1DU_1$; from LU by $U = DU'$

5.1 Triangular Matrices

In Chapter~4, we introduced the **generalized upper triangular matrix** to handle rectangular and rank-deficient matrices in QR decomposition. Here we focus on the standard case: square upper and lower triangular matrices, which have simpler structure and appear frequently in matrix decompositions and linear system solving.

5.1.1 Definition: Upper Triangular Matrix

A matrix $A \in \mathbb{R}^{N \times N}$ is **upper triangular** if $a_{ij} = 0 \quad \forall i > j$.

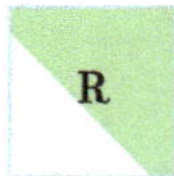

5.1.2 Definition: Lower Triangular Matrix

A matrix $A \in \mathbb{R}^{N \times N}$ is **lower triangular** if $a_{ij} = 0 \quad \forall i < j$.

5.1.3 Theorem: Transpose of Triangular Matrix

A matrix A is upper triangular **iff** $A^\top$ is lower triangular.

5.1.4 Algorithm: Inverse of an Upper Triangular Matrix

Algorithm 4 Inverse of an Upper Triangular Matrix

Input: Invertible upper triangular matrix $U \in \mathbb{R}^{N \times N}$
Output: U^{-1} (also upper triangular)
  for $n = 0, \dots, N-1$ **do**
    Solve $U\mathbf{v}_n = \mathbf{e}_n$ by back substitution ▷ $\mathbf{e}_n$ is the n-th standard basis vector
  end for
  return $U^{-1} = \begin{bmatrix} \mathbf{v}_0 & \cdots & \mathbf{v}_{N-1} \end{bmatrix}$

5.1.5 Theorem: Inverse of Upper Triangular Matrix

If U is an invertible upper triangular matrix:

1. $(U^{-1})_{nn} = \frac{1}{u_{nn}}$.
2. U^{-1} is also upper triangular.

5.1.6 Theorem: Invertibility of Upper Triangular Matrix

An upper triangular matrix U is invertible **iff** $u_{nn} \neq 0 \quad \forall n$.

This follows from the fact that $\det(U) = \prod_{n=0}^{N-1} u_{nn}$ (see equation 5.10), and a matrix is invertible **iff** its determinant is non-zero.

5.1.7 Theorem: Inverse of Lower Triangular Matrix

If L is an invertible lower triangular matrix:

1. $(L^{-1})_{nn} = \frac{1}{l_{nn}}$.
2. L^{-1} is also lower triangular.

5.1.8 Theorem: Woodbury Identity — Special Case for Ridge Regularization

Ridge regression (also known as L2 regularization or Tikhonov regularization)—fitting a linear model with a penalty term $\lambda\|\mathbf{w}\|_2^2$ in the loss—leads to a closed-form solution that involves inverting $A^\top A + \lambda I_N$, where A is the design matrix and $\lambda > 0$ is the regularization parameter. When the number of features N is large, that inverse is costly to compute. It turns out we can rewrite it in terms of the *other* Gram matrix, $AA^\top + \lambda I_M$, which is only $M \times M$ (the number of samples). If $M \ll N$, inverting the smaller matrix is much cheaper. The identity that makes this possible is a special case of the Woodbury (Sherman–Morrison–Woodbury) matrix inversion lemma.

Statement. For any $A \in \mathbb{R}^{M\times N}$, with I_N and I_M the identity matrices of size $N \times N$ and $M \times M$ respectively,

$$(A^\top A + I_N)^{-1} = I_N - A^\top(AA^\top + I_M)^{-1}A \tag{5.1}$$

(Here we have set $\lambda = 1$; for general $\lambda > 0$, replace I_N and I_M by λI_N and λI_M.) So the inverse of the $N \times N$ matrix $A^\top A + I_N$ is expressed using the inverse of the $M \times M$ matrix $AA^\top + I_M$—no $N \times N$ inversion needed.

Proof. To verify the identity, we multiply the right-hand side by $(A^\top A + I_N)$ and show that the result is I_N:

$$\begin{aligned}
&(A^\top A + I_N)\left[I_N - A^\top(AA^\top + I_M)^{-1}A\right] \\
&= (A^\top A + I_N) - (A^\top A + I_N)A^\top(AA^\top + I_M)^{-1}A \\
&= A^\top A + I_N - A^\top AA^\top(AA^\top + I_M)^{-1}A - A^\top(AA^\top + I_M)^{-1}A \\
&= A^\top A + I_N - A^\top\left[AA^\top + I_M\right](AA^\top + I_M)^{-1}A \\
&= A^\top A + I_N - A^\top A = I_N.
\end{aligned}$$

The matrix $(A^\top A + I_N)$ is positive definite (hence invertible), so the identity follows.

> Both $A^\top A + I_N$ and $AA^\top + I_M$ are positive definite: each is a positive semidefinite Gram matrix plus a positive definite identity. That is why the inverses in *5.1* exist. The same argument applies with λI in place of I for any $\lambda > 0$.

Why ridge regression handles singular design matrices. When the columns of A are linearly dependent, $A^\top A$ is singular and the normal equations $A^\top A\mathbf{w} = A^\top\mathbf{b}$ may have no solution or infinitely many. Adding λI_N with $\lambda > 0$ shifts every eigenvalue of $A^\top A$ by λ, so $A^\top A + \lambda I_N$ becomes positive definite and invertible. Ridge regression therefore always has a unique solution $\mathbf{w} = (A^\top A + \lambda I_N)^{-1}A^\top\mathbf{b}$, even when the design matrix is rank deficient. Equivalently, one minimizes $\|A\mathbf{w} - \mathbf{b}\|^2 + \lambda\|\mathbf{w}\|^2$; the Woodbury form with $(AA^\top + \lambda I_M)^{-1}$ is preferable whenever $M \ll N$.

Computational advantage. When $M \ll N$ (few samples, many features), forming and inverting $A^\top A + I_N$ costs $\mathcal{O}(N^3)$, while forming and inverting $AA^\top + I_M$ costs $\mathcal{O}(M^3)$. The same low-rank-update viewpoint is useful in online learning: when new samples arrive, one can update the smaller inverse incrementally (roughly $\mathcal{O}(M^2)$ per update) instead of refactoring an $N \times N$ inverse from scratch.

> In **kernel ridge regression**, the kernel matrix $K = AA^\top$ is $M \times M$ (one entry per sample pair), while the feature dimension N may be very large or infinite. The dual formulation inverts a matrix of size M, which is exactly the setting where *5.1* (with general λ) is the practical route.

5.2 Trace

We have already encountered the trace operation in Chapter~2, where it appeared in the equivalent forms of the Frobenius norm: $||A||_F^2 = \text{tr}(A^\top A)$. Here we formally define the trace and establish its fundamental properties.

5.2.1 Definition: Trace

For $A \in \mathbb{R}^{N\times N}$, the **trace** of A, denoted $\text{tr}(A)$, is the sum of its diagonal entries:

$$\text{tr}(A) = \sum_{n=0}^{N-1} a_{nn} \tag{5.2}$$

5.2.2 Theorem: Trace Properties

For $A, B \in \mathbb{R}^{N\times N}$ and scalar α:

1. $\text{tr}(A + B) = \text{tr}(A) + \text{tr}(B)$.
2. $\text{tr}(\alpha A) = \alpha \cdot \text{tr}(A)$.
3. $\text{tr}(AB) = \text{tr}(BA)$.
4. $\text{tr}(I_{N\times N}) = N$.

5.2.3 Theorem: Cyclic Property of Trace

For $A \in \mathbb{R}^{M\times N}$, $B \in \mathbb{R}^{N\times P}$, $C \in \mathbb{R}^{P\times M}$,

$$\text{tr}(ABC) = \text{tr}(BCA) = \text{tr}(CAB) \tag{5.3}$$

5.2.4 Theorem: Similarity Invariance of Trace

If P is invertible,

$$\text{tr}(PAP^{-1}) = \text{tr}(A) \tag{5.4}$$

Proof:

Using the cyclic property of trace (equation 5.3), we have:

$$\text{tr}(PAP^{-1}) = \text{tr}((PA)P^{-1}) = \text{tr}(P^{-1}(PA)) = \text{tr}((P^{-1}P)A) = \text{tr}(IA) = \text{tr}(A).$$

This theorem shows that trace is invariant under similarity transformations. This property is fundamental in Chapter~6, where we study eigen-decomposition and show that similar matrices share the same trace (along with eigenvalues, determinant, and rank).

5.2.5 Theorem: Matrix Derivative of Trace

For $A \in \mathbb{R}^{M\times N}$ and $X \in \mathbb{R}^{N\times M}$,

$$\frac{d}{dX}\text{tr}(AX) = A^\top \tag{5.5}$$

Note on matrix derivatives: The derivative of a scalar function with respect to a matrix is defined element-wise: the (i, j)-th entry of $\frac{d}{dX}f(X)$ equals $\frac{\partial f}{\partial x_{ij}}$. This convention is consistent with the **Frobenius inner product** (also called the Frobenius norm inner product), which for matrices $A, B \in \mathbb{R}^{M\times N}$ is defined as:

$$\langle A, B\rangle_F = \mathrm{tr}(A^\top B) = \sum_{i=0}^{M-1}\sum_{j=0}^{N-1} a_{ij}b_{ij} \tag{5.6}$$

Under this convention, the matrix derivative satisfies the chain rule and behaves like a standard gradient in the matrix space equipped with the Frobenius inner product. The full theory of matrix calculus, including gradients, Jacobians, and Hessians, is covered in Chapter~11.

5.2.6 Theorem: Trace of AA^T

For $A \in \mathbb{R}^{M\times N}$,

$$\mathrm{tr}(AA^\top) = \sum_{m=0}^{M-1}\sum_{n=0}^{N-1} a_{mn}^2 = ||A||_F^2 \tag{5.7}$$

Proof:

From the cyclic property of trace (equation 5.3), we have $\mathrm{tr}(AA^\top) = \mathrm{tr}(A^\top A)$. The result then follows directly from Chapter~2 (equation 2.11), where it was established that $\mathrm{tr}(A^\top A) = \sum_{m=0}^{M-1}\sum_{n=0}^{N-1} a_{mn}^2 = ||A||_F^2$.

Additional connections: From Chapter~2, we also know that $\mathrm{tr}(AA^\top) = \sum_{i=0}^{M-1} \lambda_i(AA^\top) = \sum_{i=0}^{\min(M,N)-1} \sigma_i(A)^2$, where $\lambda_i(AA^\top)$ are the eigenvalues of $AA^\top$ and $\sigma_i(A)$ are the singular values of A (see equation 2.11).

> The trace operator simplifies both mathematical derivation and implementation of weight regularization in deep learning. For $W \in \mathbb{R}^{M\times N}$, computing $\mathrm{tr}(W^\top W) = ||W||_F^2$ requires $O(MN)$ time (as it must access all elements). The gradient $\frac{d}{dW}\mathrm{tr}(W^\top W) = 2W$ (see Chapter~11) is mathematically compact, but evaluating it still requires $O(MN)$ time since it involves element-wise scaling. However, the matrix formulation enables vectorized execution on GPUs, which is significantly faster in practice than scalar derivatives despite the same theoretical complexity. The cyclic property $\mathrm{tr}(AB) = \mathrm{tr}(BA)$ allows optimizing matrix multiplication order (e.g., rearranging $\mathrm{tr}(ABC)$ to $\mathrm{tr}(BCA)$ may significantly reduce FLOPs if intermediate dimensions are smaller). For L layers, the total penalty $\lambda \sum_{\ell=1}^{L} \mathrm{tr}(W_\ell^\top W_\ell)$ has complexity $O(\sum_{\ell=1}^{L} M_\ell N_\ell)$ since each layer is processed independently.

5.3 Determinant

We have already encountered the determinant in earlier chapters: in Chapter~2, it appeared in the characteristic equation $\det(A - \lambda I) = 0$ for finding eigenvalues, while Chapter~3 established that orthonormal matrices have determinant ± 1. Here we formally develop the theory of determinants, exploring their geometric meaning, computational methods, and fundamental properties.

5.3.1 Definition: Determinant

For a square matrix $A \in \mathbb{R}^{N\times N}$, the **determinant** of A, denoted $\det(A)$ or $|A|$, is defined recursively as follows:

Base case ($N = 1$): For $A = [a_{00}]$, $\det(A) = a_{00}$.

Recursive case ($N > 1$): For any fixed row index i,

$$\det(A) = \sum_{j=0}^{N-1} (-1)^{i+j} a_{ij} \det(A_{ij}) \tag{5.8}$$

where A_{ij} is the $(N-1) \times (N-1)$ matrix obtained by removing row i and column j from A. This is called the **Laplace expansion** along row i. The value $\det(A)$ is independent of the choice of row i (or column, by symmetry).

Alternative definition via permutations: The determinant can also be defined as

$$\det(A) = \sum_{\sigma \in S_N} \operatorname{sgn}(\sigma) \prod_{i=0}^{N-1} a_{i,\sigma(i)},$$

where S_N is the set of all permutations of $\{0, 1, \dots, N-1\}$, and $\operatorname{sgn}(\sigma)$ is the sign of permutation σ ($+1$ for even permutations, -1 for odd permutations).

> The geometric interpretation of the determinant (signed volume of parallelepiped, area/volume scaling factor in transformations) is covered in Chapter~6 in the context of geometric transformations.

5.3.2 Definition: Singular Matrix

Matrix $A \in \mathbb{R}^{N \times N}$ is

- **singular** if $\det(A) = 0$,
- **non-singular** if $\det(A) \neq 0$.

5.3.2.1 Example: 2×2 Determinant

For $A = \begin{bmatrix} a_{00} & a_{01} \\ a_{10} & a_{11} \end{bmatrix}$, $\det(A) = a_{00}a_{11} - a_{01}a_{10}$.

5.3.3 Property: Determinant Properties

1. $\det(A) = \det(A^\top)$: The determinant is unchanged under transpose. This follows from the fact that the Laplace expansion (equation 5.8) can be applied along any row or column, and transposing swaps rows and columns without changing the algebraic structure.
2. $\det(AB) = \det(A)\det(B)$: The determinant is multiplicative. This fundamental property follows from the fact that matrix multiplication corresponds to composition of linear transformations, and the determinant measures the volume scaling factor, which is multiplicative under composition.
3. $\det(A^{-1}) = (\det A)^{-1}$: From property 2 and $AA^{-1} = I$, we have $\det(A)\det(A^{-1}) = \det(I) = 1$, so $\det(A^{-1}) = 1/\det(A)$.
4. **Scaling one row/column scales determinant**: If B is obtained from A by multiplying row i by scalar c, then $\det(B) = c\det(A)$. This follows from the multilinearity of the determinant: each term in the expansion contains exactly one element from each row, so scaling one row scales the entire determinant by that factor.
5. **Swapping rows/columns flips sign**: If B is obtained from A by swapping two rows, then $\det(B) = -\det(A)$. This follows from the alternating (antisymmetric) property of the determinant: interchanging two arguments of a multilinear alternating function changes its sign.

6. **Row addition does not change determinant**: If B is obtained from A by adding a multiple of one row to another, then $\det(B) = \det(A)$. This follows from the multilinearity and alternating property: adding a row to itself (scaled) creates a linear combination that preserves the determinant.

5.3.4 Theorem: Laplace Expansion

The determinant can be computed using **Laplace expansion** along any row or column. This is the same as the recursive definition (equation 5.8), which establishes that the expansion is independent of the choice of row or column:

Expansion along row i:

$$\det(A) = \sum_{j=0}^{N-1} (-1)^{i+j} a_{ij} \det(A_{ij}) \tag{5.9}$$

Expansion along column j:

$$\det(A) = \sum_{i=0}^{N-1} (-1)^{i+j} a_{ij} \det(A_{ij})$$

where A_{ij} is the $(N-1) \times (N-1)$ matrix obtained by removing row i and column j from A.

5.3.5 Definition: Characteristic Polynomial

For $A \in \mathbb{R}^{N \times N}$, the **characteristic polynomial** of A is defined as

$$p_A(\lambda) = \det(A - \lambda I),$$

which is a degree-N polynomial in λ.

The characteristic polynomial is fundamental for computing eigenvalues. The **characteristic equation** is $p_A(\lambda) = \det(A - \lambda I) = 0$, and the eigenvalues of A are exactly the roots of this polynomial equation (see Chapter~6). This definition establishes that:

1. There are exactly N eigenvalues (counting multiplicities; some may be complex even for real matrices), since a degree-N polynomial has N roots (by the fundamental theorem of algebra).
2. Eigenvalues can be found by solving a polynomial equation, connecting linear algebra to polynomial root-finding.
3. The coefficients of the characteristic polynomial have important relationships: the constant term is $\det(A)$ (when $\lambda = 0$), and the coefficient of λ^{N-1} is $(-1)^{N-1}\mathrm{tr}(A)$.

Example: For a 2×2 matrix $A = \begin{bmatrix} a_{00} & a_{01} \\ a_{10} & a_{11} \end{bmatrix}$,

$$p_A(\lambda) = \det(A-\lambda I) = \det \begin{bmatrix} a_{00} - \lambda & a_{01} \\ a_{10} & a_{11} - \lambda \end{bmatrix} = (a_{00}-\lambda)(a_{11}-\lambda)-a_{01}a_{10} = \lambda^2-\mathrm{tr}(A)\lambda+\det(A),$$

which is indeed a degree-2 polynomial in λ.

5.3.6 Theorem: Triangular Determinant

If A is triangular (upper or lower),

$$\det(A) = \prod_{n=0}^{N-1} a_{nn} \tag{5.10}$$

Proof:

For an upper triangular matrix, we can compute the determinant using Laplace expansion (equation 5.9) along the first row. Since all entries below the diagonal are zero, the only non-zero term in the expansion comes from a_{00}, and the corresponding minor is also upper triangular. By induction, we get $\det(A) = a_{00} \cdot a_{11} \cdots a_{N-1,N-1} = \prod_{n=0}^{N-1} a_{nn}$.

For a lower triangular matrix, the same argument applies using the first column, or we can use the fact that $\det(A) = \det(A^\top)$ and $A^\top$ is upper triangular.

Connection to invertibility: This theorem provides an alternative proof for the invertibility criterion: an upper triangular matrix U is invertible **iff** $u_{nn} \neq 0$ for all n (see Theorem: Invertibility of Upper Triangular Matrix), since $\det(U) = \prod_{n=0}^{N-1} u_{nn} \neq 0$ **iff** all diagonal elements are non-zero.

5.3.7 Theorem: Invertibility Equivalence

For $A \in \mathbb{R}^{N \times N}$, the following are equivalent:

1. $\det(A) \neq 0$
2. $\operatorname{rank}(A) = N$
3. A is invertible

5.4 LU Decomposition

5.4.1 Definition: Leading Principal Minors

Let A_n be the top-left $n \times n$ submatrix of A. $\det(A_n)$ is the n-th **leading principal minor**.

5.4.2 Theorem: LU Decomposition

If all leading principal minors are non-zero, then

$$A = LU \tag{5.11}$$

where

- L is lower triangular with $l_{ii} = 1$,
- U is upper triangular.

5.4.3 Algorithm: LU Decomposition

Let $A = [\mathbf{a}_0 \ \cdots \ \mathbf{a}_{N-1}]$ where $\mathbf{a}_n$ is the n-th column of A, and let A_n^* denote the n-th row of A (as a row vector). Similarly, let $\mathbf{u}_n$ be the n-th column of U, $\mathbf{l}_n$ be the n-th column of L, U_n^* be the n-th row of U, and L_n^* be the n-th row of L.

This algorithm computes L and U incrementally. At each step n, we use the previously computed parts of L and U to determine new rows/columns. The algorithm alternates between computing rows of U

Algorithm 5 LU Decomposition

Input: $A \in \mathbb{R}^{N \times N}$ (all leading principal minors are non-zero)
Output: L (lower triangular) and U (upper triangular) such that $A = LU$
for $n = 0, \dots, N-1$ **do**
Compute U_n^* by solving $L_n^* U = A_n^*$ ▷ Uses $L_0^*, \dots, L_{n-1}^*$
Compute $\mathbf{l}_n$ by solving $L\mathbf{u}_n = \mathbf{a}_n$ ▷ Uses $\mathbf{u}_0, \dots, \mathbf{u}_{n-1}$
end for

and columns of L, ensuring that the triangular structure is maintained throughout the process.

5.4.3.1 Example

Given the matrix

$$A = \begin{bmatrix} 2 & 4 & 6 \\ 1 & 3 & 5 \\ 4 & 8 & 10 \end{bmatrix},$$

find the LU decomposition $A = LU$.

Solution:

First, verify that all leading principal minors are non-zero (required for LU decomposition to exist):

1. **First leading principal minor**: $\det(A_1) = \det([2]) = 2 \neq 0$.
2. **Second leading principal minor**:

$$\det(A_2) = \det\left(\begin{bmatrix} 2 & 4 \\ 1 & 3 \end{bmatrix}\right) = 2 \cdot 3 - 4 \cdot 1 = 6 - 4 = 2 \neq 0.$$

3. **Third leading principal minor** (the full matrix):

$$\begin{aligned} \det(A_3) = \det(A) &= 2(3 \cdot 10 - 5 \cdot 8) - 4(1 \cdot 10 - 5 \cdot 4) + 6(1 \cdot 8 - 3 \cdot 4) \\ &= 2(30 - 40) - 4(10 - 20) + 6(8 - 12) = -20 + 40 - 24 = -4 \neq 0. \end{aligned}$$

Since all leading principal minors are non-zero, the LU decomposition exists.

Using Gaussian elimination:

1. **Step 1**: Eliminate the $(1, 0)$ entry: subtract $\frac{1}{2}\times$ row 0 from row 1:

$$L_1 = \begin{bmatrix} 1 & 0 & 0 \\ \frac{1}{2} & 1 & 0 \\ 0 & 0 & 1 \end{bmatrix}, \quad L_1 A = \begin{bmatrix} 2 & 4 & 6 \\ 0 & 1 & 2 \\ 4 & 8 & 10 \end{bmatrix}.$$

2. **Step 2**: Eliminate the $(2, 0)$ entry: subtract $2\times$ row 0 from row 2:

$$L_2 = \begin{bmatrix} 1 & 0 & 0 \\ 0 & 1 & 0 \\ 2 & 0 & 1 \end{bmatrix}, \quad L_2 L_1 A = \begin{bmatrix} 2 & 4 & 6 \\ 0 & 1 & 2 \\ 0 & 0 & -2 \end{bmatrix}.$$

3. **Final decomposition:**

$$L = (L_2 L_1)^{-1} = \begin{bmatrix} 1 & 0 & 0 \\ \frac{1}{2} & 1 & 0 \\ 2 & 0 & 1 \end{bmatrix}, \quad U = \begin{bmatrix} 2 & 4 & 6 \\ 0 & 1 & 2 \\ 0 & 0 & -2 \end{bmatrix}.$$

Verification:

$$LU = \begin{bmatrix} 1 & 0 & 0 \\ \frac{1}{2} & 1 & 0 \\ 2 & 0 & 1 \end{bmatrix} \begin{bmatrix} 2 & 4 & 6 \\ 0 & 1 & 2 \\ 0 & 0 & -2 \end{bmatrix} = \begin{bmatrix} 2 & 4 & 6 \\ 1 & 3 & 5 \\ 4 & 8 & 10 \end{bmatrix} = A.$$

Note that L has unit diagonal (all diagonal elements are 1), and U is upper triangular. The diagonal elements of U are 2, 1, and -2, which are the pivots from Gaussian elimination.

5.4.4 Theorem: Determinant via LU Decomposition

If $A = LU$,

$$\det(A) = \prod_{n=0}^{N-1} u_{nn} \tag{5.12}$$

5.4.4.1 Example

Use the LU decomposition to compute the determinant of the following matrix:

$$A = \begin{bmatrix} 4 & 3 & 1 \\ 2 & 1 & 3 \\ 6 & 7 & 8 \end{bmatrix}.$$

Solution:

Recall that $A = LU$, where

$$L = \begin{bmatrix} 1 & 0 & 0 \\ 0.5 & 1 & 0 \\ 1.5 & -5 & 1 \end{bmatrix}, \quad U = \begin{bmatrix} 4 & 3 & 1 \\ 0 & -0.5 & 2.5 \\ 0 & 0 & 19 \end{bmatrix}.$$

Using the theorem that $\det(A) = \prod_{n=0}^{N-1} u_{nn}$ (the product of the diagonal elements of U), we have:

$$\det(A) = u_{00} \cdot u_{11} \cdot u_{22} = 4 \cdot (-0.5) \cdot 19 = -38.$$

5.4.5 Theorem: Inverse via LU Decomposition

If $A = LU$,

$$A^{-1} = U^{-1} L^{-1} \tag{5.13}$$

5.4.5.1 Example

Use the LU decomposition to compute the inverse of the following matrix:

$$A = \begin{bmatrix} 4 & 3 & 1 \\ 2 & 1 & 3 \\ 6 & 7 & 8 \end{bmatrix}.$$

Solution:

Recall that $A = LU$, where

$$L = \begin{bmatrix} 1 & 0 & 0 \\ 0.5 & 1 & 0 \\ 1.5 & -5 & 1 \end{bmatrix}, \quad U = \begin{bmatrix} 4 & 3 & 1 \\ 0 & -0.5 & 2.5 \\ 0 & 0 & 19 \end{bmatrix}.$$

We compute the inverses of L and U:

1. **Inverse of L**: Since L is lower triangular with unit diagonal, we can compute L^{-1} by solving $LL^{-1} = I$ using forward substitution:

$$L^{-1} = \begin{bmatrix} 1 & 0 & 0 \\ -0.5 & 1 & 0 \\ -4 & 5 & 1 \end{bmatrix}.$$

2. **Inverse of U**: Since U is upper triangular, we can compute U^{-1} by solving $UU^{-1} = I$ using back substitution:

$$U^{-1} = \begin{bmatrix} \frac{1}{4} & \frac{3}{2} & -\frac{4}{19} \\ 0 & -2 & \frac{5}{19} \\ 0 & 0 & \frac{1}{19} \end{bmatrix}.$$

3. **Compute A^{-1}**: Using the theorem that $(LU)^{-1} = U^{-1}L^{-1}$:

$$A^{-1} = U^{-1}L^{-1} = \frac{1}{38} \begin{bmatrix} 13 & 17 & -8 \\ -2 & -26 & 10 \\ -8 & 10 & 2 \end{bmatrix}.$$

5.4.6 Solving Linear Systems

For the matrix A in above example, solve $Ax = b$ for arbitrary b.

Solution

$$x = A^{-1}b = \frac{1}{38} \begin{bmatrix} 13 & 17 & -8 \\ -2 & -26 & 10 \\ -8 & 10 & 2 \end{bmatrix} b.$$

5.5 PLU Decomposition (Pivoted LU)

LU decomposition requires that all leading principal minors are non-zero and is typically limited to square matrices. **PLU decomposition** (Pivoted LU) generalizes LU decomposition: it works for **ANY rectangular matrix** $A \in \mathbb{R}^{M \times N}$ and **ALWAYS** exists (via row permutation P for numerical stability).

5.5.1 Theorem: PLU Decomposition

For **any matrix** $A \in \mathbb{R}^{M \times N}$ (including rectangular matrices where $M \neq N$), there exists a permutation matrix $P \in \mathbb{R}^{M \times M}$, a lower triangular matrix $L \in \mathbb{R}^{M \times \min(M,N)}$ with unit diagonal, and an upper triangular matrix $U \in \mathbb{R}^{\min(M,N) \times N}$ such that

$$A = PLU \tag{5.14}$$

The permutation matrix P ensures numerical stability by reordering rows to avoid small pivots. Note that L has shape $M \times \min(M,N)$ and U has shape $\min(M,N) \times N$, reflecting the fact that PLU works for rectangular matrices.

5.5.1.1 Example

Given the matrix

$$A = \begin{bmatrix} 0 & 2 \\ 1 & 1 \\ 2 & 0 \end{bmatrix},$$

find the PLU decomposition $A = PLU$.

Solution:

Using Gaussian elimination with partial pivoting:

1. **Step 1**: The first column has pivot at row 2 (largest absolute value: $|1|$). Permute rows 1 and 2:

$$P_1 = \begin{bmatrix} 0 & 1 & 0 \\ 1 & 0 & 0 \\ 0 & 0 & 1 \end{bmatrix}, \quad P_1 A = \begin{bmatrix} 1 & 1 \\ 0 & 2 \\ 2 & 0 \end{bmatrix}.$$

2. **Step 2**: Eliminate below the pivot. Subtract $2\times$ row 1 from row 3:

$$L_1 = \begin{bmatrix} 1 & 0 & 0 \\ 0 & 1 & 0 \\ 2 & 0 & 1 \end{bmatrix}, \quad L_1 P_1 A = \begin{bmatrix} 1 & 1 \\ 0 & 2 \\ 0 & -2 \end{bmatrix}.$$

3. **Step 3**: The second column pivot is already in position. Eliminate below:

$$L_2 = \begin{bmatrix} 1 & 0 & 0 \\ 0 & 1 & 0 \\ 0 & 1 & 1 \end{bmatrix}, \quad L_2 L_1 P_1 A = \begin{bmatrix} 1 & 1 \\ 0 & 2 \\ 0 & 0 \end{bmatrix} = U.$$

4. **Final decomposition**:

$$L = (L_2 L_1)^{-1} = \begin{bmatrix} 1 & 0 & 0 \\ 0 & 1 & 0 \\ -2 & -1 & 1 \end{bmatrix}, \quad P = P_1 = \begin{bmatrix} 0 & 1 & 0 \\ 1 & 0 & 0 \\ 0 & 0 & 1 \end{bmatrix}.$$

Verification:

$$PLU = \begin{bmatrix} 0 & 1 & 0 \\ 1 & 0 & 0 \\ 0 & 0 & 1 \end{bmatrix} \begin{bmatrix} 1 & 0 & 0 \\ 0 & 1 & 0 \\ -2 & -1 & 1 \end{bmatrix} \begin{bmatrix} 1 & 1 \\ 0 & 2 \\ 0 & 0 \end{bmatrix} = \begin{bmatrix} 0 & 2 \\ 1 & 1 \\ 2 & 0 \end{bmatrix} = A.$$

Note that L has unit diagonal (all diagonal elements are 1), and U is upper triangular with shape $\min(3,2) \times 2 = 2 \times 2$.

PLU decomposition (pivoted LU) is more numerically stable than standard LU decomposition and is the default method used in numerical libraries (e.g., `np.linalg.lu`). The pivoting strategy (partial pivoting or complete pivoting) prevents numerical instability when dealing with ill-conditioned matrices. In machine learning, PLU is used in **linear system solving** (normal equations in linear regression), **matrix inversion** (computing weight updates), and **determinant computation**. The permutation matrix P can be interpreted as reordering data samples or features to improve numerical stability, which is important when working with real-world datasets that may have correlated or poorly scaled features.

5.6 LDU Decomposition

5.6.1 Theorem: LDU Decomposition

For any matrix $A \in \mathbb{R}^{N\times N}$ that admits LU decomposition, there exists a decomposition

$$A = LDU \tag{5.15}$$

where $L \in \mathbb{R}^{N\times N}$ is a lower triangular matrix with unit diagonal, $D \in \mathbb{R}^{N\times N}$ is a diagonal matrix, and $U \in \mathbb{R}^{N\times N}$ is an upper triangular matrix with unit diagonal.

This decomposition is obtained from LU decomposition by factoring $U = DU'$ where D contains the diagonal elements of U and U' is the unit upper triangular matrix obtained by dividing each row of U by its diagonal element.

5.6.1.1 Example

Given the matrix

$$A = \begin{bmatrix} 2 & 4 \\ 1 & 3 \end{bmatrix},$$

find the LDU decomposition $A = LDU$.

Solution:

First, perform LU decomposition (without pivoting, since the matrix is well-conditioned):

1. **LU decomposition**: Using Gaussian elimination:
   - Eliminate the $(2,1)$ entry: subtract $\frac{1}{2}\times$ row 1 from row 2:

     $$L = \begin{bmatrix} 1 & 0 \\ \frac{1}{2} & 1 \end{bmatrix}, \quad U = \begin{bmatrix} 2 & 4 \\ 0 & 1 \end{bmatrix}.$$

   - Verify: $LU = \begin{bmatrix} 1 & 0 \\ \frac{1}{2} & 1 \end{bmatrix} \begin{bmatrix} 2 & 4 \\ 0 & 1 \end{bmatrix} = \begin{bmatrix} 2 & 4 \\ 1 & 3 \end{bmatrix} = A.$
2. **Extract diagonal from U**: The diagonal elements of U are 2 and 1:

   $$D = \begin{bmatrix} 2 & 0 \\ 0 & 1 \end{bmatrix}.$$

3. **Form unit upper triangular U'**: Divide each row of U by its diagonal element:

   $$U' = \begin{bmatrix} 1 & \frac{4}{2} \\ 0 & 1 \end{bmatrix} = \begin{bmatrix} 1 & 2 \\ 0 & 1 \end{bmatrix}.$$

4. **Final LDU decomposition:**

$$L = \begin{bmatrix} 1 & 0 \\ \frac{1}{2} & 1 \end{bmatrix}, \quad D = \begin{bmatrix} 2 & 0 \\ 0 & 1 \end{bmatrix}, \quad U = \begin{bmatrix} 1 & 2 \\ 0 & 1 \end{bmatrix}.$$

Verification:

$$LDU = \begin{bmatrix} 1 & 0 \\ \frac{1}{2} & 1 \end{bmatrix} \begin{bmatrix} 2 & 0 \\ 0 & 1 \end{bmatrix} \begin{bmatrix} 1 & 2 \\ 0 & 1 \end{bmatrix} = \begin{bmatrix} 1 & 0 \\ \frac{1}{2} & 1 \end{bmatrix} \begin{bmatrix} 2 & 4 \\ 0 & 1 \end{bmatrix} = \begin{bmatrix} 2 & 4 \\ 1 & 3 \end{bmatrix} = A.$$

Note that both L and U have unit diagonals, and D contains the scaling factors from the original U matrix.

LDU decomposition separates the scaling (diagonal matrix D) from the triangular structure, providing a more symmetric form than LU. In **numerical linear algebra**, LDU decomposition is useful for analyzing the conditioning of matrices and for solving systems with multiple right-hand sides. The diagonal matrix D reveals the scaling factors, which is important for understanding numerical stability. In **machine learning**, LDU decomposition appears in **preconditioning** strategies for iterative solvers, where the diagonal scaling can be used to improve convergence rates. The decomposition also facilitates **block matrix operations** where diagonal blocks can be processed independently.

5.7 Mindmap

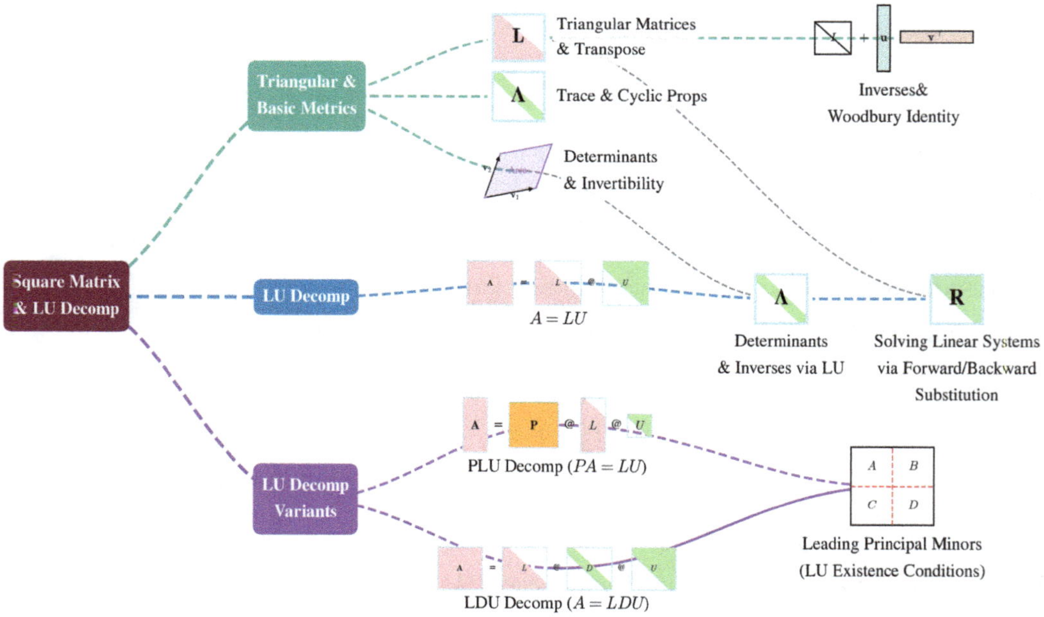

Figure 5.1: Mind map of square matrix and LU decomposition topics

5.8 Exercises

5.8.1 Triangular Matrix Inverses in Linear Systems

In solving linear systems $A\mathbf{x} = \mathbf{b}$ via LU decomposition, we solve $L\mathbf{y} = \mathbf{b}$ and $U\mathbf{x} = \mathbf{y}$, where L is lower triangular and U is upper triangular. Computing triangular matrix inverses efficiently is crucial.

1. Compute the inverse of the upper triangular matrix

$$U = \begin{bmatrix} 2 & 3 & 1 \\ 0 & 4 & 2 \\ 0 & 0 & 5 \end{bmatrix}.$$

   Verify that $UU^{-1} = I$.

2. Compute the inverse of the lower triangular matrix

$$L = \begin{bmatrix} 1 & 0 & 0 \\ 2 & 3 & 0 \\ 4 & 5 & 6 \end{bmatrix}.$$

   Verify that $LL^{-1} = I$.

3. For invertible 2×2 matrices $A = \begin{bmatrix} 1 & 2 \\ 3 & 4 \end{bmatrix}$ and $B = \begin{bmatrix} 2 & 1 \\ 1 & 2 \end{bmatrix}$, compute $(AB)^{-1}$ and $B^{-1}A^{-1}$ separately, and verify that $(AB)^{-1} = B^{-1}A^{-1}$.

4. Show that if A is invertible, then $(A^\top)^{-1} = (A^{-1})^\top$. Use this to show that if $A = LU$ is the LU decomposition, then $A^{-1} = U^{-1}L^{-1}$.

5.8.2 Trace and L2 Regularization in Deep Learning

Background: In deep learning, **L2 regularization** (also called weight decay) adds a penalty term $\lambda||W||_F^2$ to the loss function, where $\lambda > 0$ is a regularization parameter and $||W||_F^2 = \sum_{i,j} w_{ij}^2$ is the squared Frobenius norm. This penalty encourages smaller weights, which helps prevent overfitting. The trace operator provides a convenient way to express this penalty: $\lambda \text{tr}(W^\top W) = \lambda||W||_F^2$. The following exercises explore how the trace operator simplifies computations involving weight matrices.

Given weight matrices $W_1 = \begin{bmatrix} 1 & 2 \\ 3 & 4 \end{bmatrix}$ and $W_2 = \begin{bmatrix} 5 & 6 \\ 7 & 8 \end{bmatrix}$:

1. Compute $\text{tr}(W_1)$, $\text{tr}(W_2)$, and $\text{tr}(W_1 + W_2)$. Verify that $\text{tr}(W_1 + W_2) = \text{tr}(W_1) + \text{tr}(W_2)$.
2. Compute $\text{tr}(W_1 W_2)$ and $\text{tr}(W_2 W_1)$, and verify that $\text{tr}(W_1 W_2) = \text{tr}(W_2 W_1)$.
3. Compute $\text{tr}(W_1 W_1^\top)$ and verify that it equals $\sum_{i,j}(W_1)_{ij}^2$ (the sum of squares of all elements of W_1), which is $||W_1||_F^2$.
4. For a neural network with two layers, the total L2 regularization penalty is $\lambda(\text{tr}(W_1^\top W_1) + \text{tr}(W_2^\top W_2))$. Show that this equals $\lambda(||W_1||_F^2 + ||W_2||_F^2)$.

5.8.3 Determinants and Invertibility in Linear Models

The determinant of a matrix determines whether it is invertible, which is crucial for solving linear systems. In linear regression, the design matrix $X^\top X$ must be invertible for the normal equations to have a unique solution.

1. Compute the determinant of the 2×2 matrix
$$A = \begin{bmatrix} 3 & 1 \\ 2 & 4 \end{bmatrix}.$$
Determine whether A is invertible.
2. Compute the determinant of the lower triangular matrix
$$L = \begin{bmatrix} 2 & 0 & 0 \\ 1 & 3 & 0 \\ 4 & 5 & 6 \end{bmatrix}.$$
Show that $\det(L) = 2 \cdot 3 \cdot 6 = 36$ (the product of diagonal elements).
3. For a 3×3 matrix A, prove that $\det(2A) = 2^3 \det(A)$. Use this to show that if A is invertible, then $\det(A^{-1}) = 1/\det(A)$.
4. In ridge regression, we solve $(X^\top X + \lambda I)\theta = X^\top \mathbf{y}$. For $X = \begin{bmatrix} 1 & 2 \\ 2 & 4 \\ 3 & 6 \end{bmatrix}$, compute $\det(X^\top X)$ and show that $X^\top X$ is singular. Then show that $\det(X^\top X + \lambda I) > 0$ for any $\lambda > 0$, ensuring invertibility. This demonstrates why ridge regression can handle singular design matrices: even when $X^\top X$ is singular (non-invertible), adding the regularization term λI with $\lambda > 0$ makes $X^\top X + \lambda I$ positive definite and hence invertible, ensuring a unique solution.

5.8.4 Triangular Matrices and Forward/Backward Substitution

In solving linear systems via LU decomposition, triangular matrices enable efficient forward and backward substitution. This is fundamental to many numerical linear algebra algorithms.

1. Classify each of the following matrices as upper triangular, lower triangular, or neither:
$$A = \begin{bmatrix} 1 & 2 & 3 \\ 0 & 4 & 5 \\ 0 & 0 & 6 \end{bmatrix}, \quad B = \begin{bmatrix} 1 & 0 & 0 \\ 2 & 3 & 0 \\ 4 & 5 & 6 \end{bmatrix}, \quad C = \begin{bmatrix} 1 & 2 \\ 3 & 4 \end{bmatrix}.$$
2. Show that if A and B are both upper triangular $N \times N$ matrices, then AB is also upper triangular. Use this to show that if $A = LU$ is the LU decomposition, then solving $A\mathbf{x} = \mathbf{b}$ requires only forward and backward substitution.
3. Show that if A is an invertible upper triangular matrix, then A^{-1} is also upper triangular. (Hint: Use the fact that solving $A\mathbf{x} = \mathbf{e}_i$ for each standard basis vector $\mathbf{e}_i$ gives the columns of A^{-1}.)
4. For an upper triangular $N \times N$ matrix A, prove that $\det(A) = \prod_{i=0}^{N-1} a_{ii}$ (the product of diagonal elements). Use this to show that A is invertible if and only if all diagonal entries are nonzero.

5.8.5 Singular Matrices and Non-Unique Solutions

In linear regression, when the design matrix X has linearly dependent columns, $X^\top X$ is singular and the normal equations have infinitely many solutions. The determinant provides a direct test for singularity.

1. Determine whether the following matrix is singular or non-singular by computing its determinant:
$$A = \begin{bmatrix} 1 & 2 & 3 \\ 2 & 4 & 6 \\ 1 & 1 & 1 \end{bmatrix}.$$
If singular, find a nonzero vector $\mathbf{x}$ such that $A\mathbf{x} = \mathbf{0}$.
2. Show that if A is a singular $N \times N$ matrix and B is any $N \times M$ matrix, then AB is also singular. (Hint: Use the fact that if A is singular, there exists $\mathbf{v} \neq \mathbf{0}$ such that $A\mathbf{v} = \mathbf{0}$.)
3. Show that if A and B are both non-singular $N \times N$ matrices, then AB is also non-singular. Use this to show that if $A = LU$ is the LU decomposition with L and U both invertible, then A is invertible.
4. In linear regression with design matrix $X = \begin{bmatrix} 1 & 2 \\ 2 & 4 \\ 1 & 1 \end{bmatrix}$, show that $X^\top X$ is singular by computing its determinant. Explain why this means the normal equations $X^\top X\theta = X^\top \mathbf{y}$ have either no solution or infinitely many solutions.

5.8.6 PLU Decomposition and Determinant in Linear Systems

In solving linear systems $A\mathbf{x} = \mathbf{b}$, PLU decomposition $A = PLU$ (with partial pivoting) provides numerical stability. The determinant can be computed efficiently from the decomposition.

1. Prove that $\det(A) = \det(P) \cdot \det(L) \cdot \det(U)$ for PLU decomposition $A = PLU$. (Hint: Use the fact that $\det(AB) = \det(A)\det(B)$ and that $\det(P) = \pm 1$ for permutation matrices, $\det(L) = 1$ for unit lower triangular L.)
2. For the matrix
$$A = \begin{bmatrix} 0 & 1 & 2 \\ 1 & 0 & 1 \\ 2 & 1 & 0 \end{bmatrix},$$
find the PLU decomposition $A = PLU$ explicitly. Compute $\det(A)$ using the decomposition and verify by computing $\det(A)$ directly.
3. Prove that if A is invertible, then U is also invertible. What can you say about the diagonal entries of U? Use this to show that $\det(A) \neq 0$ if and only if all diagonal entries of U are nonzero.
4. Show that the number of row exchanges in the PLU decomposition equals the number of -1 factors in $\det(P)$. For the matrix in part 2, count the row exchanges and verify this relationship.

5.8.7 LDU Decomposition and Matrix Factorization

LDU decomposition $A = LDU$ (where L and U have unit diagonals) provides a normalized form of LU decomposition that is useful for theoretical analysis and numerical stability.

1. Prove that if A is invertible, then the LDU decomposition is unique (up to scaling of D). (Hint: Assume $A = L_1D_1U_1 = L_2D_2U_2$ and use the fact that L and U have unit diagonals.)
2. Given the matrix
$$A = \begin{bmatrix} 2 & 4 & 6 \\ 1 & 3 & 5 \\ 0 & 1 & 2 \end{bmatrix},$$

find the LDU decomposition explicitly. Verify that both L and U have unit diagonals and that $A = LDU$.

3. Prove that if A is symmetric and admits LDU decomposition, then $U = L^\top$. Use this to show that $A = LDL^\top$ for symmetric matrices.

4. In second-order optimization methods (e.g., Newton's method), we need to solve $H\mathbf{p} = -\mathbf{g}$ where H is the Hessian matrix. For a symmetric positive definite Hessian H with LDL decomposition $H = LDL^\top$, show that all diagonal entries of D are positive. Use this to show that the system $H\mathbf{p} = -\mathbf{g}$ has a unique solution.

5.8.8 PLU Decomposition and Numerical Stability in Linear Systems

In solving large-scale linear systems $A\mathbf{x} = \mathbf{b}$ (e.g., in finite element methods or solving normal equations in linear regression), PLU decomposition with partial pivoting provides numerical stability. Understanding the growth factor and error bounds is crucial for reliable computation.

1. Prove that partial pivoting (choosing the largest element in the current column as the pivot) ensures that all entries of L satisfy $|L_{ij}| \leq 1$ for $i > j$. Show that this bounds the growth of numerical errors during Gaussian elimination. (Hint: Consider how the pivot choice affects the elimination multipliers.)

2. The growth factor ρ is defined as $\rho = \frac{\max_{i,j} |U_{ij}|}{\max_{i,j} |A_{ij}|}$, measuring how much matrix entries can grow during elimination. Prove that for partial pivoting, $\rho \leq 2^{N-1}$ in the worst case, but in practice ρ is usually much smaller. Give an example of a 3×3 matrix that achieves $\rho = 2^2 = 4$.

3. Complete pivoting (choosing the largest element in the remaining submatrix) is more stable but more expensive ($O(N^3)$ vs $O(N^2)$ for finding pivots). Prove that with complete pivoting, the growth factor satisfies $\rho \leq \sqrt{N} \cdot 2^{N-1} \cdot \prod_{k=1}^{N-1} k^{1/(2k)}$, which grows much slower than 2^{N-1}. For $N = 4$, compare the worst-case bounds: partial pivoting gives $\rho \leq 8$, while complete pivoting gives a much smaller bound.

4. In solving the normal equations $(X^\top X)\theta = X^\top \mathbf{y}$ for linear regression, the condition number $\kappa(X^\top X)$ can be large. Using PLU decomposition, prove that the relative error in the computed solution satisfies:

$$\frac{||\theta - \hat{\theta}||}{||\theta||} \leq \kappa(X^\top X) \cdot \frac{\rho \cdot \epsilon}{1 - \kappa(X^\top X) \cdot \rho \cdot \epsilon},$$

where ρ is the growth factor, ϵ is machine precision, and $\hat{\theta}$ is the computed solution. Explain why this bound justifies the use of pivoting and why ridge regression (adding λI to $X^\top X$) can improve numerical stability.

Chapter 6

Geometric Transformation and Eigen Decomposition

Geometric transformations, linear and affine mappings, and eigen decomposition

"The mathematician's patterns, like the painter's or the poet's, must be beautiful."

— G. H. Hardy

- ▷ `Eigenvalues` $A\mathbf{x} = \lambda\mathbf{x}$, $A \in \mathbb{R}^{N\times N}$, $\lambda \in \mathbb{C}$, $\mathbf{x} \in \mathbb{C}^{N\times 1} \neq \mathbf{0}$
- ▷ `Eigenvectors` $(A - \lambda I)\mathbf{x} = 0$; alg. mult. $k \Rightarrow$ at most k lin. indep. eigenvectors; different eigenvalues $\Rightarrow$ lin. indep.
- ▷ `Characteristic equ.` $\det(A - \lambda I) = 0$; N eigenvalues (multiplicities, may be complex)
- ▷ `Eigen/Spectral decomp` $A = V\Lambda V^{-1} = \sum_{n=0}^{N-1} \lambda_n \mathbf{v}_n \mathbf{p}_n^\top$ where $V = [\mathbf{v}_0, \dots, \mathbf{v}_{N-1}]$, $\Lambda = \operatorname{diag}(\lambda_0, \dots, \lambda_{N-1})$, $\mathbf{p}_n$ (cols of $P = (V^{-1})^\top$); symmetric: $A = \sum_{n=0}^{N-1} \lambda_n \mathbf{v}_n \mathbf{v}_n^\top$ (orthonormal $\mathbf{v}_n$); requires N lin. indep. eigenvectors; $A^k = V\Lambda^k V^{-1}$
- ▷ `Similarity transformation` $B = P^{-1}AP$; same eigenvalues, det, tr, rank
- ▷ `Left eigenvectors` $A = V\Lambda V^{-1}$, $P = (V^{-1})^\top \Rightarrow A^\top = P\Lambda P^{-1}$, $A^\top \mathbf{p}_n = \lambda_n \mathbf{p}_n$; biorthogonal: $\mathbf{p}_i^\top \mathbf{v}_j = \delta_{ij}$
- ▷ `Powers of matrix` $A = V\Lambda V^{-1} \Rightarrow A^k = V\Lambda^k V^{-1}$, $k \in \mathbb{N} \cup \{0\}$
- ▷ `Neumann series` $\rho(A) < 1$ $(|\lambda_n| < 1\ \forall n) \Rightarrow (I - A)^{-1} = \sum_{k=0}^{\infty} A^k = V\operatorname{diag}(\frac{1}{1-\lambda_0}, \dots, \frac{1}{1-\lambda_{N-1}})V^{-1}$
- ▷ `Trace-eigenvalue` $\operatorname{tr}(A) = \sum_{n=0}^{N-1} \lambda_n$
- ▷ `Determinant-eigenvalue` $\det(A) = \prod_{n=0}^{N-1} \lambda_n$
- ▷ `Linear transformation` $A\mathbf{x} = V\Lambda V^{-1}\mathbf{x}$; $\mathbf{x} = V\mathbf{c}$, $\mathbf{c} = V^{-1}\mathbf{x}$; normal: $A = V\Lambda V^\top$ (rotation $\to$ scaling $\to$ rotation)
- ▷ `Power method` $|\lambda_0| > |\lambda_1| \geq \cdots$: $\mathbf{v}_{\text{new}} \leftarrow A\mathbf{v}$ (normalize), $\lambda = \frac{\mathbf{v}^\top A\mathbf{v}}{\mathbf{v}^\top \mathbf{v}}$; left: $\mathbf{p}_{\text{new}} \leftarrow A^\top \mathbf{p}$
- ▷ `Deflation` Biorthogonal $\mathbf{p}_i^\top \mathbf{v}_j = \delta_{ij}$: $A \leftarrow A - \sum_{i=0}^{n-1} \lambda_i \mathbf{v}_i \mathbf{p}_i^\top$, then power method for $\lambda_n, \mathbf{v}_n, \mathbf{p}_n$; symmetric: $\mathbf{p}_i = \mathbf{v}_i$

6.1 Geometric Transformation and Matrix

Geometric transformations in vector spaces can be broadly categorized into three types based on the properties they preserve:

1. **Linear Transformations**: Preserve the origin and keep grid lines straight, parallel, and evenly spaced. They are represented by a matrix A as $T(\mathbf{x}) = A\mathbf{x}$.
    - **Isometric**: Preserves distances and angles (e.g., rotation, reflection). Requires an **orthogonal matrix**.
    - **Similarity**: Preserves angles and shape proportions, but allows uniform scaling (e.g., $T(\mathbf{x}) = kQ\mathbf{x}$).
    - **Area-preserving**: Preserves area/volume (e.g., shear, squeeze). Requires $|\det(A)| = 1$.
    - **Orthogonal Projection**: Collapses vectors onto a subspace (e.g., $P = \mathbf{u}\mathbf{u}^\top$). It is non-invertible and singular ($\det(P) = 0$).
2. **Affine Transformations**: Linear transformations plus **translation**. They preserve lines and parallelism but can shift the origin. They are represented as $T(\mathbf{x}) = A\mathbf{x} + \mathbf{b}$.
3. **Nonlinear Transformations**: More general mappings that can bend, twist, or distort space. They cannot be represented by a single matrix and often turn straight lines into curves (e.g., activation functions in neural networks).

6.1.1 Linear Transformation

As introduced in Chapter~2, a matrix $A \in \mathbb{R}^{M\times N}$ represents a **linear transformation** (or **linear mapping**) that maps vectors from $\mathbb{R}^{N\times 1}$ to $\mathbb{R}^{M\times 1}$.

A transformation $T : \mathbb{R}^N \to \mathbb{R}^M$ is **linear** if it satisfies:

1. **Additivity**: $T(\mathbf{x}+\mathbf{y}) = T(\mathbf{x}) + T(\mathbf{y})$ for all $\mathbf{x}, \mathbf{y} \in \mathbb{R}^N$
2. **Homogeneity**: $T(c\mathbf{x}) = cT(\mathbf{x})$ for all $\mathbf{x} \in \mathbb{R}^N$ and $c \in \mathbb{R}$

These two properties together are equivalent to: $T(c_1\mathbf{x}_1 + c_2\mathbf{x}_2) = c_1T(\mathbf{x}_1) + c_2T(\mathbf{x}_2)$ for all vectors $\mathbf{x}_1, \mathbf{x}_2$ and scalars c_1, c_2.

Every linear transformation from $\mathbb{R}^N$ to $\mathbb{R}^M$ can be represented by a matrix $A \in \mathbb{R}^{M\times N}$ such that $T(\mathbf{x}) = A\mathbf{x}$.

For a square matrix $A \in \mathbb{R}^{N\times N}$, this transformation maps vectors **within the same space**: $\mathbf{x} \in \mathbb{R}^{N\times 1} \mapsto A\mathbf{x} \in \mathbb{R}^{N\times 1}$.

6.1.2 Scaling

Scaling transformations change the size or proportions of objects. When scaling is applied **along the coordinate axes**, the transformation is represented by a **diagonal matrix**.

In 2D, **Uniform scaling** scales all directions by the same factor k:

$$S = \begin{bmatrix} k & 0 \\ 0 & k \end{bmatrix} = kI$$

In this case, the object's shape is preserved, and its area changes by a factor of $\det(S) = k^2$.

Non-uniform scaling (also called **squeeze** or **stretch**) uses different factors for each axis. In 2D, this is represented by:

$$S = \begin{bmatrix} k_1 & 0 \\ 0 & k_2 \end{bmatrix}$$

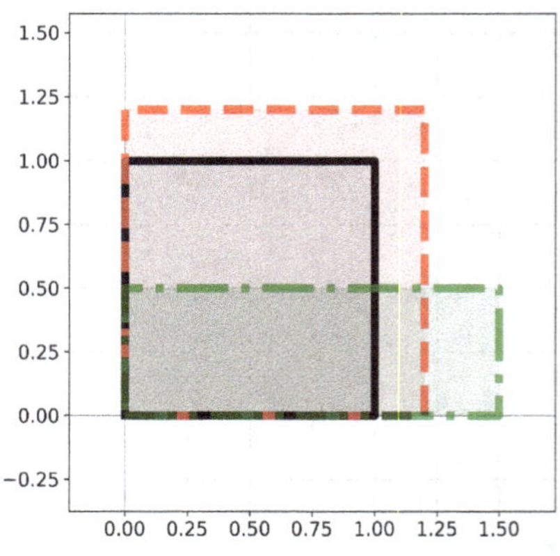

Figure 6.1: Scaling

This changes the object's aspect ratio. The area changes by a factor of $\det(S) = k_1 k_2$. A special case where $k_1 k_2 = 1$ is called a **squeeze mapping**, which is area-preserving.

As shown in Figure 6.1, 2D uniform scaling (red dashed line, $k = 1.2$) expands the square equally, while non-uniform scaling (green dash-dot line, $k_1 = 1.5, k_2 = 0.5$) stretches it horizontally and compresses it vertically. In both cases, the coordinate axes remain the principal directions of the transformation.

6.1.3 Rotation

In 2D, a **rotation** matrix rotates vectors counterclockwise by angle θ:

$$R(\theta) = \begin{bmatrix} \cos\theta & -\sin\theta \\ \sin\theta & \cos\theta \end{bmatrix}$$

As shown in Figure 6.2, a 2D rotation transformation (red dashed line) shifts the object by an angle θ around the origin. It preserves both the shape and the area of the object.

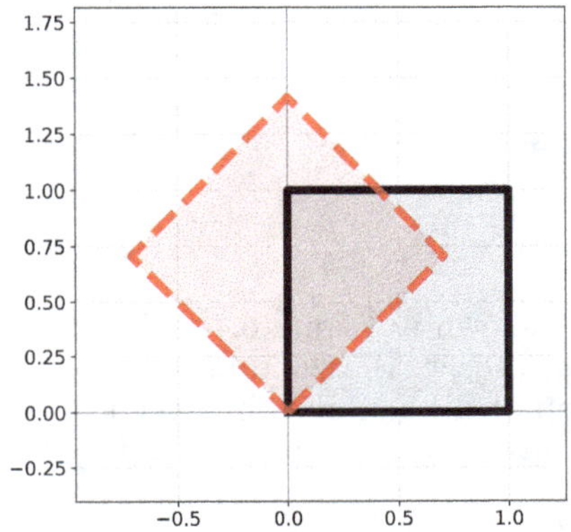

Figure 6.2: Rotation

Properties:

- $R(\theta_1)R(\theta_2) = R(\theta_1 + \theta_2)$ (rotation composition)
- $R(\theta)^{-1} = R(-\theta) = R(\theta)^\top$ (orthogonal matrix)
- $\det(R(\theta)) = 1$ (preserves orientation)

6.1.3.1 Example: Rotary Positional Embedding (RoPE)

In modern Large Language Models (like Llama), **RoPE** uses 2D rotation matrices to encode positional information. As discussed in Chapter~3 (see Example~3.8.6.1), rotations preserve inner products. By applying a rotation $R(m\theta)$ to the m-th token, the model naturally captures relative distance through the property $R(m\theta)^\top R(n\theta) = R((n-m)\theta)$.

6.1.4 Reflection

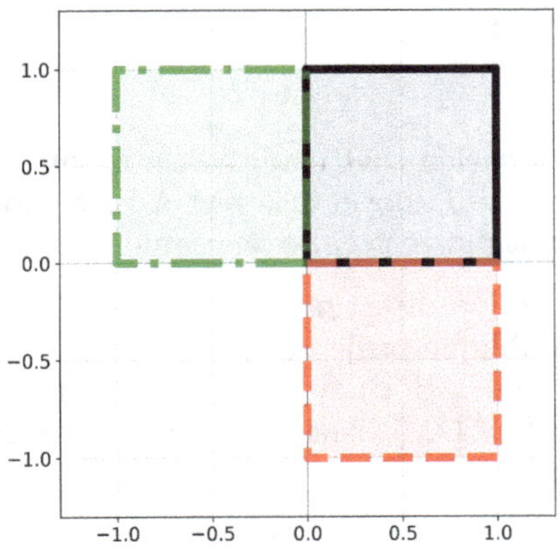

Figure 6.3: Reflection

In 2D, a **reflection** matrix reflects vectors across a line through the origin. For reflection across a line making angle θ with the x-axis:

$$F(\theta) = \begin{bmatrix} \cos(2\theta) & \sin(2\theta) \\ \sin(2\theta) & -\cos(2\theta) \end{bmatrix}$$

For reflection across the x-axis: $F = \begin{bmatrix} 1 & 0 \\ 0 & -1 \end{bmatrix}$

For reflection across the y-axis: $F = \begin{bmatrix} -1 & 0 \\ 0 & 1 \end{bmatrix}$

As shown in Figure 6.3, a 2D reflection transformation flips the object across an axis. Reflection across the x-axis (red dashed line) and y-axis (green dash-dot line) both preserve the area but reverse the orientation of the object.

Properties:

- $F^2 = I$ (reflecting twice returns to original)

- $\det(F) = -1$ (reverses orientation)

6.1.4.1 Example: Data Augmentation in Computer Vision

In deep learning, **data augmentation** frequently uses reflection transformations to increase the diversity of training data. For instance, horizontally flipping an image (reflection across the vertical axis) allows a model to learn that a "cat" remains a "cat" regardless of which way it is facing. This improves the model's **spatial invariance** and helps prevent overfitting.

6.1.5 Shear

In 2D, a **shear** transformation preserves one coordinate while shifting the other proportionally.

Horizontal shear:

$$H_h = \begin{bmatrix} 1 & k \\ 0 & 1 \end{bmatrix}$$

Vertical shear:

$$H_v = \begin{bmatrix} 1 & 0 \\ k & 1 \end{bmatrix}$$

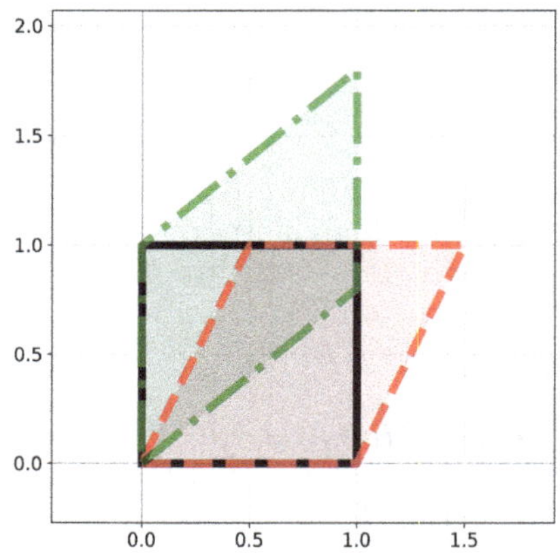

Figure 6.4: Shear

As shown in Figure 6.4, a 2D shear transformation tilts the object. Horizontal shear (red dashed line) shifts the top of the square, while vertical shear (green dash-dot line) shifts its side. Interestingly, shear transformations are area-preserving ($\det(H) = 1$).

6.1.5.1 Example: Geometric Robustness in Generative AI

In modern Generative AI models for images, such as **Diffusion Models** (e.g., Stable Diffusion), **shear** transformations are extensively used during training as part of **geometric data augmentation**. By exposing the model to sheared versions of objects (e.g., a tilted building or a slanted tree), the model learns a robust representation that is invariant to perspective distortions. This allows the AI to generate realistic objects in a vast array of poses and orientations, ensuring that the generated content remains semantically consistent even under significant geometric transformations.

6.1.6 Orthogonal Projection

An **orthogonal projection** is a special type of linear transformation that maps a vector onto its "closest point" in a subspace. Unlike the transformations discussed above, it is **not a bijection** (not invertible) because it collapses information, mapping multiple vectors to the same point.

While we focus on **orthogonal projection** (where the projection direction is perpendicular to the subspace), there also exist **oblique projections**. In an oblique projection, vectors are still collapsed onto a subspace, but the "shadow" is cast at an angle. Both types are singular and non-invertible, but orthogonal projection is far more common in AI/ML because it minimizes the "distance" (error) between the original vector and its projection.

In 2D, a projection matrix projects vectors onto a subspace (like a line). This generalizes the vector-to-vector projection discussed in Section~1.1.12.1 of Chapter~1 to matrix-based projections onto subspaces.

Projection onto the x-axis:

$$P_x = \begin{bmatrix} 1 & 0 \\ 0 & 0 \end{bmatrix}$$

Projection onto a line through the origin with unit direction vector $\mathbf{u}$:

$$P = \mathbf{u}\mathbf{u}^\top$$

This is a special case of the general projection matrix $P = X(X^\top X)^{-1}X^\top$ discussed in detail in Property~2.10.3. When projecting onto a single unit vector $\mathbf{u}$, the term $(X^\top X)^{-1}$ becomes $(\mathbf{u}^\top \mathbf{u})^{-1} = 1$, simplifying the expression to an outer product.

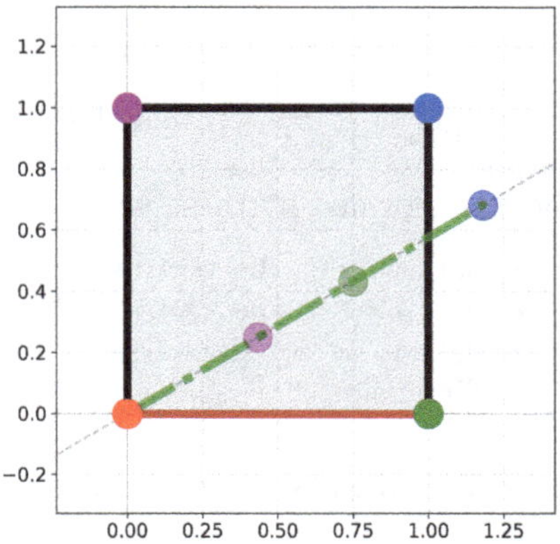

Figure 6.5: Projection

As shown in Figure 6.5, a 2D projection transformation collapses the object onto a lower-dimensional space (e.g., projecting 2D points onto a 1D line). Projection onto the x-axis (red dashed line) flattens the square into a line segment on the x-axis, while projection onto an arbitrary line (green dash-dot line) flattens it onto that line. The vertices and their corresponding projection points are highlighted; connecting any vertex to its projection yields the **residual vector R**, which is always orthogonal to the subspace of projection.

Properties (see Property~2.10.3 for algebraic proofs):

- $P^2 = P$ (projecting twice is the same as projecting once)
- $\det(P) = 0$ (singular, not invertible as it collapses dimensions)

6.1.6.1 Example: Feature Extraction in Linear Layers

In neural networks, a **linear layer** (or fully connected layer) acts as a learned projection when the output dimension is smaller than the input dimension (e.g., mapping from 1024 → 128 units). The layer projects high-dimensional input vectors onto a lower-dimensional subspace defined by its weight matrix. This process filters out redundant information and extracts the most relevant features for the task, achieving both data compression and dimensionality reduction.

6.1.7 Similarity Transformation

A **similarity transformation** is a linear transformation that preserves **angles** and **shape proportions**, but may change the absolute size (scale). It is a combination of an isometric transformation (rotation/reflection) and a uniform scaling.

Mathematically, a similarity transformation S can be written as:

$$S = kQ$$

where $k > 0$ is a uniform scaling factor and Q is an orthogonal matrix ($Q^\top Q = I$).

Geometric Meaning:

- It preserves the "shape" of objects (e.g., a square remains a square, a circle remains a circle).
- It preserves the ratio of distances between any two points.
- In 2D, the area changes by a factor of k^2.

6.1.8 Determinant and Geometric Meaning

The **determinant** of a square matrix $A \in \mathbb{R}^{N \times N}$ was formally defined in Chapter~5 (see Definition~5.3.1). Here we explore its geometric interpretation in the context of linear transformations.

6.1.8.1 Signed Volume of Parallelepiped

If $A = \begin{bmatrix} \mathbf{a}_0 & \cdots & \mathbf{a}_{N-1} \end{bmatrix}$, then $\det(A)$ is the **signed volume** (oriented volume) of the N-dimensional parallelepiped spanned by the column vectors $\mathbf{a}_0, \dots, \mathbf{a}_{N-1}$. The sign indicates the orientation of the basis formed by these vectors:

- $\det(A) > 0$: the vectors form a right-handed basis
- $\det(A) < 0$: the vectors form a left-handed basis
- $\det(A) = 0$: the vectors are linearly dependent (the parallelepiped collapses to lower dimension)

6.1.8.2 Area/Volume Scaling Factor

For a linear transformation $T(\mathbf{x}) = A\mathbf{x}$, the determinant measures how the transformation scales areas and volumes:

- **2D transformation**: $|\det(A)|$ represents the factor by which areas are scaled. If a region has area S, its image under transformation A has area $|\det(A)| \cdot S$.
- **3D transformation**: $|\det(A)|$ represents the factor by which volumes are scaled. If a region has volume V, its image has volume $|\det(A)| \cdot V$.

This scaling property follows directly from the parallelepiped interpretation: the unit cube (spanned by standard basis vectors) is transformed into a parallelepiped with volume $|\det(A)|$.

6.1.8.3 Orientation Preservation

- If $\det(A) > 0$: the transformation **preserves orientation** (e.g., rotation, scaling)
- If $\det(A) < 0$: the transformation **reverses orientation** (e.g., reflection)
- If $\det(A) = 0$: the transformation is **singular** (not invertible, collapses to lower dimension)

NOTES: For determinant properties (multiplicative property, transpose, scaling, etc.), see Chapter~5, Property: Determinant Properties.

6.1.8.4 Examples

- **Rotation**: $\det(R(\theta)) = \cos^2\theta + \sin^2\theta = 1$ (preserves area and orientation)
- **Uniform scaling by** k: $\det(kI) = k^N$ (scales area/volume by k^N)
- **Reflection**: $\det(F) = -1$ (preserves area but reverses orientation)
- **Projection**: $\det(P) = 0$ (collapses to lower dimension)

6.1.9 Inverse Transformation

For a linear transformation represented by matrix A, the **inverse transformation** (if it exists) is represented by A^{-1}:

$$\text{If } \mathbf{y} = A\mathbf{x}, \text{ then } \mathbf{x} = A^{-1}\mathbf{y}$$

A transformation is invertible if and only if:

- The matrix A is invertible ($\det(A) \neq 0$)
- The transformation is one-to-one and onto (bijective)

For composition of transformations:

$$(AB)^{-1} = B^{-1}A^{-1}$$

6.1.9.1 Common Inverse Transformations

- **Rotation**: $R(\theta)^{-1} = R(-\theta) = R(\theta)^\top$
- **Scaling**: If $S = \begin{bmatrix} k_1 & 0 \\ 0 & k_2 \end{bmatrix}$ and $k_1, k_2 \neq 0$, then $S^{-1} = \begin{bmatrix} 1/k_1 & 0 \\ 0 & 1/k_2 \end{bmatrix}$
- **Reflection**: $F^{-1} = F$ (reflecting twice returns to original)

6.1.10 Affine Transformation

An **affine transformation** is a linear transformation followed by a translation:

$$T(\mathbf{x}) = A\mathbf{x} + \mathbf{b}$$

where $A \in \mathbb{R}^{M\times N}$ is a matrix and $\mathbf{b} \in \mathbb{R}^{M\times 1}$ is a translation vector.

Affine transformations preserve:

- Parallel lines remain parallel
- Ratios of distances along parallel lines
- Convexity of sets

However, they do **not** preserve:

- Distances (unless A is orthogonal)
- Angles (unless A is a similarity transformation)

6.1.10.1 Homogeneous Coordinates

Affine transformations can be represented as linear transformations using **homogeneous coordinates**. For a 2D affine transformation:

$$\begin{bmatrix} x' \\ y' \\ 1 \end{bmatrix} = \begin{bmatrix} a_{00} & a_{01} & b_0 \\ a_{10} & a_{11} & b_1 \\ 0 & 0 & 1 \end{bmatrix} \begin{bmatrix} x \\ y \\ 1 \end{bmatrix}$$

This allows composition of affine transformations through matrix multiplication.

6.1.10.2 Translation

Translation is the simplest affine transformation (pure translation, no linear part):

$$T(\mathbf{x}) = \mathbf{x} + \mathbf{b}$$

In **homogeneous coordinates**, translating by $\mathbf{b} = [b_0, b_1]^\top$ is represented by a 3×3 matrix. This "trick" allows us to express the addition of $\mathbf{b}$ as a matrix multiplication:

$$\begin{bmatrix} 1 & 0 & b_0 \\ 0 & 1 & b_1 \\ 0 & 0 & 1 \end{bmatrix} \begin{bmatrix} x \\ y \\ 1 \end{bmatrix} = \begin{bmatrix} x + b_0 \\ y + b_1 \\ 1 \end{bmatrix}$$

For example, to translate a point $\mathbf{x} = [1, 2]^\top$ by $\mathbf{b} = [3, -1]^\top$:

$$\begin{bmatrix} 1 & 0 & 3 \\ 0 & 1 & -1 \\ 0 & 0 & 1 \end{bmatrix} \begin{bmatrix} 1 \\ 2 \\ 1 \end{bmatrix} = \begin{bmatrix} 1+3 \\ 2-1 \\ 1 \end{bmatrix} = \begin{bmatrix} 4 \\ 1 \\ 1 \end{bmatrix}$$

Note: Translation is **not** a linear transformation (it doesn't preserve the origin), but it is an affine transformation.

6.1.10.3 Example: Translation Invariance in CNNs

In Deep Learning, **Convolutional Neural Networks (CNNs)** are designed to be **translation invariant** (or more accurately, translation equivariant). This means that if an object (like a face) shifts its position in an image (a translation transformation), the network can still recognize it. The convolution operation shares weights across all spatial locations, allowing the model to detect the same features regardless of their absolute coordinates. This is a fundamental property that makes CNNs so effective for computer vision tasks.

6.1.11 Nonlinear Transformations

Not all transformations are linear. A **nonlinear transformation** does not satisfy the linearity properties above. **Crucially, nonlinear transformations cannot be represented by a matrix**—this is exactly what distinguishes them from linear transformations. While every linear transformation $T : \mathbb{R}^N \to \mathbb{R}^M$ can be written as $T(\mathbf{x}) = A\mathbf{x}$ for some matrix A, nonlinear transformations require more general functions.

Common examples of nonlinear transformations include:

- **Polynomial transformations**: $T(\mathbf{x}) = [x_0^2, x_1^2, ...]^\top$
- **Exponential transformations**: $T(\mathbf{x}) = [e^{x_0}, e^{x_1}, ...]^\top$
- **Trigonometric transformations**: $T(\mathbf{x}) = [\sin(x_0), \cos(x_1), ...]^\top$
- **Softmax function**: Applied row-wise to a matrix, $\text{softmax}(A)_{ij} = \frac{e^{a_{ij}}}{\sum_k e^{a_{ik}}}$

Figure 6.6 illustrates three examples of nonlinear transformations applied to a regular grid:

1. **Original grid** (left): A uniform 11×11 grid in the plane.
2. **Non-equidistant transformation** (middle): A polynomial-like transformation where $x' = x^2/10$ and $y' = y$. This compresses the grid near $x = 0$ and stretches it near $x = 10$, demonstrating how nonlinear transformations can create non-uniform spacing that cannot be achieved with linear transformations.
3. **Twisted transformation** (right): A trigonometric transformation where $x' = x + \sin(y/2)$ and $y' = y + \sin(x/2)$. This creates curved, wavy grid lines, showing how nonlinear transformations can introduce curvature and distortion.

Unlike linear transformations, which preserve straight lines and uniform spacing (up to scaling and rotation), nonlinear transformations can bend, compress, and distort the grid in complex ways. This flexibility is essential in machine learning, where nonlinear activation functions enable neural networks to learn complex, non-linear decision boundaries.

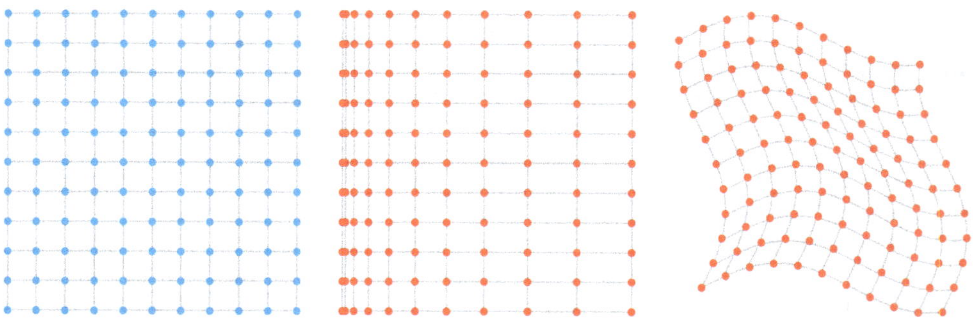

Figure 6.6: Nonlinear transformation

There is a fundamental asymmetry in mathematical theory: **linear transformations have been extensively studied** and form a well-developed branch of mathematics (linear algebra), with rich theory including eigenvalues, eigenvectors, matrix decompositions, and geometric interpretations. In contrast, **nonlinear transformations are much less unified**—they form a vast, diverse class with no single representation (like matrices) or general theory. Each nonlinear function must be analyzed individually, making systematic study more challenging. Linear transformations provide a solid, well-understood foundation that can be extended to understand nonlinear phenomena through composition (e.g., neural networks: linear layers + nonlinear activations).

6.1.11.1 Activation Functions: ReLU and Sigmoid

Two of the most important nonlinear transformations in deep learning are the **Rectified Linear Unit (ReLU)** and the **sigmoid** activation functions. These activation functions are applied element-wise after linear transformations in neural networks:

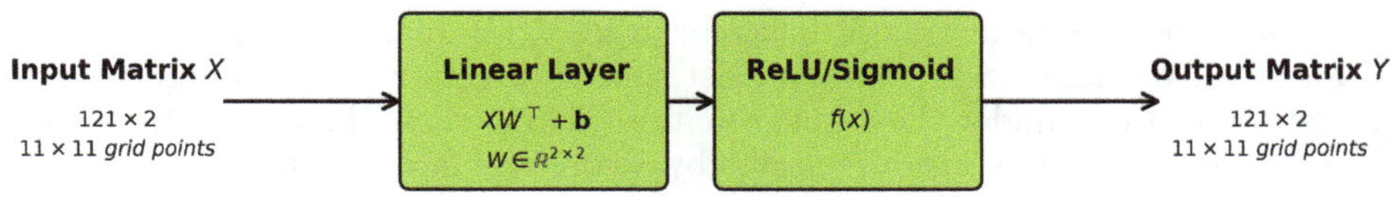

Figure 6.7: Neural network architecture

Figure 6.7 illustrates the architecture of a simple neural network layer: **Input Matrix** → **Linear Layer** ($WX + b$) → **Activation Function** (ReLU or Sigmoid) → **Output Matrix**. The linear layer performs matrix multiplication with weight matrix W and adds bias b, while the activation function introduces nonlinearity that enables the network to learn complex patterns.

- **ReLU**: $\text{ReLU}(x) = \max(0, x) = \begin{cases} x & \text{if } x \geq 0 \\ 0 & \text{if } x < 0 \end{cases}$
- **Sigmoid**: $\sigma(x) = \frac{1}{1+e^{-x}}$

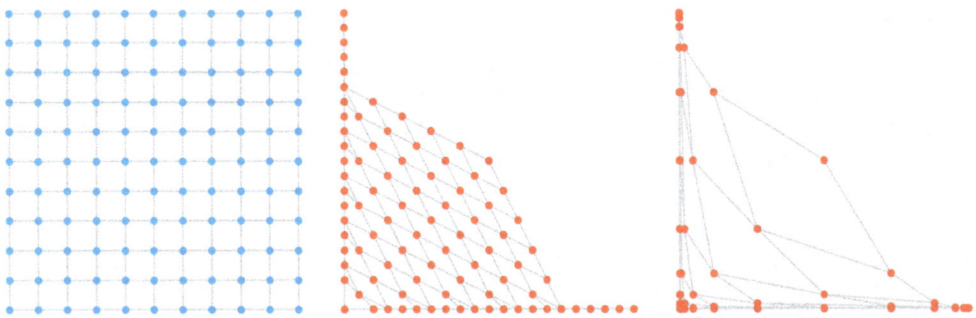

Figure 6.8: Activation function transformations

Figure 6.8 demonstrates how these activation functions transform a regular grid:

1. **Original grid** (left): A uniform 11×11 grid in the plane.
2. **ReLU transformation** (middle): The ReLU function clips all negative coordinates to zero, creating a "fold" along the axes. Points with $x < 0$ or $y < 0$ are mapped to the boundary $x = 0$ or $y = 0$, respectively. This creates a sharp discontinuity and compresses the negative half-space into the axes, demonstrating ReLU's non-differentiability at $x = 0$.

3. **Sigmoid transformation** (right): The sigmoid function maps the entire real line to the interval $(0, 1)$, creating an S-shaped compression. The grid is compressed more at the extremes and stretched in the middle, with all points converging toward the center region. This smooth, bounded transformation is useful for probability outputs and prevents gradient explosion.

The ReLU transformation introduces a **sharp boundary** (non-differentiable at zero), which is beneficial for creating piecewise-linear decision boundaries in neural networks. The sigmoid transformation provides a **smooth, bounded mapping** that is essential for binary classification and probability estimation. Both transformations are fundamentally nonlinear and cannot be represented by matrices, which is why they enable neural networks to learn complex, non-linear patterns that linear transformations alone cannot capture.

While nonlinear transformations are essential for the expressive power of neural networks, their complexity often makes them difficult to analyze directly. In contrast, linear transformations possess a deep, elegant internal structure that allows us to decompose complex mappings into simpler, more intuitive operations.

As we have seen, a general linear transformation can scale, rotate, and shear space simultaneously. However, for any given matrix, there often exist special directions—called **eigenvectors**—that remain remarkably stable under the transformation. Along these directions, the mapping does not rotate or shear; it simply scales the vector by a constant factor called the **eigenvalue**.

Uncovering these "characteristic" directions and scaling factors is fundamental to understanding the soul of a linear transformation. This leads us to one of the most powerful concepts in all of mathematics and AI: Eigenvalues and Eigenvectors.

6.2 Eigenvalue and Eigenvector

For a real-valued square matrix $A \in \mathbb{R}^{N \times N}$, we solve

$$A\mathbf{x} = \lambda \mathbf{x} \tag{6.1}$$

where

1. **Eigenvalue:** $\lambda \in \mathbb{C}$
2. **Eigenvector:** non-zero $\mathbf{x} \in \mathbb{C}^{N \times 1}$

Even though A has real entries, the eigenvalues and eigenvectors may be complex. This occurs because the characteristic polynomial $\det(A - \lambda I) = 0$ may have complex roots. For the essential complex number concepts needed here (complex conjugate, modulus, Euler's formula, and the connection to rotations), see Chapter~3.

> Eigenvalues and eigenvectors are strictly defined only for square matrices ($N \times N$). Rectangular matrices ($M \times N$, $M \neq N$) do not have eigenvalues, but they have an analogous concept: **singular values**. Since we often need to analyze the properties of rectangular matrices in AI and engineering, we use Singular Value Decomposition (SVD) instead, which we cover in Chapter~9.

6.2.1 Definition: Characteristic Equation and Eigenvalues

For a real-valued square matrix $A \in \mathbb{R}^{N \times N}$, the **characteristic equation** is

$$\det(A - \lambda I) = 0 \tag{6.2}$$

The eigenvalues of A are the solutions to this equation. A real-valued $N \times N$ matrix has exactly N eigenvalues (counting multiplicities; some may be complex).

This follows from the fact that a non-zero $\mathbf{x}$ exists for $(A-\lambda I)\mathbf{x}=0$ if and only if the columns of $A-\lambda I$ are linearly dependent, i.e., $\det(A-\lambda I)=0$.

6.2.2 Definition: Eigenvectors

For each eigenvalue $\lambda \in \mathbb{C}$ of $A \in \mathbb{R}^{N\times N}$, the **eigenvectors** corresponding to λ are the non-zero solutions to

$$(A-\lambda I)\mathbf{x}=0 \tag{6.3}$$

The set of all eigenvectors for a given eigenvalue λ (together with the zero vector) forms the **eigenspace** of λ, which is the null space of $A-\lambda I$.

6.2.3 Properties of Eigenvectors

For a matrix $A \in \mathbb{R}^{N\times N}$:

1. If λ has algebraic multiplicity k, then the number of linearly independent eigenvectors associated with λ is **at most** k.
2. Eigenvectors associated with different eigenvalues are linearly independent.
3. If the total number of linearly independent eigenvectors equals N, then the eigenvectors form a basis of $\mathbb{R}^N$ and A is diagonalizable.

6.2.4 Theorem: Eigen Decomposition

Let $A \in \mathbb{R}^{N\times N}$ have N linearly independent eigenvectors

$$V=\begin{bmatrix}\mathbf{v}_0 & \mathbf{v}_1 & \cdots & \mathbf{v}_{N-1}\end{bmatrix}$$

with eigenvalues $\lambda_0,\dots,\lambda_{N-1}$, and define

$$\Lambda=\operatorname{diag}(\lambda_0,\dots,\lambda_{N-1}).$$

Then:

1. $$A=V\Lambda V^{-1} \tag{6.4}$$

2. Let $P=(V^{-1})^\top=\begin{bmatrix}\mathbf{p}_0 & \cdots & \mathbf{p}_{N-1}\end{bmatrix}$. Then

$$A=\sum_{n=0}^{N-1}\lambda_n\mathbf{v}_n\mathbf{p}_n^\top \tag{6.5}$$

6.2.4.1 Example

Consider the matrix

$$A=\begin{bmatrix}2 & 1\\0 & 3\end{bmatrix}.$$

Step 1: Find eigenvalues

The characteristic equation is $\det(A-\lambda I)=0$:

$$\det\begin{bmatrix}2-\lambda & 1\\0 & 3-\lambda\end{bmatrix}=(2-\lambda)(3-\lambda)=0.$$

So the eigenvalues are $\lambda_0 = 2$ and $\lambda_1 = 3$.

Step 2: Find eigenvectors

For $\lambda_0 = 2$: Solve $(A - 2I)\mathbf{v}_0 = \mathbf{0}$:

$$\begin{bmatrix} 0 & 1 \\ 0 & 1 \end{bmatrix} \mathbf{v}_0 = \mathbf{0} \quad \Rightarrow \quad \mathbf{v}_0 = \begin{bmatrix} 1 \\ 0 \end{bmatrix}.$$

For $\lambda_1 = 3$: Solve $(A - 3I)\mathbf{v}_1 = \mathbf{0}$:

$$\begin{bmatrix} -1 & 1 \\ 0 & 0 \end{bmatrix} \mathbf{v}_1 = \mathbf{0} \quad \Rightarrow \quad \mathbf{v}_1 = \begin{bmatrix} 1 \\ 1 \end{bmatrix}.$$

Step 3: Construct the decomposition

$$V = \begin{bmatrix} 1 & 1 \\ 0 & 1 \end{bmatrix}, \quad \Lambda = \begin{bmatrix} 2 & 0 \\ 0 & 3 \end{bmatrix}, \quad V^{-1} = \begin{bmatrix} 1 & -1 \\ 0 & 1 \end{bmatrix}.$$

Step 4: Verification

$$V \Lambda V^{-1} = \begin{bmatrix} 1 & 1 \\ 0 & 1 \end{bmatrix} \begin{bmatrix} 2 & 0 \\ 0 & 3 \end{bmatrix} \begin{bmatrix} 1 & -1 \\ 0 & 1 \end{bmatrix} = \begin{bmatrix} 2 & 1 \\ 0 & 3 \end{bmatrix} = A.$$

6.2.5 Definition: Normal Matrix

A real square matrix A is a **normal matrix** if it commutes with its transpose:

$$A^\top A = AA^\top$$

Important examples of normal matrices include:

- **Symmetric** matrices: $A = A^\top$
- **Skew-symmetric** matrices: $A = -A^\top$
- **Orthogonal** matrices: $A^\top A = AA^\top = I$

The key property of normal matrices is that they have **orthonormal eigenvectors** (Spectral Theorem). This means for a normal matrix, the eigenvector matrix V is orthogonal, so $V^{-1} = V^\top$ and the Eigen Decomposition simplifies to $A = V \Lambda V^\top$.

In general, V is **not orthogonal**. The columns of V are linearly independent eigenvectors, but they are not necessarily orthonormal. V is orthogonal (i.e., $V^{-1} = V^\top$) only when A is a **normal matrix** (e.g., symmetric $A = A^\top$, skew-symmetric, or orthogonal). For normal matrices, the Eigen Decomposition simplifies to $A = V \Lambda V^\top$, which has the geometric interpretation of rotation → scaling → rotation. In **numerical linear algebra**, the orthogonality of V for symmetric matrices makes Eigen Decomposition more stable and efficient. In **machine learning**, understanding when V is orthogonal is crucial for methods like **Principal Component Analysis (PCA)** and **spectral clustering**, where symmetric matrices naturally arise.

6.2.6 Similarity Transformation and Similar Matrices

Two matrices A and B are said to be **similar** if there exists an invertible matrix P such that

$$B = P^{-1}AP \tag{6.6}$$

The transformation $P^{-1}AP$ is called a **similarity transformation**. Similar matrices share important properties: **they have the same eigenvalues, determinant, trace, and rank.** The similarity transformation represents a change of basis: if A represents a linear transformation in one basis, then $P^{-1}AP$ represents the same transformation in a different basis (the columns of P).

Proof that similar matrices share eigenvalues, determinant, trace, and rank:

Let $B = P^{-1}AP$ where P is invertible.

1. **Same eigenvalues**: If λ is an eigenvalue of A with eigenvector $\mathbf{v}$, then $A\mathbf{v} = \lambda\mathbf{v}$. Multiplying both sides by P^{-1} on the left and inserting $PP^{-1} = I$:

$$P^{-1}A\mathbf{v} = \lambda P^{-1}\mathbf{v} \quad \Rightarrow \quad P^{-1}APP^{-1}\mathbf{v} = \lambda P^{-1}\mathbf{v}$$

   Letting $\mathbf{w} = P^{-1}\mathbf{v}$, we have $B\mathbf{w} = \lambda\mathbf{w}$, so λ is also an eigenvalue of B with eigenvector $\mathbf{w}$. Conversely, if λ is an eigenvalue of B with eigenvector $\mathbf{w}$, then $B\mathbf{w} = \lambda\mathbf{w}$, which gives $P^{-1}AP\mathbf{w} = \lambda\mathbf{w}$. Multiplying by P on the left: $AP\mathbf{w} = \lambda P\mathbf{w}$, so λ is an eigenvalue of A with eigenvector $P\mathbf{w}$. Therefore, A and B have the same eigenvalues.

2. **Same determinant**: Using the multiplicative property of determinants (Chapter~5):

$$\begin{aligned}
\det(B) &= \det(P^{-1}AP) \\
&= \det(P^{-1})\det(A)\det(P) \\
&= \det(P^{-1})\det(P)\det(A) \\
&= \det(P^{-1}P)\det(A) \\
&= \det(I)\det(A) \\
&= \det(A)
\end{aligned}$$

3. **Same trace**: The trace invariance under similarity transformations was established in Chapter~5 (equation 5.4): $\operatorname{tr}(P^{-1}AP) = \operatorname{tr}(A)$.

4. **Same rank**: Since P and P^{-1} are invertible (full rank), they preserve the rank of matrices under multiplication. Specifically, $\operatorname{rank}(B) = \operatorname{rank}(P^{-1}AP) = \operatorname{rank}(A)$ because multiplying by invertible matrices on the left and right does not change the rank (the column space dimension is preserved under invertible transformations).

6.2.7 Diagonalization

Diagonalization is the process of transforming a matrix A into a diagonal matrix through a similarity transformation. A matrix $A \in \mathbb{R}^{N\times N}$ is **diagonalizable** if and only if it has N linearly independent eigenvectors.

The diagonalization of A is given by:

$$A = V\Lambda V^{-1} \quad \Leftrightarrow \quad \Lambda = V^{-1}AV \tag{6.7}$$

where V contains the eigenvectors of A as columns, and Λ is the diagonal matrix of eigenvalues. Here, A and Λ are **similar matrices** related by the similarity transformation $V^{-1}AV = \Lambda$.

Key properties: - Diagonalizable matrices preserve eigenvalues under similarity transformations - The diagonal form Λ makes it easy to compute powers: $A^k = V\Lambda^k V^{-1}$ - Not all matrices are diagonalizable (e.g., defective matrices)

Defective matrices: A matrix is **defective** (also called **deficient**) if it has fewer than N linearly independent eigenvectors. A defective matrix cannot be diagonalized. Note that a defective matrix can still be **full rank** (i.e., $\text{rank}(A) = N$). For example, a matrix with repeated eigenvalues where the geometric multiplicity (the dimension of the eigenspace, i.e., the number of linearly independent eigenvectors for that eigenvalue) is less than the algebraic multiplicity (multiplicity as a root of the characteristic polynomial) is defective but may still be full rank. However, if a matrix is **not full rank** (i.e., $\text{rank}(A) < N$), then it is necessarily defective and cannot be diagonalized.

Note on terminology: The term "deficient" is also used in "rank-deficient", which refers to a matrix whose rank is less than its dimension. This is a different concept from defective matrices. To avoid confusion, "defective" is more commonly used when discussing diagonalization, though the two terms are interchangeable in that context.

Example: Consider the matrix

$$A = \begin{bmatrix} 2 & 1 \\ 0 & 3 \end{bmatrix}$$

The eigenvalues are $\lambda_1 = 2$ and $\lambda_2 = 3$, with corresponding eigenvectors $\mathbf{v}_1 = \begin{bmatrix} 1 \\ 0 \end{bmatrix}$ and $\mathbf{v}_2 = \begin{bmatrix} 1 \\ 1 \end{bmatrix}$.

To construct the diagonalization, we form:

- $V = \begin{bmatrix} \mathbf{v}_1 & \mathbf{v}_2 \end{bmatrix} = \begin{bmatrix} 1 & 1 \\ 0 & 1 \end{bmatrix}$ (eigenvectors as columns)
- $\Lambda = \text{diag}(\lambda_1, \lambda_2) = \begin{bmatrix} \lambda_1 & 0 \\ 0 & \lambda_2 \end{bmatrix} = \begin{bmatrix} 2 & 0 \\ 0 & 3 \end{bmatrix}$ (eigenvalues on the diagonal in the same order as the columns of V)

Then $A = V\Lambda V^{-1}$, which can be verified by computing $V\Lambda V^{-1} = \begin{bmatrix} 2 & 1 \\ 0 & 3 \end{bmatrix} = A$.

In **numerical linear algebra**, diagonalization is fundamental for solving systems of differential equations, computing matrix functions (e.g., $\exp(A)$, $\log(A)$), and analyzing the stability of dynamical systems.

6.2.8 Spectral Decomposition

The **spectral decomposition** (also called **spectral theorem** for diagonalizable matrices) expresses a matrix as a weighted sum of rank-1 matrices formed from its eigenvectors:

$$A = \sum_{n=0}^{N-1} \lambda_n \mathbf{v}_n \mathbf{p}_n^\top \tag{6.8}$$

where $\mathbf{v}_n$ are the right eigenvectors (columns of V), $\mathbf{p}_n$ are the left eigenvectors (rows of V^{-1}, or columns of $P = (V^{-1})^\top$), and λ_n are the eigenvalues.

Each term $\lambda_n \mathbf{v}_n \mathbf{p}_n^\top$ is a rank-1 matrix that represents the contribution of the n-th eigenmode to the matrix A. The spectral decomposition reveals how A acts on vectors: it projects onto each eigenvector direction, scales by the corresponding eigenvalue, and recombines the results.

Special case for symmetric matrices: When A is symmetric, $\mathbf{p}_n = \mathbf{v}_n$ (left and right eigenvectors coincide), and the spectral decomposition simplifies to:

$$A = \sum_{n=0}^{N-1} \lambda_n \mathbf{v}_n \mathbf{v}_n^\top \tag{6.9}$$

where the eigenvectors $\mathbf{v}_n$ form an orthonormal basis. This form is particularly important in **Principal Component Analysis (PCA)**, where symmetric covariance matrices are decomposed into principal directions.

6.2.9 Theorem: Left Eigenvectors

If $A = V\Lambda V^{-1}$ and $P = (V^{-1})^\top$, then:

1. $A^\top = P\Lambda P^{-1}$.
2. $A^\top \mathbf{p}_n = \lambda_n \mathbf{p}_n, \quad n = 0, \ldots, N-1$.

The left eigenvectors $\mathbf{p}_n$ are the right eigenvectors of $A^\top$ with the same eigenvalues, forming a **biorthogonal basis** with the right eigenvectors: $\mathbf{p}_i^\top \mathbf{v}_j = \delta_{ij}$. They enable the decomposition $A = \sum_{n=0}^{N-1} \lambda_n \mathbf{v}_n \mathbf{p}_n^\top$ and are essential in **deflation methods** for computing all eigenvalues, where they remove already-computed eigencomponents. For non-symmetric matrices, left and right eigenvectors differ and both are needed; for symmetric matrices, they coincide ($\mathbf{p}_n = \mathbf{v}_n$).

6.2.10 Theorem: Powers of a Matrix

If $A = V\Lambda V^{-1}$, then for any $k \in \mathbb{N} \cup \{0\}$,

$$A^k = V\Lambda^k V^{-1} \tag{6.10}$$

6.2.11 Definition: Spectral Radius

For a square matrix $A \in \mathbb{R}^{N\times N}$ with eigenvalues $\lambda_0, \lambda_1, \ldots, \lambda_{N-1}$ (possibly complex), the **spectral radius** is defined as

$$\rho(A) = \max\{|\lambda_i|\} \tag{6.11}$$

The spectral radius is the maximum absolute value among all eigenvalues of A. It plays a crucial role in determining the convergence of iterative methods, matrix power series, and the stability of linear dynamical systems. For example, the Neumann series converges if and only if $\rho(A) < 1$.

6.2.12 Neumann Series

If $|\lambda_n| < 1$ for all eigenvalues of A (equivalently, the spectral radius $\rho(A) < 1$),

$$(I - A)^{-1} = \sum_{k=0}^{\infty} A^k = V \begin{bmatrix} \frac{1}{1-\lambda_0} & & \\ & \ddots & \\ & & \frac{1}{1-\lambda_{N-1}} \end{bmatrix} V^{-1} \tag{6.12}$$

The Neumann series provides an infinite series representation of the matrix inverse $(I - A)^{-1}$ when the spectral radius is less than 1. This is fundamental in **iterative methods** for solving linear systems, **fixed-point iterations**, and **perturbation theory**. The eigendecomposition form shows that convergence depends on all eigenvalues being inside the unit circle, and the inverse is computed by inverting each eigenvalue term $(1 - \lambda_n)^{-1}$ in the eigenbasis.

6.2.13 Remark: Linear Transformation View

Let $A = V\Lambda V^{-1}$ be the Eigen Decomposition of $A \in \mathbb{R}^{N\times N}$, where $V = \begin{bmatrix}\mathbf{v}_0 & \mathbf{v}_1 & \cdots & \mathbf{v}_{N-1}\end{bmatrix}$ contains the eigenvectors and $\Lambda = \text{diag}(\lambda_0, \lambda_1, \dots, \lambda_{N-1})$ contains the eigenvalues.

To understand how A transforms a vector $\mathbf{x}$ (i.e., compute $A\mathbf{x}$), we first express $\mathbf{x}$ in the eigenvector basis as

$$\mathbf{x} = \sum_{n=0}^{N-1} \mathbf{v}_n c_n = V\mathbf{c},$$

where $\mathbf{c} = \begin{bmatrix}c_0 & c_1 & \cdots & c_{N-1}\end{bmatrix}^\top$ are the coefficients of $\mathbf{x}$ in the eigenvector basis, given by $\mathbf{c} = V^{-1}\mathbf{x}$.

Step-by-step transformation:

1. **Change of basis**: Express $\mathbf{x}$ in the eigenvector basis:

$$\mathbf{x} = V\mathbf{c} \quad \Rightarrow \quad \mathbf{c} = V^{-1}\mathbf{x}.$$

2. **Diagonal transformation**: Apply the diagonal matrix Λ to the coefficient vector:

$$\Lambda\mathbf{c} = \begin{bmatrix} \lambda_0 c_0 \\ \lambda_1 c_1 \\ \vdots \\ \lambda_{N-1} c_{N-1} \end{bmatrix}.$$

   This scales each component c_n by its corresponding eigenvalue λ_n.

3. **Change back to standard basis**: Transform back using V:

$$A\mathbf{x} = V(\Lambda\mathbf{c}) = V\Lambda V^{-1}\mathbf{x}.$$

Result:

$$A\mathbf{x} = V\Lambda\mathbf{c} = \sum_{n=0}^{N-1} \mathbf{v}_n(\lambda_n c_n).$$

Geometric interpretation: The linear transformation A acts on $\mathbf{x}$ (i.e., $A\mathbf{x}$ is A right-multiplied by $\mathbf{x}$) by:

- Decomposing $\mathbf{x}$ into components along the eigenvector directions ($\mathbf{c} = V^{-1}\mathbf{x}$)
- Scaling each component by its corresponding eigenvalue ($\Lambda\mathbf{c}$)
- Recombining the scaled components in the eigenvector basis ($V(\Lambda\mathbf{c})$)

This view shows that **in the eigenvector basis, the transformation A becomes diagonal** (represented by Λ), making it easier to understand and compute the action of A on any vector.

Special case: Rotation–Scaling–Rotation interpretation

When A is a **normal matrix** (e.g., symmetric $A = A^\top$ or orthogonal), the eigenvectors form an orthonormal basis, so V is orthogonal and $V^{-1} = V^\top$. In this case, the Eigen Decomposition $A = V\Lambda V^\top$ can be interpreted geometrically as **Rotation+Scaling+Rotation**. Figure 6.9 illustrates this for a 2D example.

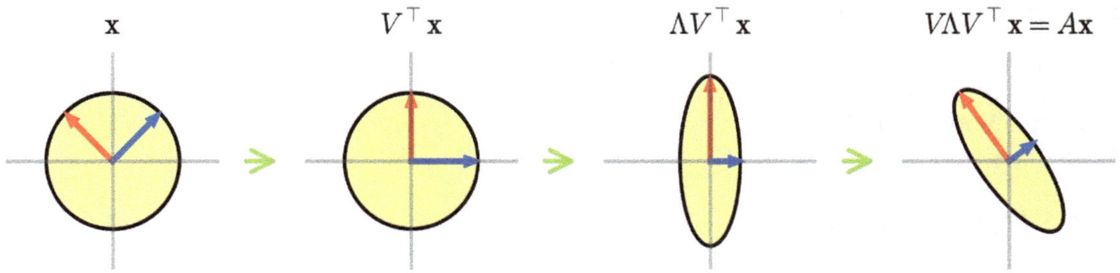

Figure 6.9: Rotation–scaling–rotation 2D View

1. **Rotate to eigenvector basis** ($V^\top$): Rotate the coordinate system to align with the eigenvector directions
2. **Scale along coordinate axes** (Λ): Scale along each coordinate axis by the corresponding eigenvalue
3. **Rotate back to original coordinate system** (V): Rotate back to the original coordinate system

This interpretation is particularly intuitive for symmetric matrices, where the transformation can be visualized as: align with principal axes → scale → realign.

> The geometric transformation order of $Ax = V\Lambda V^\top x$ is from right to left: rotation ($V^\top$) → scaling (Λ) → rotation (V). Moreover, V represents a pure rotation only when it is orthogonal and $\det(V) = 1$; otherwise, it may include reflections. This distinction is important in **computer graphics** and **robotics**, where rotation matrices must preserve orientation (no reflections).

6.2.14 Theorem: Trace–Eigenvalue Relation

For $A \in \mathbb{R}^{N\times N}$,

$$\operatorname{tr}(A) = \sum_{n=0}^{N-1} \lambda_n \tag{6.13}$$

6.2.15 Theorem: Determinant–Eigenvalue Relation

For $A \in \mathbb{R}^{N\times N}$,

$$\det(A) = \prod_{n=0}^{N-1} \lambda_n \tag{6.14}$$

> When performing Eigen Decomposition on real matrices, one may encounter **complex eigenvalues** and **complex eigenvectors**. This occurs because the characteristic polynomial $\det(A - \lambda I) = 0$ may have complex roots even when A has real entries. Complex eigenvalues and eigenvectors always appear in **conjugate pairs**: if λ is an eigenvalue with eigenvector $\mathbf{v}$, then $\bar{\lambda}$ (the complex conjugate) is also an eigenvalue with eigenvector $\bar{\mathbf{v}}$. The essential complex number concepts (complex conjugate $\bar{z}$, modulus $|z|$, Euler's formula $e^{i\theta} = \cos\theta + i\sin\theta$, and the connection between complex multiplication and 2D rotations) are introduced in Chapter~3. Complex matrices have their own algebraic structure: the transpose of a complex matrix is called the **Hermitian transpose** (denoted by superscript H, e.g., A^H), the complex analog of an orthogonal matrix is a **unitary matrix** (satisfying $U^H U = I$), and the complex analog of a symmetric matrix is a **normal matrix** (satisfying $A^H A = AA^H$). Complex matrix theory is beyond the scope of this book; interested readers may refer to specialized texts on linear algebra over complex fields.

6.2.15.1 Example: Applications

1. Markov transition probability matrices
2. Linear recurrences (e.g., Fibonacci sequence)
3. Differential equations

6.3 Power Method

While the eigenvalue equation $A\mathbf{x} = \lambda\mathbf{x}$ provides a theoretical foundation, computing eigenvalues and eigenvectors for large matrices requires efficient numerical algorithms. The **power method** is an iterative algorithm that finds the **dominant eigenvalue** (the eigenvalue with the largest absolute value) and its corresponding eigenvector. Note that the absolute value of the dominant eigenvalue equals the **spectral radius** $\rho(A)$ (see Definition~6.2.11, equation *6.11*). This method is particularly useful when:

- Only the largest eigenvalue is needed (e.g., in Markov chains, the dominant eigenvalue determines the steady-state distribution)
- The matrix is too large for direct methods (e.g., computing $\det(A - \lambda I)$ is computationally prohibitive)
- Matrix-vector products $A\mathbf{v}$ are available but the full matrix decomposition is expensive

The power method exploits the fact that repeatedly applying A to a random vector $\mathbf{v}$ will cause it to converge to the direction of the eigenvector corresponding to the dominant eigenvalue, since components along other eigenvectors decay exponentially relative to the dominant one.

Assumption: All eigenvalues are real.

Let eigenvalues satisfy

$$|\lambda_0| > |\lambda_1| \geq \cdots \geq |\lambda_{N-1}|.$$

This condition ensures that the dominant eigenvalue λ_0 is unique in magnitude, which guarantees convergence of the power method.

6.3.1 Algorithm: Dominant Eigenvalue and Eigenvectors

Algorithm 6 Dominant Eigenvalue and Eigenvectors

Input: Matrix $A \in \mathbb{R}^{N \times N}$, tolerance $\varepsilon > 0$
Output: Dominant eigenvalue λ and corresponding eigenvectors $\mathbf{v}$, $\mathbf{p}$
  Initialize random $\mathbf{v}, \mathbf{p} \in \mathbb{R}^{N \times 1}$
  $\mathbf{v} \leftarrow \frac{\mathbf{v}}{||\mathbf{v}||_2}$, $\mathbf{p} \leftarrow \frac{\mathbf{p}}{||\mathbf{p}||_2}$ ▷ Normalize
  repeat
    $\mathbf{v}_{\text{new}} \leftarrow A\mathbf{v}$, $\mathbf{p}_{\text{new}} \leftarrow A^\top \mathbf{p}$
    Normalize: $\mathbf{v}_{\text{new}} \leftarrow \frac{\mathbf{v}_{\text{new}}}{||\mathbf{v}_{\text{new}}||_2}$, $\mathbf{p}_{\text{new}} \leftarrow \frac{\mathbf{p}_{\text{new}}}{||\mathbf{p}_{\text{new}}||_2}$
    $\mathbf{v} \leftarrow \mathbf{v}_{\text{new}}$, $\mathbf{p} \leftarrow \mathbf{p}_{\text{new}}$
  until $||\mathbf{v}_{\text{new}} - \mathbf{v}||_2 < \varepsilon$ and $||\mathbf{p}_{\text{new}} - \mathbf{p}||_2 < \varepsilon$
  $\lambda \leftarrow \frac{\mathbf{v}^\top A\mathbf{v}}{\mathbf{v}^\top \mathbf{v}}$ ▷ Compute eigenvalue
  return λ, $\mathbf{v}$, $\mathbf{p}$

6.3.2 Algorithm: All Eigenvalues via Deflation

Prerequisite: A is diagonalizable, and the left and right eigenvectors are normalized such that $\mathbf{p}_i^\top \mathbf{v}_j = \delta_{ij}$ (biorthogonal condition). Alternatively, if A is symmetric, then $\mathbf{p}_i = \mathbf{v}_i$ and the condition is automatically satisfied.

Algorithm 7 All Eigenvalues via Deflation

Input: Diagonalizable matrix $A \in \mathbb{R}^{N \times N}$, tolerance $\varepsilon > 0$
Output: All eigenvalues $\lambda_0, \ldots, \lambda_{N-1}$ and eigenvectors $\mathbf{v}_0, \ldots, \mathbf{v}_{N-1}$, $\mathbf{p}_0, \ldots, \mathbf{p}_{N-1}$
for $n = 0, \ldots, N-1$ **do**
Deflate: $A \leftarrow A - \sum_{i=0}^{n-1} \lambda_i \mathbf{v}_i \mathbf{p}_i^\top$
Apply power method to A to obtain λ_n, $\mathbf{v}_n$, $\mathbf{p}_n$
end for
return $\lambda_0, \ldots, \lambda_{N-1}$, $\mathbf{v}_0, \ldots, \mathbf{v}_{N-1}$, $\mathbf{p}_0, \ldots, \mathbf{p}_{N-1}$

6.4 Matrix Functions

Eigendecomposition enables the natural extension of scalar functions to matrices. If $f : \mathbb{R} \to \mathbb{R}$ is a scalar function and $A = V\Lambda V^{-1}$ is diagonalizable, we can define:

$$f(A) = Vf(\Lambda)V^{-1} = V \begin{bmatrix} f(\lambda_0) & & \\ & \ddots & \\ & & f(\lambda_{N-1}) \end{bmatrix} V^{-1} \tag{6.15}$$

This definition is natural because it reduces to applying f to each eigenvalue, then reconstructing the matrix in the original basis.

6.4.1 Definition: Matrix Exponential

The **matrix exponential** of $A \in \mathbb{R}^{N \times N}$ is defined by the power series:

$$\exp(A) = e^A = \sum_{k=0}^{\infty} \frac{A^k}{k!} = I + A + \frac{A^2}{2!} + \frac{A^3}{3!} + \cdots \tag{6.16}$$

This series converges for all square matrices A.

For a diagonalizable matrix $A = V\Lambda V^{-1}$:

$$\exp(A) = V\exp(\Lambda)V^{-1} = V \begin{bmatrix} e^{\lambda_0} & & \\ & \ddots & \\ & & e^{\lambda_{N-1}} \end{bmatrix} V^{-1} \tag{6.17}$$

Proof

Using the power series and $A^k = V\Lambda^k V^{-1}$:

$$\exp(A) = \sum_{k=0}^{\infty} \frac{V\Lambda^k V^{-1}}{k!} = V\left(\sum_{k=0}^{\infty} \frac{\Lambda^k}{k!}\right) V^{-1} = V\exp(\Lambda)V^{-1}.$$

Since Λ is diagonal, $\exp(\Lambda)$ is diagonal with entries e^{λ_n}.

6.4.2 Properties of Matrix Exponential

1. **Identity**: $\exp(\mathbf{0}) = I$
2. **Inverse**: $\exp(A)^{-1} = \exp(-A)$
3. **Determinant**: $\det(\exp(A)) = e^{\text{tr}(A)}$

4. **Transpose**: $\exp(A^\top) = \exp(A)^\top$
5. **Commutativity condition**: $\exp(A+B) = \exp(A)\exp(B)$ if and only if $AB = BA$
6. **Similarity**: If $B = PAP^{-1}$, then $\exp(B) = P\exp(A)P^{-1}$
7. **Derivative:** $\frac{d}{dt}\exp(tA) = A\exp(tA) = \exp(tA)A$

Proof of Property 3

Using Property 8 from the eigenvalue section ($\det(A) = \prod_n \lambda_n$):

$$\det(\exp(A)) = \prod_{n=0}^{N-1} e^{\lambda_n} = e^{\sum_{n=0}^{N-1} \lambda_n} = e^{\operatorname{tr}(A)}.$$

6.4.3 Connection to Linear ODEs

The matrix exponential is the fundamental solution to linear ordinary differential equations. Consider the first-order linear ODE:

$$\frac{d\mathbf{x}(t)}{dt} = A\mathbf{x}(t), \quad \mathbf{x}(0) = \mathbf{x}_0 \tag{6.18}$$

The unique solution is:

$$\mathbf{x}(t) = \exp(tA)\mathbf{x}_0 \tag{6.19}$$

Proof

Differentiate using Property 7:

$$\frac{d\mathbf{x}(t)}{dt} = \frac{d}{dt}\exp(tA)\mathbf{x}_0 = A\exp(tA)\mathbf{x}_0 = A\mathbf{x}(t).$$

At $t = 0$: $\mathbf{x}(0) = \exp(0)\mathbf{x}_0 = I\mathbf{x}_0 = \mathbf{x}_0$.

The matrix exponential is fundamental in many areas of mathematics and physics. In **dynamical systems**, it describes the evolution of linear systems. In **Lie theory**, the matrix exponential maps elements of a Lie algebra to the corresponding Lie group (e.g., skew-symmetric matrices to rotation matrices). In **quantum mechanics**, the time evolution operator is a matrix exponential of the Hamiltonian. The connection to ODEs (equation *6.19*) shows that understanding the eigenstructure of A reveals the qualitative behavior of solutions: eigenvalues with positive real parts cause exponential growth, negative real parts cause decay, and purely imaginary eigenvalues cause oscillation. This analysis is central to **stability theory** of dynamical systems.

6.4.3.1 Example: Rotation via Matrix Exponential

Consider the skew-symmetric matrix:

$$A = \begin{bmatrix} 0 & -\theta \\ \theta & 0 \end{bmatrix}.$$

The eigenvalues are $\lambda = \pm i\theta$ (purely imaginary). Computing the matrix exponential:

$$\exp(A) = \begin{bmatrix} \cos\theta & -\sin\theta \\ \sin\theta & \cos\theta \end{bmatrix} = R(\theta).$$

This shows that the matrix exponential of a skew-symmetric matrix is a rotation matrix, connecting algebra to geometry.

6.4.4 Definition: Matrix Logarithm

For an invertible matrix A with no negative real eigenvalues, the **matrix logarithm** $\log(A)$ is defined as the matrix B such that:

$$\exp(B) = A \tag{6.20}$$

For a diagonalizable matrix $A = V\Lambda V^{-1}$ with positive eigenvalues:

$$\log(A) = V\log(\Lambda)V^{-1} = V \begin{bmatrix} \log\lambda_0 & & \\ & \ddots & \\ & & \log\lambda_{N-1} \end{bmatrix} V^{-1} \tag{6.21}$$

The matrix logarithm is not unique in general (similar to the complex logarithm), but the **principal logarithm** is uniquely defined for matrices with no eigenvalues on the closed negative real axis.

6.4.5 Definition: Matrix Square Root

For a positive semi-definite matrix A (i.e., $A = A^\top$ and all eigenvalues $\lambda_n \geq 0$), the **matrix square root** $A^{1/2}$ is defined as:

$$A^{1/2} = V\Lambda^{1/2}V^\top = V \begin{bmatrix} \sqrt{\lambda_0} & & \\ & \ddots & \\ & & \sqrt{\lambda_{N-1}} \end{bmatrix} V^\top \tag{6.22}$$

where $A = V\Lambda V^\top$ is the spectral decomposition (Chapter~7).

Properties: - $(A^{1/2})^2 = A$ - $A^{1/2}$ is symmetric and positive semi-definite - $A^{1/2}$ is unique among symmetric positive semi-definite matrices

> The matrix square root is essential in **whitening transformations**: if Σ is a covariance matrix, then $\Sigma^{-1/2}\mathbf{x}$ has identity covariance. In **natural gradient** methods, the Fisher information matrix square root is used for preconditioning. In **optimal transport**, the matrix square root appears in the Wasserstein distance formula. The Cholesky decomposition $A = LL^\top$ provides an alternative "square root" that is computationally efficient but not symmetric.

6.4.6 Theorem: Cayley-Hamilton

Every square matrix $A \in \mathbb{R}^{N\times N}$ satisfies its own characteristic equation. If the characteristic polynomial is:

$$p(\lambda) = \det(A - \lambda I) = (-1)^N(\lambda^N - c_{N-1}\lambda^{N-1} - \cdots - c_1\lambda - c_0),$$

then:

$$p(A) = A^N - c_{N-1}A^{N-1} - \dots - c_1 A - c_0 I = \mathbf{0} \tag{6.23}$$

Consequence: Any matrix power A^k with $k \geq N$ can be expressed as a linear combination of $I, A, A^2, \dots, A^{N-1}$.

6.4.6.1 Example

For a 2×2 matrix with characteristic polynomial $p(\lambda) = \lambda^2 - \text{tr}(A)\lambda + \det(A)$:

$$A^2 = \text{tr}(A) \cdot A - \det(A) \cdot I.$$

This allows computation of A^2 (and higher powers) using only A and I.

> The Cayley-Hamilton theorem has important computational applications. It implies that matrix functions (exponential, logarithm, etc.) can be computed using only the first N powers of A. For example, $\exp(A)$ for an $N \times N$ matrix can be written as a polynomial of degree at most $N-1$ in A. This is the basis for **Padé approximation** methods for computing matrix exponentials. In **control theory**, Cayley-Hamilton is used to design feedback controllers and analyze system reachability.

6.5 Mindmap

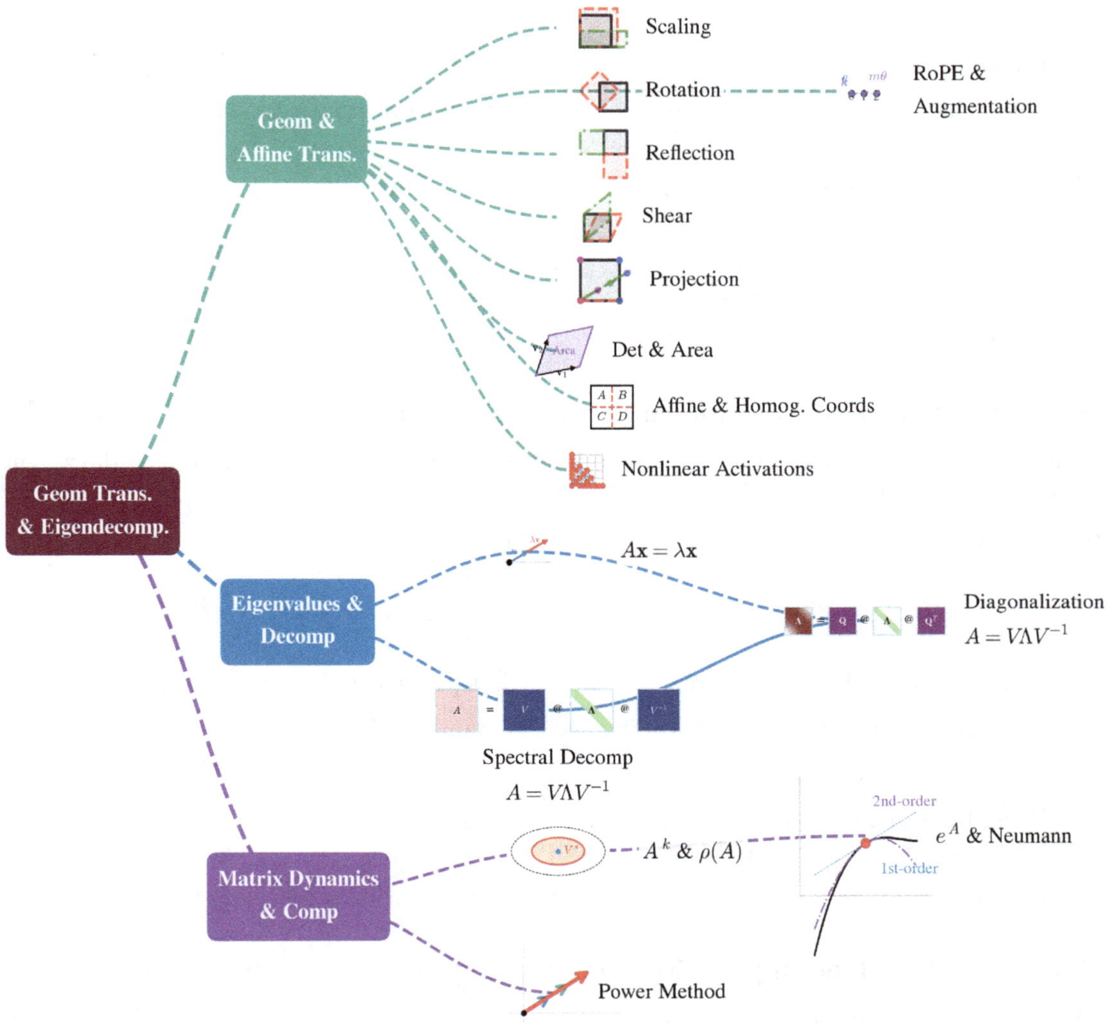

Figure 6.10: Mind map of geometric transformation and eigen decomposition topics

6.6 Exercises

6.6.1 Eigenvalues and Eigenvectors in Linear Transformations

In neural networks, weight matrices represent linear transformations. Understanding their eigenvalues and eigenvectors helps analyze how information flows through layers. Consider a weight matrix:

$$W = \begin{bmatrix} 2 & -1 & 1 \\ 0 & 1 & 0 \\ 0 & 0 & 1 \end{bmatrix}.$$

1. Compute all eigenvalues of W (including multiplicities) by solving $\det(W - \lambda I) = 0$.
2. For each eigenvalue, find a corresponding eigenvector (or a basis for the eigenspace if the eigenvalue has multiplicity greater than 1).
3. Verify that $W\mathbf{v} = \lambda\mathbf{v}$ for each eigenvalue-eigenvector pair $(\lambda, \mathbf{v})$.
4. For an input vector $\mathbf{x} = \begin{bmatrix} 1 \\ 1 \\ 1 \end{bmatrix}$, compute $W\mathbf{x}$ and express it in terms of the eigenvectors and eigenvalues.

6.6.2 Eigen Decomposition and Matrix Powers in Recurrent Networks

In recurrent neural networks (RNNs), we need to compute $W^k\mathbf{x}$ for large k, where W is the weight matrix. Eigen decomposition $W = V\Lambda V^{-1}$ makes this efficient: $W^k = V\Lambda^k V^{-1}$.

For the matrix

$$W = \begin{bmatrix} 2 & -1 & 1 \\ 0 & 1 & 0 \\ 0 & 0 & 1 \end{bmatrix},$$

1. Find the matrix V whose columns are linearly independent right eigenvectors of W.
2. Find the diagonal matrix Λ containing the eigenvalues of W.
3. Compute V^{-1} and verify that $W = V\Lambda V^{-1}$.
4. Using the eigen decomposition, compute W^3 without directly multiplying W three times. Verify your result by computing W^3 directly.
5. For what values of k does $\lim_{k\to\infty} W^k$ exist? (Hint: Consider the eigenvalues.)

6.6.3 Spectral Decomposition and Principal Component Analysis

In PCA, we decompose the covariance matrix $C = \frac{1}{N}X^\top X$ (where X is the centered data matrix) using eigen decomposition. The eigenvectors are the principal directions, and the eigenvalues indicate the variance along each direction.

Consider a 2×2 covariance matrix:

$$C = \begin{bmatrix} 5 & 2 \\ 2 & 2 \end{bmatrix}.$$

1. Compute the eigenvalues λ_1, λ_2 and corresponding eigenvectors $\mathbf{v}_1, \mathbf{v}_2$ of C. Normalize the eigenvectors to have unit length.
2. Verify that $C = V\Lambda V^\top$ where $V = \begin{bmatrix} \mathbf{v}_1 & \mathbf{v}_2 \end{bmatrix}$ and $\Lambda = \text{diag}(\lambda_1, \lambda_2)$.
3. Express C using spectral decomposition: $C = \lambda_1\mathbf{v}_1\mathbf{v}_1^\top + \lambda_2\mathbf{v}_2\mathbf{v}_2^\top$. Verify by computing the right-hand side.
4. Which principal direction (eigenvector) captures more variance? What percentage of total variance does it capture?

6.6.4 Trace and Determinant from Eigenvalues

In optimization, the trace and determinant of the Hessian matrix H (second derivative matrix) provide information about the curvature of the loss landscape. These can be computed efficiently from eigenvalues.

For the matrix

$$A = \begin{bmatrix} 3 & 1 & 0 \\ 1 & 3 & 1 \\ 0 & 1 & 3 \end{bmatrix},$$

1. Compute the eigenvalues $\lambda_1, \lambda_2, \lambda_3$ of A.
2. Verify that $\text{tr}(A) = \lambda_1 + \lambda_2 + \lambda_3$ by computing both sides.
3. Verify that $\det(A) = \lambda_1 \lambda_2 \lambda_3$ by computing both sides.
4. If A represents a Hessian matrix in optimization, what do the eigenvalues tell you about the curvature? (Hint: Positive eigenvalues indicate positive curvature, negative eigenvalues indicate negative curvature.)

6.6.5 Matrix Powers and Stability in Dynamical Systems

In analyzing the stability of linear dynamical systems $\mathbf{x}_{t+1} = A\mathbf{x}_t$ (e.g., in control theory or analyzing RNNs), we need to understand the long-term behavior, which depends on the eigenvalues of A.

Consider the matrix

$$A = \begin{bmatrix} 0.8 & 0.2 \\ 0.1 & 0.9 \end{bmatrix},$$

which represents a transition matrix in a linear system.

1. Compute the eigenvalues λ_1, λ_2 of A.
2. Find the eigen decomposition $A = V\Lambda V^{-1}$.
3. Using the eigen decomposition, compute A^{10} without directly multiplying A ten times.
4. Determine whether the system $\mathbf{x}_{t+1} = A\mathbf{x}_t$ is stable (i.e., $\lim_{t\to\infty} \mathbf{x}_t = \mathbf{0}$ for any initial $\mathbf{x}_0$). (Hint: The system is stable if and only if all eigenvalues have absolute value less than 1.)
5. For an initial state $\mathbf{x}_0 = \begin{bmatrix} 1 \\ 1 \end{bmatrix}$, compute $\mathbf{x}_t = A^t\mathbf{x}_0$ for $t = 1, 2, 3$ using the eigen decomposition.

6.6.6 Left Eigenvectors and Biorthogonality

In deflation methods for computing all eigenvalues, we need both left and right eigenvectors that satisfy the biorthogonality condition. For non-symmetric matrices, left and right eigenvectors differ.

For the matrix

$$A = \begin{bmatrix} 2 & 1 \\ 0 & 3 \end{bmatrix},$$

1. Find the right eigenvectors $\mathbf{v}_1, \mathbf{v}_2$ of A and construct $V = \begin{bmatrix} \mathbf{v}_1 & \mathbf{v}_2 \end{bmatrix}$.
2. Compute V^{-1} and verify that $A = V\Lambda V^{-1}$ where $\Lambda = \text{diag}(2, 3)$.
3. Compute $P = (V^{-1})^\top$ and verify that the columns of P are left eigenvectors of A (i.e., $A^\top \mathbf{p}_n = \lambda_n \mathbf{p}_n$ for $n = 1, 2$).

4. Verify the biorthogonality condition: $\mathbf{p}_i^\top \mathbf{v}_j = \delta_{ij}$ for $i, j \in \{1, 2\}$, where δ_{ij} is the Kronecker delta.

5. Express A using the spectral decomposition: $A = \lambda_1 \mathbf{v}_1 \mathbf{p}_1^\top + \lambda_2 \mathbf{v}_2 \mathbf{p}_2^\top$. Verify by computing the right-hand side.

6.6.7 Matching Linear Transformations with Matrices

Understanding how different transformation matrices affect geometric shapes is crucial for visualizing linear transformations. The figure below shows four plots labeled (a), (b), (c), and (d), each representing a different linear transformation applied to a square grid.

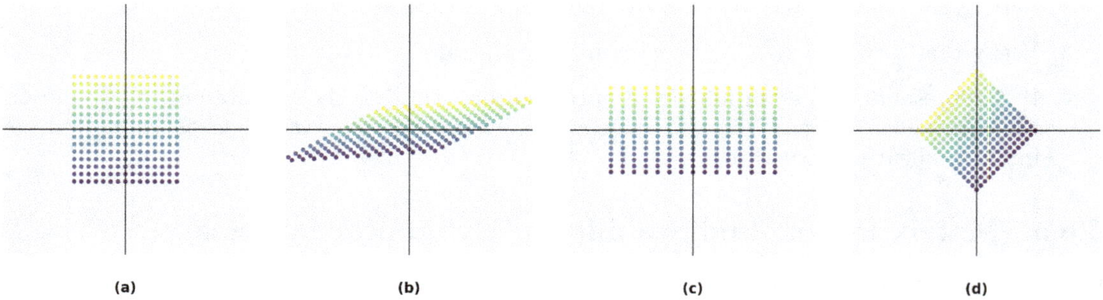

Figure 6.11: Linear Transformations

Consider the following four 2×2 transformation matrices:

$$M_1 = \begin{bmatrix} 1 & 0 \\ 0 & 1 \end{bmatrix}, \quad M_2 = \begin{bmatrix} 1.5 & 1.0 \\ 0.1 & 0.5 \end{bmatrix}, \quad M_3 = \begin{bmatrix} 2 & 0 \\ 0 & 1 \end{bmatrix}, \quad M_4 = \begin{bmatrix} \frac{\sqrt{2}}{2} & -\frac{\sqrt{2}}{2} \\ \frac{\sqrt{2}}{2} & \frac{\sqrt{2}}{2} \end{bmatrix}.$$

1. Determine which matrix corresponds to each plot by analyzing the geometric properties preserved or changed by each transformation.

2. For each transformation matrix, find its eigenvalues and eigenvectors. Using only the eigenvalues, determine which transformation preserves the area of the unit square, which preserves lengths, and which preserves orthogonality.

3. One of the matrices represents an orthogonal transformation. Identify which one and prove that it satisfies $M^\top M = I$. Show that the eigenvalues of this matrix have absolute value 1 and are complex conjugates. What does this tell you about the geometric effect?

4. For the matrix that scales along one axis, determine the scaling factor from its eigenvalues. Prove that the eigenvectors are aligned with the coordinate axes. If this transformation is applied to a unit circle, what is the area of the resulting ellipse? Express your answer in terms of the eigenvalues.

5. For the matrix with both scaling and shearing components, compute its determinant and trace. Using the relationship $\det(M) = \lambda_1 \lambda_2$ and $\text{tr}(M) = \lambda_1 + \lambda_2$, determine the eigenvalues without directly solving the characteristic equation. What does the sign of the determinant tell you about whether the transformation preserves orientation?

Chapter 7

Symmetric Matrix

Symmetric matrices: spectral decomposition, positive definiteness, and Cholesky factorization

> *"Pure mathematics is, in its way, the poetry of logical ideas."*
>
> — Albert Einstein

- ▷ `Real eigenval` All eigenvalues of symmetric $A \in \mathbb{R}^{N\times N}$ are real
- ▷ `Orthogonal eigenvec` $\mathbf{q}_i^\top \mathbf{q}_j = 0$ for eigenvectors $\mathbf{q}_i, \mathbf{q}_j$ of symmetric A with $\lambda_i \neq \lambda_j$
- ▷ `Spectral theorem` $A = Q\Lambda Q^\top = \sum_{n=0}^{N-1} \lambda_n \mathbf{q}_n \mathbf{q}_n^\top$, Q is orthonormal, $\Lambda = \text{diag}(\lambda_0, ...)$
- ▷ `Matrix power` $A^k\mathbf{x} = Q\Lambda^k\mathbf{c} = \sum_{n=0}^{N-1} \mathbf{q}_n \lambda_n^k c_n$ where $\mathbf{x} = Q\mathbf{c}$ and $A = Q\Lambda Q^\top$
- ▷ `Inverse via spectral` $A^{-1} = Q\Lambda^{-1}Q^\top$ for invertible symmetric $A = Q\Lambda Q^\top$
- ▷ `Rayleigh quotient bounds` $\lambda_{\text{min}} \leq R_A(\mathbf{x}) \leq \lambda_{\text{max}}$ for any nonzero $\mathbf{x}$
- ▷ `Eigenvalue bounds of sums` $\lambda_{\text{min}}(A) + \lambda_{\text{min}}(B) \leq \lambda_{\text{min}}(A+B) \leq \lambda_{\text{max}}(A+B) \leq \lambda_{\text{max}}(A) + \lambda_{\text{max}}(B)$ for symmetric A, B
- ▷ `Quadratic form` $Q_A(\mathbf{x}) = \mathbf{x}^\top A\mathbf{x} = \sum_{i=0}^{N-1} a_{ii}x_i^2 + 2\sum_{i<j} a_{ij}x_ix_j$ for symmetric A
- ▷ `Quadratic form via spectral` $\mathbf{x}^\top A\mathbf{x} = \sum_{n=0}^{N-1} \lambda_n c_n^2$ where $\mathbf{c} = Q^\top\mathbf{x}$ and $A = Q\Lambda Q^\top$
- ▷ `Gradient of quadratic form` $\nabla(\mathbf{x}^\top A\mathbf{x}) = 2A\mathbf{x}$ for symmetric A
- ▷ `Hessian of quadratic form` $\nabla^2(\mathbf{x}^\top A\mathbf{x}) = 2A$ for symmetric A
- ▷ `Positive (semi-)definite` $A \succ 0$ if $\mathbf{x}^\top A\mathbf{x} > 0$ for all $\mathbf{x} \neq 0$; $A \succeq 0$ if $\mathbf{x}^\top A\mathbf{x} \geq 0$
- ▷ `Positive definite criterion` $A \succ 0$ iff all eigenvalues $\lambda_i > 0$
- ▷ `Positive semi-definite criterion` $A \succeq 0$ iff all eigenvalues $\lambda_i \geq 0$
- ▷ `Cholesky decomp` $A = LL^\top$ for positive definite A; equivalently $A = R^\top R$ where R
- ▷ `Gram matrices` $AA^\top$ and $A^\top A$ are SPSD for any $A \in \mathbb{R}^{M\times N}$
- ▷ `Trace equality` $\text{tr}(AA^\top) = \text{tr}(A^\top A)$
- ▷ `Grams Eigenval/rank` $\lambda_i(AA^\top) = \lambda_i(A^\top A)$, $\text{rank}(A) = \text{rank}(AA^\top) = \text{rank}(A^\top A)$

7.1 Definition: Symmetric matrix

A real-valued matrix $A \in \mathbb{R}^{N \times N}$ is **symmetric** if

$$A = A^\top \tag{7.1}$$

i.e. $a_{ij} = a_{ji} \quad \forall i, j$

Examples of Symmetric Matrices:

1. Undirected graph adjacency matrices
2. Kernel matrices
3. Covariance matrices in statistics
4. Hessian matrices in multivariable calculus

Symmetric matrices are ubiquitous in AI/ML. **Covariance matrices** Σ for multivariate Gaussian distributions are symmetric and positive semi-definite, essential in Bayesian inference, variational autoencoders, and probabilistic models. **Kernel matrices** K in support vector machines and Gaussian processes are symmetric (since $K(x_i, x_j) = K(x_j, x_i)$), enabling efficient eigendecomposition for kernel methods. **Attention mechanisms** in transformers use symmetric similarity matrices when computing self-attention, and **graph neural networks** rely on symmetric adjacency matrices for undirected graphs. **Principal component analysis (PCA)** performs eigendecomposition on symmetric covariance matrices to find principal directions of data variation.

7.2 Spectral Decomposition

7.2.1 Theorem: Real Eigenvalues and Eigenvectors

For a real-valued symmetric matrix $A \in \mathbb{R}^{N \times N}$:

- All eigenvalues are real, and A admits a complete set of real eigenvectors.
- Equivalently, A admits a spectral decomposition (see Spectral Theorem below).

Proof

Let $\lambda \in \mathbb{C}$ and $\mathbf{x} \in \mathbb{C}^{N \times 1}$ satisfy

$$A\mathbf{x} = \lambda \mathbf{x},$$

with $\mathbf{x} \neq 0$ (where $\mathbf{x}$ may be complex). Here we use the complex conjugate notation $\overline{\lambda}$ (see Chapter~3 for complex number basics).

Consider the Hermitian form:

$$\mathbf{x}^* A\mathbf{x} = \mathbf{x}^*(\lambda \mathbf{x}) = \lambda \mathbf{x}^* \mathbf{x}.$$

Since $A = A^\top$ and A is real, we also have:

$$\mathbf{x}^* A\mathbf{x} = (A\mathbf{x})^* \mathbf{x} = (\lambda \mathbf{x})^* \mathbf{x} = \overline{\lambda} \mathbf{x}^* \mathbf{x}.$$

Because $\mathbf{x}^*\mathbf{x} > 0$ (since $\mathbf{x} \neq 0$), we conclude $\lambda = \overline{\lambda}$, hence $\lambda \in \mathbb{R}$.

Since all coefficients in $(A - \lambda I)\mathbf{x} = 0$ are real, eigenvectors can be chosen to be real.

7.2.2 Theorem: Orthogonal Eigenvectors

For a real-valued symmetric matrix $A \in \mathbb{R}^{N \times N}$, eigenvectors $\mathbf{q}_i$ and $\mathbf{q}_j$ associated with **distinct** eigenvalues $\lambda_i \neq \lambda_j$ satisfy

$$\langle \mathbf{q}_i, \mathbf{q}_j \rangle = 0.$$

Proof

Approach 1

$$\mathbf{q}_i^\top A \mathbf{q}_j = \lambda_j \mathbf{q}_i^\top \mathbf{q}_j.$$

Approach 2

$$\mathbf{q}_i^\top A \mathbf{q}_j = (A\mathbf{q}_i)^\top \mathbf{q}_j = \lambda_i \mathbf{q}_i^\top \mathbf{q}_j.$$

Thus,

$$(\lambda_j - \lambda_i)\mathbf{q}_i^\top \mathbf{q}_j = 0.$$

Since $\lambda_i \neq \lambda_j$,

$$\mathbf{q}_i^\top \mathbf{q}_j = 0.$$

> If an eigenvalue has multiplicity k, one can find k orthogonal eigenvectors (e.g., via Gram–Schmidt).

7.2.3 Theorem: Spectral Theorem

Any real-valued symmetric matrix $A \in \mathbb{R}^{N \times N}$ admits the decomposition

$$A = Q\Lambda Q^\top = \sum_{n=0}^{N-1} \lambda_n \mathbf{q}_n \mathbf{q}_n^\top \tag{7.2}$$

where

$$Q = \begin{bmatrix} \mathbf{q}_0 & \cdots & \mathbf{q}_{N-1} \end{bmatrix}$$

is a real-valued orthonormal matrix.

Proof

Eigenvectors corresponding to distinct eigenvalues are orthogonal; eigenvectors within a repeated eigenspace can be orthonormalized. Hence $Q^\top Q = I$ and $Q^{-1} = Q^\top$.

It is just eigen-decomposition of symmetric matrices. The spectral decomposition $A = Q\Lambda Q^\top$ of symmetric matrices is computationally efficient and numerically stable, making it the foundation for many dimensionality reduction and feature extraction techniques. The term **spectral** derives from the analogy between eigenvalues and the spectrum of a physical system (e.g., the frequencies of light emitted by an atom), where the eigenvalues represent the intrinsic resonant frequencies or characteristic properties of the matrix.

7.2.4 Remark: Linear Transformation View

Let $A = Q\Lambda Q^\top$. For any $\mathbf{x} \in \mathbb{R}^{N \times 1}$, represent

$$\mathbf{x} = \sum_{n=0}^{N-1} \mathbf{q}_n c_n = Q\mathbf{c} \tag{7.3}$$

Then for any $k \in \mathbb{N}$,

$$A^k\mathbf{x} = Q\Lambda^k\mathbf{c} = \sum_{n=0}^{N-1} \mathbf{q}_n \lambda_n^k c_n \tag{7.4}$$

7.2.5 Theorem: Inverse via Spectral Decomposition

If

$$A = Q\Lambda Q^\top \tag{7.5}$$

and all eigenvalues are non-zero, then

$$A^{-1} = Q\Lambda^{-1}Q^\top \tag{7.6}$$

7.2.6 Theorem: Eigenvalue Bounds of Sums

Let $A, B \in \mathbb{R}^{N\times N}$ be symmetric. All eigenvalues of $A+B$ lie in $[\lambda_{\text{min}}(A+B), \lambda_{\text{max}}(A+B)]$, and

$$\lambda_{\text{min}}(A) + \lambda_{\text{min}}(B) \le \lambda_{\text{min}}(A+B) \tag{7.7}$$

$$\lambda_{\text{max}}(A+B) \le \lambda_{\text{max}}(A) + \lambda_{\text{max}}(B) \tag{7.8}$$

Hence each eigenvalue of $A+B$ lies in

$$[\lambda_{\text{min}}(A) + \lambda_{\text{min}}(B), \lambda_{\text{max}}(A) + \lambda_{\text{max}}(B)].$$

7.3 Quadratic Forms and Definiteness

Quadratic forms serve as the fundamental link between scalar-valued calculus and linear algebra. By encoding a second-degree polynomial into the compact form $\mathbf{x}^\top A\mathbf{x}$ (where A is symmetric), we can leverage spectral theory to decode the function's high-dimensional geometry. The eigenvalues of A are not merely algebraic numbers; they represent the principal curvatures of the landscape. They rigorously characterize the critical point as a convex 'valley' (positive definite, see below), a concave 'peak' (negative definite), or a 'saddle point' (indefinite, see below). In optimization, this structure underpins second-order methods, where the Hessian matrix locally approximates the objective function as a quadratic form to determine the optimal step direction. The connection between quadratic forms and the Hessian matrix becomes explicit through the **second-order Taylor expansion** (see Chapter~11, equation (11.33)), where the quadratic term $\frac{1}{2}(\mathbf{x}-\mathbf{a})^\top H_f(\mathbf{x}-\mathbf{a})$ (with $H_f = \nabla^2 f(\mathbf{a})$ being the symmetric Hessian matrix) captures the local curvature of the function f near point $\mathbf{a}$. This explains why positive definiteness of the Hessian guarantees a local minimum: the quadratic form is positive for all directions, indicating the function curves upward in all directions (a valley).

7.3.1 Definition: Quadratic Form

For a symmetric matrix $A \in \mathbb{R}^{N\times N}$ and a vector $\mathbf{x} \in \mathbb{R}^{N\times 1}$, the expression

$$Q_A(\mathbf{x}) = \mathbf{x}^\top A\mathbf{x} \tag{7.9}$$

is called a **quadratic form** in $\mathbf{x}$.

Explicit Expansion

For $\mathbf{x} = [x_0, x_1, \dots, x_{N-1}]^\top$ and $A = [a_{ij}]$,

$$\mathbf{x}^\top A\mathbf{x} = \sum_{i=0}^{N-1}\sum_{j=0}^{N-1} a_{ij}x_ix_j.$$

Since A is symmetric ($a_{ij} = a_{ji}$), this can be written as

$$\mathbf{x}^\top A\mathbf{x} = \sum_{i=0}^{N-1} a_{ii}x_i^2 + 2\sum_{i<j} a_{ij}x_ix_j.$$

7.3.2 Definition: Positive, Negative Definite

A real-valued symmetric matrix A is:

- **Positive definite** (denoted $A \succ 0$) if

$$\mathbf{x}^\top A\mathbf{x} > 0 \quad \forall \mathbf{x} \neq 0 \tag{7.10}$$

- **Negative definite** (denoted $A \prec 0$) if

$$\mathbf{x}^\top A\mathbf{x} < 0 \quad \forall \mathbf{x} \neq 0 \tag{7.11}$$

7.3.3 Definition: Positive, Negative Semi-Definite and Indefinite

- **Positive semi-definite** (denoted $A \succeq 0$) if

$$\mathbf{x}^\top A\mathbf{x} \geq 0 \tag{7.12}$$

- **Negative semi-definite** (denoted $A \preceq 0$) if

$$\mathbf{x}^\top A\mathbf{x} \leq 0 \tag{7.13}$$

- **Indefinite** if A is neither positive semi-definite nor negative semi-definite, i.e., there exist vectors $\mathbf{x}$ and $\mathbf{y}$ such that $\mathbf{x}^\top A\mathbf{x} > 0$ and $\mathbf{y}^\top A\mathbf{y} < 0$. Equivalently, A has both positive and negative eigenvalues.

7.3.4 Theorem: Positive Definite Eigenvalue Criterion

A real-valued symmetric matrix is positive definite **iff** all eigenvalues are strictly positive.

Proof

Let $A = Q\Lambda Q^\top$ be the spectral decomposition of A, where Q is orthonormal and $\Lambda = \mathrm{diag}(\lambda_0, \dots, \lambda_{N-1})$.

($\Leftarrow$) If all eigenvalues are strictly positive:

For any $\mathbf{x} \neq 0$, let $\mathbf{c} = Q^\top\mathbf{x}$. Since Q is orthonormal, $\mathbf{c} \neq 0$ (otherwise $\mathbf{x} = Q\mathbf{c} = 0$). By the Quadratic Form via Spectral Decomposition theorem (see equation (7.14)):

$$\mathbf{x}^\top A\mathbf{x} = \sum_{n=0}^{N-1} \lambda_n c_n^2.$$

Since all $\lambda_n > 0$ and at least one $c_n \neq 0$ (because $\mathbf{c} \neq 0$), we have $\mathbf{x}^\top A\mathbf{x} > 0$ for all $\mathbf{x} \neq 0$. Therefore, A is positive definite.

($\Rightarrow$) If A is positive definite:

For each eigenvector $\mathbf{q}_n$ of A with eigenvalue λ_n, we have:

$$\mathbf{q}_n^\top A\mathbf{q}_n = \mathbf{q}_n^\top(\lambda_n\mathbf{q}_n) = \lambda_n\|\mathbf{q}_n\|_2^2 = \lambda_n > 0,$$

where the last inequality follows from the positive definiteness of A (since $\mathbf{q}_n \neq 0$) and the fact that $\|\mathbf{q}_n\|_2 = 1$ (since Q is orthonormal). Therefore, all eigenvalues $\lambda_n > 0$.

7.3.5 Theorem: Positive Semi-Definite Eigenvalue Criterion

A real-valued symmetric matrix is positive semi-definite **iff** all eigenvalues are non-negative.

Proof

The proof is analogous to that of the Positive Definite Eigenvalue Criterion theorem above, replacing strict inequalities ($>$) with non-strict inequalities ($\geq$) and the condition $\mathbf{x} \neq 0$ with $\mathbf{x}$ (allowing zero).

7.3.6 Geometric Interpretation of Quadratic Form

The quadratic form $\mathbf{x}^\top A\mathbf{x}$ has important geometric interpretations:

1. **Level Sets**: The set $\{\mathbf{x} : \mathbf{x}^\top A\mathbf{x} = \alpha\}$ (see Chapter~13 for set notation) for constant α forms:
   - An **ellipsoid** if A is positive definite (for $\alpha > 0$)or negative definite (for $\alpha < 0$)
   - A **hyperboloid** if A is indefinite
   - A **paraboloid** or **degenerate surface** if A is singular or semi-definite
2. **Principal Axes**: When A is positive definite, the eigenvectors $\mathbf{q}_n$ of A determine the **principal axes** of the ellipsoid. The eigenvalues λ_n determine the lengths of these axes.
3. **Distance Measure**: For positive definite A, the expression $\sqrt{\mathbf{x}^\top A\mathbf{x}}$ defines the **M-Norm** of $\mathbf{x}$ (see Chapter~1, equation (1.4)). When $A = \Sigma^{-1}$ where Σ is a covariance matrix, this accounts for correlations in the data (see Chapter~13, equation (13.38) for the definition of correlation). The Mahalanobis distance between two points $\mathbf{a}$ and $\mathbf{b}$ is $\sqrt{(\mathbf{a}-\mathbf{b})^\top A(\mathbf{a}-\mathbf{b})} = \|\mathbf{a}-\mathbf{b}\|_M$, which is the M-norm of the vector $(\mathbf{a}-\mathbf{b})$.

7.3.6.1 Example: Anomaly Detection with Correlated Features

Problem: A health monitoring system tracks patients' height (in cm) and weight (in kg). The training data shows that height and weight are positively correlated: taller people tend to weigh more. The mean vector is $\mu = [170, 70]^\top$ (170cm, 70kg) and the covariance matrix is

$$\Sigma = \begin{bmatrix} 100 & 40 \\ 40 & 25 \end{bmatrix}.$$

The diagonal elements of Σ are the variances: $\sigma^2_{\text{height}} = 100$ (standard deviation $\sigma_{\text{height}} = 10$ cm) and $\sigma^2_{\text{weight}} = 25$ (standard deviation $\sigma_{\text{weight}} = 5$ kg). The off-diagonal covariance $\Sigma_{12} = 40$ indicates a positive correlation.

We evaluate two new patients:
- **Patient A**: $\mathbf{x}_A = [180, 80]^\top$ (10cm taller and 10kg heavier than average)
- **Patient B**: $\mathbf{x}_B = [180, 60]^\top$ (10cm taller but 10kg lighter than average)

Using Euclidean distance: Both patients are at the exact same Euclidean distance from the mean:

$$\|\mathbf{x}_A - \mu\|_2 = \|\mathbf{x}_B - \mu\|_2 = \sqrt{10^2 + (\pm 10)^2} = \sqrt{200} \approx 14.1.$$

Euclidean distance is misleading here because it ignores scale differences between the features (adding centimeters directly to kilograms) and is blind to their correlation.

Using Mahalanobis distance: The Mahalanobis distance is $\|\mathbf{x} - \mu\|_M = \sqrt{(\mathbf{x} - \mu)^\top \Sigma^{-1} (\mathbf{x} - \mu)}$. We compute the inverse covariance matrix:

$$\Sigma^{-1} = \frac{1}{900} \begin{bmatrix} 25 & -40 \\ -40 & 100 \end{bmatrix} = \begin{bmatrix} 0.0278 & -0.0444 \\ -0.0444 & 0.1111 \end{bmatrix}.$$

For a two-dimensional Gaussian distribution, the squared Mahalanobis distance d_M^2 follows a χ_2^2 distribution, where the typical 95% anomaly threshold is $\chi^2_{2,0.95} \approx 5.99$.

- **For Patient A**: $\mathbf{x}_A - \mu = [10, 10]^\top$. Evaluating the distance:

$$d_{M,A}^2 = \begin{bmatrix} 10 & 10 \end{bmatrix} \begin{bmatrix} 0.0278 & -0.0444 \\ -0.0444 & 0.1111 \end{bmatrix} \begin{bmatrix} 10 \\ 10 \end{bmatrix} = 5.0 \quad \Rightarrow \quad d_{M,A} = \sqrt{5.0} \approx 2.24.$$

  Since $d_{M,A}^2 = 5.0 < 5.99$, Patient A falls within the 95% probability ellipse and is classified as normal. Note that if we treated the features as independent (using a diagonal covariance $\Sigma_{\text{diag}} = \text{diag}(100, 25)$), the squared distance would also be $\frac{10^2}{100} + \frac{10^2}{25} = 5.0$. For Patient A, the two metrics happen to coincide—a quirk of this particular direction, not the general case, as Patient B makes clear.

- **For Patient B**: $\mathbf{x}_B - \mu = [10, -10]^\top$. Evaluating the distance:

$$d_{M,B}^2 = \begin{bmatrix} 10 & -10 \end{bmatrix} \begin{bmatrix} 0.0278 & -0.0444 \\ -0.0444 & 0.1111 \end{bmatrix} \begin{bmatrix} 10 \\ -10 \end{bmatrix} = \frac{205}{9} \approx 22.78 \quad \Rightarrow \quad d_{M,B} \approx 4.77.$$

  Since $d_{M,B}^2 \approx 22.78 \gg 5.99$, Patient B is far outside the 95% threshold and is flagged as highly anomalous. If we ignored correlation, Patient B's independent squared distance would also be 5.0, incorrectly classifying them as normal. Because height and weight are positively correlated, a patient who is both taller and significantly lighter is extremely rare under the joint distribution. Mahalanobis distance correctly exposes this anomaly by measuring distance relative to the covariance geometry.

In **anomaly detection**, features are often correlated (e.g., height and weight in health monitoring, CPU usage and memory usage in system monitoring). Euclidean distance $\|\mathbf{a} - \mathbf{b}\|_2$ treats all dimensions independently, potentially flagging normal but correlated variations as anomalies (false positives) or failing to detect anomalous combinations that violate correlation patterns (false negatives). Mahalanobis distance $\|\mathbf{a} - \mathbf{b}\|_M$ accounts for these correlations through Σ^{-1}, measuring distance along the data's principal directions (see the Principal Axes section above). This is why **Gaussian Mixture Models (GMMs)** and **multivariate Gaussian-based anomaly detectors** use Mahalanobis distance rather than Euclidean distance.

7.3.6.2 Example: Quadratic Form in 2D

To illustrate these concepts concretely, consider the two-dimensional case. For $N = 2$ with

$$A = \begin{bmatrix} a & c \\ c & b \end{bmatrix}, \quad \mathbf{x} = \begin{bmatrix} x_0 \\ x_1 \end{bmatrix},$$

the quadratic form expands to

$$f(\mathbf{x}) = \mathbf{x}^\top A\mathbf{x} = ax_0^2 + 2cx_0x_1 + bx_1^2.$$

When A is positive definite with eigenvalues $\lambda_0, \lambda_1 > 0$ and corresponding eigenvectors $\mathbf{q}_0, \mathbf{q}_1$, the level set $\{\mathbf{x} : \mathbf{x}^\top A\mathbf{x} = 1\}$ forms an ellipse with principal axes along $\mathbf{q}_0$ and $\mathbf{q}_1$, and semi-axis lengths of $1/\sqrt{\lambda_0}$ and $1/\sqrt{\lambda_1}$.

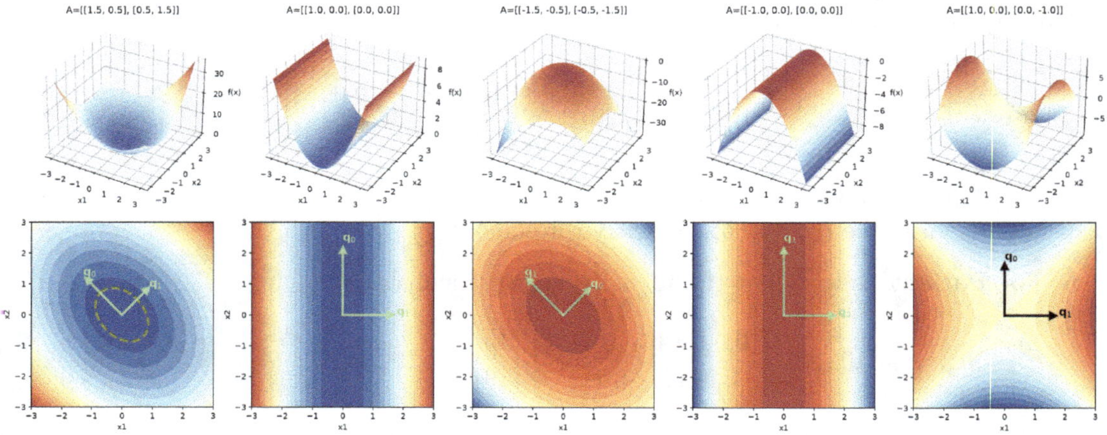

Figure 7.1: Quadratic Forms and Definiteness

Figure 7.1 visualizes quadratic forms for five different matrix types, demonstrating how the definiteness properties manifest geometrically. The top row shows 3D surface plots, and the bottom row shows 2D contour plots. The **principal axes** (eigenvectors $\mathbf{q}_0$ and $\mathbf{q}_1$) are indicated by lightgreen (or black for the indefinite case) arrows in the contour plots, with arrow lengths proportional to $1/\sqrt{|\lambda|}$ to reflect the relative importance of each axis direction. For the **positive definite** matrix (leftmost), the yellow dashed ellipse represents the level set $\{\mathbf{x} : \mathbf{x}^\top A\mathbf{x} = 1\}$, which corresponds to points at unit **Mahalanobis distance** from the origin. From left to right: (1) **Positive definite** $A = [1.5, 0.5; 0.5, 1.5]$ exhibits a bowl-shaped surface (all eigenvalues positive); (2) **Positive semi-definite** $A = [1, 0; 0, 0]$ shows a valley with a flat bottom (one eigenvalue zero, one positive); (3) **Negative definite** $A = [-1.5, -0.5; -0.5, -1.5]$ displays a peak (all eigenvalues negative); (4) **Negative semi-definite** $A = [-1, 0; 0, 0]$ shows a ridge (one eigenvalue zero, one negative); (5) **Indefinite** $A = [1, 0; 0, -1]$ exhibits a saddle point (one positive and one negative eigenvalue).

7.3.7 Theorem: Uniqueness of Symmetric Matrix Representation

For any quadratic form $Q(\mathbf{x})$, there exists a **unique symmetric matrix** A such that $Q(\mathbf{x}) = \mathbf{x}^\top A\mathbf{x}$.

Proof

Given a quadratic form $Q(\mathbf{x})$, we can always write it as $Q(\mathbf{x}) = \mathbf{x}^\top M\mathbf{x}$ for some square matrix M (which may not be symmetric). Define

$$A = \frac{M + M^\top}{2}.$$

Then A is symmetric. Moreover, for any $\mathbf{x}$:

$$\mathbf{x}^\top A\mathbf{x} = \mathbf{x}^\top \frac{M + M^\top}{2}\mathbf{x} = \frac{1}{2}(\mathbf{x}^\top M\mathbf{x} + \mathbf{x}^\top M^\top \mathbf{x}).$$

Since $\mathbf{x}^\top M\mathbf{x}$ **is a scalar**, we have $\mathbf{x}^\top M^\top \mathbf{x} = (\mathbf{x}^\top M\mathbf{x})^\top = \mathbf{x}^\top M\mathbf{x}$. Therefore,

$$\mathbf{x}^\top A\mathbf{x} = \frac{1}{2}(2\mathbf{x}^\top M\mathbf{x}) = \mathbf{x}^\top M\mathbf{x} = Q(\mathbf{x}).$$

To show uniqueness, suppose A_1 and A_2 are both symmetric matrices such that $\mathbf{x}^\top A_1\mathbf{x} = \mathbf{x}^\top A_2\mathbf{x}$ for all $\mathbf{x}$. Then $\mathbf{x}^\top(A_1 - A_2)\mathbf{x} = 0$ for all $\mathbf{x}$. Since $A_1 - A_2$ is symmetric, this implies $A_1 - A_2 = 0$, hence $A_1 = A_2$.

7.3.8 Theorem: Quadratic Form via Spectral Decomposition

Let $A = Q\Lambda Q^\top$ be the spectral decomposition of a symmetric matrix A, where Q is orthonormal and $\Lambda = \operatorname{diag}(\lambda_0, \dots, \lambda_{N-1})$.

For any $\mathbf{x} \in \mathbb{R}^{N\times 1}$, let $\mathbf{c} = Q^\top \mathbf{x}$ be the coordinates of $\mathbf{x}$ in the eigenvector basis. Then

$$\mathbf{x}^\top A\mathbf{x} = \sum_{n=0}^{N-1} \lambda_n c_n^2 \tag{7.14}$$

Proof

$$\mathbf{x}^\top A\mathbf{x} = \mathbf{x}^\top Q\Lambda Q^\top \mathbf{x} = (Q^\top \mathbf{x})^\top \Lambda (Q^\top \mathbf{x}) = \mathbf{c}^\top \Lambda \mathbf{c} = \sum_{n=0}^{N-1} \lambda_n c_n^2.$$

7.3.9 Theorem: Bounds on Quadratic Forms

For a symmetric matrix A with eigenvalues $\lambda_0 \le \lambda_1 \le \dots \le \lambda_{N-1}$:

1. For any $\mathbf{x} \neq 0$, the normalized quadratic form (see the Rayleigh Quotient section below) satisfies:

$$\lambda_0 \le \frac{\mathbf{x}^\top A\mathbf{x}}{\mathbf{x}^\top \mathbf{x}} \le \lambda_{N-1} \tag{7.15}$$

2. The minimum is achieved when $\mathbf{x}$ is an eigenvector corresponding to λ_0.
3. The maximum is achieved when $\mathbf{x}$ is an eigenvector corresponding to λ_{N-1}.

> This shows that quadratic forms are naturally bounded by the eigenvalues of the matrix. The normalized form $\frac{\mathbf{x}^\top A\mathbf{x}}{\mathbf{x}^\top \mathbf{x}}$ is called the Rayleigh quotient and is discussed in detail in the next section.

7.3.10 Theorem: Derivative of Quadratic Form

For a quadratic form $Q_A(\mathbf{x}) = \mathbf{x}^\top A\mathbf{x}$ where $A \in \mathbb{R}^{N\times N}$ is symmetric and $\mathbf{x} \in \mathbb{R}^{N\times 1}$:

1. **Gradient (First Derivative):**

$$\nabla Q_A(\mathbf{x}) = 2A\mathbf{x} \tag{7.16}$$

The gradient is a column vector in $\mathbb{R}^{N\times 1}$.

2. **Jacobian**: For a scalar-valued function $Q_A : \mathbb{R}^N \to \mathbb{R}$, the Jacobian is the row vector (transpose of the gradient):

$$\mathbf{J}_{Q_A}(\mathbf{x}) = \nabla Q_A(\mathbf{x})^\top = 2\mathbf{x}^\top A \tag{7.17}$$

For scalar-valued functions, the Jacobian is simply the transpose of the gradient. They represent the same first derivative information, just in different formats (column vector vs. row vector).

3. **Hessian (Second Derivative)**:

$$\nabla^2 Q_A(\mathbf{x}) = 2A \tag{7.18}$$

The Hessian is a symmetric matrix in $\mathbb{R}^{N\times N}$.

Proof

For the gradient, we compute the partial derivatives. Writing $Q_A(\mathbf{x}) = \sum_{i=0}^{N-1}\sum_{j=0}^{N-1} a_{ij}x_ix_j$ and using the symmetry $a_{ij} = a_{ji}$:

$$\frac{\partial Q_A}{\partial x_k} = \frac{\partial}{\partial x_k}\left(\sum_{i=0}^{N-1}\sum_{j=0}^{N-1} a_{ij}x_ix_j\right) = \sum_{j=0}^{N-1} a_{kj}x_j + \sum_{i=0}^{N-1} a_{ik}x_i = 2\sum_{j=0}^{N-1} a_{kj}x_j = 2(A\mathbf{x})_k.$$

Therefore, $\nabla Q_A(\mathbf{x}) = 2A\mathbf{x}$.

For the Hessian, we compute the second partial derivatives:

$$\frac{\partial^2 Q_A}{\partial x_k \partial x_\ell} = \frac{\partial}{\partial x_\ell}(2\sum_{j=0}^{N-1} a_{kj}x_j) = 2a_{k\ell}.$$

Therefore, $\nabla^2 Q_A(\mathbf{x}) = 2A$.

Alternatively, using matrix calculus notation: since $Q_A(\mathbf{x}) = \mathbf{x}^\top A\mathbf{x}$ and A is symmetric, we have:

$$\frac{\partial}{\partial \mathbf{x}}(\mathbf{x}^\top A\mathbf{x}) = (A + A^\top)\mathbf{x} = 2A\mathbf{x},$$

and

$$\frac{\partial^2}{\partial \mathbf{x}\partial \mathbf{x}^\top}(\mathbf{x}^\top A\mathbf{x}) = 2A.$$

This is a special case of the general gradient formula for quadratic forms (see Chapter~11, formula 2). For a general matrix W (not necessarily symmetric), the gradient is $\nabla_x(x^TWx) = (W + W^T)x$ where $x \in \mathbb{R}^{N\times 1}$. When W is symmetric ($W = W^T$), this simplifies to $\nabla_x(x^TWx) = 2Wx$, which matches the result $\nabla Q_A(\mathbf{x}) = 2A\mathbf{x}$ above (with $W = A$ and $x = \mathbf{x}$).

7.3.10.1 Example: Gradient of a 2D Quadratic Form

For $A = \begin{bmatrix} a & c \\ c & b \end{bmatrix}$ and $\mathbf{x} = \begin{bmatrix} x_0 \\ x_1 \end{bmatrix}$:

$$Q_A(\mathbf{x}) = ax_0^2 + 2cx_0x_1 + bx_1^2,$$

$$\nabla Q_A(\mathbf{x}) = \begin{bmatrix} 2ax_0 + 2cx_1 \\ 2cx_0 + 2bx_1 \end{bmatrix} = 2\begin{bmatrix} a & c \\ c & b \end{bmatrix}\begin{bmatrix} x_0 \\ x_1 \end{bmatrix} = 2A\mathbf{x}.$$

7.3.11 Why Require Symmetry in the Definition?

Technically, the expression $\mathbf{x}^\top W\mathbf{x}$ is a quadratic form for **any** square matrix W, regardless of whether W is symmetric. However, the value of $\mathbf{x}^\top W\mathbf{x}$ depends **only on the symmetric part** of W. The skew-symmetric part contributes nothing to the sum.

Mathematical explanation: Any square matrix W can be decomposed into a symmetric part and a skew-symmetric part. A matrix A is **skew-symmetric** (or **antisymmetric**) if $A = -A^\top$, i.e., $a_{ij} = -a_{ji}$ for all i, j. In particular, all diagonal entries of a skew-symmetric matrix must be zero: $a_{ii} = -a_{ii}$ implies $a_{ii} = 0$ for all i.

For any square matrix W, we can write:

$$W = W_S + W_A$$

where $W_S = \frac{W+W^\top}{2}$ is the **symmetric part** and $W_A = \frac{W-W^\top}{2}$ is the **skew-symmetric part.**

Verification that W_A is skew-symmetric:

$$W_A^\top = \left(\frac{W - W^\top}{2}\right)^\top = -\frac{W - W^\top}{2} = -W_A$$

Therefore, $W_A^\top = -W_A$, confirming that W_A is skew-symmetric. Similarly, $W_S^\top = \left(\frac{W+W^\top}{2}\right)^\top = \frac{W^\top+W}{2} = W_S$, confirming that W_S is symmetric.

When computing $\mathbf{x}^\top W\mathbf{x}$:

$$\mathbf{x}^\top W\mathbf{x} = \mathbf{x}^\top W_S\mathbf{x} + \mathbf{x}^\top W_A\mathbf{x}$$

The term involving the skew-symmetric part, $\mathbf{x}^\top W_A\mathbf{x}$, is **always zero**. This is because:

1. $\mathbf{x}^\top W_A\mathbf{x}$ is a scalar, so it equals its own transpose: $\mathbf{x}^\top W_A\mathbf{x} = (\mathbf{x}^\top W_A\mathbf{x})^\top = \mathbf{x}^\top W_A^\top\mathbf{x}$.
2. Since W_A is skew-symmetric (as defined above), we have $W_A^\top = -W_A$ by the definition of skew-symmetric matrices.
3. Therefore: $\mathbf{x}^\top W_A^\top\mathbf{x} = \mathbf{x}^\top(-W_A)\mathbf{x} = -\mathbf{x}^\top W_A\mathbf{x}$.
4. Combining steps 1 and 3: $\mathbf{x}^\top W_A\mathbf{x} = -\mathbf{x}^\top W_A\mathbf{x}$, which implies $2(\mathbf{x}^\top W_A\mathbf{x}) = 0$, hence $\mathbf{x}^\top W_A\mathbf{x} = 0$.

Why then require symmetry in the definition? There are three fundamental reasons:

1. **Uniqueness**: If we allow non-symmetric matrices, the same quadratic form can be represented by infinitely many different matrices. Requiring symmetry ensures a **one-to-one correspondence** between quadratic forms and symmetric matrices, eliminating ambiguity.

2. **Spectral properties**: Only real symmetric matrices guarantee **real eigenvalues** and **orthogonal eigenvectors** (by the Spectral Theorem). This is essential for analyzing the geometry of quadratic forms (whether they represent "valleys", "peaks", or "saddle points") and for understanding positive/negative definiteness, which are crucial in optimization.
3. **Calculus and optimization**: When A is symmetric, gradient formulas simplify: $\nabla(\mathbf{x}^\top A\mathbf{x}) = 2A\mathbf{x}$ (see Chapter~11). The Hessian matrix in optimization must be symmetric, and requiring symmetry from the start simplifies analysis in methods like Newton's method.

Therefore, while $\mathbf{x}^\top W\mathbf{x}$ is a quadratic form for any W, **the mathematical convention** requires the matrix to be symmetric to ensure uniqueness, enable spectral analysis, and simplify calculus operations.

Quadratic forms are central to many optimization and compression problems in AI/ML. In **GPTQ (GPT Quantization)**, a model compression technique, the quantization error is minimized using a quadratic form $\mathbf{e}^\top H\mathbf{e}$ where H is the Hessian matrix of the loss function. This quadratic form captures how quantization errors in different parameters interact, allowing GPTQ to achieve 4-bit quantization with minimal loss. The Hessian matrix $H = \nabla^2\mathcal{L}$ (which is symmetric) appears in second-order optimization methods like **Newton's method** and **natural gradient descent**, where the update direction is determined by $H^{-1}\nabla\mathcal{L}$. The **Mahalanobis distance** $\sqrt{\mathbf{x}^\top\Sigma^{-1}\mathbf{x}}$ (where Σ is a symmetric positive definite covariance matrix) is a quadratic form used in anomaly detection, clustering, and dimensionality reduction techniques to measure distances that account for correlations in the data.

7.4 Rayleigh Quotient

7.4.1 Definition: Rayleigh Quotient

For a symmetric matrix $A \in \mathbb{R}^{N\times N}$ and a nonzero vector $\mathbf{x} \in \mathbb{R}^{N\times 1}$, the **Rayleigh quotient** is defined as

$$R_A(\mathbf{x}) = \frac{\mathbf{x}^\top A\mathbf{x}}{\mathbf{x}^\top\mathbf{x}} = \frac{\langle \mathbf{x}, A\mathbf{x}\rangle}{\langle \mathbf{x}, \mathbf{x}\rangle} \tag{7.19}$$

The Rayleigh quotient is **scale-invariant**: $R_A(\alpha\mathbf{x}) = R_A(\mathbf{x})$ for any $\alpha \neq 0$. This follows directly from the definition:

$$R_A(\alpha\mathbf{x}) = \frac{(\alpha\mathbf{x})^\top A(\alpha\mathbf{x})}{(\alpha\mathbf{x})^\top(\alpha\mathbf{x})} = \frac{\alpha^2\mathbf{x}^\top A\mathbf{x}}{\alpha^2\mathbf{x}^\top\mathbf{x}} = \frac{\mathbf{x}^\top A\mathbf{x}}{\mathbf{x}^\top\mathbf{x}} = R_A(\mathbf{x}) \tag{7.20}$$

Due to this scale-invariance, we can assume $\|\mathbf{x}\|_2 = 1$ without loss of generality.

Geometrically, the Rayleigh quotient measures the **stretching factor** of the transformation A in the direction of $\mathbf{x}$—it answers "How much does A stretch vectors in the direction of $\mathbf{x}$?" Under the normalization $\|\mathbf{x}\|_2 = 1$, we have $R_A(\mathbf{x}) = \mathbf{x}^\top A\mathbf{x} = \langle \mathbf{x}, A\mathbf{x}\rangle$. By the vector projection formula (see Chapter~1, *1.10*), this equals the length of the projection of $A\mathbf{x}$ onto $\mathbf{x}$ (with sign), since $\langle \mathbf{x}, \mathbf{x}\rangle = 1$ when $\|\mathbf{x}\|_2 = 1$.

While the expression $\frac{\mathbf{x}^\top W\mathbf{x}}{\mathbf{x}^\top\mathbf{x}}$ is well-defined for any square matrix W, the value depends only on the symmetric part of W. To see this, decompose $W = W_S + W_A$ where $W_S = \frac{W+W^\top}{2}$ is symmetric and $W_A = \frac{W-W^\top}{2}$ is skew-symmetric. Since $\mathbf{x}^\top W_A\mathbf{x} = 0$ for any skew-symmetric matrix W_A (as shown in the "Why Require Symmetry in the Definition?" section for quadratic forms), we have:

$$\frac{\mathbf{x}^\top W\mathbf{x}}{\mathbf{x}^\top\mathbf{x}} = \frac{\mathbf{x}^\top(W_S + W_A)\mathbf{x}}{\mathbf{x}^\top\mathbf{x}} = \frac{\mathbf{x}^\top W_S\mathbf{x}}{\mathbf{x}^\top\mathbf{x}}$$

7.4.2 Theorem: Rayleigh Quotient Bounds

Let $\lambda_{\min}$ and $\lambda_{\max}$ be the smallest and largest eigenvalues of a symmetric matrix $A \in \mathbb{R}^{N\times N}$. For any nonzero vector $\mathbf{x} \in \mathbb{R}^{N\times 1}$, the Rayleigh quotient (see *7.19*) satisfies

$$\lambda_{\min} \leq R_A(\mathbf{x}) \leq \lambda_{\max} \tag{7.21}$$

Each bound is tight: the lower bound is achieved when $\mathbf{x}$ is an eigenvector corresponding to $\lambda_{\min}$, and the upper bound is achieved when $\mathbf{x}$ is an eigenvector corresponding to $\lambda_{\max}$.

7.4.3 Proof: Rayleigh Quotient Bounds

Geometric Proof via Definition: From the geometric definition, $R_A(\mathbf{x})$ equals the length of the projection of $A\mathbf{x}$ onto $\mathbf{x}$ when $\|\mathbf{x}\|_2 = 1$ (see the definition above). When $\mathbf{x}$ is an eigenvector $\mathbf{v}_n$ with eigenvalue λ_n, we have $A\mathbf{v}_n = \lambda_n \mathbf{v}_n$, so the projection length is exactly λ_n. For an arbitrary direction $\mathbf{x}$, the projection length of $A\mathbf{x}$ onto $\mathbf{x}$ must lie between the minimum and maximum projection lengths achievable over all directions, which occur when $\mathbf{x}$ aligns with the eigenvectors corresponding to $\lambda_{\min}$ and $\lambda_{\max}$. Therefore, $\lambda_{\min} \leq R_A(\mathbf{x}) \leq \lambda_{\max}$.

Algebraic Proof via Spectral Decomposition: The bounds can also be derived using the spectral decomposition. Let $A = V\Lambda V^\top$ be the eigendecomposition of a symmetric matrix A, where V is orthonormal and $\Lambda = \mathrm{diag}(\lambda_0, \lambda_1, \dots, \lambda_{N-1})$ contains the eigenvalues.

The numerator of $R_A(\mathbf{x})$ can be written as:

$$\mathbf{x}^\top A\mathbf{x} = \mathbf{x}^\top V\Lambda V^\top \mathbf{x} = (V^\top \mathbf{x})^\top \Lambda (V^\top \mathbf{x}) \tag{7.22}$$

Let $\mathbf{y} = V^\top \mathbf{x}$, which represents the coordinates of $\mathbf{x}$ in the eigenbasis. Then:

$$\mathbf{x}^\top A\mathbf{x} = \mathbf{y}^\top \Lambda \mathbf{y} = \sum_{n=0}^{N-1} \lambda_n y_n^2 \tag{7.23}$$

Similarly, the denominator becomes:

$$\mathbf{x}^\top \mathbf{x} = \mathbf{x}^\top VV^\top \mathbf{x} = (V^\top \mathbf{x})^\top (V^\top \mathbf{x}) = \mathbf{y}^\top \mathbf{y} = \sum_{n=0}^{N-1} y_n^2 \tag{7.24}$$

Therefore, the Rayleigh quotient can be expressed as a function of $\mathbf{y}$:

$$R_A(\mathbf{x}) = R(\mathbf{y}) = \frac{\sum_{n=0}^{N-1} \lambda_n y_n^2}{\sum_{n=0}^{N-1} y_n^2} = \frac{\lambda_0 y_0^2 + \lambda_1 y_1^2 + \dots + \lambda_{N-1} y_{N-1}^2}{y_0^2 + y_1^2 + \dots + y_{N-1}^2} \tag{7.25}$$

For a 2×2 matrix, this simplifies to:

$$R(\mathbf{y}) = \frac{\lambda_1 y_1^2 + \lambda_2 y_2^2}{y_1^2 + y_2^2} \tag{7.26}$$

where λ_1, λ_2 are the eigenvalues and y_1, y_2 are the components of $\mathbf{y} = V^\top \mathbf{x}$ in the eigenbasis.

This form makes it clear that $R(\mathbf{y})$ is a **weighted average** of the eigenvalues, with weights proportional to y_n^2. Since the weights are non-negative and sum to $\|\mathbf{y}\|^2$, we immediately obtain:

$$\lambda_{\min} \leq R(\mathbf{y}) \leq \lambda_{\max}$$

The bounds are achieved when $\mathbf{y}$ has only one nonzero component corresponding to the minimum or maximum eigenvalue.

7.4.4 Geometric Interpretation

The Rayleigh quotient has a geometric interpretation as a constrained optimization problem. The constraint $\|\mathbf{x}\|_2 = 1$ defines a **unit circle** (in 2D), a **unit sphere** (in 3D), or more generally a **unit hypersphere** (in higher dimensions). The objective function $f(\mathbf{x}) = \mathbf{x}^\top A\mathbf{x}$ is a **quadratic form** whose level sets form ellipses (in 2D) or ellipsoids (in higher dimensions).

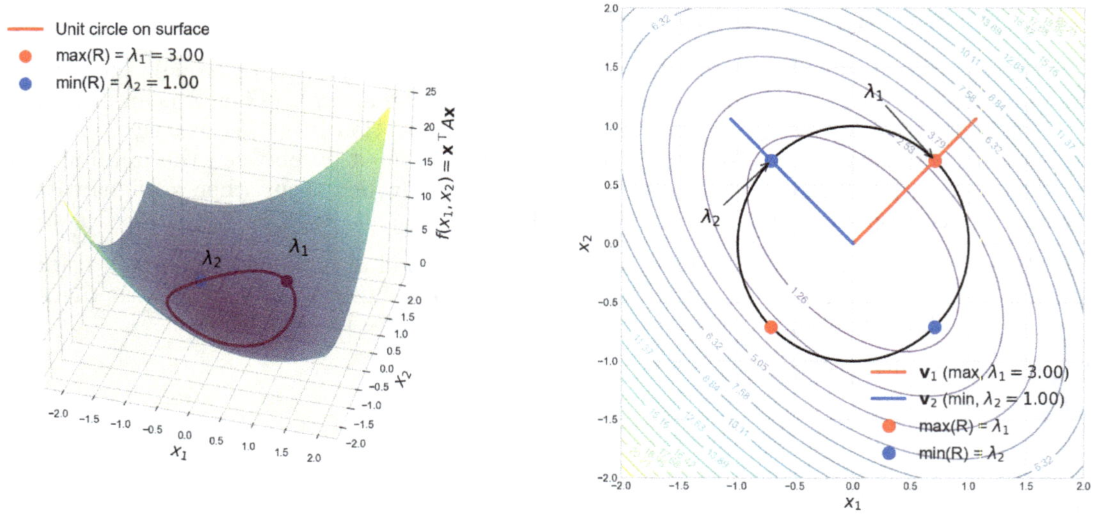

Figure 7.2: Rayleigh Quotient: Contour and 3D Surface (bowl-shaped)

Figure 7.2 shows the quadratic form $f(\mathbf{x}) = \mathbf{x}^\top A\mathbf{x}$ in two views for $\mathbf{x} \in \mathbb{R}^2$: a 3D surface plot (left) and a 2D contour plot (right). The left plot visualizes the 3D surface $z = f(x_1, x_2)$, which exhibits a **bowl-shaped** (concave upward) surface for the positive definite matrix A (see the Quadratic Forms and Definiteness section), with the unit circle mapped onto this surface (red curve), where the extremal points correspond to the maximum and minimum Rayleigh quotient values. The right plot shows contour lines with the unit circle constraint and eigenvectors. The maximum and minimum values occur at intersections of the unit circle with contour lines, which are along eigenvector directions. Each eigenvector direction appears twice on the unit circle (as $\mathbf{v}$ and $-\mathbf{v}$), and both yield the same Rayleigh quotient value due to scale-invariance.

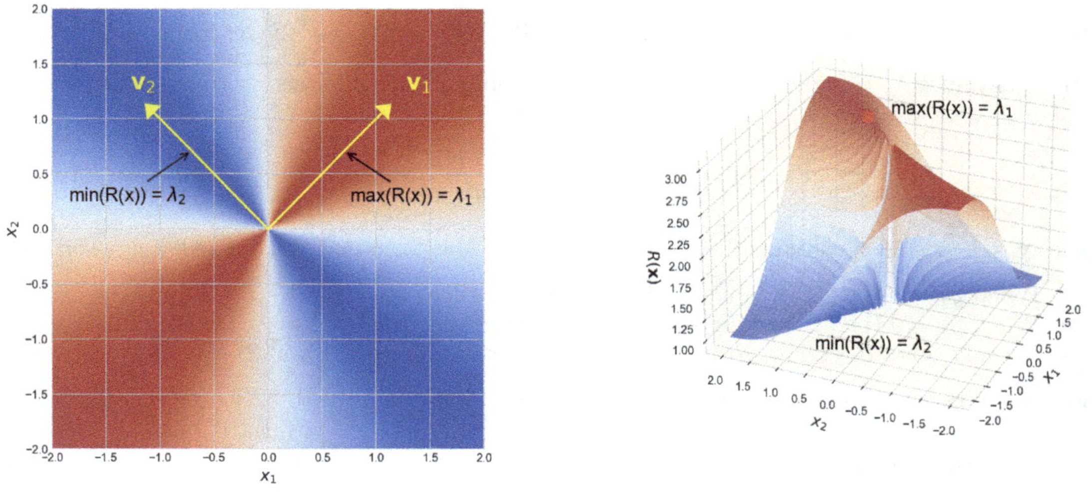

Figure 7.3: Rayleigh Quotient Surface (saddle-shaped)

Figure 7.3 shows the Rayleigh quotient $R(\mathbf{x})$ in two views: a 2D heatmap (left) and a 3D surface plot (right). Note that $R(\mathbf{x})$ is undefined at the origin. The 2D heatmap reveals the radial color gradient, while the 3D surface shows a **saddle-like shape** (hyperbolic paraboloid) characteristic of the normalized quadratic form. Both visualizations show that the maximum occurs along the eigenvector corresponding to λ_{max} (yellow arrow $\mathbf{v}_1$) and the minimum along the eigenvector corresponding to λ_{min} (yellow arrow $\mathbf{v}_2$). Since the Rayleigh quotient is scale-invariant, infinitely many points achieve the maximum value, all lying on the line spanned by $\mathbf{v}_1$.

In 3D, the constraint $\|\mathbf{x}\|_2 = 1$ defines a **unit sphere**. Figure 7.4 shows the Rayleigh quotient values mapped onto a 3D unit sphere as a color mapping, with warmer colors indicating higher values and cooler colors indicating lower values. The extremal values occur along eigenvector directions, creating a pattern that reflects the matrix's spectral structure.

The Rayleigh quotient's **scale-invariance** implies all points on a ray from the origin have the same value. This allows us to restrict attention to the unit sphere without loss of generality. Finding extremal values is a **constrained optimization problem**; using Lagrange multipliers (Chapter~12), the critical points occur exactly when $\mathbf{x}$ is an eigenvector of A.

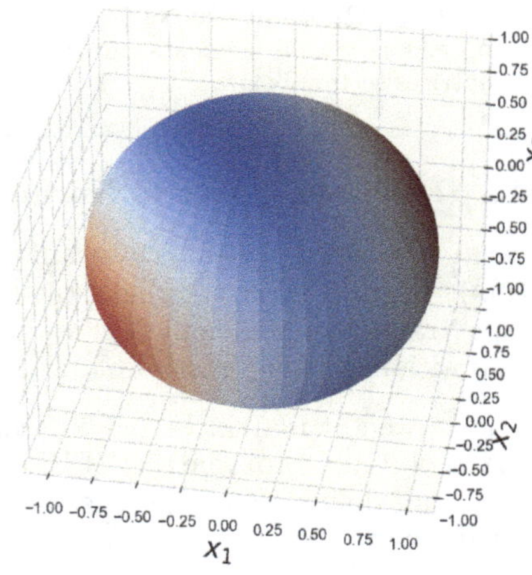

Figure 7.4: Rayleigh Quotient on 3D Unit Sphere

7.4.4.1 Examples in 2D

Consider the symmetric matrix $A = \begin{bmatrix} 2 & 1 \\ 1 & 2 \end{bmatrix}$ with eigenvalues $\lambda_{\text{min}} = 1$ and $\lambda_{\text{max}} = 3$. The corresponding unit eigenvectors are $\mathbf{q}_0 = \frac{1}{\sqrt{2}}[1, -1]^\top$ and $\mathbf{q}_1 = \frac{1}{\sqrt{2}}[1, 1]^\top$.

Verification shows that $R_A(\mathbf{q}_0) = 1 = \lambda_{\text{min}}$ and $R_A(\mathbf{q}_1) = 3 = \lambda_{\text{max}}$. For an arbitrary unit vector $\mathbf{x} = [\cos\theta, \sin\theta]^\top$, we have $R_A(\mathbf{x}) = 2 + \sin(2\theta)$, which confirms that $1 \leq R_A(\mathbf{x}) \leq 3$ for all unit vectors.

The eigenbasis representation (see *7.26*) provides additional insight. For the matrix $A = \begin{bmatrix} 1.5 & 0.5 \\ 0.5 & 1.5 \end{bmatrix}$ with eigenvalues $\lambda_1 = 2$ and $\lambda_2 = 1$, the Rayleigh quotient in eigenbasis coordinates becomes:

$$R(\mathbf{y}) = \frac{2y_1^2 + y_2^2}{y_1^2 + y_2^2}$$

where $\mathbf{y} = V^\top \mathbf{x}$ and V contains the eigenvectors of A. This form makes it immediately clear that:

- When $y_2 = 0$ (i.e., $\mathbf{x}$ is along the first eigenvector), $R(\mathbf{y}) = 2 = \lambda_1$ (maximum)
- When $y_1 = 0$ (i.e., $\mathbf{x}$ is along the second eigenvector), $R(\mathbf{y}) = 1 = \lambda_2$ (minimum)
- For any other direction, $R(\mathbf{y})$ is a weighted average between 1 and 2

Two special cases illustrate important boundary behaviors:

- **Singular matrix**: $A = \begin{bmatrix} 1 & 0 \\ 0 & 0 \end{bmatrix}$ has eigenvalues $\lambda_1 = 1, \lambda_2 = 0$. The Rayleigh quotient achieves its minimum of 0 along the null space direction, demonstrating that the lower bound can be exactly zero when the matrix is not positive definite. This case is relevant in dimensionality reduction (e.g., PCA with rank-deficient covariance matrices).

- **Indefinite matrix**: $A = \begin{bmatrix} 0 & 1 \\ 1 & 0 \end{bmatrix}$ has eigenvalues $\lambda_1 = 1, \lambda_2 = -1$. The Rayleigh quotient can be negative, with bounds $[-1, 1]$, showing that the bounds extend to negative values when the matrix has both positive and negative eigenvalues. This is important in optimization problems where the Hessian matrix may be indefinite (e.g., saddle points in non-convex optimization).

> In **Graph Neural Networks (GNNs)**, spectral convolutions are defined from the graph Laplacian; graph frequency response and smoothness are analyzed through the Laplacian's Rayleigh quotient. **Rayleigh Quotient GNNs** use abnormal values of that quotient to flag anomalous nodes, which tend to break the smoothness assumed by neighboring nodes. (Graph Laplacians return in the Graph Matrices section below.)

7.4.4.2 Example: Spectral Normalization in Generative Adversarial Networks

Spectral Normalization[1] is a technique used to stabilize **Generative Adversarial Networks (GANs)**[2] training by constraining the **spectral norm** (largest singular value) of weight matrices. The spectral norm and singular values are covered in detail in Chapter~8 (Singular Value Decomposition). This directly relates to the Rayleigh quotient bounds (see *7.21*).

Consider a weight matrix $W \in \mathbb{R}^{m\times n}$ in a neural network layer. Let $G = W^\top W$ be the **Gram matrix** (column Gram matrix) of W (see the Gram Matrices section), which is symmetric and positive semi-definite (see the theorem on positive semi-definiteness of Gram matrices). The **spectral norm** of W is defined as:

$$\sigma_{\max}(W) = \max_{\|\mathbf{x}\|_2=1} \|W\mathbf{x}\|_2 = \sqrt{\lambda_{\max}(G)}$$

[1]Miyato, T., Kataoka, T., Koyama, M., & Yoshida, Y. "Spectral Normalization for Generative Adversarial Networks." *International Conference on Machine Learning (ICML)*, 2018. https://arxiv.org/abs/1802.05957

[2]Goodfellow, I., Pouget-Abadie, J., Mirza, M., Xu, B., Warde-Farley, D., Ozair, S., Courville, A., & Bengio, Y. (2014). Generative Adversarial Networks. In *Advances in Neural Information Processing Systems* (NeurIPS), 27. https://arxiv.org/abs/1406.2661

where $\lambda_{\max}(G)$ is the largest eigenvalue of G. By the Rayleigh quotient bounds (see *7.21*), we have:

$$\lambda_{\max}(G) = \max_{\mathbf{x}\neq 0} R_G(\mathbf{x}) = \max_{\|\mathbf{x}\|_2=1} \mathbf{x}^\top G\mathbf{x} = \max_{\|\mathbf{x}\|_2=1} \mathbf{x}^\top W^\top W\mathbf{x} = \max_{\|\mathbf{x}\|_2=1} \|W\mathbf{x}\|_2^2$$

Spectral normalization constrains $\sigma_{\max}(W) \leq 1$ by normalizing the weight matrix:

$$W_{\text{SN}} = \frac{W}{\sigma_{\max}(W)}$$

This ensures that the Lipschitz constant of the layer is bounded by 1, preventing gradient explosion and stabilizing training. More generally, gradient explosion or vanishing in deep networks and RNNs tracks the spectral radius of the weight matrices—the upper end of the Rayleigh quotient for $G = W^\top W$. Spectral normalization is one practical way to keep that upper bound under control. The key insight is that by controlling the maximum Rayleigh quotient of G, we control the maximum amplification of signals passing through the layer.

In practice, $\sigma_{\max}(W)$ is estimated using the power iteration method, which iteratively computes:

$$\mathbf{v}_{t+1} = \frac{G\mathbf{v}_t}{\|G\mathbf{v}_t\|_2}$$

This converges to the eigenvector corresponding to $\lambda_{\max}(G)$, and the Rayleigh quotient $R_G(\mathbf{v}_t)$ converges to $\lambda_{\max}(G)$.

[a]: **Generative Adversarial Networks (GANs)** are a class of machine learning models where two neural networks (a generator and a discriminator) compete in a zero-sum game. The generator creates synthetic data while the discriminator tries to distinguish real from synthetic data. In **Physics-Informed Machine Learning (PINNs)**, finding the ground-state energy of an operator such as a quantum Hamiltonian is often equivalent to minimizing its Rayleigh quotient. Models can take that quotient directly as a loss and descend on it with gradient methods.

---

[a] Goodfellow, I., Pouget-Abadie, J., Mirza, M., Xu, B., Warde-Farley, D., Ozair, S., Courville, A., & Bengio, Y. (2014). Generative Adversarial Networks. In *Advances in Neural Information Processing Systems* (NeurIPS), 27. https://arxiv.org/abs/1406.2661

7.5 Cholesky Decomposition

7.5.1 Theorem: Cholesky Decomposition

If $A \in \mathbb{R}^{N\times N}$ is symmetric positive definite, then

$$A = LL^\top \tag{7.27}$$

where L is lower triangular.

Equivalently, A can be decomposed as $A = R^\top R$, where R is upper triangular($R = L^\top$).

Algorithm 8 Cholesky Decomposition

Input: Symmetric positive definite matrix $A \in \mathbb{R}^{N\times N}$
Output: Lower triangular matrix L such that $A = LL^\top$
  Initialize $L = \mathbf{0}_{N\times N}$
  $l_{00} \leftarrow \sqrt{a_{00}}$
  for $n = 0, \dots, N-1$ **do**
    Update $\mathbf{l}_n$ by solving $L\mathbf{l}_n^\top = \mathbf{a}_n$ using forward substitution
  end for
  return L

7.5.2 Algorithm: Find L in Cholesky Decomposition

where:

- $\mathbf{l}_n$ (`\mathbf{l}_n` in latex) is the n-th column vector of L, i.e., $\mathbf{l}_n = [l_{0n}, l_{1n}, \dots, l_{nn}, 0, \dots, 0]^\top \in \mathbb{R}^{N\times 1}$ (only the first $n+1$ elements are non-zero since L is lower triangular)
- $\mathbf{a}_n$ is the n-th column vector of A, i.e., $\mathbf{a}_n = [a_{0n}, a_{1n}, \dots, a_{(N-1)n}]^\top \in \mathbb{R}^{N\times 1}$
- The equation $L\mathbf{l}_n^\top = \mathbf{a}_n$ is solved using forward substitution. Since L is lower triangular and we only need the first $n+1$ elements of $\mathbf{l}_n$, we solve the system:

$$\begin{bmatrix} l_{00} & 0 & \cdots & 0 \\ l_{10} & l_{11} & \cdots & 0 \\ \vdots & \vdots & \ddots & \vdots \\ l_{n0} & l_{n1} & \cdots & l_{nn} \end{bmatrix} \begin{bmatrix} l_{0n} \\ l_{1n} \\ \vdots \\ l_{nn} \end{bmatrix} = \begin{bmatrix} a_{0n} \\ a_{1n} \\ \vdots \\ a_{nn} \end{bmatrix}.$$

7.5.2.1 Example: Cholesky Decomposition

Do Cholesky decomposition of the following matrix:

$$A = \begin{bmatrix} 4 & 2 & 2 \\ 2 & 2 & 1 \\ 2 & 1 & 3 \end{bmatrix}.$$

Solution

We compute L column by column using the algorithm above.

Step 1: Initialize: $l_{00} = \sqrt{a_{00}} = \sqrt{4} = 2$

Step 2: Compute column 1 $(n = 1)$

We solve $L\mathbf{l}_1^\top = \mathbf{a}_1$:

$$\begin{bmatrix} 2 & 0 \\ l_{10} & l_{11} \end{bmatrix} \begin{bmatrix} l_{01} \\ l_{11} \end{bmatrix} = \begin{bmatrix} 2 \\ 2 \end{bmatrix}.$$

Using the relation $a_{ij} = \sum_{k=0}^{\min(i,j)} l_{ik}l_{jk}$ from $A = LL^\top$, we can get $l_{10} = 1$, $l_{11} = 1$.

Step 3: Compute column 2 $(n = 2)$ Using $a_{ij} = \sum_{k=0}^{\min(i,j)} l_{ik}l_{jk}$:

- $a_{20} = l_{20}l_{00}$, so $2 = l_{20} \cdot 2$, hence $l_{20} = 1$
- $a_{21} = l_{20}l_{10} + l_{21}l_{11}$, so $1 = 1 \cdot 1 + l_{21} \cdot 1$, hence $l_{21} = 0$
- $a_{22} = l_{20}^2 + l_{21}^2 + l_{22}^2$, so $3 = 1^2 + 0^2 + l_{22}^2$, hence $l_{22} = \sqrt{2}$

Therefore,

$$L = \begin{bmatrix} 2 & 0 & 0 \\ 1 & 1 & 0 \\ 1 & 0 & \sqrt{2} \end{bmatrix}$$

7.5.2.2 Example: Multivariate Gaussian Generation

Generate $\mathbf{x} \sim \mathcal{N}(\mu, \Sigma)$, where $\mu \in \mathbb{R}^{N\times 1}$ and $\Sigma \in \mathbb{R}^{N\times N}$ is positive definite. (See Chapter~13 for the definition of the multivariate Gaussian distribution.)

Solution

Since $\mathbf{z} \sim \mathcal{N}(0, I)$, we have $\mathbb{E}[\mathbf{z}] = 0$ and $\mathrm{Cov}(\mathbf{z}) = I$. The linear transformation $\mathbf{y} = L\mathbf{z}$ gives:

- $\mathbb{E}[\mathbf{y}] = L \cdot 0 = 0$
- $\mathrm{Cov}(\mathbf{y}) = L \cdot I \cdot L^\top = LL^\top = \Sigma$

Therefore, $\mathbf{y} = L\mathbf{z} \sim \mathcal{N}(0, \Sigma)$. Adding the mean shift, we obtain: $\mathbf{x} = \mu + \mathbf{y} \sim \mathcal{N}(\mu, \Sigma)$. Therefore, $\mathbf{x}$ can be generated in the following steps:

1. Generate $\mathbf{z} \sim \mathcal{N}(0, I)$.
2. Do Cholesky decomposition $\Sigma = LL^\top$.
3. Set $\mathbf{x} = \mu + L\mathbf{z}$.

7.6 Optional: A Distance Between Positive Definite Matrices

The set of symmetric positive definite matrices

$$\mathcal{S}^n_{++} = \{A \in \mathbb{R}^{n\times n} : A = A^\top,\ \mathbf{x}^\top A\mathbf{x} > 0\ \forall \mathbf{x} \neq 0\}$$

is **not** a Euclidean vector space: if $A, B \in \mathcal{S}^n_{++}$, the ordinary difference $A - B$ need not be positive definite. Measuring "how far" two SPD matrices are therefore requires care.

7.6.1 Definition: Matrix Logarithm on $\mathcal{S}^n_{++}$

If $A = Q\Lambda Q^\top$ with $\Lambda = \mathrm{diag}(\lambda_1, \dots, \lambda_n)$ and $\lambda_i > 0$, the **matrix logarithm** is

$$\log(A) = Q\,\mathrm{diag}(\log\lambda_1, \dots, \log\lambda_n)\,Q^\top. \tag{7.28}$$

The result is a symmetric matrix (not necessarily definite). Thus log maps $\mathcal{S}^n_{++}$ into the linear space of symmetric matrices, where the Frobenius norm is well-defined.

7.6.2 Definition: Log-Euclidean Distance

For $A, B \in \mathcal{S}^n_{++}$, the **log-Euclidean distance** is

$$d_{\mathrm{LE}}(A, B) = \|\log(A) - \log(B)\|_F. \tag{7.29}$$

It is the Euclidean distance after mapping both matrices to the tangent (symmetric) space via log. In particular $d_{\mathrm{LE}}(A, A) = 0$, and d_{LE} is invariant under inversion in the sense that $d_{\mathrm{LE}}(A^{-1}, B^{-1}) = d_{\mathrm{LE}}(A, B)$ because $\log(A^{-1}) = -\log(A)$.

7.6.3 Example: 2×2

Let $A = I_2$ and $B = 2I_2$. Then $\log(A) = 0$ and $\log(B) = (\log 2)\, I_2$, so

$$d_{\mathrm{LE}}(A, B) = \sqrt{2}\,|\log 2|.$$

By contrast, $\|A - B\|_F = \sqrt{2}$, which does not account for the multiplicative structure of eigenvalues on $\mathcal{S}^n_{++}$.

This optional section only introduces one practical SPD distance. A fuller Riemannian theory of $\mathcal{S}^n_{++}$ (geodesics, affine-invariant metric) is beyond the scope of this chapter; the log-Euclidean construction is enough to see why **Euclidean subtraction is the wrong default** for positive definite matrices and how the spectral calculus from earlier in this chapter yields a workable alternative.[a]

[a] Arsigny, V., Fillard, P., Pennec, X., & Ayache, N. (2007). Geometric means in a novel vector space structure on symmetric positive-definite matrices. *SIAM Journal on Matrix Analysis and Applications*, 29(1), 328–347.

7.7 Gram Matrices

7.7.1 Definition: Gram Matrix

For $A \in \mathbb{R}^{M \times N}$, matrices $AA^\top$ and $A^\top A$ are **Gram matrices**:

- $A^\top A$: **Gram matrix of columns** (inner products between columns of A)
- $AA^\top$: **Gram matrix of rows** (inner products between rows of A)

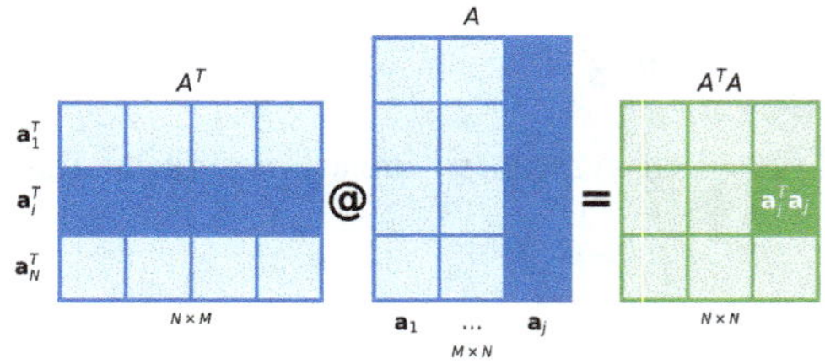

Figure 7.5: Column Gram Matrix

7.7.2 Theorem: Symmetry of Gram Matrices

$AA^\top$ and $A^\top A$ are symmetric.

7.7.3 Theorem: Trace Equality of Gram Matrices

$$\operatorname{tr}(AA^\top) = \operatorname{tr}(A^\top A) \tag{7.30}$$

7.7.4 Theorem: Positive Semi-Definiteness of Gram Matrices

$AA^\top$ and $A^\top A$ are positive semi-definite.

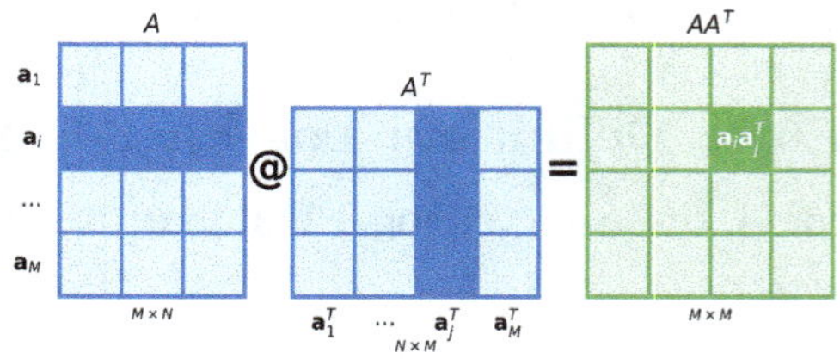

Figure 7.6: Row Gram Matrix

Proof

We show that both matrices satisfy the definition of positive semi-definiteness: for any vector $\mathbf{x}$, we have $\mathbf{x}^\top B\mathbf{x} \geq 0$ (see equation (7.12)).

For $A^\top A$:

Let $\mathbf{x} \in \mathbb{R}^{N \times 1}$ be arbitrary. Then

$$\mathbf{x}^\top(A^\top A)\mathbf{x} = (A\mathbf{x})^\top(A\mathbf{x}) = \|A\mathbf{x}\|^2 \geq 0,$$

where the first equality follows from $(A\mathbf{x})^\top = \mathbf{x}^\top A^\top$ and the second equality uses the fact that the dot product of a vector with itself equals its squared norm.

For $AA^\top$:

Let $\mathbf{y} \in \mathbb{R}^{M\times 1}$ be arbitrary. Then

$$\mathbf{y}^\top(AA^\top)\mathbf{y} = (A^\top\mathbf{y})^\top(A^\top\mathbf{y}) = \|A^\top\mathbf{y}\|^2 \geq 0,$$

where the first equality follows from $(A^\top\mathbf{y})^\top = \mathbf{y}^\top A$ and the second equality again uses the squared norm property.

Since both quadratic forms are non-negative for all vectors, $AA^\top$ and $A^\top A$ are positive semi-definite.

7.7.5 Theorem: Eigenvalue Equality of Gram Matrices

$AA^\top$ and $A^\top A$ share the same non-zero eigenvalues.

Proof

We show that if $\lambda \neq 0$ is an eigenvalue of $A^\top A$, then it is also an eigenvalue of $AA^\top$, and vice versa.

Direction 1: If $\lambda \neq 0$ is an eigenvalue of $A^\top A$, then it is an eigenvalue of $AA^\top$.

Let $\lambda \neq 0$ be an eigenvalue of $A^\top A$ with corresponding eigenvector $\mathbf{v} \neq \mathbf{0}$, so that

$$A^\top A\mathbf{v} = \lambda\mathbf{v}.$$

Multiplying both sides by A on the left:

$$AA^\top(A\mathbf{v}) = A(A^\top A\mathbf{v}) = A(\lambda\mathbf{v}) = \lambda(A\mathbf{v}).$$

Since $\lambda \neq 0$ and $\mathbf{v} \neq \mathbf{0}$, we have $A\mathbf{v} \neq \mathbf{0}$ (otherwise $A^\top A\mathbf{v} = A^\top\mathbf{0} = \mathbf{0} = \lambda\mathbf{v}$ would imply $\lambda = 0$, a contradiction). Therefore, $A\mathbf{v}$ is a non-zero eigenvector of $AA^\top$ with eigenvalue λ.

Direction 2: If $\lambda \neq 0$ is an eigenvalue of $AA^\top$, then it is an eigenvalue of $A^\top A$.

Let $\lambda \neq 0$ be an eigenvalue of $AA^\top$ with corresponding eigenvector $\mathbf{u} \neq \mathbf{0}$, so that

$$AA^\top\mathbf{u} = \lambda\mathbf{u}.$$

Multiplying both sides by $A^\top$ on the left:

$$A^\top A(A^\top\mathbf{u}) = A^\top(AA^\top\mathbf{u}) = A^\top(\lambda\mathbf{u}) = \lambda(A^\top\mathbf{u}).$$

Since $\lambda \neq 0$ and $\mathbf{u} \neq \mathbf{0}$, we have $A^\top\mathbf{u} \neq \mathbf{0}$ (otherwise $AA^\top\mathbf{u} = A\mathbf{0} = \mathbf{0} = \lambda\mathbf{u}$ would imply $\lambda = 0$, a contradiction). Therefore, $A^\top\mathbf{u}$ is a non-zero eigenvector of $A^\top A$ with eigenvalue λ.

Hence, $AA^\top$ and $A^\top A$ share the same non-zero eigenvalues.

The rank equality $\text{rank}(AA^\top) = \text{rank}(A^\top A) = \text{rank}(A)$ for Gram matrices was established in Chapter~4 (see equation 4.5).

7.7.5.1 Example: Style Transfer

In **Neural Style Transfer**[3] (Gatys et al., 2015), the style of an image is represented via the **row Gram matrix** of convolutional feature maps. A feature map $F \in \mathbb{R}^{C\times HW}$ is a matrix where rows are channels (feature detectors) and columns are spatial locations.

The Gram matrix $G = FF^\top \in \mathbb{R}^{C\times C}$ encodes channel-wise correlations: $G_{ij} = \sum_{k=1}^{HW} F_{ik}F_{jk}$ quantifies how frequently features i and j co-occur across spatial locations. The style loss is $\mathcal{L}_s = \|G_c - G_s\|_F^2$, where G_c and G_s are Gram matrices for content and style feature maps, and $\|\cdot\|_F$ is the Frobenius norm.

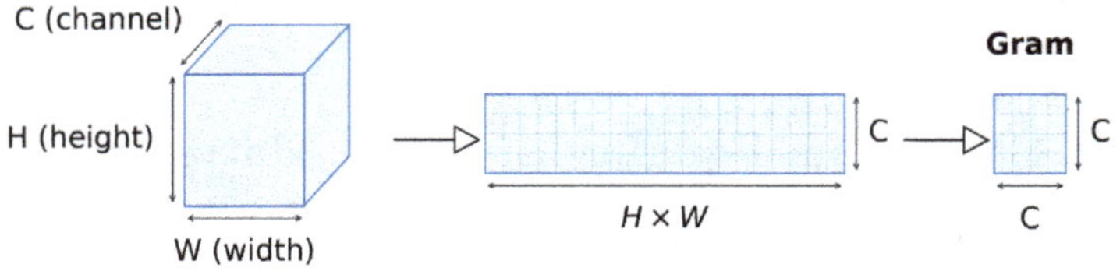

Figure 7.7: From Feature Map to Gram Matrix

Figure 7.7 illustrates the transformation from a 3D feature map tensor $(H \times W \times C)$ to a flattened matrix $F \in \mathbb{R}^{C\times HW}$, then to the Gram matrix $G = FF^\top \in \mathbb{R}^{C\times C}$. The Gram matrix discards spatial information while preserving feature co-occurrences, making it ideal for style representation.

Given feature maps $F_c, F_s \in \mathbb{R}^{3\times 4}$ (where c denotes content and s denotes style), compute the Gram matrices G_c and G_s, and evaluate the style loss $\mathcal{L}_s$.

$$F_c = \begin{bmatrix} 1 & 2 & 0 & 1 \\ 0 & 1 & 2 & 1 \\ 1 & 0 & 1 & 2 \end{bmatrix}, \quad F_s = \begin{bmatrix} 2 & 1 & 1 & 0 \\ 1 & 2 & 0 & 1 \\ 0 & 1 & 2 & 1 \end{bmatrix},$$

Solution:

Computing the Gram matrices:

$$G_c = F_cF_c^\top = \begin{bmatrix} 6 & 3 & 3 \\ 3 & 6 & 4 \\ 3 & 4 & 6 \end{bmatrix}, \quad G_s = F_sF_s^\top = \begin{bmatrix} 6 & 4 & 3 \\ 4 & 6 & 3 \\ 3 & 3 & 6 \end{bmatrix}.$$

The difference matrix is:

$$G_c - G_s = \begin{bmatrix} 0 & -1 & 0 \\ -1 & 0 & 1 \\ 0 & 1 & 0 \end{bmatrix}.$$

Therefore, the style loss is:

[3] The seminal work on neural style transfer - Gatys, L. A., Ecker, A. S., & Bethge, M. "A Neural Algorithm of Artistic Style." *Journal of Vision*, 16(12):326, 2016. https://arxiv.org/abs/1508.06576.

$$\mathcal{L}_s = \|G_c - G_s\|_F^2 = \sum_{i,j}(G_{c,ij} - G_{s,ij})^2 = 4.$$

Figure 7.8 demonstrates neural style transfer: a style image (left) is combined with a content image (middle) to produce a stylized output (right) that preserves the content structure while adopting the artistic style, involving minimization of the style loss $\mathcal{L}_s$ between Gram matrices along with a content loss.

Figure 7.8: Style Transfer via Gram

Gram matrices are fundamental in **singular value decomposition (SVD)**: the SVD $A = U\Sigma V^\top$ is computed by first finding the eigen-decomposition of $A^\top A$ (the column Gram matrix), where the eigenvalues are the squares of singular values $\sigma_i^2 = \lambda_i$ and the eigenvectors form V. In **principal component analysis (PCA)**, the covariance matrix $\frac{1}{N-1}XX^\top$ (a row Gram matrix) captures data variance, and its eigendecomposition yields principal directions. **Kernel methods** in support vector machines and Gaussian processes use Gram matrices $K_{ij} = \kappa(\mathbf{x}_i, \mathbf{x}_j)$ as similarity matrices, enabling nonlinear pattern recognition.

In **recommendation systems**, user-item interaction matrices R produce Gram matrices $R^\top R$ and $RR^\top$ that encode item-item and user-user similarities, powering collaborative filtering. **Word embeddings** in natural language processing compute Gram matrices of word vectors to measure semantic similarity, and **neural network training** often involves Gram matrices in second-order optimization methods like natural gradient descent, where $G = J^\top J$ (where J is the Jacobian) approximates the Fisher information matrix.

7.8 From Spectral Methods to Graph Neural Networks

This section is an application example that demonstrates how ***symmetric matrices*** *are used in practice. It can be skipped without affecting the understanding of the core mathematical concepts covered in this book.*

This section traces the evolution from classical spectral graph methods to modern Graph Neural Networks (GNNs)[^gnn], showing how symmetric matrix theory (particularly the eigendecomposition of the graph Laplacian) provides the mathematical foundation for graph-based learning. We begin with the fundamental graph matrices and spectral decomposition, then show how these concepts lead to principled solutions for node classification, and finally demonstrate how GNNs operationalize these spectral ideas at scale.

7.8.1 Graph Matrices: Adjacency, Degree, and Laplacian

Consider an **undirected graph** $\mathcal{G} = (\mathcal{V}, \mathcal{E})$ with N nodes (vertices) $\mathcal{V} = \{v_0, v_1, \dots, v_{N-1}\}$ and edges $\mathcal{E}$. The graph can be represented using several symmetric matrices.

The **adjacency matrix** $A \in \mathbb{R}^{N\times N}$ encodes edge connections:

$$A_{ij} = \begin{cases} 1 & \text{if } (v_i, v_j) \in \mathcal{E} \\ 0 & \text{otherwise} \end{cases}$$

For undirected graphs, edges are bidirectional ($A_{ij} = A_{ji}$), so $A = A^\top$ is **symmetric**. Directed graphs generally produce non-symmetric adjacency matrices, and weighted graphs are symmetric if weights satisfy $w_{ij} = w_{ji}$.

The **degree matrix** $D \in \mathbb{R}^{N \times N}$ is a diagonal matrix:

$$D = \text{diag}(d_0, d_1, \dots, d_{N-1}), \quad d_i = \sum_{j=0}^{N-1} A_{ij}$$

where d_i is the **degree** of node v_i (number of edges connected to it). Since D is diagonal, $D = D^\top$, so D is **always symmetric.**

The **graph Laplacian matrix** $L \in \mathbb{R}^{N \times N}$ is defined as:

$$L = D - A \tag{7.31}$$

If A is symmetric, then L is **symmetric** (since $L^\top = (D - A)^\top = D^\top - A^\top = D - A = L$). This form is called the **Combinatorial Laplacian** (or **Unnormalized Laplacian**). For directed graphs, variants like the "random walk Laplacian" are generally not symmetric unless symmetrized.

Modern GNNs often use the **symmetric normalized Laplacian**:

$$L_{\text{sym}} = D^{-1/2} L D^{-1/2} = I - D^{-1/2} A D^{-1/2} \tag{7.32}$$

where $D^{-1/2}$ is the diagonal matrix with entries $1/\sqrt{d_i}$ (assuming $d_i > 0$). The normalized Laplacian has eigenvalues bounded in $[0, 2]$, which is crucial for numerical stability in deep neural networks (preventing exploding gradients). This boundedness explains why the Graph Convolutional Network (GCN) formula uses normalized adjacency matrices.

The graph Laplacian matrix (both combinatorial and normalized) is **positive semi-definite.** For any vector $\mathbf{x} \in \mathbb{R}^{N \times 1}$:

$$\mathbf{x}^\top L \mathbf{x} = \frac{1}{2} \sum_{i,j} A_{ij} (x_i - x_j)^2 \geq 0 \tag{7.33}$$

This quadratic form measures the "smoothness" of the signal $\mathbf{x}$ on the graph: it is zero when $\mathbf{x}$ is constant, and increases when adjacent nodes have very different values.

7.8.2 Spectral Decomposition of Graph Laplacian

Since L is symmetric and positive semi-definite, it admits a spectral decomposition:

$$L = Q \Lambda Q^\top = \sum_{n=0}^{N-1} \lambda_n \mathbf{q}_n \mathbf{q}_n^\top \tag{7.34}$$

where $\Lambda = \text{diag}(\lambda_0, \lambda_1, \dots, \lambda_{N-1})$ with $0 = \lambda_0 \leq \lambda_1 \leq \dots \leq \lambda_{N-1}$, and $Q = [\mathbf{q}_0, \mathbf{q}_1, \dots, \mathbf{q}_{N-1}]$ is orthonormal. The smallest eigenvalue $\lambda_0 = 0$ has eigenvector $\mathbf{q}_0 = \frac{1}{\sqrt{N}} \mathbf{1}$ (constant vector) if the graph is connected. The number of zero eigenvalues equals the number of connected components in the graph.

The eigenvalues λ_n are called the **graph frequencies**, and the eigenvectors $\mathbf{q}_n$ are the **graph Fourier modes**. This terminology comes from signal processing. The constant eigenvector $\mathbf{q}_0$ (corresponding to $\lambda_0 = 0$) represents the **DC component** (0 Hz), which is the average value of the signal across all nodes. The Fiedler vector $\mathbf{q}_1$ (corresponding to λ_1) is the **lowest frequency**

wave, like a sine wave that crosses zero once, creating a natural bipartition of the graph. Large eigenvalues λ_{N-1} correspond to **high frequency noise**, where adjacent nodes have alternating signs $(+, -, +, -)$, representing rapid oscillations. Low frequencies correspond to smooth signals (slowly varying across the graph), while high frequencies correspond to oscillatory signals (rapidly varying).

7.8.3 Rayleigh Quotient as Graph Smoothness Objective

The Rayleigh quotient of the graph Laplacian provides a natural measure of signal smoothness:

$$R_L(\mathbf{x}) = \frac{\mathbf{x}^\top L\mathbf{x}}{\mathbf{x}^\top \mathbf{x}} = \frac{\sum_{i,j} A_{ij}(x_i - x_j)^2}{2\sum_i x_i^2} \tag{7.35}$$

Physical Interpretation: Minimizing $R_L(\mathbf{x})$ is equivalent to making connected nodes have similar values while keeping the representation normalized. The numerator $\mathbf{x}^\top L\mathbf{x} = \sum_{(i,j)\in\mathcal{E}}(x_i - x_j)^2$ penalizes differences between adjacent nodes.

By the Rayleigh quotient bounds (see *7.21*), we have:

$$0 = \lambda_0 \le R_L(\mathbf{x}) \le \lambda_{N-1}$$

The lower bound $R_L(\mathbf{x}) = 0$ is achieved when $\mathbf{x}$ is constant (perfectly smooth), while the upper bound $R_L(\mathbf{x}) = \lambda_{N-1}$ is achieved when $\mathbf{x}$ is the eigenvector corresponding to the largest eigenvalue (most oscillatory).

Optimization Viewpoint: In machine learning, the Rayleigh quotient often appears as a **graph smoothness regularizer**:

$$\mathcal{L} = \mathcal{L}_{\text{task}} + \lambda \cdot \text{tr}(\mathbf{H}^\top L\mathbf{H})$$

where $\mathbf{H}$ is a node embedding matrix and λ controls the trade-off between task performance and graph smoothness. This regularizer encourages connected nodes to have similar embeddings.

7.8.4 Spectral Methods for Node Classification

The connection between Rayleigh quotient, spectral analysis, and graph-based classification can be made explicit. Consider the optimization problem: minimize the Rayleigh quotient $R_L(\mathbf{x}) = \frac{\mathbf{x}^\top L\mathbf{x}}{\mathbf{x}^\top \mathbf{x}}$ subject to constraints. This is equivalent to finding node representations that make connected nodes similar while keeping the representation normalized.

Key Result: Under the constraints $\mathbf{x}^\top\mathbf{x} = 1$ and $\mathbf{x} \perp \mathbf{1}$ (orthogonal to the constant vector), the minimizer of the Rayleigh quotient is the **second smallest eigenvector** $\mathbf{v}_1$ of L (also called the **Fiedler vector**). This eigenvector provides a natural way to partition the graph into two groups.

Connection to Min-Cut Problem: The Fiedler vector solves the **continuous relaxation** of the **minimum cut problem**. The discrete problem asks: partition the graph into two groups S and $\bar{S}$ to minimize the number of edges between them. This is NP-hard. However, if we relax the binary constraint (nodes must be in S or $\bar{S}$) to allow real-valued assignments, the solution is exactly the Fiedler vector. This connects discrete graph geometry to continuous linear algebra: the eigenvector $\mathbf{v}_1$ provides the optimal continuous approximation to the discrete min-cut problem.

The sign of $\mathbf{v}_1$ induces a natural bipartition: nodes with positive values form one group, and nodes with negative values form another. This is the foundation of **spectral clustering** and **graph bipartition** methods.

7.8.5 Complete Example: User Classification via Spectral Bipartition

Consider a social network with $N = 5$ users: Alice (node 0), Bob (node 1), Charlie (node 2), David (node 3), and Eve (node 4). We want to classify users as "active" vs. "inactive" based solely on the graph structure.

Graph Setup:

The adjacency matrix A (friendship connections), degree matrix $D = \text{diag}(2, 3, 4, 3, 2)$, and graph Laplacian $L = D - A$ are:

$$A = \begin{bmatrix} 0 & 1 & 1 & 0 & 0 \\ 1 & 0 & 1 & 1 & 0 \\ 1 & 1 & 0 & 1 & 1 \\ 0 & 1 & 1 & 0 & 1 \\ 0 & 0 & 1 & 1 & 0 \end{bmatrix}, \quad D = \begin{bmatrix} 2 & 0 & 0 & 0 & 0 \\ 0 & 3 & 0 & 0 & 0 \\ 0 & 0 & 4 & 0 & 0 \\ 0 & 0 & 0 & 3 & 0 \\ 0 & 0 & 0 & 0 & 2 \end{bmatrix}, \quad L = \begin{bmatrix} 2 & -1 & -1 & 0 & 0 \\ -1 & 3 & -1 & -1 & 0 \\ -1 & -1 & 4 & -1 & -1 \\ 0 & -1 & -1 & 3 & -1 \\ 0 & 0 & -1 & -1 & 2 \end{bmatrix}$$

Note that $A = A^\top$ (symmetric, since friendship is mutual) and L is symmetric and positive semi-definite.

Exact Spectral Analysis:

The characteristic polynomial of L is:

$$p(\lambda) = \det(\lambda I - L) = \lambda(\lambda - 5)(\lambda - 3)(\lambda^2 - 6\lambda + 7)$$

The exact eigenvalues are:

$$\lambda_0 = 0, \quad \lambda_1 = 3 - \sqrt{2}, \quad \lambda_2 = 3, \quad \lambda_3 = 3 + \sqrt{2}, \quad \lambda_4 = 5$$

with numerical approximations:

$$\lambda_0 = 0, \quad \lambda_1 \approx 1.586, \quad \lambda_2 = 3, \quad \lambda_3 \approx 4.414, \quad \lambda_4 = 5$$

Why $\lambda_0 = 0$? For any undirected graph, $L\mathbf{1} = \mathbf{0}$, where $\mathbf{1}$ is the all-ones vector. This follows from the definition: $(L\mathbf{1})_i = d_i - \sum_j A_{ij} = 0$ for all i. If the graph is connected, $\lambda_0 = 0$ has multiplicity 1.

The **second smallest eigenvector** $\mathbf{v}_1$ (corresponding to $\lambda_1 = 3 - \sqrt{2}$) is called the **Fiedler vector** (named after Czech mathematician Miroslav Fiedler). The Fiedler vector is fundamental in spectral graph theory because it provides the optimal solution to the graph bipartition problem: it naturally separates the graph into two clusters by minimizing the number of edges between them while balancing cluster sizes. The sign of each component of the Fiedler vector indicates which cluster a node belongs to.

The Fiedler vector is approximately:

$$\mathbf{v}_1 \approx [0.63, 0.49, 0.22, -0.22, -0.63]^\top$$

Classification Result:

The sign of $\mathbf{v}_1$ provides a natural bipartition, where $(\mathbf{v}_1)_i$ denotes the i-th component (element) of the Fiedler vector $\mathbf{v}_1$:

$$\text{label}_i = \begin{cases} \text{Active} & \text{if } (\mathbf{v}_1)_i > 0 \\ \text{Inactive} & \text{if } (\mathbf{v}_1)_i < 0 \end{cases}$$

This classification emerges **purely from the graph structure**—no supervised labels were used. The method groups nodes 0–2 together (they form a more densely connected subgraph) and separates them from nodes 3–4. Table 7.1 lists the resulting sign-based classification for each node.

Table 7.1: Node classification by sign of the Fiedler vector

Node ID	User	$(\mathbf{v}_1)_i$	Classification	Color
0	Alice	+0.63	**Active**	Red
1	Bob	+0.49	**Active**	Red
2	Charlie	+0.22	**Active**	Red
3	David	−0.22	**Inactive**	Green
4	Eve	−0.63	**Inactive**	Green

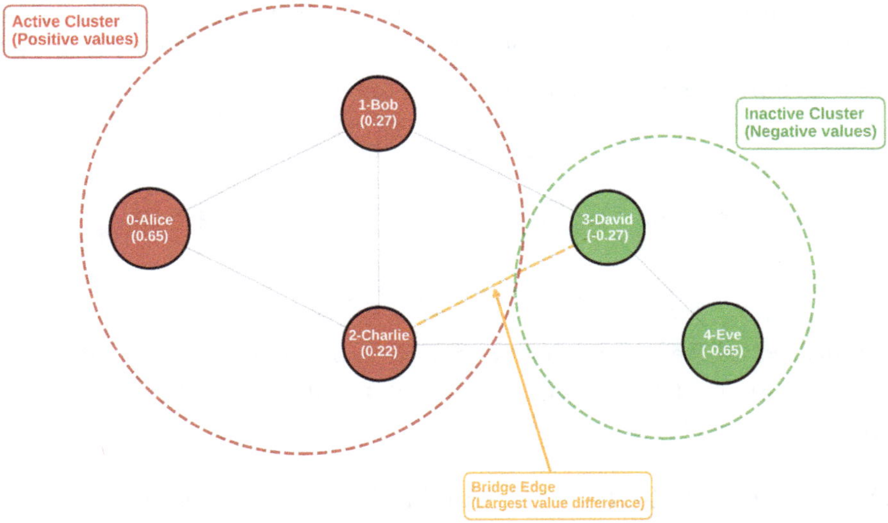

Figure 7.9: Fiedler Vector Visualization

Figure 7.9 shows the values of $\mathbf{v}_1$ plotted on the graph, making the "zero-crossing" explicit:

- **Nodes 0, 1, 2** (Alice, Bob, Charlie): Positive values (red). They form a triangle (clique) with strong internal connections.
- **Nodes 3, 4** (David, Eve): Negative values (green). They form a linear tail with weaker connections.
- **The Edge (2, 3)**: This is the "bridge" between the two clusters (highlighted in orange). Notice that the value difference $|(\mathbf{v}_1)_2 - (\mathbf{v}_1)_3| = |0.22 - (-0.22)| = 0.44$ is significant. The Laplacian quadratic form $\mathbf{v}_1^\top L\mathbf{v}_1$ "stretches" the most across this weak link, which is why the Fiedler vector naturally separates the graph at this point.

The spectral approach to node classification demonstrates how symmetric matrix theory (specifically, the eigendecomposition of the graph Laplacian) provides a principled solution to graph-based learning problems. Beyond classification, the spectral decomposition also enables **spectral graph convolution**, a technique for filtering and processing graph signals.

7.8.6 Spectral Graph Convolution

Spectral graph convolution filters graph signals in the frequency domain. In **spectral-based GNNs** (e.g., Graph Convolutional Networks, ChebNet), graph convolution is defined using the eigendecomposition of L:

1. **Graph Fourier Transform**: For a signal $\mathbf{x} \in \mathbb{R}^{N\times 1}$ on the graph, its Fourier transform is:

$$\hat{\mathbf{x}} = Q^\top \mathbf{x}$$

   where $\hat{x}_n = \mathbf{q}_n^\top \mathbf{x}$ is the coefficient of the n-th frequency mode.

2. **Spectral Filtering**: Apply a **filter** $g(\lambda)$ in the frequency domain. A filter is a function that selectively amplifies or attenuates different frequency components of the signal. The filter $g(\lambda)$ takes an eigenvalue (frequency) λ as input and returns a weight that determines how much of that frequency component to preserve:

$$\hat{\mathbf{y}} = g(\Lambda)\hat{\mathbf{x}} = \mathrm{diag}(g(\lambda_0), g(\lambda_1), \dots, g(\lambda_{N-1}))\hat{\mathbf{x}}$$

   Here, $g(\Lambda)$ applies the filter function g to each eigenvalue λ_n, creating a diagonal matrix that scales each frequency component $\hat{x}_n$ by $g(\lambda_n)$. This allows us to emphasize certain frequencies (e.g., smooth signals) while suppressing others (e.g., noisy high-frequency components).

3. **Inverse Graph Fourier Transform**: Transform back to the spatial domain:

$$\mathbf{y} = Q\hat{\mathbf{y}} = Qg(\Lambda)Q^\top \mathbf{x}$$

The filtered signal $\mathbf{y}$ is computed as:

$$\mathbf{y} = Qg(\Lambda)Q^\top \mathbf{x} = g(L)\mathbf{x} \tag{7.36}$$

where $g(L)$ is defined via the spectral decomposition. Common filters include the **low-pass filter** $g(\lambda) = e^{-\tau\lambda}$ (which smooths the signal) and the **polynomial filter** $g(\lambda) = \sum_{k=0}^{K} \theta_k \lambda^k$ (used in ChebNet with Chebyshev polynomials).

7.8.7 From Spectral Filters to Spatial Message Passing

How do we bridge the gap between the expensive spectral definition $g(L) = Qg(\Lambda)Q^\top$ and the efficient message passing of GNNs? The key is **polynomial approximation**. Instead of computing eigenvectors, we can approximate the filter function $g(\lambda)$ using a polynomial of order K:

$$g(\lambda) \approx \sum_{k=0}^{K} \theta_k \lambda^k$$

This is a game-changer because of the **local property of matrix powers**: A^1 aggregates information from 1-hop neighbors, A^2 aggregates information from 2-hop neighbors, and A^k aggregates information from k-hop neighbors. Therefore, the polynomial filter $g(L) \approx \sum_{k=0}^{K} \theta_k L^k$ can be computed using only matrix multiplications with A and D, without eigendecomposition. This explains why GNN layers look like matrix multiplications with the adjacency matrix—they are **spatially localized approximations** of spectral filters.

While spectral methods provide elegant mathematical solutions, they face computational limitations. The eigendecomposition requires $O(N^3)$ operations, making it impractical for large-scale graphs. Polynomial approximation reduces this to $O(K \cdot |\mathcal{E}|)$ where K is the polynomial order and $|\mathcal{E}|$ is the number of edges, making it scalable to large graphs.

7.8.8 Graph Neural Networks: Scalable Approximation of Spectral Methods

Classical Spectral Method: Explicitly compute the eigendecomposition $L = Q\Lambda Q^\top$ and use $\mathbf{v}_1$ for classification, or apply spectral filters $g(L)$ for signal processing. This is mathematically elegant but computationally expensive ($O(N^3)$) and does not scale to large graphs.

Graph Neural Networks: Instead of computing eigenvectors explicitly, GNNs use **iterative message passing** to approximate the low-frequency eigenspaces. The **Graph Convolutional Network (GCN)** layer can be written as:

$$\mathbf{H}^{(l+1)} = \sigma\left(\tilde{D}^{-1/2}\tilde{A}\tilde{D}^{-1/2}\mathbf{H}^{(l)}W^{(l)}\right)$$

where $\tilde{A} = A + I$ (add self-loops) and $\tilde{D}$ is the corresponding degree matrix.

Connection to Normalized Laplacian: The matrix $\tilde{D}^{-1/2}\tilde{A}\tilde{D}^{-1/2}$ is closely related to the symmetric normalized Laplacian (see *7.32*). Specifically, $\tilde{D}^{-1/2}\tilde{A}\tilde{D}^{-1/2} = I - \tilde{L}_{\text{sym}}$, where $\tilde{L}_{\text{sym}}$ is the normalized Laplacian of the graph with self-loops. This means GCN is applying a first-order polynomial filter on the normalized Laplacian, which explains why the normalization is crucial for numerical stability.

This operation performs local aggregation (message passing) via matrix multiplication with $\tilde{A}$, implicitly approximates low-frequency Laplacian eigenspaces through polynomial approximation, is learnable (via $W^{(l)}$) and nonlinear (via σ), and scales to large graphs with sparse matrix operations ($O(|\mathcal{E}|)$ per layer).

GNNs do not abandon spectral theory; they operationalize it. While eigendecomposition solves the problem explicitly, message passing approximates the same solution implicitly, but at a scale that modern hardware can sustain. The graph structure itself encodes the geometric meaning of the Rayleigh quotient (graph smoothness) into the network architecture.

The graph Laplacian's symmetry and positive semi-definiteness are crucial for GNNs. **Spectral GNNs**[4] (GCN, ChebNet, GAT) use the eigen-decomposition $L = Q\Lambda Q^\top$ to define graph convolutions in the frequency domain. **Spatial GNNs**[5] (GraphSAGE, GIN) aggregate neighbor information directly, but the underlying graph structure is still encoded by symmetric adjacency matrices. The Rayleigh quotient $R_L(\mathbf{x})$ measures signal smoothness on graphs, which is fundamental for understanding how GNNs propagate information. Work on **Rayleigh Quotient GNNs**[6] uses the Rayleigh quotient as a regularizer to encourage smooth node representations, improving generalization. The connection between graph smoothness (low $R_L(\mathbf{x})$) and node similarity is exploited in **graph-based semi-supervised learning**[7], where labels propagate along edges connecting similar nodes.

---

[4] **Spectral GNNs** define convolution in the graph Fourier domain via the eigendecomposition of the Laplacian. **GCN** (Kipf & Welling, 2017) uses a first-order approximation; **ChebNet** uses Chebyshev polynomials of the Laplacian; **GAT** (Graph Attention Networks) adds attention weights over neighbors while retaining a spectral-style aggregation.

[5] **Spatial GNNs** avoid explicit eigendecomposition by aggregating features from neighboring nodes in the graph. **GraphSAGE** samples and aggregates neighbors for scalable inductive learning; **GIN** (Graph Isomorphism Network) uses a sum aggregator and can distinguish graph structure as expressively as the WL test.

[6] **Rayleigh Quotient GNNs** incorporate the Rayleigh quotient $R_L(\mathbf{x})$ (or related smoothness terms) into the training objective or as a regularizer, encouraging node representations to be smooth with respect to the graph Laplacian and improving generalization on graphs.

[7] **Graph-based semi-supervised learning** assumes that nearby nodes in the graph tend to share labels. Algorithms such as label propagation and the method of Zhu et al. (2003) propagate labels along edges so that a node's label is influenced by its neighbors, leveraging the graph structure when only a few nodes are labeled.

7.9 Mindmap

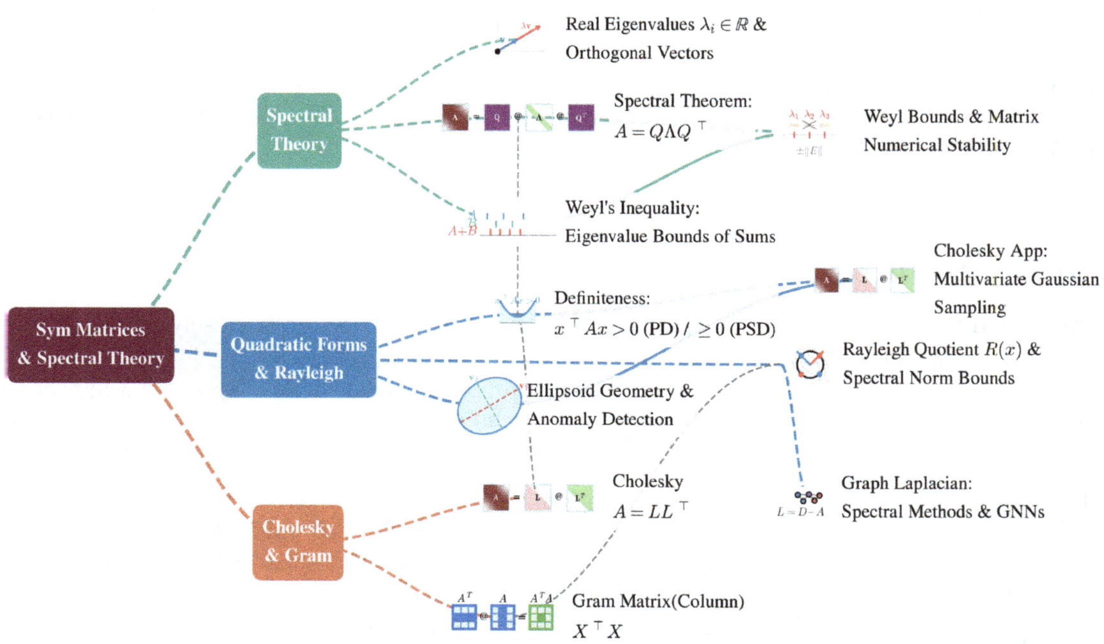

Figure 7.10: Mind map of symmetric matrix and spectral theory topics

7.10 Exercises

7.10.1 Spectral Decomposition of Symmetric Matrices

In machine learning, symmetric matrices (e.g., covariance matrices, kernel matrices) are decomposed using spectral decomposition for dimensionality reduction and feature extraction. Consider a symmetric matrix:

$$A = \begin{bmatrix} 5 & 2 & -1 \\ 2 & 0 & 2 \\ -1 & 2 & 5 \end{bmatrix}.$$

1. Find all eigenvalues $\lambda_0, \lambda_1, \lambda_2$ of A (ordered such that $\lambda_0 \leq \lambda_1 \leq \lambda_2$) by solving $\det(A - \lambda I) = 0$.
2. For each eigenvalue, find a corresponding unit eigenvector (normalized to have unit length).
3. Construct the orthonormal matrix Q whose columns are the unit eigenvectors, arranged in the same order as the eigenvalues. Verify that $Q^\top Q = I$.
4. Construct the diagonal matrix $\Lambda = \text{diag}(\lambda_0, \lambda_1, \lambda_2)$ and verify that $A = Q\Lambda Q^\top$ by computing the product.

5. Express A using spectral decomposition: $A = \sum_{n=0}^{2} \lambda_n \mathbf{q}_n \mathbf{q}_n^\top$ where $\mathbf{q}_n$ are the columns of Q. Verify by computing the sum.

7.10.2 Kernel Matrices and Positive Semi-Definiteness

In kernel methods (e.g., support vector machines, Gaussian processes), kernel matrices K with entries $\kappa_{ij} = \kappa(\mathbf{x}_i, \mathbf{x}_j)$ must be symmetric and positive semi-definite. Consider feature vectors:

$$\phi_0 = \begin{bmatrix} 1 \\ 2 \end{bmatrix}, \quad \phi_1 = \begin{bmatrix} 2 \\ 1 \end{bmatrix}, \quad \phi_2 = \begin{bmatrix} 1 \\ 1 \end{bmatrix},$$

and define the kernel matrix $K \in \mathbb{R}^{3\times 3}$ with entries $\kappa_{ij} = \phi_i^\top \phi_j$ (linear kernel).

1. Compute the kernel matrix K explicitly and show that K is symmetric, i.e., $K = K^\top$.
2. Express K in matrix form using the feature matrix $\Phi = \begin{bmatrix} \phi_0 & \phi_1 & \phi_2 \end{bmatrix} \in \mathbb{R}^{2\times 3}$.
3. Show that K is positive semi-definite by showing that for any $\mathbf{x} \in \mathbb{R}^3$, we have $\mathbf{x}^\top K \mathbf{x} \geq 0$. (Hint: Use the fact that $K = \Phi^\top \Phi$.)
4. Compute the eigenvalues of K and verify that they are all non-negative (confirming positive semi-definiteness).
5. For $\mathbf{x} = \begin{bmatrix} 1 \\ -1 \\ 1 \end{bmatrix}$, compute $\mathbf{x}^\top K \mathbf{x}$ and verify that it is non-negative.

7.10.3 Cholesky Decomposition for Multivariate Gaussian Sampling

In Bayesian inference and variational autoencoders, we need to sample from multivariate Gaussian distributions $\mathcal{N}(\mu, \Sigma)$ where Σ is a covariance matrix. The Cholesky decomposition $\Sigma = LL^\top$ enables efficient sampling via $\mathbf{x} = \mu + L\mathbf{z}$ where $\mathbf{z} \sim \mathcal{N}(\mathbf{0}, I)$.

Consider the covariance matrix:

$$\Sigma = \begin{bmatrix} 9 & 3 & -3 \\ 3 & 5 & 1 \\ -3 & 1 & 6 \end{bmatrix}.$$

1. Prove that Σ is symmetric and positive definite by showing that all leading principal minors are positive. Compute each leading principal minor explicitly.
2. Find the unique Cholesky decomposition $\Sigma = LL^\top$ where L is lower triangular with positive diagonal entries. Express all entries of L in exact form (using radicals if necessary).
3. Using the Cholesky decomposition from part (2), prove that if $\mathbf{z} \sim \mathcal{N}(\mathbf{0}, I_3)$, then $\mathbf{x} = L\mathbf{z}$ has covariance $\text{Cov}(\mathbf{x}) = \Sigma$. (Hint: Use the fact that $\text{Cov}(L\mathbf{z}) = L \cdot \text{Cov}(\mathbf{z}) \cdot L^\top$.)
4. The Cholesky decomposition is unique for positive definite matrices. Prove that if L_1 and L_2 are both lower triangular with positive diagonal entries such that $L_1 L_1^\top = L_2 L_2^\top = \Sigma$, then $L_1 = L_2$. (Hint: Consider the equation $L_1 L_1^\top = L_2 L_2^\top$ and use the uniqueness of matrix square roots.)
5. For efficient computation, we want to verify that Σ^{-1} can be computed as $(L^\top)^{-1} L^{-1}$ without explicitly inverting Σ. Prove this identity and use it to compute Σ^{-1} using the Cholesky factor L from part (2).

7.10.4 Principal Component Analysis (PCA) with Dimensionality Reduction

Given a mean-centered data matrix $X \in \mathbb{R}^{100\times 3}$ (100 samples, 3 features), the sample covariance matrix is

$$\Sigma = \frac{1}{99}X^\top X = \begin{bmatrix} 4 & 1 & -1 \\ 1 & 3 & 0 \\ -1 & 0 & 2 \end{bmatrix}.$$

1. Prove that Σ is symmetric and positive semi-definite. (Hint: Use the fact that $\Sigma = \frac{1}{N-1}X^\top X$ for mean-centered X.)
2. Find the exact spectral decomposition $\Sigma = Q\Lambda Q^\top$ where Q is orthonormal and $\Lambda = \text{diag}(\lambda_0, \lambda_1, \lambda_2)$ with $\lambda_0 \le \lambda_1 \le \lambda_2$. Express all eigenvalues and eigenvectors in exact form (using radicals if necessary).
3. Using the spectral decomposition, prove that the total variance $\text{tr}(\Sigma)$ equals the sum of eigenvalues: $\text{tr}(\Sigma) = \lambda_0 + \lambda_1 + \lambda_2$.
4. We want to reduce dimensionality while preserving at least 85% of the total variance. Determine the minimum number of principal components k needed, and compute the exact variance preserved by these k components as a fraction.
5. Let $\mathbf{x} \in \mathbb{R}^3$ be a new data point. Using the top k principal components from part (4), find the projection $\hat{\mathbf{x}}$ of $\mathbf{x}$ onto the k-dimensional subspace. Express $\hat{\mathbf{x}}$ in terms of $\mathbf{x}$, the selected eigenvectors, and verify that $\|\hat{\mathbf{x}}\|_2^2 \le \|\mathbf{x}\|_2^2$ with equality if and only if $\mathbf{x}$ lies in the selected subspace.

7.10.5 Kernel Matrix Properties and Spectral Analysis

Consider $N = 4$ data points in $\mathbb{R}^2$:

$$\mathbf{x}_1 = \begin{bmatrix} 0 \\ 0 \end{bmatrix}, \quad \mathbf{x}_2 = \begin{bmatrix} 1 \\ 0 \end{bmatrix}, \quad \mathbf{x}_3 = \begin{bmatrix} 0 \\ 1 \end{bmatrix}, \quad \mathbf{x}_4 = \begin{bmatrix} 1 \\ 1 \end{bmatrix},$$

and the RBF kernel function $\kappa(\mathbf{x}_i, \mathbf{x}_j) = \exp\left(-\frac{1}{2}\|\mathbf{x}_i - \mathbf{x}_j\|^2\right)$.

1. Compute the exact 4×4 kernel matrix K where $K_{ij} = \kappa(\mathbf{x}_i, \mathbf{x}_j)$. Express all entries in terms of $e^{-1/2}$ and e^{-1}.
2. Prove that for any kernel function $\kappa(\mathbf{x}_i, \mathbf{x}_j) = \kappa(\mathbf{x}_j, \mathbf{x}_i)$ (symmetric), the kernel matrix K is always symmetric, regardless of the data points.
3. Prove that K is positive semi-definite by showing that for any $\mathbf{v} \in \mathbb{R}^4$, we have $\mathbf{v}^\top K\mathbf{v} \ge 0$. (Hint: Use the fact that K can be written as $K = \Phi^\top\Phi$ for some feature matrix Φ.)
4. Find the exact spectral decomposition $K = Q\Lambda Q^\top$ where eigenvalues are ordered as $\lambda_0 \le \lambda_1 \le \lambda_2 \le \lambda_3$. Determine the rank of K and identify which eigenvalues are zero (if any).
5. Using the spectral decomposition, show that the kernel matrix can be written as $K = \sum_{i=0}^{r-1} \lambda_i \mathbf{q}_i \mathbf{q}_i^\top$ where $r = \text{rank}(K)$ and $\mathbf{q}_i$ are the columns of Q. Prove that this representation is unique up to the ordering of non-zero eigenvalues.

7.10.6 Optimal Quantization Error with Hessian Constraint

In GPTQ quantization, the quantization error is modeled as a quadratic form $\mathbf{e}^\top H\mathbf{e}$ where $H \in \mathbb{R}^{2\times 2}$ is the Hessian matrix of the loss function:

$$H = \begin{bmatrix} 5 & 2 \\ 2 & 3 \end{bmatrix}.$$

1. Prove that H is symmetric and positive definite by showing that all leading principal minors are positive.
2. Find the unique Cholesky decomposition $H = LL^\top$ where L is lower triangular with positive diagonal entries. Verify that L is unique.
3. The quantization error is constrained by $\|\mathbf{e}\|_2 = 1$. Using the Rayleigh quotient theorem, prove that the minimum quantization error is $\lambda_{\text{min}}(H)$ and the maximum is $\lambda_{\text{max}}(H)$, where λ_{min} and λ_{max} are the smallest and largest eigenvalues of H respectively.
4. Find the exact eigenvalues $\lambda_0 \leq \lambda_1$ of H and the corresponding unit eigenvectors $\mathbf{q}_0, \mathbf{q}_1$. Determine the unique direction $\mathbf{e}^*$ (up to sign) that minimizes $\mathbf{e}^\top H\mathbf{e}$ subject to $\|\mathbf{e}\|_2 = 1$.
5. Using the spectral decomposition $H = Q\Lambda Q^\top$, prove that for any $\mathbf{e}$ with $\|\mathbf{e}\|_2 = 1$, we have $\lambda_0 \leq \mathbf{e}^\top H\mathbf{e} = \sum_{i=0}^{1} \lambda_i c_i^2 \leq \lambda_1$ where $\mathbf{c} = Q^\top \mathbf{e}$ and $\|\mathbf{c}\|_2 = 1$. Show that the bounds are tight.

7.10.7 Derivatives of Quadratic Forms in Newton's Method

In second-order optimization methods like Newton's method, we minimize a loss function $\mathcal{L}(\mathbf{w})$ by approximating it with a quadratic form. Consider the quadratic approximation:

$$\mathcal{L}(\mathbf{w}) \approx \mathcal{L}(\mathbf{w}_0) + \nabla\mathcal{L}(\mathbf{w}_0)^\top(\mathbf{w} - \mathbf{w}_0) + \frac{1}{2}(\mathbf{w} - \mathbf{w}_0)^\top H(\mathbf{w} - \mathbf{w}_0),$$

where $H \in \mathbb{R}^{3\times 3}$ is the symmetric Hessian matrix:

$$H = \begin{bmatrix} 6 & 2 & 1 \\ 2 & 5 & 0 \\ 1 & 0 & 4 \end{bmatrix},$$

and $\mathbf{w}_0 = \begin{bmatrix} 1 \\ -1 \\ 2 \end{bmatrix}$ is the current parameter vector.

1. Define the quadratic form $Q(\mathbf{w}) = \frac{1}{2}(\mathbf{w} - \mathbf{w}_0)^\top H(\mathbf{w} - \mathbf{w}_0)$. Prove that the gradient of Q with respect to $\mathbf{w}$ is $\nabla Q(\mathbf{w}) = H(\mathbf{w} - \mathbf{w}_0)$ by:
   - Computing partial derivatives $\frac{\partial Q}{\partial w_k}$ explicitly using the expansion $Q(\mathbf{w}) = \frac{1}{2}\sum_{i,j} h_{ij}(w_i - w_{0,i})(w_j - w_{0,j})$
   - Verifying that $\nabla Q(\mathbf{w}) = H(\mathbf{w} - \mathbf{w}_0)$ using matrix calculus
2. Prove that the Hessian of Q with respect to $\mathbf{w}$ is $\nabla^2 Q(\mathbf{w}) = H$ by computing the second partial derivatives $\frac{\partial^2 Q}{\partial w_k \partial w_\ell}$.
3. In Newton's method, we find the minimum by setting $\nabla Q(\mathbf{w}) = \mathbf{0}$. Using the gradient from part (1), prove that the Newton step $\Delta\mathbf{w} = \mathbf{w} - \mathbf{w}_0$ satisfies $H\Delta\mathbf{w} = -\nabla\mathcal{L}(\mathbf{w}_0)$. If $\nabla\mathcal{L}(\mathbf{w}_0) = \begin{bmatrix} 2 \\ -3 \\ 1 \end{bmatrix}$, compute the exact Newton step $\Delta\mathbf{w}$.
4. The Jacobian of Q (as a scalar function) is $\mathbf{J}_Q(\mathbf{w}) = \nabla Q(\mathbf{w})^\top$. Prove that $\mathbf{J}_Q(\mathbf{w}) = (\mathbf{w} - \mathbf{w}_0)^\top H$ and verify this for $\mathbf{w} = \begin{bmatrix} 2 \\ 0 \\ 3 \end{bmatrix}$.

5. Using the spectral decomposition $H = Q\Lambda Q^\top$ where $\Lambda = \text{diag}(\lambda_0, \lambda_1, \lambda_2)$ with $\lambda_0 \leq \lambda_1 \leq \lambda_2$, show that the gradient can be written as $\nabla Q(\mathbf{w}) = Q\Lambda Q^\top(\mathbf{w} - \mathbf{w}_0)$. Prove that the Newton step in the eigenvector basis is $\mathbf{c} = Q^\top \Delta\mathbf{w}$ where $c_i = -\frac{1}{\lambda_i}(Q^\top \nabla \mathcal{L}(\mathbf{w}_0))_i$ for each i.

7.10.8 Mahalanobis Distance and Anomaly Detection Threshold

For a bivariate Gaussian distribution $\mathcal{N}(\mu, \Sigma)$ with $\mu = \begin{bmatrix} 0 \\ 0 \end{bmatrix}$ and covariance matrix

$$\Sigma = \begin{bmatrix} 4 & 2 \\ 2 & 3 \end{bmatrix},$$

the Mahalanobis distance is defined as $d_M(\mathbf{x}) = \sqrt{(\mathbf{x} - \mu)^\top \Sigma^{-1} (\mathbf{x} - \mu)}$.

1. Prove that Σ is symmetric and positive definite. Show that Σ^{-1} exists and is also symmetric and positive definite.
2. Find the Cholesky decomposition $\Sigma = LL^\top$ and use it to compute Σ^{-1} via $(LL^\top)^{-1} = (L^\top)^{-1} L^{-1}$.
3. For $\mathbf{x} = \begin{bmatrix} 2 \\ 1 \end{bmatrix}$, compute the exact Mahalanobis distance $d_M(\mathbf{x})$ in simplified radical form.
4. Using the spectral decomposition $\Sigma = Q\Lambda Q^\top$ where $\Lambda = \text{diag}(\lambda_0, \lambda_1)$ with $\lambda_0 \leq \lambda_1$, prove that:
   - $\Sigma^{-1} = Q\Lambda^{-1}Q^\top$ where $\Lambda^{-1} = \text{diag}(1/\lambda_0, 1/\lambda_1)$
   - $d_M(\mathbf{x})^2 = \sum_{i=0}^{1} \frac{c_i^2}{\lambda_i}$ where $\mathbf{c} = Q^\top(\mathbf{x} - \mu)$
5. For a threshold $\tau = 3$, determine the exact set of all points $\mathbf{x} \in \mathbb{R}^2$ such that $d_M(\mathbf{x}) = \tau$. Express this set in the form of an ellipse equation. Verify whether $\mathbf{x} = \begin{bmatrix} 3 \\ 2 \end{bmatrix}$ satisfies $d_M(\mathbf{x}) > \tau$.

7.10.9 Gram Matrix Rank and Spectral Properties

Given a user-item rating matrix $R \in \mathbb{R}^{5\times 4}$:

$$R = \begin{bmatrix} 5 & 4 & 0 & 2 \\ 3 & 5 & 1 & 0 \\ 0 & 2 & 4 & 5 \\ 4 & 0 & 3 & 4 \\ 2 & 3 & 5 & 1 \end{bmatrix},$$

define the Gram matrices $G_c = R^\top R$ (item-item) and $G_r = RR^\top$ (user-user).

1. Compute $G_c = R^\top R$ and $G_r = RR^\top$ explicitly. Prove that both are symmetric and positive semi-definite using the fact that $G_c = R^\top R$ and $G_r = RR^\top$.
2. Prove that $\text{rank}(G_c) = \text{rank}(G_r) = \text{rank}(R)$. Determine the exact rank of R and hence the ranks of G_c and G_r.
3. Prove that G_c and G_r share the same non-zero eigenvalues. Specifically, if $\lambda \neq 0$ is an eigenvalue of G_c with eigenvector $\mathbf{v}$, show that λ is also an eigenvalue of G_r with eigenvector $R\mathbf{v}$ (after normalization).
4. Find the spectral decomposition $G_c = Q\Lambda Q^\top$ where $\Lambda = \text{diag}(\lambda_0, \lambda_1, \lambda_2, \lambda_3)$ with $\lambda_0 \leq \lambda_1 \leq \lambda_2 \leq \lambda_3$. Identify which eigenvalues are zero.

5. For dimensionality reduction, we want to keep only the top k dimensions that capture at least 90% of the "energy" $\text{tr}(G_c)$. Determine the minimum value of k and compute the exact percentage of energy captured.

7.10.10 Graph Laplacian Spectral Properties

For an undirected graph with 4 nodes, the adjacency matrix is:

$$A = \begin{bmatrix} 0 & 1 & 1 & 0 \\ 1 & 0 & 1 & 1 \\ 1 & 1 & 0 & 1 \\ 0 & 1 & 1 & 0 \end{bmatrix}.$$

The graph Laplacian is defined as $L = D - A$ where $D = \text{diag}(d_0, d_1, d_2, d_3)$ is the degree matrix with $d_i = \sum_{j=0}^{3} A_{ij}$.

1. Compute the degree matrix D and the Laplacian $L = D - A$. Prove that L is symmetric.
2. Prove that L is positive semi-definite by showing that for any $\mathbf{x} \in \mathbb{R}^4$, we have $\mathbf{x}^\top L\mathbf{x} = \frac{1}{2}\sum_{i,j} A_{ij}(x_i - x_j)^2 \geq 0$.
3. Find the exact spectral decomposition $L = Q\Lambda Q^\top$ where $\Lambda = \text{diag}(\lambda_0, \lambda_1, \lambda_2, \lambda_3)$ with $\lambda_0 \leq \lambda_1 \leq \lambda_2 \leq \lambda_3$. Determine the smallest eigenvalue λ_0 and its geometric multiplicity. (Hint: For a connected graph, $\lambda_0 = 0$ with multiplicity 1, and the corresponding eigenvector is $\mathbf{1}$.)
4. Prove that the number of connected components in the graph equals the geometric multiplicity of eigenvalue 0. Verify this for the given graph.
5. The second smallest eigenvalue λ_1 (algebraic connectivity) measures how well-connected the graph is. Compute λ_1 and interpret its value. Prove that $\lambda_1 > 0$ if and only if the graph is connected.

7.10.11 Rayleigh Quotient in Eigenbasis Coordinates

Consider the symmetric matrix $A = \begin{bmatrix} 1.5 & 0.5 \\ 0.5 & 1.5 \end{bmatrix}$ with eigenvalues $\lambda_1 = 2$ and $\lambda_2 = 1$.

1. Find the orthonormal matrix V whose columns are the unit eigenvectors of A corresponding to λ_1 and λ_2 respectively.
2. Using the eigenbasis representation, show that the Rayleigh quotient can be written as $R(\mathbf{y}) = \frac{2y_1^2+y_2^2}{y_1^2+y_2^2}$ where $\mathbf{y} = V^\top\mathbf{x}$. Verify that when $\mathbf{x}$ is an eigenvector, the Rayleigh quotient equals the corresponding eigenvalue.
3. For a unit vector $\mathbf{x} = \begin{bmatrix} \cos\theta \\ \sin\theta \end{bmatrix}$ in the original coordinates, show that $R(\mathbf{x}) = 1.5 + 0.5\sin(2\theta)$.
4. Using the expression from part (3), prove that $1 \leq R(\mathbf{x}) \leq 2$ for all unit vectors $\mathbf{x}$, confirming that the bounds are $\lambda_2 = 1$ and $\lambda_1 = 2$.
5. Express the Rayleigh quotient $R(\mathbf{y})$ in terms of the angle ϕ where $y_1 = \cos\phi$ and $y_2 = \sin\phi$. Show that this representation makes it clear that $R(\mathbf{y})$ is a weighted average of the eigenvalues, with weights determined by the direction of $\mathbf{y}$.

Chapter 8

Singular Value Decomposition (SVD)

Singular Value Decomposition: the fundamental matrix factorization for data analysis

"The purpose of computing is insight, not numbers."

— Richard Hamming

- ▷ `Full SVD` $A = U\Sigma V^\top$, $U \in \mathbb{R}^{M\times M}$, $V \in \mathbb{R}^{N\times N}$ o.n., $\Sigma \in \mathbb{R}^{M\times N}$ $\cdots \geq \sigma_{\min(M,N)-1} \geq 0$
- ▷ `Singular values` $\sigma_i = \sqrt{\lambda_i}$ where λ_i are eigenvalues of $A^\top A$
- ▷ `Left singular vectors` $\mathbf{u}_i = \frac{1}{\sigma_i} A\mathbf{v}_i$ for $i = 0, \ldots, r-1$ where $r = \text{rank}(A)$
- ▷ `Orthonormality` $\mathbf{u}_i^\top \mathbf{u}_j = \delta_{ij}$, $\mathbf{v}_i^\top \mathbf{v}_j = \delta_{ij}$ where δ_{ij} is Kronecker delta
- ▷ `Thin` $A = U_k \Sigma_k V_k^\top$, $k = \min(M, N)$, $U_k \in \mathbb{R}^{M\times k}$, $\Sigma_k = \text{diag}(\sigma_0, \ldots, \sigma_{k-1})$, $V_k \in \mathbb{R}^{N\times k}$
- ▷ `Compact` $A = U_r \Sigma_r V_r^\top = \sum_{i=0}^{r-1} \sigma_i \mathbf{u}_i \mathbf{v}_i^\top$ where $r = \text{rank}(A)$
- ▷ `Col/Row-Gram via SVD` $A^\top A = \sum_{i=0}^{r-1} \sigma_i^2 \mathbf{v}_i \mathbf{v}_i^\top$; $AA^\top = \sum_{i=0}^{r-1} \sigma_i^2 \mathbf{u}_i \mathbf{u}_i^\top$
- ▷ `Polar decomposition` $A = QS$ where $Q = UV^\top$ is orthonormal, $S = V\Sigma V^\top$ is SPSD
- ▷ `Subspaces` U_r span $C(A)$, V_r span $C(A^\top)$, $\bar{V}_r$ span $N(A)$, $\bar{U}_r$ span $N(A^\top)$
- ▷ `Linear equations` $\mathbf{x} = V_r \Sigma_r^{-1} U_r^\top \mathbf{b} + \bar{V}_r \mathbf{c}$ for $A\mathbf{x} = \mathbf{b}$ when $U_r U_r^\top \mathbf{b} = \mathbf{b}$
- ▷ `MP pseudoinverse` $A^* = \lim_{\lambda\to 0^+} (A^\top A + \lambda I)^{-1} A^\top = V_r \Sigma_r^{-1} U_r^\top = \sum_{i=0}^{r-1} \mathbf{v}_i \sigma_i^{-1} \mathbf{u}_i^\top$
- ▷ `Least squares solution` $\hat{\mathbf{x}} = A^* \mathbf{b}$ for $\min_{\mathbf{x}} \|A\mathbf{x} - \mathbf{b}\|_2^2$
- ▷ `Energy from singular values` $\|A\|_F^2 = \sum_{i=0}^{r-1} \sigma_i^2$
- ▷ `Truncated` $A_k = \sum_{i=0}^{k-1} \sigma_i \mathbf{u}_i \mathbf{v}_i^\top$ for $k < r$; error: $\frac{\|A - A_k\|_F^2}{\|A\|_F^2} = 1 - \frac{\sum_{i=0}^{k-1} \sigma_i^2}{\sum_{i=0}^{r-1} \sigma_i^2}$
- ▷ `Rank` $\text{rank}(A) = r$ where r is the number of non-zero singular values
- ▷ `Spectral norm` $\|A\|_2 = \sigma_0$ (largest singular value)
- ▷ `Matrix-vector products` $A\mathbf{v}_i = \sigma_i \mathbf{u}_i$, $A^\top \mathbf{u}_i = \sigma_i \mathbf{v}_i$
- ▷ `Eckart-Young theorem` A_k minimizes $\|A - B\|_F$ over all $B \in \mathbb{R}^{M\times N}$ with $\text{rank}(B) \leq k$

8.1 SVD Formulae

8.1.1 Theorem: Singular Value Decomposition — Full Form

Every matrix $A \in \mathbb{R}^{M \times N}$ can be decomposed as

$$A = U\Sigma V^\top \tag{8.1}$$

where:

1. $U = \begin{bmatrix} \mathbf{u}_0 & \mathbf{u}_1 & \cdots & \mathbf{u}_{M-1} \end{bmatrix} \in \mathbb{R}^{M \times M}$ is an orthonormal matrix. Each column $\mathbf{u}_i$ is a **left singular vector**.
2. $\Sigma \in \mathbb{R}^{M \times N}, \Sigma_{mn} = \sigma_m \delta_{mn}$ where δ_{mn} is the **Kronecker delta**:

   $$\delta_{mn} = \begin{cases} 1 & \text{if } m = n \\ 0 & \text{if } m \neq n \end{cases}$$

   This means Σ is a diagonal matrix with singular values

   $$\sigma_0 \geq \sigma_1 \geq \cdots \geq \sigma_{\min(M,N)-1} \geq 0.$$

3. $V = \begin{bmatrix} \mathbf{v}_0 & \mathbf{v}_1 & \cdots & \mathbf{v}_{N-1} \end{bmatrix} \in \mathbb{R}^{N \times N}$ is an orthonormal matrix. Each column $\mathbf{v}_i$ is a **right singular vector**.

Proof

Step 1 (Eigen-decomposition)

Solve

$$A^\top A\mathbf{v} = \lambda \mathbf{v}.$$

Since $A^\top A$ is symmetric positive semi-definite, it has $r = \operatorname{rank}(A^\top A)$ positive eigenvalues

$$\lambda_0 \geq \cdots \geq \lambda_{r-1} > 0,$$

and $N - r$ zero eigenvalues, with orthonormal eigenvectors $\{\mathbf{v}_i\}_{i=0}^{N-1}$.

Step 2 (Singular values) Define

$$\sigma_i = \sqrt{\lambda_i}, \qquad i = 0, \ldots, r-1.$$

Step 3 (Left singular vectors) For $i = 0, \ldots, r-1$, define

$$\mathbf{u}_i = \frac{1}{\sigma_i} A\mathbf{v}_i.$$

Then

$$\mathbf{u}_i^\top \mathbf{u}_j = \delta_{ij}.$$

Extend $\{\mathbf{u}_i\}_{i=0}^{r-1}$ to an orthonormal basis of $\mathbb{R}^M$.

Step 4 (Verification) For all $i = 0, \ldots, N-1$,

$$A\mathbf{v}_i = U\Sigma V^\top \mathbf{v}_i.$$

Since V is orthonormal, we have $V^\top \mathbf{v}_i = \mathbf{e}_i$ (the i-th standard basis vector). Then:

- For $i < r$: $U\Sigma \mathbf{e}_i = \sigma_i \mathbf{u}_i = A\mathbf{v}_i$ (by construction in Step 3).
- For $i \geq r$: $U\Sigma \mathbf{e}_i = \mathbf{0}$ (since $\sigma_i = 0$), and $A\mathbf{v}_i = \mathbf{0}$ (since $\mathbf{v}_i$ is in the null space of $A^\top A$).

Since $\{\mathbf{v}_i\}$ spans $\mathbb{R}^N$, the decomposition holds.

The eigen-decomposition of $A^\top A$ is the foundation of SVD. Since $A^\top A$ is symmetric and positive semi-definite, all its eigenvalues are non-negative, and it has a complete set of orthonormal eigenvectors. The number of positive eigenvalues equals the rank r of $A^\top A$ (which equals the rank of A). The eigenvectors $\{\mathbf{v}_i\}$ corresponding to positive eigenvalues will become the right singular vectors, while those corresponding to zero eigenvalues span the null space of A. This eigen-decomposition can be computed efficiently using algorithms for symmetric matrices, such as the QR algorithm or divide-and-conquer methods. The relationship $\sigma_i = \sqrt{\lambda_i}$ connects the singular values of A to the eigenvalues of $A^\top A$, providing a bridge between the two decompositions.

8.2 Thin, Compact and Truncated Forms

Having constructed the full SVD $A = U\Sigma V^\top$, we rarely need every factor in practice. Unused singular vectors can be dropped, and small singular values can be discarded for approximation.

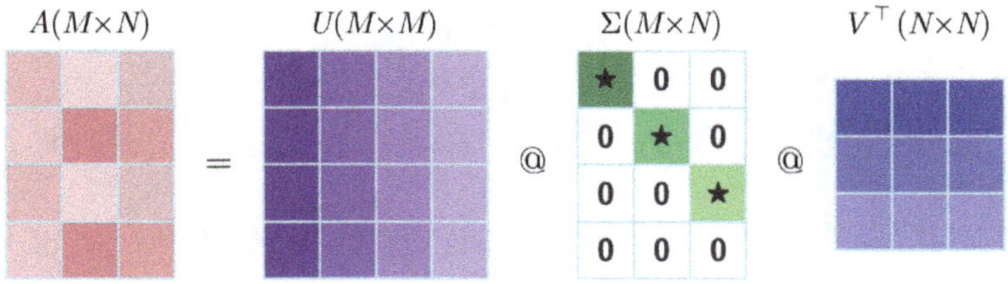

Figure 8.1: Full SVD

In the full factorization (Figure 8.1), U and V are square and Σ has the same shape as A. The next three subsections strip this down, one step at a time.

8.2.1 Definition: Thin SVD

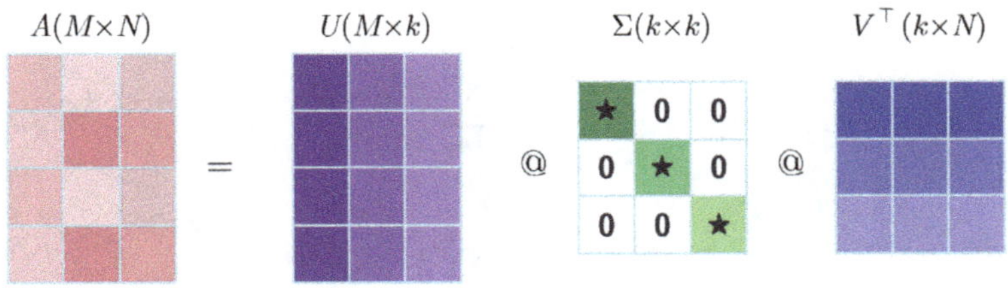

Figure 8.2: Thin SVD

Let $k = \min(M, N)$. Then

$$A = U_k \Sigma_k V_k^\top \tag{8.2}$$

where

- $U_k \in \mathbb{R}^{M\times k}$
- $\Sigma_k = \text{diag}(\sigma_0, \dots, \sigma_{k-1}) \in \mathbb{R}^{k\times k}$
- $V_k \in \mathbb{R}^{N\times k}$

This is the economy-size form in Figure 8.2: unused columns of U are gone, so Σ is square. If $\text{rank}(A) = r < k$, the diagonal still contains $k - r$ zeros. When $M \gg N$ or $N \gg M$, avoiding those unused singular vectors saves substantial storage.

8.2.2 Definition: Compact SVD

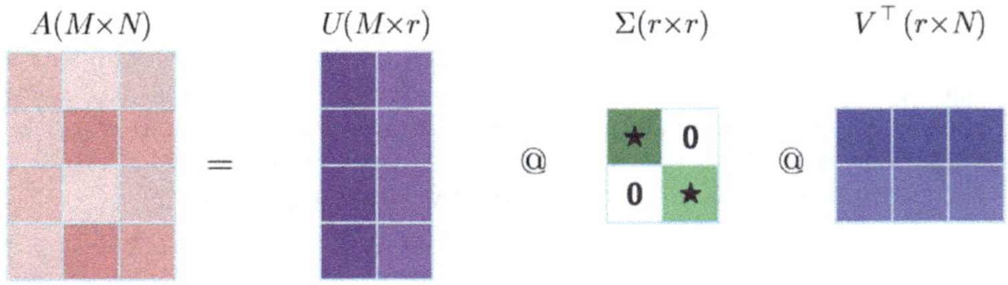

Figure 8.3: Compact SVD

If rank$(A) = r$, the **compact SVD** keeps only the positive singular values (Figure 8.3):

$$A = U_r \Sigma_r V_r^\top \tag{8.3}$$

with

- $U_r \in \mathbb{R}^{M\times r}$
- $\Sigma_r = \text{diag}(\sigma_0, \dots, \sigma_{r-1})$
- $V_r \in \mathbb{R}^{N\times r}$

Equivalently,

$$A = \sum_{i=0}^{r-1} \sigma_i \mathbf{u}_i \mathbf{v}_i^\top. \tag{8.4}$$

8.2.2.1 Example: Products of $A^\top A$ and $AA^\top$

Given $A = \sum_{i=0}^{r-1} \sigma_i \mathbf{u}_i \mathbf{v}_i^\top$, we have:

1. $$A^\top A = \sum_{i=0}^{r-1} \sigma_i^2 \mathbf{v}_i \mathbf{v}_i^\top.$$

2. $$AA^\top = \sum_{i=0}^{r-1} \sigma_i^2 \mathbf{u}_i \mathbf{u}_i^\top.$$

8.2.3 Definition: Truncated SVD

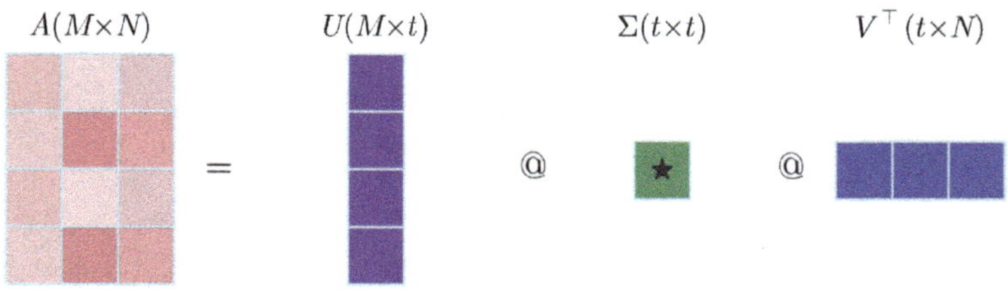

Figure 8.4: Truncated SVD

One more reduction is possible: keep only the top t singular values with $t < r$ (Figure 8.4),

$$A_t = U_t \Sigma_t V_t^\top = \sum_{i=0}^{t-1} \sigma_i \mathbf{u}_i \mathbf{v}_i^\top,$$

with

- $U_t \in \mathbb{R}^{M \times t}$
- $\Sigma_t = \mathrm{diag}(\sigma_0, \dots, \sigma_{t-1})$
- $V_t \in \mathbb{R}^{N \times t}$

Unlike Thin and Compact SVD, this product is no longer exact: it is a low-rank approximation—a **truncated SVD**. We take up its error analysis later in the chapter (@sec:truncated-svd).

8.3 Polar Decomposition

8.3.1 Theorem: Polar Decomposition

Any square matrix $A \in \mathbb{R}^{N \times N}$ can be decomposed as

$$A = QS \tag{8.5}$$

where

- Q is orthonormal
- S is SPSD(symmetric positive semi-definite)

Construction

Given the SVD $A = U\Sigma V^\top$ (where U contains the left singular vectors), we construct the polar decomposition as follows:

$$S = V\Sigma V^\top, \qquad Q = UV^\top.$$

Verification:

- $S = V\Sigma V^\top$ is SPSD because Σ is diagonal with non-negative entries and V is orthonormal
- $Q = UV^\top$ is orthonormal (product of two orthonormal matrices)
- $QS = (UV^\top)(V\Sigma V^\top) = U(V^\top V)\Sigma V^\top = U\Sigma V^\top = A$ ✓

Geometric Intuition: Analogy with Complex Numbers

To understand the geometric meaning, we can draw an analogy with complex numbers. A complex number in polar form is $z = re^{i\theta}$, where $e^{i\theta} = \cos\theta + i\sin\theta$ by Euler's formula (see Chapter~3):

- $e^{i\theta}$ represents the direction (rotation), analogous to Q in polar decomposition
- r represents the magnitude (scaling), analogous to S in polar decomposition

Polar Decomposition: Any linear transformation A can be viewed as "first apply a deformation/stretching (S), then apply a rotation (Q)". This is $A = QS$.

SVD goes further: While S represents stretching, its principal axes may not align with the coordinate axes. SVD reveals that if we first rotate the coordinate system by $V^\top$, then the stretching becomes simple scaling along the coordinate axes (given by Σ), followed by another rotation U. This is $A = U\Sigma V^\top$.

8.3.2 Spectral Norm and Applications in Optimization

The **spectral norm** (also called the **induced L2 norm** or **operator norm**) of a matrix A is defined as the largest singular value: $||A||_2 = \sigma_0(A)$. Note that while the notation $||A||_2$ looks similar to the L2 norm for vectors, for matrices it denotes the induced norm (the maximum amplification factor), not the element-wise L2 norm. The spectral norm plays a crucial role in controlling the stability and generalization of neural networks, as it bounds the Lipschitz constant of linear layers. For a weight matrix W in a neural network, the spectral norm $||W||_2$ determines the maximum amplification factor that the layer can apply to its input, directly affecting gradient flow and training stability.

The spectral norm is intimately connected to polar decomposition. In the polar decomposition $A = QS$, the symmetric positive semi-definite factor S has eigenvalues equal to the singular values of A. Therefore, the spectral norm $||A||_2 = \sigma_0(A)$ equals the largest eigenvalue of S, which represents the maximum scaling factor in the stretching component of the transformation.

The **Muon optimizer** (MomentUm Orthogonalized by Newton-Schulz)[a] demonstrates excellent performance in managing the spectral norm of weight matrices during training, specifically designed for **linear layers** (2D weight matrices) in neural networks. It applies polar decomposition to the gradient update matrix ΔW (not the weight matrix W itself), decomposing it as $\Delta W = QS$ where Q is orthogonal and S is symmetric positive semi-definite. To compute the orthogonal factor Q efficiently, Muon uses the **Newton-Schulz iteration** (a numerical algorithm that avoids expensive SVD computations), which iteratively converges to the orthogonal part of the polar decomposition using only matrix multiplications. The Newton-Schulz algorithm and its convergence properties (including the requirement that the input matrix has spectral norm less than $\sqrt{3}$) are discussed in detail in Chapter~11. By discarding the scaling component S and **retaining only the orthogonal factor** Q, Muon enforces that the update step has unit spectral norm ($||\Delta W||_2 = 1$), which is distinct from **spectral normalization** (Miyato et al.)[b] that constrains the weight matrix itself ($||W||_2 \approx 1$). This approach helps maintain the spectral norm of the update near unity, ensuring uniform gradient flow across all directions and enabling stable training with higher learning rates. While Muon encourages well-conditioned weights, it strictly enforces that the gradient update has unit spectral norm, preventing the update magnitude from growing uncontrollably and thus avoiding exploding gradients and training instability.

---

[a] **Muon optimizer**: An optimizer for 2D parameters (linear layers) in neural networks that uses Newton-Schulz iteration to orthogonalize gradient updates. See Keller Jordan et al., "Muon: An optimizer for hidden layers in neural networks," 2024, https://kellerjordan.github.io/posts/muon/; and the related paper **Muon is Scalable for LLM Training** at https://arxiv.org/abs/2502.16982.

[b] **Spectral Normalization**: A technique introduced by Takeru Miyato et al. in "Spectral Normalization for Generative Adversarial Networks" (ICML 2018) that constrains the spectral norm of weight matrices to stabilize training. Unlike Muon which constrains the update step, spectral normalization directly normalizes the weight matrix itself by dividing it by its largest singular value, ensuring $||W||_2 \approx 1$. See https://arxiv.org/abs/1802.05957.

8.4 Four Subspaces via SVD

8.4.1 Theorem: Four Subspaces via SVD

Let $A \in \mathbb{R}^{M \times N}$ have rank r.

1. $\{\mathbf{u}_i\}_{i=0}^{r-1}$ spans $C(A)$
2. $\{\mathbf{v}_i\}_{i=0}^{r-1}$ spans $C(A^\top)$
3. $\{\mathbf{v}_i\}_{i=r}^{N-1}$ spans $N(A)$
4. $\{\mathbf{u}_i\}_{i=r}^{M-1}$ spans $N(A^\top)$

The SVD provides a complete orthonormal basis for all four fundamental subspaces of a matrix. The first r left singular vectors $\{\mathbf{u}_i\}_{i=0}^{r-1}$ form an orthonormal basis for the column space $C(A)$, while the first r right singular vectors $\{\mathbf{v}_i\}_{i=0}^{r-1}$ span the row space $C(A^\top)$. The remaining right singular vectors $\{\mathbf{v}_i\}_{i=r}^{N-1}$ span the null space $N(A)$, and the remaining left singular vectors $\{\mathbf{u}_i\}_{i=r}^{M-1}$ span the left null space $N(A^\top)$. This elegant decomposition reveals that the singular vectors corresponding to non-zero singular values capture the range spaces, while those corresponding to zero singular values capture the null spaces, providing a unified geometric view of how A maps vectors between $\mathbb{R}^N$ and $\mathbb{R}^M$.

8.5 Linear Equations

8.5.1 Theorem: Linear Equations via SVD

Let $A \in \mathbb{R}^{M\times N}$ be with rank r. Suppose its SVD takes the following form:

$$A = \sum_{i=0}^{r-1} \mathbf{u}_i \sigma_i \mathbf{v}_i^\top .$$

Let $\{\mathbf{v}_i : i \in \{0, 1, \dots, N-1\}\}$ be an orthonormal basis of $\mathbb{R}^N$.

Denote

$$\bar{V}_r = \begin{bmatrix} \mathbf{v}_r & \mathbf{v}_{r+1} & \cdots & \mathbf{v}_{N-1} \end{bmatrix}.$$

Denote

$$\Sigma_r^{-1} = \operatorname{diag}\left(\sigma_0^{-1}, \sigma_1^{-1}, \dots, \sigma_{r-1}^{-1}\right).$$

For the equation

$$A\mathbf{x} = \mathbf{b}$$

with $\mathbf{b} \in \mathbb{R}^{M\times 1}$:

1. If

$$U_r U_r^\top \mathbf{b} \neq \mathbf{b},$$

then the equation $A\mathbf{x} = \mathbf{b}$ has no solution.

2. If

$$U_r U_r^\top \mathbf{b} = \mathbf{b},$$

then the solution is

$$\mathbf{x} = \sum_{i=0}^{r-1} \mathbf{v}_i \frac{\mathbf{u}_i^\top \mathbf{b}}{\sigma_i} + \sum_{i=r}^{N-1} \mathbf{v}_i c_i = V_r \Sigma_r^{-1} U_r^\top \mathbf{b} + \bar{V}_r \mathbf{c} \tag{8.6}$$

with arbitrary $\mathbf{c} = \begin{bmatrix} c_r \\ \vdots \\ c_{N-1} \end{bmatrix} \in \mathbb{R}^{(N-r)\times 1}$.

In particular, if $r = N$, then the feasible solution is unique.

Proof

1. The equation $A\mathbf{x} = \mathbf{b}$ can be written as

$$\sum_{i=0}^{r-1} \mathbf{u}_i \left(\sigma_i \mathbf{v}_i^\top \mathbf{x}\right) = \mathbf{b}.$$

Thus, the above equation has at least one solution if and only if $\mathbf{b}$ is one element of the span of $\{\mathbf{u}_i : i \in \{0, 1, \dots, r-1\}\}$. That is, the residual of the projection of $\mathbf{b}$ onto $\{\mathbf{u}_i : i \in \{0, 1, \dots, r-1\}\}$ is 0.

The residual is equal to

$$\mathbf{b} - \sum_{i=0}^{r-1} \mathbf{u}_i \left(\mathbf{u}_i^\top \mathbf{b}\right) = \mathbf{b} - U_r U_r^\top \mathbf{b}.$$

Therefore, the equation $A\mathbf{x} = \mathbf{b}$ has no solution if $U_r U_r^\top \mathbf{b} \neq \mathbf{b}$.

2. Now, we consider the condition that $U_r U_r^\top \mathbf{b} = \mathbf{b}$. That is, the equation $A\mathbf{x} = \mathbf{b}$ has at least one solution.

Recall that this equation can be written as

$$\sum_{i=0}^{r-1} \mathbf{u}_i \left(\sigma_i \mathbf{v}_i^\top \mathbf{x}\right) = \mathbf{b}.$$

Left-multiply both sides by $\mathbf{u}_j^\top$ for any $j \in \{0, 1, \dots, r-1\}$, we get

$$\sigma_j \mathbf{v}_j^\top \mathbf{x} = \mathbf{u}_j^\top \mathbf{b}.$$

Thus,

$$\mathbf{v}_j^\top \mathbf{x} = \frac{\mathbf{u}_j^\top \mathbf{b}}{\sigma_j}.$$

Because $\{\mathbf{v}_i : i \in \{0, 1, \dots, N-1\}\}$ is an orthonormal basis of $\mathbb{R}^N$, we have

$$\mathbf{x} = \sum_{i=0}^{r-1} \mathbf{v}_i \frac{\mathbf{u}_i^\top \mathbf{b}}{\sigma_i} + \sum_{i=r}^{N-1} \mathbf{v}_i c_i = V_r \Sigma_r^{-1} U_r^\top \mathbf{b} + \bar{V}_r \mathbf{c}$$

with arbitrary $\mathbf{c} = \begin{bmatrix} c_r \\ \vdots \\ c_{N-1} \end{bmatrix} \in \mathbb{R}^{(N-r)\times 1}$.

In particular, if $r = N$, then the feasible solution is unique.

8.6 Moore–Penrose Pseudoinverse

Motivation: For the linear regression problem

$$\min_{\mathbf{x}} \|A\mathbf{x} - \mathbf{b}\|_2^2,$$

where $A \in \mathbb{R}^{M\times N}$ and $\mathbf{b} \in \mathbb{R}^{M\times 1}$, the solution is

$$\hat{\mathbf{x}} = (A^\top A)^{-1} A^\top \mathbf{b}.$$

However, it is problematic if $A^\top A$ is not invertible.

8.6.1 Theorem: Moore–Penrose Pseudoinverse

Define

$$A^* = \lim_{\lambda\to 0^+} (A^\top A + \lambda I_{N\times N})^{-1} A^\top.$$

Let A be with the SVD

$$A = \sum_{i=0}^{r-1} \mathbf{u}_i \sigma_i \mathbf{v}_i^\top = U_r \Sigma_r V_r^\top.$$

Then

$$A^* = \sum_{i=0}^{r-1} \mathbf{v}_i \sigma_i^{-1} \mathbf{u}_i^\top = V_r \Sigma_r^{-1} U_r^\top. \tag{8.7}$$

Proof

We have

$$\begin{aligned}
A^* &= \lim_{\lambda \to 0^+} (A^\top A + \lambda I_{N\times N})^{-1} A^\top \\
&= \lim_{\lambda \to 0^+} \left(\sum_{i=0}^{r-1} \sigma_i^2 \mathbf{v}_i \mathbf{v}_i^\top + \lambda I_{N\times N} \right)^{-1} \left(\sum_{i=0}^{r-1} \sigma_i \mathbf{v}_i \mathbf{u}_i^\top \right) \\
&= \lim_{\lambda \to 0^+} \left(\sum_{i=0}^{r-1} \sigma_i^2 \mathbf{v}_i \mathbf{v}_i^\top + \lambda \sum_{i=0}^{r-1} \mathbf{v}_i \mathbf{v}_i^\top \right)^{-1} \left(\sum_{i=0}^{r-1} \sigma_i \mathbf{v}_i \mathbf{u}_i^\top \right) \\
&= \lim_{\lambda \to 0^+} \left(\sum_{i=0}^{r-1} (\sigma_i^2 + \lambda) \mathbf{v}_i \mathbf{v}_i^\top + \lambda \sum_{i=r}^{N-1} \mathbf{v}_i \mathbf{v}_i^\top \right)^{-1} \left(\sum_{i=0}^{r-1} \sigma_i \mathbf{v}_i \mathbf{u}_i^\top \right) \\
&= \lim_{\lambda \to 0^+} \left(\sum_{i=0}^{r-1} (\sigma_i^2 + \lambda)^{-1} \mathbf{v}_i \mathbf{v}_i^\top + \lambda^{-1} \sum_{i=r}^{N-1} \mathbf{v}_i \mathbf{v}_i^\top \right) \left(\sum_{i=0}^{r-1} \sigma_i \mathbf{v}_i \mathbf{u}_i^\top \right) \\
&= \lim_{\lambda \to 0^+} \left(\sum_{i=0}^{r-1} \sum_{j=0}^{r-1} (\sigma_i^2 + \lambda)^{-1} \mathbf{v}_i \mathbf{v}_i^\top \cdot \sigma_j \mathbf{v}_j \mathbf{u}_j^\top + \lambda^{-1} \sum_{i=r}^{N-1} \sum_{j=0}^{r-1} \mathbf{v}_i \mathbf{v}_i^\top \cdot \sigma_j \mathbf{v}_j \mathbf{u}_j^\top \right) \\
&= \lim_{\lambda \to 0^+} \sum_{i=0}^{r-1} \frac{\sigma_i}{\sigma_i^2 + \lambda} \mathbf{v}_i \mathbf{u}_i^\top \\
&= \sum_{i=0}^{r-1} \mathbf{v}_i \sigma_i^{-1} \mathbf{u}_i^\top \\
&= V_r \Sigma_r^{-1} U_r^\top.
\end{aligned}$$

8.6.2 Theorem: Least Squares Solution

In the linear regression problem

$$\min_{\mathbf{x}} \|A\mathbf{x} - \mathbf{b}\|_2^2,$$

we have

$$\hat{\mathbf{x}} = A^* \mathbf{b}. \tag{8.8}$$

Proof

Recall that $\hat{\mathbf{x}}$ is the solution to the equation:

$$(A^\top A)\hat{\mathbf{x}} = A^\top \mathbf{b}.$$

Thus, $\hat{\mathbf{x}}$ satisfies the following equation:

$$\lim_{\lambda \to 0^+} (A^\top A + \lambda I_{N\times N})\hat{\mathbf{x}} = A^\top \mathbf{b}.$$

Therefore,

$$\begin{aligned}
\hat{\mathbf{x}} &= \lim_{\lambda \to 0^+} (A^\top A + \lambda I_{N\times N})^{-1} A^\top \mathbf{b} \\
&= A^* \mathbf{b}.
\end{aligned}$$

8.7 Energy from Singular Values

8.7.1 Definition: Matrix Energy

The **energy** of a matrix $A \in \mathbb{R}^{M\times N}$ is defined as the squared Frobenius norm:

$$\|A\|_F^2 = \sum_{m=0}^{M-1}\sum_{n=0}^{N-1} a_{mn}^2.$$

This generalizes the concept of vector energy (defined in Chapter~1 as $\|\mathbf{v}\|_2^2$) to matrices. Just as vector energy measures the total squared magnitude of a vector's components, matrix energy measures the total squared magnitude of all matrix entries.

8.7.2 Theorem: Energy from Singular Values

Let $A \in \mathbb{R}^{M\times N}$ have rank $r = \text{rank}(A)$ and the following reduced form of SVD:

$$A = \sum_{i=0}^{r-1} \sigma_i \mathbf{u}_i \mathbf{v}_i^\top.$$

Then the matrix energy (squared Frobenius norm) equals the sum of squared singular values:

$$\|A\|_F^2 = \sum_{i=0}^{r-1} \left\|\sigma_i \mathbf{u}_i \mathbf{v}_i^\top\right\|_F^2 = \sum_{i=0}^{r-1} \sigma_i^2. \tag{8.9}$$

Proof

We have

$$\begin{aligned}
\|A\|_F^2 &= \sum_{m=0}^{M-1}\sum_{n=0}^{N-1} a_{mn}^2 \\
&= \sum_{m=0}^{M-1}\sum_{n=0}^{N-1} \left(\sum_{i=0}^{r-1} \sigma_i u_{im} v_{in}\right)^2 \\
&= \sum_{m=0}^{M-1}\sum_{n=0}^{N-1}\sum_{i=0}^{r-1}\sum_{j=0}^{r-1} \sigma_i u_{im} v_{in} \sigma_j u_{jm} v_{jn} \\
&= \sum_{i=0}^{r-1}\sum_{j=0}^{r-1} \sigma_i \sigma_j \left(\sum_{m=0}^{M-1} u_{im} u_{jm}\right)\left(\sum_{n=0}^{N-1} v_{in} v_{jn}\right) \\
&= \sum_{i=0}^{r-1}\sum_{j=0}^{r-1} \sigma_i \sigma_j \delta_{ij} \delta_{ij} \\
&= \sum_{i=0}^{r-1} \sigma_i^2,
\end{aligned}$$

where the fifth equality follows from the property that $\{\mathbf{u}_i : i \in \{0, 1, \dots, r-1\}\}$ are orthonormal and $\{\mathbf{v}_i : i \in \{0, 1, \dots, r-1\}\}$ are orthonormal.

8.8 Truncated SVD

8.8.1 Definition: Rank-k Truncated SVD

The rank-k truncated SVD is

$$A_k = \sum_{i=0}^{k-1} \sigma_i \mathbf{u}_i \mathbf{v}_i^\top.$$

8.8.2 Theorem: Energy loss from truncated SVD - Truncated SVD Error

$$\frac{\|A - A_k\|_F^2}{\|A\|_F^2} = 1 - \frac{\sum_{i=0}^{k-1} \sigma_i^2}{\sum_{i=0}^{r-1} \sigma_i^2}. \tag{8.10}$$

Proof

The proof is similar to the total energy one. Using the fact that

$$A - A_k = \sum_{i=k}^{r-1} \sigma_i \mathbf{u}_i \mathbf{v}_i^\top,$$

we have

$$\|A - A_k\|_F^2 = \sum_{i=k}^{r-1} \sigma_i^2.$$

Combined with $\|A\|_F^2 = \sum_{i=0}^{r-1} \sigma_i^2$ from the energy theorem, we obtain the desired result. So we omit the detailed proof.

8.8.2.1 Example: Truncated SVD Error

Consider the matrix

$$A = \begin{bmatrix} 3 & 0 \\ 0 & 2 \\ 0 & 0 \end{bmatrix}.$$

1. Compute the SVD $A = U\Sigma V^\top$ and find all singular values σ_0, σ_1.
2. Compute the rank-1 truncated SVD approximation $A_1 = \sigma_0 \mathbf{u}_0 \mathbf{v}_0^\top$.
3. Compute the relative error $\frac{\|A-A_1\|_F^2}{\|A\|_F^2}$ using the formula

$$\frac{\|A - A_1\|_F^2}{\|A\|_F^2} = 1 - \frac{\sigma_0^2}{\sigma_0^2 + \sigma_1^2}.$$

4. Verify your result by directly computing $\|A - A_1\|_F^2$ and $\|A\|_F^2$ from the matrix entries.

Solution

Step 1: Compute SVD

We compute $A^\top A$:

$$A^\top A = \begin{bmatrix} 3 & 0 & 0 \\ 0 & 2 & 0 \end{bmatrix} \begin{bmatrix} 3 & 0 \\ 0 & 2 \\ 0 & 0 \end{bmatrix} = \begin{bmatrix} 9 & 0 \\ 0 & 4 \end{bmatrix}.$$

The eigenvalues of $A^\top A$ are $\lambda_0 = 9$ and $\lambda_1 = 4$, so the singular values are $\sigma_0 = 3$ and $\sigma_1 = 2$.

The eigenvectors of $A^\top A$ are $\mathbf{v}_0 = \begin{bmatrix}1\\0\end{bmatrix}$ (for $\lambda_0 = 9$) and $\mathbf{v}_1 = \begin{bmatrix}0\\1\end{bmatrix}$ (for $\lambda_1 = 4$).

The left singular vectors are:

$$\mathbf{u}_0 = \frac{1}{\sigma_0} A\mathbf{v}_0 = \frac{1}{3}\begin{bmatrix}3\\0\\0\end{bmatrix} = \begin{bmatrix}1\\0\\0\end{bmatrix}, \quad \mathbf{u}_1 = \frac{1}{\sigma_1} A\mathbf{v}_1 = \frac{1}{2}\begin{bmatrix}0\\2\\0\end{bmatrix} = \begin{bmatrix}0\\1\\0\end{bmatrix}.$$

Therefore, the SVD is:

$$A = \begin{bmatrix}1 & 0\\0 & 1\\0 & 0\end{bmatrix}\begin{bmatrix}3 & 0\\0 & 2\end{bmatrix}\begin{bmatrix}1 & 0\\0 & 1\end{bmatrix}^\top = U\Sigma V^\top.$$

Step 2: Rank-1 Truncated SVD

$$A_1 = \sigma_0 \mathbf{u}_0 \mathbf{v}_0^\top = 3\begin{bmatrix}1\\0\\0\end{bmatrix}\begin{bmatrix}1 & 0\end{bmatrix} = \begin{bmatrix}3 & 0\\0 & 0\\0 & 0\end{bmatrix}.$$

Step 3: Relative Error Using Formula

$$\frac{\|A - A_1\|_F^2}{\|A\|_F^2} = 1 - \frac{\sigma_0^2}{\sigma_0^2 + \sigma_1^2} = 1 - \frac{9}{9+4} = 1 - \frac{9}{13} = \frac{4}{13}.$$

Step 4: Direct Verification

$$A - A_1 = \begin{bmatrix}3 & 0\\0 & 2\\0 & 0\end{bmatrix} - \begin{bmatrix}3 & 0\\0 & 0\\0 & 0\end{bmatrix} = \begin{bmatrix}0 & 0\\0 & 2\\0 & 0\end{bmatrix}.$$

$$\|A - A_1\|_F^2 = 0^2 + 0^2 + 0^2 + 2^2 + 0^2 + 0^2 = 4, \quad \|A\|_F^2 = 3^2 + 0^2 + 0^2 + 2^2 + 0^2 + 0^2 = 9 + 4 = 13.$$

Therefore,

$$\frac{\|A - A_1\|_F^2}{\|A\|_F^2} = \frac{4}{13},$$

which matches the result from Step 3.

8.9 Randomized SVD

For large matrices $A \in \mathbb{R}^{M\times N}$ where M, N are very large, computing the full SVD or even a truncated SVD can be computationally expensive (typically $O(MN\min(M,N))$). **Randomized SVD** is a randomized algorithm that computes an approximate rank-k truncated SVD with significantly lower computational cost, especially when $k \ll \min(M,N)$. It is important to emphasize that randomized SVD is not a new model or a new decomposition—the mathematical meaning of SVD remains unchanged: we still seek $A = U\Sigma V^\top$ where U and V are orthonormal and Σ is diagonal. Rather, randomized SVD uses random projections at the algorithmic level to approximate the computation of this same decomposition, making it computationally efficient for large matrices. The output is still a standard SVD decomposition $A \approx U_k\Sigma_k V_k^\top$; only the method of computation differs.

8.9.1 Algorithm: Randomized SVD

Algorithm 9 Randomized SVD

Input: Matrix $A \in \mathbb{R}^{M\times N}$, target rank k, oversampling parameter p (typically $p = 10$)
Output: Rank-k approximation $A \approx U_k\Sigma_k V_k^\top$
Generate random matrix $\Omega \in \mathbb{R}^{N\times(k+p)}$ ▷ Random sampling: columns from standard Gaussian
$Y \leftarrow A\Omega$ ▷ Range finding: project onto $(k+p)$-dimensional subspace
Compute QR decomposition: $Y = QR$ where $Q \in \mathbb{R}^{M\times(k+p)}$ ▷ Orthonormalization
$B \leftarrow Q^\top A$ ▷ Project A onto subspace spanned by Q
Compute SVD: $B = \hat{U}\Sigma_k\hat{V}^\top$ where $\hat{U} \in \mathbb{R}^{(k+p)\times k}$, $\Sigma_k \in \mathbb{R}^{k\times k}$, $\hat{V} \in \mathbb{R}^{N\times k}$
$U_k \leftarrow Q\hat{U}$, $V_k \leftarrow \hat{V}$ ▷ Reconstruction
return U_k, Σ_k, V_k such that $A \approx U_k\Sigma_k V_k^\top$

Optional power iteration: For better accuracy, especially when the singular values decay slowly, we can apply power iterations: replace step 2 with $Y = (AA^\top)^q A\Omega$ for $q \geq 1$ iterations, which helps capture the dominant singular vectors more accurately.

8.9.1.1 Example: Computational Complexity in Recommendation Systems

Consider a recommendation system with a user-item interaction matrix $A \in \mathbb{R}^{10^6\times 10^4}$, where 10^6 users have interacted with 10^4 items (e.g., products, movies, or articles). We wish to compute a rank-k truncated SVD with $k = 100$ to discover latent factors for collaborative filtering:

- **Full SVD**: Requires $O(MN\min(M,N)) = O(10^6 \times 10^4 \times 10^4) = O(10^{14})$ operations, which is computationally prohibitive for real-time applications.
- **Truncated SVD** (iterative methods): Still requires multiple passes over the matrix, roughly $O(MNk) = O(10^6 \times 10^4 \times 100) = O(10^{12})$ operations, which may take hours or days on large datasets.
- **Randomized SVD**:
  - Random sampling: $O(MN(k+p)) = O(10^6 \times 10^4 \times 110) = O(10^{12})$ for matrix multiplication
  - QR on Y: $O(M(k+p)^2) = O(10^6 \times 110^2) = O(10^{10})$
  - SVD on B: $O(N(k+p)^2) = O(10^4 \times 110^2) = O(10^8)$
  - Total: Dominated by $O(MN(k+p))$, but with better constants and often only one pass over the data, making it feasible to compute in minutes rather than hours.

The key advantage is that randomized SVD can work with **streaming data** or **matrices stored in distributed systems** where only matrix-vector products Av are available, without requiring the full matrix to be stored in memory.

Randomized SVD is the standard method for computing approximate SVD of very large matrices in modern machine learning and data science. It is used in **dimensionality reduction** (PCA on large datasets), **recommendation systems** (collaborative filtering with millions of users and items), **image and video processing** (compressing large media files), and **natural language processing** (computing embeddings for large text corpora). The algorithm provides a rank-k approximation with error guarantees: with high probability, the approximation error satisfies $||A - U_k\Sigma_k V_k^\top||_2 \leq (1+\epsilon)\sigma_{k+1}$ for small ϵ, where σ_{k+1} is the $(k+1)$-th singular value. The oversampling parameter p controls the trade-off between accuracy and computational cost: larger p improves accuracy but increases computation. Power iterations ($q \geq 1$) are particularly important when the singular values decay slowly, as they help the algorithm capture the dominant subspace more accurately. Randomized SVD is the default or strongly recommended method in major machine learning libraries: **scikit-learn** uses `sklearn.decomposition.PCA(svd_solver="randomized")` as the default for large datasets, **Spark MLlib** employs randomized SVD for distributed PCA, and **cuML** (RAPIDS machine learning library) defaults to randomized SVD for GPU-accelerated dimensionality reduction. The underlying implementation is available via `sklearn.utils.extmath.randomized_svd` for direct use.

8.10 Latent Representation, Embedding

8.10.1 Remark: Latent Representation, Embedding

Let $A = \begin{bmatrix} \mathbf{a}_0^\top \\ \vdots \\ \mathbf{a}_{M-1}^\top \end{bmatrix} \in \mathbb{R}^{M\times N}$, $A = U\Sigma V^\top$.

1. Right-singular-coordinate representation:

$$AV = U\Sigma.$$

2. Top-k latent representation/embedding:

$$AV_{[:,0:k]} = U_{[:,0:k]}\Sigma_{0:k,0:k} \in \mathbb{R}^{M\times k}.$$

3. Row-wise kernel similarity of any pair of two row vecctors in A:

$$(AV)(AV)^\top = U\Sigma\Sigma^\top U^\top.$$

8.11 Transformation Perspective

Let $A \in \mathbb{R}^{M\times N}$ be with rank r and the following reduced SVD:

$$A = U_r\Sigma_r V_r^\top.$$

Then for any $\mathbf{x} \in \mathbb{R}^{N\times 1}$, $A\mathbf{x}$ can be interpreted as a linear transformation mapping $\mathbf{x} \in \text{span}(V_r)$ in $\mathbb{R}^N$ to $\text{span}(U_r)$ in $\mathbb{R}^M$.

Specifically,

$$A\mathbf{x} = U_r\Sigma_r(V_r^\top \mathbf{x}),$$

which decomposes into: (1) projection $V_r^\top\mathbf{x}$ onto $\text{span}(V_r)$ (the row space of A), (2) scaling by singular values Σ_r, and (3) mapping to $\text{span}(U_r)$ (the column space of A) via U_r.

The transformation A acts as a rank-r linear map, compressing $\mathbb{R}^N$ to an r-dimensional intermediate space, then expanding to $\mathbb{R}^M$.

8.12 Mindmap

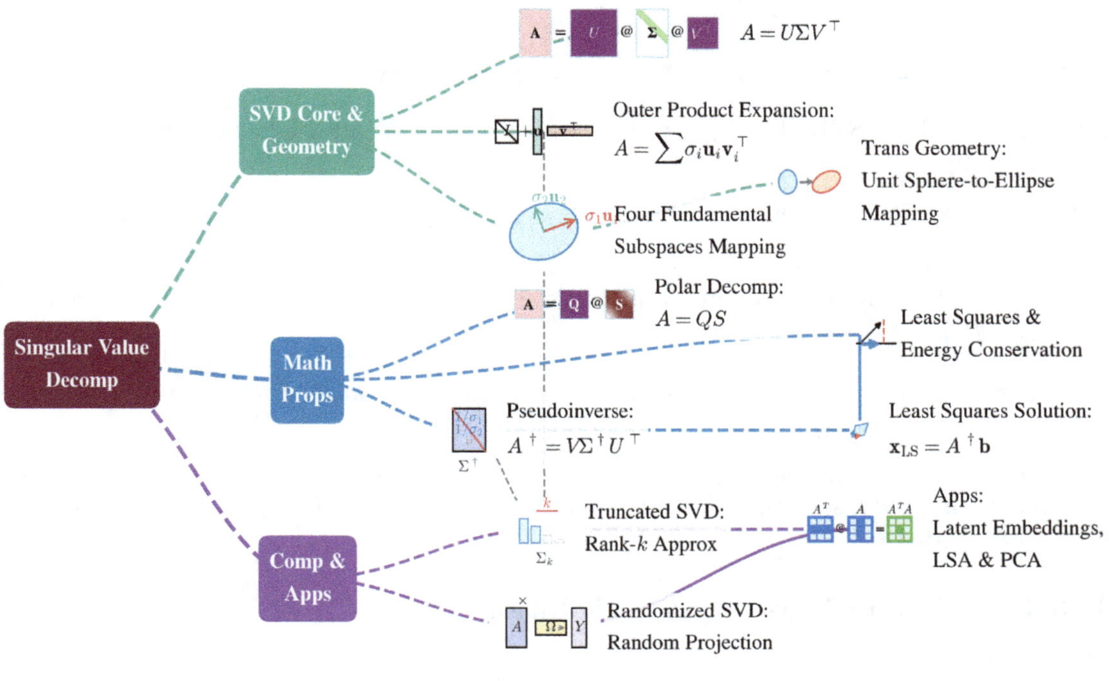

Figure 8.5: Mind map of Singular Value Decomposition topics

8.13 Exercises

8.13.1 Computing $A^\top A$ and $AA^\top$ from SVD

Given a matrix $A \in \mathbb{R}^{M\times N}$ with SVD $A = \sum_{i=0}^{r-1} \sigma_i \mathbf{u}_i \mathbf{v}_i^\top$, where $\{\mathbf{u}_i\}$ and $\{\mathbf{v}_i\}$ are orthonormal sets:

1. Prove that $A^\top A = \sum_{i=0}^{r-1} \sigma_i^2 \mathbf{v}_i \mathbf{v}_i^\top$.
2. Prove that $AA^\top = \sum_{i=0}^{r-1} \sigma_i^2 \mathbf{u}_i \mathbf{u}_i^\top$.
3. Using the results from parts (1) and (2), prove that each column $\mathbf{v}_i$ of V is an eigenvector of $A^\top A$ with eigenvalue σ_i^2.
4. Prove that each column $\mathbf{u}_i$ of U is an eigenvector of $AA^\top$ with eigenvalue σ_i^2.

8.13.2 Properties of Singular Values and Matrix Norms

Let $A \in \mathbb{R}^{M\times N}$ have SVD $A = U\Sigma V^\top$ with singular values $\sigma_0 \geq \sigma_1 \geq \cdots \geq \sigma_{r-1} > 0$ where $r = \text{rank}(A)$.

1. Prove that $\|A\|_F^2 = \sum_{i=0}^{r-1} \sigma_i^2$ using the fact that $\|A\|_F^2 = \text{tr}(A^\top A)$ and the spectral decomposition of $A^\top A$.
2. Prove that $\|A\|_2 = \sigma_0$ (the largest singular value) by showing that $\|A\|_2 = \max_{\|\mathbf{x}\|_2=1} \|A\mathbf{x}\|_2 = \sigma_0$.

3. For a square matrix $A \in \mathbb{R}^{N \times N}$ with SVD $A = U\Sigma V^\top$, prove that $|\det(A)| = \prod_{i=0}^{N-1} \sigma_i$ if A is invertible. (Hint: Use the fact that $\det(A) = \det(U)\det(\Sigma)\det(V^\top)$ and the orthonormality of U and V.)

8.13.3 Rank-k Truncated SVD and Optimality

Let $A \in \mathbb{R}^{M \times N}$ have SVD $A = \sum_{i=0}^{r-1} \sigma_i \mathbf{u}_i \mathbf{v}_i^\top$ with $r = \text{rank}(A)$ and singular values $\sigma_0 \geq \sigma_1 \geq \cdots \geq \sigma_{r-1} > 0$.

1. For $k < r$, define the rank-k truncated SVD as $A_k = \sum_{i=0}^{k-1} \sigma_i \mathbf{u}_i \mathbf{v}_i^\top$. Prove that $\text{rank}(A_k) = k$.
2. Prove that A_k minimizes $\|A - B\|_F$ over all matrices $B \in \mathbb{R}^{M \times N}$ with $\text{rank}(B) \leq k$. (This is the Eckart–Young theorem. Hint: Use the fact that for any rank-k matrix B, we have $\|A - B\|_F^2 \geq \sum_{i=k}^{r-1} \sigma_i^2$.)
3. Express the relative error $\frac{\|A - A_k\|_F^2}{\|A\|_F^2}$ in terms of the singular values. Prove your formula using the energy theorem.
4. Given a matrix with singular values $\sigma_0 = 10$, $\sigma_1 = 5$, $\sigma_2 = 1$, $\sigma_3 = 0.1$, determine the minimum rank k such that the relative error is less than 5%. Compute the exact relative error for this k.

8.13.4 SVD and Least Squares Solution

For the matrix

$$A = \begin{bmatrix} 1 & 1 \\ 0 & 1 \\ 1 & 0 \end{bmatrix}, \quad \mathbf{b} = \begin{bmatrix} 1 \\ 1 \\ 1 \end{bmatrix},$$

1. Compute the SVD $A = U\Sigma V^\top$ explicitly. Find all singular values and the corresponding left and right singular vectors.
2. Using the SVD from part (1), compute the Moore–Penrose pseudoinverse $A^* = V\Sigma^{-1}U^\top$ where Σ^{-1} has entries $1/\sigma_i$ for non-zero singular values.
3. Find the least squares solution $\hat{\mathbf{x}} = A^*\mathbf{b}$ to $A\mathbf{x} = \mathbf{b}$ and verify that it minimizes $\|A\mathbf{x} - \mathbf{b}\|_2^2$.

8.13.5 Moore-Penrose Pseudoinverse Properties

Let $A \in \mathbb{R}^{M \times N}$ have SVD $A = U\Sigma V^\top$ with singular values $\sigma_0 \geq \sigma_1 \geq \cdots \geq \sigma_{r-1} > 0$ and $\sigma_i = 0$ for $i \geq r$, where $r = \text{rank}(A)$.

1. Define the Moore-Penrose pseudoinverse as $A^+ = V\Sigma^+U^\top$, where Σ^+ is a diagonal matrix with entries $(\Sigma^+)_{ii} = 1/\sigma_i$ for $i < r$ and $(\Sigma^+)_{ii} = 0$ for $i \geq r$. Prove that this definition is well-defined and unique.
2. Using the SVD form, prove the four Penrose conditions:
   - $AA^+A = A$
   - $A^+AA^+ = A^+$
   - $(AA^+)^\top = AA^+$
   - $(A^+A)^\top = A^+A$
3. If A is an invertible square matrix, prove that $A^+ = A^{-1}$.
4. For the least squares problem $\min_{\mathbf{x} \in \mathbb{R}^N} \|A\mathbf{x} - \mathbf{b}\|_2^2$, prove that $\mathbf{x} = A^+\mathbf{b}$ is a solution. Show that if $\text{rank}(A) = N$, then this solution is unique.

8.13.6 Image Compression with Truncated SVD

Consider a 64×64 grayscale image represented as a matrix $A \in \mathbb{R}^{64\times 64}$ with singular values $\sigma_0 = 100$, $\sigma_1 = 80$, $\sigma_2 = 60$, $\sigma_3 = 40$, $\sigma_4 = 20$, and $\sigma_i = 0.1$ for $i \geq 5$ (representing noise).

1. Compute the SVD $A = U\Sigma V^\top$ and determine the rank r of A. For a rank-k truncated SVD approximation $A_k = \sum_{i=0}^{k-1} \sigma_i \mathbf{u}_i \mathbf{v}_i^\top$, prove that the storage requirement is $k(64 + 64 + 1) = 129k$ elements compared to 4096 for the original matrix.

2. Compute the compression ratio $\frac{129k}{4096}$ for $k = 5, 10, 20$. Determine the minimum rank k needed to achieve a compression ratio of at least 10:1 (i.e., $\frac{4096}{129k} \geq 10$).

3. Using the energy loss formula, determine the minimum rank k needed to preserve at least 90% of the image energy (i.e., $\frac{\|A_k\|_F^2}{\|A\|_F^2} \geq 0.9$). Express your answer in terms of the singular values and compute it explicitly.

4. For $k = 4$, compute the relative error $\frac{\|A-A_4\|_F^2}{\|A\|_F^2}$ and explain the trade-off between compression and reconstruction quality.

8.13.7 Low-Rank Adaptation (LoRA) with Truncated SVD

In fine-tuning large language models, LoRA approximates weight updates using low-rank matrices. Consider a weight update matrix $\Delta W \in \mathbb{R}^{4096\times 4096}$ with SVD $\Delta W = U\Sigma V^\top$ and singular values $\sigma_0 = 50$, $\sigma_1 = 30$, $\sigma_2 = 20$, $\sigma_3 = 10$, $\sigma_4 = 5$, and $\sigma_i < 1$ for $i \geq 5$.

1. Prove that the rank-k approximation $\Delta W_k = \sum_{i=0}^{k-1} \sigma_i \mathbf{u}_i \mathbf{v}_i^\top$ can be factorized as $\Delta W_k = BA$ where $B \in \mathbb{R}^{4096\times k}$ and $A \in \mathbb{R}^{k\times 4096}$. Find explicit expressions for B and A in terms of the SVD components.

2. Prove that the number of trainable parameters is reduced from $4096^2 = 16,777,216$ (for full ΔW) to $2 \cdot 4096 \cdot k = 8192k$ (for the factorized form BA). Compute the parameter reduction ratio for $k = 8$ and $k = 16$.

3. If we want to preserve at least 95% of the update energy (i.e., $\frac{\|\Delta W_k\|_F^2}{\|\Delta W\|_F^2} \geq 0.95$), determine the minimum rank k required. Express your answer in terms of the singular values and compute it explicitly.

4. Using the energy loss formula, prove that for $k = 8$, the relative error is less than 5% if $\sum_{i=8}^{r-1} \sigma_i^2 < 0.05 \sum_{i=0}^{r-1} \sigma_i^2$. Verify this condition for the given singular values.

8.13.8 Noise Removal with Truncated SVD

A noisy signal matrix $A \in \mathbb{R}^{100\times 100}$ can be decomposed as $A = A_{\text{signal}} + A_{\text{noise}}$, where A_{signal} has rank $r = 5$ and A_{noise} represents noise. The SVD of A yields singular values $\sigma_0 = 100$, $\sigma_1 = 50$, $\sigma_2 = 10$, $\sigma_3 = 5$, $\sigma_4 = 2$, $\sigma_5 = 0.5$, $\sigma_6 = 0.1$, and $\sigma_i < 0.05$ for $i \geq 7$.

1. Prove that if the noise components correspond to small singular values (i.e., $\sigma_i \ll \sigma_0$ for $i \geq r$), then the rank-k truncated SVD $A_k = \sum_{i=0}^{k-1} \sigma_i \mathbf{u}_i \mathbf{v}_i^\top$ (where $k \leq r$) filters out noise while preserving the signal. (Hint: Show that $\|A_k - A_{\text{signal}}\|_F^2$ is small when $k = r$.)

2. Determine an appropriate rank k for noise removal. Justify your choice by computing the energy distribution: $\frac{\sum_{i=0}^{k-1} \sigma_i^2}{\sum_{i=0}^{r-1} \sigma_i^2}$ for $k = 3, 4, 5, 6$.

3. Prove that the signal-to-noise ratio (SNR) improvement can be quantified as $\frac{\|A_k\|_F^2}{\|A-A_k\|_F^2}$ when A_k captures the signal and $A - A_k$ contains mostly noise. Compute the SNR improvement for $k = 5$ using the given singular values.

4. For $k = 5$, compute the relative error $\frac{\|A-A_5\|_F^2}{\|A\|_F^2}$ and analyze the trade-off between noise removal and signal preservation.

8.13.9 Newton-Schulz Iteration for Polar Decomposition

The **Newton-Schulz iteration** is a numerical method to compute the orthogonal factor Q of the polar decomposition $A = QS$ without performing SVD. The iteration is:

$$Q_{k+1} = \frac{1}{2} Q_k (3I - Q_k^\top Q_k)$$

1. **Fixed point analysis (scalar case):** Consider the scalar iteration $x_{k+1} = \frac{1}{2} x_k (3 - x_k^2)$.
   - Prove that $x = 1$ is a fixed point (i.e., if $x_k = 1$, then $x_{k+1} = 1$).
   - For $x_0 = 0.5$, compute x_1 and x_2 by hand. Prove that the sequence converges to 1.
   - For $x_0 = 1.5$, compute x_1 and x_2. Determine whether the iteration converges or diverges.
2. **Singular value perspective:** Let $G \in \mathbb{R}^{n \times n}$ be a square matrix with SVD $G = U_G \Sigma_G V_G^\top$ with singular values $\sigma_0, \sigma_1, \dots, \sigma_{r-1}$.
   - Prove that if we apply the Newton-Schulz iteration to G, each iteration acts on the singular values independently. (Hint: Use the fact that $G = U_G \Sigma_G V_G^\top$ and analyze $G^\top G$.)
   - Prove that if G has singular values σ_i, then after one iteration, the singular values become $\frac{1}{2}\sigma_i(3 - \sigma_i^2)$.
3. **Convergence condition and spectral normalization:**
   - For the scalar iteration $x_{k+1} = \frac{1}{2} x_k (3 - x_k^2)$, find the exact range of initial values x_0 for which the iteration converges to 1. (Hint: Consider when $|x_{k+1} - 1| < |x_k - 1|$ and solve the inequality.)
   - In practice, **spectral normalization** is applied before Newton-Schulz iteration: given a matrix G with largest singular value $\sigma_{\max}$, we normalize it as $\tilde{G} = G/\sigma_{\max}$. Prove that this normalization is necessary for convergence. What happens if $\sigma_{\max} > \sqrt{3}$? Prove your answer.
4. **Derivation using Taylor expansion:** The Newton-Schulz iteration can be derived using Taylor expansion. Our goal is to compute the orthogonal factor Q of the polar decomposition $G = QS$, which can be written as $Q = G(G^\top G)^{-1/2}$.
   - Suppose Q_k is our current estimate. Define the error term $E_k = I - Q_k^\top Q_k$. If Q_k is close to orthogonal, then E_k is small. Prove that $Q_k^\top Q_k = I - E_k$.
   - The next iteration should be $Q_{k+1} = Q_k (Q_k^\top Q_k)^{-1/2} = Q_k (I - E_k)^{-1/2}$. Use the Taylor expansion $(I - E)^{-1/2} = I + \frac{1}{2}E + \frac{3}{8}E^2 + O(E^3)$ (valid for small E) to approximate this.
   - Keeping terms up to first order in E_k (i.e., $E_k^2 \approx 0$), prove that $(I - E_k)^{-1/2} \approx \frac{3}{2}I - \frac{1}{2}Q_k^\top Q_k$.
   - Substitute this approximation into $Q_{k+1} = Q_k (I - E_k)^{-1/2}$ and prove that you obtain the Newton-Schulz iteration: $Q_{k+1} = \frac{1}{2} Q_k (3I - Q_k^\top Q_k)$.
5. **Connection to polar decomposition:** Given a matrix G with polar decomposition $G = QS$, prove that the Newton-Schulz iteration, when applied to a properly normalized G (i.e., $\tilde{G} = G/\sigma_{\max}$), converges to the orthogonal factor Q of the polar decomposition of $\tilde{G}$.

8.13.10 Low-Rank Adaptation (LoRA) with Truncated SVD

In fine-tuning large language models, LoRA approximates weight updates using low-rank matrices. Consider a weight matrix $W \in \mathbb{R}^{d \times d}$ and its update ΔW.

1. Show that if ΔW has SVD $\Delta W = U \Sigma V^\top$ with singular values $\sigma_0 \geq \sigma_1 \geq \dots \geq \sigma_{r-1}$, then the rank-$k$ approximation $\Delta W_k = \sum_{i=0}^{k-1} \sigma_i \mathbf{u}_i \mathbf{v}_i^\top$ can be factorized as $\Delta W_k = BA$ where

$B \in \mathbb{R}^{d \times k}$ and $A \in \mathbb{R}^{k \times d}$.

2. Prove that the number of trainable parameters is reduced from d^2 (for full ΔW) to $2kd$ (for the factorized form BA). Compute the parameter reduction ratio when $k = 8$ and $d = 4096$.

3. If we want to preserve at least 95% of the update energy (i.e., $\frac{\|\Delta W_k\|_F^2}{\|\Delta W\|_F^2} \geq 0.95$), determine the minimum rank k required. Express your answer in terms of the singular values.

4. In practice, LoRA often uses $k = 8$ or $k = 16$. Using the energy loss formula, explain why these small ranks are often sufficient for effective fine-tuning.

8.13.11 Noise Removal with Truncated SVD

A noisy signal matrix $A \in \mathbb{R}^{M \times N}$ can be decomposed as $A = A_{\text{signal}} + A_{\text{noise}}$, where A_{signal} has rank r and A_{noise} represents noise.

1. Prove that if the noise components correspond to small singular values, then the rank-k truncated SVD $A_k = \sum_{i=0}^{k-1} \sigma_i \mathbf{u}_i \mathbf{v}_i^\top$ (where $k \leq r$) filters out noise while preserving the signal.

2. Given a matrix A with singular values $\sigma_0 = 100$, $\sigma_1 = 50$, $\sigma_2 = 10$, $\sigma_3 = 0.5$, $\sigma_4 = 0.1$, determine an appropriate rank k for noise removal. Justify your choice using the energy distribution.

3. Show that the signal-to-noise ratio (SNR) improvement can be quantified as $\frac{\|A_k\|_F^2}{\|A - A_k\|_F^2}$ when A_k captures the signal and $A - A_k$ contains mostly noise.

4. For a specific noisy matrix, compute the rank-k approximation for different values of k and analyze the trade-off between noise removal and signal preservation using the energy loss formula.

Chapter 9

Matrix Decompositions and Beyond

Specialized decompositions: LDL, rank-revealing QR, interpolative decomposition, CUR, NMF, LQ, Schur, low-rank factorization (LoRA), and tensor decompositions

"The purpose of computing is insight, not numbers."

— Richard Hamming

- ▷ `LDL` $A = L_1 D L_1^\top$ for symmetric A; if $A \succ 0$ then $A = (L_1\sqrt{D})(L_1\sqrt{D})^\top$
- ▷ `LDL to Cholesky` If $A = L_1 D L_1^\top \succ 0$ then $A = (L_1\sqrt{D})(L_1\sqrt{D})^\top$
- ▷ `RRQR` $A = PQR$, P is permutation, $Q \in \mathbb{R}^{M\times r}$ o.n., R_{11} well-conditioned, $\|R_{22}\|$ small
- ▷ `Interpolative Decomposition` $A = A(:,\mathcal{J})C$, $\mathcal{J}$ is r column indices, $C \in \mathbb{R}^{r\times N}$
- ▷ `CUR` $A = A(:,\mathcal{C})\cdot A(\mathcal{R},\mathcal{C})^{-1}\cdot A(\mathcal{R},:)$, $\mathcal{R},\mathcal{C}$ index sets of size r, $A(\mathcal{R},\mathcal{C}) \in \mathbb{R}^{r\times r}$ invertible
- ▷ `CUR form` $A = CUR$, $C = A(:,\mathcal{C})$, $U = A(\mathcal{R},\mathcal{C})^{-1}$, $R = A(\mathcal{R},:)$
- ▷ `NMF` $A \approx WH$ for $A \in \mathbb{R}_{\geq 0}^{M\times N}$, $W \in \mathbb{R}_{\geq 0}^{M\times r}$, $H \in \mathbb{R}_{\geq 0}^{r\times N}$, minimizing $\|A - WH\|_F^2$
- ▷ `NMF updates` $H \leftarrow H \odot \frac{W^\top A}{W^\top WH}$, $W \leftarrow W \odot \frac{AH^\top}{WHH^\top}$, $\odot$ is element-wise
- ▷ `LQ` $A = LQ$, $L \in \mathbb{R}^{M\times r}$ is lower triangular, $Q \in \mathbb{R}^{r\times N}$ has orthonormal rows ($QQ^\top = I$)
- ▷ `LQ from QR` If $A^\top = Q_R R$ then $A = LQ$, $L = R^\top$, $Q = Q_R^\top$
- ▷ `Schur` $A = QUQ^\top$, Q is orthonormal, U is upper triangular
- ▷ `Schur for normal` If $AA^\top = A^\top A$ then U is diagonal and $A = Q\Lambda Q^\top$ (spectral)
- ▷ `QR algorithm` $A_0 = A$, $A_{k+1} = R_k Q_k$, $A_k = Q_k R_k$; A_k converges to Schur form
- ▷ `Rank from RRQR` $\text{rank}(A) = r$, r is size of well-conditioned R_{11} block
- ▷ `ID selection` Skeleton columns $\mathcal{J}$ selected so $A(:,\mathcal{J})$ has full column rank r and is well-conditioned
- ▷ `CUR selection` Skeleton $A(\mathcal{R},\mathcal{C})$ must be invertible and well-conditioned
- ▷ `LQ row space` Rows of Q in $A = LQ$ form o.n. basis for row space; $\text{rank}(A) = \text{rank}(Q)$
- ▷ `Schur eigenvalues` Diagonal entries of U in $A = QUQ^\top$ are eigenvalues of A
- ▷ `LoRA` $\mathbf{y} = W_0\mathbf{x} + \frac{\alpha}{r}BA\mathbf{x}$, $\Delta W = \frac{\alpha}{r}BA$, $\text{rank}(\Delta W) \leq r$; train $r(m+n)$ params vs mn for full fine-tuning
- ▷ `LoRA merge` $W_{\text{merged}} = W_0 + \frac{\alpha}{r}BA$ at inference (no extra latency)
- ▷ `Tensor Train` $\mathcal{A}(i_1,\ldots,i_d) = \sum_{r_1,\ldots,r_{d-1}} \mathcal{G}_1(i_1,r_1)\cdots\mathcal{G}_d(r_{d-1},i_d)$; $O(dnR^2)$ vs $O(n^d)$

We have explored various matrix decompositions in previous chapters: **QR decomposition** in Chapter~4, **LU, PLU, and LDU decompositions** in Chapter~5, **Spectral decomposition** and **Cholesky decomposition** in Chapter~7, and **SVD** and **Polar decomposition** in Chapter~8. This chapter covers additional specialized decompositions that are particularly useful in machine learning and numerical computing.

9.1 LDL Decomposition

9.1.1 Theorem: LDL Decomposition

For a **symmetric** matrix $A \in \mathbb{R}^{N\times N}$ that admits LU decomposition, there exists a decomposition

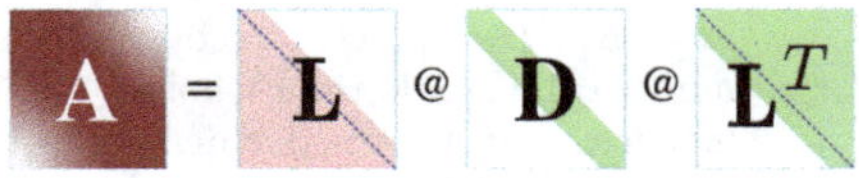

$$A = L_1 D L_1^\top \tag{9.1}$$

where $L_1 \in \mathbb{R}^{N\times N}$ is a lower triangular matrix with unit diagonal, and $D \in \mathbb{R}^{N\times N}$ is a diagonal matrix. We use the notation L_1 (with subscript 1) to emphasize that it has unit diagonal, distinguishing it from L used in other decompositions (e.g., Cholesky) where the diagonal may not be 1.

This is the symmetric version of LDU decomposition, where the symmetry of A implies $U = L_1^\top$.

9.1.1.1 Example

Given the symmetric matrix

$$A = \begin{bmatrix} 4 & 2 \\ 2 & 3 \end{bmatrix},$$

find the LDL decomposition $A = L_1 D L_1^\top$.

Solution:

Since A is symmetric, we can use the fact that $U = L^\top$ from LDU decomposition.

1. **LDU decomposition**: First, perform LU decomposition:
   - Eliminate the $(2,1)$ entry: subtract $\frac{1}{2}\times$ row 1 from row 2:
   $$L = \begin{bmatrix} 1 & 0 \\ \frac{1}{2} & 1 \end{bmatrix}, \quad U = \begin{bmatrix} 4 & 2 \\ 0 & 2 \end{bmatrix}.$$
   - Extract diagonal: $D = \begin{bmatrix} 4 & 0 \\ 0 & 2 \end{bmatrix}$.
   - Form unit upper triangular: $U' = \begin{bmatrix} 1 & \frac{1}{2} \\ 0 & 1 \end{bmatrix}$.
   - So $A = LDU'$ where $L = \begin{bmatrix} 1 & 0 \\ \frac{1}{2} & 1 \end{bmatrix}$, $D = \begin{bmatrix} 4 & 0 \\ 0 & 2 \end{bmatrix}$, $U' = \begin{bmatrix} 1 & \frac{1}{2} \\ 0 & 1 \end{bmatrix}$.
2. **Use symmetry**: Since A is symmetric, we have $A = A^\top = (LDU')^\top = (U')^\top D^\top L^\top = (U')^\top D L^\top$. But also $A = LDU'$, so by uniqueness (up to scaling), we must have $(U')^\top = L$, i.e., $U' = L^\top$.
3. **Final LDL decomposition**:
   $$L_1 = \begin{bmatrix} 1 & 0 \\ \frac{1}{2} & 1 \end{bmatrix}, \quad D = \begin{bmatrix} 4 & 0 \\ 0 & 2 \end{bmatrix}, \quad L_1^\top = \begin{bmatrix} 1 & \frac{1}{2} \\ 0 & 1 \end{bmatrix}.$$

Verification:

$$L_1 D L_1^\top = \begin{bmatrix} 1 & 0 \\ \frac{1}{2} & 1 \end{bmatrix} \begin{bmatrix} 4 & 0 \\ 0 & 2 \end{bmatrix} \begin{bmatrix} 1 & \frac{1}{2} \\ 0 & 1 \end{bmatrix} = \begin{bmatrix} 1 & 0 \\ \frac{1}{2} & 1 \end{bmatrix} \begin{bmatrix} 4 & 2 \\ 0 & 2 \end{bmatrix} = \begin{bmatrix} 4 & 2 \\ 2 & 3 \end{bmatrix} = A.$$

Note that L_1 has unit diagonal, and D contains the scaling factors. Since all diagonal entries of D are positive, A is positive definite, and we could also compute the Cholesky decomposition $A = (L_1\sqrt{D})(L_1\sqrt{D})^\top$.

> LDL decomposition is a generalization of Cholesky decomposition that does not require positive definiteness. If A is positive definite, then all diagonal entries of D are positive, and we can write $A = (L_1\sqrt{D})(L_1\sqrt{D})^\top$, recovering the Cholesky decomposition. LDL decomposition is more numerically stable than Cholesky for matrices that are symmetric but not positive definite, or for matrices that are positive definite but near singular. In **machine learning**, LDL decomposition is used in **second-order optimization methods** (e.g., quasi-Newton methods) where the Hessian approximation may not be positive definite, in **Gaussian process regression** for covariance matrices that may be ill-conditioned, and in **Kalman filtering** for state estimation. The diagonal matrix D reveals the eigenvalues of A (when A is diagonalizable), making LDL useful for **eigenvalue analysis** and **conditioning assessment**.

9.2 Rank-Revealing QR (RRQR)

9.2.1 Theorem: Rank-Revealing QR Decomposition

For a matrix $A \in \mathbb{R}^{M\times N}$ with $\text{rank}(A) = r$, there exists a permutation matrix $P \in \mathbb{R}^{M\times M}$ such that

$$A = PQR = PQ \begin{bmatrix} R_{11} & R_{12} \\ 0 & R_{22} \end{bmatrix} \tag{9.2}$$

where $P \in \mathbb{R}^{M\times M}$ is a permutation matrix (reordering rows), $Q \in \mathbb{R}^{M\times r}$ has orthonormal columns, $R_{11} \in \mathbb{R}^{r\times r}$ is upper triangular and well-conditioned, and $||R_{22}||$ is small (indicating that the last $N-r$ columns are nearly linearly dependent on the first r columns).

9.2.1.1 Example

Given the matrix

$$A = \begin{bmatrix} 1 & 2 & 3 \\ 2 & 4 & 6 \\ 1 & 1 & 2 \end{bmatrix},$$

find the Rank-Revealing QR decomposition $A = PQR$.

Solution:

Note that the third column is a linear combination of the first two: $\begin{bmatrix} 3 \\ 6 \\ 2 \end{bmatrix} = \begin{bmatrix} 1 \\ 2 \\ 1 \end{bmatrix} + 2 \begin{bmatrix} 1 \\ 1 \\ 0.5 \end{bmatrix}$ (approximately). The rank is $r = 2$.

1. **Perform QR decomposition with column pivoting:**

   - Start with A and identify the column with largest norm (column 2: $|| \begin{bmatrix} 2 \\ 4 \\ 1 \end{bmatrix} ||_2 = \sqrt{21}$).

- Permute columns to bring the most important ones first (if needed).
- Apply Householder reflections or Gram-Schmidt to get Q and R.

2. **Result** (after row permutation for numerical stability):

$$P = \begin{bmatrix} 0 & 1 & 0 \\ 1 & 0 & 0 \\ 0 & 0 & 1 \end{bmatrix}, \quad Q = \begin{bmatrix} -0.408 & -0.707 \\ -0.816 & 0.707 \\ -0.408 & 0 \end{bmatrix}, \quad R = \begin{bmatrix} -2.449 & -4.899 & -7.348 \\ 0 & 0.707 & 1.414 \\ 0 & 0 & 0 \end{bmatrix}.$$

3. **Block structure**:

$$R = \begin{bmatrix} R_{11} & R_{12} \\ 0 & R_{22} \end{bmatrix} = \begin{bmatrix} -2.449 & -4.899 & -7.348 \\ 0 & 0.707 & 1.414 \\ 0 & 0 & 0 \end{bmatrix},$$

where $R_{11} = \begin{bmatrix} -2.449 & -4.899 \\ 0 & 0.707 \end{bmatrix}$ is upper triangular and well-conditioned, $R_{12} = \begin{bmatrix} -7.348 \\ 1.414 \end{bmatrix}$, and $R_{22} = \begin{bmatrix} 0 \end{bmatrix}$ (a 1×1 block with small norm, indicating the third column is linearly dependent).

4. **Verification**:

$$PQR = \begin{bmatrix} 0 & 1 & 0 \\ 1 & 0 & 0 \\ 0 & 0 & 1 \end{bmatrix} \begin{bmatrix} -0.408 & -0.707 \\ -0.816 & 0.707 \\ -0.408 & 0 \end{bmatrix} \begin{bmatrix} -2.449 & -4.899 & -7.348 \\ 0 & 0.707 & 1.414 \\ 0 & 0 & 0 \end{bmatrix} \approx \begin{bmatrix} 1 & 2 & 3 \\ 2 & 4 & 6 \\ 1 & 1 & 2 \end{bmatrix} = A.$$

The small R_{22} block confirms that the matrix has rank $r = 2$, and the first two columns form a well-conditioned basis for the column space.

> Rank-Revealing QR (RRQR) explicitly identifies the rank of a matrix and provides a well-conditioned basis for the column space. This is crucial in **feature selection** (identifying the most important features), **dimensionality reduction** (determining the intrinsic dimensionality of data), and **low-rank approximation** (finding the best rank-r approximation). In machine learning, RRQR is used in **subspace learning**, **anomaly detection** (identifying low-rank structure), and **matrix completion** problems. The permutation P reorders rows to ensure numerical stability, while the structure of R reveals which columns (features) are most important, enabling interpretable feature selection and efficient low-rank representations.

9.3 Interpolative Decomposition (ID)

9.3.1 Theorem: Interpolative Decomposition

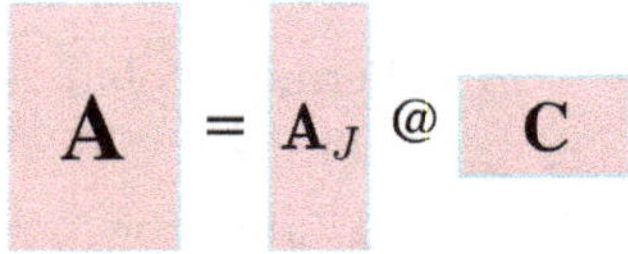

For a matrix $A \in \mathbb{R}^{M \times N}$ with $\text{rank}(A) = r$, there exists a decomposition

$$A = A(:, \mathcal{J})C \tag{9.3}$$

where $\mathcal{J} \subset \{0, 1, \ldots, N-1\}$ is a set of r column indices (the "skeleton" columns), $A(:, \mathcal{J}) \in \mathbb{R}^{M \times r}$ consists of r selected columns of A, and $C \in \mathbb{R}^{r \times N}$ is an interpolation matrix with $||C|| \leq \sqrt{r(N-r)+1}$.

9.3.1.1 Example

Given the matrix

$$A = \begin{bmatrix} 1 & 2 & 3 & 4 \\ 2 & 4 & 6 & 8 \\ 1 & 1 & 2 & 3 \end{bmatrix},$$

find the Interpolative Decomposition $A = A(:, \mathcal{J})C$.

Solution:

Note that the rank of A is $r = 2$ (columns 3 and 4 are linear combinations of columns 1 and 2).

1. **Select skeleton columns**: Using column pivoting or other selection methods, we identify the most important columns. Here, columns 1 and 2 form a well-conditioned basis:

$$\mathcal{J} = \{0, 1\}, \quad A(:, \mathcal{J}) = \begin{bmatrix} 1 & 2 \\ 2 & 4 \\ 1 & 1 \end{bmatrix}.$$

2. **Construct interpolation matrix** C: For each column j of A, find coefficients c_j such that $A(:, j) = A(:, \mathcal{J})c_j$:

   - Column 0: $A(:, 0) = \begin{bmatrix} 1 \\ 2 \\ 1 \end{bmatrix} = A(:, \mathcal{J}) \begin{bmatrix} 1 \\ 0 \end{bmatrix}$
   - Column 1: $A(:, 1) = \begin{bmatrix} 2 \\ 4 \\ 1 \end{bmatrix} = A(:, \mathcal{J}) \begin{bmatrix} 0 \\ 1 \end{bmatrix}$
   - Column 2: $A(:, 2) = \begin{bmatrix} 3 \\ 6 \\ 2 \end{bmatrix} = A(:, \mathcal{J}) \begin{bmatrix} 1 \\ 1 \end{bmatrix}$ (since column 2 = column 0 + column 1)
   - Column 3: $A(:, 3) = \begin{bmatrix} 4 \\ 8 \\ 3 \end{bmatrix} = A(:, \mathcal{J}) \begin{bmatrix} 1 \\ 1.5 \end{bmatrix}$ (approximately, since column 3 ≈ column 0 + 1.5 × column 1)

3. **Final decomposition**:

$$C = \begin{bmatrix} 1 & 0 & 1 & 1 \\ 0 & 1 & 1 & 1.5 \end{bmatrix}.$$

4. **Verification**:

$$A(:, \mathcal{J})C = \begin{bmatrix} 1 & 2 \\ 2 & 4 \\ 1 & 1 \end{bmatrix} \begin{bmatrix} 1 & 0 & 1 & 1 \\ 0 & 1 & 1 & 1.5 \end{bmatrix} = \begin{bmatrix} 1 & 2 & 3 & 4 \\ 2 & 4 & 6 & 8 \\ 1 & 1 & 2 & 2.5 \end{bmatrix} \approx A.$$

The skeleton columns $\mathcal{J} = \{0, 1\}$ are actual columns from A, making this decomposition interpretable: we can express all columns of A as linear combinations of these two representative columns.

Interpolative Decomposition (ID) provides a **column-skeleton** representation where the matrix is expressed in terms of a small subset of its columns. This is highly interpretable: the selected columns (skeleton columns) are actual data points or features from the original matrix.

> In machine learning, ID is used for **data summarization** (selecting representative samples), **feature selection** (identifying important features), **matrix compression** (storing only r columns instead of N), and **interpretable dimensionality reduction**. Unlike SVD which produces abstract principal components, ID produces interpretable representations using actual columns of the data matrix, making it valuable for **explainable AI** and **data analysis**.

9.4 Skeleton Decomposition (Cross Approximation / CUR)

9.4.1 Theorem: Skeleton Decomposition

For a matrix $A \in \mathbb{R}^{M \times N}$ with $\text{rank}(A) = r$, there exist index sets $\mathcal{R} \subset \{0, 1, \dots, M-1\}$ and $\mathcal{C} \subset \{0, 1, \dots, N-1\}$ of size r such that

$$A = A(:, \mathcal{C}) \cdot A(\mathcal{R}, \mathcal{C})^{-1} \cdot A(\mathcal{R}, :) \tag{9.4}$$

which can also be written as the **CUR decomposition**

$$A = CUR \tag{9.5}$$

where $C = A(:, \mathcal{C})$ (selected columns), $U = A(\mathcal{R}, \mathcal{C})^{-1}$ (inverse of the skeleton), and $R = A(\mathcal{R}, :)$ (selected rows). Here, $A(\mathcal{R}, \mathcal{C}) \in \mathbb{R}^{r \times r}$ is the submatrix formed by rows $\mathcal{R}$ and columns $\mathcal{C}$ (the "skeleton"), and $A(\mathcal{R}, \mathcal{C})$ is invertible.

9.4.1.1 Example

Given the matrix

$$A = \begin{bmatrix} 1 & 2 & 3 & 4 \\ 2 & 4 & 6 & 8 \\ 1 & 1 & 2 & 3 \\ 3 & 5 & 7 & 9 \end{bmatrix},$$

find the Skeleton Decomposition $A = A(:, \mathcal{C}) \cdot A(\mathcal{R}, \mathcal{C})^{-1} \cdot A(\mathcal{R}, :)$.

Solution:

Note that the rank of A is $r = 2$ (rows 2 and 3 are linear combinations of rows 0 and 1, and columns 2 and 3 are linear combinations of columns 0 and 1).

1. **Select skeleton indices**: Choose $r = 2$ rows and columns such that $A(\mathcal{R}, \mathcal{C})$ is invertible. Here, rows 0 and 2, and columns 0 and 1:

$$\mathcal{R} = \{0, 2\}, \quad \mathcal{C} = \{0, 1\}.$$

2. **Extract skeleton components**:

   - $A(\mathcal{R}, :) = \begin{bmatrix} 1 & 2 & 3 & 4 \\ 1 & 1 & 2 & 3 \end{bmatrix}$ (selected rows)

- $A(:, \mathcal{C}) = \begin{bmatrix} 1 & 2 \\ 2 & 4 \\ 1 & 1 \\ 3 & 5 \end{bmatrix}$ (selected columns)
- $A(\mathcal{R}, \mathcal{C}) = \begin{bmatrix} 1 & 2 \\ 1 & 1 \end{bmatrix}$ (skeleton submatrix, invertible)

3. **Compute inverse of skeleton:**

$$A(\mathcal{R}, \mathcal{C})^{-1} = \begin{bmatrix} 1 & 2 \\ 1 & 1 \end{bmatrix}^{-1} = \begin{bmatrix} -1 & 2 \\ 1 & -1 \end{bmatrix}.$$

4. **Verification:**

$$A(:, \mathcal{C}) \cdot A(\mathcal{R}, \mathcal{C})^{-1} \cdot A(\mathcal{R}, :) = \begin{bmatrix} 1 & 2 \\ 2 & 4 \\ 1 & 1 \\ 3 & 5 \end{bmatrix} \begin{bmatrix} -1 & 2 \\ 1 & -1 \end{bmatrix} \begin{bmatrix} 1 & 2 & 3 & 4 \\ 1 & 1 & 2 & 3 \end{bmatrix}$$

$$= \begin{bmatrix} 1 & 2 \\ 2 & 4 \\ 1 & 1 \\ 3 & 5 \end{bmatrix} \begin{bmatrix} 1 & 0 & 1 & 2 \\ 0 & 1 & 1 & 1 \end{bmatrix} = \begin{bmatrix} 1 & 2 & 3 & 4 \\ 2 & 4 & 6 & 8 \\ 1 & 1 & 2 & 3 \\ 3 & 5 & 7 & 9 \end{bmatrix} = A.$$

The skeleton $A(\mathcal{R}, \mathcal{C}) = \begin{bmatrix} 1 & 2 \\ 1 & 1 \end{bmatrix}$ is the intersection of the selected rows and columns, and its inverse enables reconstructing the entire matrix from just r rows and r columns.

Skeleton Decomposition (also called **cross approximation** or **CUR decomposition**, $A \approx CUR$) expresses a matrix using a small subset of its rows and columns. This provides **interpretable low-rank approximation**: the selected rows and columns are actual data points/features from the original matrix.

In **Deep Neural Networks (DNNs)**, skeleton decomposition is used to **compress large, redundant weight matrices** by approximating them using a small subset of their original rows and columns, reducing parameter space, lowering computational complexity (FLOPs), and helping address **matrix completion** problems. In machine learning, skeleton decomposition is also used in **recommendation systems** (selecting representative users and items), **data summarization** (finding representative samples and features), and **interpretable dimensionality reduction**. Unlike SVD which produces abstract factors, skeleton decomposition uses actual rows and columns, making it valuable for **explainable AI** where interpretability is crucial. The decomposition $A = CUR$ (where C are selected columns, R are selected rows, and U is the inverse of the intersection) enables efficient computation and storage while maintaining interpretability.

9.5 Non-Negative Matrix Factorization (NMF)

9.5.1 Definition: Non-Negative Matrix Factorization

For a non-negative matrix $A \in \mathbb{R}_{\geq 0}^{M \times N}$ (i.e., $A_{ij} \geq 0$ for all i, j), there exist non-negative matrices $W \in \mathbb{R}_{\geq 0}^{M \times r}$ and $H \in \mathbb{R}_{\geq 0}^{r \times N}$ such that

$$A \approx WH \tag{9.6}$$

where the approximation minimizes $||A - WH||_F^2$ subject to $W \geq 0$ and $H \geq 0$.

9.5.1.1 Example

Given the non-negative matrix

$$A = \begin{bmatrix} 1 & 2 & 3 \\ 2 & 4 & 6 \\ 1 & 1 & 2 \end{bmatrix},$$

find a Non-Negative Matrix Factorization $A \approx WH$ with rank $r = 2$.

Solution:

We seek non-negative matrices $W \in \mathbb{R}_{\geq 0}^{3 \times 2}$ and $H \in \mathbb{R}_{\geq 0}^{2 \times 3}$ such that $A \approx WH$.

1. **Initialization**: Start with random non-negative matrices:

$$W^{(0)} = \begin{bmatrix} 1 & 0.5 \\ 2 & 1 \\ 0.5 & 0.5 \end{bmatrix}, \quad H^{(0)} = \begin{bmatrix} 1 & 2 & 3 \\ 0 & 0 & 0 \end{bmatrix}.$$

2. **Iterative optimization**: Using multiplicative update rules (alternating optimization):
   - Update H while keeping W fixed:

$$H \leftarrow H \odot \frac{W^\top A}{W^\top WH} \tag{9.7}$$

   (element-wise multiplication and division)
   - Update W while keeping H fixed:

$$W \leftarrow W \odot \frac{AH^\top}{WHH^\top} \tag{9.8}$$

   - Repeat until convergence
3. **Result** (after convergence):

$$W = \begin{bmatrix} 0.707 & 0.707 \\ 1.414 & 1.414 \\ 0.577 & 0.577 \end{bmatrix}, \quad H = \begin{bmatrix} 0.707 & 1.414 & 2.121 \\ 0.707 & 1.414 & 2.121 \end{bmatrix}.$$

4. **Verification**:

$$WH = \begin{bmatrix} 0.707 & 0.707 \\ 1.414 & 1.414 \\ 0.577 & 0.577 \end{bmatrix} \begin{bmatrix} 0.707 & 1.414 & 2.121 \\ 0.707 & 1.414 & 2.121 \end{bmatrix} = \begin{bmatrix} 1 & 2 & 3 \\ 2 & 4 & 6 \\ 0.816 & 1.633 & 2.449 \end{bmatrix} \approx A.$$

Note that all entries in W and H are non-negative, and the factorization provides an interpretable parts-based representation: W can be interpreted as basis vectors (e.g., user preferences or topics), and H as coefficients (e.g., item characteristics or document-topic distributions).

> Non-Negative Matrix Factorization (NMF) is fundamental in many machine learning applications where data and factors must be non-negative. In **recommendation systems**, user-item rating matrices are factorized via NMF to discover latent factors (e.g., user preferences, item characteristics) that are interpretable and non-negative. In **topic modeling** (e.g., LDA alternatives), document-term matrices are factorized to discover topics, where each topic is a non-negative combination of words. In **image processing**, image data (non-negative pixel values) is factorized to discover parts-based representations (e.g., facial features, object parts). In **audio signal processing**, spectrograms are factorized to separate sources. The non-negativity constraint enables **additive models** where components are combined additively (rather than subtractively as in PCA), leading to **parts-based representations** that are more interpretable than holistic representations. NMF is solved using alternating optimization algorithms (e.g., multiplicative updates, coordinate descent) that maintain non-negativity throughout the optimization.

9.6 LQ Decomposition

9.6.1 Theorem: LQ Decomposition

For any matrix $A \in \mathbb{R}^{M\times N}$ with $\text{rank}(A) = r$, there exists a decomposition

$$A = LQ \tag{9.9}$$

where $L \in \mathbb{R}^{M\times r}$ is lower triangular, and $Q \in \mathbb{R}^{r\times N}$ has orthonormal rows ($QQ^\top = I_{r\times r}$).

9.6.1.1 Example

Given the matrix

$$A = \begin{bmatrix} 1 & 2 & 3 \\ 2 & 4 & 6 \\ 1 & 1 & 2 \end{bmatrix},$$

find the LQ decomposition $A = LQ$.

Solution:

Note that LQ decomposition can be obtained by applying QR decomposition to $A^\top$ and then transposing. Alternatively, we can use Gram-Schmidt on the rows of A.

1. **Using QR on $A^\top$**: Compute QR decomposition of $A^\top$:

$$A^\top = \begin{bmatrix} 1 & 2 & 1 \\ 2 & 4 & 1 \\ 3 & 6 & 2 \end{bmatrix} = Q_R R,$$

   where Q_R has orthonormal columns and R is upper triangular.

2. **Transpose to get LQ**: Since $A = (A^\top)^\top = (Q_R R)^\top = R^\top Q_R^\top$, we have:

$$L = R^\top \text{ (lower triangular)}, \quad Q = Q_R^\top \text{ (orthonormal rows)} \tag{9.10}$$

3. **Result:**

$$L = \begin{bmatrix} \sqrt{6} & 0 \\ 2\sqrt{6} & \sqrt{2} \\ \sqrt{6} & \sqrt{2} \end{bmatrix}, \quad Q = \begin{bmatrix} \frac{1}{\sqrt{6}} & \frac{2}{\sqrt{6}} & \frac{1}{\sqrt{6}} \\ 0 & \frac{1}{\sqrt{2}} & \frac{1}{\sqrt{2}} \end{bmatrix}.$$

4. **Verification:**
   - L is lower triangular (zeros above the diagonal)
   - $QQ^\top = I_{2\times 2}$ (orthonormal rows)
   - $LQ = A$ (approximately, up to numerical precision)

Note that Q has orthonormal rows, providing an orthonormal basis for the row space of A, which is the key difference from QR decomposition (which provides a basis for the column space).

LQ decomposition is the "transpose" of QR decomposition: if $A = LQ$, then $A^\top = Q^\top L^\top$ is the QR decomposition of $A^\top$. LQ decomposition provides an orthonormal basis for the **row space** of A (as opposed to QR which provides a basis for the column space).

> In machine learning, LQ is used when working with **transposed data matrices** (samples as columns, features as rows), in **dual formulations** of optimization problems, and in algorithms that require row-space operations. The decomposition is particularly useful in **kernel methods** where Gram matrices $K = XX^\top$ are decomposed, and in **online learning** where new samples (rows) are added incrementally.

9.7 Schur Decomposition

9.7.1 Theorem: Schur Decomposition

For any square matrix $A \in \mathbb{R}^{N\times N}$, there exists an orthonormal matrix $Q \in \mathbb{R}^{N\times N}$ and an upper triangular matrix $U \in \mathbb{R}^{N\times N}$ such that

$$A = QUQ^\top \tag{9.11}$$

The diagonal entries of U are the eigenvalues of A.

9.7.1.1 Example

Given the non-symmetric matrix

$$A = \begin{bmatrix} 1 & 2 \\ 0 & 3 \end{bmatrix},$$

find the Schur decomposition $A = QUQ^\top$.

Solution:

1. **Find eigenvalues**: The characteristic polynomial is $\det(A - \lambda I) = (1-\lambda)(3-\lambda) = 0$, so $\lambda_1 = 1$ and $\lambda_2 = 3$.

2. **Find eigenvectors and construct Q**:

   - For $\lambda_1 = 1$: $(A - I)\mathbf{v}_1 = 0$ gives $\mathbf{v}_1 = \begin{bmatrix} 1 \\ 0 \end{bmatrix}$.
   - For $\lambda_2 = 3$: $(A - 3I)\mathbf{v}_2 = 0$ gives $\mathbf{v}_2 = \begin{bmatrix} 1 \\ 1 \end{bmatrix}$.

- Normalize and orthogonalize (using Gram-Schmidt if needed):

$$Q = \begin{bmatrix} 1 & 0 \\ 0 & 1 \end{bmatrix}.$$

3. **Construct U**: Since A is already upper triangular, we can take $Q = I$ and $U = A$:

$$U = \begin{bmatrix} 1 & 2 \\ 0 & 3 \end{bmatrix}.$$

4. **Verification**:

$$QUQ^\top = \begin{bmatrix} 1 & 0 \\ 0 & 1 \end{bmatrix} \begin{bmatrix} 1 & 2 \\ 0 & 3 \end{bmatrix} \begin{bmatrix} 1 & 0 \\ 0 & 1 \end{bmatrix} = \begin{bmatrix} 1 & 2 \\ 0 & 3 \end{bmatrix} = A.$$

Note that the diagonal entries of U (1 and 3) are the eigenvalues of A. Since A is already upper triangular, the Schur decomposition is trivial with $Q = I$. For general non-symmetric matrices, U is upper triangular (not necessarily diagonal), and Q is an orthonormal matrix that triangularizes A.

> Schur decomposition is the most general form of triangularization for square matrices and is fundamental in eigenvalue computation algorithms (e.g., QR algorithm). Unlike spectral decomposition which requires A to be symmetric, Schur decomposition exists for any square matrix. In machine learning, Schur decomposition appears in **dynamical systems analysis** (analyzing stability of recurrent neural networks), **control theory** (state-space representations), and **eigenvalue computation** for large matrices. The upper triangular form U reveals the eigenvalue structure, and the orthonormal matrix Q provides a change of basis that triangularizes A. For normal matrices (including symmetric matrices), Schur decomposition reduces to spectral decomposition with U being diagonal.

9.8 Low-Rank Matrix Factorization and Tensor Decompositions

While matrix decompositions are fundamental, many modern applications involve higher-order tensors (multi-dimensional arrays). However, in engineering practice, **low-rank matrix factorization** (including methods like **LoRA** - Low-Rank Adaptation) has largely replaced the role of tensor decompositions in many deep learning applications. This section briefly introduces the key tensor decomposition methods, with emphasis on those most relevant to current practice.

9.8.1 Low-Rank Matrix Factorization (LoRA)

Low-Rank Adaptation (LoRA)[1] is the dominant approach for parameter-efficient fine-tuning of large language models. It applies the low-rank factorization ideas from this chapter—and truncated SVD (Chapter~8)—to the weight *update* ΔW rather than to the full matrix W_0.

For a frozen pre-trained weight $W_0 \in \mathbb{R}^{m \times n}$, the adapted layer computes

$$\mathbf{y} = W_0\mathbf{x} + \frac{\alpha}{r} BA\mathbf{x},$$

with

[1] Hu, E. J., Shen, Y., Wallis, P., Allen-Zhu, Z., Li, Y., Wang, S., Wang, L., & Chen, W. "LoRA: Low-Rank Adaptation of Large Language Models." ICLR 2022. https://arxiv.org/abs/2106.09685

$$W = W_0 + \Delta W, \qquad \Delta W = \frac{\alpha}{r} BA, \qquad \text{rank}(\Delta W) \leq r. \tag{9.12}$$

Here $B \in \mathbb{R}^{m \times r}$, $A \in \mathbb{R}^{r \times n}$, $r \ll \min(m, n)$, and $\alpha > 0$ is a scaling hyperparameter. Only B and A are trained, reducing trainable parameters from mn to $r(m+n)$. The base weights W_0 stay fixed (and can be shared across tasks); B is initialized to zero so $\Delta W = 0$ at the start of fine-tuning. After training, the factors can be merged as $W_{\text{merged}} = W_0 + \frac{\alpha}{r} BA$ with no extra inference cost. LoRA learns B and A by gradient descent instead of explicitly truncating an SVD of ΔW. In Transformers, it is typically applied to the query and value projections W_Q and W_V in each attention block.

9.8.1.1 Example: Parameter Count

For one attention projection with $m = n = 4096$ and $r = 8$, full fine-tuning trains $mn = 16{,}777{,}216$ parameters, whereas LoRA trains $r(m+n) = 65{,}536$—a reduction of roughly $256\times$.

Low-rank matrix factorization has become the engineering standard for parameter-efficient fine-tuning in modern deep learning. **LoRA** and its variants (e.g., QLoRA, AdaLoRA) are widely used because they offer an excellent trade-off between model capacity and parameter efficiency. The key insight is that weight updates during fine-tuning often have low intrinsic rank, making low-rank factorization both theoretically justified and practically effective. This approach has largely supplanted traditional tensor decomposition methods in many deep learning applications, as matrix factorization is simpler, more numerically stable, and better supported by existing frameworks.

9.8.2 Tensor Train (TT) Decomposition

For a d-dimensional tensor $\mathcal{A} \in \mathbb{R}^{n_1 \times n_2 \times \cdots \times n_d}$, the **Tensor Train (TT) decomposition** represents it as a chain of low-rank core tensors:

$$\mathcal{A}(i_1, i_2, \ldots, i_d) = \sum_{r_1=1}^{R_1} \sum_{r_2=1}^{R_2} \cdots \sum_{r_{d-1}=1}^{R_{d-1}} \mathcal{G}_1(i_1, r_1) \mathcal{G}_2(r_1, i_2, r_2) \cdots \mathcal{G}_{d-1}(r_{d-2}, i_{d-1}, r_{d-1}) \mathcal{G}_d(r_{d-1}, i_d),$$

where $\mathcal{G}_k$ are core tensors with ranks $R_k \ll n_k$. The storage requirement is $O(dnR^2)$ instead of $O(n^d)$, providing exponential compression for high-dimensional tensors.

9.8.2.1 Example

For a 3-dimensional tensor $\mathcal{A} \in \mathbb{R}^{10 \times 10 \times 10}$ with TT ranks $R_1 = R_2 = 5$, the TT decomposition requires storing three core tensors: $\mathcal{G}_1 \in \mathbb{R}^{10 \times 5}$, $\mathcal{G}_2 \in \mathbb{R}^{5 \times 10 \times 5}$, and $\mathcal{G}_3 \in \mathbb{R}^{5 \times 10}$, totaling $50 + 250 + 50 = 350$ parameters instead of 1000 for the full tensor, achieving significant compression.

Tensor Train decomposition is currently one of the most important tensor decomposition methods in deep learning compression, particularly for **large-scale embedding compression** and **high-dimensional scientific computing**. TT decomposition excels at compressing very high-dimensional tensors (e.g., embedding tables in recommendation systems, quantum state representations) where the exponential storage reduction is crucial. In practice, TT is used for compressing embedding layers in large language models, where embedding tables can have billions of parameters. The chain structure of TT cores enables efficient computation and makes TT particularly suitable for very high-dimensional problems where other tensor decompositions become impractical.

9.8.3 Other Tensor Decompositions

Tucker decomposition (including HOSVD - Higher-Order Singular Value Decomposition) decomposes a tensor into a core tensor and factor matrices for each mode, providing flexibility for medium-scale tensor modeling in image processing and multi-view data analysis. **CP decomposition** (CANDECOMP/PARAFAC) represents a tensor as a sum of rank-1 components, historically important in signal processing and interpretable modeling, though less commonly used in modern deep learning practice.

9.9 Mindmap

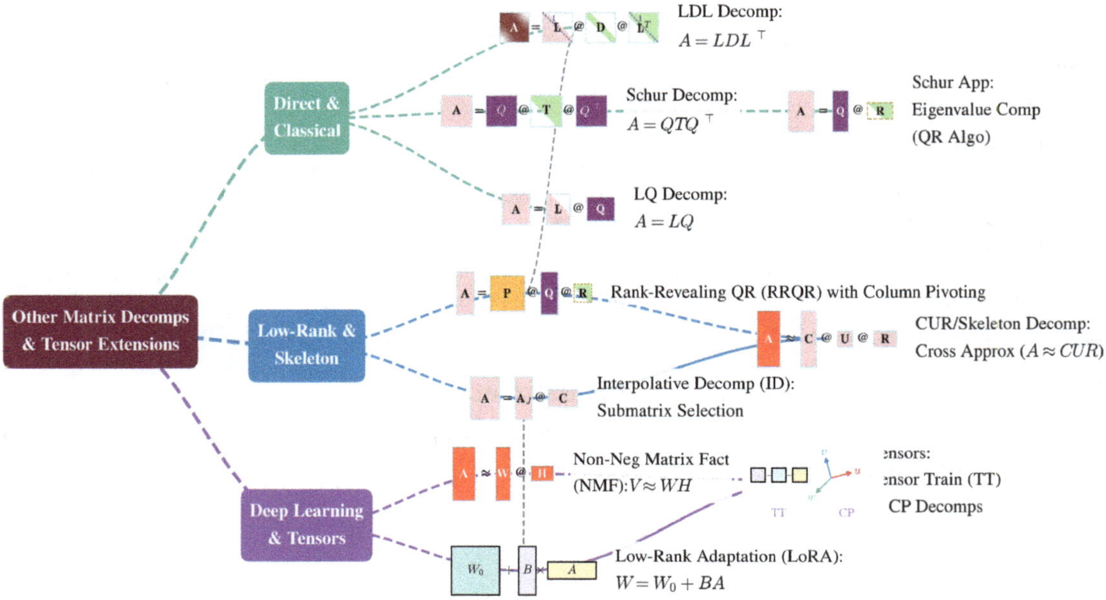

Figure 9.1: Mind map of other matrix decompositions

9.10 Exercises

9.10.1 LDL Decomposition and Positive Definiteness in Optimization

In second-order optimization methods (e.g., quasi-Newton methods), the Hessian approximation H may be symmetric but not positive definite. LDL decomposition $H = LDL^\top$ provides a numerically stable way to check positive definiteness and solve linear systems.

1. Prove that a symmetric matrix A is positive definite if and only if all diagonal entries of D in its LDL decomposition $A = LDL^\top$ are positive. (Hint: Use the fact that $\mathbf{x}^\top A\mathbf{x} = \mathbf{x}^\top LDL^\top\mathbf{x} = (L^\top\mathbf{x})^\top D(L^\top\mathbf{x})$.)

2. Given the symmetric matrix

$$H = \begin{bmatrix} 4 & 2 & 2 \\ 2 & 3 & 1 \\ 2 & 1 & 3 \end{bmatrix},$$

which represents a Hessian approximation in optimization, find the LDL decomposition $H = LDL^\top$. Determine whether H is positive definite by examining the diagonal entries of D.

3. Prove that if A is positive definite with LDL decomposition $A = LDL^\top$, then the Cholesky decomposition $A = (L\sqrt{D})(L\sqrt{D})^\top$ exists, where $\sqrt{D}$ is the diagonal matrix with entries $\sqrt{D_{ii}}$. Compute the Cholesky factor for the matrix in part 2.
4. In Gaussian process regression, the covariance matrix K may be near singular (ill-conditioned). Show that LDL decomposition is more numerically stable than Cholesky decomposition when A is near singular (i.e., when some diagonal entries of D are very small but positive). Explain why this matters for solving $A\mathbf{x} = \mathbf{b}$.

9.10.2 Rank-Revealing QR and Feature Selection

For a matrix $A \in \mathbb{R}^{M\times N}$ with rank r, consider the Rank-Revealing QR decomposition $A = PQR$ where $R = \begin{bmatrix} R_{11} & R_{12} \\ 0 & R_{22} \end{bmatrix}$:

1. Prove that $\text{rank}(A) = r$ if and only if $R_{11} \in \mathbb{R}^{r\times r}$ is invertible and $||R_{22}||$ is small (ideally zero). (Hint: Use the fact that the rank equals the number of linearly independent columns.)
2. Given the matrix
$$A = \begin{bmatrix} 1 & 2 & 3 & 4 \\ 2 & 4 & 6 & 8 \\ 1 & 1 & 2 & 3 \\ 0 & 1 & 1 & 2 \end{bmatrix},$$
determine the rank r and find a Rank-Revealing QR decomposition. Identify which columns of A form a well-conditioned basis.
3. Prove that the first r columns of AP (after permutation) form a basis for the column space of A. Show that these columns correspond to the columns with largest norms in the QR decomposition with column pivoting.
4. In the context of feature selection, explain why Rank-Revealing QR can identify the most important features (columns) of a data matrix. How does the structure of R reveal feature importance?

9.10.3 Interpolative Decomposition and Interpretable Low-Rank Approximation

For a matrix $A \in \mathbb{R}^{M\times N}$ with rank r, consider the Interpolative Decomposition $A = A(:, \mathcal{J})C$:

1. Prove that the interpolation matrix C has the property that $C(:, \mathcal{J}) = I_r$ (the $r \times r$ identity matrix), where $\mathcal{J}$ is the set of skeleton column indices. (Hint: Use the fact that $A(:, \mathcal{J}) = A(:, \mathcal{J}) \cdot I_r$.)
2. Given the matrix
$$A = \begin{bmatrix} 1 & 2 & 3 & 5 \\ 2 & 4 & 6 & 10 \\ 1 & 1 & 2 & 3 \end{bmatrix},$$
find an Interpolative Decomposition with skeleton columns $\mathcal{J} = \{0, 1\}$. Construct the interpolation matrix C explicitly.
3. Prove that the norm bound $||C|| \leq \sqrt{r(N-r)+1}$ is tight in the worst case. (Hint: Consider a matrix where the non-skeleton columns are nearly orthogonal to the skeleton columns.)

4. Show that Interpolative Decomposition provides an interpretable low-rank approximation: unlike SVD which produces abstract principal components, ID uses actual columns from the original matrix. Explain why this is advantageous for explainable AI.

9.10.4 Skeleton Decomposition (CUR) and Matrix Completion

For a matrix $A \in \mathbb{R}^{M\times N}$ with rank r, consider the Skeleton Decomposition $A = A(:,\mathcal{C})\cdot A(\mathcal{R},\mathcal{C})^{-1}\cdot A(\mathcal{R},:)$:

1. Prove that the skeleton submatrix $A(\mathcal{R},\mathcal{C})$ must be invertible for the decomposition to exist. Show that if $\text{rank}(A) = r$, then there always exist index sets $\mathcal{R}$ and $\mathcal{C}$ of size r such that $A(\mathcal{R},\mathcal{C})$ is invertible.
2. Given the matrix
$$A = \begin{bmatrix} 1 & 2 & 3 & 4 \\ 2 & 4 & 6 & 8 \\ 1 & 1 & 2 & 3 \\ 3 & 5 & 7 & 9 \\ 0 & 1 & 1 & 2 \end{bmatrix},$$
find a Skeleton Decomposition. Verify that $A = CUR$ where $C = A(:,\mathcal{C})$, $U = A(\mathcal{R},\mathcal{C})^{-1}$, and $R = A(\mathcal{R},:)$.
3. Prove that the storage requirement for the CUR decomposition is $Mr + r^2 + rN$ compared to MN for the full matrix. For what values of r is CUR more storage-efficient than the full matrix?
4. In the context of matrix completion, suppose we observe only a subset of entries of A. Show that if we can identify a skeleton $A(\mathcal{R},\mathcal{C})$ from the observed entries, we can reconstruct the entire matrix using the CUR decomposition. What conditions must the observed entries satisfy?

9.10.5 Non-Negative Matrix Factorization and Topic Modeling

For a non-negative matrix $A \in \mathbb{R}^{M\times N}_{\geq 0}$, consider the NMF $A \approx WH$ where $W \in \mathbb{R}^{M\times r}_{\geq 0}$ and $H \in \mathbb{R}^{r\times N}_{\geq 0}$:

1. Prove that NMF is generally not unique: if $A = WH$, then for any invertible matrix $T \in \mathbb{R}^{r\times r}$ with $T \geq 0$ and $T^{-1} \geq 0$, we have $A = (WT)(T^{-1}H)$. Show that such matrices T exist (e.g., permutation matrices or positive diagonal scaling matrices).
2. The multiplicative update rules for NMF are:
$$H \leftarrow H \odot \frac{W^\top A}{W^\top WH}, \quad W \leftarrow W \odot \frac{AH^\top}{WHH^\top},$$
where $\odot$ denotes element-wise multiplication and division is element-wise. Prove that these updates maintain non-negativity: if $W \geq 0$ and $H \geq 0$, then the updated matrices are also non-negative.
3. Prove that the multiplicative updates decrease (or maintain) the Frobenius norm error $||A - WH||_F^2$ at each iteration. (Hint: Use the fact that the updates can be derived from gradient descent with a specific step size that ensures non-negativity.)
4. In topic modeling, a document-term matrix $A \in \mathbb{R}^{M\times N}_{\geq 0}$ (where A_{ij} is the frequency of word j in document i) is factorized as $A \approx WH$. Interpret the columns of W and rows of H in this context. Explain why NMF is more suitable than SVD for this application.

9.10.6 LQ Decomposition and Kernel Methods

For a matrix $A \in \mathbb{R}^{M \times N}$ with LQ decomposition $A = LQ$:

1. Prove that LQ decomposition can be obtained by applying QR decomposition to $A^\top$ and then transposing. Specifically, if $A^\top = Q_R R$ is the QR decomposition of $A^\top$, then $A = LQ$ where $L = R^\top$ and $Q = Q_R^\top$.
2. Prove that the rows of Q form an orthonormal basis for the row space of A. Show that $\text{rank}(A) = \text{rank}(Q)$.
3. Given the matrix
$$A = \begin{bmatrix} 1 & 2 & 3 \\ 2 & 4 & 6 \\ 1 & 1 & 2 \\ 0 & 1 & 1 \end{bmatrix},$$
find the LQ decomposition explicitly. Verify that Q has orthonormal rows.
4. In kernel methods, we often work with Gram matrices $K = XX^\top$ where $X \in \mathbb{R}^{N \times d}$ is a data matrix. Show that if $X = LQ$ is the LQ decomposition, then $K = LL^\top$. Explain how this can be used to reduce computational complexity when $N \gg d$.

9.10.7 Schur Decomposition and Eigenvalue Computation

For a square matrix $A \in \mathbb{R}^{N \times N}$, consider the Schur decomposition $A = QUQ^\top$:

1. Prove that the Schur decomposition always exists for any square matrix (over $\mathbb{C}$). Show that the diagonal entries of U are the eigenvalues of A. (Hint: Use the fact that A and U are similar: $A = QUQ^\top$ implies A and U have the same eigenvalues.)
2. Prove that if A is normal (i.e., $AA^\top = A^\top A$), then the Schur decomposition reduces to the spectral decomposition: U is diagonal and Q contains the eigenvectors. Show that symmetric matrices are normal.
3. The QR algorithm for computing eigenvalues iteratively constructs a sequence of matrices:
$$A_0 = A, \quad A_{k+1} = R_k Q_k \text{ where } A_k = Q_k R_k \text{ (QR decomposition).}$$
Prove that $A_{k+1} = Q_k^\top A_k Q_k$, so all A_k are similar to A and have the same eigenvalues. Show that under certain conditions, A_k converges to an upper triangular matrix (the Schur form).
4. For a 2×2 matrix
$$A = \begin{bmatrix} a & b \\ c & d \end{bmatrix},$$
find the Schur decomposition explicitly. Express Q and U in terms of a, b, c, d and the eigenvalues λ_1, λ_2 of A. (Hint: Consider the case when A has distinct eigenvalues and when it has a repeated eigenvalue.)

9.10.8 Rank-Revealing QR and Low-Rank Approximation

For a matrix $A \in \mathbb{R}^{M \times N}$ with Rank-Revealing QR decomposition $A = PQR$:

1. Prove that the Rank-Revealing QR decomposition provides a rank-k approximation $A_k = PQ_k R_k$ (where Q_k consists of the first k columns of Q and R_k consists of the first k rows of R) that satisfies:
$$||A - A_k||_2 \leq \sigma_{k+1} \cdot \sqrt{1 + ||R_{12}R_{22}^{-1}||_2^2},$$

where σ_{k+1} is the $(k+1)$-th singular value of A and $R = \begin{bmatrix} R_{11} & R_{12} \\ 0 & R_{22} \end{bmatrix}$. Compare this bound to the optimal rank-k approximation error σ_{k+1} from SVD.

2. The Strong Rank-Revealing QR (SRRQR) decomposition ensures that $||R_{12}R_{22}^{-1}||_2 \leq f(r, N)$ for some function f. Prove that if $||R_{12}R_{22}^{-1}||_2 \leq c$ for a constant c, then the rank-k approximation error is at most $(1+c)\sigma_{k+1}$, which is within a constant factor of optimal.

3. In randomized algorithms for low-rank approximation, we compute $Y = A\Omega$ where $\Omega \in \mathbb{R}^{N\times k}$ is a random matrix, then perform QR decomposition on Y to get Q. Prove that if Ω has certain properties (e.g., Gaussian random matrix), then with high probability, the column space of Q captures most of the range of A. (Hint: Use the fact that $Y = A\Omega$ projects the range of A onto a random subspace.)

4. Design an algorithm that uses Rank-Revealing QR to compute a rank-k approximation of A with error guarantee $||A - A_k||_2 \leq (1+\epsilon)\sigma_{k+1}$ for a given tolerance $\epsilon > 0$. Analyze the computational complexity of your algorithm and compare it to computing the full SVD.

9.10.9 Skeleton Decomposition and Neural Network Compression

For a matrix $A \in \mathbb{R}^{M\times N}$ with Skeleton Decomposition $A = CUR$:

1. Prove that if A has exact rank r, then the CUR decomposition is exact: $A = A(:, \mathcal{C}) \cdot A(\mathcal{R}, \mathcal{C})^{-1} \cdot A(\mathcal{R}, :)$. Show that the choice of skeleton indices $\mathcal{R}$ and $\mathcal{C}$ affects the conditioning of $A(\mathcal{R}, \mathcal{C})$.

2. For a matrix A with approximate rank r (i.e., A has r large singular values and the rest are small), prove that the CUR approximation error satisfies:

$$||A - CUR||_2 \leq \sigma_{r+1} \cdot (1 + ||A(\mathcal{R}, \mathcal{C})^{-1}||_2 \cdot ||A(\mathcal{R}, \mathcal{C}^c)||_2),$$

where $\mathcal{C}^c$ denotes the complement of $\mathcal{C}$ and σ_{r+1} is the $(r+1)$-th singular value.

3. The problem of selecting optimal skeleton indices $\mathcal{R}$ and $\mathcal{C}$ to minimize the approximation error is NP-hard. However, greedy algorithms based on maximizing the volume (absolute determinant) of $A(\mathcal{R}, \mathcal{C})$ work well in practice. Prove that if $A(\mathcal{R}, \mathcal{C})$ has maximum volume among all $r \times r$ submatrices, then $||A(\mathcal{R}, \mathcal{C})^{-1}||_2$ is minimized.

4. In the context of compressing neural network weight matrices, suppose we want to approximate a weight matrix $W \in \mathbb{R}^{4096\times 4096}$ using CUR decomposition with $r = 64$ skeleton rows and columns. Compute the compression ratio and the number of parameters saved compared to the full matrix. If W has rank approximately r, estimate the approximation error using the bound from part (2).

9.10.10 Schur Decomposition and Matrix Functions in Optimization

For a square matrix $A \in \mathbb{R}^{N\times N}$ with Schur decomposition $A = QUQ^\top$:

1. Prove that for any analytic function f (e.g., $f(x) = e^x$, $f(x) = \log(x)$, $f(x) = x^p$), we can define $f(A) = Qf(U)Q^\top$, where $f(U)$ is applied to the upper triangular matrix U in a specific way. Show that if U is diagonal, then $f(U)$ is simply the diagonal matrix with $f(\lambda_i)$ on the diagonal.

2. For the matrix exponential e^A, prove that if $A = QUQ^\top$ is the Schur decomposition, then $e^A = Qe^UQ^\top$. Show that e^U can be computed efficiently when U is upper triangular using the fact that $e^U = I + U + \frac{U^2}{2!} + \cdots$ and that powers of upper triangular matrices have a specific structure.

3. The matrix logarithm $\log(A)$ is defined as the inverse of the matrix exponential: if $e^B = A$, then $B = \log(A)$. Prove that if A has Schur decomposition $A = QUQ^\top$ and all eigenvalues

of A are positive, then $\log(A) = Q \log(U) Q^\top$. Show that $\log(U)$ can be computed when U is upper triangular.

4. In the context of optimization, the matrix square root $A^{1/2}$ (satisfying $(A^{1/2})^2 = A$) is used in second-order methods. Prove that if A is positive definite with Schur decomposition $A = QUQ^\top$, then $A^{1/2} = QU^{1/2}Q^\top$ where $U^{1/2}$ is the upper triangular square root of U. Design an algorithm to compute $U^{1/2}$ when U is upper triangular with positive diagonal entries.

Chapter 10

Matrix Block Partitioning

Block matrices and Kronecker products: structural tools for scalable computation

"The purpose of computing is insight, not numbers."

— Richard Hamming

▷ `Kronecker product` $A \otimes B = [a_{ij}B]_{m \times n} \in \mathbb{R}^{mp \times nq}$ for $A \in \mathbb{R}^{m \times n}$, $B \in \mathbb{R}^{p \times q}$

▷ `Kronecker mixed product` $(A \otimes B)(C \otimes D) = (AC) \otimes (BD)$

▷ `Kronecker property` $(A \otimes B)^{-1} = A^{-1} \otimes B^{-1}$; $(A \otimes B)^\top = A^\top \otimes B^\top$ for invertible A, B

▷ `Kronecker eigenvalues` If $A\mathbf{u} = \lambda\mathbf{u}$ and $B\mathbf{v} = \mu\mathbf{v}$, then $(A \otimes B)(\mathbf{u} \otimes \mathbf{v}) = \lambda\mu(\mathbf{u} \otimes \mathbf{v})$

▷ `Vector Kronecker` $\mathbf{a} \otimes \mathbf{b} = [a_i\mathbf{b}]_{i=0}^{m-1} \in \mathbb{R}^{mn \times 1}$ for $\mathbf{a} \in \mathbb{R}^{m \times 1}$, $\mathbf{b} \in \mathbb{R}^{n \times 1}$

▷ `Vec operator` $\mathrm{vec}(ABC) = (C^\top \otimes A)\mathrm{vec}(B)$

▷ `Block multiplication` $(AB)_{ij} = \sum_{k=0}^{r-1} A_{ik}B_{kj}$

▷ `Schur complement` $S = A_{11} - A_{10}A_{00}^{-1}A_{01}$ for $A = \begin{bmatrix} A_{00} & A_{01} \\ A_{10} & A_{11} \end{bmatrix}$ with invertible A_{00}

▷ `BlockInverse` $A^{-1} = \begin{bmatrix} A_{00}^{-1} + A_{00}^{-1}A_{01}S^{-1}A_{10}A_{00}^{-1} & -A_{00}^{-1}A_{01}S^{-1} \\ -S^{-1}A_{10}A_{00}^{-1} & S^{-1} \end{bmatrix}$ where S is Schur complement

▷ `BlockLU` $A = \begin{bmatrix} L_{00} & 0 \\ L_{10} & L_{11} \end{bmatrix} \begin{bmatrix} U_{00} & U_{01} \\ 0 & U_{11} \end{bmatrix}$ where $L_{00}U_{00} = A_{00}$, $L_{00}U_{01} = A_{01}$, $L_{10}U_{00} = A_{10}$, $L_{11}U_{11} = S$

▷ `BlockCholesky` $A = \begin{bmatrix} L_{00} & 0 \\ L_{10} & L_{11} \end{bmatrix} \begin{bmatrix} L_{00}^\top & L_{10}^\top \\ 0 & L_{11}^\top \end{bmatrix}$ for SPD A where $L_{00}L_{00}^\top = A_{00}$, $L_{10} = A_{10}(L_{00}^\top)^{-1}$, $L_{11}L_{11}^\top = S$

▷ `BlockQR` $A = QR$ where $A = [A_0 \cdots A_{k-1}]$, $Q = [Q_0 \cdots Q_{k-1}]$ has orthonormal columns, R is block upper triangular

10.1 Kronecker Product (Matrix Tensor Product)

The Kronecker product (also called tensor product) is a fundamental operation that creates block-structured matrices from smaller matrices.

Kronecker Product vs. Kronecker Delta: The **Kronecker product** $A \otimes B$ (computed via `np.kron`) is a matrix operation that creates block-structured matrices. This is distinct from the **Kronecker delta** function δ_{ij}, which is a scalar indicator function defined as $\delta_{ij} = 1$ if $i = j$ and $\delta_{ij} = 0$ if $i \neq j$. The Kronecker delta is used to express diagonal matrices (e.g., $\Sigma_{ij} = \sigma_i \delta_{ij}$ for a diagonal matrix Σ with diagonal entries σ_i) and appears in many mathematical contexts (e.g., orthonormality conditions $\mathbf{u}_i^\top \mathbf{u}_j = \delta_{ij}$). Despite sharing the name "Kronecker," these are completely different concepts: Kronecker product is a matrix operation, while Kronecker delta is a scalar function.

10.1.1 Definition: Kronecker Product

For matrices $A \in \mathbb{R}^{m \times n}$ and $B \in \mathbb{R}^{p \times q}$, the **Kronecker product** $A \otimes B$ is defined as the $mp \times nq$ block matrix:

$$A \otimes B = \begin{bmatrix} a_{00}B & a_{01}B & \cdots & a_{0(n-1)}B \\ a_{10}B & a_{11}B & \cdots & a_{1(n-1)}B \\ \vdots & \vdots & \ddots & \vdots \\ a_{(m-1)0}B & a_{(m-1)1}B & \cdots & a_{(m-1)(n-1)}B \end{bmatrix} \tag{10.1}$$

Each element a_{ij} of A is multiplied by the entire matrix B, creating a block structure.

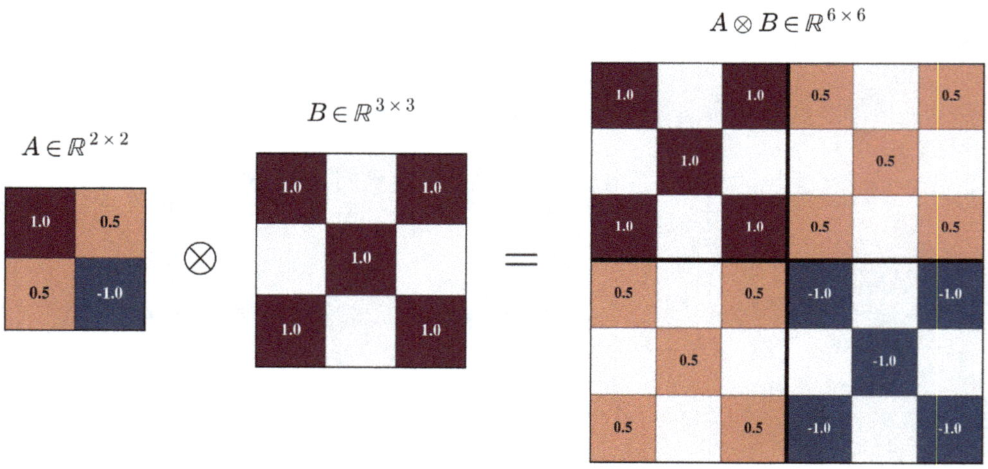

Figure 10.1: Kronecker Product Structure

Figure 10.1 illustrates the Kronecker product $A \otimes B$ for a 2×2 matrix A and a 3×3 matrix B, resulting in a 6×6 block matrix. The visualization shows how each element a_{ij} of A scales the entire matrix B, creating a block-structured pattern. Notice how the pattern of B is repeated and scaled by the values in A: positive elements of A (like $a_{00} = 1$) preserve the pattern of B, while negative elements (like $a_{11} = -1$) invert it. The black separator lines in the result highlight the 2×2 block structure corresponding to the dimensions of A. This visual representation makes clear why the Kronecker product creates matrices of size $mp \times nq$ when $A \in \mathbb{R}^{m \times n}$ and $B \in \mathbb{R}^{p \times q}$.

10.1.2 Properties of Kronecker Product

1. **Bilinearity**: $(A + C) \otimes B = A \otimes B + C \otimes B$ and $A \otimes (B + D) = A \otimes B + A \otimes D$.
2. **Scalar Multiplication**: $(kA) \otimes B = A \otimes (kB) = k(A \otimes B)$ for any scalar k.
3. **Transpose**: $(A \otimes B)^\top = A^\top \otimes B^\top$.
4. **Mixed Product Rule**:

$$(A \otimes B)(C \otimes D) = (AC) \otimes (BD) \tag{10.2}$$

   provided the dimensions are compatible.
5. **Inverse**: If A and B are invertible, then

$$(A \otimes B)^{-1} = A^{-1} \otimes B^{-1} \tag{10.3}$$

6. **Determinant**: If $A \in \mathbb{R}^{n \times n}$ and $B \in \mathbb{R}^{m \times m}$, then

$$\det(A \otimes B) = (\det A)^m (\det B)^n \tag{10.4}$$

7. **Trace**: If $A \in \mathbb{R}^{n \times n}$ and $B \in \mathbb{R}^{m \times m}$, then

$$\operatorname{tr}(A \otimes B) = \operatorname{tr}(A) \cdot \operatorname{tr}(B) \tag{10.5}$$

8. **Eigenvalues**: If λ is an eigenvalue of A with eigenvector $\mathbf{u}$, and μ is an eigenvalue of B with eigenvector $\mathbf{v}$, then $\lambda\mu$ is an eigenvalue of $A \otimes B$ with eigenvector $\mathbf{u} \otimes \mathbf{v}$.

10.1.3 Vector Kronecker Product

For vectors $\mathbf{a} \in \mathbb{R}^{m \times 1}$ and $\mathbf{b} \in \mathbb{R}^{n \times 1}$, the Kronecker product is:

$$\mathbf{a} \otimes \mathbf{b} = \begin{bmatrix} a_0 \mathbf{b} \\ a_1 \mathbf{b} \\ \vdots \\ a_{m-1} \mathbf{b} \end{bmatrix} \in \mathbb{R}^{mn \times 1} \tag{10.6}$$

This is equivalent to the outer product $\mathbf{a}\mathbf{b}^\top$ when reshaped appropriately.

10.1.4 Distinction from Other Operations

Kronecker Product vs. Other Tensor Operations:

1. `np.kron(A, B)` computes the Kronecker product $A \otimes B$, creating a block matrix where each element of A is multiplied by the entire matrix B.
2. `np.tensordot(A, B, axes=...)` computes tensor contractions along specified axes, which is different from Kronecker product. For example, `np.tensordot(A, B, axes=1)` computes $\sum_k A_{ik} B_{kj}$ (matrix multiplication), not Kronecker product.
3. **Outer Product** (`np.outer(a, b)` or `a @ b.T` for vectors): For vectors $\mathbf{a} \in \mathbb{R}^{m \times 1}$ and $\mathbf{b} \in \mathbb{R}^{n \times 1}$, the outer product $\mathbf{a}\mathbf{b}^\top \in \mathbb{R}^{m \times n}$ is a matrix, while the Kronecker product $\mathbf{a} \otimes \mathbf{b} \in \mathbb{R}^{mn \times 1}$ is a vector. They are related by reshaping: $\operatorname{vec}(\mathbf{a}\mathbf{b}^\top) = \mathbf{a} \otimes \mathbf{b}$.

4. **np.einsum**: Einstein summation notation can express various tensor operations, but Kronecker product is specifically the block-structured operation defined above.

5. **Vector Tensor Product**: For vectors, the Kronecker product $\mathbf{a} \otimes \mathbf{b}$ produces a column vector, while the outer product $\mathbf{a}\mathbf{b}^\top$ produces a matrix. The relationship: $\mathbf{a} \otimes \mathbf{b} = \text{vec}(\mathbf{a}\mathbf{b}^\top)$.

Note on Terminology: The **Kronecker delta** δ_{ij} (a scalar function: $\delta_{ij} = 1$ if $i = j$, 0 otherwise) is unrelated to the **Kronecker product** $A \otimes B$ (a matrix operation). The Kronecker delta is used to express diagonal matrices and orthonormality conditions, while the Kronecker product creates block-structured matrices.

> The Kronecker product is fundamental in many areas of machine learning and numerical computation. In **convolutional neural networks**, the Kronecker product appears when flattening and reshaping weight matrices for efficient computation. In **tensor networks** and **quantum machine learning**, Kronecker products are used to represent high-dimensional tensor operations. In **system identification** and **control theory**, Kronecker products enable compact representation of matrix equations. The **vec operator** (which stacks matrix columns into a vector) satisfies $\text{vec}(ABC) = (C^\top \otimes A)\text{vec}(B)$, which is crucial for solving matrix equations. In **distributed optimization**, Kronecker products enable efficient representation of block-structured Hessian matrices. The mixed product rule $(A \otimes B)(C \otimes D) = (AC) \otimes (BD)$ is particularly important for efficient computation, as it allows large Kronecker products to be computed without explicitly forming the full matrix. Understanding the distinction between `np.kron` (Kronecker product) and `np.tensordot` (tensor contraction) is crucial: Kronecker product creates block-structured matrices, while tensor contraction performs summation over specified indices, serving different purposes in numerical computation and machine learning.

10.2 Matrix Partitioning (Block Matrices)

10.2.1 Definition: Block Matrix

A matrix $A \in \mathbb{R}^{M \times N}$ can be **partitioned** into blocks (submatrices) as

$$A = \begin{bmatrix} A_{00} & A_{01} & \cdots & A_{0(q-1)} \\ A_{10} & A_{11} & \cdots & A_{1(q-1)} \\ \vdots & \vdots & \ddots & \vdots \\ A_{(p-1)0} & A_{(p-1)1} & \cdots & A_{(p-1)(q-1)} \end{bmatrix} \tag{10.7}$$

where each block $A_{ij} \in \mathbb{R}^{m_i \times n_j}$ with $\sum_{i=0}^{p-1} m_i = M$ and $\sum_{j=0}^{q-1} n_j = N$.

10.2.1.1 Example: Block Matrix Partitioning

Consider the matrix

$$A = \begin{bmatrix} 1 & 2 & 3 & 0 & 0 \\ 4 & 5 & 6 & 0 & 0 \\ 0 & 0 & 0 & -1 & 0 \\ 0 & 0 & 0 & 0 & 1 \end{bmatrix} \in \mathbb{R}^{4 \times 5}.$$

We can partition A into blocks as:

$$A = \begin{bmatrix} A_{00} & A_{01} \\ A_{10} & A_{11} \end{bmatrix},$$

where

$$A_{00} = \begin{bmatrix} 1 & 2 & 3 \\ 4 & 5 & 6 \end{bmatrix} \in \mathbb{R}^{2\times 3}, \quad A_{01} = \begin{bmatrix} 0 & 0 \\ 0 & 0 \end{bmatrix} \in \mathbb{R}^{2\times 2},$$

$$A_{10} = \begin{bmatrix} 0 & 0 & 0 \\ 0 & 0 & 0 \end{bmatrix} \in \mathbb{R}^{2\times 3}, \quad A_{11} = \begin{bmatrix} -1 & 0 \\ 0 & 1 \end{bmatrix} \in \mathbb{R}^{2\times 2}.$$

This partitioning reveals that A has a block structure where A_{00} and A_{11} contain the non-zero entries, while A_{01} and A_{10} are zero blocks.

10.2.2 Block Matrix Operations

10.2.2.1 Transpose of Block Matrix

For a block matrix A partitioned as in the definition above, the transpose is:

$$A^\top = \begin{bmatrix} A_{00}^\top & A_{10}^\top & \cdots & A_{(p-1)0}^\top \\ A_{01}^\top & A_{11}^\top & \cdots & A_{(p-1)1}^\top \\ \vdots & \vdots & \ddots & \vdots \\ A_{0(q-1)}^\top & A_{1(q-1)}^\top & \cdots & A_{(p-1)(q-1)}^\top \end{bmatrix}.$$

The transpose operation has two layers: first transpose the block structure (rows and columns of blocks), then transpose each individual block.

10.2.2.2 Scalar Multiplication

For a scalar k and block matrix A:

$$kA = \begin{bmatrix} kA_{00} & kA_{01} & \cdots & kA_{0(q-1)} \\ kA_{10} & kA_{11} & \cdots & kA_{1(q-1)} \\ \vdots & \vdots & \ddots & \vdots \\ kA_{(p-1)0} & kA_{(p-1)1} & \cdots & kA_{(p-1)(q-1)} \end{bmatrix}.$$

10.2.2.3 Addition and Subtraction

For block matrices A and B with the same partition structure:

$$A \pm B = \begin{bmatrix} A_{00} \pm B_{00} & A_{01} \pm B_{01} & \cdots & A_{0(q-1)} \pm B_{0(q-1)} \\ A_{10} \pm B_{10} & A_{11} \pm B_{11} & \cdots & A_{1(q-1)} \pm B_{1(q-1)} \\ \vdots & \vdots & \ddots & \vdots \\ A_{(p-1)0} \pm B_{(p-1)0} & A_{(p-1)1} \pm B_{(p-1)1} & \cdots & A_{(p-1)(q-1)} \pm B_{(p-1)(q-1)} \end{bmatrix}.$$

10.2.3 Theorem: Block Matrix Multiplication

Let $A \in \mathbb{R}^{M\times d}$ and $B \in \mathbb{R}^{d\times N}$ be partitioned into blocks:

$$A = \begin{bmatrix} A_{00} & \cdots & A_{0(r-1)} \\ \vdots & \ddots & \vdots \\ A_{(p-1)0} & \cdots & A_{(p-1)(r-1)} \end{bmatrix}, \quad B = \begin{bmatrix} B_{00} & \cdots & B_{0(q-1)} \\ \vdots & \ddots & \vdots \\ B_{(r-1)0} & \cdots & B_{(r-1)(q-1)} \end{bmatrix},$$

where the column partition of A matches the row partition of B. Then

$$(AB)_{ij} = \sum_{k=0}^{r-1} A_{ik}B_{kj} \tag{10.8}$$

where $(AB)_{ij}$ denotes the (i,j)-th block of AB.

Proof

The (i,j)-th block of AB corresponds to rows i of A and columns j of B. Since matrix multiplication is row-by-column, we have:

$$(AB)_{ij} = A_{i*}B_{*j} = \begin{bmatrix} A_{i0} & \cdots & A_{i(r-1)} \end{bmatrix} \begin{bmatrix} B_{0j} \\ \vdots \\ B_{(r-1)j} \end{bmatrix} = \sum_{k=0}^{r-1} A_{ik}B_{kj}.$$

10.2.3.1 Example: Block Matrix Multiplication

Consider matrices

$$A = \begin{bmatrix} 1 & 2 & 3 & 4 \\ 5 & 6 & 7 & 8 \\ 9 & 10 & 11 & 12 \end{bmatrix} \in \mathbb{R}^{3\times 4}, \quad B = \begin{bmatrix} 1 & 2 \\ 3 & 4 \\ 5 & 6 \\ 7 & 8 \end{bmatrix} \in \mathbb{R}^{4\times 2}.$$

Partition A into 2×2 blocks and B into 2×1 blocks:

$$A = \begin{bmatrix} A_{00} & A_{01} \\ A_{10} & A_{11} \end{bmatrix} = \begin{bmatrix} \begin{bmatrix} 1 & 2 \\ 5 & 6 \end{bmatrix} & \begin{bmatrix} 3 & 4 \\ 7 & 8 \end{bmatrix} \\ \begin{bmatrix} 9 & 10 \end{bmatrix} & \begin{bmatrix} 11 & 12 \end{bmatrix} \end{bmatrix},$$

$$B = \begin{bmatrix} B_{00} \\ B_{10} \end{bmatrix} = \begin{bmatrix} \begin{bmatrix} 1 & 2 \\ 3 & 4 \end{bmatrix} \\ \begin{bmatrix} 5 & 6 \\ 7 & 8 \end{bmatrix} \end{bmatrix}.$$

Then the block matrix product is:

$$AB = \begin{bmatrix} A_{00}B_{00} + A_{01}B_{10} \\ A_{10}B_{00} + A_{11}B_{10} \end{bmatrix} = \begin{bmatrix} \begin{bmatrix} 1 & 2 \\ 5 & 6 \end{bmatrix} \begin{bmatrix} 1 & 2 \\ 3 & 4 \end{bmatrix} + \begin{bmatrix} 3 & 4 \\ 7 & 8 \end{bmatrix} \begin{bmatrix} 5 & 6 \\ 7 & 8 \end{bmatrix} \\ \begin{bmatrix} 9 & 10 \end{bmatrix} \begin{bmatrix} 1 & 2 \\ 3 & 4 \end{bmatrix} + \begin{bmatrix} 11 & 12 \end{bmatrix} \begin{bmatrix} 5 & 6 \\ 7 & 8 \end{bmatrix} \end{bmatrix}.$$

Computing each block:

$$A_{00}B_{00} = \begin{bmatrix} 1 & 2 \\ 5 & 6 \end{bmatrix} \begin{bmatrix} 1 & 2 \\ 3 & 4 \end{bmatrix} = \begin{bmatrix} 7 & 10 \\ 23 & 34 \end{bmatrix},$$

$$A_{01}B_{10} = \begin{bmatrix} 3 & 4 \\ 7 & 8 \end{bmatrix} \begin{bmatrix} 5 & 6 \\ 7 & 8 \end{bmatrix} = \begin{bmatrix} 43 & 50 \\ 91 & 106 \end{bmatrix},$$

$$A_{10}B_{00} = \begin{bmatrix} 9 & 10 \end{bmatrix} \begin{bmatrix} 1 & 2 \\ 3 & 4 \end{bmatrix} = \begin{bmatrix} 39 & 58 \end{bmatrix},$$

$$A_{11}B_{10} = \begin{bmatrix} 11 & 12 \end{bmatrix} \begin{bmatrix} 5 & 6 \\ 7 & 8 \end{bmatrix} = \begin{bmatrix} 139 & 162 \end{bmatrix}.$$

Therefore,

$$AB = \begin{bmatrix} \begin{bmatrix} 7 & 10 \\ 23 & 34 \end{bmatrix} + \begin{bmatrix} 43 & 50 \\ 91 & 106 \end{bmatrix} \\ \begin{bmatrix} 39 & 58 \end{bmatrix} + \begin{bmatrix} 139 & 162 \end{bmatrix} \end{bmatrix} = \begin{bmatrix} 50 & 60 \\ 114 & 140 \\ 178 & 220 \end{bmatrix}.$$

This matches the direct matrix multiplication AB computed element-wise.

10.2.4 Block Matrix Multiplication: Key Partitioning Patterns

For block matrix multiplication, the most important patterns are when both matrices are partitioned, enabling parallel computation and distributed processing.

10.2.4.1 Pattern 1: Both Matrices Fully Partitioned

When both A and B are fully partitioned into blocks:

$$A = \begin{bmatrix} A_{00} & A_{01} \\ A_{10} & A_{11} \end{bmatrix}, \quad B = \begin{bmatrix} B_{00} & B_{01} \\ B_{10} & B_{11} \end{bmatrix},$$

where the column partition of A matches the row partition of B, then

$$AB = \begin{bmatrix} A_{00}B_{00} + A_{01}B_{10} & A_{00}B_{01} + A_{01}B_{11} \\ A_{10}B_{00} + A_{11}B_{10} & A_{10}B_{01} + A_{11}B_{11} \end{bmatrix}.$$

This is the most general block multiplication form, combining elements from both the scalar product expansion (row-by-column) and the outer product expansion (column-by-row) perspectives.

10.2.4.2 Pattern 2: A Partitioned Vertically, B Partitioned Horizontally

When A is partitioned vertically and B is partitioned horizontally:

$$A = \begin{bmatrix} A_0 \\ A_1 \end{bmatrix}, \quad B = \begin{bmatrix} B_0 & B_1 \end{bmatrix},$$

where $A_0 \in \mathbb{R}^{k \times D}$, $A_1 \in \mathbb{R}^{(n-k) \times D}$, $B_0 \in \mathbb{R}^{D \times r}$, $B_1 \in \mathbb{R}^{D \times (m-r)}$, then

$$AB = \begin{bmatrix} A_0 \\ A_1 \end{bmatrix} \begin{bmatrix} B_0 & B_1 \end{bmatrix} = \begin{bmatrix} A_0B_0 & A_0B_1 \\ A_1B_0 & A_1B_1 \end{bmatrix}.$$

This pattern enables **parallel computation** by allowing different processors to work on different blocks simultaneously. These block partitioning patterns are essential for **distributed computing** (including **tensor parallelism**, **pipeline parallelism**, and **data parallelism**) and **memory-efficient**

algorithms where large matrices are processed in chunks. **Pattern 1** (complete block structure) is the most general form and is used in distributed matrix multiplication where matrices are partitioned across multiple devices. **Pattern 2** enables parallel computation by allowing different processors to compute different blocks (e.g., A_0B_0 and A_1B_1 can be computed in parallel). Understanding these patterns helps choose the most efficient computational strategy for a given problem and hardware architecture. For basic matrix multiplication perspectives (column-wise, row-wise, outer product), see Chapter~2.

10.2.5 Definition: Block Diagonal Matrix

A matrix $A \in \mathbb{R}^{N\times N}$ is **block diagonal** if it can be partitioned as

$$A = \begin{bmatrix} A_0 & & & \\ & A_1 & & \\ & & \ddots & \\ & & & A_{k-1} \end{bmatrix},$$

where $A_i \in \mathbb{R}^{n_i\times n_i}$ are square blocks on the diagonal, and all off-diagonal blocks are zero matrices. We denote this as $A = \text{blkdiag}(A_0, A_1, \dots, A_{k-1})$.

10.2.6 Definition: Block Upper/Lower Triangular Matrix

A matrix $A \in \mathbb{R}^{N\times N}$ is **block upper triangular** if all blocks below the block diagonal are zero:

$$A = \begin{bmatrix} A_{00} & A_{01} & \cdots & A_{0(k-1)} \\ & A_{11} & \cdots & A_{1(k-1)} \\ & & \ddots & \vdots \\ & & & A_{(k-1)(k-1)} \end{bmatrix}.$$

Similarly, A is **block lower triangular** if all blocks above the block diagonal are zero.

10.2.7 Theorem: Schur Complement

Let $A \in \mathbb{R}^{N\times N}$ be partitioned as

$$A = \begin{bmatrix} A_{00} & A_{01} \\ A_{10} & A_{11} \end{bmatrix},$$

where $A_{00} \in \mathbb{R}^{m\times m}$ is invertible. The **Schur complement** of A_{00} in A is defined as

$$S = A_{11} - A_{10}A_{00}^{-1}A_{01} \tag{10.9}$$

If A is invertible, then

$$A^{-1} = \begin{bmatrix} A_{00}^{-1} + A_{00}^{-1}A_{01}S^{-1}A_{10}A_{00}^{-1} & -A_{00}^{-1}A_{01}S^{-1} \\ -S^{-1}A_{10}A_{00}^{-1} & S^{-1} \end{bmatrix} \tag{10.10}$$

provided that S is invertible.

Proof

We verify by direct multiplication:

$$\begin{bmatrix} A_{00} & A_{01} \\ A_{10} & A_{11} \end{bmatrix} \begin{bmatrix} A_{00}^{-1} + A_{00}^{-1}A_{01}S^{-1}A_{10}A_{00}^{-1} & -A_{00}^{-1}A_{01}S^{-1} \\ -S^{-1}A_{10}A_{00}^{-1} & S^{-1} \end{bmatrix} = I_{N\times N}.$$

Expanding the top-left block:

$$A_{00}(A_{00}^{-1} + A_{00}^{-1}A_{01}S^{-1}A_{10}A_{00}^{-1}) - A_{01}S^{-1}A_{10}A_{00}^{-1} = I + A_{01}S^{-1}A_{10}A_{00}^{-1} - A_{01}S^{-1}A_{10}A_{00}^{-1} = I.$$

Expanding the top-right block:

$$-A_{00}A_{00}^{-1}A_{01}S^{-1} + A_{01}S^{-1} = -A_{01}S^{-1} + A_{01}S^{-1} = 0.$$

The bottom blocks follow similarly.

The Schur complement is fundamental in many areas of machine learning and optimization. In **Gaussian elimination**, the Schur complement appears naturally when eliminating variables block-wise. In **Gaussian processes**, the Schur complement formula enables efficient computation of conditional distributions: if $\mathbf{x} = [\mathbf{x}_0, \mathbf{x}_1]^\top$ follows a multivariate Gaussian with covariance $\Sigma = \begin{bmatrix} \Sigma_{00} & \Sigma_{01} \\ \Sigma_{10} & \Sigma_{11} \end{bmatrix}$, then the conditional covariance of $\mathbf{x}_1$ given $\mathbf{x}_0$ is the Schur complement $\Sigma_{11} - \Sigma_{10}\Sigma_{00}^{-1}\Sigma_{01}$. In **optimization**, block-structured Hessian matrices can be inverted efficiently using Schur complements, which is crucial for second-order methods like Newton's method and natural gradient descent.

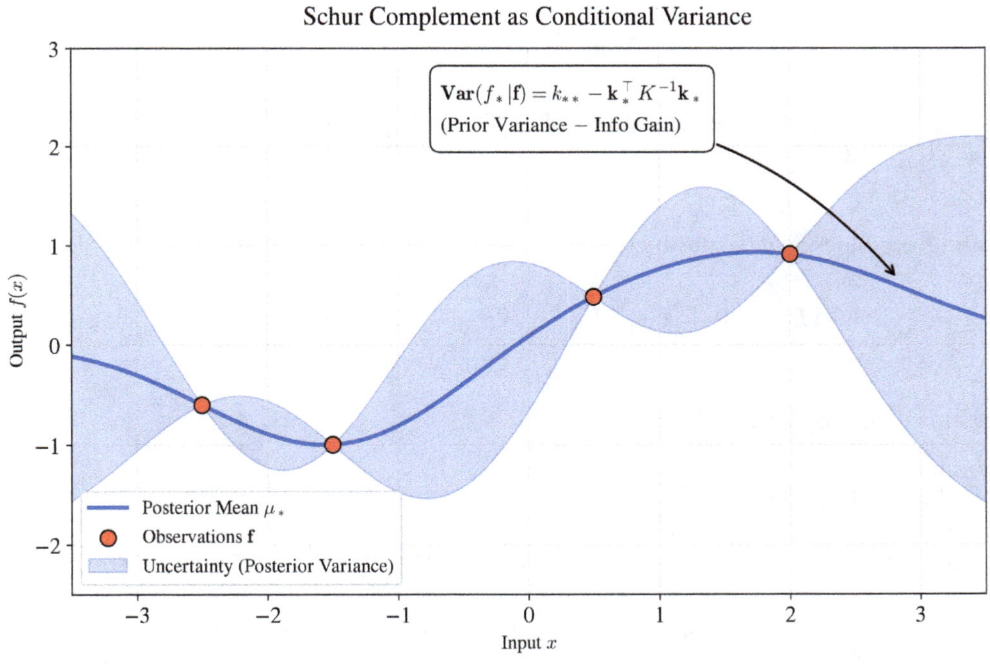

Figure 10.2: Schur Complement in Gaussian Processes

Figure 10.2 demonstrates the Schur complement in the context of Gaussian process regression. The plot shows training observations (red points) and the posterior mean prediction (blue line) with a 95% confidence interval (shaded blue region). The key insight is that **the posterior variance at any test point is exactly the Schur complement** $k_{**} - \mathbf{k}_*^\top K^{-1}\mathbf{k}_*$, where k_{**} is the prior variance, K is the covariance matrix of training points, and $\mathbf{k}_*$ is the covariance between training and test points. The shaded region represents the reduction in uncertainty (information gain) compared to the prior: near training points, the variance is small (high confidence), while far from training data, the variance approaches the prior variance. This visualization connects the abstract linear algebra concept of the Schur complement to the intuitive notion of "uncertainty reduction" in probabilistic modeling, making clear why the Schur complement is fundamental in Gaussian processes, Bayesian inference, and conditional probability computations.

10.3 Block Matrix Decompositions

Block decompositions extend classical decompositions to partitioned matrices, enabling scalable and parallel algorithms.

10.3.1 Block LU Decomposition

10.3.2 Theorem: Block LU Decomposition

Let $A \in \mathbb{R}^{N \times N}$ be partitioned as

$$A = \begin{bmatrix} A_{00} & A_{01} \\ A_{10} & A_{11} \end{bmatrix},$$

where $A_{00} \in \mathbb{R}^{m \times m}$ is invertible. Then A can be decomposed as

$$A = \begin{bmatrix} L_{00} & 0 \\ L_{10} & L_{11} \end{bmatrix} \begin{bmatrix} U_{00} & U_{01} \\ 0 & U_{11} \end{bmatrix} \tag{10.11}$$

where $L_{00}U_{00} = A_{00}$ (LU decomposition of A_{00}), $L_{00}U_{01} = A_{01}$, $L_{10}U_{00} = A_{10}$, and $L_{11}U_{11} = A_{11} - L_{10}U_{01}$ (the Schur complement).

Proof

Expanding the block matrix product:

$$\begin{bmatrix} L_{00} & 0 \\ L_{10} & L_{11} \end{bmatrix} \begin{bmatrix} U_{00} & U_{01} \\ 0 & U_{11} \end{bmatrix} = \begin{bmatrix} L_{00}U_{00} & L_{00}U_{01} \\ L_{10}U_{00} & L_{10}U_{01} + L_{11}U_{11} \end{bmatrix}.$$

Setting this equal to A and solving:

- $L_{00}U_{00} = A_{00}$ (standard LU of A_{00})
- $L_{00}U_{01} = A_{01}$ (solve for U_{01})
- $L_{10}U_{00} = A_{10}$ (solve for L_{10})
- $L_{10}U_{01} + L_{11}U_{11} = A_{11}$, so $L_{11}U_{11} = A_{11} - L_{10}U_{01}$ (LU of Schur complement)

Block LU decomposition enables **parallel computation** by allowing different processors to work on different blocks simultaneously. In large-scale optimization, Hessian matrices are often block-structured (e.g., blocks corresponding to different layers in a neural network), and block LU decomposition enables efficient Newton step computation. The recursive application of block LU decomposition leads to **divide-and-conquer algorithms** that are highly parallelizable and cache-efficient, making them suitable for modern GPU architectures and distributed computing systems.

10.3.3 Block Cholesky Decomposition

10.3.4 Theorem: Block Cholesky Decomposition

Let $A \in \mathbb{R}^{N\times N}$ be symmetric positive definite, partitioned as

$$A = \begin{bmatrix} A_{00} & A_{01} \\ A_{10} & A_{11} \end{bmatrix},$$

where $A_{00} \in \mathbb{R}^{m\times m}$ is positive definite. Then A can be decomposed as

$$A = \begin{bmatrix} L_{00} & 0 \\ L_{10} & L_{11} \end{bmatrix} \begin{bmatrix} L_{00}^\top & L_{10}^\top \\ 0 & L_{11}^\top \end{bmatrix} \tag{10.12}$$

where $L_{00}L_{00}^\top = A_{00}$ (Cholesky of A_{00}), $L_{10} = A_{10}(L_{00}^\top)^{-1}$, and $L_{11}L_{11}^\top = A_{11} - L_{10}L_{10}^\top$ (Cholesky of the Schur complement).

> Block Cholesky decomposition is essential for large-scale covariance matrix computations in Gaussian processes, Bayesian neural networks, and multivariate statistics. When covariance matrices have block structure (e.g., spatial or temporal blocks), block Cholesky enables efficient sampling and inference. The decomposition also appears in **preconditioning** for iterative solvers, where block-structured preconditioners are constructed to accelerate convergence in optimization algorithms.

10.3.5 Block QR Decomposition

10.3.6 Theorem: Block QR Decomposition

Let $A \in \mathbb{R}^{M\times N}$ be partitioned column-wise as

$$A = \begin{bmatrix} A_0 & A_1 & \cdots & A_{k-1} \end{bmatrix},$$

where each $A_i \in \mathbb{R}^{M\times n_i}$ with $\sum_{i=0}^{k-1} n_i = N$. Then A can be decomposed as

$$A = QR = \begin{bmatrix} Q_0 & Q_1 & \cdots & Q_{k-1} \end{bmatrix} R \tag{10.13}$$

where Q has orthonormal columns and R is block upper triangular.

> Block QR decomposition enables **incremental QR factorization**, where new columns are added to the decomposition without recomputing from scratch. This is crucial in **online learning** algorithms, **streaming PCA**, and **adaptive filtering**. In neural networks, block QR can be used for **orthonormal weight initialization** and **gradient normalization** across parameter groups, maintaining training stability while enabling efficient block-wise updates.

10.4 Geometric and Computational Interpretations

10.4.1 Decomposition as Rotation and Scaling

Matrix decompositions reveal geometric structure:

1. **SVD**: $A = U\Sigma V^\top$ decomposes any linear transformation into:
   - Rotation/reflection ($V^\top$): maps input to principal axes

  - Scaling (Σ): stretches along principal axes
  - Rotation/reflection (U): maps to output space

2. **Polar Decomposition**: $A = QS$ separates:
  - Orthogonal part (Q): preserves angles and distances
  - Symmetric part (S): scales along eigenvector directions

3. **Spectral Decomposition**: $A = Q\Lambda Q^\top$ (for symmetric A) shows:
  - Scaling along eigenvector directions
  - No rotation (since Q is the same for input and output)

These geometric interpretations are fundamental for understanding how neural network layers transform data. In **convolutional neural networks**, the geometric structure of weight matrices (learned through training) determines how features are extracted and combined. In **attention mechanisms**, the decomposition of attention matrices reveals how information flows between tokens. The scaling factors (singular values, eigenvalues) determine the **conditioning** of the transformation, which affects gradient flow, optimization stability, and generalization in deep learning.

10.4.2 Blocks as Subspaces and Conditional Structures

Block partitioning reveals structural relationships:

1. **Block Diagonal**: $A = \text{blkdiag}(A_0, A_1, \dots, A_{k-1})$ indicates that the transformation acts **independently** on k subspaces. This appears in:
  - Multi-task learning (different tasks have separate parameter blocks)
  - Modular neural networks (different modules process different inputs)
  - Distributed systems (different processors handle different blocks)

2. **Block Triangular**: Indicates **hierarchical dependencies**:
  - Lower blocks depend on upper blocks
  - Enables sequential computation (process blocks in order)
  - Appears in causal models, autoregressive models, and sequential neural networks

3. **Schur Complement**: Represents **conditional structure**:
  - $S = A_{11} - A_{10}A_{00}^{-1}A_{01}$ is the covariance of A_{11} conditioned on A_{00}
  - Fundamental in Gaussian processes, Bayesian networks, and probabilistic graphical models

Block structures in neural networks often correspond to architectural choices. In **transformer architectures**, attention matrices can be block-sparse (local attention, sparse attention patterns) to reduce computational cost. In **residual networks**, the block structure of weight matrices enables skip connections and gradient flow. In **mixture-of-experts models**, different experts correspond to different blocks, and routing mechanisms determine which blocks are activated. Understanding block structure is crucial for **model interpretability**, **efficient inference**, and **distributed training**. In **tensor parallel training**, weight matrices are partitioned into blocks across multiple devices: for a weight matrix $W \in \mathbb{R}^{m\times n}$, if partitioned column-wise as $W = [W_0, W_1, \dots, W_{k-1}]$ across k devices, then the forward pass computes $W\mathbf{x} = [W_0, W_1, \dots, W_{k-1}]\mathbf{x} = [W_0\mathbf{x}, W_1\mathbf{x}, \dots, W_{k-1}\mathbf{x}]$ where each device computes its block-vector product independently, and results are concatenated. This corresponds to **Perspective 1** (column-wise partitioning) of block matrix multiplication, enabling efficient parallel computation of large matrix operations.

10.4.3 Whitening, Decoupling, and Conditioning

Matrix decompositions enable data transformations:

1. **Whitening**: For a covariance matrix Σ, Cholesky decomposition $\Sigma = LL^\top$ enables whitening: $\mathbf{z} = L^{-1}\mathbf{x}$ has identity covariance. This is used in:
   - Data preprocessing (whitening layers in CNNs)
   - Normalization techniques (batch normalization, layer normalization)
   - Feature decorrelation
2. **Decoupling**: Block diagonalization decouples different components:
   - Independent parameter groups in optimization
   - Separate feature spaces in multi-modal learning
   - Modular computation in distributed systems
3. **Conditioning**: The condition number $\kappa(A) = \sigma_{\text{max}}(A)/\sigma_{\text{min}}(A)$ (from SVD) measures numerical stability:
   - Well-conditioned matrices ($\kappa \approx 1$) enable stable computation
   - Ill-conditioned matrices ($\kappa \gg 1$) cause numerical errors
   - Regularization (e.g., ridge regression) improves conditioning

These concepts are central to modern deep learning practice. **Whitening** is implicitly performed by normalization layers (batch norm, layer norm) that decorrelate and standardize activations, improving training stability and convergence. **Decoupling** appears in **multi-task learning** where different tasks have separate parameter blocks, and in **modular architectures** where different modules process different inputs. **Conditioning** is crucial for understanding why some optimization problems are difficult (ill-conditioned Hessians) and how regularization (weight decay, dropout) improves optimization landscapes. In **second-order optimization**, preconditioning matrices are constructed to improve conditioning and accelerate convergence.

10.5 Applications in AI and Optimization

10.5.1 Hessian Block Structure in Optimization

In large-scale optimization, the Hessian matrix $H = \nabla^2 \mathcal{L}(\theta)$ (formally defined in Chapter~11) often exhibits block structure:

$$H = \begin{bmatrix} H_{00} & H_{01} & \cdots \\ H_{10} & H_{11} & \cdots \\ \vdots & \vdots & \ddots \end{bmatrix},$$

where blocks H_{ij} correspond to interactions between parameter groups (e.g., different layers, different modules).

Block-structured Hessians enable **block-wise second-order methods** where different parameter groups are updated using different step sizes or different optimization algorithms. In **natural gradient descent**, the Fisher information matrix (a block-structured approximation of the Hessian) is used to compute update directions. Block structure also enables **distributed optimization** where different processors update different parameter blocks in parallel.

10.5.1.1 GPTQ: Block-wise Hessian for Quantization

GPTQ (GPT Quantization) is a post-training quantization method that uses block-wise Hessian decomposition to quantize weight matrices with minimal accuracy loss. The key insight is that computing and inverting the full Hessian matrix is computationally infeasible for large weight matrices, but block-wise decomposition makes it tractable.

Problem Setup:

For a Transformer linear layer with weight matrix $W \in \mathbb{R}^{d_{\text{out}} \times d_{\text{in}}}$, quantization seeks to find a quantized weight matrix $\hat{W}$ that minimizes the reconstruction error:

$$\min_{\hat{W}} (\hat{W} - W)^\top H (\hat{W} - W),$$

where H is the Hessian matrix of the loss with respect to the weights. For a full weight matrix, $H \in \mathbb{R}^{(d_{\text{out}} \cdot d_{\text{in}}) \times (d_{\text{out}} \cdot d_{\text{in}})}$, which is computationally intractable for large models (e.g., $d_{\text{out}} = 4096$, $d_{\text{in}} = 4096$ gives $H \in \mathbb{R}^{16,777,216 \times 16,777,216}$).

Block-wise Decomposition (The Key Innovation):

GPTQ partitions the weight matrix W into blocks, typically column-wise (or row-wise), with block size b (commonly $b = 32$ to 128). For column-wise partitioning:

$$W = \begin{bmatrix} W_0 & W_1 & \cdots & W_{k-1} \end{bmatrix},$$

where each block $W_i \in \mathbb{R}^{d_{\text{out}} \times b}$ and $k = \lceil d_{\text{in}} / b \rceil$.

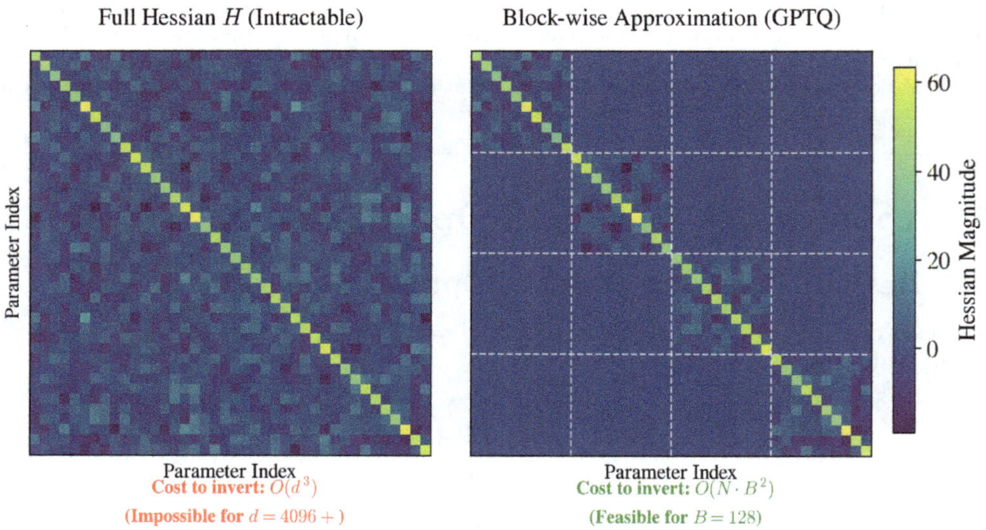

Figure 10.3: Hessian Structure for GPTQ

Figure 10.3 compares the full Hessian matrix (left) with the block-wise approximation used in GPTQ (right). The full Hessian $H \in \mathbb{R}^{(d_{\text{out}} \cdot d_{\text{in}}) \times (d_{\text{out}} \cdot d_{\text{in}})}$ is dense and computationally intractable to invert for large models (e.g., $O(d^3)$ cost for $d = 4096+$). GPTQ's key insight is to approximate the Hessian as block-diagonal, where only diagonal blocks $H_{\text{block}} \in \mathbb{R}^{b \times b}$ are retained (with $b = 128$ typically). This reduces the inversion cost to $O(N \cdot B^2)$ where N is the number of blocks and B is the block size, making quantization feasible for large language models. The white grid lines in the right panel emphasize the block structure, showing how the problem is decomposed into independent, tractable subproblems. This block-wise approach enables GPTQ to achieve 4-bit quantization with minimal accuracy loss by processing weight matrices in manageable chunks rather than attempting to optimize the entire matrix at once.

Making the Hessian Computable:

Instead of computing the full Hessian $H \in \mathbb{R}^{(d_{\text{out}} \cdot d_{\text{in}}) \times (d_{\text{out}} \cdot d_{\text{in}})}$, GPTQ computes block-wise Hessians:

$$H_{\text{block}} \in \mathbb{R}^{b \times b}, \quad b \ll d_{\text{in}}.$$

For each block i, the optimization problem becomes:

$$\min_{\hat{W}_i} (\hat{W}_i - W_i)^\top H_{\text{block}} (\hat{W}_i - W_i),$$

where H_{block} is the Hessian for the i-th block. This reduction from $O(d_{\text{in}}^2)$ to $O(b^2)$ makes the Hessian computation and inversion (or Cholesky decomposition) computationally feasible.

Algorithm Steps:

1. **Block-wise Processing**: Process blocks sequentially (or in parallel for independent blocks).
2. **Hessian Computation**: For each block, compute H_{block} using calibration data.
3. **Cholesky/Inverse**: Compute H_{block}^{-1} or the Cholesky decomposition L where $H_{\text{block}} = LL^\top$ (since H_{block} is positive definite).
4. **Quantization with Error Propagation**: Quantize each block while propagating quantization errors to subsequent blocks to maintain global accuracy.

Why Block-wise Works:

- **Computational Feasibility**: Block size $b = 128$ gives $H_{\text{block}} \in \mathbb{R}^{128 \times 128}$, which is manageable, compared to full Hessian $H \in \mathbb{R}^{16M \times 16M}$.
- **Preserves Structure**: Block-wise processing respects the column-wise structure of weight matrices, enabling efficient implementation.
- **Error Compensation**: Error propagation between blocks ensures that quantization errors in early blocks are compensated in later blocks, maintaining overall model accuracy.

GPTQ demonstrates how block structure transforms an intractable problem (full Hessian inversion) into a tractable one (block-wise Hessian inversion). This principle applies broadly: whenever a large matrix operation is infeasible, block-wise decomposition can make it computationally viable. The block size b is a trade-off: smaller blocks reduce memory and computation per block but may lose some cross-block correlations; larger blocks capture more correlations but require more computation. In practice, $b = 128$ is a common choice that balances accuracy and efficiency. GPTQ achieves 4-bit quantization with minimal accuracy loss (often $< 1\%$ degradation) by leveraging this block-wise Hessian approach, making it a standard method for quantizing large language models.

10.5.2 Covariance and PCA: Primal vs Dual

Principal Component Analysis (PCA) can be formulated in two ways:

1. **Primal (Covariance)**: Decompose $C = \frac{1}{N-1} X^\top X$ (covariance matrix) via eigendecomposition: $C = Q \Lambda Q^\top$.
2. **Dual (Gram)**: Decompose $K = \frac{1}{N-1} X X^\top$ (Gram matrix) via eigendecomposition: $K = U \Lambda U^\top$.

The relationship: $X = U \Sigma Q^\top$ (SVD), where Q contains principal components and U contains dual representations.

The primal-dual perspective is fundamental in kernel methods. In **kernel PCA**, we work with the Gram matrix $K_{ij} = \kappa(\mathbf{x}_i, \mathbf{x}_j)$ instead of the covariance matrix, enabling nonlinear dimensionality reduction. In **support vector machines**, the dual formulation uses the Gram matrix to find support vectors. The block structure of covariance matrices (e.g., spatial blocks in image data, temporal blocks in time series) enables efficient PCA computation via block decompositions. In **large-scale PCA**, block-wise SVD is used to handle datasets that don't fit in memory.

10.5.3 Attention Matrices and Low-Rank Structure

Attention matrices in transformers have the form

$$\text{Attention} = \text{softmax}\left(\frac{QK^\top}{\sqrt{d_k}}\right)V,$$

where $Q, K, V \in \mathbb{R}^{N\times d}$ are query, key, and value matrices for N tokens.

Attention matrices are often **low-rank** or have **block-sparse structure. Sparse attention**[1] patterns (local attention, strided attention, random attention) create block-sparse attention matrices, reducing computational cost from $O(N^2)$ to $O(N\sqrt{N})$ or $O(N \log N)$. **Low-rank attention** approximations (e.g., using truncated SVD) enable efficient computation for long sequences. **Block attention**[2] (dividing sequences into blocks and applying attention within and between blocks) is used in models like Longformer[3] and BigBird[4] to handle very long sequences.

Understanding the block and low-rank structure of attention matrices is crucial for designing efficient transformer architectures.

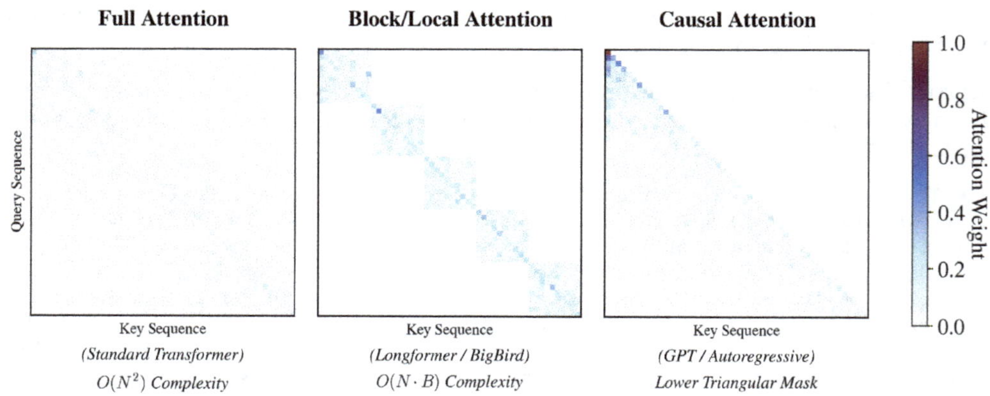

Figure 10.4: Attention Matrix Patterns

[1]Child, R., Gray, S., Radford, A., & Sutskever, I. "Generating Long Sequences with Sparse Transformers." OpenAI, arXiv 2019. https://arxiv.org/abs/1904.10509

[2]Block (local/window) attention is used in Longformer (https://arxiv.org/abs/2004.05150) and BigBird (https://arxiv.org/abs/2007.14062) for long sequences.

[3]Beltagy, I., Peters, M. E., & Cohan, A. "Longformer: The Long-Document Transformer." arXiv 2020. https://arxiv.org/abs/2004.05150

[4]Zaheer, M., Guruganesh, G., Dubey, K. A., et al. "Big Bird: Transformers for Longer Sequences." NeurIPS 2020. https://arxiv.org/abs/2007.14062

Figure 10.4 visualizes three common block structures in transformer attention matrices. **Panel A (Full Attention)** shows the dense attention pattern used in standard transformers, where every token attends to every other token, resulting in $O(N^2)$ complexity. **Panel B (Block/Local Attention)** demonstrates the block-sparse pattern used in models like Longformer and BigBird, where attention is restricted to local blocks along the diagonal, reducing complexity to $O(N \cdot B)$ where B is the block size. **Panel C (Causal Attention)** shows the lower triangular mask used in autoregressive models like GPT, where tokens can only attend to previous tokens in the sequence. The white regions represent masked (zero) attention weights, while colored regions show active attention. These block structures enable efficient computation for long sequences by exploiting sparsity, demonstrating how block matrix concepts translate directly to practical architectural choices in modern transformer models.

10.5.4 Tensor Parallelism and Distributed Matrix Operations

In **tensor parallel training**[5], large weight matrices are partitioned across multiple devices (GPUs) to enable distributed computation.

10.5.4.1 Column-wise Partitioning (Tensor Parallel)

For a weight matrix $W \in \mathbb{R}^{m \times n}$ partitioned column-wise across k devices:

$$W = \begin{bmatrix} W_0 & W_1 & \cdots & W_{k-1} \end{bmatrix},$$

where $W_i \in \mathbb{R}^{m \times (n/k)}$ is stored on device i. The forward pass computes:

$$W\mathbf{x} = \begin{bmatrix} W_0 & W_1 & \cdots & W_{k-1} \end{bmatrix} \mathbf{x} = \begin{bmatrix} W_0\mathbf{x} \\ W_1\mathbf{x} \\ \vdots \\ W_{k-1}\mathbf{x} \end{bmatrix},$$

where each device computes $W_i\mathbf{x}$ independently, and results are concatenated. This corresponds to **Perspective 1** (column-wise partitioning) of block matrix multiplication.

10.5.4.2 Row-wise Partitioning (Tensor Parallel)

Alternatively, W can be partitioned row-wise:

$$W = \begin{bmatrix} W_0 \\ W_1 \\ \vdots \\ W_{k-1} \end{bmatrix},$$

where $W_i \in \mathbb{R}^{(m/k) \times n}$ is stored on device i. The forward pass computes:

[5] Fuheng Wu, *Distributed AI Systems: A Practical Guide to Building Scalable Training, Inference, and Serving Systems for Production AI.* https://www.amazon.com/dp/B0GN8NRVQX/

$$W\mathbf{x} = \begin{bmatrix} W_0 \\ W_1 \\ \vdots \\ W_{k-1} \end{bmatrix} \mathbf{x} = \begin{bmatrix} W_0\mathbf{x} \\ W_1\mathbf{x} \\ \vdots \\ W_{k-1}\mathbf{x} \end{bmatrix},$$

where each device computes $W_i\mathbf{x}$ independently. This corresponds to **Perspective 4** (row-wise partitioning).

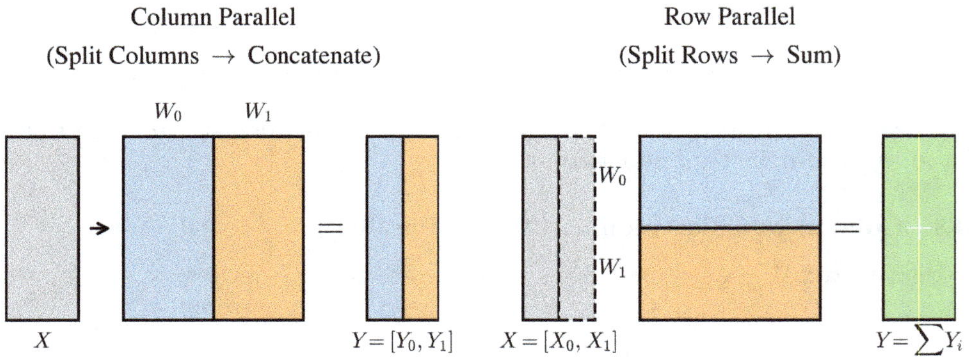

Figure 10.5: Tensor Parallel Partitioning

Figure 10.5 illustrates the two primary partitioning strategies used in tensor parallelism. **Panel 1 (Column Parallel)** shows column-wise partitioning where the weight matrix W is split vertically into blocks W_0 and W_1. The input X is broadcast to all devices, each device computes W_iX independently, and the outputs are concatenated: $Y = [Y_0, Y_1]$. This corresponds to **Perspective 1** of block matrix multiplication. **Panel 2 (Row Parallel)** shows row-wise partitioning where W is split horizontally and the input X is partitioned column-wise. Each device computes W_iX_i on its local input partition, and the outputs are summed via all-reduce: $Y = \sum Y_i$. This corresponds to **Perspective 4** of block matrix multiplication. The different colors (light blue and light orange) distinguish the blocks assigned to different devices, while the green output in Panel 2 indicates the summed result. These visualizations clarify the geometric and computational differences between the two strategies, showing how block matrix structure directly maps to distributed computation patterns in modern deep learning systems.

Tensor parallelism is a fundamental distributed training strategy for large language models and transformer architectures. By partitioning weight matrices into blocks across devices, tensor parallel enables training models that are too large to fit on a single device. The block matrix multiplication perspectives (especially **Perspective 1** for column-wise and **Perspective 4** for row-wise) directly correspond to how matrix operations are distributed in tensor parallel implementations (e.g., Megatron-LM[6], PyTorch's tensor parallel[7]). The communication pattern (all-reduce[8], all-gather[9]) required to combine results from different devices follows naturally from

---

[6] NVIDIA Megatron-LM: https://github.com/NVIDIA/Megatron-LM

[7] PyTorch tensor parallel (torch.distributed.tensor): https://pytorch.org/docs/stable/distributed.tensor.html

[8] All-reduce: each rank holds a tensor; all ranks receive the reduction (e.g., sum) across ranks. NCCL collective operations: https://docs.nvidia.com/deeplearning/nccl/user-guide/docs/api/colls.html

[9] All-gather: each rank holds a shard; all ranks receive the concatenation of all shards. NCCL collective

the block structure. Understanding block matrix multiplication is essential for implementing and optimizing tensor parallel algorithms, as it reveals the communication and computation trade-offs in distributed matrix operations.

10.6 Mindmap

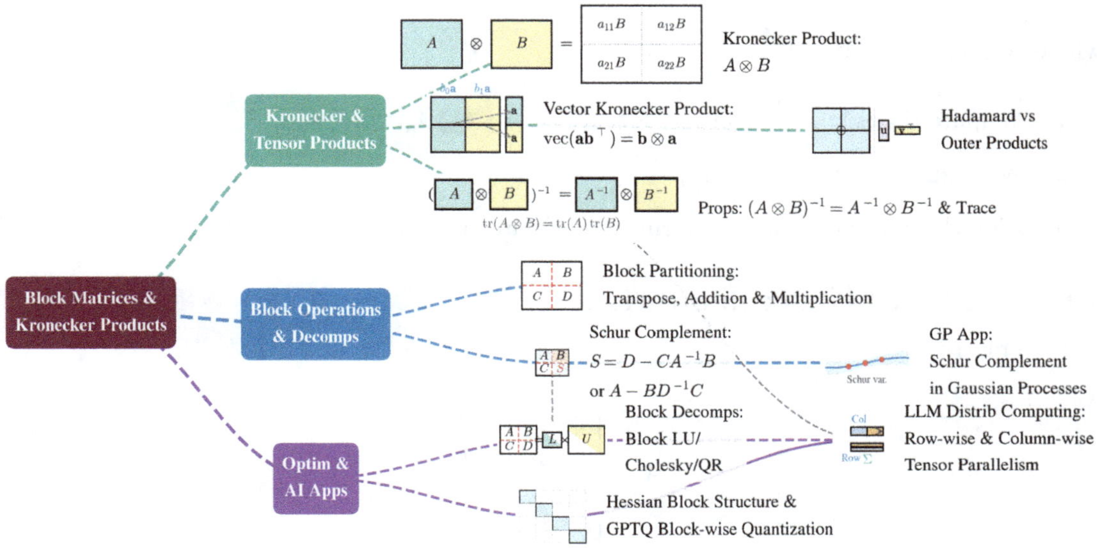

Figure 10.6: Mind map of block matrix and Kronecker product topics

10.7 Exercises

10.7.1 Block Matrix Multiplication

Given block matrices

$$A = \begin{bmatrix} A_{00} & A_{01} \\ A_{10} & A_{11} \end{bmatrix}, \quad B = \begin{bmatrix} B_{00} & B_{01} \\ B_{10} & B_{11} \end{bmatrix},$$

where $A_{ij}, B_{ij} \in \mathbb{R}^{n \times n}$:

1. Compute the block representation of AB.
2. Show that if A and B are both block diagonal (i.e., $A_{01} = A_{10} = 0$ and $B_{01} = B_{10} = 0$), then AB is also block diagonal.
3. Prove that $(AB)_{11} = A_{10}B_{01} + A_{11}B_{11}$ using block multiplication rules.

operations: https://docs.nvidia.com/deeplearning/nccl/user-guide/docs/api/colls.html This directly applies block matrix multiplication perspectives.

10.7.2 Schur Complement and Matrix Inversion

For the partitioned matrix

$$A = \begin{bmatrix} A_{00} & A_{01} \\ A_{10} & A_{11} \end{bmatrix} = \begin{bmatrix} 2 & 1 & 0 \\ 1 & 2 & 1 \\ 0 & 1 & 2 \end{bmatrix},$$

where $A_{00} = \begin{bmatrix} 2 & 1 \\ 1 & 2 \end{bmatrix}$:

1. Compute the Schur complement $S = A_{11} - A_{10}A_{00}^{-1}A_{01}$.
2. Verify that $S = 2 - \begin{bmatrix} 0 & 1 \end{bmatrix} A_{00}^{-1} \begin{bmatrix} 0 \\ 1 \end{bmatrix}$.
3. Using the Schur complement formula, compute A^{-1} and verify that $AA^{-1} = I$.

10.7.3 Block LU Decomposition

For the matrix

$$A = \begin{bmatrix} 4 & 2 & 1 & 0 \\ 2 & 3 & 0 & 1 \\ 1 & 0 & 2 & 1 \\ 0 & 1 & 1 & 2 \end{bmatrix},$$

partitioned as $A = \begin{bmatrix} A_{00} & A_{01} \\ A_{10} & A_{11} \end{bmatrix}$ where $A_{00} \in \mathbb{R}^{2\times 2}$:

1. Compute the LU decomposition of A_{00}: $A_{00} = L_{00}U_{00}$.
2. Solve $L_{00}U_{01} = A_{01}$ to find U_{01}.
3. Solve $L_{10}U_{00} = A_{10}$ to find L_{10}.
4. Compute the Schur complement $S = A_{11} - L_{10}U_{01}$.
5. Compute the LU decomposition of S: $S = L_{11}U_{11}$.
6. Verify that $A = \begin{bmatrix} L_{00} & 0 \\ L_{10} & L_{11} \end{bmatrix} \begin{bmatrix} U_{00} & U_{01} \\ 0 & U_{11} \end{bmatrix}$.

10.7.4 Block Cholesky Decomposition

For the symmetric positive definite matrix

$$A = \begin{bmatrix} 9 & 3 & 0 & 0 \\ 3 & 5 & 0 & 0 \\ 0 & 0 & 4 & 2 \\ 0 & 0 & 2 & 3 \end{bmatrix},$$

1. Verify that A is block diagonal with blocks $A_0 = \begin{bmatrix} 9 & 3 \\ 3 & 5 \end{bmatrix}$ and $A_1 = \begin{bmatrix} 4 & 2 \\ 2 & 3 \end{bmatrix}$.
2. Compute the Cholesky decomposition of A_0: $A_0 = L_0L_0^\top$.
3. Compute the Cholesky decomposition of A_1: $A_1 = L_1L_1^\top$.
4. Show that $A = \text{blkdiag}(L_0, L_1) \cdot \text{blkdiag}(L_0^\top, L_1^\top)$.
5. Prove that for any block diagonal positive definite matrix $A = \text{blkdiag}(A_0, A_1, \dots, A_{k-1})$, the Cholesky decomposition is $L = \text{blkdiag}(L_0, L_1, \dots, L_{k-1})$ where $A_i = L_iL_i^\top$.

10.7.5 Schur Complement in Gaussian Processes

In Gaussian process regression, the joint distribution of training points $\mathbf{f}$ and test point f_* is

$$\begin{bmatrix} \mathbf{f} \\ f_* \end{bmatrix} \sim \mathcal{N}\left(\mathbf{0}, \begin{bmatrix} K & \mathbf{k}_* \\ \mathbf{k}_*^\top & k_{**} \end{bmatrix}\right),$$

where $K \in \mathbb{R}^{N\times N}$ is the covariance matrix of training points, $\mathbf{k}_* \in \mathbb{R}^N$ is the covariance between training and test points, and $k_{**} \in \mathbb{R}$ is the variance of the test point.

1. Using the Schur complement formula, show that the conditional distribution $f_*|\mathbf{f}$ has:
   - Mean: $\mathbf{k}_*^\top K^{-1}\mathbf{f}$
   - Variance: $k_{**} - \mathbf{k}_*^\top K^{-1}\mathbf{k}_*$ (this is the Schur complement)
2. Prove that the conditional variance is always less than or equal to the marginal variance k_{**}.
3. Interpret the Schur complement $k_{**} - \mathbf{k}_*^\top K^{-1}\mathbf{k}_*$ as the reduction in uncertainty about f_* given observations $\mathbf{f}$.

10.7.6 Block Structure in Hessian Matrices

Consider a neural network with two layers. The loss function $\mathcal{L}(\theta)$ depends on parameters $\theta = [\theta_0, \theta_1]^\top$ where θ_0 are first-layer parameters and θ_1 are second-layer parameters.

1. Show that the Hessian matrix $H = \nabla^2\mathcal{L}(\theta)$ has block structure:

$$H = \begin{bmatrix} H_{00} & H_{01} \\ H_{10} & H_{11} \end{bmatrix},$$

   where $H_{00} = \frac{\partial^2\mathcal{L}}{\partial\theta_0\partial\theta_0^\top}$, $H_{11} = \frac{\partial^2\mathcal{L}}{\partial\theta_1\partial\theta_1^\top}$, and $H_{01} = H_{10}^\top = \frac{\partial^2\mathcal{L}}{\partial\theta_0\partial\theta_1^\top}$.
2. Prove that if the network has no skip connections and θ_1 does not directly affect the computation involving θ_0, then $H_{01} = 0$ (the Hessian is block diagonal).
3. In practice, even with skip connections, H_{01} is often small. Using the Schur complement, show that the block-diagonal approximation $H \approx \text{blkdiag}(H_{00}, H_{11})$ is reasonable when $||H_{01}||_F \ll ||H_{00}||_F, ||H_{11}||_F$.

10.7.7 Low-Rank Approximation of Attention Matrices

For an attention matrix $A \in \mathbb{R}^{N\times N}$ with SVD $A = U\Sigma V^\top$ and singular values $\sigma_0 \geq \sigma_1 \geq \cdots \geq \sigma_{N-1}$:

1. Prove that the rank-k truncated SVD $A_k = \sum_{i=0}^{k-1} \sigma_i \mathbf{u}_i \mathbf{v}_i^\top$ minimizes $||A - B||_F$ over all rank-k matrices B.
2. Given that attention matrices in transformers often have rapidly decaying singular values (e.g., $\sigma_i \approx \sigma_0 \cdot 0.9^i$), determine the minimum rank k needed to capture 90% of the Frobenius norm energy: $\frac{||A_k||_F^2}{||A||_F^2} \geq 0.9$.
3. Show that computing $A_k\mathbf{v}$ (for any vector $\mathbf{v}$) requires only $O(kN)$ operations instead of $O(N^2)$ for the full matrix-vector product, demonstrating the computational advantage of low-rank approximations.

10.7.8 Block QR for Incremental Factorization

Consider a matrix $A = \begin{bmatrix} A_0 & A_1 \end{bmatrix}$ where $A_0 \in \mathbb{R}^{M \times N_0}$ and $A_1 \in \mathbb{R}^{M \times N_1}$.

1. Suppose we have the QR decomposition $A_0 = Q_0 R_0$ where $Q_0 \in \mathbb{R}^{M \times N_0}$ has orthonormal columns and $R_0 \in \mathbb{R}^{N_0 \times N_0}$ is upper triangular. Show that to compute the QR decomposition of $A = \begin{bmatrix} A_0 & A_1 \end{bmatrix}$, we can:
   - Compute $\tilde{A}_1 = (I - Q_0 Q_0^\top) A_1$ (project A_1 onto the orthogonal complement of $\text{col}(A_0)$)
   - Compute QR of $\tilde{A}_1$: $\tilde{A}_1 = Q_1 R_1$
   - Then $A = \begin{bmatrix} Q_0 & Q_1 \end{bmatrix} \begin{bmatrix} R_0 & Q_0^\top A_1 \\ 0 & R_1 \end{bmatrix}$
2. Prove that this incremental approach is more efficient than recomputing QR from scratch when $N_1 \ll N_0$.
3. This incremental QR is used in online learning algorithms. Explain why maintaining an orthonormal basis incrementally is crucial for streaming PCA and adaptive filtering applications.

II

Differentiation and Optimization

Chapters 11–12

Learning = structured search in parameter space

Learning through structured search: from gradients to knowledge

Learning in AI is fundamentally an optimization problem defined over high-dimensional parameter spaces. Calculus supplies the tools to quantify sensitivity, direction, and curvature, enabling principled updates through gradients and higher-order information.

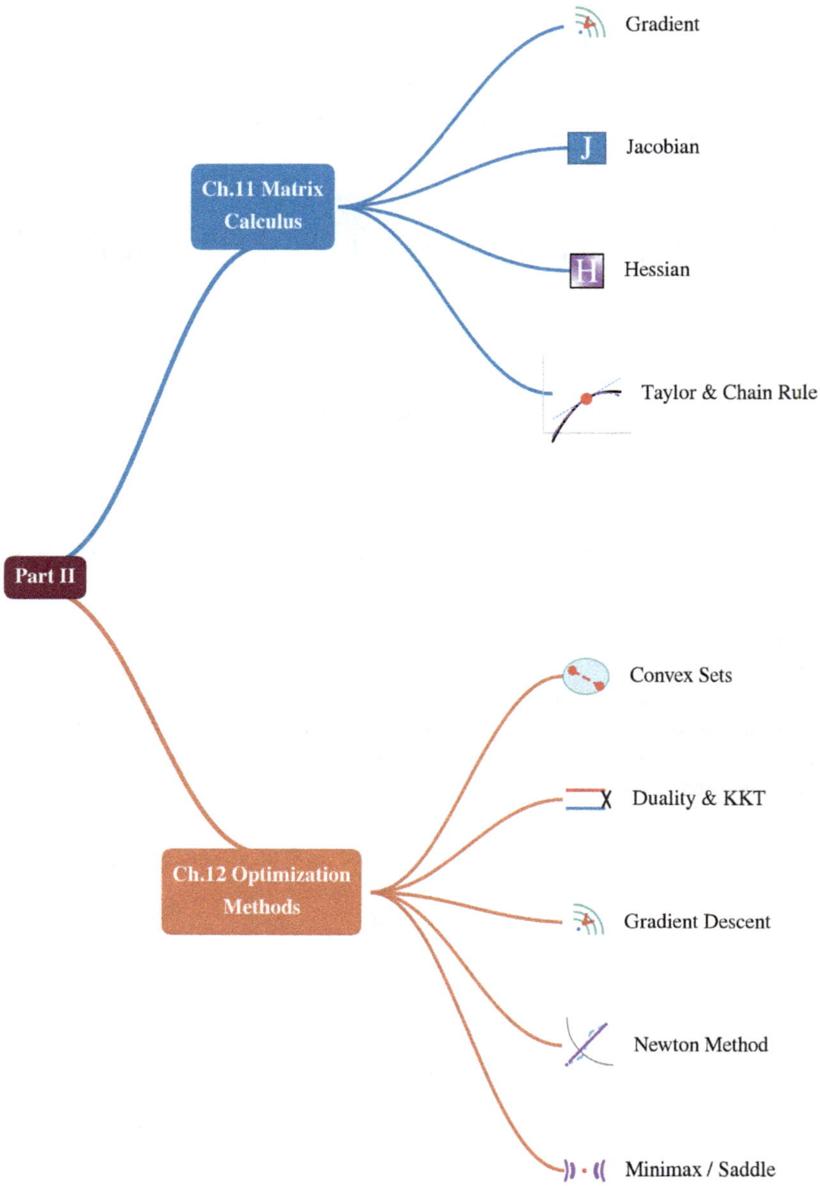
Part II
Ch.11 Matrix Calculus
Gradient
Jacobian
Hessian
Taylor & Chain Rule
Ch.12 Optimization Methods
Convex Sets
Duality & KKT
Gradient Descent
Newton Method
Minimax / Saddle

Chapter 11

Matrix Calculus

Matrix calculus: gradients, Jacobians, and Hessians for machine learning

> *"Mathematics is the art of giving the same name to different things."*
>
> — Henri Poincaré

- ▷ `Derivative/Partial` $\frac{df(x)}{dx} = \lim_{h \to 0} \frac{f(x+h)-f(x)}{h}$; $\frac{\partial f(x)}{\partial x_n} = \lim_{h \to 0} \frac{f(x_{-n}, x_n+h)-f(x)}{h}$
- ▷ `Product/Quotient` $(f(x)g(x))' = f'(x)g(x) + f(x)g'(x)$; $\left(\frac{f(x)}{g(x)}\right)' = \frac{f'(x)g(x)-f(x)g'(x)}{(g(x))^2}$
- ▷ `Chain` $(g(f(x)))' = g'(f(x))f'(x)$; $\nabla_{\mathbf{t}} z = (\nabla_{\mathbf{t}} \mathbf{x})^T (\nabla_{\mathbf{x}} z) \in \mathbb{R}^{N \times 1}$ for $z = f(\mathbf{x})$, $\mathbf{x} = \mathbf{g}(\mathbf{t})$
- ▷ `Sigmoid'` $\frac{d\sigma(x)}{dx} = \sigma(x)(1-\sigma(x)) \in [0, 0.25]$ where $\sigma(x) = \frac{1}{1+e^{-x}} \in [0,1]$
- ▷ `Tanh'` $\frac{dt(x)}{dx} = 1 - t^2(x) = 4\sigma(2x) \cdot (1-\sigma(2x)) \in [0,1]$ where $t(x) = \tanh(x) \in [-1,1]$
- ▷ `Gradient(v/m)` $\nabla_{\mathbf{x}} f(\mathbf{x}) = [\frac{\partial f}{\partial x_0}, \dots, \frac{\partial f}{\partial x_{N-1}}]^T \in \mathbb{R}^{N \times 1}$; $\nabla_X f(X) = [\frac{\partial f}{\partial x_{mn}}]_{M \times N} \in \mathbb{R}^{M \times N}$
- ▷ `Gradient` $\nabla_{\mathbf{x}}(\mathbf{x}^T \mathbf{a}) = \mathbf{a} \in \mathbb{R}^{N \times 1}$ for $\mathbf{x}, \mathbf{a} \in \mathbb{R}^{N \times 1}$
- ▷ `Gradient` $\nabla_{\mathbf{x}}(\mathbf{x}^T W \mathbf{x}) = (W + W^T)\mathbf{x} \in \mathbb{R}^{N \times 1}$ for $\mathbf{x} \in \mathbb{R}^{N \times 1}$, $W \in \mathbb{R}^{N \times N}$
- ▷ `Gradient` $\nabla_{\mathbf{x}}((A\mathbf{x}-\mathbf{b})^T W (A\mathbf{x}-\mathbf{b})) = A^T(W+W^T)(A\mathbf{x}-\mathbf{b}) \in \mathbb{R}^{N \times 1}$
- ▷ `Gradient` $\nabla_X(a^T X b) = ab^T \in \mathbb{R}^{M \times N}$ for $a \in \mathbb{R}^{M \times 1}$, $X \in \mathbb{R}^{M \times N}$, $b \in \mathbb{R}^{N \times 1}$
- ▷ `Gradient` $\nabla_U \|U\|_F^2 = 2U \in \mathbb{R}^{M \times N}$ where $\|U\|_F = \sqrt{\sum_{i,j} U_{ij}^2}$
- ▷ `Gradient` $\nabla_U \|UV^T - A\|_F^2 = 2(UV^T - A)V \in \mathbb{R}^{M \times K}$
- ▷ `Jacobian matrix` $\frac{\partial \mathbf{f}}{\partial \mathbf{x}} = \mathbf{J}_{\mathbf{f}}(\mathbf{x}) = [\frac{\partial f_m}{\partial x_n}]_{M \times N}$ for $\mathbf{f}: \mathbb{R}^N \to \mathbb{R}^M$
- ▷ `Jacobian of Ax` $\frac{\partial (A\mathbf{x})}{\partial \mathbf{x}} = A \in \mathbb{R}^{M \times N}$ for $A \in \mathbb{R}^{M \times N}$, $\mathbf{x} \in \mathbb{R}^{N \times 1}$
- ▷ `Softmax Jacobian` $\frac{\partial S(\mathbf{x})}{\partial \mathbf{x}} = \operatorname{diag}(S(\mathbf{x})) - (S(\mathbf{x}))(S(\mathbf{x}))^T \in \mathbb{R}^{N \times N}$
- ▷ `Hessian matrix` $\mathbf{H}_f = \nabla_{\mathbf{x}}^2 f(\mathbf{x}) = \left[\frac{\partial^2 f}{\partial x_i \partial x_j}\right]_{N \times N}$ for $f: \mathbb{R}^N \to \mathbb{R}$; $\mathbf{H}_f = \mathbf{H}_f^T$
- ▷ `Sigmoid Hessian` $\nabla_{\mathbf{x}}^2 \sigma(\mathbf{w}^T\mathbf{x}) = \sigma(\mathbf{w}^T\mathbf{x})(1-\sigma(\mathbf{w}^T\mathbf{x}))(1-2\sigma(\mathbf{w}^T\mathbf{x}))\mathbf{w}\mathbf{w}^T \in \mathbb{R}^{N \times N}$
- ▷ `Taylor` $f(\mathbf{x}) = f(\mathbf{a}) + (\nabla_{\mathbf{x}} f(\mathbf{a}))^T(\mathbf{x}-\mathbf{a}) + \frac{1}{2}(\mathbf{x}-\mathbf{a})^T \nabla_{\mathbf{x}}^2 f(\mathbf{a})(\mathbf{x}-\mathbf{a}) + o(\|\mathbf{x}-\mathbf{a}\|_2^2)$; $f(\mathbf{x}) = f(\mathbf{a}) + J\mathbf{e} + \frac{1}{2}\mathbf{e}^T H \mathbf{e} + o(\|\mathbf{e}\|_2^2)$ where $\mathbf{e} = \mathbf{x} - \mathbf{a}$, $J = (\nabla_{\mathbf{x}} f(\mathbf{a}))^T \in \mathbb{R}^{1 \times N}$, $H = \nabla_{\mathbf{x}}^2 f(\mathbf{a}) \in \mathbb{R}^{N \times N}$
- ▷ `Newton-Schulz` $X_{t+1} = 2X_t - X_t A X_t \Rightarrow A^{-1}$; $Q_{t+1} = \frac{1}{2}Q_t(3I - Q_t^\top Q_t)$(polar decomp)

11.1 Calculus Conceptual Hierarchy

The field of calculus can be organized into a conceptual hierarchy:

$$\text{Single-Variable Calculus} \subset \text{Multivariable Calculus} \supset \begin{cases} \text{Vector Calculus} \\ \text{Matrix Calculus} \end{cases}$$

Single-Variable Calculus deals with functions $f : \mathbb{R} \to \mathbb{R}$ and their derivatives $\frac{df}{dx}$.

Multivariable Calculus extends this to functions of multiple variables. It branches into two specialized areas:

- **Vector Calculus** studies scalar and vector fields over continuous spaces, emphasizing geometric intuition (gradient, divergence, curl). It is used in physics, PDEs, and continuous dynamics.
- **Matrix Calculus** focuses on derivatives of functions with respect to vectors and matrices, emphasizing algebraic computation (gradients, Jacobians, Hessians). It is the primary tool for optimization and machine learning.

In machine learning, **matrix calculus** is essential for computing gradients during training (backpropagation, gradient descent), while **vector calculus** re-emerges in continuous-time models (diffusion models, score matching, ODEs/SDEs). Both are needed, and understanding their relationship provides a significant conceptual advantage.

This chapter focuses on **matrix calculus**, which builds upon the foundations of single-variable and multivariable calculus. We begin by reviewing the essential concepts, starting with single-variable derivatives and progressively building up to gradients, Jacobians, and Hessians for vector and matrix functions.

11.2 Scalar and Vector Fields

Before diving into matrix calculus, we introduce the fundamental concepts of **scalar fields** and **vector fields**, which are central to vector calculus and will be important in later chapters on score functions and diffusion models.

11.2.1 Definition: Scalar Field

A **scalar field** is a function $f : \mathbb{R}^d \to \mathbb{R}$ that assigns a real number (scalar) to each point in space. At each point $\mathbf{x} \in \mathbb{R}^d$, the field takes a value $f(\mathbf{x}) \in \mathbb{R}$.

Examples:

- **Temperature field**: $T(\mathbf{x})$ gives the temperature at location $\mathbf{x}$ in a room
- **Probability density**: $p(\mathbf{x})$ gives the probability density at point $\mathbf{x}$ (see Chapter~13)
- **Log-probability**: $\log p(\mathbf{x})$ is a scalar field used in score-based generative models (see Chapter~17)
- **Energy function**: $E(\mathbf{x})$ gives the energy at configuration $\mathbf{x}$ in energy-based models

11.2.2 Definition: Vector Field

A **vector field** is a function $\mathbf{F} : \mathbb{R}^d \to \mathbb{R}^d$ that assigns a vector to each point in space. At each point $\mathbf{x} \in \mathbb{R}^d$, the field takes a value $\mathbf{F}(\mathbf{x}) \in \mathbb{R}^d$.

Examples:

- **Velocity field**: $\mathbf{v}(\mathbf{x})$ gives the velocity of a fluid at location $\mathbf{x}$
- **Force field**: $\mathbf{F}(\mathbf{x})$ gives the force acting on a particle at position $\mathbf{x}$

- **Gradient field**: $\nabla f(\mathbf{x})$ gives the gradient of a scalar field f at point $\mathbf{x}$
- **Score field**: $\nabla_{\mathbf{x}} \log p(\mathbf{x})$ is a vector field used in score-based generative models (see Chapter~17)

11.3 Single-variable Derivative

We start with the fundamental concept of a derivative for functions of a single variable, which forms the foundation for all subsequent generalizations.

11.3.1 Definition: Derivative

The derivative of a function $f(x)$ at a point x is defined as the limit:

$$\frac{df(x)}{dx} = \lim_{h \to 0} \frac{f(x+h) - f(x)}{h} \tag{11.1}$$

when this limit exists. The derivative represents the instantaneous rate of change of f with respect to x.

Figure 11.1 shows a function $f(x)$ with its tangent line at a point $(x_0, f(x_0))$. The slope of the tangent line equals the derivative $f'(x_0)$, representing the instantaneous rate of change at that point.

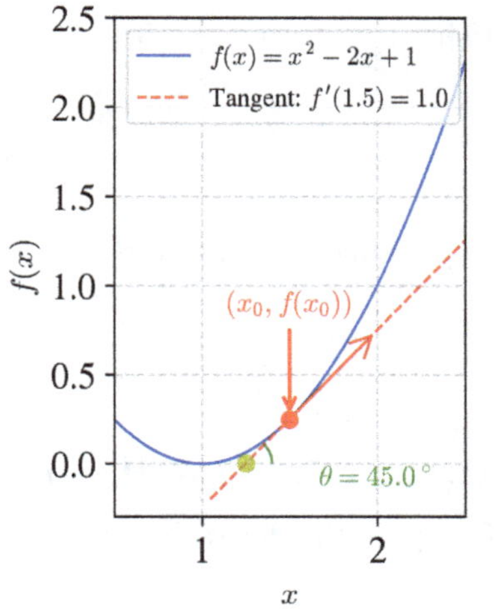

Figure 11.1: Single-Variable Derivative

11.3.2 Theorem: Differentiation Rules

The following rules allow us to compute derivatives of composite functions:

1. **Sum rule**: $(f(x) + g(x))' = f'(x) + g'(x)$
2. **Product rule**: $(f(x)g(x))' = f'(x)g(x) + f(x)g'(x)$
3. **Quotient rule**: $\left(\frac{f(x)}{g(x)}\right)' = \frac{f'(x)g(x) - f(x)g'(x)}{(g(x))^2}$
4. **Chain rule**: For composite functions $g(f(x))$, we have

$$(g \circ f)'(x) = g'(f(x)) \cdot f'(x) \tag{11.2}$$

The chain rule is particularly important as it extends naturally to multivariable and matrix calculus.

11.3.2.1 Example: Chain Rule Application

Compute the derivative of $f(x) = (2x+1)^4$.

Solution

Let $u = 2x + 1$, so $f(x) = u^4$. By the chain rule:

$$\frac{df(x)}{dx} = \frac{d}{du}(u^4) \cdot \frac{du}{dx} = 4u^3 \cdot 2 = 8(2x+1)^3$$

11.3.2.2 Example: Sigmoid Derivative

Compute the derivative of the sigmoid function $\sigma(x)$, where

$$\sigma(x) = \frac{1}{1+e^{-x}}$$

Solution

Using the chain rule with $u = 1 + e^{-x}$, we have $\sigma(x) = u^{-1}$:

$$\frac{d\sigma(x)}{dx} = \frac{d}{du}(u^{-1}) \cdot \frac{du}{dx} = -u^{-2} \cdot (-e^{-x}) = \frac{e^{-x}}{(1+e^{-x})^2}$$

Expressing in terms of $\sigma(x)$:

$$\frac{d\sigma(x)}{dx} = \frac{1+e^{-x}-1}{(1+e^{-x})^2} = \frac{1}{1+e^{-x}} - \frac{1}{(1+e^{-x})^2} = \sigma(x) - \sigma^2(x) = \sigma(x)(1-\sigma(x)) \tag{11.3}$$

11.3.2.3 Example: Hyperbolic Tangent Derivative

Compute the derivative of

$$\tanh(x) = \frac{e^x - e^{-x}}{e^x + e^{-x}}$$

Solution

Using the quotient rule with $f(x) = e^x - e^{-x}$ and $g(x) = e^x + e^{-x}$:

- $f'(x) = e^x + e^{-x} = g(x)$
- $g'(x) = e^x - e^{-x} = f(x)$

Applying the quotient rule:

$$\frac{d}{dx}\tanh(x) = \frac{f'(x)g(x) - f(x)g'(x)}{(g(x))^2} = \frac{(g(x))^2 - (f(x))^2}{(g(x))^2} = \frac{4}{(e^x + e^{-x})^2} = 1 - \tanh^2(x) \tag{11.4}$$

11.3.2.4 Activation Functions and the Derivatives

The following table summarizes commonly used activation functions in neural networks and their derivatives, which are essential for backpropagation and gradient-based optimization.

Activation	Formula	Derivative
Sigmoid	$\sigma(z) = \frac{1}{1+e^{-z}}$	$\sigma'(z) = \sigma(z)(1-\sigma(z))$
Tanh	$\tanh(z) = \frac{e^z - e^{-z}}{e^z + e^{-z}} = 2\sigma(2 \cdot z) - 1$	$\tanh'(z) = 1 - \tanh^2(z) = \text{sech}^2(z)$
ReLU	$\text{ReLU}(z) = \max(0, z)$	$\text{ReLU}'(z) = 1$ if $z > 0$, 0 if $z \leq 0$
Leaky ReLU	$\text{LeakyReLU}(z) = \max(\alpha z, z)$	$\text{LeakyReLU}'(z) = 1$ if $z > 0$, α if $z \leq 0$
ELU	$\text{ELU}(z) = z$ if $z > 0$, $\alpha(e^z - 1)$ if $z \leq 0$	$\text{ELU}'(z) = 1$ if $z > 0$, αe^z if $z \leq 0$
GELU	$\text{GELU}(z) = z \cdot \Phi(z)$	$\text{GELU}'(z) = \Phi(z) + z \cdot \phi(z)$
Swish	$\text{Swish}(z) = z \cdot \sigma(z)$	$\text{Swish}'(z) = \sigma(z) + z \cdot \sigma(z)(1-\sigma(z))$
Mish	$\text{Mish}(z) = z \cdot \tanh(\text{Softplus}(z)) = z \cdot \tanh(\ln(1+e^z))$	$\text{Mish}'(z) = \frac{e^z(4(z+1)+4e^{2z}+e^{3z}+e^z(4z+6))}{(1+e^z)^2(1+e^{2z})}$

Activation	Formula	Derivative
GEGLU	$\text{GEGLU}(z) = z \odot \text{GELU}(z)$	$\text{GEGLU}'(z) = \text{GELU}(z) + z \cdot \text{GELU}'(z) = \text{GELU}(z) + z(\Phi(z) + z \cdot \phi(z))$
ReGLU	$\text{ReGLU}(z) = z \odot \text{ReLU}(z)$	$\text{ReGLU}'(z) = 2z$ if $z > 0$, 0 if $z \leq 0$
SwiGLU	$\text{SwiGLU}(z) = z \odot \text{Swish}(z) = z^2 \cdot \sigma(z)$	$\text{SwiGLU}'(z) = 2z \cdot \sigma(z) + z^2 \cdot \sigma(z)(1 - \sigma(z))$
Softplus	$\text{Softplus}(z) = \ln(1 + e^z)$	$\text{Softplus}'(z) = \sigma(z) = \frac{1}{1+e^{-z}}$
Softmax	$\text{Softmax}(\mathbf{z})_i = \frac{e^{z_i}}{\sum_{j=1}^{n} e^{z_j}}$	$\nabla_{\mathbf{z}}\text{Softmax}(\mathbf{z})_{ij} = \text{Softmax}(\mathbf{z})_i(1 - \text{Softmax}(\mathbf{z})_i)$ if $i = j$, $-\text{Softmax}(\mathbf{z})_i \cdot \text{Softmax}(\mathbf{z})_j$ if $i \neq j$

Notation notes:

- **Variable** z **vs** x: In activation functions, z typically denotes the **pre-activation** value (the input to the activation function), which is usually the output of a linear layer: $z = \mathbf{w}^\top \mathbf{x} + b$ or $z = W\mathbf{x} + \mathbf{b}$. The symbol x is typically reserved for the **raw input data**. This convention distinguishes between the original input (x) and the transformed value (z) that enters the activation function.
- **GELU symbols**: For the Gaussian Error Linear Unit (GELU), $\Phi(z)$ denotes the **cumulative distribution function (CDF)** of the standard normal distribution: $\Phi(z) = \int_{-\infty}^{z} \phi(t)dt = \frac{1}{2}\left(1 + \text{erf}\left(\frac{z}{\sqrt{2}}\right)\right)$, and $\phi(z)$ denotes the **probability density function (PDF)** of the standard normal distribution: $\phi(z) = \frac{1}{\sqrt{2\pi}}e^{-z^2/2}$. Here, $\text{erf}(z) = \frac{2}{\sqrt{\pi}}\int_0^z e^{-t^2}dt$ is the **error function** (also called the Gauss error function), which is a special function related to the normal distribution. The error function satisfies $\text{erf}(-z) = -\text{erf}(z)$ and $\lim_{z\to\infty} \text{erf}(z) = 1$.
- **Element-wise product**: In the formulas above, $\odot$ denotes the **element-wise product** (Hadamard product). For vectors $\mathbf{z}, \mathbf{w} \in \mathbb{R}^{N\times 1}$, $[\mathbf{z} \odot \mathbf{w}]_i = z_i w_i$ (see Chapter~2 for details). When applied element-wise to a vector, $\mathbf{z} \odot f(\mathbf{z})$ means each element z_i is multiplied by the corresponding element $f(\mathbf{z})_i$.

11.4 Partial Differentiation and Gradients

11.4.1 Definition: Partial Derivative

Let $f : \mathbb{R}^N \to \mathbb{R}$. The partial derivative with respect to x_n is

$$\frac{\partial f(x)}{\partial x_n} = \lim_{h\to 0} \frac{f(x_{-n}, x_n + h) - f(x)}{h} \tag{11.5}$$

where $f(x_{-n}, x_n + h)$ means to evaluate f at the point where only the n-th coordinate is perturbed by h, and all other coordinates stay exactly the same. Here, x_{-n} denotes the vector $\mathbf{x}$ with the n-th coordinate removed (or held fixed). The $-$ is an index-removal operator, not arithmetic subtraction. The same pattern appears in X_{-i} in statistics for dataset without sample i, θ_{-k} in optimization for all parameters except block k, and S_{-j} in combinatorics for set with element j removed.

Many texts use the equivalent "holding constant" rule: treat all other variables as constants and differentiate with respect to x_n. While this computational approach is often more practical, the explicit limit-based definition with notation $f(x_{-n}, x_n + h)$ is necessary for generalizing to matrix calculus and Fréchet derivatives, where we must precisely specify which variables are being perturbed.

Figure 11.2 illustrates $\frac{\partial f}{\partial x_1}$ for $f(x_1, x_2) = x_1^2 + x_2^2$ at a point $(x_1, x_2) = (2, 0.5)$: the white curve is the slice with x_2 fixed; the pink line is the tangent (slope $2x_1 = 4$); the value $\frac{\partial f}{\partial x_1} = 4$ is shown on the floor.

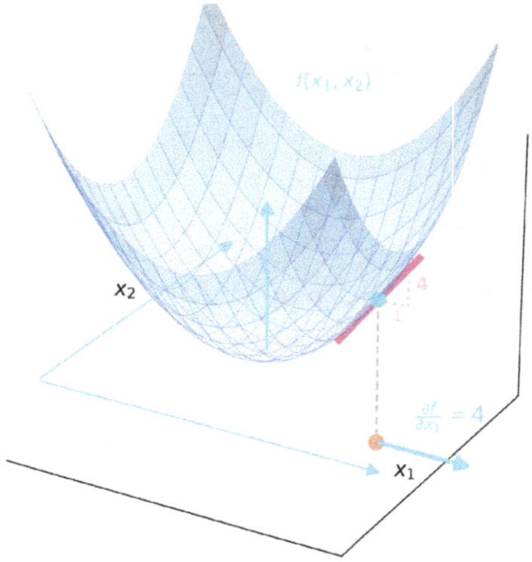

Figure 11.2: Partial derivative

11.4.2 Definition: Gradient

The symbol ∇ (pronounced "nabla") is a **vector differential operator**. For a function $f : \mathbb{R}^N \to \mathbb{R}$, the operator $\nabla_{\mathbf{x}}$ measures the rate of change (slope) of f in each coordinate direction separately, and then synthesizes these partial derivatives into a single vector. This resulting vector is the **gradient**.

Formally, for $\mathbf{x} \in \mathbb{R}^N$ and $f : \mathbb{R}^N \to \mathbb{R}$, the gradient is defined as:

$$\nabla_{\mathbf{x}} f(\mathbf{x}) = \begin{bmatrix} \frac{\partial f}{\partial x_0} \\ \frac{\partial f}{\partial x_1} \\ \vdots \\ \frac{\partial f}{\partial x_{N-1}} \end{bmatrix} \in \mathbb{R}^{N \times 1} \tag{11.6}$$

For matrix-valued inputs, if $X \in \mathbb{R}^{M \times N}$ and $f : \mathbb{R}^{M \times N} \to \mathbb{R}$, the gradient is:

$$\nabla_X f(X) = \left[\frac{\partial f}{\partial x_{mn}} \right]_{M \times N} \in \mathbb{R}^{M \times N} \tag{11.7}$$

Both $\nabla_{\mathbf{x}} f(\mathbf{x})$ and $\nabla_{\mathbf{x}} f$ are commonly used notations for the gradient. The form $\nabla_{\mathbf{x}} f(\mathbf{x})$ explicitly indicates that the gradient is evaluated at the point $\mathbf{x}$, while $\nabla_{\mathbf{x}} f$ emphasizes that the gradient is an operator acting on the function f. Both notations are equivalent and acceptable; the choice is often a matter of style or emphasis.

11.4.3 Connection: Gradient as a Vector Field

The **gradient** of a scalar field $f : \mathbb{R}^d \to \mathbb{R}$ is itself a vector field $\nabla f : \mathbb{R}^d \to \mathbb{R}^d$ that assigns to each point $\mathbf{x}$ the gradient vector $\nabla f(\mathbf{x})$ defined in equation 11.6. The gradient field points in the direction of steepest increase of the scalar field at each point, and its magnitude indicates the rate of change.

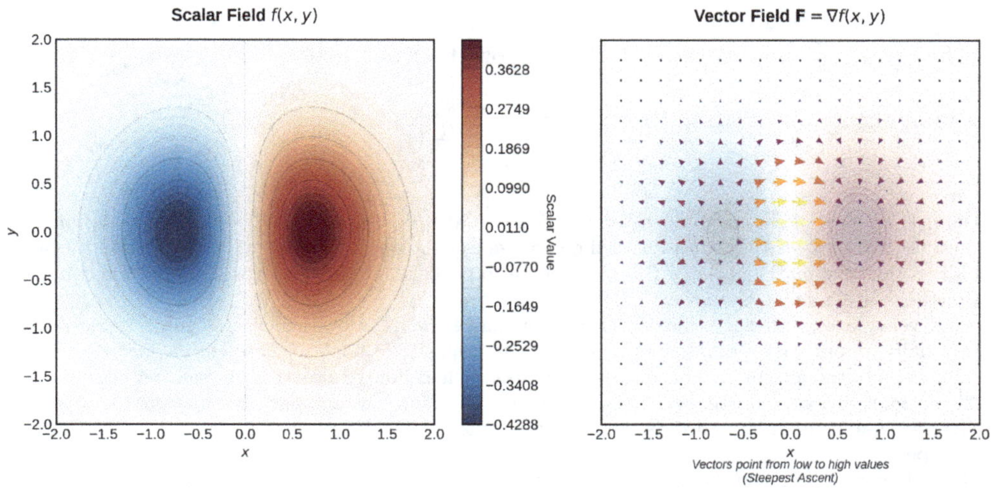

Figure 11.3: Scalar and Vector Fields

Figure 11.3 illustrates the fundamental relationship between a scalar field and its gradient vector field using a **dipole function** $f(x, y) = xe^{-x^2-y^2}$. The left panel shows the scalar field $f(x, y)$, which assigns a scalar value to each point in the plane. This particular function creates a positive peak (source) on the right and a negative valley (sink) on the left. The right panel shows the corresponding gradient vector field $\nabla f(x, y)$, which assigns a vector to each point. The key insight demonstrated here is that **the gradient vector field points in the direction of steepest increase** of the scalar field at each point. Notice how vectors flow outward from the positive peak (source) and inward toward the negative valley (sink), creating a dipole pattern. This relationship—that every scalar field has an associated gradient vector field—is fundamental to understanding how score functions work in generative modeling (Chapter~17), where the score field $\nabla_{\mathbf{x}} \log p(\mathbf{x})$ is the gradient of the log-probability scalar field.

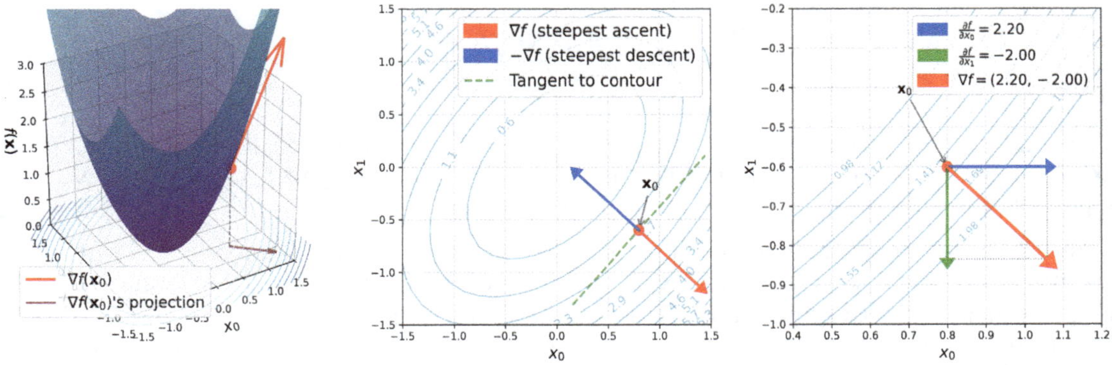

Figure 11.4: Gradient Visualization

Figure 11.4 illustrates the geometric interpretation of the gradient for a 2D function $f : \mathbb{R}^2 \to \mathbb{R}$, where $\mathbf{x} = [x_0, x_1]^\top$ following the general definition in equation 11.6. The left panel shows the 3D surface of the function with contour lines projected onto the bottom plane, providing a complete view of how the gradient relates to the function's landscape. The middle panel focuses on a single point $\mathbf{x}_0$, showing that the gradient $\nabla_{\mathbf{x}} f(\mathbf{x}_0)$ points in the direction of steepest ascent (red arrow), while $-\nabla_{\mathbf{x}} f(\mathbf{x}_0)$ points in the direction of steepest descent (blue arrow). The gradient is perpendicular to the contour line (green dashed line) at that point, meaning it points in the direction where the function increases most rapidly. The right panel shows a zoomed-in view of the same point, demonstrating that the gradient vector $\nabla_{\mathbf{x}} f(\mathbf{x}_0)$ (red arrow) is the vector sum of its partial derivative components: $\frac{\partial f}{\partial x_0}$ (blue arrow, horizontal component) and $\frac{\partial f}{\partial x_1}$ (green arrow, vertical component), illustrating that $\nabla_{\mathbf{x}} f(\mathbf{x}) = \begin{bmatrix} \frac{\partial f}{\partial x_0} \\ \frac{\partial f}{\partial x_1} \end{bmatrix}$.

The development of the nabla operator ∇ has a rich historical context. When dealing with functions of multiple variables, writing out all partial derivative calculations becomes cumbersome and inelegant. To address this, the mathematician Hamilton chose to package all partial derivative operations in all directions into a single operator.
A few years later, Tait gave this operator a new name: nabla. However, at this point, nabla remained essentially an ordinary derivative operator. In Maxwell's 1873 book, *A Treatise on Electricity and Magnetism*, although he introduced the ∇ symbol, it had limited practical impact. In actual calculations, expressions still had to be expanded into numerous component equations. It wasn't until later that Gibbs, often regarded as the father of modern vector analysis, recognized the full potential of this operator.
Through spatial intuition, Gibbs observed that the algebraic form of ∇ resembled a vector. He discovered that allowing nabla to participate in algebraic operations as a vector produced valid results. This insight led him to treat the nabla operator as a vector, despite a crucial difference: while the components of an ordinary vector are numerical, the components of nabla are differential operations. This dual nature—inheriting vector properties to sense direction while retaining differentiation capabilities to perceive change—makes nabla a powerful tool for understanding how fields change in space. The versatility of nabla in describing diverse physical phenomena (heat, light, electricity) and modern applications (including AI) stems from its ability to decompose complex changes into three fundamental forms: gradient, divergence, and curl. This book focuses on gradient.

11.4.3.1 Alternative Geometric Interpretation: Gradient as Normal Vector Projection

From a **Vector Calculus** perspective, the gradient has an important geometric interpretation that complements the 2D contour view shown in the previous figure. Consider a function $f : \mathbb{R}^N \to \mathbb{R}$ and define the level surface $F(\mathbf{x}, z) = f(\mathbf{x}) - z = 0$, which represents the graph of f in $\mathbb{R}^{N+1}$ (where $z = f(\mathbf{x})$). The gradient of F with respect to $(\mathbf{x}, z)$ is:

$$\nabla_{(\mathbf{x},z)} F = \begin{bmatrix} \nabla_{\mathbf{x}} f(\mathbf{x}) \\ -1 \end{bmatrix}$$

This gradient vector is **normal** (perpendicular) to the tangent plane of the surface $F(\mathbf{x}, z) = 0$ at any point. When projected onto the horizontal plane (the $\mathbf{x}$-space), the gradient $\nabla_{\mathbf{x}} f(\mathbf{x})$ represents the direction of steepest ascent of the function f.

Figure 11.5 illustrates this relationship. The left panel shows the 3D surface $z = f(\mathbf{x})$ with the **normal vector** $\nabla_{(\mathbf{x},z)} F = [\nabla_{\mathbf{x}} f(\mathbf{x}_0), -1]^\top$ (red arrow) at point $\mathbf{x}_0$. This normal vector is perpendicular to the **tangent plane** (yellow surface) at that point. When this normal vector is projected onto the input space (x_0, x_1)-plane (dark red arrow), we obtain the gradient $\nabla_{\mathbf{x}} f(\mathbf{x}_0)$, which points in the direction of steepest ascent in the 2D input space. The right panel shows the same gradient vector in the 2D contour view, demonstrating that it is perpendicular to the contour line (green dashed line) at that point.

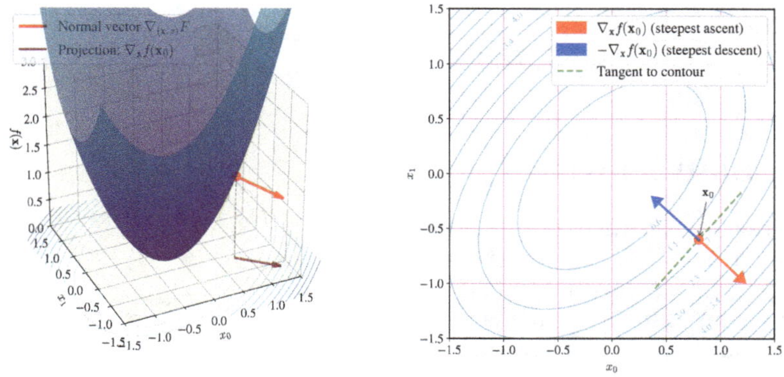

Figure 11.5: Gradient as Normal Vector Projection

The normal vector[1] $[\nabla_{\mathbf{x}} f(\mathbf{x}_0), -1]^\top$ is **different** from the gradient direction along the surface (shown in the previous figure). While both vectors have the same projection $[\nabla_{\mathbf{x}} f(\mathbf{x}_0), 0]^\top$ onto the (x_0, x_1)-plane, they differ in their z-component: the normal vector has $z = -1$ (perpendicular to the surface), while the gradient direction along the surface has $z = \|\nabla_{\mathbf{x}} f(\mathbf{x}_0)\|$ (following the steepest ascent path on the surface). This difference reflects that the normal vector points **out of** the surface, while the gradient direction lies **on** the surface.

These two geometric interpretations are **equivalent**:

- **2D view** (contour plot): The gradient is perpendicular to contour lines, pointing in the direction of steepest ascent.
- **3D view** (surface plot): The gradient is the projection of the surface's normal vector onto the input space $\mathbb{R}^2$ the (x_0, x_1)-plane.

This geometric view connects the algebraic definition of the gradient to its role in optimization: following the gradient direction increases the function value most rapidly, while following the negative gradient (gradient descent) decreases it most rapidly.

This geometric interpretation belongs to **Vector Calculus**, which emphasizes spatial intuition. In **Matrix Calculus** (the focus of this chapter), we primarily work with the algebraic properties of gradients for computational purposes in machine learning.

11.4.3.2 Notation

Throughout this chapter, we adopt the partial derivative notation $\frac{\partial f}{\partial \mathbf{x}}$ (which is equivalent to $\nabla_{\mathbf{x}} f$ for gradients, and $\mathbf{J}_{\mathbf{f}}(\mathbf{x})$ for Jacobians) because it makes the chain rule structure explicit through symbolic cancellation of intermediate variables. The diagram below illustrates how different combinations of scalar and vector inputs and outputs lead to different derivative shapes.

11.4.3.3 Example: Gradient Computation

Compute the gradient of

$$f(\mathbf{x}) = (x_0 + 2x_1^2)^3$$

Solution

[1] In this case, originating from a single point, there are actually two normal vectors pointing in opposite directions, both lying along the same line. When we refer to the normal vector and the gradient having the same projection, we are considering only one of the two normal vectors.

$$\nabla_{\mathbf{x}} f(\mathbf{x}) = \begin{bmatrix} 3(x_0 + 2x_1^2)^2 \\ 12x_1(x_0 + 2x_1^2)^2 \end{bmatrix}$$

11.5 Jacobian Matrix

The gradient provides derivatives for scalar-valued functions $f : \mathbb{R}^N \rightarrow \mathbb{R}$, giving us a column vector of partial derivatives. However, many important functions in machine learning and optimization are **vector-valued**, mapping $\mathbb{R}^N \rightarrow \mathbb{R}^M$ where $M > 1$. For example, neural network layers output vectors, and optimization algorithms often work with vector-valued objectives. The **Jacobian matrix** naturally extends the gradient concept to vector-valued functions by organizing the gradients of each output component as rows, creating an $M \times N$ matrix that captures how all M outputs change with respect to all N inputs simultaneously. This matrix representation is essential for the chain rule in vector calculus, backpropagation in neural networks, and understanding how transformations affect multidimensional spaces.

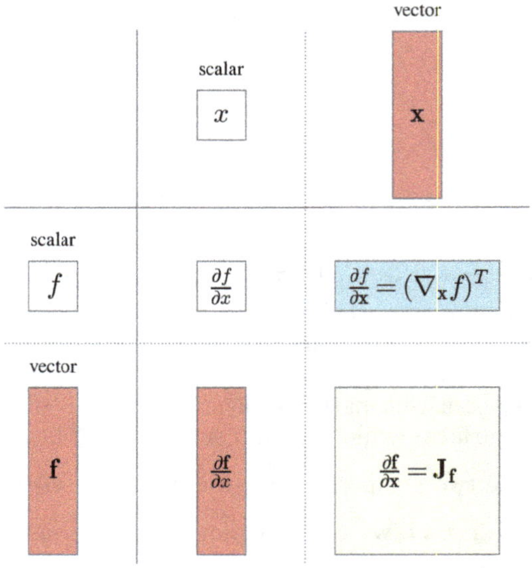

Figure 11.6: Types of derivatives: scalar and vector functions

11.5.1 Definition: Jacobian Matrix

Let $\mathbf{f} : \mathbb{R}^N \rightarrow \mathbb{R}^M$ be a vector-valued function. For $\mathbf{x} \in \mathbb{R}^N$, the **Jacobian matrix** is denoted by:

$$\frac{\partial \mathbf{f}}{\partial \mathbf{x}} = \mathbf{J_f}(\mathbf{x}) = \begin{bmatrix} \frac{\partial f_m}{\partial x_n} \end{bmatrix}_{M \times N} \in \mathbb{R}^{M \times N} \tag{11.8}$$

Alternatively, the Jacobian can be expressed in terms of the gradients of each component:

$$\frac{\partial \mathbf{f}}{\partial \mathbf{x}} = \begin{bmatrix} (\nabla_{\mathbf{x}} f_0(\mathbf{x}))^T \\ (\nabla_{\mathbf{x}} f_1(\mathbf{x}))^T \\ \vdots \\ (\nabla_{\mathbf{x}} f_{M-1}(\mathbf{x}))^T \end{bmatrix}_{M \times N} \tag{11.9}$$

where each row is the transpose of the gradient of the corresponding scalar component function. The matrix shape is $M \times N$.

Relationship with Gradient: For a scalar function $f : \mathbb{R}^N \rightarrow \mathbb{R}$ (i.e., $M = 1$), the Jacobian reduces to a $1 \times N$ row vector, which is the transpose of the gradient (a column vector). That is:

$$\frac{\partial f}{\partial \mathbf{x}} = (\nabla_{\mathbf{x}} f(\mathbf{x}))^T \in \mathbb{R}^{1 \times N}$$

where $\nabla_{\mathbf{x}} f(\mathbf{x}) \in \mathbb{R}^{N \times 1}$ is the gradient.

Matrix-to-Matrix Functions: For a function $f : \mathbb{R}^{M\times N} \to \mathbb{R}^{P\times Q}$, the Jacobian is a 4-dimensional tensor. However, by vectorizing both input and output matrices using the vec operator (which stacks columns), we can represent it as a $(PQ) \times (MN)$ matrix:

$$\nabla_X f(X) = \left[\frac{\partial \text{vec}(f(X))_i}{\partial \text{vec}(X)_j}\right]_{(PQ)\times(MN)}$$

where $\text{vec}(X) \in \mathbb{R}^{MN\times 1}$ and $\text{vec}(f(X)) \in \mathbb{R}^{PQ\times 1}$ are the vectorized forms of X and $f(X)$, respectively.

Neural Network Hidden Layers: In neural networks, hidden layers often process batches where both input and output are matrices. For example, a linear layer with batch processing takes $X \in \mathbb{R}^{B\times N}$ (batch size B, feature dimension N) and outputs $Y = XW^\top + b \in \mathbb{R}^{B\times M}$ (batch size B, output dimension M), where $W \in \mathbb{R}^{M\times N}$ and $b \in \mathbb{R}^{M\times 1}$. While the full Jacobian $\nabla_X Y$ would be a $(BM) \times (BN)$ matrix (after vectorization), in practice we typically compute gradients with respect to the parameters W and b rather than the full Jacobian with respect to the input X. However, the Jacobian concept is still fundamental: during backpropagation, we compute $\nabla_X L$ (gradient of loss with respect to input) by multiplying the upstream gradient with the Jacobian of the layer transformation. This operation, computing $\mathbf{v}^T J$ for an upstream gradient vector $\mathbf{v}$ and Jacobian J, is called the **Vector-Jacobian Product (VJP)**.

Connection to Automatic Differentiation Frameworks: In practice, frameworks like PyTorch (via `backward()`) and JAX (via `grad()` and `vjp()`) compute VJPs efficiently without explicitly forming the full Jacobian matrix, which is crucial for computational efficiency in deep learning. When you call `loss.backward()` in PyTorch, the framework does not construct the full Jacobian matrix $J \in \mathbb{R}^{M\times N}$ (which would require $O(MN)$ memory). Instead, it computes the VJP $\mathbf{v}^T J$ directly using reverse-mode automatic differentiation, where $\mathbf{v}$ is the upstream gradient (typically $\mathbf{v} = \nabla_Y L$). This is why we care about the **operator** $f'(X)[\Delta]$ rather than the explicit matrix representation: the operator can be applied efficiently without materializing the full Jacobian. The same principle applies to JAX's `jax.vjp()` function, which explicitly returns a VJP function that computes $\mathbf{v}^T J$ without forming J. This computational efficiency is what makes backpropagation feasible for deep neural networks with millions or billions of parameters.

For a single sample (i.e., when X is a row vector $\mathbf{x}^\top \in \mathbb{R}^{1\times N}$ and Y is a row vector $\mathbf{y}^\top \in \mathbb{R}^{1\times M}$), the transformation is $\mathbf{y}^\top = \mathbf{x}^\top W^\top + b^\top$, and the Jacobian $\nabla_{\mathbf{x}} \mathbf{y} = W \in \mathbb{R}^{M\times N}$. For the batched case, this can be extended through the vectorized representation.

Convolutional Neural Networks (CNNs): In CNNs, inputs are often 4D tensors in NCHW format: $X \in \mathbb{R}^{N\times C\times H\times W}$ where N is batch size, C is number of channels, and $H \times W$ are spatial dimensions. A convolutional layer transforms this to $Y \in \mathbb{R}^{N\times C'\times H'\times W'}$. The Jacobian $\nabla_X Y$ would be an 8-dimensional tensor, but by vectorizing all dimensions, we can represent it as a $(NC'H'W') \times (NCHW)$ matrix. In practice, we use the **convolution transpose** (also called deconvolution or transposed convolution) operation during backpropagation to efficiently compute $\nabla_X L$ without explicitly forming the full Jacobian. The key insight is that the Jacobian of a convolution operation has a structured sparsity pattern that can be exploited through convolution transpose, making backpropagation through CNNs computationally efficient despite the high dimensionality of the input and output tensors.

11.5.2 Theorem: Chain Rule for Vector Functions

The chain rule is fundamental to computing derivatives of composite functions. When dealing with vector-valued functions, the chain rule takes on a matrix form that depends on the dimensions of the inputs and outputs. The general form for the composition $\mathbf{f}(\mathbf{g}(\mathbf{x}))$ is:

$$\frac{\partial \mathbf{f}}{\partial \mathbf{x}} = \frac{\partial \mathbf{f}}{\partial \mathbf{g}} \frac{\partial \mathbf{g}}{\partial \mathbf{x}} \tag{11.10}$$

where the matrix dimensions must align correctly for the multiplication to be valid.

11.5.2.1 Chain Rule Cases: Scalar and Vector Combinations

The chain rule can be applied to various combinations of scalar and vector functions. The diagram below illustrates all six possible combinations, showing how the derivative shapes change based on whether the input x or g (and the output f) are scalars or vectors. Each cell in the grid represents a different case, with color coding: white for scalars, light red for column vectors, light blue for row vectors, and light yellow for matrices.

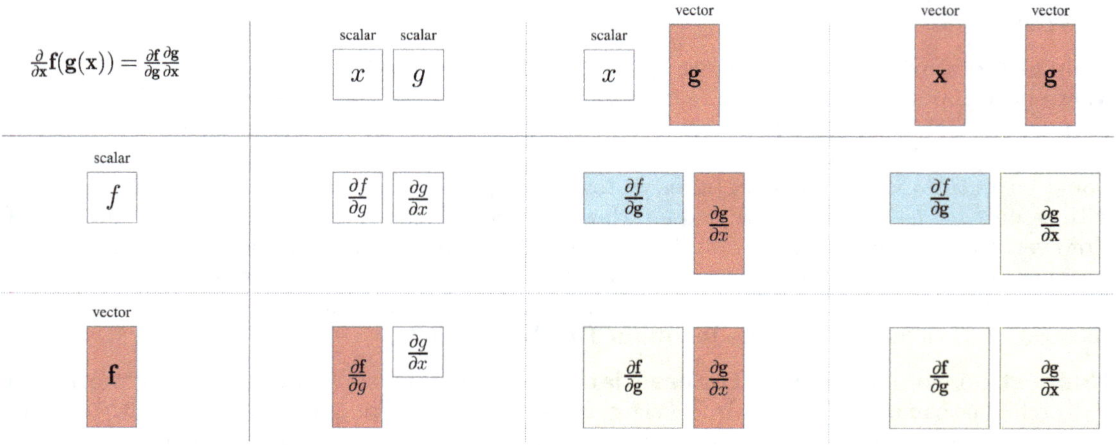

Figure 11.7: Chain rule cases for scalar and vector functions

Row 1: Scalar output f

1. **Case 1 (Column 1)**: $f(g(x))$ where $f : \mathbb{R} \to \mathbb{R}$, $g : \mathbb{R} \to \mathbb{R}$, $x \in \mathbb{R}$
   - $\frac{\partial f}{\partial g} \in \mathbb{R}^{1\times 1}$ (scalar); $\frac{\partial g}{\partial x} \in \mathbb{R}^{1\times 1}$ (scalar)
   - Chain rule: $\frac{\partial f}{\partial x} = \frac{\partial f}{\partial g}\frac{\partial g}{\partial x}$ (scalar multiplication)
2. **Case 2 (Column 2)**: $f(\mathbf{g}(x))$ where $f : \mathbb{R}^M \to \mathbb{R}$, $\mathbf{g} : \mathbb{R} \to \mathbb{R}^M$, $x \in \mathbb{R}$
   - $\frac{\partial f}{\partial \mathbf{g}} \in \mathbb{R}^{1\times M}$ (row vector, gradient transpose)
   - $\frac{\partial \mathbf{g}}{\partial x} \in \mathbb{R}^{M\times 1}$ (vector function derivative)
   - Chain rule: $\frac{\partial f}{\partial x} = \frac{\partial f}{\partial \mathbf{g}}\frac{\partial \mathbf{g}}{\partial x} \in \mathbb{R}$ (row vector times column vector gives scalar)
3. **Case 3 (Column 3)**: $f(\mathbf{g}(\mathbf{x}))$ where $f : \mathbb{R}^M \to \mathbb{R}$, $\mathbf{g} : \mathbb{R}^N \to \mathbb{R}^M$, $\mathbf{x} \in \mathbb{R}^N$
   - $\frac{\partial f}{\partial \mathbf{g}} \in \mathbb{R}^{1\times M}$ (row vector, gradient transpose); $\frac{\partial \mathbf{g}}{\partial \mathbf{x}} \in \mathbb{R}^{M\times N}$ (Jacobian matrix)
   - Chain rule: $\frac{\partial f}{\partial \mathbf{x}} = \frac{\partial f}{\partial \mathbf{g}}\frac{\partial \mathbf{g}}{\partial \mathbf{x}} \in \mathbb{R}^{1\times N}$ (row vector times matrix gives row vector)

Row 2: Vector output $\mathbf{f}$

4. **Case 4 (Column 1)**: $\mathbf{f}(g(x))$ where $\mathbf{f} : \mathbb{R} \to \mathbb{R}^M$, $g : \mathbb{R} \to \mathbb{R}$, $x \in \mathbb{R}$
   - $\frac{\partial \mathbf{f}}{\partial g} \in \mathbb{R}^{M\times 1}$ (vector function derivative); $\frac{\partial g}{\partial x} \in \mathbb{R}^{1\times 1}$ (scalar)
   - Chain rule: $\frac{\partial \mathbf{f}}{\partial x} = \frac{\partial \mathbf{f}}{\partial g}\frac{\partial g}{\partial x} \in \mathbb{R}^{M\times 1}$ (column vector times scalar gives column vector)
5. **Case 5 (Column 2)**: $\mathbf{f}(\mathbf{g}(x))$ where $\mathbf{f} : \mathbb{R}^M \to \mathbb{R}^P$, $\mathbf{g} : \mathbb{R} \to \mathbb{R}^M$, $x \in \mathbb{R}$
   - $\frac{\partial \mathbf{f}}{\partial \mathbf{g}} \in \mathbb{R}^{P\times M}$ (Jacobian matrix); $\frac{\partial \mathbf{g}}{\partial x} \in \mathbb{R}^{M\times 1}$ (vector function derivative)
   - Chain rule: $\frac{\partial \mathbf{f}}{\partial x} = \frac{\partial \mathbf{f}}{\partial \mathbf{g}}\frac{\partial \mathbf{g}}{\partial x} \in \mathbb{R}^{P\times 1}$ (matrix times column vector gives column vector)
6. **Case 6 (Column 3)**: $\mathbf{f}(\mathbf{g}(\mathbf{x}))$ where $\mathbf{f} : \mathbb{R}^M \to \mathbb{R}^P$, $\mathbf{g} : \mathbb{R}^N \to \mathbb{R}^M$, $\mathbf{x} \in \mathbb{R}^N$
   - $\frac{\partial \mathbf{f}}{\partial \mathbf{g}} \in \mathbb{R}^{P\times M}$ (Jacobian matrix); $\frac{\partial \mathbf{g}}{\partial \mathbf{x}} \in \mathbb{R}^{M\times N}$ (Jacobian matrix)
   - Chain rule: $\frac{\partial \mathbf{f}}{\partial \mathbf{x}} = \frac{\partial \mathbf{f}}{\partial \mathbf{g}}\frac{\partial \mathbf{g}}{\partial \mathbf{x}} \in \mathbb{R}^{P\times N}$ (matrix multiplication gives Jacobian matrix)

Case 6 is the most general form and is the standard chain rule for vector functions used in backpropagation and automatic differentiation. In Cases 2, 4, and 5, the derivatives $\frac{\partial \mathbf{v}}{\partial s} \in \mathbb{R}^{M\times 1}$

are called **vector function derivatives** (derivative of a vector function w.r.t. a scalar). Although they have the same shape as a gradient, they are **not gradients**: a gradient is $\nabla_{\mathbf{x}} f$ where $f : \mathbb{R}^N \to \mathbb{R}$ (scalar function w.r.t. vector), while a vector function derivative is $\frac{\partial \mathbf{v}}{\partial s}$ where $\mathbf{v} : \mathbb{R} \to \mathbb{R}^M$ (vector function w.r.t. scalar). *A vector function derivative can be viewed as a degenerate Jacobian when the input dimension is 1.*

Although Case 6 represents the most general form of the chain rule, we provide a detailed proof for Case 3 (scalar function through vector intermediate) for several reasons. First, Case 3 demonstrates the essential matrix multiplication structure of the chain rule in a simpler setting, making the key ideas more accessible. Second, this case is frequently encountered in optimization problems, where we compute gradients of scalar loss functions with respect to parameters. Third, the general Case 6 follows the same pattern as Case 3, but with full matrix multiplication instead of vector-matrix multiplication. The remaining cases (1, 2, 4, and 5) can be derived as special cases by setting appropriate dimensions to 1.

11.5.2.2 Proof: Case 3 — Scalar Function Through Vector Intermediate

This proof establishes the chain rule for **Case 3**: $f(\mathbf{g}(\mathbf{x}))$ where $f : \mathbb{R}^M \to \mathbb{R}$ is a scalar function, $\mathbf{g} : \mathbb{R}^N \to \mathbb{R}^M$ is a vector function, and $\mathbf{x} \in \mathbb{R}^N$. The chain rule states:

$$\frac{\partial f}{\partial \mathbf{x}} = \frac{\partial f}{\partial \mathbf{g}} \frac{\partial \mathbf{g}}{\partial \mathbf{x}} \in \mathbb{R}^{1 \times N}$$

Note on notation: In this proof, we use the partial derivative notation $\frac{\partial}{\partial}$ (as in the case descriptions) rather than the gradient notation ∇, because it makes the chain rule structure explicit. The relationship is: $\frac{\partial f}{\partial \mathbf{x}} = (\nabla_{\mathbf{x}} f)^T$, where $\nabla_{\mathbf{x}} f \in \mathbb{R}^{N \times 1}$ is the gradient (column vector) and $\frac{\partial f}{\partial \mathbf{x}} \in \mathbb{R}^{1 \times N}$ is its transpose (row vector).

Proof

Since f depends on $\mathbf{g}$ and $\mathbf{g}$ depends on $\mathbf{x}$, we have the composition $f(\mathbf{g}(\mathbf{x}))$.

For each component x_n of $\mathbf{x}$ (where $n = 0, 1, \ldots, N-1$), by the scalar chain rule:

$$\frac{\partial f}{\partial x_n} = \sum_{m=0}^{M-1} \frac{\partial f}{\partial g_m} \frac{\partial g_m}{\partial x_n}$$

This can be written in matrix form. Note that:

- $\frac{\partial f}{\partial \mathbf{g}} \in \mathbb{R}^{1 \times M}$ is a row vector with entries $\frac{\partial f}{\partial g_m}$ (this is the transpose of the gradient $\nabla_{\mathbf{g}} f$)
- $\frac{\partial \mathbf{g}}{\partial \mathbf{x}} \in \mathbb{R}^{M \times N}$ is the Jacobian matrix with entries $\left[\frac{\partial \mathbf{g}}{\partial \mathbf{x}}\right]_{m,n} = \frac{\partial g_m}{\partial x_n}$

Therefore, the n-th component of $\frac{\partial f}{\partial \mathbf{x}}$ is:

$$\left[\frac{\partial f}{\partial \mathbf{x}}\right]_n = \sum_{m=0}^{M-1} \left[\frac{\partial f}{\partial \mathbf{g}}\right]_m \left[\frac{\partial \mathbf{g}}{\partial \mathbf{x}}\right]_{m,n} = \left[\frac{\partial f}{\partial \mathbf{g}} \frac{\partial \mathbf{g}}{\partial \mathbf{x}}\right]_n$$

Since this holds for all $n = 0, 1, \ldots, N-1$, we have:

$$\frac{\partial f}{\partial \mathbf{x}} = \frac{\partial f}{\partial \mathbf{g}} \frac{\partial \mathbf{g}}{\partial \mathbf{x}} \in \mathbb{R}^{1 \times N}$$

where the matrix dimensions are: $\frac{\partial f}{\partial \mathbf{g}} \in \mathbb{R}^{1 \times M}$ and $\frac{\partial \mathbf{g}}{\partial \mathbf{x}} \in \mathbb{R}^{M \times N}$, giving the result $\in \mathbb{R}^{1 \times N}$.

Equivalence with gradient notation: In gradient notation, this corresponds to $\nabla_{\mathbf{x}} f = \left(\frac{\partial \mathbf{g}}{\partial \mathbf{x}}\right)^T \nabla_{\mathbf{g}} f \in \mathbb{R}^{N \times 1}$, where $\nabla_{\mathbf{g}} f \in \mathbb{R}^{M \times 1}$ is the gradient of f with respect to $\mathbf{g}$.

11.5.2.3 Proof: Case 6 — General Vector Chain Rule

This proof establishes the chain rule for **Case 6**, the most general form: a vector function that depends on another vector function. For $\mathbf{y} = \mathbf{f}(\mathbf{g}(\mathbf{x}))$ where $\mathbf{f} : \mathbb{R}^M \to \mathbb{R}^P$ and $\mathbf{g} : \mathbb{R}^N \to \mathbb{R}^M$ are vector functions, the chain rule becomes:

$$\frac{\partial \mathbf{f}}{\partial \mathbf{x}} = \frac{\partial \mathbf{f}}{\partial \mathbf{g}} \frac{\partial \mathbf{g}}{\partial \mathbf{x}} \in \mathbb{R}^{P \times N}$$

This is the standard chain rule for vector functions used in backpropagation and automatic differentiation.

Proof

Since $\mathbf{y} = \mathbf{f}(\mathbf{g})$ and $\mathbf{g} = \mathbf{g}(\mathbf{x})$, we have the composition $\mathbf{y} = \mathbf{f}(\mathbf{g}(\mathbf{x}))$.

For each output component y_p (where $p = 0, 1, \dots, P-1$) and each input component x_n (where $n = 0, 1, \dots, N-1$), by the scalar chain rule:

$$\frac{\partial y_p}{\partial x_n} = \sum_{m=0}^{M-1} \frac{\partial y_p}{\partial g_m} \frac{\partial g_m}{\partial x_n}$$

This can be written in matrix form. Note that:

- $\frac{\partial \mathbf{f}}{\partial \mathbf{g}} \in \mathbb{R}^{P \times M}$ is the Jacobian matrix with entries $\left[\frac{\partial \mathbf{f}}{\partial \mathbf{g}}\right]_{p,m} = \frac{\partial y_p}{\partial g_m}$
- $\frac{\partial \mathbf{g}}{\partial \mathbf{x}} \in \mathbb{R}^{M \times N}$ is the Jacobian matrix with entries $\left[\frac{\partial \mathbf{g}}{\partial \mathbf{x}}\right]_{m,n} = \frac{\partial g_m}{\partial x_n}$

Therefore, the (p, n)-th entry of $\frac{\partial \mathbf{f}}{\partial \mathbf{x}}$ is:

$$\left[\frac{\partial \mathbf{f}}{\partial \mathbf{x}}\right]_{p,n} = \sum_{m=0}^{M-1} \left[\frac{\partial \mathbf{f}}{\partial \mathbf{g}}\right]_{p,m} \left[\frac{\partial \mathbf{g}}{\partial \mathbf{x}}\right]_{m,n} = \left[\frac{\partial \mathbf{f}}{\partial \mathbf{g}} \frac{\partial \mathbf{g}}{\partial \mathbf{x}}\right]_{p,n}$$

Since this holds for all $p = 0, 1, \dots, P-1$ and $n = 0, 1, \dots, N-1$, we have:

$$\frac{\partial \mathbf{f}}{\partial \mathbf{x}} = \frac{\partial \mathbf{f}}{\partial \mathbf{g}} \frac{\partial \mathbf{g}}{\partial \mathbf{x}} \in \mathbb{R}^{P \times N}$$

This is standard matrix multiplication, where the Jacobian $\frac{\partial \mathbf{f}}{\partial \mathbf{g}} \in \mathbb{R}^{P \times M}$ multiplies the Jacobian $\frac{\partial \mathbf{g}}{\partial \mathbf{x}} \in \mathbb{R}^{M \times N}$ to produce the result $\in \mathbb{R}^{P \times N}$.

11.5.2.4 Example: Chain Rule Application (Case 3)

This example demonstrates **Case 3** of the chain rule. Let $z = f(\mathbf{x}) = x_0^2 + x_1^2$ be a scalar function where $\mathbf{x} = \begin{bmatrix} x_0 \\ x_1 \end{bmatrix} \in \mathbb{R}^2$, and let $\mathbf{x} = \mathbf{g}(\mathbf{t}) = A\mathbf{t}$ where $A = \begin{bmatrix} 1 & 2 \\ 3 & 4 \end{bmatrix}$ and $\mathbf{t} = \begin{bmatrix} t_0 \\ t_1 \end{bmatrix} \in \mathbb{R}^2$.

Notation correspondence: In this example, $z = f(\mathbf{x})$ corresponds to Case 3's $f(\mathbf{g}(\mathbf{x}))$, where $f : \mathbb{R}^2 \to \mathbb{R}$ is the scalar function, $\mathbf{g} : \mathbb{R}^2 \to \mathbb{R}^2$ is the vector function (linear transformation $A\mathbf{t}$), and $\mathbf{t}$ plays the role of $\mathbf{x}$ in Case 3. We compute $\nabla_{\mathbf{t}} z$ (the gradient) or equivalently $\frac{\partial f}{\partial \mathbf{t}}$ (the row vector, transpose of the gradient).

Solution

Step 1: Compute the gradient $\nabla_{\mathbf{x}} f$

$$\nabla_{\mathbf{x}} f = \nabla_{\mathbf{x}} z = \begin{bmatrix} \frac{\partial f}{\partial x_0} \\ \frac{\partial f}{\partial x_1} \end{bmatrix} = \begin{bmatrix} 2x_0 \\ 2x_1 \end{bmatrix} \in \mathbb{R}^{2\times 1}$$

Step 2: Compute the Jacobian $\frac{\partial \mathbf{g}}{\partial \mathbf{t}}$

Since $\mathbf{x} = \mathbf{g}(\mathbf{t}) = A\mathbf{t}$ and $\mathbf{t} = \begin{bmatrix} t_0 \\ t_1 \end{bmatrix}$, we have:

- $x_0 = g_0(\mathbf{t}) = A_{00}t_0 + A_{01}t_1 = t_0 + 2t_1$
- $x_1 = g_1(\mathbf{t}) = A_{10}t_0 + A_{11}t_1 = 3t_0 + 4t_1$

The Jacobian matrix is:

$$\frac{\partial \mathbf{g}}{\partial \mathbf{t}} = \begin{bmatrix} \frac{\partial g_0}{\partial t_0} & \frac{\partial g_0}{\partial t_1} \\ \frac{\partial g_1}{\partial t_0} & \frac{\partial g_1}{\partial t_1} \end{bmatrix} = \begin{bmatrix} 1 & 2 \\ 3 & 4 \end{bmatrix} = A \in \mathbb{R}^{2\times 2}$$

Step 3: Apply the chain rule (Case 3)

Using the chain rule from Case 3: $\frac{\partial f}{\partial \mathbf{t}} = \frac{\partial f}{\partial \mathbf{g}}\frac{\partial \mathbf{g}}{\partial \mathbf{t}}$, where $\frac{\partial f}{\partial \mathbf{g}} = (\nabla_{\mathbf{x}} f)^T \in \mathbb{R}^{1\times 2}$:

$$\frac{\partial f}{\partial \mathbf{t}} = (\nabla_{\mathbf{x}} f)^T \frac{\partial \mathbf{g}}{\partial \mathbf{t}} = \begin{bmatrix} 2x_0 & 2x_1 \end{bmatrix} \begin{bmatrix} 1 & 2 \\ 3 & 4 \end{bmatrix} = \begin{bmatrix} 2x_0 + 6x_1 & 4x_0 + 8x_1 \end{bmatrix} \in \mathbb{R}^{1\times 2}$$

Taking the transpose to get the gradient:

$$\nabla_{\mathbf{t}} z = \nabla_{\mathbf{t}} f = \left(\frac{\partial f}{\partial \mathbf{t}}\right)^T = \begin{bmatrix} 2x_0 + 6x_1 \\ 4x_0 + 8x_1 \end{bmatrix} \in \mathbb{R}^{2\times 1}$$

Substituting $\mathbf{x} = A\mathbf{t}$:

$$\nabla_{\mathbf{t}} z = \begin{bmatrix} 2(t_0 + 2t_1) + 6(3t_0 + 4t_1) \\ 4(t_0 + 2t_1) + 8(3t_0 + 4t_1) \end{bmatrix} = \begin{bmatrix} 20t_0 + 28t_1 \\ 28t_0 + 40t_1 \end{bmatrix} \in \mathbb{R}^{2\times 1}$$

Verification: We can verify by direct computation. Since $z = (t_0 + 2t_1)^2 + (3t_0 + 4t_1)^2 = t_0^2 + 4t_0t_1 + 4t_1^2 + 9t_0^2 + 24t_0t_1 + 16t_1^2 = 10t_0^2 + 28t_0t_1 + 20t_1^2$, we have:

$$\nabla_{\mathbf{t}} z = \begin{bmatrix} 20t_0 + 28t_1 \\ 28t_0 + 40t_1 \end{bmatrix}$$

which matches our result from the chain rule.

11.5.3 Common Gradient Formulas

Having established the fundamental concepts of gradients, Jacobians, and the chain rule, we now present a collection of commonly used gradient formulas. These formulas are frequently used in machine learning, optimization, and matrix calculus. **It is highly recommended to memorize these formulas** as they appear repeatedly in gradient computations, backpropagation, and optimization algorithms.

Most formulas below are **gradients** (for scalar-valued functions), except formula 3 which is a **Jacobian** (for vector-valued functions). For scalar functions $f : \mathbb{R}^N \to \mathbb{R}$, the gradient $\nabla_{\mathbf{x}} f(\mathbf{x}) \in \mathbb{R}^{N\times 1}$ is a column vector. For vector functions $\mathbf{f} : \mathbb{R}^N \to \mathbb{R}^M$, the Jacobian $\frac{\partial \mathbf{f}}{\partial \mathbf{x}} = \mathbf{J}_{\mathbf{f}}(\mathbf{x}) \in \mathbb{R}^{M\times N}$ is a matrix. For scalar functions with matrix inputs $f : \mathbb{R}^{M\times N} \to \mathbb{R}$, the gradient $\nabla_X f(X) \in \mathbb{R}^{M\times N}$ is a matrix.

1. Let $\mathbf{x}, \mathbf{a} \in \mathbb{R}^{N\times 1}$, then

$$\boxed{\nabla_{\mathbf{x}}(\mathbf{x}^T\mathbf{a}) = \mathbf{a} \in \mathbb{R}^{N\times 1}} \tag{11.11}$$

Proof

Since $\mathbf{x}^T\mathbf{a} = \sum_{i=0}^{N-1} x_i a_i$, and $\nabla_{\mathbf{x}}(\mathbf{x}^T\mathbf{a}) \in \mathbb{R}^{N\times 1}$ is a column vector (with N rows and 1 column), the notation $[\nabla_{\mathbf{x}}(\mathbf{x}^T\mathbf{a})]_j$ refers to the j-th element (the element in the j-th row, since there is only one column). We have:

$$[\nabla_{\mathbf{x}}(\mathbf{x}^T\mathbf{a})]_j = \frac{\partial}{\partial x_j}\sum_{i=0}^{N-1} x_i a_i = \sum_{i=0}^{N-1} a_i \frac{\partial x_i}{\partial x_j} = \sum_{i=0}^{N-1} a_i \delta_{ij} = a_j$$

Since this holds for all $j = 0, \dots, N-1$, we conclude that $\nabla_{\mathbf{x}}(\mathbf{x}^T\mathbf{a}) = \mathbf{a}$.

2. Let $\mathbf{x} \in \mathbb{R}^{N\times 1}$, $W \in \mathbb{R}^{N\times N}$, then

$$\boxed{\nabla_{\mathbf{x}}(\mathbf{x}^T W\mathbf{x}) = (W + W^T)\mathbf{x} \in \mathbb{R}^{N\times 1}} \tag{11.12}$$

Proof

Step 1: Write $\mathbf{x}^T W\mathbf{x}$ in index form.
The product $\mathbf{x}^T W$ is a $1 \times N$ row; its j-th entry is $(\mathbf{x}^T W)_j = \sum_{i=0}^{N-1} x_i W_{ij}$. Then the scalar $\mathbf{x}^T W\mathbf{x}$ is that row times the column $\mathbf{x}$:

$$\mathbf{x}^T W\mathbf{x} = \sum_{j=0}^{N-1}(\mathbf{x}^T W)_j x_j = \sum_{j=0}^{N-1}\left(\sum_{i=0}^{N-1} x_i W_{ij}\right) x_j = \sum_{i=0}^{N-1}\sum_{j=0}^{N-1} x_i W_{ij} x_j.$$

Step 2: Compute the k-th component of the gradient.
We differentiate the double sum with respect to x_k. For each term $x_i W_{ij} x_j$, the product rule gives two contributions (since both x_i and x_j may depend on x_k). Using the Kronecker delta $\frac{\partial x_i}{\partial x_k} = \delta_{ik}$ and $\frac{\partial x_j}{\partial x_k} = \delta_{jk}$:

$$\frac{\partial}{\partial x_k}(x_i W_{ij} x_j) = \frac{\partial x_i}{\partial x_k} W_{ij} x_j + x_i W_{ij}\frac{\partial x_j}{\partial x_k} = \delta_{ik} W_{ij} x_j + x_i W_{ij}\delta_{jk}.$$

Summing over i and j:

$$[\nabla_{\mathbf{x}}(\mathbf{x}^T W\mathbf{x})]_k = \frac{\partial}{\partial x_k}\sum_{i,j} x_i W_{ij} x_j = \sum_{i,j}\left(\delta_{ik} W_{ij} x_j + x_i W_{ij}\delta_{jk}\right).$$

Step 3: Collapse the sums using the Kronecker delta.
The identity $\sum_i \delta_{ik} F(i,j) = F(k,j)$ (and similarly $\sum_j \delta_{jk} G(i,j) = G(i,k)$) gives:

$$\sum_{i,j}\delta_{ik} W_{ij} x_j = \sum_j W_{kj} x_j = [W\mathbf{x}]_k, \qquad \sum_{i,j} x_i W_{ij}\delta_{jk} = \sum_i x_i W_{ik} = \sum_i W^T_{ik} x_i = [W^T\mathbf{x}]_k.$$

Therefore:

$$[\nabla_{\mathbf{x}}(\mathbf{x}^T W\mathbf{x})]_k = [W\mathbf{x}]_k + [W^T\mathbf{x}]_k = [(W + W^T)\mathbf{x}]_k.$$

Since this holds for all $k = 0, \dots, N-1$, we conclude:

$$\nabla_{\mathbf{x}}(\mathbf{x}^T W\mathbf{x}) = (W + W^T)\mathbf{x}$$

$\mathbf{x}^T W\mathbf{x}$ is a **generalized quadratic form** (see Chapter~7 for the standard definition with symmetric matrices). If W is symmetric ($W = W^T$), then the gradient simplifies to $\nabla_{\mathbf{x}}(\mathbf{x}^T W\mathbf{x}) = 2W\mathbf{x}$.

3. Let $A \in \mathbb{R}^{M\times N}$, $\mathbf{x} \in \mathbb{R}^{N\times 1}$, then

$$\boxed{\frac{\partial(A\mathbf{x})}{\partial\mathbf{x}} = A \in \mathbb{R}^{M\times N}} \tag{11.13}$$

This is a **Jacobian** (not a gradient) because $A\mathbf{x} : \mathbb{R}^N \to \mathbb{R}^M$ is a vector-valued function. In gradient notation, this is sometimes written as $\nabla_{\mathbf{x}}(A\mathbf{x}) = A$, but we prefer the partial derivative notation $\frac{\partial}{\partial\mathbf{x}}$ for Jacobians to maintain consistency with the chain rule notation.

Proof

Since $A\mathbf{x} \in \mathbb{R}^{M\times 1}$ with $[A\mathbf{x}]_m = \sum_{n=0}^{N-1} A_{mn}x_n$, the Jacobian matrix has entries:

$$\left[\frac{\partial(A\mathbf{x})}{\partial\mathbf{x}}\right]_{m,n} = \frac{\partial[A\mathbf{x}]_m}{\partial x_n} = A_{mn}$$

Therefore, $\frac{\partial(A\mathbf{x})}{\partial\mathbf{x}} = A$.

4. Let $A \in \mathbb{R}^{M\times N}$, $\mathbf{x} \in \mathbb{R}^{N\times 1}$, $\mathbf{b} \in \mathbb{R}^{M\times 1}$, $W \in \mathbb{R}^{M\times M}$, then

$$\boxed{\nabla_{\mathbf{x}}((A\mathbf{x}-\mathbf{b})^T W(A\mathbf{x}-\mathbf{b})) = A^T(W+W^T)(A\mathbf{x}-\mathbf{b}) \in \mathbb{R}^{N\times 1}} \tag{11.14}$$

Proof

Let $\mathbf{y} = A\mathbf{x} - \mathbf{b}$, so we need $\nabla_{\mathbf{x}}(\mathbf{y}^T W\mathbf{y})$. Using the chain rule (Theorem: Chain Rule for Vector Functions):

$$\nabla_{\mathbf{x}}(\mathbf{y}^T W\mathbf{y}) = \left(\frac{\partial\mathbf{y}}{\partial\mathbf{x}}\right)^T \nabla_{\mathbf{y}}(\mathbf{y}^T W\mathbf{y})$$

From Proof 3, $\frac{\partial\mathbf{y}}{\partial\mathbf{x}} = \frac{\partial(A\mathbf{x}-\mathbf{b})}{\partial\mathbf{x}} = A \in \mathbb{R}^{M\times N}$ (this is the Jacobian of the vector-valued function $\mathbf{y} = A\mathbf{x} - \mathbf{b}$). From Proof 2, $\nabla_{\mathbf{y}}(\mathbf{y}^T W\mathbf{y}) = (W + W^T)\mathbf{y} = (W + W^T)(A\mathbf{x} - \mathbf{b}) \in \mathbb{R}^{M\times 1}$.

Therefore:

$$\nabla_{\mathbf{x}}((A\mathbf{x}-\mathbf{b})^T W(A\mathbf{x}-\mathbf{b})) = A^T(W+W^T)(A\mathbf{x}-\mathbf{b})$$

5. Let $a \in \mathbb{R}^{M\times 1}$, $X \in \mathbb{R}^{M\times N}$, $b \in \mathbb{R}^{N\times 1}$, then

$$\boxed{\nabla_X(a^T Xb) = ab^T \in \mathbb{R}^{M\times N}} \tag{11.15}$$

Proof

We have $a^T Xb = \sum_{i,j} a_i X_{ij} b_j$. The (m,n)-th entry of the gradient is:

$$[\nabla_X(a^T Xb)]_{m,n} = \frac{\partial}{\partial X_{mn}} \sum_{i,j} a_i X_{ij} b_j = a_m b_n$$

Therefore, $\nabla_X(a^T Xb) = ab^T$.

6. Let $U \in \mathbb{R}^{M \times N}$, then

$$\boxed{\nabla_U \|U\|_F^2 = 2U \in \mathbb{R}^{M \times N}} \tag{11.16}$$

where $\|U\|_F$ is the Frobenius norm, defined as $\|U\|_F = \sqrt{\sum_{i,j} U_{ij}^2} = \sqrt{\text{tr}(U^T U)}$.

Proof

Since $\|U\|_F^2 = \sum_{i,j} U_{ij}^2$, the (m, n)-th entry of the gradient is:

$$[\nabla_U \|U\|_F^2]_{m,n} = \frac{\partial}{\partial U_{mn}} \sum_{i,j} U_{ij}^2 = 2U_{mn}$$

Therefore, $\nabla_U \|U\|_F^2 = 2U$.

11.5.3.1 Example: Gradient of Frobenius Norm Squared

Compute the gradient of $\|U\|_F^2$ where $U = \begin{bmatrix} 1 & 2 \\ 3 & 4 \end{bmatrix} \in \mathbb{R}^{2\times 2}$.

Solution

First, compute $\|U\|_F^2$:

$$\|U\|_F^2 = \sum_{i,j} U_{ij}^2 = 1^2 + 2^2 + 3^2 + 4^2 = 1 + 4 + 9 + 16 = 30$$

Using formula 6, the gradient is:

$$\nabla_U \|U\|_F^2 = 2U = 2 \begin{bmatrix} 1 & 2 \\ 3 & 4 \end{bmatrix} = \begin{bmatrix} 2 & 4 \\ 6 & 8 \end{bmatrix}$$

Verification: We can verify by computing the gradient element-wise:

- $\frac{\partial \|U\|_F^2}{\partial U_{00}} = \frac{\partial}{\partial U_{00}}(U_{00}^2 + U_{01}^2 + U_{10}^2 + U_{11}^2) = 2U_{00} = 2(1) = 2$
- $\frac{\partial \|U\|_F^2}{\partial U_{01}} = 2U_{01} = 2(2) = 4$
- $\frac{\partial \|U\|_F^2}{\partial U_{10}} = 2U_{10} = 2(3) = 6$
- $\frac{\partial \|U\|_F^2}{\partial U_{11}} = 2U_{11} = 2(4) = 8$

This confirms that $\nabla_U \|U\|_F^2 = \begin{bmatrix} 2 & 4 \\ 6 & 8 \end{bmatrix} = 2U$.

7. Let $U \in \mathbb{R}^{M \times K}$, $V \in \mathbb{R}^{N \times K}$, $A \in \mathbb{R}^{M \times N}$, then

$$\boxed{\nabla_U \|UV^T - A\|_F^2 = 2(UV^T - A)V \in \mathbb{R}^{M \times K}} \tag{11.17}$$

Proof

Let $E = UV^T - A$, so $\|E\|_F^2 = \sum_{i,j} E_{ij}^2$ where $E_{ij} = \sum_k U_{ik} V_{jk} - A_{ij}$.

The (m, n)-th entry of the gradient is:

$$[\nabla_U \|UV^T - A\|_F^2]_{m,n} = \frac{\partial}{\partial U_{mn}} \sum_{i,j} E_{ij}^2$$

Using the chain rule:

$$= \sum_{i,j} 2E_{ij} \frac{\partial E_{ij}}{\partial U_{mn}}$$

Now, since $E_{ij} = \sum_k U_{ik} V_{jk} - A_{ij}$, we compute the partial derivative:

$$\frac{\partial E_{ij}}{\partial U_{mn}} = \frac{\partial}{\partial U_{mn}} \left(\sum_k U_{ik} V_{jk} - A_{ij} \right) = \sum_k \frac{\partial U_{ik}}{\partial U_{mn}} V_{jk} = \sum_k \delta_{im} \delta_{kn} V_{jk} = \delta_{im} V_{jn}$$

where δ_{im} selects row $i = m$ and δ_{kn} selects column $k = n$ from V, giving V_{jn}. Substituting back:

$$= \sum_{i,j} 2E_{ij} \delta_{im} V_{jn} = 2 \sum_j E_{mj} V_{jn}$$

where δ_{im} is the Kronecker delta. This gives:

$$[\nabla_U \|UV^T - A\|_F^2]_{m,n} = 2[EV]_{m,n} = 2[(UV^T - A)V]_{m,n}$$

Therefore, $\nabla_U \|UV^T - A\|_F^2 = 2(UV^T - A)V$.

11.5.3.2 Example: Linear Transformation Gradients

Let $y = Wx + b$, where

- $W \in \mathbb{R}^{M \times N}$
- $b \in \mathbb{R}^{M \times 1}$
- $x \in \mathbb{R}^{N \times 1}$

Compute the following gradients:

1. $\frac{\partial \mathbf{y}}{\partial \mathbf{x}}$. The final answer should be in a matrix form.
2. $\nabla_W y$. The final answer should be in an element-wise form.
3. $\nabla_b y$. The final answer should be in a matrix form.

Solution

1. Since $\mathbf{y} \in \mathbb{R}^{M \times 1}$ and $\mathbf{x} \in \mathbb{R}^{N \times 1}$, $\frac{\partial \mathbf{y}}{\partial \mathbf{x}}$ is the Jacobian matrix $\in \mathbb{R}^{M \times N}$.

We have $\mathbf{y} = W\mathbf{x} + \mathbf{b}$, so $y_m = \sum_{n=0}^{N-1} W_{mn} x_n + b_m$. Therefore:

$$\frac{\partial y_m}{\partial x_n} = W_{mn}$$

Hence, $\frac{\partial \mathbf{y}}{\partial \mathbf{x}} = W \in \mathbb{R}^{M \times N}$.

2. Since $\mathbf{y} \in \mathbb{R}^{M \times 1}$ and $W \in \mathbb{R}^{M \times N}$, $\nabla_W \mathbf{y} \in \mathbb{R}^{M \times M \times N}$ (a 3D tensor).

We have $y_m = \sum_{i=0}^{N-1} w_{mi} x_i + b_m$. To compute $\nabla_W \mathbf{y}$, we need $\frac{\partial y_m}{\partial w_{kn}}$ for all m, k, n, where:

- m is the output index (which element of y)
- k is the row index of W (we use k instead of m to avoid confusion with the output index)
- n is the column index of W (which column of W)

Therefore:

$$\frac{\partial y_m}{\partial w_{kn}} = \frac{\partial}{\partial w_{kn}} \left(\sum_{i=0}^{N-1} w_{mi} x_i + b_m \right) = x_n \delta_{mk}$$

where δ_{mk} is the Kronecker delta (equal to 1 if $m = k$, and 0 otherwise). The derivative is non-zero only when $k = m$ because y_m only depends on row m of W (i.e., elements w_{mi} for $i = 0, \ldots, N-1$). When we take the derivative with respect to w_{kn}, it's zero unless $k = m$. In element-wise form: $\frac{\partial y_m}{\partial w_{kn}} = x_n \delta_{mk}$.

3. For $\nabla_{\mathbf{b}} \mathbf{y}$, we have $\mathbf{y} = W\mathbf{x} + \mathbf{b}$, so $y_m = \sum_{n=0}^{N-1} W_{mn} x_n + b_m$.

Therefore:

$$\frac{\partial y_m}{\partial b_k} = \delta_{mk}$$

where δ_{mk} is the Kronecker delta. Hence, $\nabla_{\mathbf{b}} \mathbf{y} = I_M \in \mathbb{R}^{M \times M}$, where I_M is the $M \times M$ identity matrix.

11.5.3.3 Example: Softmax Gradient

Define $\text{Softmax} : \mathbb{R}^N \to \mathbb{R}^N$, where

$$(\text{Softmax}(x))_m = \frac{\exp(x_m)}{\sum_{i=0}^{N-1} \exp(x_i)}$$

Compute $\frac{\partial \text{Softmax}(\mathbf{x})}{\partial \mathbf{x}}$.

Solution

We have

$$\frac{\partial (\text{Softmax}(x))_m}{\partial x_n} = \frac{\partial}{\partial x_n} \left(\frac{\exp(x_m)}{\sum_{i=0}^{N-1} \exp(x_i)} \right)$$

Using the quotient rule:

$$= \frac{\exp(x_m)\delta_{mn} \sum_{i=0}^{N-1} \exp(x_i) - \exp(x_m)\exp(x_n)}{\left(\sum_{i=0}^{N-1} \exp(x_i)\right)^2}$$

where δ_{mn} is the **Kronecker delta**. This simplifies to:

$$= \frac{\exp(x_m)\delta_{mn}}{\sum_{i=0}^{N-1} \exp(x_i)} - \frac{\exp(x_m)\exp(x_n)}{\left(\sum_{i=0}^{N-1} \exp(x_i)\right)^2}$$

$$= (\text{Softmax}(x))_m \delta_{mn} - (\text{Softmax}(x))_m (\text{Softmax}(x))_n$$

Therefore, the Jacobian matrix $\frac{\partial \text{Softmax}(\mathbf{x})}{\partial \mathbf{x}} \in \mathbb{R}^{N \times N}$ has entries:

$$\left[\frac{\partial \text{Softmax}(\mathbf{x})}{\partial \mathbf{x}}\right]_{m,n} = (\text{Softmax}(\mathbf{x}))_m \delta_{mn} - (\text{Softmax}(\mathbf{x}))_m (\text{Softmax}(\mathbf{x}))_n$$

In matrix form:

$$\frac{\partial \text{Softmax}(\mathbf{x})}{\partial \mathbf{x}} = \text{diag}(\text{Softmax}(\mathbf{x})) - (\text{Softmax}(\mathbf{x}))(\text{Softmax}(\mathbf{x}))^T \in \mathbb{R}^{N \times N}$$

where $\text{diag}(\text{Softmax}(x))$ is the diagonal matrix with $(\text{Softmax}(x))_m$ on the diagonal, and the second term is the outer product of $\text{Softmax}(x)$ with itself.

11.6 Non-Differentiable Operations in Chain Rule

Backpropagation[2] in neural networks is fundamentally an application of the chain rule: gradients flow backward through the computational graph by repeatedly applying the chain rule at each layer. The chain rule and gradient formulas presented thus far assume that all functions are differentiable. However, many operations commonly used in neural networks are not differentiable at certain points, or are fundamentally discrete in nature. To enable gradient-based optimization through backpropagation, these operations require special handling. This section examines three important classes of such operations: piecewise linear activations with non-differentiable points, pooling operations with discontinuous gradients, and discrete quantization operations.

11.6.1 ReLU and Subgradients

The **Rectified Linear Unit (ReLU)** activation function is defined as:

$$\text{ReLU}(x) = \max(0, x) = \begin{cases} x & \text{if } x > 0 \\ 0 & \text{if } x \leq 0 \end{cases}$$

For $x \neq 0$, the derivative exists and is given by:

$$\frac{d}{dx}\text{ReLU}(x) = \begin{cases} 1 & \text{if } x > 0 \\ 0 & \text{if } x < 0 \end{cases}$$

At $x = 0$, the function is not differentiable in the classical sense. The left derivative is 0 and the right derivative is 1, so the derivative does not exist. However, for automatic differentiation to proceed, a convention must be established. The subdifferential of ReLU at $x = 0$ is the interval $[0, 1]$, and any value in this interval is a valid subgradient. Common conventions are:

- $\frac{d}{dx}\text{ReLU}(0) = 0$ (treating the point as part of the non-active region)
- $\frac{d}{dx}\text{ReLU}(0) = 1$ (treating the point as part of the active region)

The choice of 0 is frequently adopted as it promotes sparsity in the gradient flow, which can aid in regularization.

For element-wise ReLU applied to a vector $\mathbf{x} \in \mathbb{R}^N$, the Jacobian is a diagonal matrix:

$$\frac{\partial \text{ReLU}(\mathbf{x})}{\partial \mathbf{x}} = \text{diag}\left(\mathbf{1}_{\mathbf{x}>0}\right) \in \mathbb{R}^{N \times N}$$

where $\mathbf{1}_{\mathbf{x}>0}$ is an indicator vector with entries 1 where $x_i > 0$ and 0 (or the chosen convention value) where $x_i \leq 0$.

11.6.2 Pooling Operations

Pooling layers in convolutional neural networks perform downsampling operations that exhibit discontinuous gradients. We examine two fundamental types:

[2] Backpropagation, short for "backward propagation of errors," is an algorithm for computing gradients in neural networks. It was independently developed by multiple researchers in the 1960s and 1970s, with key contributions from Seppo Linnainmaa (1970), Paul Werbos (1974), and popularized by Rumelhart, Hinton, and Williams (1986). The algorithm efficiently computes gradients by applying the chain rule in reverse order through the computational graph, avoiding redundant calculations that would occur in naive forward-mode differentiation.

Max Pooling: For a pooling window containing values $\{x_0, x_1, \dots, x_{k-1}\}$, the output is $\max(x_0, x_1, \dots, x_{k-1})$. The derivative with respect to each input is:

$$\frac{\partial}{\partial x_i} \max(x_0, x_1, \dots, x_{k-1}) = \begin{cases} 1 & \text{if } x_i = \max(x_0, x_1, \dots, x_{k-1}) \\ 0 & \text{otherwise} \end{cases}$$

The gradient flows only to the position that achieved the maximum. If multiple positions share the maximum value, the gradient is typically distributed uniformly among them. For example, if k positions share the maximum, each receives gradient $\frac{1}{k}$.

Average Pooling: This operation computes the mean over the pooling window and is fully differentiable:

$$\frac{\partial}{\partial x_i} \frac{1}{k} \sum_{j=0}^{k-1} x_j = \frac{1}{k}$$

The gradient is uniform across all positions in the pooling window.

For a 2D max pooling operation over a spatial region, the Jacobian with respect to the input is sparse: only the positions selected as maxima receive non-zero gradients during backpropagation.

11.6.3 Straight-Through Estimator

Discrete operations, such as quantization in Vector Quantized Variational Autoencoders (VQ-VAE), present a fundamental challenge: they are not differentiable anywhere in their domain. The **Straight-Through Estimator (STE)** is a gradient approximation technique that enables backpropagation through such operations.

Consider a quantization operation $Q : \mathbb{R} \to \mathcal{Q}$ that maps continuous values to a discrete codebook $\mathcal{Q} = \{q_0, q_1, \dots, q_{k-1}\}$:

$$\hat{x} = Q(x) = \arg\min_{q \in \mathcal{Q}} \|x - q\|$$

The forward pass uses the quantized value $\hat{x}$. During backpropagation, the STE approximates the gradient by treating the quantization operation as the identity function:

$$\frac{\partial L}{\partial x} \approx \frac{\partial L}{\partial \hat{x}}$$

This corresponds to setting $\frac{\partial \hat{x}}{\partial x} = 1$ in the chain rule, even though the true derivative is zero (or undefined) everywhere.

The STE is motivated by the observation that, while quantization is discrete, the loss function $\mathcal{L}(\hat{x})$ is typically smooth with respect to the quantized values. The gradient $\frac{\partial \mathcal{L}}{\partial \hat{x}}$ provides directional information for updating x to reduce the loss, despite the non-differentiability of the quantization step itself. This heuristic has proven effective in practice for training models with discrete bottlenecks.

For a vector quantization operation $\mathbf{Q} : \mathbb{R}^N \to \mathcal{Q}^N$, the STE approximation extends to:

$$\frac{\partial L}{\partial \mathbf{x}} \approx \frac{\partial L}{\partial \hat{\mathbf{x}}} \cdot I_N = \frac{\partial L}{\partial \hat{\mathbf{x}}}$$

where I_N is the $N \times N$ identity matrix.

> The STE is a biased gradient estimator. More sophisticated alternatives exist, such as Gumbel-Softmax relaxation and REINFORCE-based methods, but the STE remains widely used due to its simplicity and empirical effectiveness.

11.7 Fréchet Derivative - Derivatives of Matrix-to-Matrix Functions

For matrix-valued functions $f : \mathbb{R}^{M\times N} \to \mathbb{R}^{P\times Q}$, we have seen that the Jacobian can be represented as a $(PQ) \times (MN)$ matrix via vectorization. However, there is an alternative and often more elegant viewpoint: treating the derivative as a **linear operator** acting directly on matrices, without vectorization.

The **Fréchet derivative** (also simply called the derivative) of a matrix-valued function f at a point X is a linear operator $f'(X)$ that maps matrices to matrices. It satisfies the first-order approximation:

$$f(X + \Delta) = f(X) + f'(X)[\Delta] + o(\|\Delta\|) \tag{11.18}$$

where $\Delta \in \mathbb{R}^{M\times N}$ is a perturbation matrix, and $o(\|\Delta\|)$ denotes terms that vanish faster than $\|\Delta\|$ as $\|\Delta\| \to 0$. The notation $f'(X)[\Delta]$ means applying the linear operator $f'(X)$ to the matrix Δ. When the context is clear (i.e., there is only one matrix variable), we use Δ without a subscript. When multiple matrices are involved and clarity is needed, we use subscripts such as Δ_X, Δ_A, etc., to indicate which matrix is being perturbed. This notation is consistent with the perturbation notation used in Chapter~4 for rank-1 perturbation analysis (e.g., backward error analysis where $A + \Delta_A = QR$)[3].

Relationship to Vectorization: The Fréchet derivative and the vectorized Jacobian are equivalent representations. The vectorized Jacobian $\nabla_X f(X) \in \mathbb{R}^{(PQ)\times(MN)}$ is the matrix representation of the linear operator $f'(X)$ with respect to the standard basis after vectorization. However, the operator viewpoint often provides more insight into the structure of derivatives, especially for functions like matrix inverses, matrix exponentials, and other matrix operations.

11.7.1 Properties of Fréchet Derivative

The Fréchet derivative satisfies several important properties:

1. **Linearity**: The Fréchet derivative is a linear operator. For matrices Δ_1, Δ_2 and scalars α, β:

$$f'(X)[\alpha\Delta_1 + \beta\Delta_2] = \alpha f'(X)[\Delta_1] + \beta f'(X)[\Delta_2] \tag{11.19}$$

2. **Chain Rule**: For composite functions $h(X) = g(f(X))$ where $f : \mathbb{R}^{M\times N} \to \mathbb{R}^{P\times Q}$ and $g : \mathbb{R}^{P\times Q} \to \mathbb{R}^{R\times S}$, the Fréchet derivative satisfies:

$$h'(X)[\Delta] = g'(f(X))[f'(X)[\Delta]] \tag{11.20}$$

[3] Some textbooks use the notation dX (or dA, dB, etc.) for perturbation matrices, following the differential notation from single-variable calculus. We prefer Δ (or Δ_X, Δ_A when needed) because it is visually clearer in expressions like $X\Delta + \Delta X$ compared to $X\,dX + dX\,X$, where the spacing between dX and X can be ambiguous. Both notations are mathematically equivalent and represent the same concept: an infinitesimal change in the matrix variable.

This expresses the chain rule as composition of linear operators.

3. **Sum Rule**: For functions $f, g : \mathbb{R}^{M\times N} \to \mathbb{R}^{P\times Q}$:

$$(f+g)'(X)[\Delta] = f'(X)[\Delta] + g'(X)[\Delta] \tag{11.21}$$

4. **Product Rule**: For matrix-valued functions $f(X) = A(X)B(X)$ where $A : \mathbb{R}^{M\times N} \to \mathbb{R}^{P\times R}$ and $B : \mathbb{R}^{M\times N} \to \mathbb{R}^{R\times Q}$:

$$f'(X)[\Delta] = A'(X)[\Delta]B(X) + A(X)B'(X)[\Delta] \tag{11.22}$$

5. **Inverse Rule**: For the matrix inverse function $f(X) = X^{-1}$ where $X \in \mathbb{R}^{N\times N}$ is invertible:

$$f'(X)[\Delta] = -X^{-1}\Delta X^{-1} \tag{11.23}$$

Proof: Consider the perturbation $X + \epsilon\Delta$ where ϵ is a small scalar. We have the identity $(X+\epsilon\Delta)(X+\epsilon\Delta)^{-1} = I$. Differentiating both sides with respect to ϵ at $\epsilon = 0$:

$$\frac{d}{d\epsilon}\left[(X+\epsilon\Delta)(X+\epsilon\Delta)^{-1}\right]\Big|_{\epsilon=0} = 0$$

Using the product rule:

$$\Delta X^{-1} + X\frac{d}{d\epsilon}(X+\epsilon\Delta)^{-1}\Big|_{\epsilon=0} = 0$$

Solving for the derivative:

$$\frac{d}{d\epsilon}(X+\epsilon\Delta)^{-1}\Big|_{\epsilon=0} = -X^{-1}\Delta X^{-1}$$

This gives the first-order term in the expansion $(X+\epsilon\Delta)^{-1} = X^{-1} - \epsilon X^{-1}\Delta X^{-1} + O(\epsilon^2)$, which matches the Fréchet derivative definition.

Inverse of the Derivative Operator: When $f'(X)$ is invertible, the inverse operator $f'^{-1}(X)$ satisfies $f'^{-1}(X)[f'(X)[\Delta]] = \Delta$. For the inverse function $f(X) = X^{-1}$, to find the inverse operator, we set $\Delta_f = f'(X)[\Delta] = -X^{-1}\Delta X^{-1}$ and solve for Δ:

$$\begin{aligned}\Delta_f &= -X^{-1}\Delta X^{-1}\\ X\Delta_f X &= -X(X^{-1}\Delta X^{-1})X = -\Delta\\ \Delta &= -X\Delta_f X.\end{aligned}$$

Therefore, the inverse operator is:

$$f'^{-1}(X)[\Delta_f] = -X\Delta_f X \tag{11.24}$$

where Δ_f is a perturbation of $f(X)$.

6. **Determinant Rule**: For the scalar-valued determinant function $f(X) = \det(X)$ where $X \in \mathbb{R}^{N\times N}$ is invertible, the Fréchet derivative (which maps matrices to scalars) is:

$$f'(X)[\Delta] = \det(X)\cdot \operatorname{tr}(X^{-1}\Delta) \tag{11.25}$$

The gradient (Jacobian) representation is $\nabla_X \det(X) = \det(X)X^{-T} \in \mathbb{R}^{N\times N}$, which satisfies $f'(X)[\Delta] = \langle \nabla_X \det(X), \Delta\rangle_F$ where $\langle\cdot,\cdot\rangle_F$ is the Frobenius inner product.

7. **Trace Rule**: For the scalar-valued trace function $f(X) = \text{tr}(X)$ where $X \in \mathbb{R}^{N\times N}$, the Fréchet derivative is:

$$f'(X)[\Delta] = \text{tr}(\Delta) \tag{11.26}$$

The gradient is $\nabla_X \text{tr}(X) = I_N \in \mathbb{R}^{N\times N}$ (identity matrix). For the function $f(X) = \text{tr}(AX)$ where $A \in \mathbb{R}^{N\times N}$ is constant:

$$f'(X)[\Delta] = \text{tr}(A\Delta) = \langle A^\top, \Delta\rangle_F$$

The gradient is $\nabla_X \text{tr}(AX) = A^\top$.

8. **Spectral Norm Rule**: For the scalar-valued spectral norm function $f(X) = \|X\|_2 = \sigma_{\max}(X)$ where $\sigma_{\max}(X)$ is the largest singular value of $X \in \mathbb{R}^{M\times N}$, the Fréchet derivative is:

$$f'(X)[\Delta] = \mathbf{u}_{\max}^\top \Delta \mathbf{v}_{\max} \tag{11.27}$$

where $\mathbf{u}_{\max} \in \mathbb{R}^M$ and $\mathbf{v}_{\max} \in \mathbb{R}^N$ are the left and right singular vectors corresponding to $\sigma_{\max}(X)$, normalized such that $\|\mathbf{u}_{\max}\|_2 = \|\mathbf{v}_{\max}\|_2 = 1$ and $\mathbf{u}_{\max}^\top X \mathbf{v}_{\max} = \sigma_{\max}(X)$. The gradient is $\nabla_X \|X\|_2 = \mathbf{u}_{\max}\mathbf{v}_{\max}^\top \in \mathbb{R}^{M\times N}$. This assumes $\sigma_{\max}(X)$ is a simple (non-repeated) singular value.

9. **Spectral Radius Rule**: For the scalar-valued spectral radius function $f(X) = \rho(X) = \max_i |\lambda_i(X)|$ where $\lambda_i(X)$ are the eigenvalues of $X \in \mathbb{R}^{N\times N}$, the Fréchet derivative is:

$$f'(X)[\Delta] = \frac{\text{Re}\left(\overline{\lambda}_{\max} \mathbf{w}_{\max}^* \Delta \mathbf{v}_{\max}\right)}{\mathbf{w}_{\max}^* \mathbf{v}_{\max}} \tag{11.28}$$

where $\lambda_{\max}$ is the eigenvalue with maximum modulus, $\mathbf{v}_{\max}$ is the corresponding right eigenvector, and $\mathbf{w}_{\max}$ is the corresponding left eigenvector, normalized such that $\mathbf{w}_{\max}^* \mathbf{v}_{\max} = 1$. Even for real matrices, eigenvalues can be complex (e.g., rotation matrices), so we use the general form with complex conjugation ($\mathbf{w}^*$ denotes the conjugate transpose). The gradient is $\nabla_X \rho(X) = \mathbf{w}_{\max}\overline{\mathbf{v}}_{\max}^\top$. This assumes $\lambda_{\max}$ is a simple eigenvalue. For real matrices with real dominant eigenvalues, this simplifies to $\mathbf{v}_{\max}^\top \Delta \mathbf{v}_{\max}/\|\mathbf{v}_{\max}\|_2^2$, with gradient $\nabla_X \rho(X) = \mathbf{v}_{\max}\mathbf{v}_{\max}^\top/\|\mathbf{v}_{\max}\|_2^2$.

Example: For the function $f(X) = X^2$ where $X \in \mathbb{R}^{N\times N}$, the Fréchet derivative at X is:

$$f'(X)[\Delta] = X\Delta + \Delta X$$

This operator view makes it clear that the derivative depends on both left and right multiplication by X, which would be less obvious in the vectorized representation.

11.7.2 Example: Newton-Schulz Iteration for Matrix Inversion

The Newton-Schulz method is an iterative algorithm for computing matrix inverses using Newton's method. It demonstrates how matrix calculus (specifically, derivatives of matrix-valued functions) can be applied to derive practical numerical algorithms.

We seek the zero of the function $f : \mathbb{R}^{N\times N} \to \mathbb{R}^{N\times N}$ defined as:

$$f(X) = X^{-1} - A,$$

where $A \in \mathbb{R}^{N\times N}$ is the matrix whose inverse we wish to compute, and $X \in \mathbb{R}^{N\times N}$ is our current estimate.

For the function $f(X) = X^{-1} - A$, the Fréchet derivative at X (see the Inverse Rule in equation 11.23) applied to matrix Δ operates as follows:

$$f'(X)[\Delta] = -X^{-1}\Delta X^{-1}.$$

The inverse of f' at X, applied to matrix Δ_f (a perturbation of $f(X)$), operates as (see equation 11.24):

$$f'^{-1}(X)[\Delta_f] = -X\Delta_f X.$$

The Newton method for root finding has at each iterate:

$$\begin{aligned}
X_{t+1} &= X_t - \underbrace{f'^{-1}(X_t)\Big[f(X_t)\Big]}_{\text{Application of } f'^{-1}} \\
&= X_t - f'^{-1}(X_t)\Big[X_t^{-1} - A\Big] \\
&= X_t - \underbrace{\Big(-X_t(X_t^{-1} - A)X_t\Big)}_{\text{Using } f'^{-1}(X)[\Delta_f] = -X\Delta_f X} \\
&= X_t + X_t(X_t^{-1} - A)X_t \\
&= X_t + \underbrace{X_t X_t^{-1}}_{I} X_t - X_t A X_t \\
&= X_t + X_t - X_t A X_t \\
&= 2X_t - X_t A X_t.
\end{aligned}$$

Crucial Observation: Notice that the explicit inverse term X_t^{-1} (which appeared in the evaluation of $f(X_t) = X_t^{-1} - A$) is perfectly cancelled by the pre-multiplication in the inverse operator f'^{-1}. This cancellation is crucial because it allows us to update the estimate using only matrix multiplication, without ever computing X_t^{-1} explicitly. This yields the **Newton-Schulz iteration**:

$$X_{t+1} = 2X_t - X_t A X_t = X_t(2I - AX_t).$$

The method has several important properties: (1) **Quadratic convergence**: Under appropriate conditions (e.g., when X_0 is sufficiently close to A^{-1}), the method converges quadratically, meaning the number of correct digits roughly doubles at each iteration. (2) **No explicit matrix inversion**: The iteration formula only involves matrix multiplications, avoiding the need to compute X_t^{-1} explicitly at each step. (3) **Initialization**: The method requires a good initial guess X_0. Common choices include $X_0 = \alpha A^\top$ for some scalar α, or $X_0 = \frac{A^\top}{\|A\|_F^2}$.

Connection to Neumann Series: The Neumann series (see Chapter~6, equation 6.12) provides an alternative iterative method for matrix inversion. The following table compares different methods for computing matrix inverses:

Property	**Direct** (LU/GE)	**Neumann Series**	**Newton-Schulz**
Convergence	N/A (exact)	Linear	Quadratic
Reqs	Matrix must be invertible	$\rho(I - A) < 1$ (or $\rho(A) < 1$ for $(I - A)^{-1}$)	Good initial guess X_0
Cost	$O(N^3)$	$O(N^3)$ per term	$O(N^3)$ per iteration
Stability	Moderate	Good	Excellent
Use Case	Small to medium matrices when exact inverse is needed	When preconditioning ensures convergence, simple implementation needed	Large matrices, high accuracy required, when matrix inverses are needed repeatedly

While Neumann series converges linearly and requires $\rho(A) < 1$, Newton-Schulz achieves quadratic convergence and is more numerically robust, making it the preferred choice in most practical applications.

11.8 Hessian Matrix

11.8.1 Definition: Hessian Matrix

For $f : \mathbb{R}^N \to \mathbb{R}$, the Hessian matrix is

$$\mathbf{H}_f = \nabla^2_{\mathbf{x}} f(\mathbf{x}) = \left[\frac{\partial^2 f}{\partial x_i \partial x_j}\right]_{N\times N} \tag{11.29}$$

11.8.2 Theorem: Symmetry of Hessian

If $f : \mathbb{R}^N \to \mathbb{R}$ is sufficiently smooth, then

$$\mathbf{H}_f = \mathbf{H}_f^T \tag{11.30}$$

11.8.2.1 Example: Sigmoid Hessian

Compute the Hessian of the sigmoid composition $\sigma(\mathbf{w}^T\mathbf{x})$, where:

- $\mathbf{x} \in \mathbb{R}^{N\times 1}$ is the input vector
- $\mathbf{w} \in \mathbb{R}^{N\times 1}$ is a weight vector (e.g., parameters in logistic regression or a neural network layer)
- $\sigma(z) = \frac{1}{1+e^{-z}}$ is the sigmoid function
- $\mathbf{w}^T\mathbf{x} = \sum_{n=0}^{N-1} w_n x_n$ is the inner product (linear combination)

This function appears frequently in machine learning, particularly in logistic regression and neural network activation functions.

Solution

Step 1: Compute the gradient (with respect to $\mathbf{x}$, treating $\mathbf{w}$ as a constant):

Let $z = \mathbf{w}^T\mathbf{x}$. The derivative of the sigmoid function is $\frac{d\sigma}{dz} = \sigma(z)(1-\sigma(z))$. Using the chain rule:

$$\nabla_{\mathbf{x}}\sigma(\mathbf{w}^T\mathbf{x}) = \sigma(\mathbf{w}^T\mathbf{x})\left(1-\sigma(\mathbf{w}^T\mathbf{x})\right)\nabla_{\mathbf{x}}(\mathbf{w}^T\mathbf{x}) = \sigma(\mathbf{w}^T\mathbf{x})\left(1-\sigma(\mathbf{w}^T\mathbf{x})\right)\mathbf{w}$$

Step 2: Compute the Hessian (with respect to $\mathbf{x}$, treating $\mathbf{w}$ as a constant):

The Hessian is the Jacobian of the gradient: $\mathbf{H}_f = \nabla^2_{\mathbf{x}}\sigma(\mathbf{w}^T\mathbf{x}) = \frac{\partial(\nabla_{\mathbf{x}}\sigma(\mathbf{w}^T\mathbf{x}))}{\partial\mathbf{x}}$.

Let $g(\mathbf{x}) = \sigma(\mathbf{w}^T\mathbf{x})(1-\sigma(\mathbf{w}^T\mathbf{x}))$ be the scalar coefficient. Then the gradient is $\nabla_{\mathbf{x}}\sigma(\mathbf{w}^T\mathbf{x}) = g(\mathbf{x})\mathbf{w}$, where the i-th component is $[\nabla_{\mathbf{x}}\sigma(\mathbf{w}^T\mathbf{x})]_i = g(\mathbf{x})w_i$.

To compute the Hessian, we need $\frac{\partial}{\partial x_j}[\nabla_{\mathbf{x}}\sigma(\mathbf{w}^T\mathbf{x})]_i = \frac{\partial}{\partial x_j}[g(\mathbf{x})w_i] = w_i\frac{\partial g(\mathbf{x})}{\partial x_j}$.

First, compute $\frac{\partial g(\mathbf{x})}{\partial x_j}$:

$$\begin{aligned}
\frac{\partial g(\mathbf{x})}{\partial x_j} &= \frac{\partial}{\partial x_j}[\sigma(\mathbf{w}^T\mathbf{x})(1-\sigma(\mathbf{w}^T\mathbf{x}))] \\
&= \frac{\partial \sigma(\mathbf{w}^T\mathbf{x})}{\partial x_j}(1-\sigma(\mathbf{w}^T\mathbf{x})) + \sigma(\mathbf{w}^T\mathbf{x})\frac{\partial(1-\sigma(\mathbf{w}^T\mathbf{x}))}{\partial x_j} \\
&= \sigma(\mathbf{w}^T\mathbf{x})(1-\sigma(\mathbf{w}^T\mathbf{x}))w_j(1-\sigma(\mathbf{w}^T\mathbf{x})) + \sigma(\mathbf{w}^T\mathbf{x})(-\sigma(\mathbf{w}^T\mathbf{x})(1-\sigma(\mathbf{w}^T\mathbf{x}))w_j) \\
&= \sigma(\mathbf{w}^T\mathbf{x})(1-\sigma(\mathbf{w}^T\mathbf{x}))w_j[(1-\sigma(\mathbf{w}^T\mathbf{x})) - \sigma(\mathbf{w}^T\mathbf{x})] \\
&= \sigma(\mathbf{w}^T\mathbf{x})(1-\sigma(\mathbf{w}^T\mathbf{x}))(1-2\sigma(\mathbf{w}^T\mathbf{x}))w_j
\end{aligned}$$

Therefore, the (i,j)-th entry of the Hessian is:

$$[\mathbf{H}_f]_{ij} = \frac{\partial}{\partial x_j}[\nabla_{\mathbf{x}}\sigma(\mathbf{w}^T\mathbf{x})]_i = w_i \cdot \sigma(\mathbf{w}^T\mathbf{x})(1-\sigma(\mathbf{w}^T\mathbf{x}))(1-2\sigma(\mathbf{w}^T\mathbf{x}))w_j$$

Writing this in matrix form:

$$\mathbf{H}_f = \nabla^2_{\mathbf{x}}\sigma(\mathbf{w}^T\mathbf{x}) = \sigma(\mathbf{w}^T\mathbf{x})\left(1-\sigma(\mathbf{w}^T\mathbf{x})\right)\left(1-2\sigma(\mathbf{w}^T\mathbf{x})\right)\mathbf{w}\mathbf{w}^T$$

where $ww^T \in \mathbb{R}^{N\times N}$ is the outer product of w with itself, giving $[ww^T]_{ij} = w_i w_j$.

11.8.3 Geometric Interpretation: Ill-Conditioned Curvature

The Hessian matrix provides crucial geometric information about the loss landscape. In machine learning optimization, the **eigenvalue distribution** of the Hessian determines the difficulty of optimization. Consider a loss function $\mathcal{L}(\mathbf{w})$ where $\mathbf{w}$ are the model parameters.

Well-conditioned curvature: When the Hessian $\mathbf{H}_L$ has eigenvalues that are all positive and of similar magnitude, the loss landscape is approximately isotropic (like a bowl). Gradient descent converges efficiently because the gradient points toward the minimum in all directions.

Ill-conditioned curvature: When the Hessian has eigenvalues with very different magnitudes (high condition number $\kappa = \lambda_{\max}/\lambda_{\min} \gg 1$), the loss landscape has a **"valley"** or **"ravine"** shape. This creates several problems:

1. **Slow convergence**: Gradient descent makes rapid progress along directions with large curvature (large eigenvalues) but very slow progress along directions with small curvature (small eigenvalues), leading to zigzagging behavior.
2. **Sensitivity to learning rate**: A learning rate that works well for high-curvature directions may be too large for low-curvature directions, causing oscillations or divergence.
3. **Quantization sensitivity**: In weight quantization (e.g., GPTQ, AWQ), the Hessian eigenvalues indicate which weight directions are most sensitive to perturbations. Directions with large eigenvalues (high curvature) require more careful quantization, as small errors in these directions cause large loss increases. The Hessian guides how to allocate quantization error: minimize error in high-sensitivity directions while allowing larger errors in low-sensitivity directions.

Visualization: A 2D loss function with ill-conditioned Hessian has elliptical contour lines that are highly elongated (like a narrow valley). The major axis corresponds to the eigenvector with the smallest eigenvalue (shallow direction), while the minor axis corresponds to the eigenvector with the largest eigenvalue (steep direction). Optimization algorithms like Adam, which adapt learning rates per dimension, and second-order methods like Newton's method, which use the Hessian inverse, are designed to handle such ill-conditioned landscapes.

Figure 11.8 compares the loss landscapes for well-conditioned (left) and ill-conditioned (right) Hessians. The well-conditioned case shows nearly circular contours (isotropic), while the ill-conditioned case shows highly elongated elliptical contours (anisotropic valley). The eigenvectors are shown as arrows: red arrows indicate steep directions (large eigenvalues), while orange arrows indicate shallow directions (small eigenvalues). In the ill-conditioned case, the valley shape makes optimization challenging because gradient descent progresses slowly along the shallow direction (major axis) while oscillating along the steep direction (minor axis).

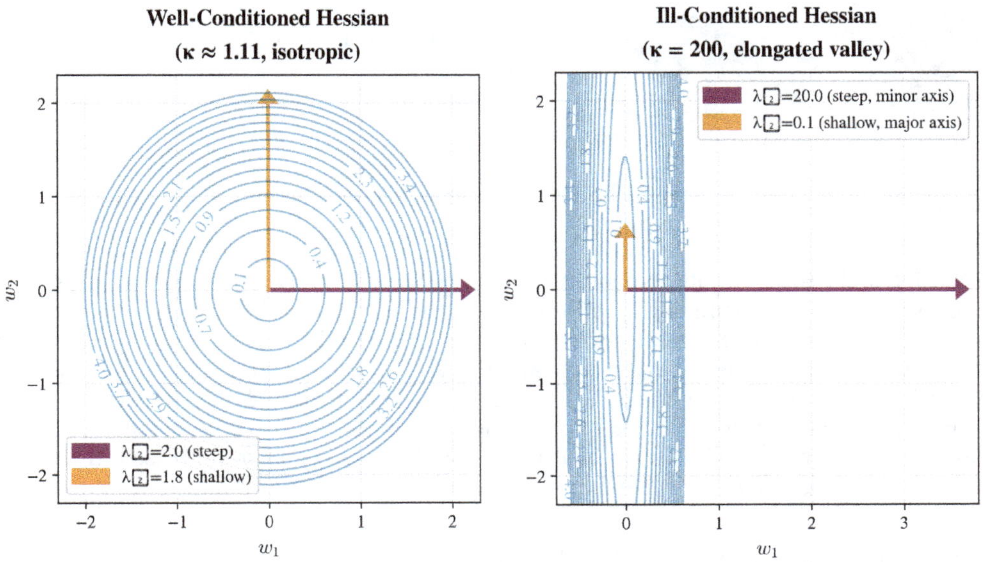

Figure 11.8: Well-conditioned vs Ill-conditioned Hessian

11.9 Taylor Expansion

Taylor expansion provides a way to approximate a function near a point using its derivatives. This section progresses from the simple single-variable case to the more general multi-variable case, focusing on **second-order expansions** which are fundamental in optimization algorithms like Newton's method.

11.9.1 Single-Variable Taylor Expansion

For a function $f : \mathbb{R} \to \mathbb{R}$ that is differentiable at a point $a \in \mathbb{R}$, the **first-order Taylor expansion** around a is:

$$f(x) = f(a) + f'(a)(x - a) + o((x - a)^2) \tag{11.31}$$

where $a \in \mathbb{R}$ is the **expansion point** (the point around which the function is approximated), $f'(a)$ is the first derivative (gradient in 1D) evaluated at a, and the remainder term $o((x - a)^2)$ satisfies $\lim_{x \to a} \frac{o((x-a)^2)}{(x-a)^2} = 0$. This gives a linear approximation: the function is approximated by its value at a plus a linear correction term.

For a function $f : \mathbb{R} \to \mathbb{R}$ that is twice differentiable at a point $a \in \mathbb{R}$, the **second-order Taylor expansion** adds curvature information:

$$f(x) = f(a) + f'(a)(x - a) + \frac{1}{2} f''(a)(x - a)^2 + o((x - a)^3) \tag{11.32}$$

where $f''(a)$ is the second derivative (Hessian in 1D, which is just a scalar) evaluated at a, and the remainder term $o((x-a)^3)$ satisfies $\lim_{x\to a} \frac{o((x-a)^3)}{(x-a)^3} = 0$. The quadratic term $\frac{1}{2}f''(a)(x-a)^2$ captures the local curvature: if $f''(a) > 0$, the function is locally convex (curves upward), while if $f''(a) < 0$, it is locally concave (curves downward).

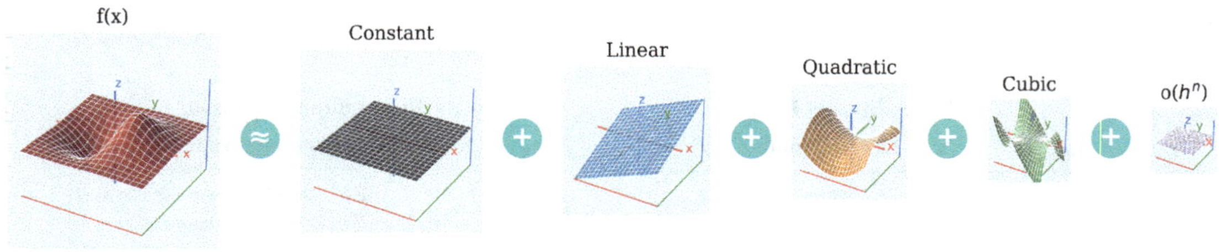

Figure 11.9: Taylor Expansion Visualization

Figure 11.9 visualizes the Taylor expansion decomposition for a 2D function $f(x, y)$, showing how the function can be approximated by progressively adding terms: the **constant term** $f(\mathbf{a})$ (flat plane), the **linear term** (gradient correction), the **quadratic term** (Hessian curvature), the **cubic term**, and the **higher-order remainder** $o(h^n)$. Each panel shows one component of the expansion, with the "+" symbols indicating addition of terms. This geometric visualization helps build intuition for how Taylor expansions approximate functions locally, and the same decomposition principle applies to the multi-variable Taylor expansion discussed in the following sections, where the gradient and Hessian become vector and matrix quantities respectively.

11.9.2 Multi-Variable Taylor Expansion

For functions $f : \mathbb{R}^N \to \mathbb{R}$ with vector inputs, the Taylor expansion generalizes naturally. The first-order term uses the **gradient** (vector of first partial derivatives), and the second-order term uses the **Hessian matrix** (matrix of second partial derivatives).

11.9.3 Theorem: Second-order Taylor Expansion

Let $f : \mathbb{R}^N \to \mathbb{R}$ be a function that is twice differentiable at a point $\mathbf{a} \in \mathbb{R}^N$. Then for $\mathbf{x} \in \mathbb{R}^N$ in a neighborhood of $\mathbf{a}$,

$$f(\mathbf{x}) = f(\mathbf{a}) + (\nabla_{\mathbf{x}} f(\mathbf{a}))^T(\mathbf{x}-\mathbf{a}) + \frac{1}{2}(\mathbf{x}-\mathbf{a})^T(\nabla^2_{\mathbf{x}} f(\mathbf{a}))(\mathbf{x}-\mathbf{a}) + o(\|\mathbf{x}-\mathbf{a}\|_2^2) \tag{11.33}$$

where $\mathbf{a} \in \mathbb{R}^N$ is the **expansion point** (the point around which the function is approximated), and the remainder term $o(\|\mathbf{x}-\mathbf{a}\|_2^2)$ satisfies $\lim_{\mathbf{x}\to\mathbf{a}} \frac{o(\|\mathbf{x}-\mathbf{a}\|_2^2)}{\|\mathbf{x}-\mathbf{a}\|_2^2} = 0$.

Alternative notation: If we let $\mathbf{e} = \mathbf{x} - \mathbf{a}$ denote the error vector, $\mathbf{J}_f = \mathbf{J}_f(\mathbf{a}) = (\nabla_{\mathbf{x}} f(\mathbf{a}))^\top \in \mathbb{R}^{1\times N}$ denote the Jacobian (transpose of the gradient), and $\mathbf{H}_f = \nabla^2_{\mathbf{x}} f(\mathbf{a}) \in \mathbb{R}^{N\times N}$ denote the Hessian, then the second-order Taylor expansion can be written more compactly as:

$$f(\mathbf{x}) = f(\mathbf{a}) + \mathbf{J}_f\mathbf{e} + \frac{1}{2}\mathbf{e}^T\mathbf{H}_f\mathbf{e} + o(\|\mathbf{e}\|_2^2) \tag{11.34}$$

This notation emphasizes that the expansion is a function of the error vector $\mathbf{e}$, with the first-order term being linear in $\mathbf{e}$ (via the Jacobian $\mathbf{J}_f$) and the second-order term being quadratic in $\mathbf{e}$ (via the Hessian $\mathbf{H}_f$).

Connection to Jacobian and Hessian:

- **First-order term**: $(\nabla_{\mathbf{x}} f(\mathbf{a}))^T(\mathbf{x}-\mathbf{a})$ uses the **gradient** $\nabla_{\mathbf{x}} f(\mathbf{a}) \in \mathbb{R}^{N\times 1}$. For a scalar function, the Jacobian is the transpose of the gradient: $\mathbf{J}_f(\mathbf{a}) = (\nabla_{\mathbf{x}} f(\mathbf{a}))^\top \in \mathbb{R}^{1\times N}$, so this term can also be written as $\mathbf{J}_f(\mathbf{a})(\mathbf{x}-\mathbf{a})$.
- **Second-order term**: $\frac{1}{2}(\mathbf{x}-\mathbf{a})^T(\nabla^2_{\mathbf{x}} f(\mathbf{a}))(\mathbf{x}-\mathbf{a})$ uses the **Hessian matrix** $\mathbf{H}_f = \nabla^2_{\mathbf{x}} f(\mathbf{a}) \in \mathbb{R}^{N\times N}$, which contains all second-order partial derivatives. The Hessian captures the local curvature of the function: positive eigenvalues indicate directions of positive curvature (valleys), while negative eigenvalues indicate negative curvature (peaks).

The expansion shows that near $\mathbf{a}$, the function $f(\mathbf{x})$ is approximated by a constant term $f(\mathbf{a})$, plus a linear term (gradient/Jacobian) capturing the slope, plus a quadratic term (Hessian) capturing the curvature.

11.9.3.1 Example: Weight Pruning and Quantization

The second-order Taylor expansion is fundamental to understanding **weight pruning** and **quantization** in neural networks. In **Optimal Brain Damage** and **Optimal Brain Surgeon**[4] (LeCun et al., 1990; Hassibi & Stork, 1993), the change in loss when removing or quantizing a weight w_i is approximated using the second-order expansion:

$$\Delta L \approx \frac{\partial L}{\partial w_i}\Delta w_i + \frac{1}{2}\frac{\partial^2 L}{\partial w_i^2}(\Delta w_i)^2$$

where Δw_i is the change in weight i. The Hessian diagonal $\frac{\partial^2 L}{\partial w_i^2}$ (or full Hessian for Optimal Brain Surgeon) determines which weights can be safely removed or quantized with minimal loss increase. This principle underlies modern quantization methods like **GPTQ**[5] (Frantar et al., 2022) and **AWQ**[6] (Lin et al., 2023), which use Hessian information to guide quantization error allocation, ensuring that weights with large Hessian values (high sensitivity) are quantized more carefully than those with small Hessian values (low sensitivity). The Taylor expansion provides the theoretical foundation: the quadratic term $\frac{1}{2}\mathbf{e}^T\mathbf{H}_f\mathbf{e}$ quantifies how quantization errors $\mathbf{e}$ propagate through the loss landscape, with the Hessian $\mathbf{H}_f$ encoding the sensitivity of the loss to perturbations in different weight directions.

---

[4] **Optimal Brain Damage** (LeCun et al., 1990) and **Optimal Brain Surgeon** (Hassibi & Stork, 1993) use second-order Taylor expansion to estimate weight importance for pruning. OBD uses the Hessian diagonal, while OBS uses the full Hessian matrix for more accurate estimation. See https://proceedings.neurips.cc/paper/1989/hash/6c9882bbac1c7093bd25041881277658-Abstract.html and https://proceedings.neurips.cc/paper/1992/hash/303ed4c69846ab36c2904d3ba8573050-Abstract.html.

[5] **GPTQ**: Elias Frantar, Saleh Ashkboos, Torsten Hoefler, and Dan Alistarh. "GPTQ: Accurate Post-Training Quantization for Generative Pre-trained Transformers." arXiv preprint arXiv:2210.17323 (2022). https://arxiv.org/abs/2210.17323. GPTQ uses Hessian information to guide quantization error allocation, quantizing weights one at a time while updating remaining weights to compensate for quantization error, minimizing the overall loss increase.

[6] **AWQ**: Lin, Ji, et al. "AWQ: Activation-aware Weight Quantization for LLMs." arXiv preprint arXiv:2306.00978 (2023). https://arxiv.org/abs/2306.00978. AWQ (Activation-aware Weight Quantization) identifies and preserves important weights (those with large activation magnitudes) during quantization, using a similar Hessian-guided approach to minimize quantization error in critical weight directions.

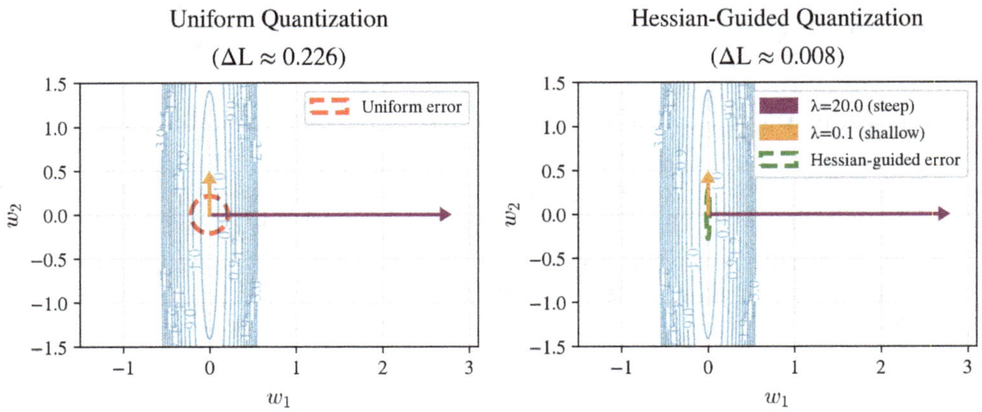

Figure 11.10: Hessian-Guided Quantization Error Allocation

Figure 11.10 illustrates how Hessian eigenvalues guide quantization error allocation. In the left panel, weights are quantized uniformly (naive approach), leading to large loss increases in high-sensitivity directions. In the right panel, quantization error is allocated based on Hessian eigenvalues: more quantization bits (smaller error) are allocated to directions with large eigenvalues (steep, sensitive directions), while fewer bits (larger error) are allocated to directions with small eigenvalues (shallow, insensitive directions). This Hessian-guided approach minimizes the overall loss increase from quantization.

11.9.3.2 Example: Another Newton-Schulz Iteration for Orthogonal Matrix (Polar Decomposition)

Beyond matrix inversion, Newton-Schulz iteration has a powerful application in computing the **orthogonal factor** of the **polar decomposition** (see Chapter~8). This is particularly useful in contexts where we need to find the "closest" orthogonal matrix to a given matrix, such as in certain QR decomposition algorithms or in optimization methods like the Muon optimizer.

Given a matrix $G \in \mathbb{R}^{N\times N}$ with polar decomposition $G = QS$ where Q is orthogonal and S is positive semidefinite, we seek to compute Q without performing SVD. We can formulate this as finding the zero of the function $f : \mathbb{R}^{N\times N} \to \mathbb{R}^{N\times N}$ defined as:

$$f(X) = X^\top X - I,$$

where $X \in \mathbb{R}^{N\times N}$ is our current estimate of the orthogonal matrix Q, and we want $X^\top X = I$ (orthogonality condition).

However, a more direct approach uses the fact that $Q = G(G^\top G)^{-1/2}$ in the polar decomposition. The Newton-Schulz iteration for computing the orthogonal factor is:

$$Q_{t+1} = \frac{1}{2}Q_t(3I - Q_t^\top Q_t). \tag{11.35}$$

Derivation: Starting from Q_t, we want to "correct" it to be more orthogonal. Define the error $E_t = I - Q_t^\top Q_t$. If Q_t is close to orthogonal, then E_t is small. The correction factor $(Q_t^\top Q_t)^{-1/2} = (I - E_t)^{-1/2}$ can be approximated using the Taylor expansion:

$$(I - E_t)^{-1/2} = I + \frac{1}{2}E_t + \frac{3}{8}E_t^2 + o(\|E_t\|^2),$$

where $o(\|E_t\|^2)$ denotes the remainder term that satisfies $\lim_{\|E_t\|\to 0} \frac{o(\|E_t\|^2)}{\|E_t\|^2} = 0$, meaning the remainder grows slower than $\|E_t\|^2$ as $E_t \to 0$.

Keeping only first-order terms (assuming E_t is small):

$$(I - E_t)^{-1/2} \approx I + \frac{1}{2}E_t = I + \frac{1}{2}(I - Q_t^\top Q_t) = \frac{3}{2}I - \frac{1}{2}Q_t^\top Q_t.$$

Multiplying by Q_t gives the iteration:

$$Q_{t+1} = Q_t(I - E_t)^{-1/2} \approx Q_t\left(\frac{3}{2}I - \frac{1}{2}Q_t^\top Q_t\right) = \frac{1}{2}Q_t(3I - Q_t^\top Q_t).$$

The iteration converges **quadratically** to an orthogonal matrix when the initial guess Q_0 has spectral norm $\|Q_0\|_2 < \sqrt{3}$. In practice, we typically initialize with $Q_0 = G/\|G\|_2$, which normalizes the input matrix G by its spectral norm, ensuring that $\|Q_0\|_2 = 1$ and thus satisfying the convergence condition.

Connection to QR Decomposition: While QR decomposition typically uses Householder reflections or Gram-Schmidt, the Newton-Schulz iteration provides an alternative method to compute the orthogonal factor. Given a matrix A, if we apply Newton-Schulz to a normalized version of A, we obtain an orthogonal matrix that can be used in certain QR-like factorizations. This approach is particularly valuable when:

1. **SVD is expensive**: For large matrices, SVD computation is costly, while Newton-Schulz only requires matrix multiplications.
2. **GPU-friendly**: Matrix multiplications are highly parallelizable on GPUs, making Newton-Schulz efficient for modern hardware.
3. **Iterative refinement**: The iteration can be used to refine an approximate orthogonal matrix obtained from other methods.

> The Muon optimizer uses this iteration to compute orthogonal update directions for neural network training. By extracting the orthogonal factor from the gradient matrix via polar decomposition, Muon optimizer eliminates anisotropic scaling effects, leading to more stable optimization.

11.10 Summary: Unified View of Derivatives

This chapter has introduced a progression of derivative concepts, from single-variable derivatives to Fréchet derivatives for matrix-to-matrix functions. Throughout this progression, a unifying theme emerges: **all derivatives are linear operators that provide first-order approximations.** The Fréchet derivative provides the most general framework that encompasses all previous concepts. The following table summarizes how different function types lead to different derivative representations, all unified under the Fréchet derivative framework:

Function Type	In/Output	Derivative	Notation	Relationship
$f : \mathbb{R} \to \mathbb{R}$	S → S	Derivative	$f'(x)$ (scalar)	$f'(x) = f'(x)$
$\mathbf{f} : \mathbb{R} \to \mathbb{R}^M$	S → V	Vector func derivative	$\frac{\partial \mathbf{f}}{\partial x}$ (vector)	$f'(x)[h] = \frac{\partial \mathbf{f}}{\partial x} \cdot h$
$f : \mathbb{R}^N \to \mathbb{R}$	V → S	Gradient	$\nabla f(\mathbf{x})$ (vector)	$f'(\mathbf{x})[\mathbf{h}] = \nabla f(\mathbf{x})^T\mathbf{h}$
$\mathbf{f} : \mathbb{R}^N \to \mathbb{R}^M$	V → V	Jacobian	$\mathbf{J}_\mathbf{f}(\mathbf{x})$ (matrix)	$f'(\mathbf{x})[\mathbf{h}] = \mathbf{J}_\mathbf{f}(\mathbf{x})\,\mathbf{h}$
$f : \mathbb{R}^N \to \mathbb{R}$ (2nd ord)	V → S	Hessian	$\mathbf{H}_f = \nabla^2 f(\mathbf{x})$	$\mathbf{H}_f = \mathbf{J}_{\nabla f}$
$f : \mathbb{R}^{M\times N} \to \mathbb{R}^{P\times Q}$	M → M	Fréchet	$f'(X)[\Delta]$	General case

In the table above, S denotes Scalar, V denotes Vector, and M denotes Matrix.

11.10.1 First-Order Derivatives

All first-order derivatives share the same fundamental structure: they are linear operators that provide first-order approximations:

- **Scalar function**:
$$f(x+h) \approx f(x) + f'(x) \cdot h$$
- **Gradient**:
$$f(\mathbf{x}+\mathbf{h}) \approx f(\mathbf{x}) + \nabla f(\mathbf{x})^T \mathbf{h}$$
- **Jacobian**:
$$\mathbf{f}(\mathbf{x}+\mathbf{h}) \approx \mathbf{f}(\mathbf{x}) + \mathbf{J}_{\mathbf{f}}(\mathbf{x})\,\mathbf{h}$$
- **Fréchet derivative**:
$$f(X+\Delta) \approx f(X) + f'(X)[\Delta]$$

11.10.2 Relationships to Fréchet Derivative

The Fréchet derivative provides a unified perspective on all derivative types:

- **Vector function derivative** ($\mathbb{R} \to \mathbb{R}^M$): The derivative $\frac{\partial \mathbf{f}}{\partial x} \in \mathbb{R}^{M\times 1}$ is a column vector, which can be viewed as a degenerate Jacobian (when input dimension is 1). The Fréchet derivative acts as:
$$f'(x)[h] = \frac{\partial \mathbf{f}}{\partial x} \cdot h \tag{11.36}$$
- **Gradient** ($\mathbb{R}^N \to \mathbb{R}$): The gradient ∇f is the coordinate representation of the Fréchet derivative with respect to the standard inner product:
$$f'(\mathbf{x})[\mathbf{h}] = \langle \nabla f(\mathbf{x}), \mathbf{h} \rangle = \nabla f(\mathbf{x})^T \mathbf{h} \tag{11.37}$$
- **Jacobian** ($\mathbb{R}^N \to \mathbb{R}^M$): The Jacobian matrix $\mathbf{J}_{\mathbf{f}}$ is the matrix representation of the Fréchet derivative:
$$f'(\mathbf{x})[\mathbf{h}] = \mathbf{J}_{\mathbf{f}}(\mathbf{x})\,\mathbf{h} \tag{11.38}$$
- **Matrix-to-scalar functions** ($\mathbb{R}^{M\times N} \to \mathbb{R}$): Functions with matrix inputs and scalar outputs use the gradient notation $\nabla_X f(X) \in \mathbb{R}^{M\times N}$, which is a matrix. This relates to the Fréchet derivative via the Frobenius inner product:
$$f'(X)[\Delta] = \langle \nabla_X f(X), \Delta \rangle_F = \mathrm{tr}(\nabla_X f(X)^T \Delta) \tag{11.39}$$
- **Hessian** (second-order): The Hessian is the Jacobian of the gradient, representing the second Fréchet derivative: $\mathbf{H}_f = \frac{\partial(\nabla f)}{\partial \mathbf{x}} = \mathbf{J}_{\nabla f}$.

11.10.3 The Unified Framework

The Fréchet derivative provides a unified framework for understanding derivatives across all function types. The operator viewpoint is particularly natural for matrix-to-matrix functions, where it preserves structural insights that would be obscured by vectorization. This unified perspective connects the fundamental concepts of single-variable calculus, multivariable calculus, and matrix calculus into one coherent framework, which is essential for understanding optimization algorithms, backpropagation in neural networks, and numerical methods in machine learning.

11.11 Mindmap

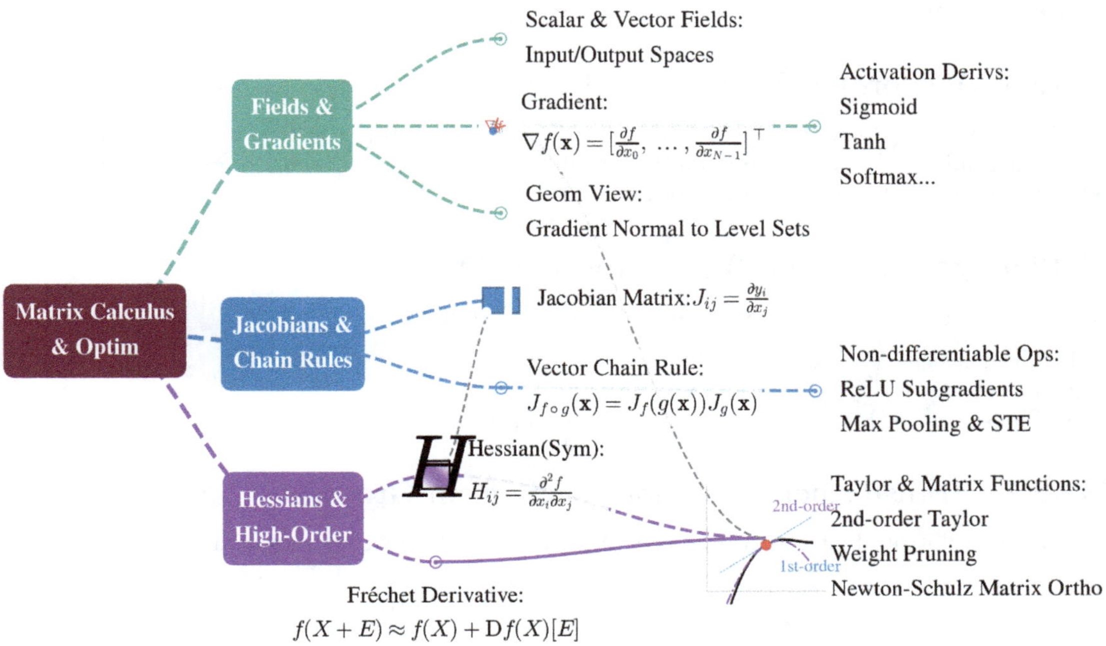

Figure 11.11: Mind map of matrix calculus topics

11.12 Exercises

11.12.1 Gradient of Quadratic Form

Compute the gradient $\nabla_{\mathbf{x}} f(\mathbf{x})$ where $f(\mathbf{x}) = \mathbf{x}^T A\mathbf{x} + \mathbf{b}^T\mathbf{x} + c$, with $A \in \mathbb{R}^{N\times N}$ symmetric, $\mathbf{b} \in \mathbb{R}^N$, and $c \in \mathbb{R}$.

11.12.2 Jacobian of Element-wise Operations

Let $\mathbf{f} : \mathbb{R}^N \to \mathbb{R}^N$ be defined as $\mathbf{f}(\mathbf{x}) = [\exp(x_0), \exp(x_1), \dots, \exp(x_{N-1})]^T$. Compute the Jacobian matrix $\frac{\partial \mathbf{f}}{\partial \mathbf{x}}$.

11.12.3 Chain Rule for Neural Network Layer

Consider a neural network layer: $\mathbf{y} = \sigma(W\mathbf{x} + \mathbf{b})$, where:

- $W \in \mathbb{R}^{M\times N}$ is a weight matrix
- $\mathbf{b} \in \mathbb{R}^{M\times 1}$ is a bias vector
- σ is the sigmoid function applied element-wise
- $\mathbf{x} \in \mathbb{R}^{N\times 1}$ is the input

Compute $\frac{\partial \mathbf{y}}{\partial \mathbf{x}}$, $\nabla_W \mathbf{y}$, and $\nabla_{\mathbf{b}} \mathbf{y}$.

11.12.4 Hessian of Loss Function

For the loss function $\mathcal{L}(\mathbf{w}) = \frac{1}{2}||\mathbf{y} - X\mathbf{w}||_2^2$ where $X \in \mathbb{R}^{K\times N}$, $\mathbf{y} \in \mathbb{R}^{K\times 1}$, and $\mathbf{w} \in \mathbb{R}^{N\times 1}$, compute the Hessian matrix $\nabla^2_{\mathbf{w}}\mathcal{L}(\mathbf{w})$.

11.12.5 Vector-Jacobian Product

Given $f : \mathbb{R}^N \to \mathbb{R}^M$ with Jacobian $J \in \mathbb{R}^{M\times N}$, and a vector $v \in \mathbb{R}^M$, compute the vector-Jacobian product $v^T J$ without explicitly forming the full Jacobian matrix. Explain why this is computationally efficient for backpropagation in deep learning.

11.12.6 Second-order Taylor Approximation

For $f(\mathbf{x}) = \log(1 + \exp(x_0) + \exp(x_1))$ where $\mathbf{x} \in \mathbb{R}^2$, compute:

1. The gradient $\nabla_{\mathbf{x}} f(\mathbf{x})$
2. The Hessian $\nabla^2_{\mathbf{x}} f(\mathbf{x})$
3. The second-order Taylor expansion around $\mathbf{x}_0 = [0, 0]^T$

11.12.7 Transformer Attention Mechanism Gradients

In the scaled dot-product attention mechanism, the attention weights are computed as:

$$\text{Attention}(Q, K, V) = \text{softmax}\left(\frac{QK^T}{\sqrt{d_k}}\right)V$$

where $Q, K, V \in \mathbb{R}^{N\times d_k}$ are query, key, and value matrices, and d_k is the dimension.

Compute:

1. $\nabla_Q \text{Attention}(Q, K, V)$ (gradient with respect to matrix Q)
2. $\nabla_K \text{Attention}(Q, K, V)$ (gradient with respect to matrix K)
3. $\nabla_V \text{Attention}(Q, K, V)$ (gradient with respect to matrix V)

These gradients are essential for training transformer models via backpropagation through the attention mechanism.

11.12.8 Diffusion Model Score Function

In diffusion models, the score function $\nabla_{x_t} \log p(x_t)$ is learned to reverse the forward diffusion process. Consider a simplified case where the score function is parameterized as $s_\theta(x_t, t) = W_\theta \phi(x_t) + b_\theta$, where:

- $W_\theta \in \mathbb{R}^{D\times d}$ is a learnable weight matrix
- $\mathbf{b}_\theta \in \mathbb{R}^{D\times 1}$ is a bias vector
- $\phi(\mathbf{x}_t) \in \mathbb{R}^{d\times 1}$ is a feature extractor (e.g., neural network output)
- $\mathbf{x}_t \in \mathbb{R}^{D\times 1}$ is the noisy data at time t

Compute:

1. $\nabla_{W_\theta} s_\theta(\mathbf{x}_t, t)$ (gradient with respect to matrix W_θ)
2. $\nabla_{\mathbf{b}_\theta} s_\theta(\mathbf{x}_t, t)$
3. $\nabla_{\mathbf{x}_t} s_\theta(\mathbf{x}_t, t)$ (needed for the reverse SDE)

These gradients enable training diffusion models to learn the score function that guides the denoising process.

11.12.9 Quantization-Aware Training (GPTQ/AWQ)

In post-training quantization methods like GPTQ[7] and AWQ[8], we optimize quantized weights to minimize reconstruction error. Consider the objective:

$$L(W_q) = ||XW - XW_q||_F^2$$

where:

- $X \in \mathbb{R}^{N \times d}$ is the input data
- $W \in \mathbb{R}^{d \times m}$ is the original full-precision weight matrix
- $W_q \in \mathbb{R}^{d \times m}$ is the quantized weight matrix (with constraints)

If $W_q = \text{quantize}(W + \Delta)$ where $\Delta \in \mathbb{R}^{d \times m}$ is a small perturbation, compute:

1. $\nabla_\Delta \mathcal{L}(W_q)$ (gradient with respect to matrix Δ, ignoring quantization constraints)
2. The optimal Δ that minimizes L in the unconstrained case
3. Explain how this relates to the Hessian-based optimization used in GPTQ

Understanding these gradients is crucial for developing efficient quantization algorithms that preserve model accuracy while reducing memory and computation costs.

[7] **GPTQ**: Elias Frantar, Saleh Ashkboos, Torsten Hoefler, and Dan Alistarh. "GPTQ: Accurate Post-Training Quantization for Generative Pre-trained Transformers." arXiv preprint arXiv:2210.17323 (2022). https://arxiv.org/abs/2210.17323. GPTQ uses Hessian information to guide quantization error allocation, quantizing weights one at a time while updating remaining weights to compensate for quantization error, minimizing the overall loss increase.

[8] **AWQ**: Lin, Ji, et al. "AWQ: Activation-aware Weight Quantization for LLMs." arXiv preprint arXiv:2306.00978 (2023). https://arxiv.org/abs/2306.00978. AWQ (Activation-aware Weight Quantization) identifies and preserves important weights (those with large activation magnitudes) during quantization, using a similar Hessian-guided approach to minimize quantization error in critical weight directions.

Chapter 12

Optimization Methods

Convex optimization, gradient descent, proximal methods, and constrained / minimax optimization for machine learning

"The best way to have a good idea is to have lots of ideas."

— Linus Pauling

- ▷ `Convex set` S is convex if $\lambda x + (1-\lambda)y \in S$ for all $x, y \in S$ and $\lambda \in (0,1)$
- ▷ `Convex` $f(\lambda x + (1-\lambda)y) \leq (<) \lambda f(x) + (1-\lambda) f(y)$ for all $x, y \in S$ $(x \neq y)$, $\lambda \in (0,1)$
- ▷ `Concave function` f is concave iff $-f$ is convex
- ▷ `Subgradient` f is convex iff $f(y) \geq f(x) + \nabla f(x)^\top (y-x)$ for all $x, y \in S$
- ▷ `Hessian test` f is convex iff $\nabla^2 f(x) \succeq 0$ for all $x \in S$
- ▷ `Jensen's inequality` If f is convex, then $f(\mathbb{E}[X]) \leq \mathbb{E}[f(X)]$
- ▷ `KL divergence` $\mathrm{KL}(P||Q) = \mathbb{E}_{X\sim P}[\log \frac{P(X)}{Q(X)}] \geq 0$
- ▷ `Convex prog` $\min_{x \in D} f(x)$ s.t. $g_i(x) \leq 0$, $h_i(x) = 0$, D convex, f, g_i convex, h_i affine
- ▷ `Global optimality` In a convex program, any local minimum is a global minimum
- ▷ `Lagrangian` $L(x, \eta, \nu) = f(x) + \sum_i \eta_i g_i(x) + \sum_i \nu_i h_i(x)$
- ▷ `Weak duality` $d^* \leq p^*$ where d^* is dual optimal value, p^* is primal optimal value
- ▷ `Strong duality` If strict feasibility holds (Slater's condition), then $d^* = p^*$
- ▷ `KKT` $g_i(x) \leq 0$, $h_i(x) = 0$; $\eta \geq 0$; $\eta_i g_i(x) = 0$; $\nabla f(x) + \sum_i \eta_i \nabla g_i(x) + \sum_i \nu_i \nabla h_i(x) = 0$
- ▷ `Necessary conditions` For $\min_x f(x)$, global minimizer $\nabla f(x^*) = 0$ and $\nabla^2 f(x^*) \succeq 0$
- ▷ `Gradient descent` $x^{(t+1)} = x^{(t)} - \alpha \nabla f(x^{(t)})$; $0 < \alpha < 2/L$, L Lipschitz const
- ▷ `Newton's method` $x^{(t+1)} = x^{(t)} - \alpha H(x^{(t)})^{-1} \nabla f(x^{(t)})$ where $H = \nabla^2 f$
- ▷ `Proj. GD` $x^{(t+1)} = \mathrm{Proj}_D(x^{(t)} - \alpha \nabla f(x^{(t)}))$ where $\mathrm{Proj}_D(x) = \arg\min_{y \in D} \|x - y\|_2$
- ▷ `Ridge regression solution` $x^* = (A^\top A + \lambda I)^{-1} A^\top b$ for $\min_x \|Ax - b\|_2^2 + \lambda \|x\|_2^2$
- ▷ `Probability simplex` $\Delta^d = \{p \in \mathbb{R}^d : p_i \geq 0, \sum_{i=0}^{d-1} p_i = 1\}$

12.1 Convex Sets

12.1.1 Definition: Convex Set

A set S is **convex** if for any $x, y \in S$ and any $\lambda \in (0,1)$,

$$\lambda x + (1-\lambda)y \in S \tag{12.1}$$

Figure 12.1 illustrates the geometric definition of convexity. The left panel shows a **convex set** (ellipse) where the line segment between any two points x and y (red dashed line) remains entirely inside the set. This demonstrates that for any $\lambda \in (0,1)$, the point $\lambda x + (1-\lambda)y$ lies within the set. The right panel shows a **non-convex set** (crescent shape) where the line segment between two points crosses the boundary, violating the convexity condition. This visualization makes concrete the definition: a set is convex if and only if it contains all line segments connecting any pair of its points. This geometric property is fundamental to convex optimization, as it ensures that any local minimum in a convex set is also a global minimum.

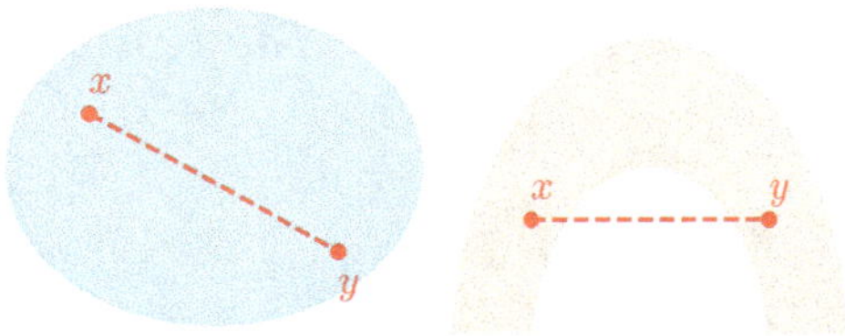

Figure 12.1: Convex vs Non-Convex Sets

12.1.1.1 Example: Euclidean Ball Convexity

Define

$$S = \{x \in \mathbb{R}^d : ||x||_2 \leq r\}.$$

Is this set convex?

Solution

Let $x, y \in S$ and $\lambda \in (0,1)$. Then

$$||\lambda x + (1-\lambda)y||_2^2 = \sum_{i=0}^{d-1} (\lambda x_i + (1-\lambda)y_i)^2.$$

Expanding and applying Cauchy–Schwarz,

$$\leq \lambda^2 ||x||_2^2 + (1-\lambda)^2 ||y||_2^2 + 2\lambda(1-\lambda)||x||_2||y||_2 \leq r^2.$$

Thus $\lambda x + (1-\lambda)y \in S$, so S is convex.

12.1.1.2 Example: Positive Semidefinite Cone Convexity

Is the set of all $N \times N$ positive semidefinite matrices convex?

Solution

Let $X, Y \succeq 0$ and $\lambda \in (0,1)$. For any $v \in \mathbb{R}^N$,

$$v^\top(\lambda X + (1-\lambda)Y)v = \lambda v^\top X v + (1-\lambda)v^\top Y v \geq 0.$$

Hence $\lambda X + (1-\lambda)Y \succeq 0$, and the set is convex.

12.1.1.3 Example: Probability Simplex Convexity

Define the probability simplex

$$\Delta^d = \left\{ p \in \mathbb{R}^d : p_i \geq 0, \sum_{i=0}^{d-1} p_i = 1 \right\}.$$

Is Δ^d convex?

Solution

Let $x, y \in \Delta^d$ and $\lambda \in (0,1)$. Then

$$\lambda x_i + (1-\lambda) y_i \geq 0,$$

and

$$\sum_{i=0}^{d-1} (\lambda x_i + (1-\lambda) y_i) = 1.$$

Thus $\lambda x + (1-\lambda) y \in \Delta^d$, so Δ^d is convex.

12.2 Convex and Concave Functions

12.2.1 Definition: Convex and Concave Function

A function $f : S \to \mathbb{R}$ defined on a convex set S is

- **strictly convex** if

$$f(\lambda x + (1-\lambda) y) < \lambda f(x) + (1-\lambda) f(y) \tag{12.2}$$

- **weakly convex** if the inequality is $\leq$:

$$f(\lambda x + (1-\lambda) y) \leq \lambda f(x) + (1-\lambda) f(y) \tag{12.3}$$

- **Concave**: f is (strictly) concave iff $-f$ is (strictly) convex.

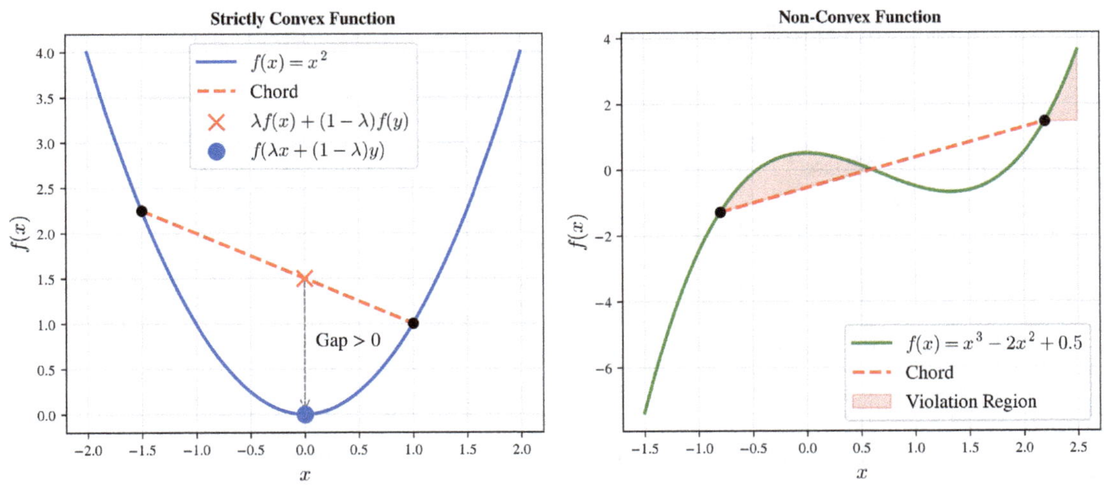

Figure 12.2: Convex vs Non-Convex Functions

Figure 12.2 visualizes the difference between strictly convex and non-convex functions. The left panel shows a **strictly convex function** $f(x) = x^2$ where the chord (red dashed line) connecting any two points lies entirely above the curve. At the midpoint $\lambda x + (1-\lambda)y$, the function value $f(\lambda x + (1-\lambda)y)$ (blue circle) is strictly less than the chord value $\lambda f(x) + (1-\lambda)f(y)$ (red cross), demonstrating the strict inequality. The gap between these values is highlighted with an arrow. The right panel shows a **non-convex function** $f(x) = x^3 - 2x^2 + 0.5$ where the chord crosses the curve, violating the convexity condition. The red shaded region highlights where the function lies above the chord, indicating non-convexity. This visualization makes concrete why convex functions are easier to optimize: the function always lies below any chord, ensuring that local minima are global minima.

12.2.1.1 Example: Strictly Convex Functions

- $f(x) = \log(1 + e^{-x})$ (logistic loss; binary classification)
- $f(x) = x^2$ (squared loss; regression, L2 regularization)
- $f(x) = -\log x$ for $x > 0$ (negative log-likelihood, MLE)

The top row of Figure 12.3 plots these three functions: each is strictly convex on its domain (the Hessian test below confirms $\nabla^2 f \succeq 0$), and they appear throughout supervised learning (classification, regression) and maximum-likelihood estimation.

12.2.1.2 Example: Strictly Concave Functions

- $f(x) = \log x$ ($x > 0$; log-likelihood, log utility)
- $f(x) = \log(1 + x)$ ($x > 0$; log-utility in bandits and RL)
- $f(p) = H(p) = -p \log p - (1-p)\log(1-p)$ on $p \in (0, 1)$ (binary entropy; classification, information theory)

In Figure 12.3, the second row plots the strictly concave functions: $f(x) = \log x$ ($x > 0$), $f(x) = \log(1 + x)$ ($x > 0$), and $f(p) = H(p)$ (binary entropy). These functions are common in AI/ML (classification, regression, MLE, information theory).

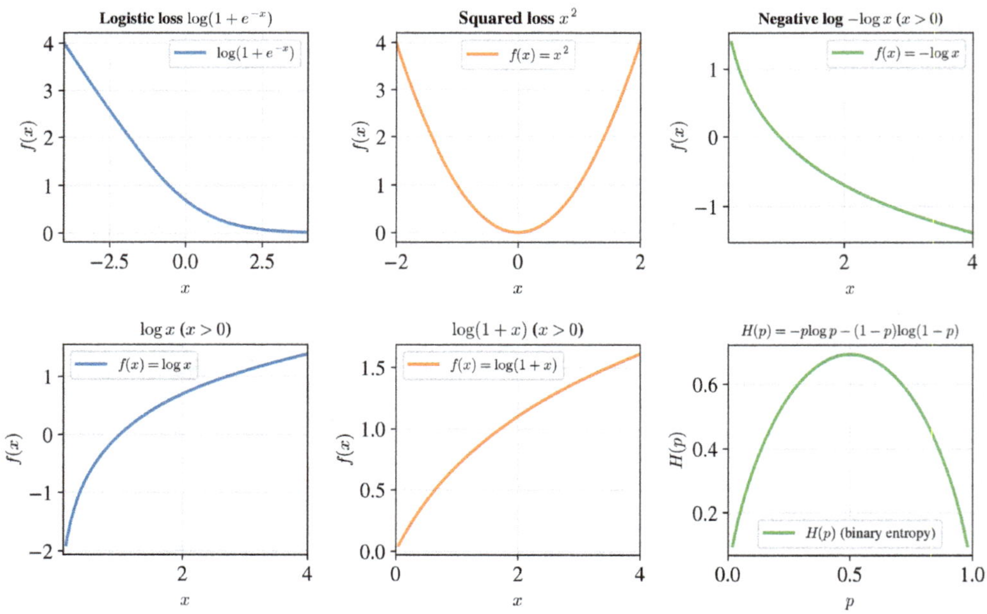

Figure 12.3: Strictly convex and strictly concave examples

12.2.2 Theorem: Subgradient Characterization

$f : S \to \mathbb{R}$ is convex **iff** for any $x, y \in S$,

$$f(y) \geq f(x) + \nabla f(x)^\top (y - x) \tag{12.4}$$

12.2.3 Theorem: Hessian Test

f is convex **iff**

$$\nabla^2 f(x) \succeq 0 \tag{12.5}$$

12.2.3.1 Example: Least Squares Loss Convexity

Show that $f(x) = (w^\top x - y)^2$ is convex.

Solution

Gradient:

$$\nabla_x f(x) = 2(w^\top x - y)w.$$

Hessian:

$$\nabla_x^2 f(x) = 2ww^\top \succeq 0.$$

Hence f is convex.

12.2.3.2 Example: Logistic Loss Convexity

Show that $f(x) = \log(1 + e^{-w^\top x})$ is convex.

Solution

Gradient:

$$\nabla f(x) = \frac{w}{1 + e^{w^\top x}}.$$

Hessian:

$$\nabla^2 f(x) = \frac{e^{w^\top x}}{(1 + e^{w^\top x})^2} ww^\top \succeq 0.$$

Thus f is convex.

12.2.3.3 Example: Euclidean Norm Convexity

Show that $f(x) = \|x\|_2$ is convex.

Solution Sketch

The Hessian is

$$\nabla^2 f(x) = \frac{1}{\|x\|_2} I - \frac{1}{\|x\|_2^3} xx^\top.$$

For any v, Cauchy–Schwarz gives $v^\top \nabla^2 f(x)\, v \geq 0$, so f is convex.

12.2.3.4 Example: Log-Sum-Exp Convexity

Show that $f(x) = \log \sum_{n=0}^{N-1} e^{x_n}$ is convex.

Solution

Hessian entries are $H_{ij} = \text{softmax}(x)_i \delta_{ij} - \text{softmax}(x)_i \text{softmax}(x)_j$. For any v, $v^\top H v \geq 0$ by Cauchy–Schwarz. Hence f is convex.

12.2.4 Theorem: Jensen's Inequality

For convex functions, the graph lies below any chord: the value of f at a weighted average of points is at most the same weighted average of the values of f. So when X is a random variable taking values in the domain of f, the expectation of $f(X)$ cannot be less than f evaluated at the expectation of X. Formally, if f is convex,

$$f(\mathbb{E}[X]) \leq \mathbb{E}[f(X)] \tag{12.6}$$

with equality if and only if f is affine on the support of X or X is degenerate. The inequality is reversed when f is concave (e.g. $f = \log$), which is used in deriving the ELBO in variational inference.

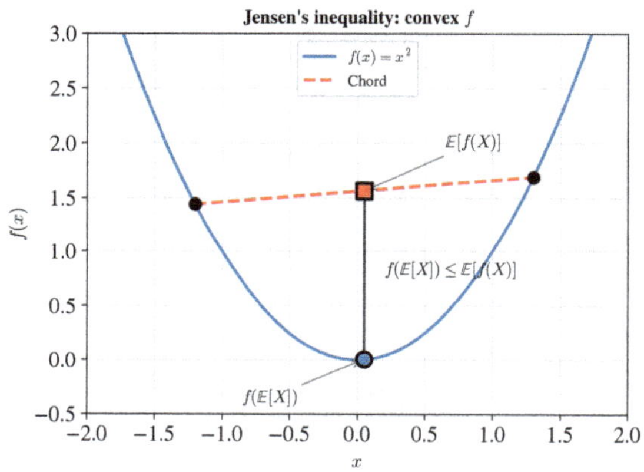

Figure 12.4: Jensen's inequality

Figure 12.4 shows the two-point case: take $X \in \{x_1, x_2\}$ with $\mathbb{P}(X = x_1) = \lambda$ and $\mathbb{P}(X = x_2) = 1 - \lambda$. Then $\mathbb{E}[X] = \lambda x_1 + (1-\lambda)x_2$ lies on the segment between x_1 and x_2, and $\mathbb{E}[f(X)] = \lambda f(x_1) + (1-\lambda) f(x_2)$ is the corresponding point on the chord. Because the chord lies above the curve, $f(\mathbb{E}[X])$ (on the curve) is below $\mathbb{E}[f(X)]$ (on the chord).

KL divergence non-negativity: The KL divergence $\text{KL}(P \| Q) = \mathbb{E}_{X \sim P}[\log \frac{P(X)}{Q(X)}] \geq 0$, with equality iff $P = Q$ a.e., follows from Jensen's inequality (using convexity of $-\log$). See Chapter~14, Section~14.6 for the definition, proof, and applications.

12.3 Convex Optimization Problems

12.3.1 Definition: Convex Program

$$\begin{aligned} \min_{x \in D} \quad & f(x) \\ \text{s.t.} \quad & g_i(x) \leq 0, \quad i = 0, \dots, M-1 \\ & h_i(x) = 0, \quad i = 0, \dots, N-1 \end{aligned} \tag{12.7}$$

where D is convex, f, g_i convex, h_i affine.

12.3.1.1 Example: Support Vector Machine

The Support Vector Machine (SVM) is a binary classification method that finds the optimal separating hyperplane. Given training data $\{(\mathbf{x}^{(m)}, y^{(m)})\}_{m=0}^{M-1}$ where $\mathbf{x}^{(m)} \in \mathbb{R}^d$ and $y^{(m)} \in \{-1, +1\}$, the soft-margin SVM optimization problem is:

$$\min_{\mathbf{w},\xi} \quad \frac{1}{2}||\mathbf{w}||_2^2 + \frac{C}{2M}\sum_{m=0}^{M-1} \xi_m$$

subject to

$$y^{(m)}\mathbf{w}^\top \mathbf{x}^{(m)} \geq 1 - \xi_m, \quad \xi_m \geq 0, \quad m = 0, \dots, M-1.$$

Here:

- $\mathbf{w} \in \mathbb{R}^d$ is the weight vector defining the hyperplane
- $\xi = [\xi_0, \dots, \xi_{M-1}]^\top$ are slack variables allowing misclassification
- $C > 0$ is a regularization parameter controlling the trade-off between margin maximization and training error
- The constraint $y^{(m)}\mathbf{w}^\top \mathbf{x}^{(m)} \geq 1 - \xi_m$ ensures correct classification (with margin) when $\xi_m = 0$, and allows violations when $\xi_m > 0$

Why is this a convex program?

1. **Objective function**: The term $\frac{1}{2}||\mathbf{w}||_2^2$ is a quadratic function (convex), and $\frac{C}{2M}\sum_m \xi_m$ is linear (convex). The sum of convex functions is convex.
2. **Constraints**: The constraint $y^{(m)}\mathbf{w}^\top \mathbf{x}^{(m)} \geq 1 - \xi_m$ is affine (linear) in $(\mathbf{w}, \xi)$, and $\xi_m \geq 0$ defines a half-space (convex set).
3. **Domain**: The domain is $\mathbb{R}^d \times \mathbb{R}^M$ (convex).

Therefore, this is a convex program, and any local minimum is a global minimum.

12.3.2 Theorem: Global Optimality in Convex Programs

In a convex program, any local minimum is a global minimum.

12.4 Duality

Constrained optimization can be studied by turning it into an unconstrained object that encodes both the objective and the constraints. The **dual problem** maximizes a lower bound on the primal optimum; when that bound is tight, we obtain strong duality and the Karush–Kuhn–Tucker (KKT) conditions characterize optimality.

12.4.1 Lagrangian

The **Lagrangian** folds the constraints into the objective using multipliers $\eta_i \geq 0$ for inequalities $g_i(x) \leq 0$ and ν_i for equalities $h_i(x) = 0$:

$$L(x, \eta, \nu) = f(x) + \sum_i \eta_i g_i(x) + \sum_i \nu_i h_i(x) \tag{12.8}$$

For fixed (η, ν) with $\eta \geq 0$, minimizing $L(x, \eta, \nu)$ over x yields a function that is always a lower bound on the primal optimal value p^* when the primal constraints are satisfied. The **dual problem** is to maximize this lower bound over $\eta \geq 0$ and ν, giving the dual optimal value d^*.

12.4.2 Weak Duality

For any feasible primal x and any $\eta \geq 0$, ν, we have $f(x) \geq L(x, \eta, \nu)$ (because $g_i(x) \leq 0$ and $h_i(x) = 0$). Maximizing over dual variables and minimizing over primal variables preserves the inequality, so the dual optimum can never exceed the primal:

$$d^* \leq p^* \tag{12.9}$$

Thus the dual always provides a **certificate**: if we find dual feasible (η, ν) with dual value equal to some primal feasible $f(x)$, then x is primal optimal.

12.4.3 Strong Duality (Slater's Condition)

In general $d^* < p^*$ is possible (a **duality gap**). For convex programs, **Slater's condition** (strict feasibility: there exists x with $g_i(x) < 0$ for all inequalities and $h_i(x) = 0$ for all equalities) removes the gap: then

$$d^* = p^* \tag{12.10}$$

So the dual bound is tight and, under mild regularity, optimal primal-dual pairs satisfy the KKT conditions below.

12.5 KKT Conditions

12.5.1 Theorem: Karush–Kuhn–Tucker Conditions

If strong duality holds:

1. **Primal feasibility**: $g_i(x) \leq 0$, $h_i(x) = 0$
2. **Dual feasibility**: $\eta \geq 0$
3. **Complementary slackness**: $\eta_i g_i(x) = 0$
4. **Stationarity**:

$$\nabla f(x) + \sum_i \eta_i \nabla g_i(x) + \sum_i \nu_i \nabla h_i(x) = 0 \tag{12.11}$$

12.6 Unconstrained Optimization

12.6.1 Theorem: Necessary Conditions for Optimality

For

$$\min_x f(x),$$

a global minimizer satisfies

$$\nabla f(x^*) = 0, \quad \nabla^2 f(x^*) \succeq 0 \tag{12.12}$$

Figure 12.5 illustrates a **saddle point** for the function $f(x, y) = x^2 - y^2$ at $(0, 0)$. The 3D surface plot shows that while $\nabla f(0, 0) = 0$ (the gradient is zero), the Hessian $\nabla^2 f$ is **indefinite** (not positive semidefinite). The blue curve along $y = 0$ shows the convex direction (minimum along the x-axis), while the red curve along $x = 0$ shows the concave direction (maximum along the y-axis). The gray transparent plane at $z = 0$ represents the zero gradient level. This visualization makes concrete why the Hessian condition $\nabla^2 f(x^*) \succeq 0$ is necessary for optimality: a point with zero gradient can be a saddle point (not a minimum) if the Hessian has both positive and negative eigenvalues. In deep learning, saddle points are common in high-dimensional optimization landscapes, and second-order methods or techniques like momentum are used to escape them.

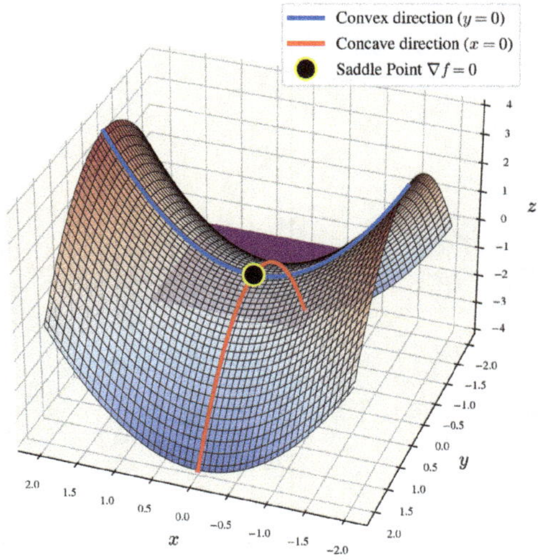

Figure 12.5: Saddle Point Surface

12.7 Gradient Descent

12.7.1 Algorithm: Gradient Descent

Algorithm 10 Gradient Descent

Input: Objective function f, initial point $x^{(0)}$, step size α with $0 < \alpha < 2/L$ where L is the Lipschitz constant
Output: Sequence $\{x^{(t)}\}$ converging to minimizer x^*

$t \leftarrow 0$
repeat
  $x^{(t+1)} \leftarrow x^{(t)} - \alpha \nabla f(x^{(t)})$
  $t \leftarrow t + 1$
until convergence
return $x^{(t)}$

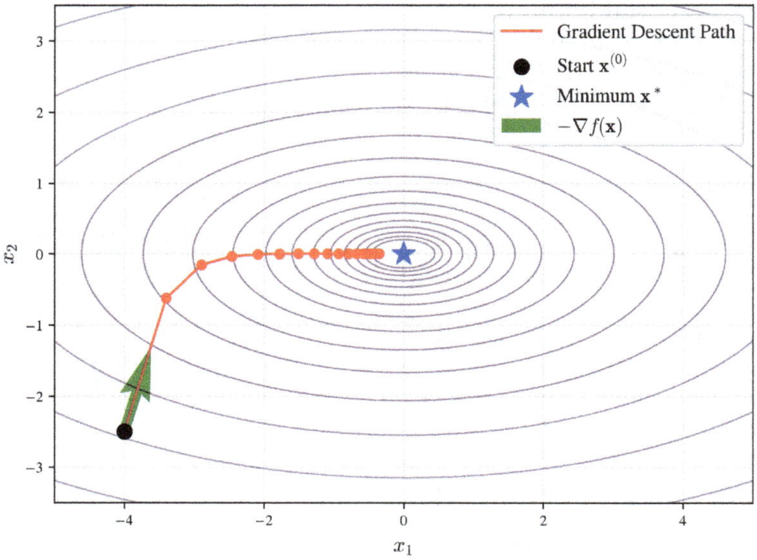

Figure 12.6: Gradient Descent Trajectory

Figure 12.6 visualizes the path taken by gradient descent on a convex 2D loss surface $f(x_1, x_2) = 0.5x_1^2 + 2.5x_2^2$.

The contour lines (background) show level sets of the function, with darker regions indicating lower values. The red path shows the iterative updates $\mathbf{x}^{(t+1)} = \mathbf{x}^{(t)} - \alpha \nabla f(\mathbf{x}^{(t)})$, zig-zagging towards the global minimum at the origin.

The green arrow at the starting point shows the negative gradient direction $-\nabla f(\mathbf{x})$, which is orthogonal to the contour lines. This visualization makes concrete how gradient descent follows the steepest descent direction at each step, moving perpendicular to the level sets. The zig-zag pattern arises from the elongated shape of the loss surface (ill-conditioning), where the curvature differs significantly along different directions. Despite the zig-zagging, gradient descent converges to the global minimum for convex functions.

12.8 Newton's Method

12.8.1 Algorithm: Newton's Method

Algorithm 11 Newton's Method

Input: Objective function f with Hessian H, initial point $x^{(0)}$, step size α
Output: Sequence $\{x^{(t)}\}$ converging to minimizer x^*

  $t \leftarrow 0$
  repeat
    $x^{(t+1)} \leftarrow x^{(t)} - \alpha H(x^{(t)})^{-1} \nabla f(x^{(t)})$
    $t \leftarrow t + 1$
  until convergence
  return $x^{(t)}$

Figure 12.7 compares the convergence trajectories of gradient descent and Newton's method on the same convex quadratic surface $f(x_1, x_2) = 0.5x_1^2 + 2.5x_2^2$. The red path shows **gradient descent**, which zig-zags due to ill-conditioning (the loss surface is elongated).

Gradient descent takes many steps because it ignores the curvature information encoded in the Hessian. The green path shows **Newton's method**, which uses the inverse Hessian H^{-1} to account for curvature and jumps directly to the minimum in a single step (for quadratic functions).

The annotations highlight that gradient descent "zig-zags" because it ignores curvature, while Newton's method "jumps directly" by using Hessian information. This visualization makes concrete why Newton's method converges faster: it adapts the step direction and size based on the local geometry of the loss surface, effectively "preconditioning" the gradient to account for the function's curvature. However, Newton's method is computationally expensive ($O(N^3)$ for matrix inversion) compared to gradient descent ($O(N^2)$ for matrix-vector products), making it impractical for large-scale problems in deep learning.

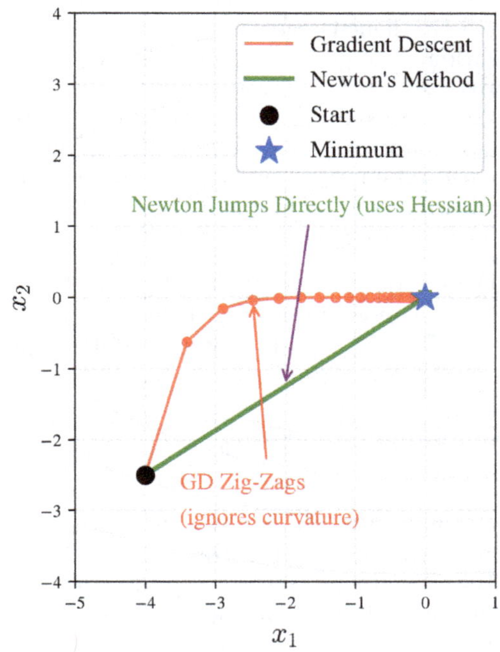

Figure 12.7: Newton's Method vs Gradient Descent

12.9 Constrained Optimization

12.9.1 Algorithm: Projected Gradient Descent

Algorithm 12 Projected Gradient Descent

Input: Objective function f, constraint set D, initial point $x^{(0)} \in D$, step size α
Output: Sequence $\{x^{(t)}\}$ in D converging to constrained minimizer
  $t \leftarrow 0$
  repeat
    $x^{(t+1)} \leftarrow \text{Proj}_D\left(x^{(t)} - \alpha \nabla f(x^{(t)})\right)$ ▷ Project onto constraint set
    $t \leftarrow t + 1$
  until convergence
  return $x^{(t)}$

where $\text{Proj}_D(x) = \arg\min_{y \in D} ||x - y||_2$ is the projection operator.

Figure 12.8 visualizes Algorithm 12.3 (Projected Gradient Descent) for constrained optimization. The green rectangle represents the constraint set $D = [0, 2] \times [0, 2]$. The gray contour lines show the unconstrained objective function $f(\mathbf{x}) = 0.5||\mathbf{x} - \mathbf{c}||^2$ with the unconstrained minimum (gold star) located outside the feasible set. The algorithm performs a two-step process at each iteration:

(1) **Gradient step** (red dashed arrow): take an unconstrained gradient step $x^{(t)} - \alpha\nabla f(x^{(t)})$, which may move outside the constraint set.

(2) **Projection step** (blue solid arrow): project the point back onto the constraint set using Proj_D.

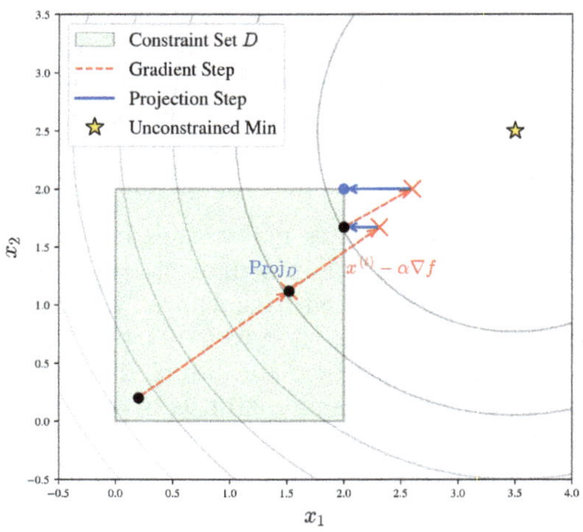

Figure 12.8: Projected Gradient Descent

The visualization shows three iterations, with each iteration following this pattern. This makes concrete how projected gradient descent ensures feasibility: even when the gradient step would violate constraints, the projection operator maps the point back to the nearest feasible point, maintaining constraint satisfaction while still making progress toward the constrained optimum. Projected gradient descent is fundamental to constrained optimization problems in machine learning, such as optimization on the probability simplex or with box constraints.

12.10 Nonsmooth Optimization and Proximal Methods

Gradient descent and Newton's method assume a differentiable objective function. Many regularizers used in machine learning—most notably the ℓ_1 penalty—are convex but **not differentiable** at zero. Ordinary gradients are undefined there, so we need a different algorithmic primitive.

12.10.1 Composite Objectives

A common template is the **composite problem**

$$\min_{x \in \mathbb{R}^n} F(x) = f(x) + g(x), \tag{12.13}$$

where f is convex and **smooth** (continuously differentiable with Lipschitz gradient) and g is convex and **nonsmooth** but **simple** in a sense made precise by the proximal operator below. Typical choices:

- $f(x) = \frac{1}{2}\|Ax - b\|_2^2$ (least squares) or a smooth logistic loss
- $g(x) = \lambda\|x\|_1$ (Lasso / ℓ_1 regularization)

The smooth part admits ordinary gradients; the nonsmooth part is handled by a closed-form **proximal mapping**.

12.10.2 Definition: Proximal Operator

For a convex function $g : \mathbb{R}^n \to \mathbb{R} \cup \{+\infty\}$ and step size $\gamma > 0$, the **proximal operator** of γg at y is

$$\mathrm{prox}_{\gamma g}(y) = \arg\min_{x \in \mathbb{R}^n} \left(g(x) + \frac{1}{2\gamma}\|x - y\|_2^2 \right). \tag{12.14}$$

Intuitively, $\mathrm{prox}_{\gamma g}(y)$ is a compromise between minimizing g and staying close to y. When $g \equiv 0$, one recovers prox = Id. When g is the indicator of a closed convex set D, the proximal operator reduces to Euclidean projection onto D (cf. projected gradient descent above).

12.10.3 Theorem: Soft-Thresholding for ℓ_1

For $g(x) = \lambda\|x\|_1$ with $\lambda \geq 0$, the proximal operator acts **coordinatewise**:

$$\left(\mathrm{prox}_{\gamma\lambda\|\cdot\|_1}(y)\right)_i = \mathrm{sign}(y_i)\,\max(|y_i| - \gamma\lambda,\, 0). \tag{12.15}$$

This map is called **soft-thresholding**: entries with $|y_i| \leq \gamma\lambda$ are set exactly to zero; larger entries are shrunk toward zero by $\gamma\lambda$. Exact zeros are the mechanism behind sparse solutions of the Lasso.

12.10.4 Algorithm: Proximal Gradient Descent (ISTA)

Algorithm 13 Proximal Gradient Descent (ISTA)

Input: Smooth f, nonsmooth g, $x^{(0)}$, step size $\alpha > 0$
Output: Sequence $\{x^{(t)}\}$ for $\min_x f(x) + g(x)$

```
t ← 0
repeat
    y^(t) ← x^(t) − α∇f(x^(t))            ▷ Gradient step on smooth part
    x^(t+1) ← prox_{αg}(y^(t))            ▷ Proximal step on nonsmooth part
    t ← t + 1
until convergence
return x^(t)
```

For $g = \lambda\|\cdot\|_1$, the proximal step is soft-thresholding (*12.15*). The method is also called **ISTA** (Iterative Shrinkage-Thresholding Algorithm). With box constraints $x \in [0, 1]^n$, one may compose soft-thresholding with projection onto the box (projected proximal gradient).

12.10.5 Example: Soft-Thresholding in One Dimension

Take $g(x) = \lambda|x|$ with $\lambda = 1$ and $\gamma = 1$, so $\mathrm{prox}(y) = \mathrm{sign}(y)\max(|y| - 1, 0)$.

- If $y = 2.5$, then $\mathrm{prox}(y) = 1.5$.
- If $y = 0.4$, then $\mathrm{prox}(y) = 0$.
- If $y = -3$, then $\mathrm{prox}(y) = -2$.

Thus values inside $[-1, 1]$ collapse to the nonsmooth kink at 0, while large values are merely shrunk.

> Proximal methods are the standard way to optimize **composite convex objectives** in sparse learning (Lasso, elastic net, total variation). The same template extends to **group sparsity** (block soft-thresholding) and to constrained problems via indicator functions. Accelerated variants (FISTA) improve the convergence rate under the same proximal primitive. See Parikh and Boyd for a systematic survey of proximal operators.[a]
>
> [a]Parikh, N., & Boyd, S. (2014). Proximal algorithms. *Foundations and Trends in Optimization*, 1(3),

127–239.

12.11 Minimax Optimization

12.11.1 Definition: Minimax Problem

A **minimax optimization problem** is an optimization problem of the form:

$$\min_{\theta\in\Theta}\max_{\phi\in\Phi}\mathcal{V}(\theta,\phi) \tag{12.16}$$

where $\mathcal{V}:\Theta\times\Phi\to\mathbb{R}$ is the **value function**, and θ and ϕ are the optimization variables.

The **minimizer** (the player choosing θ) seeks to reduce $\mathcal{V}$, and the **maximizer** (the player choosing ϕ) seeks to increase it. One player's gain is the other's loss, so the setup is a **zero-sum game**.

12.11.2 Nash Equilibrium

A point (θ^*,ϕ^*) is a **Nash equilibrium** (or **saddle point**) if:

$$\mathcal{V}(\theta^*,\phi)\le\mathcal{V}(\theta^*,\phi^*)\le\mathcal{V}(\theta,\phi^*)$$

for all $\theta\in\Theta$ and $\phi\in\Phi$.

At a Nash equilibrium, neither player can do better by changing strategy alone. With ϕ^* fixed, the minimizer cannot lower $\mathcal{V}$ by moving θ; with θ^* fixed, the maximizer cannot raise $\mathcal{V}$ by moving ϕ.

12.11.3 Application: Generative Adversarial Networks (GANs)

Generative Adversarial Networks (GANs)[1] provide a concrete application of minimax optimization in machine learning. GANs learn to generate samples from a data distribution by training two neural networks in an adversarial game.

12.11.3.1 The GAN Minimax Objective

The GAN training objective is:

$$\min_{\theta}\max_{\phi}\mathcal{V}(G_\theta,D_\phi) \tag{12.17}$$

Here G_θ is the **generator** (a neural network that maps noise $\mathbf{z}$ to fake samples $G_\theta(\mathbf{z})$, the minimizer) and D_ϕ is the **discriminator** (a neural network that takes a sample and outputs a probability that it is real, the maximizer). The value function is:

$$\mathcal{V}(G_\theta,D_\phi)=\mathbb{E}_{\mathbf{x}\sim p_{\text{data}}}[\log D_\phi(\mathbf{x})]+\mathbb{E}_{\mathbf{z}\sim p(\mathbf{z})}[\log(1-D_\phi(G_\theta(\mathbf{z})))] \tag{12.18}$$

Components:

[1] Goodfellow, I., Pouget-Abadie, J., Mirza, M., Xu, B., Warde-Farley, D., Ozair, S., Courville, A., & Bengio, Y. (2014). Generative Adversarial Networks. In *Advances in Neural Information Processing Systems* (NeurIPS), 27. https://arxiv.org/abs/1406.2661

- **Generator** G_θ: A neural network that maps random noise $\mathbf{z} \sim p(\mathbf{z})$ to generated samples $G_\theta(\mathbf{z})$. The generator is the **minimizer**.
- **Discriminator** D_ϕ: A neural network that classifies samples as real ($D_\phi(\mathbf{x}) \to 1$) or fake ($D_\phi(G_\theta(\mathbf{z})) \to 0$). The discriminator is the **maximizer**.
- **Data distribution** $p_{\text{data}}(\mathbf{x})$: The true distribution of real data samples.

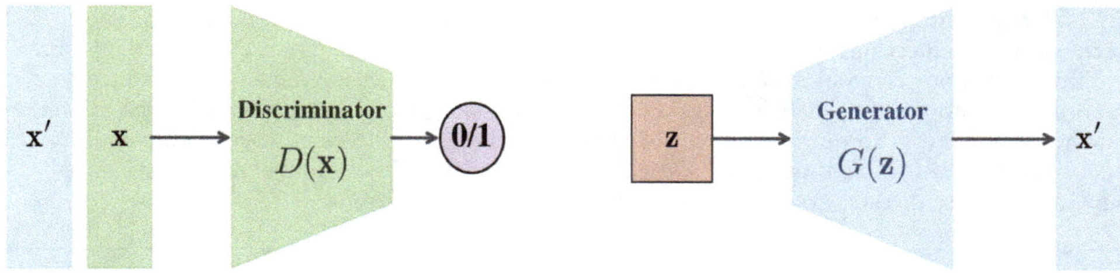

Figure 12.9: GAN Architecture

Figure 12.9 illustrates the GAN architecture. On the left, the discriminator $D(\mathbf{x})$ (green trapezoid) receives both real data $\mathbf{x}$ (green rectangle) and fake data $\mathbf{x}'$ (blue rectangle) as inputs, and outputs a probability score (0/1 in the purple circle) indicating whether the input is real or fake. On the right, the generator $G(\mathbf{z})$ (blue trapezoid) takes random noise $\mathbf{z}$ (red square) from the latent space and generates fake data $\mathbf{x}'$ (blue rectangle). The generated data $\mathbf{x}'$ is then fed to the discriminator, creating the adversarial training loop. This diagram shows the minimax game: the generator tries to produce realistic samples that fool the discriminator, while the discriminator tries to correctly distinguish real from fake samples.

The discriminator aims to maximize $\mathcal{V}$ by telling real and fake samples apart; the generator aims to minimize $\mathcal{V}$ by making fakes look real so that $D_\phi(G_\theta(\mathbf{z})) \to 1$.

12.11.3.2 Optimal Discriminator

For a fixed generator G_θ, the optimal discriminator D^*_ϕ maximizes the value function. Taking the derivative with respect to $D_\phi(\mathbf{x})$ and setting it to zero:

$$\frac{\partial \mathcal{V}}{\partial D_\phi(\mathbf{x})} = \frac{p_{\text{data}}(\mathbf{x})}{D_\phi(\mathbf{x})} - \frac{p_g(\mathbf{x})}{1 - D_\phi(\mathbf{x})} = 0$$

where $p_g(\mathbf{x})$ is the distribution induced by the generator: $p_g(\mathbf{x}) = \int p(\mathbf{z})\delta(\mathbf{x} - G_\theta(\mathbf{z}))\, d\mathbf{z}$.

Solving for $D_\phi(\mathbf{x})$:

$$D^*_\phi(\mathbf{x}) = \frac{p_{\text{data}}(\mathbf{x})}{p_{\text{data}}(\mathbf{x}) + p_g(\mathbf{x})} \tag{12.19}$$

Interpretation: The optimal discriminator outputs the probability that a sample is real, given by the ratio of the real data density to the total density (real + generated).

12.11.3.3 Connection to Jensen-Shannon Divergence

Substituting the optimal discriminator D^*_ϕ back into the value function, the generator's objective becomes:

$$\mathcal{V}(G_\theta, D^*_\phi) = 2 \cdot D_{\text{JS}}(p_{\text{data}} || p_g) - \log 4 \tag{12.20}$$

where D_{JS} is the **Jensen-Shannon divergence** (see Chapter 14, Section~14.7).

Key insight: When the discriminator is optimal, minimizing the GAN objective is equivalent to minimizing JS divergence between the real and generated distributions. At the global optimum (Nash equilibrium), $p_g = p_{data}$ and $D_{JS}(p_{data}||p_g) = 0$, meaning the generator perfectly matches the data distribution.

> GANs demonstrate how minimax optimization can be applied to generative modeling. The mathematical connection to Jensen-Shannon divergence shows how different divergence measures lead to different training dynamics. **Wasserstein GANs (WGANs)** use Wasserstein distance (see Chapter 14, Section~14.9) instead of JS divergence, providing more stable training. The minimax framework illustrates how optimization objectives shape the properties of learned distributions. In practice, GANs are trained using alternating gradient updates: first update the discriminator (maximization step), then update the generator (minimization step).

12.11.4 Application: Sharpness-Aware Minimization (SAM)

Standard gradient methods seek a parameter w with small training loss $L(w)$. Empirically, **flat** local minima—regions where L stays small under small perturbations of w—often generalize better than sharp ones. Sharpness-Aware Minimization (SAM)[2] encodes this preference as a local **minimax** problem of the same form as *12.16*.

12.11.4.1 The Sharpness-Aware Objective

Fix a radius $\rho > 0$. The **sharpness-aware** objective is

$$\min_{w} \max_{\|\epsilon\|_2 \leq \rho} L(w + \epsilon). \tag{12.21}$$

The inner maximization looks for the worst-case loss in a Euclidean ball around w; the outer minimization chooses w so that even that worst-case neighbor has small loss. Flat minima are exactly those w for which the inner maximum remains close to $L(w)$.

12.11.4.2 First-Order Approximation

For small ρ, a linearization $L(w + \epsilon) \approx L(w) + \nabla L(w)^\top \epsilon$ yields the approximate maximizer

$$\hat{\epsilon}(w) \approx \rho \frac{\nabla L(w)}{\|\nabla L(w)\|_2}.$$

A practical update then descends on $L(w+\hat{\epsilon}(w))$ rather than on $L(w)$ alone: one gradient evaluation constructs the adversarial perturbation, and a second evaluates the loss at the perturbed point.

> SAM illustrates how the **minimax** viewpoint of this section applies beyond adversarial games between two networks: the "adversary" can be a local perturbation of the parameters themselves, and the resulting geometry of the loss landscape is what the optimizer seeks to control. Flatness is a geometric reading of generalization that complements the convexity and duality tools earlier in the chapter. In deep learning practice, SAM and its variants are used as drop-in replacements for ordinary SGD or Adam when one wants solutions that are robust to small parameter noise. The same min-max template

[2]Foret, P., Kleiner, A., Mobahi, H., & Neyshabur, B. (2021). Sharpness-Aware Minimization for Efficiently Improving Generalization. In *International Conference on Learning Representations* (ICLR). https://arxiv.org/abs/2010.01412.

also appears in adversarial training, where the perturbation acts on inputs rather than weights.

12.12 Mindmap

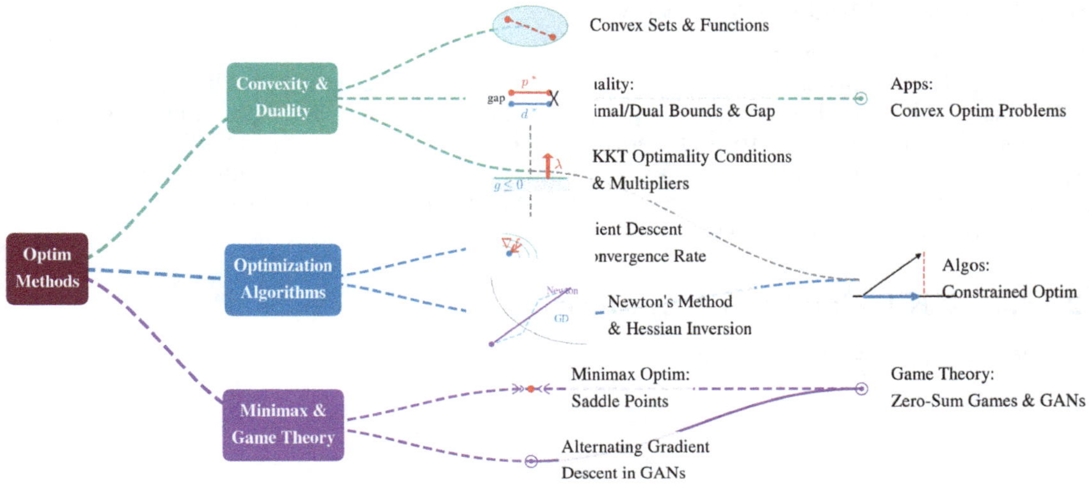

Figure 12.10: Mind map of optimization methods topics

12.13 Exercises

12.13.1 Ridge Regression Solution

Consider the ridge regression optimization problem:

$$\min_{x \in \mathbb{R}^N} \quad ||Ax - b||_2^2 + \lambda ||x||_2^2$$

where:

- $A \in \mathbb{R}^{M \times N}$ is the design matrix
- $x \in \mathbb{R}^{N \times 1}$ is the parameter vector to be optimized
- $b \in \mathbb{R}^{M \times 1}$ is the target vector
- $\lambda > 0$ is the regularization parameter

1. Prove that the solution is
$$x^* = (A^\top A + \lambda I)^{-1} A^\top b.$$
2. Ridge regression is a regularized linear regression method that prevents overfitting. Prove that $(A^\top A + \lambda I)$ is always positive definite (and hence invertible) even when $A^\top A$ is singular. (Hint: Show that for any $v \neq 0$, $v^\top (A^\top A + \lambda I) v > 0$.)

12.13.2 Lagrange Multipliers for Constrained Optimization

Solve the constrained optimization problem:

$$\min_{x \in \mathbb{R}^2} \quad x_0^2 + x_1^2$$
$$\text{s.t.} \quad x_0 + x_1 = 1$$

using the method of Lagrange multipliers.

1. Write down the Lagrangian function.
2. Compute the optimal solution x^* and the corresponding Lagrange multiplier ν^*.
3. This problem models finding the point on a line closest to the origin; similar constrained optimization arises in **support vector machines (SVMs)** and **portfolio optimization.** Compute the optimal value $f(x^*) = (x_0^*)^2 + (x_1^*)^2$. If the constraint changes to $x_0 + x_1 = 1 + \delta$ for small δ, show that the change in optimal value is approximately $\nu^* \cdot \delta$ (sensitivity interpretation of the Lagrange multiplier).

12.13.3 KKT Conditions Application

Consider the optimization problem:

$$\min_{x \in \mathbb{R}} \quad x^2$$
$$\text{s.t.} \quad x \geq 0, \quad x \leq 2$$

1. Write down the KKT conditions for this problem (primal feasibility, dual feasibility, complementary slackness, and stationarity). Let $\eta_1 \geq 0$ be the multiplier for $x \geq 0$ and $\eta_2 \geq 0$ be the multiplier for $x \leq 2$.
2. Find all points that satisfy the KKT conditions. (There should be exactly one point that satisfies all conditions.)
3. Identify the global minimizer x^* and verify it satisfies the KKT conditions.
4. KKT conditions are fundamental to **support vector machines (SVMs)** and **Lasso regression**. State the theorem: for a convex program with Slater's condition, a point x^* is optimal if and only if it satisfies the KKT conditions. Verify that your answer in part 3 satisfies this theorem.

12.13.4 Gradient Descent Convergence

For the function $f(x) = \frac{1}{2}x^T Ax$ where $A \in \mathbb{R}^{N \times N}$ is positive definite with largest eigenvalue $\lambda_{\text{max}}(A) = L$ and smallest eigenvalue $\lambda_{\text{min}}(A) = \mu$:

1. Prove that gradient descent with step size $\alpha = \frac{1}{L}$ converges to the global minimum $x^* = 0$.
2. Derive the convergence rate and express it in terms of the condition number $\kappa = \frac{L}{\mu}$. Show that $||x^{(t)}||_2 \leq \left(1 - \frac{1}{\kappa}\right)^t ||x^{(0)}||_2$.
3. In neural network training, the Hessian matrix determines convergence speed and a large condition number κ indicates ill-conditioning. Prove that if $\alpha = \frac{2}{L}$, then $||x^{(t+1)}||_2 \geq ||x^{(t)}||_2$ (divergence). (Hint: Consider the update $x^{(t+1)} = x^{(t)} - \alpha A x^{(t)}$ and analyze the eigenvalues.)

12.13.5 Newton's Method vs Gradient Descent

For the quadratic function $f(x) = \frac{1}{2}(x - x^*)^T H(x - x^*)$ where $H \in \mathbb{R}^{N \times N}$ is positive definite and x^* is the global minimizer:

1. Prove that Newton's method with step size $\alpha = 1$ converges to x^* in exactly one step, regardless of the initial point $x^{(0)}$.

2. For gradient descent with optimal step size $\alpha = \frac{1}{L}$ where $L = \lambda_{\max}(H)$, show that $||x^{(t)} - x^*||_2 \leq \left(1 - \frac{1}{\kappa}\right)^t ||x^{(0)} - x^*||_2$ where $\kappa = \frac{\lambda_{\max}(H)}{\lambda_{\min}(H)}$. Derive the number of iterations t needed to achieve $||x^{(t)} - x^*||_2 \leq \epsilon||x^{(0)} - x^*||_2$. Express your answer as $t \geq \frac{\log(1/\epsilon)}{\log(\kappa/(\kappa-1))}$ (or an equivalent form).
3. Compare the computational cost per iteration: gradient descent requires $O(N^2)$ operations (matrix-vector product), while Newton's method requires $O(N^3)$ operations (matrix inversion). In deep learning, full Newton's method is rarely used due to the $O(N^3)$ cost; for quadratic functions, however, Newton's method converges in 1 iteration while gradient descent needs $O(\kappa \log(1/\epsilon))$ iterations. Compute the total cost (iterations × cost per iteration) for both methods when $N = 1000$, $\kappa = 100$, and $\epsilon = 10^{-6}$. Which method is more efficient?

12.13.6 Projected Gradient Descent on Probability Simplex

Consider optimizing $f(x) = \frac{1}{2}||x - c||_2^2$ subject to $x \in \Delta^d$ (the probability simplex), where $\Delta^d = \{x \in \mathbb{R}^d : x_i \geq 0, \sum_{i=0}^{d-1} x_i = 1\}$ and $c \in \mathbb{R}^d$ is given.

1. Derive the projection operator $\text{Proj}_{\Delta^d}(c)$ that projects c onto the probability simplex. (Hint: The projection involves sorting and thresholding.)
2. Write down the projected gradient descent update rule for minimizing $f(x)$ subject to $x \in \Delta^d$.
3. The probability simplex constraint appears in **softmax output layers**, **policy optimization in reinforcement learning**, and **mixture models**. Prove that if we use standard gradient descent $x^{(t+1)} = x^{(t)} - \alpha \nabla f(x^{(t)})$ without projection, then $x^{(t+1)}$ may not satisfy $\sum_i x_i^{(t+1)} = 1$ or $x_i^{(t+1)} \geq 0$. Verify that the projection Proj_{Δ^d} ensures both constraints are satisfied.

12.13.7 Soft-Thresholding and Proximal Gradient

Let $g(x) = \lambda|x|$ on $\mathbb{R}$ with $\lambda > 0$, and let $\gamma > 0$.

1. Derive $\text{prox}_{\gamma g}(y) = \text{sign}(y) \max(|y| - \gamma\lambda, 0)$ by solving the scalar minimization

$$\min_{x \in \mathbb{R}} \lambda|x| + \frac{1}{2\gamma}(x - y)^2.$$

(Hint: treat the cases $x > 0$, $x < 0$, and $x = 0$ separately.)

2. For the composite objective $F(x) = \frac{1}{2}(x - a)^2 + \lambda|x|$ with $a = 3$ and $\lambda = 1$, write one proximal-gradient step starting from $x^{(0)} = 0$ with step size $\alpha = 1$. What is $x^{(1)}$?

3. Explain in one or two sentences why soft-thresholding produces **exact zeros**, whereas a smooth penalty such as $\frac{\lambda}{2}x^2$ (ridge) typically does not.

III

Probability and Information for Learning

Chapters 13–16

From "fitting functions" to "modeling uncertainty"

Modeling uncertainty, capturing reality: the language of intelligence

Real-world data is uncertain, incomplete, and often generated by latent mechanisms. Probability theory provides a principled framework for modeling uncertainty and reasoning about distributions rather than individual samples.

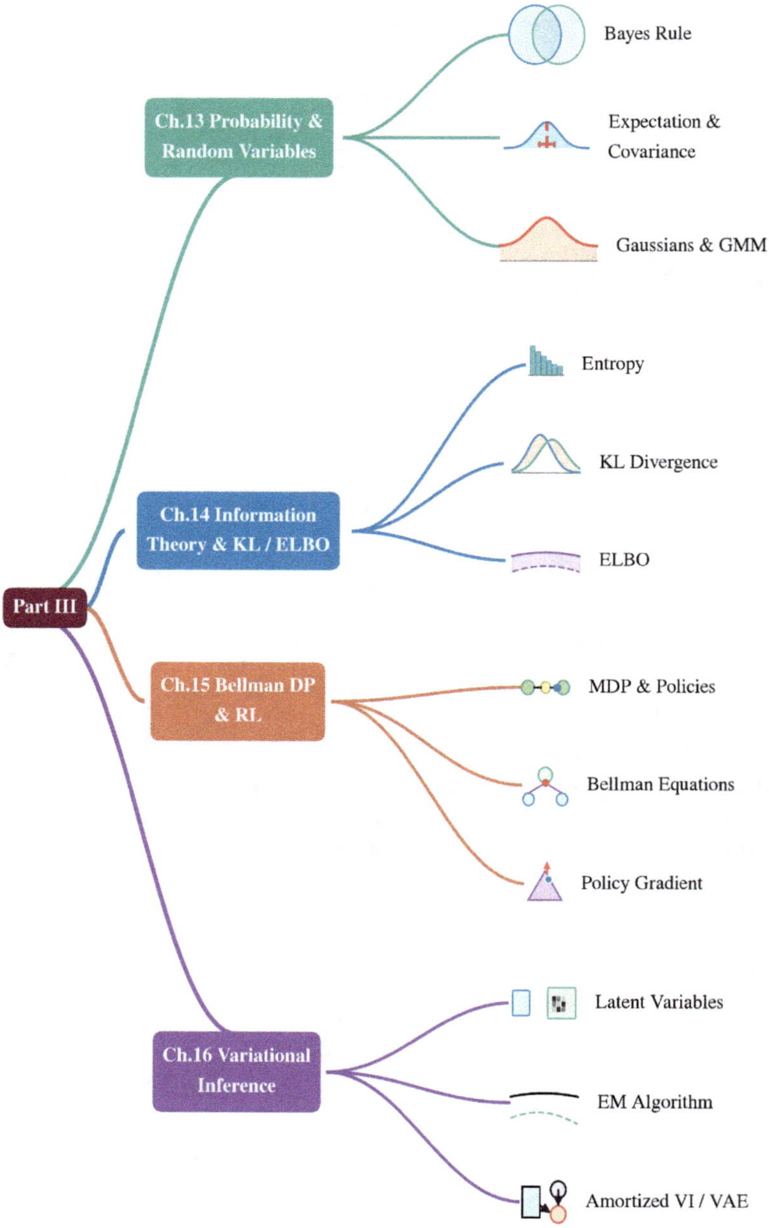
Part III
Ch.13 Probability & Random Variables
Bayes Rule
Expectation & Covariance
Gaussians & GMM
Ch.14 Information Theory & KL / ELBO
Entropy
KL Divergence
ELBO
Ch.15 Bellman DP & RL
MDP & Policies
Bellman Equations
Policy Gradient
Ch.16 Variational Inference
Latent Variables
EM Algorithm
Amortized VI / VAE

Chapter 13

Probability and Random Variables

From certainty to uncertainty: probability distributions and the mathematics of chance

"Probability is the very guide of life."

— Cicero

▷ `Joint P` $P(X,Y)$; CDF: $P(X \leq x, Y \leq y)$; PDF: $f(x,y)$ with $\int_{-\infty}^{\infty}\int_{-\infty}^{\infty} f(x,y)dxdy = 1$, $f(x,y) = \frac{\partial^2}{\partial x \partial y}P(X \leq x, Y \leq y)$

▷ `Marginal P` $P(X=x) = \sum_y P(X=x, Y=y)$; $f(x) = \int f(x,y)\,dy$

▷ `Independence` $P(X,Y) = P(X)P(Y)$; $f(x,y) = f(x)f(y)$; $P_{XY} = \mathbf{p}_X\mathbf{p}_Y^\top$ (rank-1) iff $X \perp Y$

▷ `Bayes' rule` $p(x|y) = \frac{p(y|x)p(x)}{p(y)}$; conditional: $q(x|y,z) = \frac{p(y|x,z)q(x|z)}{p(y|z)}$

▷ `Variance` $\mathrm{Var}[X] = \mathbb{E}[(X - \mathbb{E}[X])^2] = \mathbb{E}[X^2] - (\mathbb{E}[X])^2 = \mathrm{Cov}[X,X] \succeq 0$; $\hat{\sigma}^2 = \frac{1}{K-1}\sum_{k=0}^{K-1}(X^{(k)} - \hat{\mu})^2$

▷ `Covar` $\mathrm{Cov}[X,Y] = \mathbb{E}[(X-\mathbb{E}[X])(Y-\mathbb{E}[Y])] = \mathbb{E}[XY] - \mathbb{E}[X]\mathbb{E}[Y]$; $\hat{\Sigma} = \frac{1}{K-1}\sum_{k=0}^{K-1}(X^{(k)} - \hat{\mu}_X)(Y^{(k)} - \hat{\mu}_Y)^\top$

▷ `Affine trans` For $Y = AX + b$: $\mathbb{E}[Y] = A\mathbb{E}[X] + b$, $\mathrm{Var}[Y] = A\mathrm{Var}[X]A^\top$

▷ `Change of vars (1D)` $Y = g(X)$, $f_Y(y) = f_X(g^{-1}(y)) \cdot |\frac{d}{dy}g^{-1}(y)|$

▷ `Change of vars (multid)` $\mathbf{Y} = g(\mathbf{X})$, $f_{\mathbf{Y}}(\mathbf{y}) = f_{\mathbf{X}}(g^{-1}(\mathbf{y})) \cdot |\det(\mathbf{J}_{g^{-1}}(\mathbf{y}))|$

▷ `Hoeffding` $P(\hat{\mu} - \mathbb{E}[\hat{\mu}] \geq \epsilon) \leq \exp(-\frac{2N^2\epsilon^2}{\sum_{n=0}^{N-1}(b_n - a_n)^2})$ (same for $\leq -\epsilon$); $\mathrm{U}[a,b]$: $\exp(-\frac{2N\epsilon^2}{(b-a)^2})$

▷ `Univariate Gaussian` $X \sim \mathcal{N}(\mu, \sigma^2)$; $f(x|\mu,\sigma^2) = \frac{1}{\sqrt{2\pi\sigma^2}}\exp(-\frac{(x-\mu)^2}{2\sigma^2})$

▷ `Multivariate Gaussian` $X \sim \mathcal{N}(\mu, \Sigma)$; $f(\mathbf{x}|\mu,\Sigma) = \frac{1}{\sqrt{(2\pi)^d|\Sigma|}}\exp(-\frac{1}{2}(\mathbf{x}-\mu)^\top\Sigma^{-1}(\mathbf{x}-\mu))$ for $X \in \mathbb{R}^{d\times 1}$

▷ `Gaussian linear comb` If $X^{(0)}, \ldots, X^{(N-1)}$ are jointly normal, $Y = \sum_{n=0}^{N-1} c_n X^{(n)}$ is normal with $\mathbb{E}[Y] = \sum_{n=0}^{N-1} c_n \mathbb{E}[X^{(n)}]$ and $\mathrm{Var}[Y] = \sum_{n,m=0}^{N-1} c_n c_m \,\mathrm{Cov}[X^{(n)}, X^{(m)}]$ $(d \times d)$

▷ `Gaussian marginal` If $\begin{bmatrix} X \\ Y \end{bmatrix} \sim \mathcal{N}\left(\begin{bmatrix} \mu_X \\ \mu_Y \end{bmatrix}, \begin{bmatrix} \Sigma_{XX} & \Sigma_{XY} \\ \Sigma_{YX} & \Sigma_{YY} \end{bmatrix}\right)$, then $X \sim \mathcal{N}(\mu_X, \Sigma_{XX})$

▷ `Von Mises-Fisher` $\mathbf{x} \sim \mathrm{vMF}(\mu, \kappa)$ on $\mathbb{S}^{d-1}$

13.1 Sets and Cardinality

13.1.1 Definition: Set

A **set** is a collection of distinct objects, called **elements**. A set is typically denoted by a capital letter (e.g., S, A, $\mathcal{X}$) and its elements are listed within curly braces, e.g., $S = \{s_1, s_2, \dots, s_n\}$.

13.1.2 Set Notation

- $x \in S$: element x belongs to set S
- $x \notin S$: element x does not belong to set S
- $A \subseteq B$: A is a **subset** of B (every element of A is also in B)
- $A \subset B$: A is a **proper subset** of B ($A \subseteq B$ and $A \neq B$)
- $\emptyset$: the **empty set** (set with no elements)
- Ω: the **universal set** (the set containing all elements under consideration)

13.1.3 Set Operations

For sets A and B:

1. **Union**: $A \cup B = \{x : x \in A \text{ or } x \in B\}$ — elements in A or B (or both)
2. **Intersection**: $A \cap B = \{x : x \in A \text{ and } x \in B\}$ — elements in both A and B
3. **Complement**: For a universal set Ω, the complement of A is $A^c = \Omega \setminus A = \{x \in \Omega : x \notin A\}$ — elements not in A
4. **Set difference**: $A \setminus B = \{x : x \in A \text{ and } x \notin B\}$ — elements in A but not in B

13.1.4 Definition: Cardinality

For a finite set S, the **cardinality** of S, denoted by $|S|$, is the number of elements in S.

For example:

- If $S = \{a, b, c\}$, then $|S| = 3$
- If $A = \{0, 1, 2, \dots, N-1\}$, then $|A| = N$
- The empty set $\emptyset$ has cardinality $|\emptyset| = 0$

For an infinite set, cardinality is defined using bijections, but we focus on finite sets.

Sets are fundamental mathematical objects used throughout probability theory and machine learning. In **probability**, sample spaces and event spaces are sets. In **machine learning**, feature spaces, label spaces, and state spaces in reinforcement learning are all sets. The cardinality notation $|S|$ is standard in mathematics and will be used throughout this book to denote the size of sets. Understanding sets and their cardinality is essential for working with discrete probability distributions, where probabilities are defined over finite sets of outcomes.

13.2 Probability Basics

13.2.1 Definition: Sample Space and Events

A **sample space** Ω is the set of all possible outcomes of a random experiment. An **event** A is a subset of the sample space, $A \subseteq \Omega$.

13.2.2 Definition: Probability Measure

A **probability measure** P assigns to each event A a number $P(A) \in [0,1]$ such that:

1. **Non-negativity**: $P(A) \geq 0$ for all events A
2. **Normalization**: $P(\Omega) = 1$
3. **Countable additivity**: For any countable collection of mutually exclusive events $A_1, A_2, ...$ (i.e., $A_i \cap A_j = \emptyset$ for $i \neq j$),

$$P\left(\bigcup_{i=1}^{\infty} A_i\right) = \sum_{i=1}^{\infty} P(A_i). \tag{13.1}$$

The triple $(\Omega, \mathcal{F}, P)$ is called a **probability space**, where $\mathcal{F}$ is the collection of events on which P is defined.

13.2.3 Definition: Random Variable

A **random variable** X assigns a real number to each outcome in the sample space Ω:

$$X : \Omega \to \mathbb{R}.$$

We write $P(X \leq x)$ and $P(X = x)$ for the probabilities of the events determined by X. Several random variables $X, Y, Z, ...$ may share the same sample space; then $X + Y$ and XY are the random variables obtained by adding and multiplying their values.

Random variables are often classified by their range:

- **Discrete**: X takes values in a countable set (e.g., $\{0, 1, 2, ...\}$ or $\{a, b, c\}$). Here a, b, c denote real numbers (or labels after a fixed encoding to $\mathbb{R}$).
- **Continuous**: X takes values in an uncountable set (e.g., $\mathbb{R}$ or an interval). For continuous random variables, we describe the distribution with a probability density function (PDF) rather than point masses $P(X = x)$.

> The axiomatic definition of probability (Kolmogorov axioms) provides the mathematical foundation for probability theory. In **machine learning**, we often work directly with random variables and their distributions without explicitly constructing the underlying sample space. The sample space Ω can be thought of as the set of all possible data points, and events as subsets of that space. Random variables map outcomes to numerical values, making them suitable for mathematical analysis and computation.

13.3 Probability Distributions of a Single Random Variable

13.3.1 Definition: Discrete random variable, PMF

For a discrete random variable X, the **probability mass function (PMF)** is defined as

$$P(X = x) \tag{13.2}$$

which gives the probability that X takes the value x. The PMF outputs values in $[0,1]$ (since it represents probabilities), and $\sum_x P(X = x) = 1$ (where the sum is over all possible values of X).

13.3.2 Definition: Continuous random variable, CDF

For a continuous random variable X, the **cumulative distribution function (CDF)** is defined as

$$P(X \leq x) \tag{13.3}$$

which gives the probability that X is less than or equal to x. The CDF is a non-decreasing function with $P(X \leq -\infty) = 0$ and $P(X \leq \infty) = 1$.

13.3.3 Definition: Continuous random variable, PDF

For a continuous random variable X, the **probability density function (PDF)** is denoted as $f(x)$ and satisfies

$$\int_{-\infty}^{\infty} f(x)dx = 1. \tag{13.4}$$

The PDF outputs values in $[0, \infty)$ (it can be greater than 1). The PDF does not represent probabilities directly; rather, $f(x)$ is a probability density. The probability that X falls in an interval $[a, b]$ is given by the integral $\int_a^b f(x)dx$. For an infinitesimal interval $[x, x + dx]$ around x, the probability is approximately $f(x)dx$ (in the limit as $dx \to 0$).

For a continuous random variable, $P(X = c) = 0$ for any specific value c, since the probability of a single point is the integral over a set of measure zero. This leads to the concept of "almost surely": an event A occurs **almost surely** (abbreviated as **a.s.**) if $P(A) = 1$. For example, "$X \neq c$ almost surely" is true for any continuous random variable X and any constant c, since $P(X \neq c) = 1 - P(X = c) = 1$.

13.3.4 Definition: CDF and PDF Relationship

For a continuous random variable X with PDF $f(x)$ and CDF $P(X \leq x)$:

$$P(X \leq x) = \int_{-\infty}^{x} f(u)du \tag{13.5}$$

and

$$f(x) = \frac{d}{dx} P(X \leq x) \tag{13.6}$$

almost everywhere (wherever the derivative exists).

13.4 Joint Probabilities of Multiple Variables

13.4.1 Definition: Discrete variables, PMF

For discrete random variables X, Y, the joint probability mass function is defined as

$$P(X = x, Y = y). \tag{13.7}$$

Similar to the single-variable case, the joint PMF outputs values in $[0, 1]$ and satisfies $\sum_x \sum_y P(X = x, Y = y) = 1$.

13.4.2 Definition: Confusion Matrix

Suppose the output is one of N labels. The prediction vs. ground truth can be summarized in an $N \times N$ matrix C, where c_{ij} is the number of instances whose ground-truth label is i and predicted label is j.

The confusion matrix provides a complete summary of classification performance. For a binary classification problem ($N = 2$), the confusion matrix is typically organized as:

Table 13.1: Binary confusion matrix

Predicted / Actual	Positive	Negative
Positive	True Positive (TP)	False Negative (FN)
Negative	False Positive (FP)	True Negative (TN)

For a multiclass classification problem with $N \geq 3$ classes, the confusion matrix is an $N \times N$ table. For example, with three classes (e.g., "gold", "silver", "bronze"), the confusion matrix might look like:

Table 13.2: Multiclass confusion matrix

Predicted / Actual	**gold**	**silver**	**bronze**
gold	c_{11}	c_{12}	c_{13}
silver	c_{21}	c_{22}	c_{23}
bronze	c_{31}	c_{32}	c_{33}

Here, c_{ii} represents the number of **correctly** classified instances for class i (true positives for that class), while c_{ij} for $i \neq j$ represents misclassifications (e.g., c_{12} is the number of instances that are actually "gold" but predicted as "silver").

From the confusion matrix, we can compute joint probabilities: $P(\text{Actual} = i, \text{Predict} = j) = \frac{c_{ij}}{\sum_{i,j} c_{ij}}$, where the denominator is the total number of instances.

> Classification metrics are fundamental to evaluating machine learning models. The **confusion matrix** provides a complete picture of classification errors, revealing which classes are commonly confused. In **multiclass** settings the same metrics can be computed per class (**macro**-averaging) or pooled globally (**micro**-averaging).

13.4.3 Definition: Precision

For a classification problem with N classes, the **precision** for class i is defined as the conditional probability:

$$\text{Precision}_i = P(\text{Actual} = i \mid \text{Predict} = i) = \frac{P(\text{Actual} = i, \text{Predict} = i)}{P(\text{Predict} = i)}.$$

In terms of the confusion matrix C with entries c_{ij}:

$$\text{Precision}_i = \frac{c_{ii}}{\sum_{j=0}^{N-1} c_{ji}} = \frac{\text{True Positives}_i}{\text{True Positives}_i + \text{False Positives}_i}.$$

Precision measures: *Of all instances the model predicted as class i, how many were actually class i?* High precision means the model makes few false positive errors for that class.

13.4.4 Definition: Recall (Sensitivity)

For a classification problem with N classes, the **recall** (also called **sensitivity**) for class i is defined as the conditional probability:

$$\text{Recall}_i = P(\text{Predict} = i \mid \text{Actual} = i) = \frac{P(\text{Actual} = i, \text{Predict} = i)}{P(\text{Actual} = i)}.$$

In terms of the confusion matrix C with entries c_{ij}:

$$\text{Recall}_i = \frac{c_{ii}}{\sum_{j=0}^{N-1} c_{ij}} = \frac{\text{True Positives}_i}{\text{True Positives}_i + \text{False Negatives}_i}.$$

Recall measures: *Of all actual instances of class i, how many did the model correctly identify?* High recall means the model finds most of the positive instances (makes few false negative errors).

13.4.5 Definition: F1 Score

For a classification problem with N classes, the **F1 score** for class i is defined as the harmonic mean of precision and recall:

$$\text{F1}_i = \frac{2 \cdot \text{Precision}_i \cdot \text{Recall}_i}{\text{Precision}_i + \text{Recall}_i} = \frac{2}{\frac{1}{\text{Precision}_i} + \frac{1}{\text{Recall}_i}}.$$

The F1 score provides a single metric that balances precision and recall. It is the harmonic mean (rather than arithmetic mean) because: 1. **Penalizes extreme values**: If either precision or recall is very low, the F1 score will be low, even if the other metric is high. 2. **Balanced evaluation**: The harmonic mean better reflects overall performance when one metric is poor. For example, if precision $= 0.9$ and recall $= 0.1$, the arithmetic mean is 0.5, but the harmonic mean is approximately 0.18, which better reflects the poor overall performance. 3. **Equal weighting**: The F1 score gives equal weight to precision and recall, ensuring that improving the lower metric has a greater impact.

13.4.6 Definition: Accuracy

For a classification problem with N classes, the **accuracy** is defined as the probability of correct prediction:

$$\text{Accuracy} = P(\text{Predict} = \text{Actual}) = \sum_{i=0}^{N-1} P(\text{Actual} = i, \text{Predict} = i) = \frac{\sum_{i=0}^{N-1} c_{ii}}{\sum_{i,j=0}^{N-1} c_{ij}}.$$

Accuracy measures the overall proportion of correct predictions. However, accuracy can be misleading when classes are imbalanced, as it may be dominated by the majority class.

13.4.7 Definition: Specificity (True Negative Rate)

For binary classification, **specificity** (also called **true negative rate**) is defined as the conditional probability:

$$\begin{aligned}\text{Specificity} &= P(\text{Predict} = \text{Negative} \mid \text{Actual} = \text{Negative}) \\ &= \frac{\text{True Negatives}}{\text{True Negatives} + \text{False Positives}} \\ &= \frac{\text{TN}}{\text{TN} + \text{FP}}.\end{aligned}$$

Specificity measures: *Of all actual negative instances, how many did the model correctly identify as negative?* High specificity means the model makes few false positive errors. Specificity is the complement of the false positive rate: Specificity $= 1 -$ False Positive Rate.

For multiclass classification, specificity for class i can be defined as the probability of correctly predicting "not class i" when the actual class is not i:

$$\text{Specificity}_i = P(\text{Predict} \neq i \mid \text{Actual} \neq i) = \frac{\sum_{j\neq i}\sum_{k\neq i} c_{jk}}{\sum_{j\neq i}\sum_{k=0}^{N-1} c_{jk}}.$$

> **Precision** and **recall** are conditional probabilities providing complementary views: precision measures the quality of positive predictions, while recall measures the completeness of positive detection. **Specificity** complements recall by measuring how well negative instances are identified. The **precision-recall trade-off** is central to model evaluation: increasing the decision threshold typically raises precision but lowers recall. The **F1 score** provides a balanced single metric, but the choice between precision and recall depends on the application. In **medical diagnosis**, high sensitivity and specificity are both critical; in **fraud detection**, high precision and specificity help avoid blocking legitimate transactions.

13.4.8 Definition: ROC Curve

For binary classification, the **Receiver Operating Characteristic (ROC) curve** plots the **True Positive Rate (TPR)** (which equals recall/sensitivity) against the **False Positive Rate (FPR)** at various classification thresholds.

- **True Positive Rate (TPR)**: $\text{TPR} = \text{Recall} = \frac{\text{TP}}{\text{TP+FN}}$
- **False Positive Rate (FPR)**: $\text{FPR} = 1 - \text{Specificity} = \frac{\text{FP}}{\text{FP+TN}}$

The ROC curve is generated by varying the decision threshold from the most conservative (classifying everything as negative, giving point $(0, 0)$) to the most liberal (classifying everything as positive, giving point $(1, 1)$). Each point on the curve represents a different trade-off between TPR and FPR.

A perfect classifier would have a ROC curve that goes from $(0, 0)$ to $(0, 1)$ to $(1, 1)$, achieving TPR $= 1$ with FPR $= 0$. A random classifier would have a ROC curve along the diagonal line from $(0, 0)$ to $(1, 1)$.

13.4.9 Definition: AUC (Area Under the ROC Curve)

The **Area Under the ROC Curve (AUC)** is defined as the area under the ROC curve:

$$\text{AUC} = \int_0^1 \text{TPR}(\text{FPR})\, d(\text{FPR}),$$

where TPR(FPR) is the true positive rate as a function of the false positive rate.

The AUC takes values in $[0, 1]$:

- **AUC = 1**: Perfect classifier (all positive instances ranked higher than all negative instances)
- **AUC = 0.5**: Random classifier (no discriminative ability)
- **AUC < 0.5**: Worse than random (can be improved by inverting predictions)

13.4.10 Interpretation of AUC

The AUC has a probabilistic interpretation: it equals the probability that a randomly chosen positive instance will be ranked higher (have a higher predicted score) than a randomly chosen negative instance:

$$\text{AUC} = P(\text{Score}(\text{Positive}) > \text{Score}(\text{Negative})),$$

where $\text{Score}(\cdot)$ is the model's prediction score (before thresholding).

> This interpretation makes AUC particularly useful for evaluating ranking quality and model discrimination ability, independent of the specific threshold chosen. The **ROC curve** visualizes the trade-off between true positive rate (sensitivity) and false positive rate (1 - specificity) across all thresholds. The **AUC** provides a threshold-independent measure of model discrimination ability, evaluating how well the model separates positive and negative instances. Use it when ranking quality matters more than any single operating point.

13.4.11 Definition: Calibration of Probabilistic Predictions

Discrimination (e.g. AUC) asks whether scores **rank** positives above negatives. **Calibration** asks whether a reported probability $f \in [0, 1]$ matches the empirical frequency of the event:

$$\mathbb{P}(Y = 1 \mid f) \approx f.$$

A model can have excellent AUC yet be poorly calibrated (e.g. systematically reporting 0.99 when the true rate is near 0.6).

13.4.12 Definition: Reliability Diagram

Partition predicted probabilities into bins $B_1, \dots, B_M$. For each bin B_m, plot the average prediction against the average outcome:

$$\Big(\tfrac{1}{|B_m|} \sum_{i \in B_m} f_i,\ \tfrac{1}{|B_m|} \sum_{i \in B_m} y_i\Big).$$

Perfect calibration lies on the diagonal $y = x$.

13.4.13 Definition: Brier Score

For binary labels $y_t \in \{0, 1\}$ and forecasts $f_t \in [0, 1]$,

$$\text{BS} = \frac{1}{N} \sum_{t=1}^{N} (f_t - y_t)^2. \tag{13.8}$$

The Brier score is the mean squared error between probabilities and outcomes: lower is better. It jointly penalizes miscalibration and lack of sharpness.

13.4.14 Definition: Temperature Scaling

Let $z \in \mathbb{R}^C$ be the logit vector of a C-class classifier and

$$\hat{p}_i = \frac{e^{z_i/T}}{\sum_{j=1}^{C} e^{z_j/T}}, \qquad T > 0. \tag{13.9}$$

The scalar **temperature** T is typically fit on a held-out validation set (e.g. by minimizing negative log-likelihood) while the classifier weights stay fixed. Because T scales all logits equally, $\arg\max_i \hat{p}_i = \arg\max_i z_i$: **hard predictions are unchanged**, only the probability levels are recalibrated. $T > 1$ softens (flattens) the distribution; $0 < T < 1$ sharpens it.[a] **Calibration** is complementary: it asks whether probabilities are numerically meaningful, not only whether rankings are correct. The **Brier score** and **reliability diagrams** quantify this. **Temperature scaling** is a one-parameter post-hoc adjustment that preserves arg max decisions.

[a] Guo, C., Pleiss, G., Sun, Y., & Weinberger, K. Q. (2017). On Calibration of Modern Neural Networks. In *International Conference on Machine Learning* (ICML), pp. 1321–1330. https://arxiv.org/abs/1706.04599.

13.4.15 Definition: Continuous variables, CDF

For continuous random variables X, Y, the joint cumulative distribution function is defined as

$$P(X \leq x, Y \leq y). \tag{13.10}$$

13.4.16 Definition: Continuous variables, PDF

Continuous random variables X, Y are **jointly continuous** if there exists a joint probability density function $f(x, y)$ such that

$$\int_{-\infty}^{\infty} \int_{-\infty}^{\infty} f(x, y) dx dy = 1. \tag{13.11}$$

Similar to the single-variable case, the joint PDF outputs values in $[0, \infty)$ (it can be greater than 1) and represents a probability density. The probability that (X, Y) falls in a region A is given by the double integral $\iint_A f(x, y) dx dy$. For an infinitesimal region $[x, x + dx] \times [y, y + dy]$ around (x, y), the probability is $f(x, y) dx dy$ (in the limit as $dx, dy \to 0$).

13.4.17 Definition: CDF and PDF Relationship

For continuous random variables X, Y with joint PDF $f(x, y)$ and joint CDF $P(X \leq x, Y \leq y)$:

$$P(X \leq x, Y \leq y) = \int_{-\infty}^{x} \int_{-\infty}^{y} f(u, v) dv du \tag{13.12}$$

and

$$f(x, y) = \frac{\partial^2}{\partial x \partial y} P(X \leq x, Y \leq y) \tag{13.13}$$

almost everywhere (wherever the mixed partial exists).

13.5 Marginal Probabilities

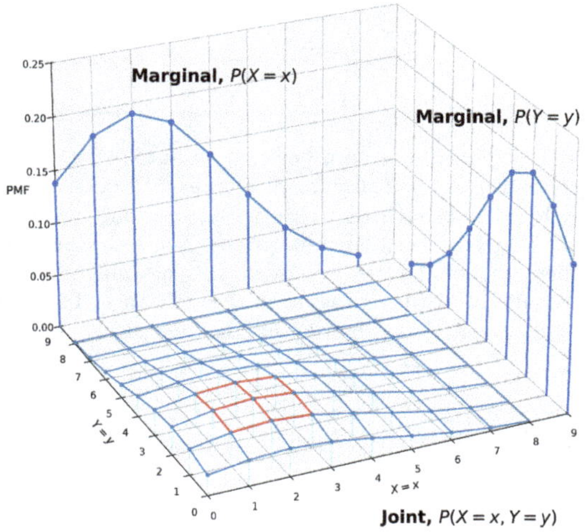

Figure 13.1: Joint PMF and its Marginals

13.5.1 Definition: Discrete variables, marginal PMF

For discrete random variables X, Y with joint PMF $P(X = x, Y = y)$, the marginal PMF of X is

$$P(X = x) = \sum_{y} P(X = x, Y = y). \tag{13.14}$$

13.5.2 Definition: Continuous variables, marginal PDF

For continuous random variables X, Y with joint PDF $f(x, y)$, the marginal PDF of X is

$$f(x) = \int_{-\infty}^{\infty} f(x, y)dy. \tag{13.15}$$

13.6 Conditional Probabilities

13.6.1 Definition: Conditional Probability / PMF (Discrete)

For discrete random variables X, Y with joint PMF $P(X = x, Y = y)$ and marginal PMF $P(Y = y) > 0$, the **conditional probability** of X given $Y = y$ is

$$P(X = x \mid Y = y) = \frac{P(X = x, Y = y)}{P(Y = y)}. \tag{13.16}$$

As a function of x (with y fixed), this is the **conditional PMF** of X given $Y = y$, also written $p_{X|Y}(x \mid y)$.

13.6.2 Definition: Conditional PDF (Continuous)

For continuous random variables X, Y with joint PDF $f(x, y)$ and marginal PDF $f_Y(y) > 0$, the **conditional probability density function** of X given $Y = y$ is

$$f_{X|Y}(x \mid y) = \frac{f(x, y)}{f_Y(y)} \tag{13.17}$$

where:

- $f(x, y)$ is the joint PDF of X and Y
- $f_X(x) = \int_{-\infty}^{\infty} f(x, y)dy$ is the marginal PDF of X
- $f_Y(y) = \int_{-\infty}^{\infty} f(x, y)dx$ is the marginal PDF of Y

Conditional probability is fundamental to understanding how information about one variable updates our beliefs about another.

> In **machine learning**, conditional probabilities appear everywhere: in classification ($P(\text{class} \mid \text{features})$), in language models ($P(\text{next word} \mid \text{context})$), and in reinforcement learning ($P(\text{next state} \mid \text{current state}, \text{action})$). Conditional probability also connects to **information theory**: the conditional entropy $H(X \mid Y)$ measures the uncertainty in X after observing Y, and the mutual information $I(X; Y) = H(X) - H(X \mid Y)$ measures how much information Y provides about X.

13.6.3 Theorem: Chain Rule of Probability

For discrete random variables X, Y:

$$P(X = x, Y = y) = P(X = x \mid Y = y)P(Y = y) = P(Y = y \mid X = x)P(X = x). \tag{13.18}$$

For continuous random variables X, Y:

$$f(x, y) = f_{X|Y}(x \mid y)f_Y(y) = f_{Y|X}(y \mid x)f_X(x). \tag{13.19}$$

> The chain rule of probability is the foundation for **probabilistic graphical models** and **Bayesian networks**, where complex joint distributions are factorized into products of conditional probabilities.

13.6.4 Theorem: Law of Total Probability (Discrete)

For discrete random variables X, Y:

$$P(X = x) = \sum_y P(X = x \mid Y = y)P(Y = y). \tag{13.20}$$

13.6.5 Theorem: Law of Total Probability (Continuous)

For continuous random variables X, Y:

$$f_X(x) = \int_{-\infty}^{\infty} f_{X|Y}(x \mid y)f_Y(y)dy. \tag{13.21}$$

The law of total probability is essential for **marginalization** in probabilistic inference, allowing us to compute marginal distributions from conditional and joint distributions.

13.6.6 Properties of Conditional Probability

1. **Normalization**: For discrete X given $Y = y$:

$$\sum_x P(X = x \mid Y = y) = 1.$$

For continuous X given $Y = y$:

$$\int_{-\infty}^{\infty} f_{X|Y}(x \mid y)dx = 1.$$

2. **Conditional expectation**: The conditional expectation of X given $Y = y$ is

$$\mathbb{E}[X \mid Y = y] = \begin{cases} \sum_x x \cdot P(X = x \mid Y = y) & \text{(discrete)} \\ \int_{-\infty}^{\infty} x \cdot f_{X|Y}(x \mid y)dx & \text{(continuous)} \end{cases}$$

13.7 Independence

13.7.1 Definition: Independence (Discrete)

Discrete random variables X and Y are **independent** if

$$P(X = x, Y = y) = P(X = x)P(Y = y) \tag{13.22}$$

for all values of x and y.

13.7.2 Definition: Independence (Continuous)

Continuous random variables X and Y are **independent** if

$$f(x, y) = f(x)f(y) \tag{13.23}$$

for all values of x and y, where $f(x, y)$ is the joint PDF and $f(x)$, $f(y)$ are the marginal PDFs.

13.7.3 Linear Algebra Perspective: Independence as Outer Product

For discrete random variables X and Y taking values in finite sets, we can view probability distributions from a linear algebra perspective:

- **Marginal distributions as vectors**: If X takes N distinct values and Y takes M distinct values, then $p_X(x)$ can be represented as a vector $\mathbf{p}_X \in \mathbb{R}^{N\times 1}$ where $[\mathbf{p}_X]_i = P(X = x_i)$, and $p_Y(y)$ can be represented as a vector $\mathbf{p}_Y \in \mathbb{R}^{M\times 1}$ where $[\mathbf{p}_Y]_j = P(Y = y_j)$.
- **Joint distribution as a matrix**: The joint PMF $p_{X,Y}(x, y)$ can be represented as a matrix $P_{XY} \in \mathbb{R}^{N\times M}$ where $[P_{XY}]_{ij} = P(X = x_i, Y = y_j)$.
- **Independence as outer product**: When X and Y are independent, the joint probability matrix equals the outer product of the marginal probability vectors:

$$P_{XY} = \mathbf{p}_X \mathbf{p}_Y^\top \tag{13.24}$$

or equivalently, $[P_{XY}]_{ij} = [\mathbf{p}_X]_i [\mathbf{p}_Y]_j = P(X = x_i)P(Y = y_j)$.

This shows that independence corresponds to a **rank-1** structure in the joint probability matrix, where the matrix can be factorized as the outer product of two vectors (the marginal distributions).

The linear algebra perspective on independence provides deep insights into the structure of probability distributions.

> In **machine learning**, this rank-1 structure is exploited in matrix factorization methods for recommendation systems, where user-item interaction matrices are factorized into user and item preference vectors. The outer product representation also connects to **tensor decompositions** in higher-dimensional probability distributions. When variables are not independent, the joint probability matrix has higher rank, requiring more complex factorizations. This perspective is fundamental in **probabilistic graphical models**, where independence assumptions correspond to low-rank structures in probability tensors.

13.8 Bayes' Rule

13.8.1 Intuition

In AI/ML we often treat X as a learnable parameter (or latent variable) and Y as an observation, and ask how Y updates our belief about X — that is, compute $P(X = x \mid Y = y)$.

13.8.2 Definition: Notation for Bayes' Rule

When both variables are discrete, we use two PMF letters to separate the **belief about** X from the **observation model for** Y:

- $q(x)$: **prior** of X, $q(x) = P(X = x)$
- $q(x \mid y)$: **posterior** of X given $Y = y$, $q(x \mid y) = P(X = x \mid Y = y)$
- $p(y \mid x)$: **likelihood**, $p(y \mid x) = P(Y = y \mid X = x)$
- $p(y)$: **evidence**, $p(y) = P(Y = y) = \sum_x p(y \mid x)q(x)$

When a variable is continuous, the corresponding factor is a PDF written with f (for example $f(y \mid x)$ or $f(x)$); the same four roles still apply.

13.8.3 Definition: Likelihood

The likelihood $p(y \mid x)$ (or $f(y \mid x)$ when Y is continuous) has the same mathematical form as a conditional probability, but a different interpretation: a **conditional probability** treats x as fixed and describes the distribution of y, whereas the **likelihood** treats the observed y as fixed and views the expression as a function of the parameter x. Maximum likelihood estimation maximizes this function of x for fixed data.

13.8.4 Theorem: Bayes' Rule, Discrete X, Discrete Y

Let X and Y be discrete random variables. With the notation above,

$$q(x \mid y) = \frac{p(y \mid x)q(x)}{p(y)}. \tag{13.25}$$

13.8.5 Theorem: Conditional Bayes' Rule

Let X, Y, Z be discrete random variables. Then

$$q(x \mid y, z) = \frac{p(y \mid x, z)q(x \mid z)}{p(y \mid z)}.$$

13.8.5.1 Example: Markov Model

Consider a transition matrix $P \in \mathbb{R}^{N \times N}$ with

$$p_{ij} = P(Y = j \mid X = i)$$

for all $i, j \in \{0, 1, \dots, N-1\}$.

Thus, $p_{ij} \geq 0$ and $\sum_{j=0}^{N-1} p_{ij} = 1$.

Suppose we are given a prior of X with $P(X = i) = q_i$ for all $i \in \{0, \dots, N-1\}$. Compute $P(X = i \mid Y = j)$.

Solution

$$P(X = i \mid Y = j) = \frac{q_i p_{ij}}{\sum_{k=0}^{N-1} q_k p_{kj}}.$$

> This Markov model is closely related to the Markov Decision Process (MDP) in reinforcement learning. In an MDP, the transition probabilities p_{ij} represent the probability of moving from state i to state j given an action, and the prior q_i can be thought of as the initial state distribution. The posterior $P(X = i \mid Y = j)$ corresponds to the belief state update in partially observable MDPs (POMDPs) or the state distribution update in policy evaluation.

13.8.6 Theorem: Bayes' Rule, Discrete X, Continuous Y

Let X be a discrete random variable and Y be a continuous random variable. Then

$$p(x \mid y) = \frac{f(y \mid x)p(x)}{f(y)}. \tag{13.26}$$

13.8.6.1 Example: Gaussian Mixture Model

Here we use the Gaussian (normal) distribution, denoted as $\mathcal{N}(\mu, \sigma^2)$, which represents a continuous distribution with mean μ and variance σ^2. For $\mathcal{N}(\mu, 1)$ (variance 1), the probability density function is $\phi(x - \mu) = \frac{1}{\sqrt{2\pi}} \exp\left(-\frac{(x-\mu)^2}{2}\right)$, where $\phi(z) = \frac{1}{\sqrt{2\pi}} e^{-z^2/2}$ is the standard normal density. (A formal definition of the Gaussian distribution is given in the "Gaussian Distribution" section later.)

Let $X \in \{0, 1\}$, $Y \sim \mathcal{N}(X, 1)$, and $P(X = 1) = q$. Compute $p(1 \mid y)$.

Solution

$$p(1 \mid y) = \frac{q\phi(y-1)}{q\phi(y-1) + (1-q)\phi(y)},$$

where $\phi(z) = \frac{1}{\sqrt{2\pi}} e^{-z^2/2}$.

This Gaussian Mixture Model (GMM) example is directly related to the `GaussianMixture` class in scikit-learn. In sklearn, a GMM is a probabilistic model that assumes data points are generated from a mixture of several Gaussian distributions. The posterior probability $p(1 \mid y)$ computed here corresponds to the `predict_proba()` method in sklearn's `GaussianMixture`, which returns the probability of each sample belonging to each component. The prior $q = P(X = 1)$ corresponds to the mixture weights (`weights_` attribute), and the Gaussian components with means X and variance 1 correspond to the `means_` and `covariances_` attributes in the fitted model.

13.8.7 Theorem: Bayes' Rule, Continuous X, Discrete Y

Let X be a continuous random variable and Y be a discrete random variable. Then

$$f(x \mid y) = \frac{p(y \mid x) f(x)}{p(y)}. \tag{13.27}$$

13.8.7.1 Example: Classification

Suppose X is a standard normal random variable. Let

$$Y = \begin{cases} 1 & \text{with probability } \sigma(3X + 2) \\ 0 & \text{with probability } 1 - \sigma(3X + 2) \end{cases},$$

where σ is the sigmoid function. Compute $f(x \mid 1)$.

Solution

We have

$$f(x \mid 1) = \frac{p(1 \mid x) f(x)}{p(1)} = \frac{p(1 \mid x) f(x)}{\int_{-\infty}^{\infty} p(1 \mid z) f(z) dz}.$$

Since $X \sim \mathcal{N}(0, 1)$, we have $f(x) = \frac{1}{\sqrt{2\pi}} \exp\left(-\frac{x^2}{2}\right)$ and $p(1 \mid x) = \sigma(3x + 2)$. Therefore,

$$f(x \mid 1) = \frac{\sigma(3x + 2) \cdot \frac{1}{\sqrt{2\pi}} \exp\left(-\frac{x^2}{2}\right)}{\int_{-\infty}^{\infty} \sigma(3z + 2) \cdot \frac{1}{\sqrt{2\pi}} \exp\left(-\frac{z^2}{2}\right) dz}.$$

13.8.8 Theorem: Bayes' Rule, Continuous X, Continuous Y

Let X and Y be jointly continuous with $f(y) > 0$. Then

$$f(x \mid y) = \frac{f(y \mid x) f(x)}{f(y)}. \tag{13.28}$$

13.8.8.1 Example: Linear Regression

Let X and ε be two independent standard normal random variables. Define $Y = X + \varepsilon$. Compute $f(x \mid y)$.

Solution

By Bayes' rule,

$$f(x \mid y) = \frac{f(y \mid x) f(x)}{f(y)} = \frac{f(y \mid x) f(x)}{\int_{-\infty}^{\infty} f(y \mid z) f(z)\, dz}.$$

Since $X \sim \mathcal{N}(0, 1)$ and $\varepsilon \sim \mathcal{N}(0, 1)$ are independent, and $Y = X + \varepsilon$, we have:

- $f(x) = \frac{1}{\sqrt{2\pi}} \exp\left(-\frac{x^2}{2}\right)$ (prior for X)
- $f(y \mid x) = \frac{1}{\sqrt{2\pi}} \exp\left(-\frac{(y-x)^2}{2}\right)$ (likelihood, since $Y \mid X = x \sim \mathcal{N}(x, 1)$)

Substituting into the formula:

$$f(x \mid y) = \frac{\frac{1}{\sqrt{2\pi}} \exp\left(-\frac{(y-x)^2}{2}\right) \cdot \frac{1}{\sqrt{2\pi}} \exp\left(-\frac{x^2}{2}\right)}{\int_{-\infty}^{\infty} \frac{1}{\sqrt{2\pi}} \exp\left(-\frac{(y-z)^2}{2}\right) \cdot \frac{1}{\sqrt{2\pi}} \exp\left(-\frac{z^2}{2}\right) dz}.$$

The exponent in the numerator is

$$-\frac{(y-x)^2}{2} - \frac{x^2}{2} = -\left(x - \frac{y}{2}\right)^2 - \frac{y^2}{4},$$

and the denominator equals $(2\pi)^{-1/2} \cdot 2^{-1/2} \exp(-y^2/4)$ (using $Y \sim \mathcal{N}(0, 2)$). Therefore

$$f(x \mid y) = \frac{1}{\sqrt{\pi}} \exp\left(-\left(x - \frac{y}{2}\right)^2\right),$$

that is, $X \mid Y = y \sim \mathcal{N}(y/2,\ 1/2)$.

13.9 Statistics Basics

13.9.1 Definition: Mean (Expected Value)

The **mean** (also called **expected value** or **expectation**) of a random variable X is

- **Discrete case**: if X has PMF $P(X = x)$,

$$\mathbb{E}[X] = \sum_x x \cdot P(X = x), \tag{13.29}$$

  and $\mathbb{E}[|X|] = \sum_x |x| \, P(X = x)$;

- **Continuous case**: if X has PDF $f(x)$,

$$\mathbb{E}[X] = \int_{-\infty}^{\infty} x \cdot f(x) \, dx, \tag{13.30}$$

  and $\mathbb{E}[|X|] = \int_{-\infty}^{\infty} |x| \, f(x) \, dx$.

The mean $\mathbb{E}[X]$ is the average value of X under its probability distribution.

> Throughout this section we assume $\mathbb{E}[|X|] < \infty$ and $\mathbb{E}[|Y|] < \infty$, so that means are finite real numbers.

For constants $a, b, c \in \mathbb{R}$, expectation has the following properties:

1. **Linearity**: $\mathbb{E}[aX + bY + c] = a\mathbb{E}[X] + b\mathbb{E}[Y] + c$
2. **Constant**: $\mathbb{E}[c] = c$ (a constant random variable has expectation equal to itself)
3. **Tower property**: $\mathbb{E}[\mathbb{E}[X]] = \mathbb{E}[X]$ (the expectation of an expectation is the expectation itself)
4. **Monotonicity**: If $X \leq Y$ almost surely, then $\mathbb{E}[X] \leq \mathbb{E}[Y]$
5. **Non-negativity**: If $X \geq 0$ almost surely, then $\mathbb{E}[X] \geq 0$
6. **Independence**: If X and Y are independent, then $\mathbb{E}[XY] = \mathbb{E}[X]\mathbb{E}[Y]$

The linearity property extends to sums of any finite number of random variables: $\mathbb{E}[\sum_{i=0}^{n-1} a_i X_i + c] = \sum_{i=0}^{n-1} a_i \mathbb{E}[X_i] + c$.

13.9.2 Definition: Mean Vector

For $X \in \mathbb{R}^{N\times 1}$:

$$\mathbb{E}[X] = \begin{bmatrix} \mathbb{E}[X_0] \\ \vdots \\ \mathbb{E}[X_{N-1}] \end{bmatrix} \in \mathbb{R}^{N\times 1} \tag{13.31}$$

The mean vector $\mathbb{E}[X]$ is a theoretical quantity that depends on the probability distribution of X. In practice, we often estimate this mean from observed data using the empirical mean.

13.9.3 Definition: Empirical Mean

Let $X^{(0)}, \dots, X^{(K-1)} \in \mathbb{R}$ be i.i.d. observations of a random variable X. Then the empirical mean (also called sample mean) is defined as

$$\hat{\mu} = \frac{1}{K} \sum_{k=0}^{K-1} X^{(k)}.$$

The terms "empirical" and "sample" are used interchangeably to refer to quantities computed from observed data, as opposed to their theoretical population counterparts. This applies to all empirical/sample quantities throughout this chapter.

13.9.4 Definition: Empirical Mean Vector

Let $X^{(0)}, \dots, X^{(K-1)} \in \mathbb{R}^{N\times 1}$ be i.i.d. observations of a random vector X. Then the empirical mean vector is defined as

$$\hat{\mu} = \frac{1}{K} \sum_{k=0}^{K-1} X^{(k)} \in \mathbb{R}^{N\times 1}.$$

13.9.5 Definition: Variance

For a random variable $X \in \mathbb{R}$ with $\mathbb{E}[X^2] < \infty$, the variance is defined as

$$\mathrm{Var}[X] = \mathbb{E}[(X - \mathbb{E}[X])^2]. \tag{13.32}$$

An equivalent formula (often more convenient for computation) is:

$$\mathrm{Var}[X] = \mathbb{E}[X^2] - (\mathbb{E}[X])^2. \tag{13.33}$$

Throughout the remainder of this section we assume $\mathbb{E}[X^2] < \infty$ and $\mathbb{E}[Y^2] < \infty$, so that variances and covariances are finite.

Proof of equivalence

We have

$$\begin{aligned}\mathrm{Var}[X] &= \mathbb{E}[(X - \mathbb{E}[X])^2] \\ &= \mathbb{E}[X^2 - 2X \cdot \mathbb{E}[X] + (\mathbb{E}[X])^2] \\ &= \mathbb{E}[X^2] - 2\mathbb{E}[X] \cdot \mathbb{E}[X] + (\mathbb{E}[X])^2 \\ &= \mathbb{E}[X^2] - (\mathbb{E}[X])^2.\end{aligned}$$

13.9.6 Properties of Variance

For a random variable $X \in \mathbb{R}$ and constants $a, b \in \mathbb{R}$:

1. **Non-negativity**: $\mathrm{Var}[X] \geq 0$ (variance is always non-negative)
2. **Constant random variable**: If $X = c$ (a constant) with probability 1, then $\mathrm{Var}[X] = 0$
3. **Scaling and shifting**: $\mathrm{Var}[aX + b] = a^2\mathrm{Var}[X]$

The third property follows from the definition: $\mathrm{Var}[aX + b] = \mathbb{E}[(aX + b - \mathbb{E}[aX + b])^2] = \mathbb{E}[(aX + b - a\mathbb{E}[X] - b)^2] = \mathbb{E}[a^2(X - \mathbb{E}[X])^2] = a^2\mathrm{Var}[X]$.

13.9.7 Definition: Empirical Variance

Let $X^{(0)}, \ldots, X^{(K-1)} \in \mathbb{R}$ be i.i.d. observations of a random variable X, and let $\hat{\mu} = \frac{1}{K}\sum_{k=0}^{K-1} X^{(k)}$ be the empirical mean. Then the empirical variance is defined as

$$\hat{\sigma}^2 = \frac{1}{K-1}\sum_{k=0}^{K-1}(X^{(k)} - \hat{\mu})^2. \tag{13.34}$$

The divisor $K - 1$ (instead of K) makes this an unbiased estimator of the population variance. This correction is necessary because $\hat{\mu}$ is an estimate of the true mean μ, not the true mean itself. When we use $\hat{\mu}$ in place of μ, we lose one degree of freedom, leaving $K - 1$ independent pieces of information for estimating the variance.

13.9.8 Definition: Variance Matrix

For a random vector $X \in \mathbb{R}^{N\times 1}$, the variance matrix (also called covariance matrix of X with itself) is defined as

$$\mathrm{Var}[X] = \mathbb{E}[(X - \mathbb{E}[X])(X - \mathbb{E}[X])^\top] \in \mathbb{R}^{N\times N}.$$

Note that $\mathrm{Var}[X] = \mathrm{Cov}[X, X]$, so the variance matrix is the covariance matrix of X with itself. The diagonal elements are the variances of the individual components: $(\mathrm{Var}[X])_{ii} = \mathrm{Var}[X_i]$. The off-diagonal elements are the covariances between different components: $(\mathrm{Var}[X])_{ij} = \mathrm{Cov}[X_i, X_j]$ for $i \neq j$.

13.9.9 Theorem: Variance Matrix is Positive Semidefinite

For a random vector $X \in \mathbb{R}^{N\times 1}$, the variance matrix $\mathrm{Var}[X]$ is positive semidefinite, i.e.,

$$\mathrm{Var}[X] \succeq 0.$$

Proof

Recall that $\mathrm{Var}[X] = \mathbb{E}[(X - \mathbb{E}[X])(X - \mathbb{E}[X])^\top]$. For any vector $v \in \mathbb{R}^{N\times 1}$, we have

$$\begin{aligned} v^\top \mathrm{Var}[X] v &= v^\top \mathbb{E}[(X - \mathbb{E}[X])(X - \mathbb{E}[X])^\top] v \\ &= \mathbb{E}[v^\top (X - \mathbb{E}[X])(X - \mathbb{E}[X])^\top v] \\ &= \mathbb{E}[(v^\top (X - \mathbb{E}[X]))^2] \geq 0, \end{aligned}$$

where the last inequality follows from the fact that the square of any real number is non-negative, and the expectation of a non-negative random variable is non-negative. Therefore, $\mathrm{Var}[X] \succeq 0$.

13.9.10 Definition: Empirical Variance Matrix

Let $X^{(0)}, \ldots, X^{(K-1)} \in \mathbb{R}^{N \times 1}$ be i.i.d. observations of a random vector X, and let $\hat{\mu}$ be the empirical mean. Then the empirical variance matrix is defined as

$$\hat{\Sigma} = \frac{1}{K-1} \sum_{k=0}^{K-1} (X^{(k)} - \hat{\mu})(X^{(k)} - \hat{\mu})^\top \in \mathbb{R}^{N \times N}.$$

13.9.10.1 Example: Scaled Dot-Product Attention in Transformers

In the **scaled dot-product attention** mechanism used in Transformers, attention scores are computed as:

$$\mathrm{Attention}(Q, K, V) = \mathrm{softmax}\left(\frac{QK^\top}{\sqrt{d_k}}\right) V,$$

where $Q, K \in \mathbb{R}^{n \times d_k}$ are query and key matrices (obtained by applying linear transformations to input embeddings, not the weight matrices themselves), and d_k is the dimension. The scaling factor $1/\sqrt{d_k}$ is crucial for maintaining stable variance.

To understand why, consider a single attention score $\mathbf{q}^\top \mathbf{k}$ where $\mathbf{q}, \mathbf{k} \in \mathbb{R}^{d_k \times 1}$ are query and key vectors. In the matrix $QK^\top$, each element $(QK^\top)_{ij}$ corresponds to the inner product between the i-th row of Q and the j-th row of K. For mathematical convenience, we represent these row vectors as column vectors $\mathbf{q}, \mathbf{k} \in \mathbb{R}^{d_k \times 1}$ (so $\mathbf{q}^\top$ is the row vector from Q). Note that $\mathbf{q}^\top \mathbf{k}$ is a scalar (inner product), which corresponds to one element of the attention score matrix $QK^\top$; this is different from $\mathbf{q}\mathbf{k}^\top$, which would be a $d_k \times d_k$ matrix (outer product). If the elements of $\mathbf{q}$ and $\mathbf{k}$ are independent random variables with mean 0 and variance σ^2, then:

$$\mathbf{q}^\top \mathbf{k} = \sum_{i=0}^{d_k - 1} q_i k_i.$$

Since q_i and k_i are independent for each i, and assuming q_i and k_j are independent for $i \neq j$, we have:

$$\begin{aligned} \mathrm{Var}[\mathbf{q}^\top \mathbf{k}] &= \mathrm{Var}\left[\sum_{i=0}^{d_k-1} q_i k_i\right] = \sum_{i=0}^{d_k-1} \mathrm{Var}[q_i k_i] \quad \text{(by independence)} \\ &= \sum_{i=0}^{d_k-1} (\mathbb{E}[q_i^2 k_i^2] - (\mathbb{E}[q_i k_i])^2) \\ &= \sum_{i=0}^{d_k-1} (\mathbb{E}[q_i^2]\mathbb{E}[k_i^2] - 0) \quad \text{(by independence, } \mathbb{E}[q_i] = \mathbb{E}[k_i] = 0) \\ &= \sum_{i=0}^{d_k-1} \sigma^2 \cdot \sigma^2 = d_k \sigma^4. \end{aligned}$$

The variance grows linearly with d_k. Without scaling, as d_k increases, the attention scores become very large, causing the softmax to saturate (producing extremely sharp distributions) and making gradients vanish during backpropagation.

By dividing by $\sqrt{d_k}$, we scale the variance:

$$\mathrm{Var}\left[\frac{\mathbf{q}^\top \mathbf{k}}{\sqrt{d_k}}\right] = \frac{1}{d_k}\mathrm{Var}[\mathbf{q}^\top \mathbf{k}] = \frac{1}{d_k} \cdot d_k \sigma^4 = \sigma^4,$$

which remains constant regardless of d_k. This ensures stable gradients and prevents the softmax from saturating, making the attention mechanism work effectively across different dimensions.

13.9.10.2 Example: Weight Initialization in Neural Networks

Proper weight initialization is crucial for training deep neural networks. The goal is to maintain stable variance of activations and gradients across layers. Consider a linear layer with ReLU activation:

$$\mathbf{y} = \mathrm{ReLU}(\mathbf{W}\mathbf{x} + \mathbf{b}),$$

where $\mathbf{W} \in \mathbb{R}^{d_{\text{out}} \times d_{\text{in}}}$ is the weight matrix, $\mathbf{x} \in \mathbb{R}^{d_{\text{in}}}$ is the input, and $\mathbf{b} \in \mathbb{R}^{d_{\text{out}}}$ is the bias. For a single output $y_i = \mathrm{ReLU}(\sum_{j=0}^{d_{\text{in}}-1} W_{ij} x_j + b_i)$.

To maintain stable variance through the network, we want $\mathrm{Var}[y_i] = \mathrm{Var}[x_j]$ (assuming inputs and outputs have the same variance). If the weights W_{ij} are initialized independently with mean 0 and variance σ_W^2, and inputs x_j are independent with mean 0 and variance σ_x^2, then:

$$\begin{aligned}
\mathrm{Var}\left[\sum_{j=0}^{d_{\text{in}}-1} W_{ij} x_j\right] &= \sum_{j=0}^{d_{\text{in}}-1} \mathrm{Var}[W_{ij} x_j] \quad \text{(by independence)} \\
&= \sum_{j=0}^{d_{\text{in}}-1} \left(\mathbb{E}[W_{ij}^2 x_j^2] - (\mathbb{E}[W_{ij} x_j])^2\right) \\
&= \sum_{j=0}^{d_{\text{in}}-1} \mathbb{E}[W_{ij}^2]\mathbb{E}[x_j^2] \quad (\text{by independence, } \mathbb{E}[W_{ij}] = \mathbb{E}[x_j] = 0) \\
&= d_{\text{in}} \cdot \sigma_W^2 \cdot \sigma_x^2.
\end{aligned}$$

For **ReLU activation**, since $\mathrm{ReLU}(z) = \max(0, z)$ and approximately half the pre-activations are positive (assuming symmetric distribution around 0), we have $\mathrm{Var}[\mathrm{ReLU}(z)] \approx \frac{1}{2}\mathrm{Var}[z]$ (the ReLU "kills" half the variance). To maintain $\mathrm{Var}[y_i] = \sigma_x^2$, we need:

$$\frac{1}{2} \cdot d_{\text{in}} \cdot \sigma_W^2 \cdot \sigma_x^2 = \sigma_x^2,$$

which gives $\sigma_W^2 = \frac{2}{d_{\text{in}}}$. This is the **He initialization** (also called Kaiming initialization), where weights are sampled from $\mathcal{N}(0, \frac{2}{d_{\text{in}}})$ or $\mathrm{Uniform}(-\sqrt{\frac{6}{d_{\text{in}}}}, \sqrt{\frac{6}{d_{\text{in}}}})$.

For **symmetric activations** like tanh (approximately variance-preserving near zero), the same forward-pass argument gives $d_{\text{in}} \cdot \sigma_W^2 = 1$, hence $\sigma_W^2 = \frac{1}{d_{\text{in}}}$. This is **LeCun initialization**: sample from $\mathcal{N}(0, \frac{1}{d_{\text{in}}})$ or $\mathrm{Uniform}(-\sqrt{\frac{3}{d_{\text{in}}}}, \sqrt{\frac{3}{d_{\text{in}}}})$. The same $\sigma_W^2 = \frac{1}{d_{\text{in}}}$ is also the usual choice for **SELU** (self-normalizing networks).

Xavier initialization (also called Glorot initialization) additionally matches gradient variance in the backward pass. Balancing d_{in} (forward) and d_{out} (backward) yields

$$\sigma_W^2 = \frac{2}{d_{\text{in}} + d_{\text{out}}},$$

with sampling from $\mathcal{N}(0, \frac{2}{d_{\text{in}}+d_{\text{out}}})$ or Uniform$(-\sqrt{\frac{6}{d_{\text{in}}+d_{\text{out}}}}, \sqrt{\frac{6}{d_{\text{in}}+d_{\text{out}}}})$. When $d_{\text{in}} = d_{\text{out}}$, Xavier reduces to LeCun.

The key insight is that proper initialization keeps activation (and gradient) variance stable across layers, preventing vanishing or exploding gradients during training.

13.9.11 Definition: Covariance

For random variables $X \in \mathbb{R}$ and $Y \in \mathbb{R}$, the covariance is defined as

$$\text{Cov}[X, Y] = \mathbb{E}[(X - \mathbb{E}[X])(Y - \mathbb{E}[Y])]. \tag{13.35}$$

13.9.12 Theorem: Covariance Formula

For random variables $X \in \mathbb{R}$ and $Y \in \mathbb{R}$:

$$\text{Cov}[X, Y] = \mathbb{E}[XY] - \mathbb{E}[X]\mathbb{E}[Y]. \tag{13.36}$$

13.9.13 Properties of Covariance

For random variables $X, Y, Z \in \mathbb{R}$ and constants $a, b, c, d \in \mathbb{R}$:

1. **Symmetry**: $\text{Cov}[X, Y] = \text{Cov}[Y, X]$
2. **Variance connection**: $\text{Cov}[X, X] = \text{Var}[X]$
3. **Bilinearity**:
   - $\text{Cov}[aX + bY, Z] = a\text{Cov}[X, Z] + b\text{Cov}[Y, Z]$
   - $\text{Cov}[X, cY + dZ] = c\text{Cov}[X, Y] + d\text{Cov}[X, Z]$
4. **Scaling**: $\text{Cov}[aX, bY] = ab\text{Cov}[X, Y]$
5. **Independence**: If X and Y are independent, then $\text{Cov}[X, Y] = 0$ (the converse is not necessarily true)
6. **Cauchy-Schwarz inequality**: $|\text{Cov}[X, Y]| \leq \sqrt{\text{Var}[X] \cdot \text{Var}[Y]}$

13.9.14 Definition: Covariance Matrix

For random vectors $X \in \mathbb{R}^{M\times 1}$ and $Y \in \mathbb{R}^{N\times 1}$, the covariance matrix is defined as

$$\text{Cov}[X, Y] = \mathbb{E}[(X - \mathbb{E}[X])(Y - \mathbb{E}[Y])^\top] \in \mathbb{R}^{M\times N}. \tag{13.37}$$

13.9.15 Definition: Correlation Coefficient

For random variables $X \in \mathbb{R}$ and $Y \in \mathbb{R}$ with $\text{Var}[X] > 0$ and $\text{Var}[Y] > 0$, the **correlation coefficient** (also called Pearson correlation coefficient) is defined as:

$$\rho_{XY} = \frac{\text{Cov}[X, Y]}{\sqrt{\text{Var}[X] \cdot \text{Var}[Y]}} = \frac{\text{Cov}[X, Y]}{\sigma_X \sigma_Y} \tag{13.38}$$

where $\sigma_X = \sqrt{\text{Var}[X]}$ and $\sigma_Y = \sqrt{\text{Var}[Y]}$ are the standard deviations of X and Y respectively.

The correlation coefficient is a normalized measure of linear dependence between X and Y, taking values in $[-1, 1]$. When $\rho_{XY} = 1$, X and Y are perfectly positively linearly related; when $\rho_{XY} = -1$, they are perfectly negatively linearly related; when $\rho_{XY} = 0$, they are uncorrelated (though not necessarily independent).

13.9.16 Definition: Empirical Covariance

Let $(X^{(0)}, Y^{(0)}), \dots, (X^{(K-1)}, Y^{(K-1)}) \in \mathbb{R} \times \mathbb{R}$ be i.i.d. observations of a pair of random variables (X, Y), and let $\hat{\mu}_X = \frac{1}{K}\sum_{k=0}^{K-1} X^{(k)}$ and $\hat{\mu}_Y = \frac{1}{K}\sum_{k=0}^{K-1} Y^{(k)}$ be their empirical means. Then the empirical covariance is defined as

$$\widehat{\text{Cov}}[X, Y] = \frac{1}{K-1}\sum_{k=0}^{K-1}(X^{(k)} - \hat{\mu}_X)(Y^{(k)} - \hat{\mu}_Y). \tag{13.39}$$

The divisor $K - 1$ (instead of K) makes this an unbiased estimator of the population covariance.

13.9.17 Definition: Empirical Correlation Coefficient

Let $(X^{(0)}, Y^{(0)}), \dots, (X^{(K-1)}, Y^{(K-1)}) \in \mathbb{R} \times \mathbb{R}$ be i.i.d. observations of a pair of random variables (X, Y), and let $\hat{\sigma}_X^2$ and $\hat{\sigma}_Y^2$ be their empirical variances. Then the empirical correlation coefficient is defined as:

$$\hat{\rho}_{XY} = \frac{\widehat{\text{Cov}}[X, Y]}{\sqrt{\hat{\sigma}_X^2 \cdot \hat{\sigma}_Y^2}} = \frac{\widehat{\text{Cov}}[X, Y]}{\hat{\sigma}_X \hat{\sigma}_Y},$$

where $\hat{\sigma}_X = \sqrt{\hat{\sigma}_X^2}$ and $\hat{\sigma}_Y = \sqrt{\hat{\sigma}_Y^2}$ are the empirical standard deviations.

13.9.18 Properties of Covariance Matrix

Let $X \in \mathbb{R}^{M\times 1}$ and $Y \in \mathbb{R}^{N\times 1}$ be random vectors, and let A and B be constant matrices of compatible size:

1. **Transpose relationship**: $\text{Cov}[Y, X] = (\text{Cov}[X, Y])^\top \in \mathbb{R}^{N\times M}$
2. **Variance matrix connection**: $\text{Cov}[X, X] = \text{Var}[X] \in \mathbb{R}^{M\times M}$ (the variance matrix)
3. **Bilinearity**:
   - $\text{Cov}[AX + b, Y] = A\text{Cov}[X, Y]$ for constant vector b of compatible size
   - $\text{Cov}[X, BY + c] = \text{Cov}[X, Y]B^\top$ for constant vector c of compatible size
4. **Positive semidefiniteness**: For the variance matrix $\text{Var}[X] = \text{Cov}[X, X]$, we have $\text{Var}[X] \succeq 0$ (see Theorem: Variance Matrix is Positive Semidefinite)

13.9.19 Definition: Empirical Covariance Matrix

Let $(X^{(0)}, Y^{(0)}), \dots, (X^{(K-1)}, Y^{(K-1)}) \in \mathbb{R}^{M\times 1} \times \mathbb{R}^{N\times 1}$ be i.i.d. observations of a pair of random vectors (X, Y). Then the empirical covariance matrix is defined as

$$\hat{\Sigma} = \frac{1}{K-1}\sum_{k=0}^{K-1}(X^{(k)} - \hat{\mu}_X)(Y^{(k)} - \hat{\mu}_Y)^\top \in \mathbb{R}^{M\times N},$$

where $\hat{\mu}_X$ and $\hat{\mu}_Y$ are the empirical means of X and Y respectively.

13.9.20 Theorem: Affine Transformations

Let $X \in \mathbb{R}^{N\times 1}$ be a random vector, $A \in \mathbb{R}^{M\times N}$ be a matrix, and $b \in \mathbb{R}^{M\times 1}$ be a vector. Then

$$\mathbb{E}[AX + b] = A\mathbb{E}[X] + b \tag{13.40}$$

$$\text{Var}[AX + b] = A\text{Var}[X]A^\top. \tag{13.41}$$

When $b = 0$, the transformation is linear (not just affine). In machine learning, the term "linear layer" or "linear transformation" is often used to refer to affine transformations $AX + b$, even when $b \neq 0$, as the bias term b is typically included in the layer parameters.

13.9.21 Definition: Estimator

An **estimator** is a function of a sample $X^{(0)}, \dots, X^{(K-1)}$ used to estimate an unknown parameter. If θ is a parameter (which may be a scalar, vector, or function), then an estimator of θ is typically denoted by $\hat{\theta}$. Before the data are observed, $\hat{\theta}$ is a random variable; after observation it yields a numerical **estimate**.

13.9.22 Definition: Bias of an Estimator

Let θ be a parameter (which may be a scalar, vector, or function) and let $\hat{\theta}$ be an estimator of θ. The **bias** of the estimator $\hat{\theta}$ is defined as:

$$\text{Bias}(\hat{\theta}) = \mathbb{E}[\hat{\theta}] - \theta.$$

An estimator is called **unbiased** if $\text{Bias}(\hat{\theta}) = 0$, i.e., if $\mathbb{E}[\hat{\theta}] = \theta$. Otherwise, it is **biased.**

For example, the empirical mean $\hat{\mu} = \frac{1}{K}\sum_{k=0}^{K-1} X^{(k)}$ is an unbiased estimator of the population mean $\mu = \mathbb{E}[X]$, since $\mathbb{E}[\hat{\mu}] = \mu$. Similarly, the empirical variance $\hat{\sigma}^2 = \frac{1}{K-1}\sum_{k=0}^{K-1}(X^{(k)} - \hat{\mu})^2$ and the empirical covariance $\widehat{\text{Cov}}[X, Y] = \frac{1}{K-1}\sum_{k=0}^{K-1}(X^{(k)} - \hat{\mu}_X)(Y^{(k)} - \hat{\mu}_Y)$ are unbiased estimators of $\sigma^2 = \text{Var}[X]$ and $\text{Cov}[X, Y]$ respectively (**the $K - 1$ divisor ensures unbiasedness**).

13.9.22.1 Example: Bias-Variance Decomposition of Mean Squared Error

The bias-variance decomposition illustrates how variance and expectation concepts apply to machine learning. Let $X \in \mathbb{R}^{d\times 1}$ be an input/feature vector, $Y \in \mathbb{R}$ be the true target value (which depends on X), and $f(X) = \mathbb{E}[Y \mid X]$ be the true regression function. Let $\hat{Y} \in \mathbb{R}$ be a prediction (which depends on training data and possibly X), and $\varepsilon = Y - f(X)$ be the irreducible error (noise) with $\mathbb{E}[\varepsilon \mid X] = 0$ and $\text{Var}[\varepsilon \mid X] = \sigma^2$.

In this context, $\hat{Y}$ is an estimator of the parameter $f(X) = \mathbb{E}[Y \mid X]$, not of the random variable Y itself. The expectation $\mathbb{E}[\cdot]$ in the decomposition is taken over both the randomness in training data (affecting $\hat{Y}$) and the randomness in Y given X. The expected mean squared error decomposes as:

$$\text{MSE} = \mathbb{E}[(\hat{Y} - Y)^2] = \text{Bias}(\hat{Y})^2 + \text{Var}[\hat{Y}] + \sigma^2,$$

where $\text{Bias}(\hat{Y}) = \mathbb{E}[\hat{Y}] - f(X) = \mathbb{E}[\hat{Y}] - \mathbb{E}[Y \mid X]$ is the bias of $\hat{Y}$ as an estimator of $f(X)$.

The three terms are:

- **Bias²**: $\text{Bias}(\hat{Y})^2 = \left(\mathbb{E}[\hat{Y}] - \mathbb{E}[Y \mid X]\right)^2$ — the squared estimator bias. The bias appears squared in the decomposition because we are decomposing the mean squared error.
- **Variance**: $\text{Var}[\hat{Y}]$ — the variability of predictions across different training sets
- **Irreducible Error**: $\sigma^2 = \text{Var}[\varepsilon]$ — the inherent noise in the data that cannot be reduced

The bias-variance decomposition is fundamental to understanding **model complexity** and **overfitting** in machine learning. **High bias** (underfitting) occurs when the model is too simple to capture the underlying patterns (e.g., linear models for nonlinear data). **High variance** (overfitting) occurs when the model is too complex and memorizes training data (e.g., deep trees with many leaves). The **bias-variance trade-off** guides model selection: simpler models reduce variance but increase bias, while complex models reduce bias but increase variance.

Proof

We have $Y = f(X) + \varepsilon$ where $f(X) = \mathbb{E}[Y \mid X]$ and $\mathbb{E}[\varepsilon \mid X] = 0$, $\text{Var}[\varepsilon \mid X] = \sigma^2$. Then:

$$\begin{aligned}
\mathbb{E}[(\hat{Y} - Y)^2] &= \mathbb{E}[(\hat{Y} - f(X) + f(X) - Y)^2] \\
&= \mathbb{E}[(\hat{Y} - f(X))^2 + 2(\hat{Y} - f(X))(f(X) - Y) + (f(X) - Y)^2] \\
&= \mathbb{E}[(\hat{Y} - f(X))^2] + 2\mathbb{E}[(\hat{Y} - f(X))(f(X) - Y)] + \mathbb{E}[(f(X) - Y)^2].
\end{aligned}$$

The middle term simplifies:

$$\begin{aligned}
\mathbb{E}[(\hat{Y} - f(X))(f(X) - Y)] &= \mathbb{E}[(\hat{Y} - f(X))(-\varepsilon)] \\
&= -\mathbb{E}[(\hat{Y} - f(X))\varepsilon].
\end{aligned}$$

If $\hat{Y}$ is constructed from training data that is independent of the test point's noise ε (which is standard in supervised learning), then $\hat{Y}$ and ε are independent, so $\mathbb{E}[(\hat{Y} - f(X))\varepsilon] = \mathbb{E}[\hat{Y} - f(X)] \cdot \mathbb{E}[\varepsilon] = \mathbb{E}[\hat{Y} - f(X)] \cdot 0 = 0$. Therefore, the middle term is zero.

For the first term:

$$\begin{aligned}
\mathbb{E}[(\hat{Y} - f(X))^2] &= \mathbb{E}[(\hat{Y} - \mathbb{E}[\hat{Y}] + \mathbb{E}[\hat{Y}] - f(X))^2] \\
&= \mathbb{E}[(\hat{Y} - \mathbb{E}[\hat{Y}])^2] + 2\mathbb{E}[(\hat{Y} - \mathbb{E}[\hat{Y}])(\mathbb{E}[\hat{Y}] - f(X))] + \mathbb{E}[(\mathbb{E}[\hat{Y}] - f(X))^2] \\
&= \text{Var}[\hat{Y}] + 0 + \text{Bias}(\hat{Y})^2,
\end{aligned}$$

where the cross term is zero because $\mathbb{E}[\hat{Y} - \mathbb{E}[\hat{Y}]] = 0$.

For the third term:

$$\mathbb{E}[(f(X) - Y)^2] = \mathbb{E}[\varepsilon^2] = \text{Var}[\varepsilon] = \sigma^2,$$

since $\mathbb{E}[\varepsilon] = 0$.

Combining all terms:

$$\mathbb{E}[(\hat{Y} - Y)^2] = \text{Bias}(\hat{Y})^2 + \text{Var}[\hat{Y}] + \sigma^2.$$

In **ensemble methods** like bagging, averaging multiple models reduces variance while maintaining low bias. In **regularization** (e.g., L2 regularization in neural networks), we explicitly trade bias for variance by constraining model complexity. The irreducible error σ^2 represents the fundamental limit of prediction accuracy, independent of the model choice.

13.9.23 Theorem: Hoeffding's Inequality

Let $X^{(0)}, \dots, X^{(N-1)}$ be independent random variables wherein $X^{(n)}$ is supported on $[a_n, b_n]$. Then for any $\epsilon > 0$,

$$P(\hat{\mu} - \mathbb{E}[\hat{\mu}] \geq \epsilon) \leq \exp\left(-\frac{2N^2\epsilon^2}{\sum_{n=0}^{N-1}(b_n - a_n)^2}\right),$$

$$P(\hat{\mu} - \mathbb{E}[\hat{\mu}] \leq -\epsilon) \leq \exp\left(-\frac{2N^2\epsilon^2}{\sum_{n=0}^{N-1}(b_n - a_n)^2}\right).$$

In the special case that $X^{(0)}, \dots, X^{(N-1)}$ are supported on $[a, b]$, we have

$$P(\hat{\mu} - \mathbb{E}[\hat{\mu}] \geq \epsilon) \leq \exp\left(-\frac{2N\epsilon^2}{(b-a)^2}\right),$$

$$P(\hat{\mu} - \mathbb{E}[\hat{\mu}] \leq -\epsilon) \leq \exp\left(-\frac{2N\epsilon^2}{(b-a)^2}\right).$$

The concentration inequalities presented here (Hoeffding's inequality and its variants) are fundamental to many algorithms in reinforcement learning. The **Upper Confidence Bound (UCB)** algorithm for multi-armed bandits uses Hoeffding's inequality to construct confidence intervals for the expected rewards of each arm. The UCB algorithm selects the arm with the highest upper confidence bound, balancing exploration (trying uncertain arms) and exploitation (choosing the best-known arm). The confidence bound formula $Q(a) + c\sqrt{\frac{\ln t}{N(a)}}$ directly relies on concentration inequalities to guarantee that the true mean lies within the confidence interval with high probability.

13.9.23.1 Example: Application of Hoeffding's Inequality

Consider a sample space Ω with a probability measure defined on it. Consider a function $f : \Omega \to [a, b]$. Let $X^{(0)}, \dots, X^{(N-1)}$ be independently and randomly drawn from Ω. Consider any subspace $A \subseteq \Omega$ and any $\epsilon > 0$.

Use Hoeffding's inequality to bound:

1.
$$P\left(\left|\frac{1}{N}\sum_{n=0}^{N-1} \mathbf{1}\{X^{(n)} \in A\} - P(A)\right| > \epsilon\right).$$

2.
$$P\left(\left|\frac{1}{N}\sum_{n=0}^{N-1} f(X^{(n)})\mathbf{1}\{X^{(n)} \in A\} - \mathbb{E}[f(X)\mathbf{1}\{X \in A\}]\right| > \epsilon\right).$$

Solution

1. We notice that
$$\mathbb{E}[\mathbf{1}\{X^{(n)} \in A\}] = P(A)$$
and $\mathbf{1}\{X^{(n)} \in A\} \in \{0, 1\}$.

Therefore, following from Hoeffding's inequality, we get

$$P\left(\left|\frac{1}{N}\sum_{n=0}^{N-1} \mathbf{1}\{X^{(n)} \in A\} - P(A)\right| > \epsilon\right) \leq 2\exp(-2N\epsilon^2).$$

2. Let $Z^{(n)} = f(X^{(n)})\mathbf{1}\{X^{(n)} \in A\}$. Although f takes values in $[a, b]$, the factor $\mathbf{1}\{X^{(n)} \in A\}$ forces $Z^{(n)} = 0$ whenever $X^{(n)} \notin A$, so $Z^{(n)}$ is supported on $[\min(a, 0), \max(b, 0)]$, not necessarily on $[a, b]$. (If $f \equiv 1$, then $[a, b] = [1, 1]$ but $Z^{(n)} \in \{0, 1\}$.) Hoeffding's inequality therefore gives

$$P\left(\left|\frac{1}{N}\sum_{n=0}^{N-1} Z^{(n)} - \mathbb{E}[Z]\right| > \epsilon\right) \leq 2\exp\left(-\frac{2N\epsilon^2}{(\max(b, 0) - \min(a, 0))^2}\right),$$

where $Z = f(X)\mathbf{1}\{X \in A\}$.

Similar principles apply to other RL algorithms like Thompson Sampling, which uses Bayesian posterior distributions, and epsilon-greedy methods that use probability bounds for exploration strategies.

13.10 Transformation of Random Variables

When we apply a function to a random variable, we obtain a new random variable with a different distribution. The **change of variables formula** (also called the **transformation of density formula**) allows us to compute the probability density function of the transformed random variable from the original density.

13.10.1 Theorem: Change of Variables

Let $\mathbf{X}$ be a continuous random variable (one-dimensional) or random vector (multidimensional) with probability density function $f_{\mathbf{X}}(\mathbf{x})$. Let $\mathbf{Y} = g(\mathbf{X})$ where g is a differentiable, invertible transformation with inverse g^{-1}.

- **One-dimensional case**: $X \in \mathbb{R}$, $Y \in \mathbb{R}$, and $g : \mathbb{R} \to \mathbb{R}$ is strictly monotonic.
- **Multidimensional case**: $\mathbf{X} \in \mathbb{R}^{d\times 1}$, $\mathbf{Y} \in \mathbb{R}^{d\times 1}$, and $g : \mathbb{R}^d \to \mathbb{R}^d$ with Jacobian matrix $\mathbf{J}_{g^{-1}}(\mathbf{y}) = \frac{\partial g^{-1}}{\partial \mathbf{y}} \in \mathbb{R}^{d\times d}$.

Then the probability density function of $\mathbf{Y}$ is:

$$f_{\mathbf{Y}}(\mathbf{y}) = f_{\mathbf{X}}(g^{-1}(\mathbf{y})) \cdot |\det(\mathbf{J}_{g^{-1}}(\mathbf{y}))|. \tag{13.42}$$

For the one-dimensional case, the Jacobian determinant reduces to the absolute value of the derivative: $|\det(\mathbf{J}_{g^{-1}}(y))| = \left|\frac{d}{dy}g^{-1}(y)\right|$, giving:

$$f_Y(y) = f_X(g^{-1}(y)) \cdot \left|\frac{d}{dy}g^{-1}(y)\right|. \tag{13.43}$$

The factor $|\det(\mathbf{J}_{g^{-1}}(\mathbf{y}))|$ (or $\left|\frac{d}{dy}g^{-1}(y)\right|$ in one dimension) accounts for how the transformation stretches, compresses, and rotates (in multidimensions) the probability density. It measures the local volume change: if the transformation expands a region, the determinant is greater than 1 and the density decreases; if it contracts, the determinant is less than 1 and the density increases.

Proof Sketch: The formula follows from the substitution rule for integrals. For an event A in the space of $\mathbf{Y}$, we have $P(\mathbf{Y} \in A) = P(g(\mathbf{X}) \in A) = P(\mathbf{X} \in g^{-1}(A)) = \int_{g^{-1}(A)} f_{\mathbf{X}}(\mathbf{x})d\mathbf{x}$. Making the substitution $\mathbf{x} = g^{-1}(\mathbf{y})$ and using the Jacobian determinant gives $P(\mathbf{Y} \in A) = \int_A f_{\mathbf{X}}(g^{-1}(\mathbf{y}))|\det(\mathbf{J}_{g^{-1}}(\mathbf{y}))|d\mathbf{y}$, from which the formula follows. In one dimension, for a strictly increasing function g, we have $P(Y \leq y) = P(g(X) \leq y) = P(X \leq g^{-1}(y))$. Differentiating both sides with respect to y gives $f_Y(y) = f_X(g^{-1}(y)) \cdot \frac{d}{dy}g^{-1}(y)$. For a strictly decreasing function, we get a negative derivative, so we take the absolute value to ensure the density is non-negative.

The change of variables formula is fundamental in probability theory and machine learning. In **normalizing flows**, we use invertible transformations to map a simple base distribution (e.g., standard Gaussian) to a complex target distribution. Each transformation layer applies a change of variables, and the overall density is computed by chaining these transformations, with the Jacobian determinant tracking how each layer modifies the density. The requirement that g be invertible (strictly monotonic in one dimension) ensures that the inverse g^{-1} exists and is unique. For non-monotonic functions in one dimension, we need to partition the domain into regions where the function is monotonic and sum over all preimages. In normalizing flows, the log-determinant of the Jacobian is computed efficiently for each transformation layer, and the overall log-likelihood includes the sum of these log-determinants. This is crucial for training flow-based generative models, where we need to evaluate the density of generated samples.

13.10.2 Example: Change of Variables

13.10.2.1 One-Dimensional Case: Uniform to Exponential

Let $X \sim \mathcal{U}(0,1)$ be uniformly distributed on $[0,1]$, and consider the transformation $Y = -\log(X)$ (which is strictly decreasing on $(0,1)$).

The inverse transformation is $X = g^{-1}(Y) = e^{-Y}$, and its derivative is $\frac{d}{dy}g^{-1}(y) = -e^{-y}$. Since g is strictly decreasing, we take the absolute value: $\left|\frac{d}{dy}g^{-1}(y)\right| = e^{-y}$.

The PDF of X is $f_X(x) = 1$ for $x \in [0,1]$ and 0 otherwise. Since $X \in (0,1)$ implies $Y \in (0,\infty)$, applying the change of variables formula:

$$f_Y(y) = f_X(e^{-y}) \cdot e^{-y} = 1 \cdot e^{-y} = e^{-y}, \quad y > 0.$$

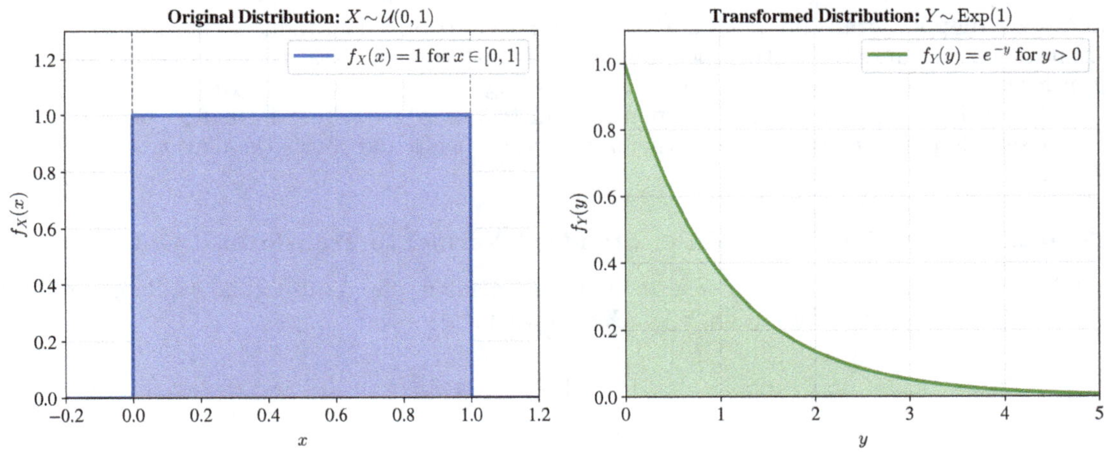

Figure 13.2: Original and Transformed Distributions

This confirms that $Y \sim \mathrm{Exp}(1)$ (exponential distribution with rate parameter 1). The factor e^{-y} accounts for how the logarithmic transformation compresses the uniform distribution: regions near 0 in X (high density) map to large values in Y (low density after transformation). The transformation is strictly decreasing, so small values of X (near 0) map to large values of Y (near $+\infty$), while values of X near 1 map to values of Y near 0. The derivative factor e^{-y} ensures that the probability mass is properly redistributed: when the transformation compresses a region (mapping a small interval in X to a large interval in Y), the density decreases proportionally.

Figure 13.2 compares the original uniform distribution $f_X(x) = 1$ for $x \in [0,1]$ (left) with the transformed exponential distribution $f_Y(y) = e^{-y}$ for $y > 0$ (right). The left panel shows the constant density of the uniform distribution on $[0, 1]$, while the right panel shows the decaying exponential density. This visual comparison illustrates how the transformation redistributes probability mass: the uniform distribution's constant density becomes a decaying exponential, with higher density near $y = 0$ (corresponding to x near 1) and lower density for large y (corresponding to x near 0).

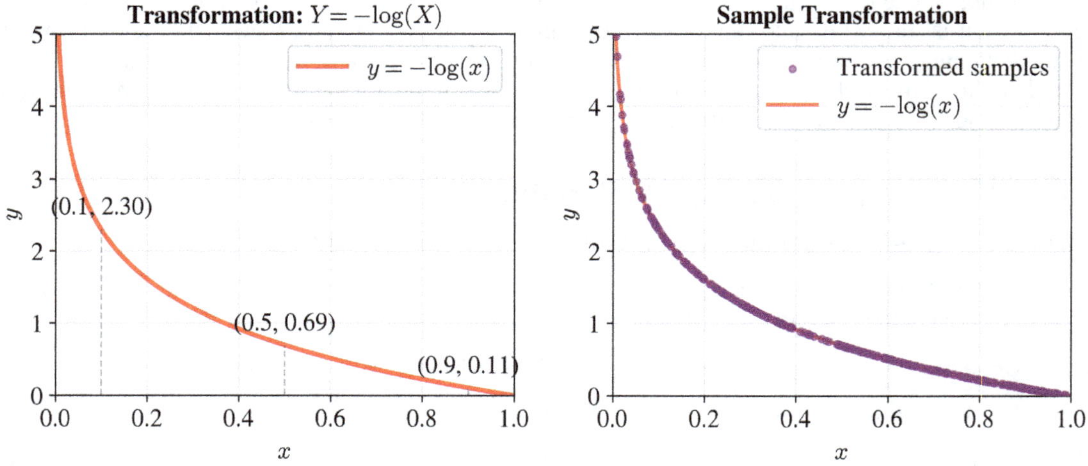

Figure 13.3: Transformation Mapping and Samples

Figure 13.3 illustrates the transformation process in detail. The left panel shows the strictly decreasing transformation $y = -\log(x)$ with key mapping points annotated, demonstrating how values are mapped from the domain $[0, 1]$ to the range $(0, \infty)$. The right panel displays sample points from the uniform distribution and their transformed values, clearly showing how the transformation compresses density: small x values (high density in the uniform distribution) map to large y values (low density in the exponential distribution). The transformation curve overlaid on the sample points confirms the mathematical relationship and shows the density compression effect.

13.10.2.2 Multidimensional Case: Standard Normal to Transformed Gaussian

Let $\mathbf{X} = [X_1, X_2]^\top$ where $X_1, X_2 \sim \mathcal{N}(0, 1)$ are independent standard normal random variables. Consider the linear transformation $\mathbf{Y} = A\mathbf{X} + \mathbf{b}$ where:

$$A = \begin{bmatrix} 2 & 1 \\ 0 & 1 \end{bmatrix}, \quad \mathbf{b} = \begin{bmatrix} 1 \\ 2 \end{bmatrix}.$$

The inverse transformation is $\mathbf{X} = A^{-1}(\mathbf{Y} - \mathbf{b})$, where:

$$A^{-1} = \begin{bmatrix} \frac{1}{2} & -\frac{1}{2} \\ 0 & 1 \end{bmatrix}.$$

The Jacobian matrix is $\mathbf{J}_{g^{-1}}(\mathbf{y}) = A^{-1}$, and its determinant is $\det(A^{-1}) = \frac{1}{2} \cdot 1 - 0 \cdot (-\frac{1}{2}) = \frac{1}{2}$.

Since X_1, X_2 are independent, the joint PDF is $f_{\mathbf{X}}(\mathbf{x}) = \frac{1}{2\pi} \exp\left(-\frac{1}{2}(x_1^2 + x_2^2)\right)$. Applying the change of variables formula:

$$f_{\mathbf{Y}}(\mathbf{y}) = f_{\mathbf{X}}(A^{-1}(\mathbf{y} - \mathbf{b})) \cdot |\det(A^{-1})| = \frac{1}{2\pi} \exp\left(-\frac{1}{2}\|A^{-1}(\mathbf{y} - \mathbf{b})\|^2\right) \cdot \frac{1}{2}.$$

The factor $\frac{1}{2}$ accounts for the volume change: the transformation A has determinant 2, so it expands the volume by a factor of 2, and the density decreases accordingly. The mean shifts from $\mathbf{0}$ to $\mathbf{b} = [1, 2]^\top$, and the covariance changes from $\mathbf{I}$ to $AA^\top = \begin{bmatrix} 5 & 1 \\ 1 & 1 \end{bmatrix}$. The Jacobian determinant $|\det(A^{-1})| = \frac{1}{2}$ ensures that probability mass is conserved: when the transformation expands volume (determinant > 1), the density must decrease proportionally. This example demonstrates how linear transformations of Gaussian random variables preserve the Gaussian form but change the mean and covariance, which is fundamental to understanding multivariate Gaussian distributions and their applications in machine learning.

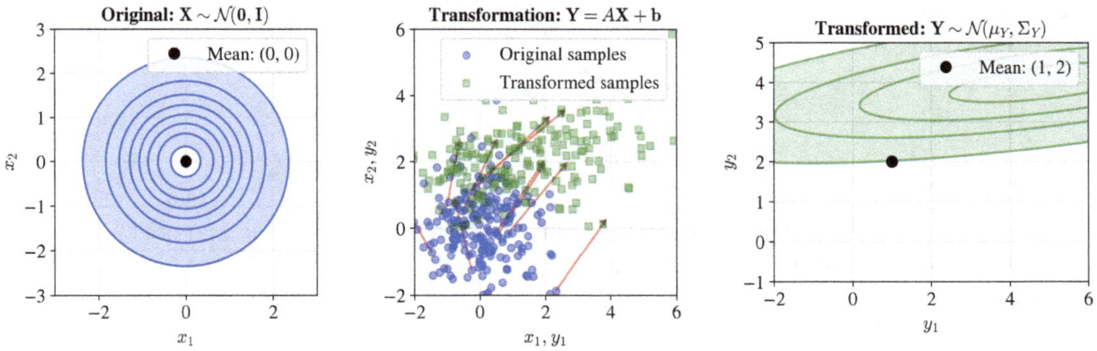

Figure 13.4: Multidimensional Change of Variables

Figure 13.4 demonstrates the multidimensional change of variables formula through a linear transformation of a standard bivariate normal distribution. The left panel shows the original standard normal distribution $\mathbf{X} \sim \mathcal{N}(\mathbf{0}, \mathbf{I})$ with circular contours indicating independent components and equal variance in all directions. The middle panel displays sample points from both distributions with arrows showing the transformation mapping, along with the transformation matrix A and vector $\mathbf{b}$ annotated. The right panel shows the transformed Gaussian distribution with elliptical contours, a shifted mean at $(1, 2)$, and rotated/scaled covariance. The elliptical contours illustrate how the transformation rotates and scales the distribution, while the Jacobian determinant factor $\frac{1}{2}$ accounts for the volume change, ensuring that probability mass is properly conserved despite the expansion.

13.10.3 Application: Normalizing Flows

Normalizing flows[1] provide a concrete application of the change of variables formula in generative modeling. They represent a complex data distribution by mapping observations to a simple base distribution through a sequence of invertible transformations.

13.10.3.1 Definition: Normalizing Flow

A **normalizing flow** is a generative model that learns a complex distribution on data $\mathbf{x}$ by an invertible map to a latent variable $\mathbf{z}$. The **base density** q_Z is a simple PDF on this latent space, typically the standard Gaussian $\mathcal{N}(\mathbf{0}, \mathbf{I})$.

[1]Rezende, D., & Mohamed, S. (2015). Variational inference with normalizing flows. In *International Conference on Machine Learning* (ICML), pp. 1530–1538. https://arxiv.org/abs/1505.05770. See also Dinh, L., Sohl-Dickstein, J., & Bengio, S. (2016). Density estimation using Real NVP. In *International Conference on Learning Representations* (ICLR 2017). https://arxiv.org/abs/1605.08803.

Let $f = f_L \circ f_{L-1} \circ \cdots \circ f_1$ be the composition of L invertible transformations. The symbol $\circ$ denotes function composition: apply the right-hand map first, then the left-hand map, so $(g \circ h)(\mathbf{u}) = g(h(\mathbf{u}))$. Thus $f(\mathbf{x}) = f_L(f_{L-1}(\cdots f_1(\mathbf{x}) \cdots))$.

The **forward** map sends data to latent space by setting $\mathbf{z}_0 = \mathbf{x}$ and applying the layers in order,

$$\mathbf{z}_0 = \mathbf{x}, \quad \mathbf{z}_1 = f_1(\mathbf{z}_0), \quad \mathbf{z}_2 = f_2(\mathbf{z}_1), \quad \ldots, \quad \mathbf{z}_L = f_L(\mathbf{z}_{L-1}), \tag{13.44}$$

and writing $\mathbf{z}$ for the final latent $\mathbf{z}_L$, so $\mathbf{z} = f(\mathbf{x})$. The **inverse** f^{-1} generates data from the base: sample $\mathbf{z} \sim q_Z$ and set $\mathbf{x}' = f^{-1}(\mathbf{z})$.

The data density follows from the **change of variables formula**:

$$q_X(\mathbf{x}) = q_Z(f(\mathbf{x})) \prod_{\ell=1}^{L} \left| \det \left(\frac{\partial f_\ell}{\partial \mathbf{z}_{\ell-1}} \right) \right|. \tag{13.45}$$

Key properties:

- **Invertibility**: Each transformation f_ℓ must be bijective (one-to-one and onto)
- **Tractable Jacobian**: The Jacobian determinant $\det(\partial f_\ell / \partial \mathbf{z}_{\ell-1})$ must be computable efficiently
- **Composition**: The overall transformation $f = f_L \circ f_{L-1} \circ \cdots \circ f_1$ is also invertible

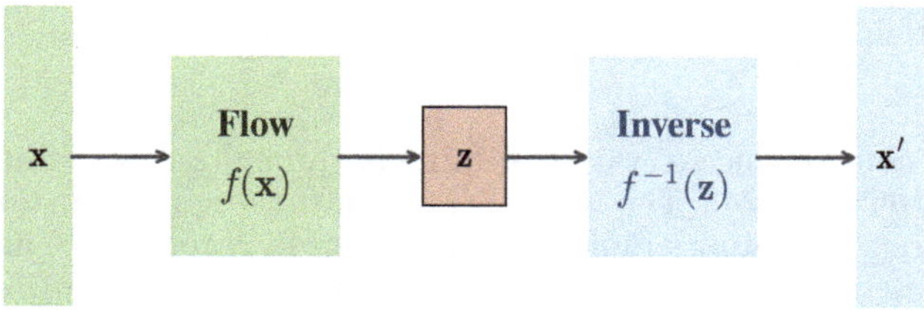

Figure 13.5: Normalizing Flow Architecture

Figure 13.5 illustrates the normalizing flow architecture. The input $\mathbf{x}$ (green rectangle) flows through the forward transformation $f(\mathbf{x})$ (green square) to the latent variable $\mathbf{z}$ (red square). The inverse transformation $f^{-1}(\mathbf{z})$ (blue square) maps $\mathbf{z}$ back to the output $\mathbf{x}'$ (blue rectangle). The forward pass enables exact density evaluation via change of variables; the inverse pass samples new data from the base distribution.

13.10.3.2 Normalizing Flows for Variational Inference

In variational inference (see Chapter 16), normalizing flows are used to create an **expressive variational family** $\mathcal{Q}$ that can approximate complex posteriors $p(\mathbf{z} \mid \mathbf{x})$.

Flow-based variational posterior: An $\mathbf{x}$-conditioned flow $f = f_L \circ \cdots \circ f_1$ maps $\mathbf{z}_0 \sim q_0$ to $\mathbf{z} = f(\mathbf{z}_0; \mathbf{x}, \phi)$, with density

$$q(\mathbf{z} \mid \mathbf{x}, \phi) = q_0(f^{-1}(\mathbf{z}; \mathbf{x}, \phi)) \prod_{\ell=1}^{L} \left| \det \left(\frac{\partial f_\ell^{-1}}{\partial \mathbf{z}_\ell} \right) \right|. \tag{13.46}$$

Training: The **Evidence Lower Bound (ELBO)** (see Chapter 14, Section~14.11) is maximized with respect to both the flow parameters ϕ and the generative model parameters θ:

$$\mathcal{L}(\theta, \phi) = \mathbb{E}_{q(\mathbf{z}|\mathbf{x},\phi)}[\log p(\mathbf{x} \mid \mathbf{z}, \theta)] - D_{\mathrm{KL}}(q(\mathbf{z} \mid \mathbf{x}, \phi) || p(\mathbf{z})) \tag{13.47}$$

The KL divergence term can be computed exactly because $q(\mathbf{z} \mid \mathbf{x}, \phi)$ has a tractable density.

13.10.3.3 Advantages of Normalizing Flows

1. **Expressiveness**: By composing many invertible transformations, normalizing flows can represent arbitrarily complex distributions (universal approximation property). The variational family $\mathcal{Q}$ can be made arbitrarily large.
2. **Exact density evaluation**: Unlike GANs (which have no explicit likelihood), normalizing flows provide exact likelihood computation, enabling:
   - Exact KL divergence computation in ELBO (see Chapter 14, Section~14.11)
   - Model comparison via log-likelihood
   - Uncertainty quantification
3. **No approximation gap (in limit)**: As the flow becomes more expressive, the optimal $q^* \in \mathcal{Q}$ can get arbitrarily close to the true posterior $p(\mathbf{z} \mid \mathbf{x})$. In the limit of infinite capacity, $q^* = p(\mathbf{z} \mid \mathbf{x})$, eliminating the approximation gap present in standard VI.
4. **Bi-directional**: The invertibility enables both:
   - **Forward pass**: Map data $\mathbf{x}$ to latent $\mathbf{z} = f(\mathbf{x})$ for exact density evaluation
   - **Inverse pass**: Sample $\mathbf{z} \sim q_Z$ and set $\mathbf{x}' = f^{-1}(\mathbf{z})$ to generate data

Normalizing flows demonstrate how the **change of variables formula** enables exact density computation for complex transformations. The key mathematical requirement is that transformations must be invertible with tractable Jacobian determinants, which constrains the design of flow architectures but enables exact likelihood computation.

> Normalizing flows illustrate how the change of variables formula enables exact density computation for complex transformations. The mathematical requirement of invertibility with tractable Jacobians connects flows to the matrix calculus concepts in Chapter 11. Flows demonstrate that by composing invertible transformations, we can create expressive variational families while maintaining tractability. **Continuous Normalizing Flows (CNFs)** extend this idea using ODEs (see Chapter 20), showing how continuous-time dynamics can define infinite-depth flows.

13.10.4 Example: Linear Transformation

Let $\mathbf{X} \in \mathbb{R}^{d\times 1}$ be a random vector with PDF $f_{\mathbf{X}}(\mathbf{x})$, and let $\mathbf{Y} = A\mathbf{X} + \mathbf{b}$ where $A \in \mathbb{R}^{d\times d}$ is an invertible matrix and $\mathbf{b} \in \mathbb{R}^{d\times 1}$ is a vector.

The inverse transformation is $\mathbf{X} = A^{-1}(\mathbf{Y} - \mathbf{b})$, and the Jacobian matrix is $\mathbf{J}_{g^{-1}}(\mathbf{y}) = A^{-1}$. Therefore, the PDF of $\mathbf{Y}$ is:

$$f_{\mathbf{Y}}(\mathbf{y}) = f_{\mathbf{X}}(A^{-1}(\mathbf{y} - \mathbf{b})) \cdot |\det(A^{-1})| = f_{\mathbf{X}}(A^{-1}(\mathbf{y} - \mathbf{b})) \cdot \frac{1}{|\det(A)|}.$$

This formula is used in the derivation of the multivariate Gaussian PDF: starting with independent standard normal random variables $\mathbf{Z} \sim \mathcal{N}(\mathbf{0}, I_d)$ and applying the linear transformation $\mathbf{X} = A\mathbf{Z} + \mu$ where $\Sigma = AA^\top$, we obtain $\mathbf{X} \sim \mathcal{N}(\mu, \Sigma)$.

13.10.5 Example: Standardization

Let $X \sim \mathcal{N}(\mu, \sigma^2)$ be a univariate Gaussian random variable. The standardization transformation is $Z = \frac{X-\mu}{\sigma}$.

The inverse transformation is $X = \sigma Z + \mu$, and $\frac{d}{dz}(\sigma z + \mu) = \sigma$. Therefore, using the change of variables formula:

$$f_Z(z) = f_X(\sigma z + \mu) \cdot \sigma = \frac{1}{\sqrt{2\pi\sigma^2}} \exp\left(-\frac{(\sigma z + \mu - \mu)^2}{2\sigma^2}\right) \cdot \sigma = \frac{1}{\sqrt{2\pi}} \exp\left(-\frac{z^2}{2}\right),$$

which confirms that $Z \sim \mathcal{N}(0,1)$.

13.10.6 Definition: Whitening Transformations

Standardization (Example~13.10.5) rescales a **scalar** to unit variance. For a random vector $X \in \mathbb{R}^d$ with mean μ and positive definite covariance $\Sigma = \mathrm{Cov}[X]$, a **whitening** map produces $Z = W(X - \mu)$ with

$$\mathrm{Cov}[Z] = I_d.$$

Any matrix square root of Σ^{-1} works: if $\Sigma = U\Lambda U^\top$ is the spectral decomposition (Chapter~7), then $\Sigma^{-1/2} = U\Lambda^{-1/2}U^\top$ satisfies $\Sigma^{-1/2}\Sigma(\Sigma^{-1/2})^\top = I$.

13.10.7 Definition: PCA Whitening vs ZCA Whitening

Two common choices of W are:

$$W_{\text{PCA}} = \Lambda^{-1/2}U^\top, \qquad W_{\text{ZCA}} = \Sigma^{-1/2} = U\Lambda^{-1/2}U^\top. \tag{13.48}$$

- **PCA whitening** first rotates into the eigenbasis of Σ, then rescales; the coordinates of Z are aligned with principal axes.
- **ZCA whitening** (zero-phase component analysis) uses the symmetric square root $\Sigma^{-1/2}$. Among linear whitening maps, it minimizes $\|WX - X\|_F$ in an average sense: whitened vectors stay as close as possible (in squared error) to the original centered vectors, while enforcing $\mathrm{Cov}[Z] = I$.

Both remove second-order correlations. They differ in how they orient the whitened coordinates relative to the original axes.

13.10.8 Example: 2 × 2 ZCA Whitening

Let $\mu = 0$ and

$$\Sigma = \begin{bmatrix} 2 & 1 \\ 1 & 2 \end{bmatrix} = U\Lambda U^\top, \quad \Lambda = \mathrm{diag}(3,1), \quad U = \frac{1}{\sqrt{2}}\begin{bmatrix} 1 & -1 \\ 1 & 1 \end{bmatrix}.$$

Then

$$W_{\text{ZCA}} = U\Lambda^{-1/2}U^\top = \frac{1}{\sqrt{2}}U\,\mathrm{diag}(3^{-1/2},1)\,U^\top.$$

For any sample x, $z = W_{\text{ZCA}}x$ has empirical covariance approaching I as the sample size grows, while remaining a rotated-and-scaled version that prefers alignment with x. Estimate Σ on training data only; adding a ridge $\Sigma + \varepsilon I$ before inversion stabilizes small eigenvalues.

13.11 Uniform Distribution

13.11.1 Definition: Discrete Uniform Distribution

A random variable X follows a **discrete uniform distribution** over the set $\{a, a+1, \ldots, b-1, b\}$ where $a, b \in \mathbb{Z}$ and $a < b$, denoted as $X \sim \mathrm{Uniform}(a,b)$ or $X \sim U(a,b)$, if its probability mass function is

$$P(X = x) = \frac{1}{b - a + 1}, \quad \text{for } x \in \{a, a + 1, \dots, b\}.$$

The mean and variance are:

$$\mathbb{E}[X] = \frac{a+b}{2}, \quad \text{Var}[X] = \frac{(b - a + 1)^2 - 1}{12}.$$

13.11.2 Definition: Continuous Uniform Distribution

A random variable X follows a **continuous uniform distribution** on the interval $[a, b]$ where $a, b \in \mathbb{R}$ and $a < b$, denoted as $X \sim \text{Uniform}(a, b)$ or $X \sim U(a, b)$, if its probability density function is

$$f(x) = \begin{cases} \frac{1}{b-a} & \text{if } a \leq x \leq b, \\ 0 & \text{otherwise.} \end{cases}$$

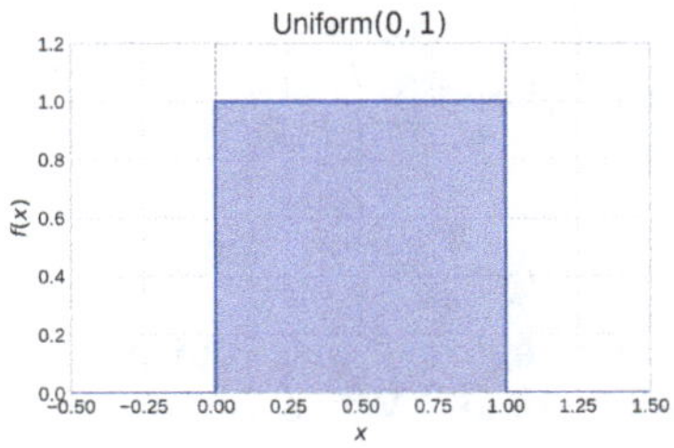

Figure 13.6: Uniform Distribution

The mean and variance are:

$$\mathbb{E}[X] = \frac{a+b}{2}, \quad \text{Var}[X] = \frac{(b-a)^2}{12}.$$

The cumulative distribution function is:

$$P(X \leq x) = \begin{cases} 0 & \text{if } x < a, \\ \frac{x-a}{b-a} & \text{if } a \leq x \leq b, \\ 1 & \text{if } x > b. \end{cases}$$

The uniform distribution is fundamental in probability theory and serves as the basis for generating random samples from other distributions. Among all distributions defined on the same interval (for continuous) or the same set (for discrete), the uniform distribution has the maximum entropy (see Chapter~14 for details on entropy).

13.12 Bernoulli Distribution

13.12.1 Definition: Bernoulli Distribution

A random variable X follows a **Bernoulli distribution** with parameter $p \in [0, 1]$, denoted as $X \sim \text{Bernoulli}(p)$ or $X \sim \text{Ber}(p)$, if it takes values in $\{0, 1\}$ with probability mass function:

$$P(X = x) = \begin{cases} p & \text{if } x = 1, \\ 1 - p & \text{if } x = 0. \end{cases}$$

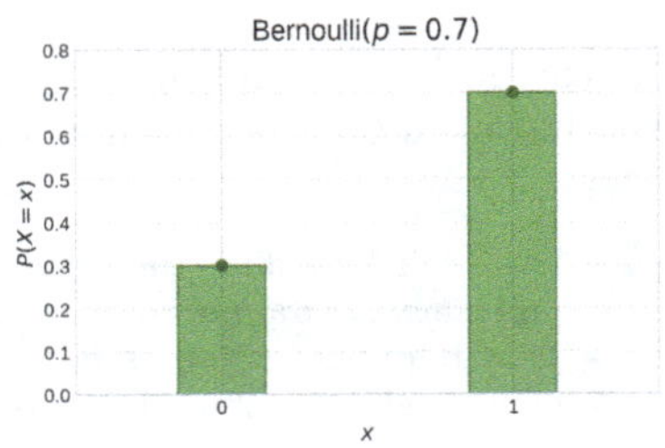

Figure 13.7: Bernoulli Distribution

Equivalently, $P(X = x) = p^x(1 - p)^{1-x}$ for $x \in \{0, 1\}$.

The mean and variance are:

$$\mathbb{E}[X] = p, \quad \text{Var}[X] = p(1 - p).$$

The Bernoulli distribution is fundamental in machine learning, particularly for **binary classification** problems. In logistic regression and neural networks with binary outputs, the output layer often uses a sigmoid activation function to model $P(Y = 1 \mid X)$, which follows a Bernoulli distribution. The parameter p represents the probability of "success" (typically $X = 1$), and $1-p$ represents the probability of "failure" ($X = 0$).

13.13 Categorical and Multinomial Distributions

13.13.1 Definition: Categorical Distribution

A random variable X follows a **categorical distribution** (also called **multinoulli distribution**) over K categories $\{0, 1, \dots, K-1\}$ with probability vector $\mathbf{p} = (p_0, p_1, \dots, p_{K-1})^\top$ where $p_k \geq 0$ for all k and $\sum_{k=0}^{K-1} p_k = 1$, denoted as $X \sim \text{Categorical}(\mathbf{p})$, if its probability mass function is:

$$P(X = k) = p_k, \quad \text{for } k \in \{0, 1, \dots, K-1\}.$$

The categorical distribution is a generalization of the Bernoulli distribution from 2 outcomes to K outcomes. When $K = 2$, the categorical distribution reduces to the Bernoulli distribution.

The mean vector and covariance matrix are:

$$\mathbb{E}[X] = \mathbf{p}, \quad \text{Var}[X] = \text{diag}(\mathbf{p}) - \mathbf{p}\mathbf{p}^\top,$$

where X is represented as a one-hot vector in $\mathbb{R}^{K \times 1}$ (i.e., $X_k = 1$ if $X = k$, and $X_k = 0$ otherwise).

13.13.2 Definition: Multinomial Distribution

A random vector $\mathbf{X} = (X_0, X_1, \dots, X_{K-1})^\top$ follows a **multinomial distribution** with n trials and probability vector $\mathbf{p} = (p_0, p_1, \dots, p_{K-1})^\top$ where $p_k \geq 0$ for all k and $\sum_{k=0}^{K-1} p_k = 1$, denoted as $\mathbf{X} \sim \text{Multinomial}(n, \mathbf{p})$, if its probability mass function is:

$$P(X_0 = x_0, X_1 = x_1, \dots, X_{K-1} = x_{K-1}) = \frac{n!}{x_0! x_1! \cdots x_{K-1}!} p_0^{x_0} p_1^{x_1} \cdots p_{K-1}^{x_{K-1}},$$

where $x_k \in \{0, 1, \dots, n\}$ for all k and $\sum_{k=0}^{K-1} x_k = n$. Each X_k represents the count of outcomes in category k after n independent trials.

The mean vector and covariance matrix are:

$$\mathbb{E}[\mathbf{X}] = n\mathbf{p}, \quad \text{Var}[\mathbf{X}] = n(\text{diag}(\mathbf{p}) - \mathbf{p}\mathbf{p}^\top).$$

The categorical distribution models a single trial with K possible outcomes, while the multinomial distribution models n independent trials.

> In machine learning, the categorical distribution is fundamental for **multiclass classification** problems. Neural networks with K output classes use a softmax activation function to produce a probability vector $\mathbf{p}$ over K categories, and the predicted class follows a categorical distribution. The multinomial distribution is used when modeling counts or frequencies across multiple categories.

13.14 Gaussian Distribution

13.14.1 Definition: Univariate Gaussian

A random variable X follows a univariate Gaussian (normal) distribution with mean $\mu \in \mathbb{R}$ and variance $\sigma^2 > 0$, denoted as $X \sim \mathcal{N}(\mu, \sigma^2)$, if its probability density function is

$$f(x \mid \mu, \sigma^2) = \frac{1}{\sqrt{2\pi\sigma^2}} \exp\left(-\frac{(x-\mu)^2}{2\sigma^2}\right).$$

The normalization constant $\frac{1}{\sqrt{2\pi\sigma^2}}$ in the Gaussian PDF is derived from the Gaussian integral $\int_{-\infty}^{\infty} e^{-x^2} dx = \sqrt{\pi}$. To compute $\int_{-\infty}^{\infty} \exp\left(-\frac{(x-\mu)^2}{2\sigma^2}\right) dx$, we use a change of variables: let $u = \frac{x-\mu}{\sigma\sqrt{2}}$, so $x = \sigma\sqrt{2}u + \mu$ and $dx = \sigma\sqrt{2}du$. Then

$$\int_{-\infty}^{\infty} \exp\left(-\frac{(x-\mu)^2}{2\sigma^2}\right) dx = \int_{-\infty}^{\infty} e^{-u^2} \sigma\sqrt{2}du = \sigma\sqrt{2} \int_{-\infty}^{\infty} e^{-u^2} du = \sigma\sqrt{2}\sqrt{\pi} = \sqrt{2\pi\sigma^2}.$$

This ensures that the PDF integrates to 1. The normalization constant is fundamental to probability density functions: it ensures that the total probability over the entire real line equals 1, satisfying the requirement that $\int_{-\infty}^{\infty} f(x)dx = 1$ for any valid PDF.

13.14.1.1 Definition: Partition Function

The **partition function** is a general term for the normalization constant of a probability distribution. For any probability distribution defined by an unnormalized density $\tilde{p}(\mathbf{x})$, the partition function Z is:

$$Z = \int \tilde{p}(\mathbf{x})\, d\mathbf{x}$$

This allows the normalized density to be written as:

$$p(\mathbf{x}) = \frac{\tilde{p}(\mathbf{x})}{Z}$$

Historical Context The concept of the partition function originated in statistical physics, where it was introduced to normalize the Boltzmann distribution over possible states of a physical system. However, its mathematical foundation was established much earlier through the work of mathematicians like Carl Friedrich Gauss.

Gauss's derivation of the Gaussian integral $\int_{-\infty}^{\infty} e^{-x^2} dx = \sqrt{\pi}$ provided the key mathematical tool needed to compute partition functions for exponential-family distributions. The Gaussian distribution itself can be viewed as an early example of a model with a partition function:

$$Z(\mu, \sigma^2) = \int_{-\infty}^{\infty} \exp\left(-\frac{(x-\mu)^2}{2\sigma^2}\right) dx = \sqrt{2\pi\sigma^2}$$

Here, the partition function $Z(\mu, \sigma^2)$ depends on the distribution's parameters μ and σ^2, but not on the random variable x. This is a general property: partition functions are functions of the model parameters, not the random variables.

Significance in Machine Learning In modern machine learning, the partition function plays a central role in energy-based models and other probabilistic models. However, for high-dimensional data like images, computing the partition function becomes intractable because it requires integration over the entire data space. This is why many modern generative models focus on learning unnormalized densities or their gradients rather than the full normalized density.

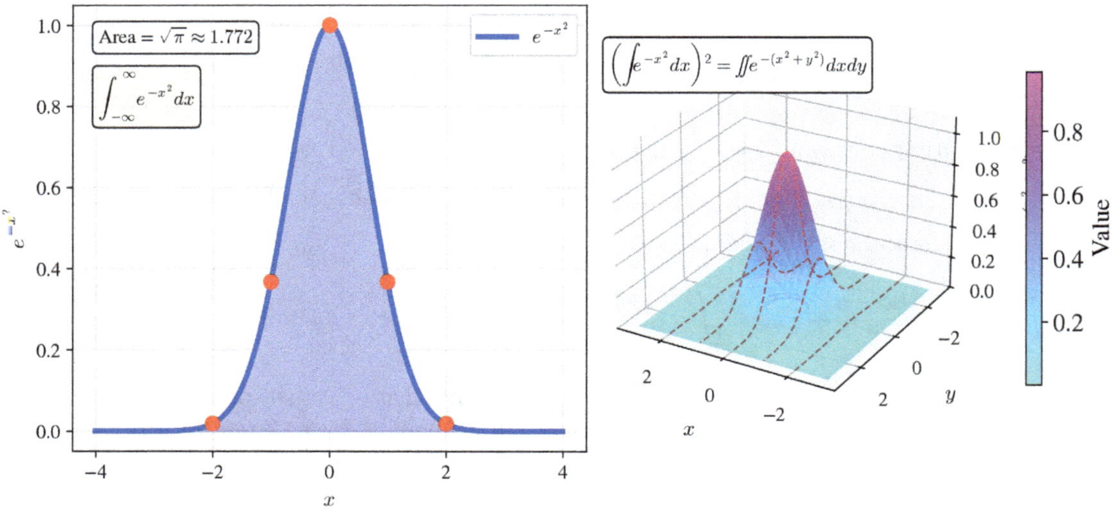

Figure 13.8: Gaussian integral: 1D curve and 2D surface.

The fundamental Gaussian integral $\int_{-\infty}^{\infty} e^{-x^2} dx = \sqrt{\pi}$ is computed using a clever trick with double integrals. Consider the square of the integral:

$$\left(\int_{-\infty}^{\infty} e^{-x^2} dx\right)^2 = \int_{-\infty}^{\infty} e^{-x^2} dx \int_{-\infty}^{\infty} e^{-y^2} dy = \int_{-\infty}^{\infty} \int_{-\infty}^{\infty} e^{-(x^2+y^2)} dxdy.$$

Figure 13.8 shows the 1D integrand e^{-x^2} (area $\sqrt{\pi}$) and the 2D surface $e^{-(x^2+y^2)}$ that appears in the double integral.

Converting to polar coordinates with $x = r\cos\theta$, $y = r\sin\theta$, and $dxdy = rdrd\theta$, we have:

$$\int_{-\infty}^{\infty} \int_{-\infty}^{\infty} e^{-(x^2+y^2)} dxdy = \int_0^{2\pi} \int_0^{\infty} e^{-r^2} rdrd\theta = 2\pi \int_0^{\infty} e^{-r^2} rdr = 2\pi \left[-\frac{1}{2}e^{-r^2}\right]_0^{\infty} = \pi.$$

Therefore, $\int_{-\infty}^{\infty} e^{-x^2} dx = \sqrt{\pi}$.

> The Gaussian integral $\int_{-\infty}^{\infty} e^{-x^2} dx = \sqrt{\pi}$ is often considered the second most beautiful formula in mathematics, after Euler's identity $e^{i\pi} + 1 = 0$. Its elegance lies in the surprising connection between the exponential function, the square root, and the fundamental constant π, achieved through the elegant technique of converting a seemingly intractable one-dimensional integral into a tractable two-dimensional integral using polar coordinates!

13.14.2 Definition: Standard Normal Distribution

A **standard normal distribution** is a Gaussian distribution with mean $\mu = 0$ and variance $\sigma^2 = 1$, denoted as $\mathcal{N}(0, 1)$. In classical probability theory, a random variable following the standard normal distribution is conventionally denoted by Z, i.e., $Z \sim \mathcal{N}(0, 1)$.

The probability density function of the standard normal distribution is:

$$\phi(z) = \frac{1}{\sqrt{2\pi}} \exp\left(-\frac{z^2}{2}\right),$$

and the cumulative distribution function is denoted by $\Phi(z) = P(Z \leq z)$.

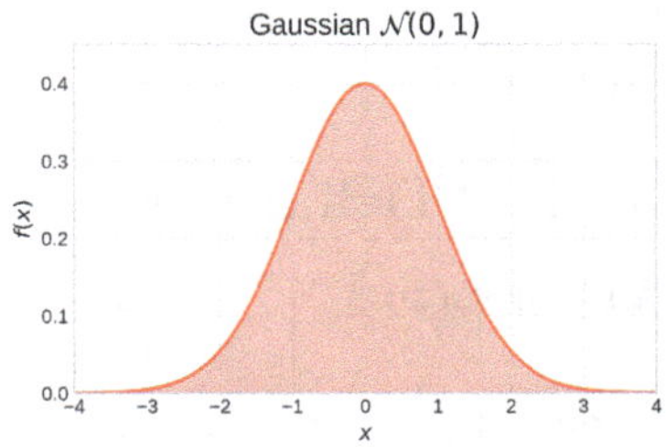

Figure 13.9: Standard Gaussian Distribution

Any Gaussian random variable $X \sim \mathcal{N}(\mu, \sigma^2)$ can be standardized to a standard normal random variable via the transformation $Z = \frac{X-\mu}{\sigma}$, so that $Z \sim \mathcal{N}(0, 1)$ (Example~13.10.5).

In classical probability theory, Z conventionally denotes a standard normal random variable. However, in machine learning contexts, Z is also commonly used to denote latent variables (as in mixture models and variational inference). The meaning of Z should therefore be determined from context. For example, in mixture models, Z typically represents the component indicator (a discrete latent variable), while in the context of standardization, Z represents a standardized Gaussian random variable.

13.14.3 Definition: Multivariate Gaussian

A random vector $X \in \mathbb{R}^{d\times 1}$ follows a multivariate Gaussian (normal) distribution with mean $\mu \in \mathbb{R}^{d\times 1}$ and covariance matrix $\Sigma \in \mathbb{R}^{d\times d}$ (positive definite), denoted as $X \sim \mathcal{N}(\mu, \Sigma)$, if its probability density function is

$$f(x \mid \mu, \Sigma) = \frac{1}{\sqrt{(2\pi)^d |\Sigma|}} \exp\left(-\frac{1}{2}(x-\mu)^\top \Sigma^{-1}(x-\mu)\right).$$

The multivariate Gaussian PDF can be derived from the univariate case. Starting with d independent standard normal random variables $Z_1, \dots, Z_d \sim \mathcal{N}(0, 1)$ with joint PDF $f(\mathbf{z}) = \prod_{i=1}^d \frac{1}{\sqrt{2\pi}} e^{-z_i^2/2} = \frac{1}{(2\pi)^{d/2}} \exp\left(-\frac{1}{2}\mathbf{z}^\top \mathbf{z}\right)$, we apply a linear transformation $\mathbf{X} = A\mathbf{Z} + \mu$, where A is a $d \times d$ matrix. Since Σ is positive definite, we can use the Cholesky decomposition $\Sigma = AA^\top$ (or eigendecomposition). By the change of variables formula (see Section "Transformation of Random Variables"), the PDF of $\mathbf{X}$ is $f(\mathbf{x}) = f(\mathbf{z})|\det(A^{-1})| = f(\mathbf{z})/|\det(A)|$. Since $|\det(A)| = \sqrt{|\Sigma|}$ and $\mathbf{z}^\top \mathbf{z} = (\mathbf{x}-\mu)^\top \Sigma^{-1}(\mathbf{x}-\mu)$, we obtain the multivariate Gaussian PDF. The normalization constant $\frac{1}{\sqrt{(2\pi)^d|\Sigma|}}$ ensures that the PDF integrates to 1 over $\mathbb{R}^d$.

13.14.4 Definition: Standard Multivariate Normal Distribution

A **standard multivariate normal distribution** is a multivariate Gaussian distribution with mean $\mu = \mathbf{0}_{d\times 1}$ (zero vector) and covariance matrix $\Sigma = I_d$ (identity matrix), denoted as $\mathcal{N}(\mathbf{0}, I_d)$.

The probability density function of the standard multivariate normal distribution is:

$$f(\mathbf{z}) = \frac{1}{(2\pi)^{d/2}} \exp\left(-\frac{1}{2}\mathbf{z}^\top \mathbf{z}\right) = \prod_{i=1}^d \frac{1}{\sqrt{2\pi}} \exp\left(-\frac{z_i^2}{2}\right),$$

which shows that the components $Z_1, \dots, Z_d$ are independent standard normal random variables.

Any multivariate Gaussian random vector $\mathbf{X} \sim \mathcal{N}(\mu, \Sigma)$ can be standardized to a standard multivariate normal random vector via the transformation $\mathbf{Z} = \Sigma^{-1/2}(\mathbf{X} - \mu)$, where $\Sigma^{-1/2}$ is the matrix square root of Σ^{-1} (e.g., from Cholesky decomposition or eigendecomposition), so that $\mathbf{Z} \sim \mathcal{N}(\mathbf{0}, I_d)$.

13.14.5 Definition: Jointly Normal Random Vectors

Random vectors $X^{(0)}, \ldots, X^{(N-1)} \in \mathbb{R}^{d\times 1}$ are said to be **jointly normal** (or **jointly Gaussian**) if their concatenated vector $\begin{bmatrix} X^{(0)} \\ \vdots \\ X^{(N-1)} \end{bmatrix} \in \mathbb{R}^{Nd\times 1}$ follows a multivariate Gaussian distribution.

Equivalently, $X^{(0)}, \ldots, X^{(N-1)}$ are jointly normal if any linear combination $\sum_{n=0}^{N-1} c_n X^{(n)}$ is a Gaussian random vector for all scalars $c_0, \ldots, c_{N-1} \in \mathbb{R}$.

13.14.6 Theorem: Linear Combinations of Normal Random Vectors

Let $X^{(0)}, \ldots, X^{(N-1)} \in \mathbb{R}^{d\times 1}$ be jointly normal. Define $Y = \sum_{n=0}^{N-1} c_n X^{(n)} \in \mathbb{R}^{d\times 1}$. Then

1. Y is a normal random vector.
2. $\mathbb{E}[Y] = \sum_{n=0}^{N-1} c_n \mathbb{E}[X^{(n)}]$.
3. $\operatorname{Var}[Y] = \sum_{n=0}^{N-1}\sum_{m=0}^{N-1} c_n c_m \operatorname{Cov}[X^{(n)}, X^{(m)}] \in \mathbb{R}^{d\times d}$, where for $n \neq m$, $\operatorname{Cov}[X^{(n)}, X^{(m)}]$ is the cross-covariance matrix of $X^{(n)}$ and $X^{(m)}$.

13.14.7 Theorem: Marginal Gaussians are Gaussians

If $\begin{bmatrix} X \\ Y \end{bmatrix} \sim \mathcal{N}\left(\begin{bmatrix} \mu_X \\ \mu_Y \end{bmatrix}, \begin{bmatrix} \Sigma_{XX} & \Sigma_{XY} \\ \Sigma_{YX} & \Sigma_{YY} \end{bmatrix}\right)$, where $\Sigma_{ij} = \operatorname{Cov}[i,j]$ for $i,j \in \{X, Y\}$, then

$$X \sim \mathcal{N}(\mu_X, \Sigma_{XX}).$$

13.14.8 Theorem: Conditional Gaussians are Gaussians

If $\begin{bmatrix} X \\ Y \end{bmatrix} \sim \mathcal{N}\left(\begin{bmatrix} \mu_X \\ \mu_Y \end{bmatrix}, \begin{bmatrix} \Sigma_{XX} & \Sigma_{XY} \\ \Sigma_{YX} & \Sigma_{YY} \end{bmatrix}\right)$, where $\Sigma_{ij} = \operatorname{Cov}[i,j]$ for $i,j \in \{X, Y\}$, then conditional on $Y = y$, the conditional distribution of X is Gaussian:

$$X \mid Y = y \sim \mathcal{N}(\mu_{X|y}, \Sigma_{X|y}),$$

with

$$\mu_{X|y} = \mu_X + \Sigma_{XY}\Sigma_{YY}^{-1}(y - \mu_Y);\ \Sigma_{X|y} = \Sigma_{XX} - \Sigma_{XY}\Sigma_{YY}^{-1}\Sigma_{YX}.$$

13.14.8.1 Example: Conditional Gaussian Distribution

Let $X_i \sim \mathcal{N}(\mu_i, \sigma_i^2)$ with $i \in \{0, 1\}$. These random variables are independent. Define

$$Y = X_0 + 3X_1 \quad \text{and} \quad Z = 2X_0 - X_1.$$

Then $Y \mid (Z = z)$ is a normal random variable. Compute its mean and variance.

Solution

We have

$$\begin{bmatrix} Y \\ Z \end{bmatrix} \sim \mathcal{N}\left(\begin{bmatrix} \mu_Y \\ \mu_Z \end{bmatrix}, \begin{bmatrix} \Sigma_{YY} & \Sigma_{YZ} \\ \Sigma_{ZY} & \Sigma_{ZZ} \end{bmatrix}\right),$$

where

$$\mu_Y = \mathbb{E}[X_0 + 3X_1] = \mu_0 + 3\mu_1, \qquad \mu_Z = \mathbb{E}[2X_0 - X_1] = 2\mu_0 - \mu_1,$$

$$\Sigma_{YY} = \text{Var}[X_0 + 3X_1] = \sigma_0^2 + 9\sigma_1^2, \qquad \Sigma_{ZZ} = \text{Var}[2X_0 - X_1] = 4\sigma_0^2 + \sigma_1^2,$$

and, using independence of X_0 and X_1,

$$\Sigma_{YZ} = \text{Cov}[X_0 + 3X_1, 2X_0 - X_1] = \text{Cov}[X_0, 2X_0] + \text{Cov}[3X_1, -X_1] = 2\sigma_0^2 - 3\sigma_1^2 = \Sigma_{ZY}.$$

Therefore,

$$\mu_{Y|z} = \mu_Y + \Sigma_{YZ}\Sigma_{ZZ}^{-1}(z - \mu_Z) = \mu_0 + 3\mu_1 + (2\sigma_0^2 - 3\sigma_1^2)(4\sigma_0^2 + \sigma_1^2)^{-1}(z - 2\mu_0 + \mu_1)$$

and

$$\Sigma_{Y|z} = \Sigma_{YY} - \Sigma_{YZ}\Sigma_{ZZ}^{-1}\Sigma_{ZY} = \sigma_0^2 + 9\sigma_1^2 - (2\sigma_0^2 - 3\sigma_1^2)(4\sigma_0^2 + \sigma_1^2)^{-1}(2\sigma_0^2 - 3\sigma_1^2).$$

13.14.8.2 Example: Bayesian Linear Regression

Consider the following linear regression model:

$$y^{(n)} = \theta^\top \mathbf{x}^{(n)} + \epsilon^{(n)}, \quad \forall n \in \{0, 1, \dots, N-1\},$$

where $\mathbf{x}^{(n)} \in \mathbb{R}^{d\times 1}$ and $\epsilon^{(n)} \sim \mathcal{N}(0, \sigma^2)$ are i.i.d. Let the prior of θ be $\mathcal{N}(\mu, \Sigma)$.

Conditional on $\{(\mathbf{x}^{(n)}, y^{(n)})\}_{n=0}^{N-1}$, the posterior of θ is still Gaussian, denoted $\mathcal{N}(\mu', \Sigma')$.

1. Compute μ' and Σ'.
2. Suppose $\mu = \mathbf{0}_{d\times 1}$ and $\Sigma = \mathbf{I}_{d\times d}$. Simplify Σ'.

Solution

Denote

$$X = \begin{bmatrix} \mathbf{x}^{(0)\top} \\ \mathbf{x}^{(1)\top} \\ \vdots \\ \mathbf{x}^{(N-1)\top} \end{bmatrix}, \quad \mathbf{y} = \begin{bmatrix} y^{(0)} \\ y^{(1)} \\ \vdots \\ y^{(N-1)} \end{bmatrix}, \quad \epsilon = \begin{bmatrix} \epsilon^{(0)} \\ \epsilon^{(1)} \\ \vdots \\ \epsilon^{(N-1)} \end{bmatrix},$$

so that $\mathbf{y} = X\theta + \epsilon$ with $\epsilon \sim \mathcal{N}(\mathbf{0}, \sigma^2\mathbf{I}_{N\times N})$. Then

$$\begin{bmatrix} \theta \\ \mathbf{y} \end{bmatrix} \mid X \sim \mathcal{N}\left(\begin{bmatrix} \mu_\theta \\ \mu_Y \end{bmatrix}, \begin{bmatrix} \Sigma_{\theta\theta} & \Sigma_{\theta Y} \\ \Sigma_{Y\theta} & \Sigma_{YY} \end{bmatrix} \right),$$

where

$$\mu_\theta = \mathbb{E}[\theta \mid X] = \mu, \qquad \mu_Y = \mathbb{E}[\mathbf{y} \mid X] = X\mu,$$
$$\Sigma_{\theta\theta} = \text{Var}[\theta \mid X] = \Sigma,$$
$$\Sigma_{\theta Y} = \text{Cov}[\theta, X\theta + \epsilon \mid X] = \Sigma X^\top, \qquad \Sigma_{Y\theta} = X\Sigma,$$
$$\Sigma_{YY} = \text{Var}[X\theta + \epsilon \mid X] = X\Sigma X^\top + \sigma^2\mathbf{I}_{N\times N}.$$

Therefore,

$$\mu' = \mu + \Sigma X^\top(X\Sigma X^\top + \sigma^2\mathbf{I}_{N\times N})^{-1}(\mathbf{y} - X\mu)$$

and

$$\Sigma' = \Sigma - \Sigma X^\top(X\Sigma X^\top + \sigma^2\mathbf{I}_{N\times N})^{-1}X\Sigma.$$

By Woodbury's matrix identity,

$$\Sigma' = \left(\Sigma^{-1} + \frac{1}{\sigma^2}X^\top X\right)^{-1}$$

and

$$\mu' = \Sigma' \left(\Sigma^{-1}\mu + \frac{1}{\sigma^2} X^\top \mathbf{y} \right).$$

When $\mu = \mathbf{0}_{d\times 1}$ and $\Sigma = \mathbf{I}_{d\times d}$,

$$\Sigma' = \left(\mathbf{I}_{d\times d} + \frac{1}{\sigma^2} X^\top X \right)^{-1}.$$

13.15 Beta Distribution

13.15.1 Definition: Beta Distribution

A random variable X follows a **Beta distribution** with shape parameters $\alpha > 0$ and $\beta > 0$, denoted as $X \sim \mathrm{Beta}(\alpha, \beta)$, if its probability density function is:

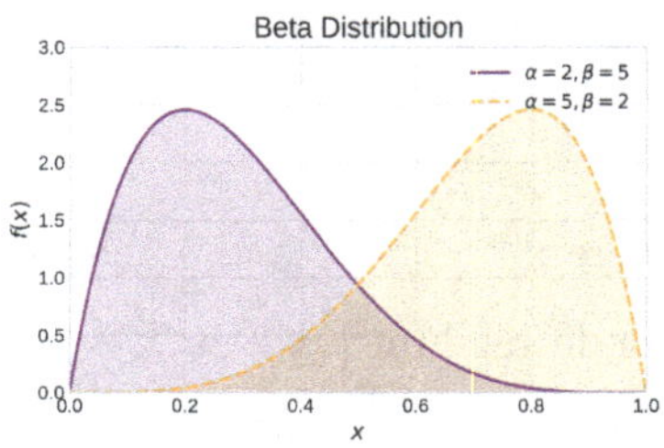

Figure 13.10: Beta Distribution

$$f(x \mid \alpha, \beta) = \frac{\Gamma(\alpha+\beta)}{\Gamma(\alpha)\Gamma(\beta)} x^{\alpha-1}(1-x)^{\beta-1}, \quad \text{for } x \in [0,1],$$

where $\Gamma(\cdot)$ is the gamma function. The normalization constant $\frac{\Gamma(\alpha+\beta)}{\Gamma(\alpha)\Gamma(\beta)}$ ensures that the PDF integrates to 1 over $[0,1]$.

The mean and variance are:

$$\mathbb{E}[X] = \frac{\alpha}{\alpha+\beta}, \quad \mathrm{Var}[X] = \frac{\alpha\beta}{(\alpha+\beta)^2(\alpha+\beta+1)}.$$

The Beta distribution is the **conjugate prior** for the Bernoulli distribution in Bayesian inference. If $X \sim \mathrm{Bernoulli}(p)$ and the prior on p is $p \sim \mathrm{Beta}(\alpha, \beta)$, then after observing data, the posterior distribution of p is also a Beta distribution. This conjugacy property makes Bayesian updating computationally tractable. The Beta distribution is also used to model probabilities and proportions, as it is naturally defined on $[0,1]$. The parameters α and β can be interpreted as "pseudo-counts" of successes and failures, respectively.

13.16 Dirichlet Distribution

13.16.1 Definition: Dirichlet Distribution

A random vector $\mathbf{X} = (X_0, X_1, \dots, X_{K-1})^\top$ follows a **Dirichlet distribution** with concentration parameters $\alpha = (\alpha_0, \alpha_1, \dots, \alpha_{K-1})^\top$ where $\alpha_k > 0$ for all k, denoted as $\mathbf{X} \sim \mathrm{Dirichlet}(\alpha)$, if its probability density function is:

$$f(\mathbf{x} \mid \alpha) = \frac{\Gamma(\sum_{k=0}^{K-1} \alpha_k)}{\prod_{k=0}^{K-1} \Gamma(\alpha_k)} \prod_{k=0}^{K-1} x_k^{\alpha_k - 1}, \quad \text{for } \mathbf{x} \in \Delta^{K-1},$$

where $\Delta^{K-1} = \{\mathbf{x} \in [0,1]^K : \sum_{k=0}^{K-1} x_k = 1\}$ is the $(K-1)$-dimensional simplex. The normalization constant ensures that the PDF integrates to 1 over the simplex.

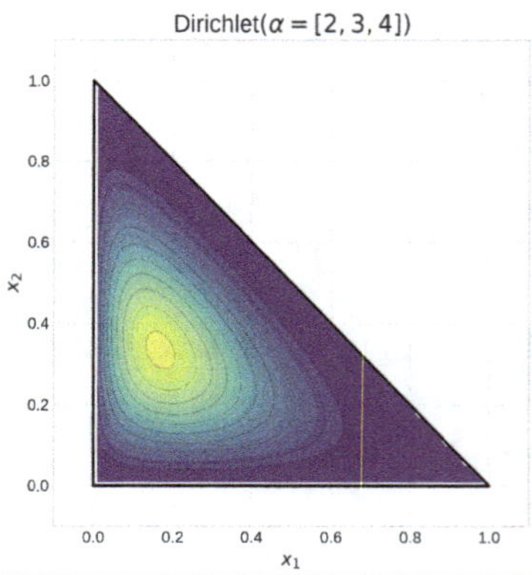

Figure 13.11: Dirichlet Distribution

The mean vector and covariance matrix are:

$$\mathbb{E}[\mathbf{X}] = \frac{\alpha}{\sum_{k=0}^{K-1} \alpha_k}, \quad \mathrm{Var}[\mathbf{X}] = \frac{1}{(\sum_{k=0}^{K-1} \alpha_k)^2(\sum_{k=0}^{K-1} \alpha_k + 1)} \left(\mathrm{diag}(\alpha) \sum_{k=0}^{K-1} \alpha_k - \alpha\alpha^\top \right).$$

When $K = 2$, the Dirichlet distribution reduces to the Beta distribution.

> The Dirichlet distribution is the **conjugate prior** for the categorical and multinomial distributions in Bayesian inference. If $\mathbf{X} \sim \mathrm{Categorical}(\mathbf{p})$ and the prior on $\mathbf{p}$ is $\mathbf{p} \sim \mathrm{Dirichlet}(\alpha)$, then after observing data, the posterior distribution of $\mathbf{p}$ is also a Dirichlet distribution. This conjugacy property is fundamental in **topic modeling** (e.g., Latent Dirichlet Allocation, LDA), where documents are modeled as mixtures of topics, and each topic is a distribution over words. The Dirichlet distribution is also used as a prior for probability vectors in Bayesian neural networks and other probabilistic models.

13.17 Mixture Distributions

13.17.1 Definition: Mixture Distribution

A **mixture distribution** is a convex combination of K component distributions. Let $Z \in \{0, 1, \dots, K-1\}$ be a discrete random variable with PMF $P(Z = k) = \pi_k$, where $\pi_k \geq 0$ for all k and $\sum_{k=0}^{K-1} \pi_k = 1$. The parameters π_k are called **mixture weights** or **component weights**.

For discrete random variables, if $X \mid Z = k$ has PMF $p_k(x)$, then the mixture distribution of X has PMF:

$$p(x) = \sum_{k=0}^{K-1} \pi_k p_k(x).$$

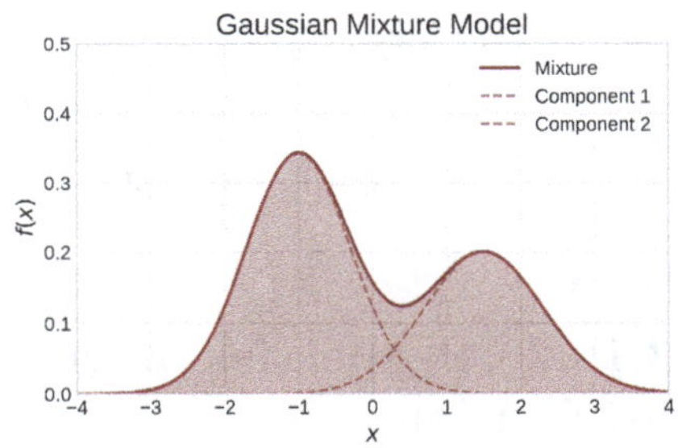

Figure 13.12: Gaussian Mixture Model

For continuous random variables, if $X \mid Z = k$ has PDF $f_k(x)$, then the mixture distribution of X has PDF:

$$f(x) = \sum_{k=0}^{K-1} \pi_k f_k(x).$$

13.17.2 Definition: Gaussian Mixture Model (GMM)

A **Gaussian Mixture Model (GMM)** is a mixture distribution where each component is a Gaussian distribution. For a GMM with K components, the PDF is:

$$f(x) = \sum_{k=0}^{K-1} \pi_k \mathcal{N}(x \mid \mu_k, \Sigma_k),$$

where $\pi_k \geq 0$, $\sum_{k=0}^{K-1} \pi_k = 1$, and $\mathcal{N}(x \mid \mu_k, \Sigma_k)$ is the PDF of a multivariate Gaussian with mean μ_k and covariance matrix Σ_k.

Mixture distributions are fundamental in **unsupervised learning** and **density estimation**. The Gaussian Mixture Model (GMM) is one of the most widely used mixture models, implemented in scikit-learn as `GaussianMixture`. GMMs can model complex, multi-modal distributions by combining simpler Gaussian components. The mixture weights π_k represent the prior probability of each component, and the posterior probability $P(Z = k \mid X = x)$ (computed via Bayes' rule) indicates which component is most likely to have generated a given observation. GMMs are used in **clustering** (where each component represents a cluster), **density estimation**, and as building blocks for more complex probabilistic models. The Expectation-Maximization (EM) algorithm is commonly used to estimate GMM parameters from data.

13.18 Von Mises-Fisher Distribution

13.18.1 Definition: Von Mises-Fisher Distribution

A random vector $\mathbf{x} \in \mathbb{R}^{d\times 1}$ follows a **von Mises-Fisher (vMF) distribution** on the $(d-1)$-dimensional unit sphere $\mathbb{S}^{d-1} = \{\mathbf{x} \in \mathbb{R}^d : \|\mathbf{x}\|_2 = 1\}$ with mean direction $\mu \in \mathbb{S}^{d-1}$ and concentration parameter $\kappa \geq 0$, denoted as $\mathbf{x} \sim \text{vMF}(\mu, \kappa)$, if its probability density function is

$$f(\mathbf{x} \mid \mu, \kappa) = C_d(\kappa) \exp(\kappa \mu^\top \mathbf{x}),$$

where $C_d(\kappa)$ is the normalization constant:

$$C_d(\kappa) = \frac{\kappa^{d/2-1}}{(2\pi)^{d/2} I_{d/2-1}(\kappa)},$$

and $I_\nu(\kappa)$ is the modified Bessel function of the first kind of order ν.

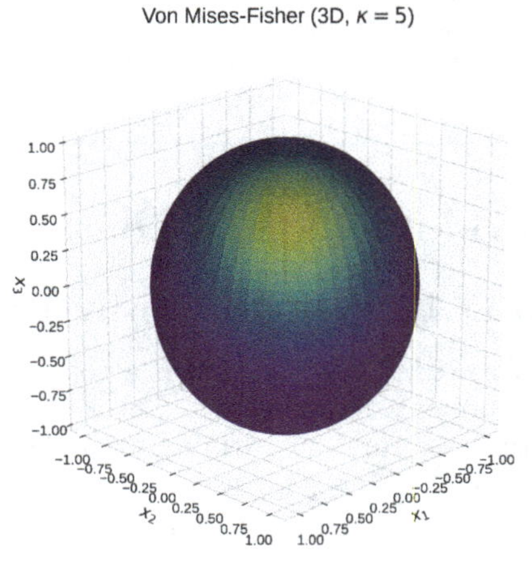

Figure 13.13: Von Mises-Fisher Distribution

13.18.2 Properties of Von Mises-Fisher Distribution

1. **Support**: The vMF distribution is defined on the unit sphere $\mathbb{S}^{d-1}$, meaning $\|\mathbf{x}\|_2 = 1$ for all $\mathbf{x}$ in the support.
2. **Mean direction**: μ is the mean direction (mode) of the distribution, and $\|\mu\|_2 = 1$.
3. **Concentration parameter**: κ controls the concentration of the distribution:
   - $\kappa = 0$: uniform distribution on the sphere
   - $\kappa \to \infty$: distribution concentrates around μ (becomes a point mass)
4. **Symmetry**: The distribution is rotationally symmetric around the mean direction μ.
5. **Expected value**: For $\mathbf{x} \sim \text{vMF}(\mu, \kappa)$, the expected value is

   $$\mathbb{E}[\mathbf{x}] = A_d(\kappa)\mu,$$

   where $A_d(\kappa) = \frac{I_{d/2}(\kappa)}{I_{d/2-1}(\kappa)}$ is the ratio of modified Bessel functions.

13.18.3 Special Case: Von Mises Distribution (2D)

For $d = 2$, the vMF distribution reduces to the **von Mises distribution** on the unit circle. If we parameterize a point on the circle by angle $\theta \in [0, 2\pi)$, the von Mises distribution has density:

$$f(\theta \mid \mu, \kappa) = \frac{1}{2\pi I_0(\kappa)} \exp(\kappa \cos(\theta - \mu)),$$

where $\mu \in [0, 2\pi)$ is the mean angle and $\kappa \geq 0$ is the concentration parameter.

13.18.4 Relationship to Other Distributions

1. **Gaussian distribution**: The vMF distribution can be viewed as a Gaussian distribution restricted to the unit sphere. Specifically, if $\mathbf{y} \sim \mathcal{N}(\kappa\mu, \mathbf{I}_{d\times d})$, then $\mathbf{x} = \frac{\mathbf{y}}{\|\mathbf{y}\|_2}$ follows $\text{vMF}(\mu, \kappa)$ approximately for large κ.
2. **Uniform distribution**: When $\kappa = 0$, the vMF distribution becomes uniform on the sphere.
3. **Watson distribution**: The vMF distribution is a special case of the Watson distribution when the concentration is symmetric.

13.18.5 Maximum Likelihood Estimation

Given observations $\mathbf{x}^{(1)}, \dots, \mathbf{x}^{(N)} \in \mathbb{S}^{d-1}$, the maximum likelihood estimates are:

- **Mean direction**: $\hat{\mu} = \frac{\bar{\mathbf{x}}}{\|\bar{\mathbf{x}}\|_2}$, where $\bar{\mathbf{x}} = \frac{1}{N}\sum_{n=1}^{N} \mathbf{x}^{(n)}$ is the sample mean.
- **Concentration parameter**: $\hat{\kappa}$ satisfies $A_d(\hat{\kappa}) = \|\bar{\mathbf{x}}\|_2$, which can be solved numerically.

13.18.6 Sampling from Von Mises-Fisher Distribution

Efficient sampling algorithms exist for the vMF distribution. One common approach uses the fact that if $\mathbf{x} \sim \text{vMF}(\mu, \kappa)$, then $\mathbf{x}$ can be generated by:

1. Sampling a direction component along μ
2. Sampling a uniform direction in the orthogonal complement
3. Combining them to form a unit vector

The von Mises-Fisher distribution is fundamental for modeling **directional data** and **unit vectors** in machine learning[2].

In **natural language processing**, word embeddings and sentence embeddings[3] are often normalized to unit vectors, and the vMF distribution is used to model the distribution of these embeddings. In **computer vision**, normalized feature vectors (e.g., SIFT descriptors[4])

[2]Gopal, S., & Yang, Y. (2014). Von Mises-Fisher clustering models. In *ICML*, pp. 154–162. https://proceedings.mlr.press/v32/gopal14.html. Davidson, T. R., Falorsi, L., De Cao, N., Kipf, T., & Tomczak, J. M. (2018). Hyperspherical variational auto-encoders. In *UAI*. https://arxiv.org/abs/1804.00891.

[3]**Word embeddings** and **sentence embeddings**: dense vector representations of words or sentences learned from text so that semantically similar items have similar vectors; often L2-normalized for cosine similarity. Canonical work: Mikolov, T., Chen, K., Corrado, G., & Dean, J. (2013). Efficient estimation of word representations in vector space (word2vec). https://arxiv.org/abs/1301.3781.

[4]**SIFT** (Scale-Invariant Feature Transform): local image descriptors formed from histograms of gradient directions around keypoints, typically 128-dimensional and L2-normalized to unit length. They are scale- and rotation-invariant and widely used for image matching and 3D reconstruction. Lowe, D. G. (2004). Distinctive image features from scale-invariant keypoints. *IJCV*, 60(2), 91–110. https://www.cs.ubc.ca/~lowe/keypoints/.

follow distributions on the unit sphere, making vMF distributions suitable for modeling. In **recommendation systems**[5], user and item embeddings are often constrained to the unit sphere, and vMF distributions can model user preferences as directions. In **spherical generative models**, vMF distributions are used as building blocks for generating data on hyperspheres. In **contrastive learning**[6], normalized representations lie on the unit sphere, and vMF distributions can model the distribution of positive and negative pairs. The concentration parameter κ provides a natural way to control the spread of embeddings: high κ means embeddings are tightly clustered around a mean direction (useful for representing coherent concepts), while low κ means embeddings are more spread out (useful for representing diverse concepts). The vMF distribution is also used in **Bayesian inference** for directional data, where the mean direction μ can be treated as a parameter with a prior distribution.

13.19 Distribution Summary

This chapter has covered probability distributions widely used in AI/ML. Figure 13.14 provides a visual overview of these distributions, showing probability density functions (PDFs) for continuous distributions, probability mass functions (PMFs) for discrete distributions, and a special non-parametric distribution.

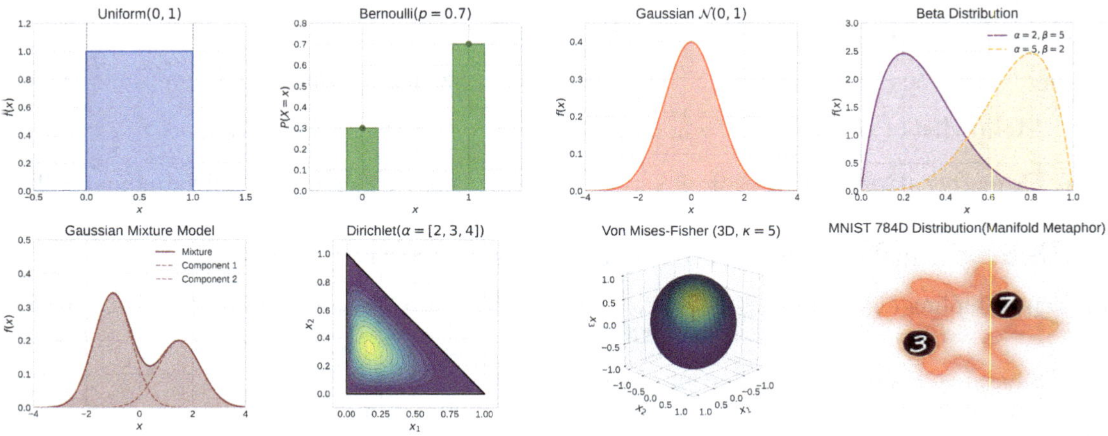

Figure 13.14: Distribution PDFs and PMFs

The final distribution is special: **MNIST 784D handwriting distribution**—a manifold metaphor representing the unknown high-dimensional probability distribution of handwritten digits, where the 784-dimensional pixel space (28×28 images) is visualized as a low-dimensional manifold structure. Unlike the parametric distributions above, the true MNIST distribution is never known in closed form; we can only approximate it from observed data. This highlights

[5]**Recommendation systems** often represent users and items as vectors; when L2-normalized these lie on the unit sphere and similarity is measured by inner product (cosine similarity). Spherical or angular metric learning is used to improve recommendation and retrieval (e.g. maximum inner product search). See e.g. https://arxiv.org/abs/1507.05910 (Bayesian personalized ranking and latent factor models).

[6]**Contrastive learning**: learning representations by pulling positive pairs together and pushing negative pairs apart; representations are often L2-normalized and thus lie on the unit sphere. Chen, T., Kornblith, S., Norouzi, M., & Hinton, G. (2020). A simple framework for contrastive learning of visual representations (SimCLR). https://arxiv.org/abs/2002.05709. Wang, T., & Isola, P. (2020). Understanding contrastive representation learning through alignment and uniformity on the hypersphere. *ICML*. https://arxiv.org/abs/2005.10242.

a fundamental principle in machine learning: **real-world data distributions often lie on low-dimensional manifolds embedded in high-dimensional spaces**, making dimensionality reduction and generative modeling possible. We will explore how to learn and approximate such unknown distributions using variational inference in Chapter 16.

13.20 Mindmap

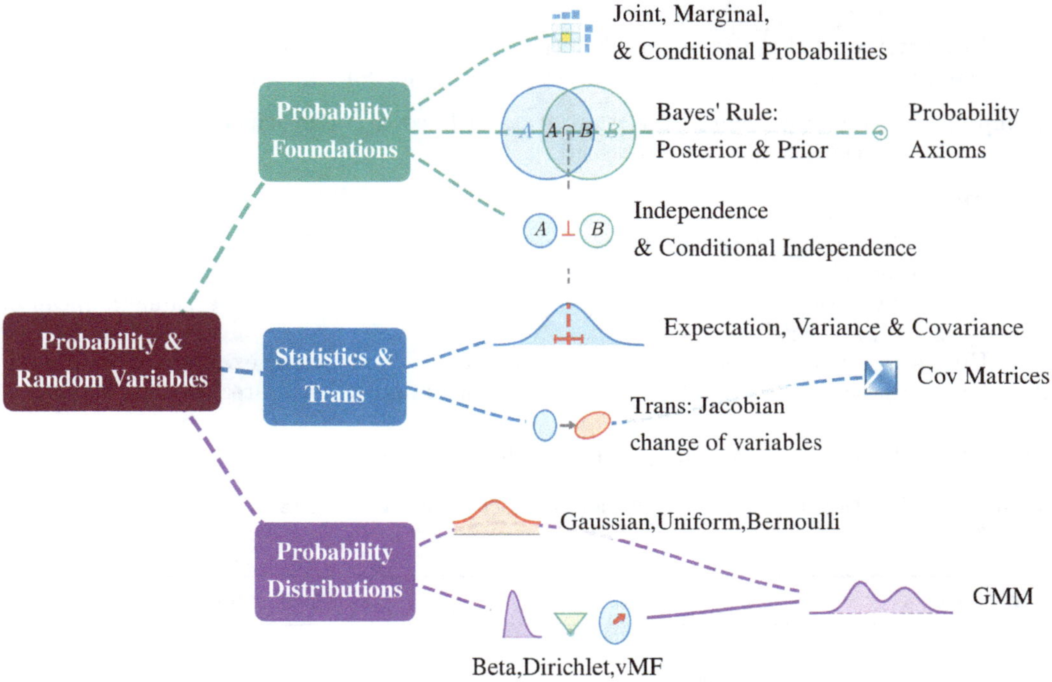

Figure 13.15: Mind map of probability and random variables topics

13.21 Exercises

13.21.1 Classification Metrics

13.21.1.1 Part 1: Multiclass Classification

A machine learning model is trained to classify images into three categories: **dog**, **cat**, and **fish**. Given the confusion matrix for this 3-class classification problem:

Actual \ Predicted	Dog	Cat	Fish
Dog	30	5	2
Cat	3	40	7
Fish	1	6	35

1. Compute the joint probabilities $P(\text{Actual} = \text{dog}, \text{Predict} = \text{dog})$, $P(\text{Actual} = \text{dog}, \text{Predict} = \text{cat})$, etc., for all combinations of actual and predicted classes.

2. Compute the overall **accuracy** of the classifier.

3. For each class (dog, cat, fish), compute:

   - **Precision**: $P(\text{Actual} = \text{class} \mid \text{Predict} = \text{class})$ — Of all images predicted as this class, how many were actually this class?
   - **Recall** (Sensitivity): $P(\text{Predict} = \text{class} \mid \text{Actual} = \text{class})$ — Of all actual images of this class, how many did the model correctly identify?
   - **F1 score**: $\text{F1} = \frac{2 \cdot \text{Precision} \cdot \text{Recall}}{\text{Precision} + \text{Recall}}$
   - **Specificity**: $P(\text{Predict} \neq \text{class} \mid \text{Actual} \neq \text{class})$ — Of all images that are not this class, how many did the model correctly identify as not this class?

4. Compute the **macro-averaged** precision, recall, F1 score, and specificity (i.e., the average across all three classes: dog, cat, and fish).

5. **Interpretation**:

   - Which class (dog, cat, or fish) has the best performance? Which class is most challenging for the model?
   - Explain why the F1 score uses the harmonic mean rather than the arithmetic mean of precision and recall.
   - How would you use these metrics to improve the model? For example, if the model has low precision for "cat", what does this mean and what strategies could help?

13.21.1.2 Part 2: Binary Classification and ROC Curve

For a binary classification problem, a classifier produces the following prediction scores and true labels:

Instance	True Label	Prediction Score
1	Positive	0.95
2	Positive	0.87
3	Positive	0.76
4	Positive	0.65
5	Positive	0.45
6	Negative	0.82
7	Negative	0.71
8	Negative	0.58
9	Negative	0.33
10	Negative	0.21

1. Sort the instances by prediction score in descending order and compute the confusion matrix for threshold $t = 0.5$. Then compute precision, recall (sensitivity), specificity, and F1 score at this threshold.

2. Compute the **True Positive Rate (TPR)** and **False Positive Rate (FPR)** for thresholds $t \in \{0.2, 0.4, 0.5, 0.6, 0.8\}$. Plot these points to sketch the ROC curve.

3. Compute the **AUC (Area Under the ROC Curve)** using the probabilistic interpretation: $\text{AUC} = P(\text{Score(Positive)} > \text{Score(Negative)})$. (Hint: Count the number of pairs where a positive instance has a higher score than a negative instance, then divide by the total number of positive-negative pairs.)

4. Interpret the AUC value: What does it tell you about the classifier's ability to distinguish

between positive and negative instances?

13.21.2 Joint CDF from PDF

Given

$$f(x, y) = \begin{cases} 3y, & 0 \le x \le y \le 1 \\ 0, & \text{otherwise} \end{cases}$$

compute $P(X \le x, Y \le y)$.

13.21.3 Marginal Probability from Confusion Matrix

Using the confusion matrix from Part 1 above, compute the probability that the prediction is **dog**.

13.21.4 Marginal PDFs

Given

$$f(x, y) = \begin{cases} 3y, & 0 \le x \le y \le 1 \\ 0, & \text{otherwise} \end{cases}$$

compute $f(x)$ and $f(y)$.

13.21.5 Von Mises-Fisher Distribution

1. Show that the von Mises-Fisher distribution $f(\mathbf{x} \mid \mu, \kappa) = C_d(\kappa) \exp(\kappa \mu^\top \mathbf{x})$ is properly normalized, i.e., $\int_{\mathbb{S}^{d-1}} f(\mathbf{x} \mid \mu, \kappa)\, d\mathbf{x} = 1$, where the integral is over the unit sphere.

2. For $\mathbf{x} \sim \text{vMF}(\mu, \kappa)$ with $\mu = [1, 0, 0]^\top$ and $\kappa = 2$, compute the probability density at $\mathbf{x} = [1, 0, 0]^\top$ and at $\mathbf{x} = [0, 1, 0]^\top$. Which point has higher density? Explain why.

3. Show that when $\kappa = 0$, the von Mises-Fisher distribution reduces to a uniform distribution on the unit sphere. (Hint: What is the value of $I_\nu(0)$?)

4. For the von Mises distribution (2D case) with mean angle $\mu = \pi/2$ and concentration $\kappa = 1$, compute the probability density at angles $\theta = \pi/2$ and $\theta = 0$. Interpret the results.

5. Given observations $\mathbf{x}^{(1)}, \dots, \mathbf{x}^{(N)} \in \mathbb{S}^{d-1}$ from a vMF distribution, show that the maximum likelihood estimate of the mean direction is $\hat{\mu} = \frac{\bar{\mathbf{x}}}{\|\bar{\mathbf{x}}\|_2}$, where $\bar{\mathbf{x}} = \frac{1}{N} \sum_{n=1}^{N} \mathbf{x}^{(n)}$.

6. Explain why the von Mises-Fisher distribution is suitable for modeling normalized embeddings in machine learning. What advantages does it have over using a Gaussian distribution for normalized vectors?

13.21.6 Calibration, Brier Score, and Temperature Scaling

1. A model predicts $f = 0.9$ on ten examples, of which six have label $y = 1$. Is this bin calibrated? Compute the contribution of these ten examples to the Brier score $\frac{1}{10} \sum_i (f_i - y_i)^2$ (with $f_i = 0.9$ for all i).

2. Let logits $z = (2, 0)$ for a binary softmax (two classes). Compute the probabilities for temperatures $T = 1$ and $T = 2$. Verify that arg max is unchanged. Which T is more "confident"?

3. Show that for binary forecasts, the Brier score admits the decomposition

$$\text{BS} = \frac{1}{N} \sum_m |B_m| \, (\bar{f}_m - \bar{y}_m)^2 + \frac{1}{N} \sum_m |B_m| \, \bar{y}_m (1 - \bar{y}_m),$$

where bins B_m partition the predictions, $\bar{f}_m$ and $\bar{y}_m$ are bin averages of f and y (Murphy decomposition into calibration and refinement terms; you may assume a fixed binning).

13.21.7 Whitening

Let $X \in \mathbb{R}^2$ have mean 0 and covariance $\Sigma = \begin{bmatrix} 2 & 0 \\ 0 & 1/2 \end{bmatrix}$.

1. Compute $W_{\text{ZCA}} = \Sigma^{-1/2}$ and verify $W_{\text{ZCA}} \Sigma W_{\text{ZCA}}^\top = I_2$.
2. For PCA whitening $W_{\text{PCA}} = \Lambda^{-1/2} U^\top$ with the eigenbasis of Σ, show that $\text{Cov}[W_{\text{PCA}} X] = I_2$ as well. How do W_{PCA} and W_{ZCA} differ in this diagonal case?

Chapter 14

Information Theory

Entropy, mutual information, KL divergence, and the evidence lower bound

> *"Information is the resolution of uncertainty."*
>
> — Claude Shannon

- ▷ `Info` $I(x) = -\log p(x)$,bits(2),nats(e); Gain: $\mathrm{IG}(Y;X) = H(Y) - H(Y|X) = I(Y;X)$
- ▷ `Entropy` $H(X) = -\sum_x p(x)\log p(x) \geq 0$; $H(X) = -\int f(x)\log f(x)dx$
- ▷ `Cond-entropy` $H(X|Y) = H(X,Y) - H(Y) \leq H(X)$ with equality iff X and Y indp.
- ▷ `Cross-entropy` $H(p,q) = -\sum_x p(x)\log q(x) = H(p) + D_{\mathrm{KL}}(p\|q)$
- ▷ `Perplexity` $\mathrm{Perplexity}(p) = 2^{H(p)}$ (bits) or $e^{H(p)}$ (nats); effective number of outcomes
- ▷ `KL Div` $d = D_{\mathrm{KL}}(p\|q) = \sum_x p(x)\log\frac{p(x)}{q(x)} \geq 0$; $D_{\mathrm{KL}}(f\|g) = \int f(x)\log\frac{f(x)}{g(x)}dx$
- ▷ `KL Div` $d(p\|q) \neq d(q\|p)$; $d(p(x,y)\|q(x,y)) = d(p(x)\|q(x)) + \mathbb{E}_{p(x)}[d(p(y|x)\|q(y|x))]$
- ▷ `JS Div` $D_{\mathrm{JS}}(p\|q) = \frac{1}{2}D_{\mathrm{KL}}(p\|\frac{p+q}{2}) + \frac{1}{2}D_{\mathrm{KL}}(q\|\frac{p+q}{2})$; symm. $0 \leq D_{\mathrm{JS}}(p\|q) \leq \ln 2$ nats (≤ 1 bit)
- ▷ `f-Div` $\sum_x q(x)f(\frac{p(x)}{q(x)})$;convex, $f(1) = 0$; KL ($t\log t$), rKL ($-\log t$), TV ($\frac{1}{2}|t-1|$)
- ▷ `Wasserstein` $W_p(p,q) = (\inf_{\gamma\in\Gamma(p,q)}\int d(x,y)^p d\gamma(x,y))^{1/p}$, $\Gamma(p,q)$ is set of couplings
- ▷ `Wasserstein-1 dual` $W_1(p,q) = \sup_{f:\|f\|_L\leq 1}[\mathbb{E}_{x\sim p}[f(x)] - \mathbb{E}_{x\sim q}[f(x)]]$ (Kantorovich-Rubinstein duality)
- ▷ `Fisher` $F(\theta)_{ij} = \mathbb{E}[\frac{\partial\log p}{\partial\theta_i}\frac{\partial\log p}{\partial\theta_j}] = -\mathbb{E}[\frac{\partial^2\log p}{\partial\theta_i\partial\theta_j}]$; SPSD. $D_{\mathrm{KL}}(p(\theta)\|p(\theta+\delta)) \approx \frac{1}{2}\delta^\top F(\theta)\delta$
- ▷ `ELBO` $\mathrm{ELBO}(q,\theta) = \mathbb{E}_{q(\mathbf{z})}[\log p(\mathbf{x},\mathbf{z}|\theta)] - \mathbb{E}_{q(\mathbf{z})}[\log q(\mathbf{z})] = \log p(\mathbf{x}|\theta) - D_{\mathrm{KL}}(q(\mathbf{z})\|p(\mathbf{z}|\mathbf{x},\theta))$
- ▷ `ELBO alt` $\mathrm{ELBO}(q,\theta) = \mathbb{E}_{q(\mathbf{z})}[\log p(\mathbf{x}|\mathbf{z},\theta)] - D_{\mathrm{KL}}(q(\mathbf{z})\|p(\mathbf{z}|\theta))$ (rec + reg)
- ▷ `Markov chain` $P(X_n = x_n|X_{n-1} = x_{n-1},\dots,X_0 = x_0) = P(X_n = x_n|X_{n-1} = x_{n-1})$
- ▷ `Sto-Mat` $P \in \mathbb{R}^{N\times N}$ with $[P]_{ij} \geq 0\ \forall i,j$ and $\sum_j [P]_{ij} = 1\ \forall i$; doubly: $\sum_i [P]_{ij} = 1\ \forall j$
- ▷ `Stationary distribution` $\pi^\top P = \pi^\top$; ergodic: $\lim_{k\to\infty} P^k = \mathbf{1}_N\pi^\top$
- ▷ `Mutual info` $I(X;Y) = \sum_{x,y} p(x,y)\log\frac{p(x,y)}{p(x)p(y)} = D_{\mathrm{KL}}(p(x,y)\|p(x)p(y)) = H(X) - H(X|Y) = H(Y) - H(Y|X)$
- ▷ `Info Bottleneck` $\min_{p(z|x)} I(X;Z) - \beta I(Z;Y)$ for $\beta > 0$ (compression vs. preservation)

14.1 Information Content

14.1.1 Definition: Information Content (Self-Information)

For a discrete random variable X with probability mass function $p(x) = P(X = x)$, the **information content** (or **self-information**) of an outcome x is defined as

$$I(x) = -\log p(x). \tag{14.1}$$

The information content $I(x)$ measures the minimum number of bits needed to encode outcome x under an optimal code, or equivalently, the surprise of the outcome. The base of the logarithm determines the unit: base 2 gives bits, base e gives nats, and base 10 gives dits (or hartleys).

14.1.1.1 Example: Fair Coin Toss

Consider a fair coin toss where X is the outcome, with $P(X = \text{head}) = P(X = \text{tail}) = 0.5$. The information content of observing a head is:

$$I(\text{head}) = -\log_2 P(X = \text{head}) = -\log_2 0.5 = -\log_2 \frac{1}{2} = 1 \text{ bit.}$$

Similarly, $I(\text{tail}) = 1$ bit. Under an optimal code, each outcome requires 1 bit (e.g., 0 for head, 1 for tail). For a biased coin with $P(X = \text{head}) = 0.9$, we have $I(\text{head}) \approx 0.152$ bits and $I(\text{tail}) \approx 3.32$ bits—optimal codes assign shorter codes to more probable outcomes.

14.1.2 Intuition

Information content provides a remarkable transformation: it converts any event (a qualitative outcome) into a number (bits needed to encode it). This numerical measure captures both the encoding length and the surprise of the outcome. Rare events ($p(x)$ small) have high information content; certain events ($p(x) = 1$) have zero information content. The information content is additive for independent events: $I(x, y) = I(x) + I(y)$ when X and Y are independent.

Information content is the fundamental building block of information theory, measuring the minimum number of bits needed to encode or transmit an outcome. In **data compression**, optimal codes (e.g., Huffman codes[a]) assign code lengths approximately equal to $I(x)$ for each outcome x, achieving compression rates close to the entropy. In machine learning, information content appears in **loss functions** (negative log-likelihood is the information content of the data under the model), **decision trees** (information gain measures how much information is gained by splitting on a feature), and **communication theory** (channel capacity and coding theorems). The logarithmic form ensures that information content is additive for independent events, which is essential for optimal coding and information transmission.

[a] **Huffman codes**: variable-length prefix codes for lossless compression. Codeword lengths are chosen from symbol frequencies (more frequent symbols get shorter codes), so average code length is close to the source entropy. The algorithm builds an optimal prefix-free code via a bottom-up merging of the two least-frequent symbols (Huffman, 1952).

14.2 Entropy

14.2.1 Definition: Entropy (Discrete)

For a discrete random variable X with probability mass function $p(x) = P(X = x)$, the **entropy** is defined as the expected information content:

$$H(X) = \mathbb{E}[I(X)] = \mathbb{E}[-\log p(X)] = -\sum_{x} p(x) \log p(x) \tag{14.2}$$

where the sum is over all possible values of X, and we use the convention $0 \log 0 = 0$.

14.2.2 Definition: Entropy (Continuous)

For a continuous random variable X with probability density function $f(x)$, the **differential entropy** is defined as

$$H(X) = -\int_{-\infty}^{\infty} f(x) \log f(x)\, dx \tag{14.3}$$

provided the integral exists.

14.2.3 Properties of Entropy

1. **Non-negativity**: $H(X) \geq 0$ for discrete X (differential entropy can be negative for continuous X).
2. **Maximum entropy**: For a discrete random variable with N possible outcomes, entropy is maximized when $p(x) = 1/N$ for all x (uniform distribution), giving $H(X) = \log N$.
3. **Additivity**: For independent random variables X and Y, $H(X,Y) = H(X) + H(Y)$.
4. **Conditional entropy**: $H(X|Y) = H(X,Y) - H(Y) \leq H(X)$, with equality if and only if X and Y are independent.

14.2.3.1 Example: Entropy Change in Scaled Dot-Product Attention

In Chapter~13, we showed that in **scaled dot-product attention**, dividing attention scores by $\sqrt{d_k}$ maintains constant variance: $\mathrm{Var}[\mathbf{q}^\top \mathbf{k}/\sqrt{d_k}] = \sigma^4$ regardless of d_k. However, **entropy behaves differently**.

For a continuous random variable $X \sim \mathcal{N}(\mu, \sigma^2)$ (Gaussian distribution), the differential entropy is:

$$H(X) = \frac{1}{2}\log(2\pi e \sigma^2) = \frac{1}{2}\log(2\pi e) + \frac{1}{2}\log(\sigma^2).$$

Derivation

The probability density function of a Gaussian distribution is:

$$f(x) = \frac{1}{\sqrt{2\pi\sigma^2}} \exp\left(-\frac{(x-\mu)^2}{2\sigma^2}\right).$$

Using the definition of differential entropy $H(X) = -\int_{-\infty}^{\infty} f(x) \log f(x)\, dx$, we have:

$$\begin{aligned} H(X) &= -\int_{-\infty}^{\infty} f(x)\log f(x)\,dx \\ &= -\int_{-\infty}^{\infty} f(x)\log\left[\frac{1}{\sqrt{2\pi\sigma^2}}\exp\left(-\frac{(x-\mu)^2}{2\sigma^2}\right)\right]dx \\ &= -\int_{-\infty}^{\infty} f(x)\left[-\frac{1}{2}\log(2\pi\sigma^2)-\frac{(x-\mu)^2}{2\sigma^2}\right]dx \\ &= \frac{1}{2}\log(2\pi\sigma^2)\int_{-\infty}^{\infty} f(x)\,dx+\frac{1}{2\sigma^2}\int_{-\infty}^{\infty}(x-\mu)^2 f(x)\,dx. \end{aligned}$$

Since $\int_{-\infty}^{\infty} f(x)\,dx = 1$ and $\int_{-\infty}^{\infty}(x-\mu)^2 f(x)\,dx = \mathrm{Var}[X] = \sigma^2$, we get:

$$H(X) = \frac{1}{2}\log(2\pi\sigma^2) + \frac{1}{2\sigma^2}\cdot\sigma^2 = \frac{1}{2}\log(2\pi\sigma^2) + \frac{1}{2} = \frac{1}{2}\log(2\pi e\sigma^2).$$

Consider the attention score before scaling: $S = \mathbf{q}^\top\mathbf{k}$ with $\mathrm{Var}[S] = d_k\sigma^4$. Assuming S is approximately Gaussian (which is reasonable when d_k is large, as S is a sum of d_k independent terms), its entropy is:

$$H(S) = \frac{1}{2}\log(2\pi e\cdot d_k\sigma^4) = \frac{1}{2}\log(2\pi e) + \frac{1}{2}\log(d_k) + \log(\sigma^2).$$

After scaling by $1/\sqrt{d_k}$, the scaled score $S' = S/\sqrt{d_k}$ has variance $\mathrm{Var}[S'] = \sigma^4$ and entropy:

$$H(S') = \frac{1}{2}\log(2\pi e\cdot\sigma^4) = \frac{1}{2}\log(2\pi e) + \log(\sigma^2).$$

The **entropy reduction** is:

$$H(S) - H(S') = \frac{1}{2}\log(d_k).$$

This shows that while scaling maintains variance (preventing gradient issues), it **reduces entropy by $\frac{1}{2}\log(d_k)$ bits**. This entropy reduction reflects the fact that the scaled distribution is more concentrated (has lower uncertainty) relative to the unscaled distribution, even though both have the same variance structure after proper initialization. The scaling factor $1/\sqrt{d_k}$ effectively "compresses" the information content, making the attention scores more predictable while maintaining numerical stability.

14.2.4 Information Gain

For random variables Y and X, the **information gain** is defined as the reduction in entropy of Y after observing X:

$$\mathrm{IG}(Y;X) = H(Y) - H(Y|X) = I(Y;X) \tag{14.4}$$

where $I(Y;X)$ is the mutual information between Y and X (see the Mutual Information section below). Information gain measures how much information about Y is gained by knowing X.

Entropy measures the **uncertainty** or **information content** of a random variable.

In machine learning, entropy is fundamental to many concepts: **decision trees** use entropy (or Gini impurity) to select optimal splits, maximizing information gain $\text{IG}(Y;X) = H(Y) - H(Y|X)$. At each node, the decision tree algorithm selects the feature X that maximizes information gain, effectively choosing the feature that most reduces the uncertainty about the target Y. This greedy approach builds trees that progressively reduce entropy, leading to pure leaf nodes. **Maximum entropy principle** is used in probabilistic modeling to find distributions that make minimal assumptions while satisfying constraints. **Entropy regularization** encourages model outputs to be more uniform, preventing overconfident predictions. In **generative models**, entropy measures the diversity of generated samples. **Differential entropy** (for continuous variables) appears in variational inference and diffusion models, where we work with continuous probability distributions.

14.3 Mutual Information

14.3.1 Definition: Mutual Information

For two random variables X and Y with joint distribution $p(x, y)$ and marginals $p(x)$ and $p(y)$, the **mutual information** is

$$I(X;Y) = \sum_{x,y} p(x,y) \log \frac{p(x,y)}{p(x)p(y)} = D_{\text{KL}}(p(x,y)||p(x)p(y)). \tag{14.5}$$

For continuous random variables, replace sums with integrals.

Notation: We use semicolon (;) for mutual information $I(X;Y)$ and comma (,) for joint entropy $H(X,Y)$. See the Preface for the full notation convention.

14.3.2 Intuition and Interpretation

Mutual information measures the **amount of information** shared between two random variables. It answers the question: "How much does knowing X reduce our uncertainty about Y (and vice versa)?"

- $I(X;Y) = 0$: X and Y are independent. Knowing X provides no information about Y.
- $I(X;Y) > 0$: X and Y are dependent. The larger the value, the stronger the dependence.
- **Maximum value**: For discrete variables, $I(X;Y) \leq \min(H(X), H(Y))$. When $I(X;Y) = H(X) = H(Y)$, one variable completely determines the other.

14.3.3 Properties of Mutual Information

1. **Symmetry**: $I(X;Y) = I(Y;X)$. Mutual information is symmetric—the information X provides about Y equals the information Y provides about X.
2. **Non-negativity**: $I(X;Y) \geq 0$, with equality if and only if X and Y are independent. This follows from the non-negativity of KL divergence.
3. **Relationship to entropy**:

$$I(X;Y) = H(X) - H(X|Y) = H(Y) - H(Y|X) = H(X) + H(Y) - H(X,Y). \tag{14.6}$$

This shows that mutual information is the reduction in uncertainty about one variable after observing the other. It can also be viewed as the "overlap" between the entropies of X and Y.

4. **Data processing inequality**: For a Markov chain $X \to Y \to Z$, we have $I(X;Z) \leq I(X;Y)$. Information can only be lost (never gained) through processing. This is fundamental to understanding information flow in neural networks.
5. **Chain rule**: For multiple variables, $I(X_1, \dots, X_n; Y) = \sum_{i=1}^{n} I(X_i; Y|X_1, \dots, X_{i-1})$.

14.3.4 Conditional Mutual Information

The **conditional mutual information** between X and Y given Z is defined as

$$I(X;Y|Z) = H(X|Z) - H(X|Y,Z) = \mathbb{E}_{z\sim p(z)}[I(X;Y|Z=z)] \tag{14.7}$$

where $I(X;Y|Z=z)$ is the mutual information between X and Y under the conditional distribution $p(x,y|z)$.

Conditional mutual information measures the information shared between X and Y that is not already contained in Z. It satisfies: $I(X;Y|Z) \geq 0$, with equality if and only if X and Y are conditionally independent given Z.

> The symbol I is used for related but distinct concepts: $I(x)$ (lowercase) denotes information content of a specific outcome x; $I(X)$ (uppercase, no semicolon) denotes the random variable $I(x)$ where x is a realization of X, with $\mathbb{E}[I(X)] = H(X)$; $I(X;Y)$ (uppercase with semicolon) denotes mutual information between random variables X and Y. The context (presence of semicolon and whether arguments are specific values or random variables) makes the distinction clear.

14.3.5 Mutual Information in Different Scenarios

Consider the mutual information $I(X;Y)$ between two random variables X and Y in three different scenarios:

1. **Perfect correlation**: If $Y = X$ (deterministic relationship), then $I(X;Y) = H(X) = H(Y)$. All information in X is shared with Y, achieving the maximum possible mutual information.
2. **Independence**: If X and Y are independent, then $p(x,y) = p(x)p(y)$, so $I(X;Y) = 0$. No information is shared between the variables.
3. **Noisy channel**: If $Y = X + \varepsilon$ where ε is independent noise, then $0 < I(X;Y) < H(X)$. The noise reduces the mutual information compared to the perfect correlation case, but some information is still preserved.

14.3.5.1 Example: Feature Selection in Decision Trees

Consider a binary classification problem with target $Y \in \{0,1\}$ and a binary feature $X \in \{0,1\}$. Given the following joint distribution:

- $P(X=0, Y=0) = 0.3$
- $P(X=0, Y=1) = 0.2$
- $P(X=1, Y=0) = 0.1$
- $P(X=1, Y=1) = 0.4$

Compute the mutual information $I(Y;X)$ to determine if X is a good feature for predicting Y.

Solution

First, compute the marginal distributions:

- $P(X=0) = P(X=0, Y=0) + P(X=0, Y=1) = 0.3 + 0.2 = 0.5$
- $P(X=1) = P(X=1, Y=0) + P(X=1, Y=1) = 0.1 + 0.4 = 0.5$
- $P(Y=0) = P(X=0, Y=0) + P(X=1, Y=0) = 0.3 + 0.1 = 0.4$
- $P(Y=1) = P(X=0, Y=1) + P(X=1, Y=1) = 0.2 + 0.4 = 0.6$

Now compute the mutual information. Using the definition:

$$\begin{aligned}
I(Y;X) &= \sum_{x,y} P(X=x,Y=y)\log_2 \frac{P(X=x,Y=y)}{P(X=x)P(Y=y)} \\
&= 0.3\log_2 \frac{0.3}{0.5\times 0.4} + 0.2\log_2 \frac{0.2}{0.5\times 0.6} \\
&\quad + 0.1\log_2 \frac{0.1}{0.5\times 0.4} + 0.4\log_2 \frac{0.4}{0.5\times 0.6} \\
&\approx 0.125 \text{ bits}
\end{aligned}$$

Alternatively, using $I(Y;X) = H(Y) - H(Y|X)$:

- $H(Y) \approx 0.971$ bits
- $H(Y|X) = 0.5 \times 0.971 + 0.5 \times 0.722 \approx 0.846$ bits
- Therefore, $I(Y;X) \approx 0.971 - 0.846 = 0.125$ bits

The mutual information $I(Y;X) \approx 0.125$ bits indicates that feature X provides some information about the target Y, but not a strong relationship. In decision tree construction, this would be compared with mutual information values for other features to select the best split.

Mutual information is one of the most important concepts in information theory and machine learning. In **feature selection**, we select features X that have high mutual information with the target Y, i.e., $\arg\max_X I(Y;X)$. In **decision trees**, information gain equals mutual information: $\text{IG}(Y;X) = I(Y;X)$, making mutual information the fundamental criterion for split selection. In **representation learning**, we learn representations Z that maximize mutual information with the data: $\max_Z I(X;Z)$ or $\max_Z I(Z;Y)$ for downstream tasks. The **information bottleneck principle** seeks representations that maximize $I(Z;Y)$ while minimizing $I(X;Z)$, balancing compression and preservation of relevant information. In **graph neural networks (GNNs)**, mutual information measures information flow between nodes: message passing aggregates information from neighbors, and mutual information quantifies how much information each node gains from its neighborhood. In **contrastive learning** (e.g., SimCLR, MoCo), mutual information is maximized between different views of the same data (e.g., different augmentations): $\max_\theta I(f_\theta(X_1); f_\theta(X_2))$ where X_1 and X_2 are augmented views. In **generative models**, mutual information between latent variables Z and generated samples X is used to ensure meaningful latent representations: $\max I(X;Z)$ prevents mode collapse and encourages diverse generations. In **variational autoencoders (VAEs)**, the ELBO can be written as $\log p(x) \geq \mathbb{E}_{q(z|x)}[\log p(x|z)] - D_{\text{KL}}(q(z|x)||p(z)) = I(X;Z) - H(Z|X) - D_{\text{KL}}(q(z|x)||p(z))$, showing the connection between mutual information and the reconstruction-regularization trade-off. In **transformers**, attention mechanisms can be viewed through an information-theoretic lens: attention weights determine how much information flows from each token to others, and mutual information measures the information shared between query and key representations. The data processing inequality shows that information can only be lost (never gained) through processing, which is fundamental to understanding information flow in neural networks, including **CNNs** where information flows through convolutional layers and pooling operations. In **causal inference**, mutual information is used to measure causal relationships and identify confounders. In **neuroscience**, mutual information quantifies how much information neural responses carry about stimuli.

14.4 Cross-Entropy

14.4.1 Definition: Cross-Entropy (Discrete)

For two discrete probability distributions $p(x)$ and $q(x)$ over the same set of outcomes, the **cross-entropy** is defined as

$$H(p,q) = -\sum_x p(x)\log q(x). \tag{14.8}$$

14.4.2 Relationship to Entropy

Cross-entropy can be decomposed as

$$H(p, q) = H(p) + D_{\text{KL}}(p||q) \tag{14.9}$$

where $H(p)$ is the entropy of p and $D_{\text{KL}}(p||q)$ is the KL divergence from p to q (defined below).

14.4.3 Relationship to Negative Log-Likelihood

For a probabilistic model with observed data $\mathcal{D} = \{x^{(n)}\}_{n=0}^{N-1}$ and true labels $y^{(n)}$, the **negative log-likelihood** $\text{NLL}(\theta) = -\sum_{n=0}^{N-1} \log p_\theta(y^{(n)}|\mathbf{x}^{(n)})$ equals the cross-entropy $H(p_{\text{true}}, p_{\text{pred}})$ when averaged over samples, showing that minimizing cross-entropy is equivalent to maximum likelihood estimation.

Cross-entropy is the standard loss function for **classification tasks** in machine learning. When p is the true distribution (one-hot encoded labels) and q is the predicted distribution (model outputs), minimizing cross-entropy is equivalent to maximizing the log-likelihood of the data. In **neural network classification**, the softmax output q represents predicted class probabilities, and cross-entropy loss $H(p, q) = -\sum_i p_i \log q_i$ measures how well the predictions match the true labels. In **gradient boosting** (e.g., XGBoost, LightGBM, CatBoost), cross-entropy loss is commonly used for classification tasks. The boosting algorithm iteratively adds weak learners (typically decision trees) that minimize the gradient of the cross-entropy loss, effectively reducing the KL divergence between true and predicted distributions at each iteration. **Histogram-based gradient boosting** (e.g., LightGBM) uses histogram binning for efficient feature splitting, and when the weak learners are decision trees, they may use information gain (mutual information) for split selection, connecting gradient boosting to information-theoretic feature selection. The relationship $H(p, q) = H(p) + D_{\text{KL}}(p||q)$ shows that cross-entropy equals entropy (constant for fixed labels) plus KL divergence, so minimizing cross-entropy is equivalent to minimizing KL divergence between true and predicted distributions. The connection to negative log-likelihood shows that classification can be viewed as maximum likelihood estimation: minimizing cross-entropy maximizes the log-likelihood of the true labels under the predicted distribution.

14.5 Perplexity

14.5.1 Definition: Perplexity

Perplexity can be defined in two ways, depending on the context:

1. **Based on entropy**: For a discrete probability distribution $p(x)$,

   $$\text{Perplexity}(p) = 2^{H(p)} \tag{14.10}$$

   where $H(p) = -\sum_x p(x) \log_2 p(x)$ is the entropy (in bits). Equivalently, using natural logarithms (nats), $\text{Perplexity}(p) = e^{H(p)}$.

2. **Based on cross-entropy**: When comparing a true distribution p_t with a model's predicted distribution p_m,

   $$\text{Perplexity} = 2^{H(p_t, p_m)} \tag{14.11}$$

   where $H(p_t, p_m) = -\sum_x p_t(x) \log_2 p_m(x)$ is the cross-entropy.

- **When p_t is one-hot** (as in standard language modeling): $H(p_t, p_m) = -\log_2 p_m(x^*)$ where x^* is the true outcome. This simplifies the calculation since only one term contributes.
- **When p_t is not one-hot** (e.g., soft labels, multi-label classification, or label smoothing): all non-zero terms in p_t contribute to the cross-entropy, requiring the full sum $-\sum_x p_t(x) \log_2 p_m(x)$.

14.5.2 Interpretation

For perplexity based on entropy, it can be interpreted as the **effective number of outcomes** that the distribution is spread over:

- A uniform distribution over N outcomes has entropy $H = \log_2 N$ and perplexity $2^H = N$
- A deterministic distribution (one outcome with probability 1) has entropy $H = 0$ and perplexity $2^0 = 1$
- Higher perplexity indicates more uncertainty (more spread-out distribution)

For perplexity based on cross-entropy (Perplexity $= 2^{H(p_t,p_m)}$), it measures how well the model predicts the true distribution. Lower perplexity indicates better prediction quality: the model assigns higher probability to the true outcomes, requiring fewer bits to encode them.

14.5.3 Perplexity in Language Modeling

In language modeling, we use **perplexity based on cross-entropy** (context 2 above). For a language model that predicts the next token given previous tokens, the perplexity on a sequence $\mathbf{x} = (x_0, x_1, \dots, x_{T-1})$ is

$$\text{Perplexity} = \exp\left(-\frac{1}{T}\sum_{t=0}^{T-1} \log p(x_t|x_0, \dots, x_{t-1})\right) = \exp\left(\frac{1}{T}\text{NLL}\right) = 2^{H(p_t,p_m)},$$

where NLL is the negative log-likelihood of the sequence. Here, $p(x_t|x_0, \dots, x_{t-1})$ is the model's predicted probability distribution (denoted as p_m), p_t is the true distribution (one-hot encoding of the actual token), and $H(p_t, p_m)$ is the cross-entropy between them. Since p_t is one-hot, this simplifies to $H(p_t, p_m) = -\log_2 p_m(x_t|x_0, \dots, x_{t-1})$ at each position t.

14.5.3.1 Example: Computing Perplexity for a Language Model

Consider a language model predicting the next token in a sequence. At a given position, the model assigns probabilities to 4 possible tokens: $p(\text{token}_0|\text{context}) = 0.5$, $p(\text{token}_1|\text{context}) = 0.25$, $p(\text{token}_2|\text{context}) = 0.15$, $p(\text{token}_3|\text{context}) = 0.1$. The true token is token_0. Compute the cross-entropy and perplexity.

Solution

The true distribution p_t is one-hot: $p_t(\text{token}_0) = 1$, $p_t(\text{token}_i) = 0$ for $i \neq 0$. The model's predicted distribution is $p_m(\text{token}_i|\text{context}) = p(\text{token}_i|\text{context})$.

Compute the cross-entropy:

$$\begin{aligned} H(p_t, p_m) &= -\sum_{i=0}^{3} p_t(\text{token}_i) \log_2 p_m(\text{token}_i|\text{context}) \\ &= -p_t(\text{token}_0) \log_2 p_m(\text{token}_0|\text{context}) \\ &= -\log_2 0.5 = 1 \text{ bit} \end{aligned}$$

Now compute the perplexity:

$$\text{Perplexity} = 2^{H(p_t,p_m)} = 2^1 = 2$$

The perplexity of 2 means the model is as uncertain as a uniform distribution over 2 tokens. Since the model assigned probability 0.5 to the correct token, it needs 1 bit to encode the outcome, resulting in a perplexity of 2. Lower perplexity (closer to 1) would indicate the model is more confident and assigns higher probability to the correct token.

If we compute the entropy of the model's predicted distribution itself (without considering the true token), we get $H(p_m) = -0.5\log_2 0.5 - 0.25\log_2 0.25 - 0.15\log_2 0.15 - 0.1\log_2 0.1 \approx 1.743$ bits, giving perplexity $2^{1.743} \approx 3.35$. This measures the model's overall uncertainty across all tokens. However, in language modeling, we use cross-entropy (which equals $-\log_2 p_m(\text{true token})$) because the true distribution is one-hot, focusing on how well the model predicts the actual token.

Perplexity is a standard evaluation metric for **language models** and **sequence models**. Lower perplexity indicates a better model that is less "perplexed" by the data (i.e., assigns higher probability to the observed sequences). In **natural language processing**, perplexity is used to compare different language models: a model with perplexity P is as uncertain as a uniform distribution over P words. For example, a perplexity of 50 means the model is as uncertain as choosing uniformly from 50 words. Perplexity is directly related to cross-entropy: Perplexity $= 2^{H(p_t,p_m)}$ where $H(p_t, p_m)$ is the cross-entropy between true distribution p_t (one-hot) and model's predicted distribution p_m. In **machine translation** and **text generation**, perplexity is used to evaluate model quality, with state-of-the-art language models achieving perplexities in the range of 10-50 on standard benchmarks.

14.6 Kullback-Leibler (KL) Divergence

14.6.1 Definition: KL Divergence

For two probability distributions over the same space, the **Kullback-Leibler (KL) divergence** is defined as follows. For discrete distributions with probability mass functions $p(x)$ and $q(x)$, and for continuous distributions with probability density functions $f(x)$ and $g(x)$:

$$D_{\text{KL}}(p||q) = \begin{cases} \sum_x p(x)\log\frac{p(x)}{q(x)} & \text{(discrete, PMFs } p \text{ and } q) \\ \int_{-\infty}^{\infty} f(x)\log\frac{f(x)}{g(x)}\,dx & \text{(continuous, PDFs } f \text{ and } g) \end{cases} \tag{14.12}$$

KL divergence measures the **expected extra information** (or "surprise") when using distribution q to encode events that actually follow distribution p, or equivalently, the **amount of information lost** when using q to replace p. In machine learning, p typically represents the **true data distribution** and q represents the **model distribution**. Minimizing $D_{\text{KL}}(p||q)$ makes the model q match the data distribution p.

Following standard convention, we use lowercase p and q as generic notation for probability distributions (PMFs for discrete, PDFs for continuous). When the context is clear, we write $D_{\text{KL}}(p||q)$ for both cases.

14.6.2 Properties of KL Divergence

1. **Non-negativity**: $D_{\text{KL}}(p||q) \geq 0$ for all p, q, with equality if and only if $p = q$ almost everywhere.
2. **Asymmetry**: $D_{\text{KL}}(p||q) \neq D_{\text{KL}}(q||p)$ in general (KL divergence is not a metric).
3. **Chain rule**: For joint distributions, $D_{\text{KL}}(p(x,y)||q(x,y)) = D_{\text{KL}}(p(x)||q(x)) + \mathbb{E}_{p(x)}[D_{\text{KL}}(p(y|x)||q(y|x))]$.
4. **Convexity**: $D_{\text{KL}}(p||q)$ is convex in both p and q.

14.6.3 Why KL Divergence is Asymmetric

Unlike a metric, KL divergence is **asymmetric**: $D_{\mathrm{KL}}(p||q) \neq D_{\mathrm{KL}}(q||p)$ in general. This asymmetry has profound consequences in machine learning. We examine it from three perspectives: intuitive interpretation, mathematical mechanism, and practical consequences.

14.6.3.1 Intuitive Interpretation: The Cost of Using the Wrong Distribution

The asymmetry arises because the cost depends on which distribution is the "truth" and which is the "approximation."

Terminology:

- **Forward KL** $D_{\mathrm{KL}}(p||q)$: Divergence from true distribution p to approximation q (eg. used in maximum likelihood estimation).
- **Reverse KL** $D_{\mathrm{KL}}(q||p)$: Divergence from approximation q to true distribution p (eg. used in variational inference).

Consider a simple example: the true distribution p is a fair 6-sided die (uniform over $\{1,2,3,4,5,6\}$), but you mistakenly use a fair coin q (uniform over $\{\text{heads}, \text{tails}\}$) to model it.

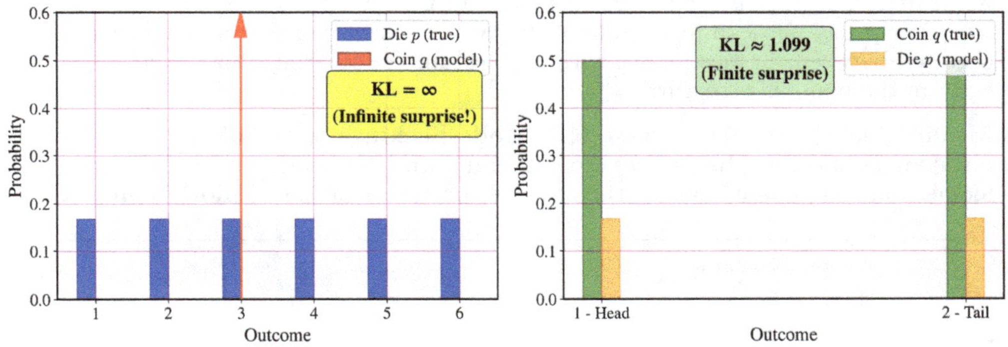

Figure 14.1: KL Divergence: Die vs Coin Example

Figure 14.1 visualizes this example. The left panel shows **forward KL** $D_{\mathrm{KL}}(p||q)$: the die p (blue bars) assigns probability 1/6 to each outcome, but the coin model q (red bars) assigns probability 0 to all die outcomes. When the die shows "3", your coin model assigns probability 0—you experience **infinite surprise** because an impossible event occurred. The KL divergence is infinite.

The right panel shows **reverse KL** $D_{\mathrm{KL}}(q||p)$: the coin q (green bars) assigns probability 1/2 to each outcome, while the die model p (orange bars) assigns 1/6 (mapping heads/tails to outcomes 1/2). When the coin shows "heads", your die model assigns 1/6—you experience only **moderate surprise** because the event was still possible. The KL divergence is finite (approximately 1.099 nats).

This illustrates the core asymmetry: **using a restrictive model to explain a rich distribution is catastrophic** (forward KL), while **using a rich model to explain a restrictive distribution is merely inefficient** (reverse KL).

14.6.3.2 Mathematical Perspective: Different Weighting Mechanisms

The asymmetry stems from which distribution weights the expectation. For continuous distributions: $D_{\mathrm{KL}}(p||q) = \int p(x) \log \frac{p(x)}{q(x)}\, dx = \mathbb{E}_{p(x)}\left[\log \frac{p(x)}{q(x)}\right]$. The expectation is taken with respect to $p(x)$, meaning $p(x)$ determines how much each point x contributes to the overall divergence—points where $p(x)$ is large contribute more, while points where $p(x)$ is small contribute less. We say p is the **weighting distribution** because it weights the contributions from different regions of the space. For **Forward KL** $D_{\text{KL}}(p||q)$, weighted by $p(x)$, we care about regions where p has high probability. If $p(x)$ is large but $q(x)$ is small, the penalty $\log(p(x)/q(x))$ becomes very large. This forces q to **avoid zeros** wherever p has mass—a "zero-avoiding" behavior. The optimal q must spread out to cover all modes of p, even if this makes q wide and blurry.

For **Reverse KL** $D_{\text{KL}}(q||p)$, the formula is $D_{\text{KL}}(q||p) = \int q(x) \log \frac{q(x)}{p(x)} dx = \mathbb{E}_{q(x)}\left[\log \frac{q(x)}{p(x)}\right]$, weighted by $q(x)$. We only care about regions where q places probability mass. If $q(x)$ is large but $p(x)$ is small, we get penalized. However, if $q(x)$ is small (or zero) where $p(x)$ is large, the penalty is minimal because $q(x)\log(q(x)/p(x)) \to 0$ as $q(x) \to 0$. This allows q to **force zeros** wherever p is small—a "zero-forcing" behavior. The optimal q can focus on a single mode of p and ignore others.

14.6.3.3 Practical Consequences: Mode-Covering vs. Mode-Seeking

The asymmetry manifests dramatically when fitting a simple model (e.g., a unimodal Gaussian) to a complex distribution (e.g., a bimodal distribution). Consider a bimodal true distribution $p(x) = 0.5 \cdot \mathcal{N}(x|-2, 0.8) + 0.5 \cdot \mathcal{N}(x|2, 0.8)$ and a unimodal Gaussian approximation $q(x) = \mathcal{N}(x|\mu, \sigma^2)$.

- **Minimizing** $D_{\text{KL}}(p||q)$ **(Forward KL, Mode-Covering)**: q must place probability mass wherever p has mass to avoid infinite penalties. Result: q becomes **wide and blurry**, centered between the two modes with high variance, "covering" both modes but failing to capture the bimodal structure.
- **Minimizing** $D_{\text{KL}}(q||p)$ **(Reverse KL, Mode-Seeking)**: q can safely ignore regions where p has low probability. Result: q becomes **sharp and focused**, locking onto one of the two modes and completely ignoring the other—a phenomenon called **mode collapse**.

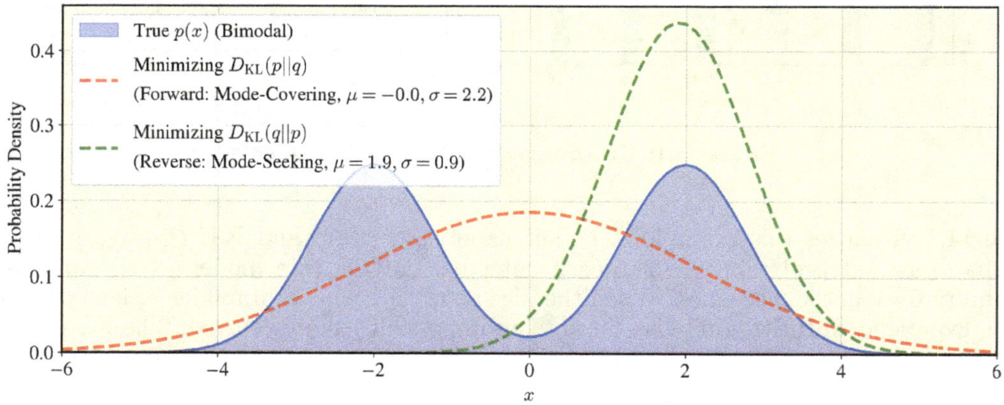

Figure 14.2: KL Divergence Asymmetry

Figure 14.2 visualizes this asymmetry: the blue curve shows the true bimodal distribution $p(x)$, the red dashed curve shows the optimal $q(x)$ when minimizing forward KL (wide and blurry, mode-covering), and the green dashed curve shows the optimal $q(x)$ when minimizing reverse KL (sharp and focused on one mode, mode-seeking).

The choice between forward and reverse KL depends on the application: use forward KL when you need to capture all modes (e.g., density estimation), and reverse KL when you prefer sharp, focused approximations (e.g., variational inference in VAEs).

14.6.4 Relationship to Cross-Entropy and Entropy

$$D_{\text{KL}}(p||q) = H(p, q) - H(p) \tag{14.13}$$

where $H(p,q)$ is cross-entropy and $H(p)$ is entropy.

KL divergence is fundamental to many machine learning algorithms: **variational inference** minimizes $D_{\text{KL}}(q(\mathbf{z})||p(\mathbf{z}|\mathbf{x}))$ (reverse KL) to approximate posterior distributions. **Variational autoencoders (VAEs)** use reverse KL as a regularization term to encourage the latent distribution to match a prior. **Knowledge distillation** uses KL divergence to transfer knowledge from a large teacher model to a small student model. **Domain adaptation** uses KL divergence to align source and target distributions. **Natural gradient descent** uses the Fisher information matrix (related to KL divergence) to compute update directions that respect the geometry of the parameter space. In **diffusion models**, KL divergence appears in the ELBO objective for training the reverse diffusion process. In **normalizing flows** and **flow matching**, KL divergence is used to match the learned distribution to the target distribution.

14.7 Jensen-Shannon (JS) Divergence

14.7.1 Definition: Jensen-Shannon Divergence

For two probability distributions $p(x)$ and $q(x)$, the **Jensen-Shannon (JS) divergence** is defined as

$$D_{\text{JS}}(p||q) = \frac{1}{2}D_{\text{KL}}\left(p||\frac{p+q}{2}\right) + \frac{1}{2}D_{\text{KL}}\left(q||\frac{p+q}{2}\right) \tag{14.14}$$

where $\frac{p+q}{2}$ is the mixture distribution with equal weights.

14.7.2 Properties of JS Divergence

1. **Symmetry**: $D_{\text{JS}}(p||q) = D_{\text{JS}}(q||p)$ (unlike KL divergence).
2. **Boundedness**: $0 \leq D_{\text{JS}}(p||q) \leq \ln 2$ nats (equivalently, ≤ 1 bit), with equality on the right when p and q have disjoint supports.
3. **Metric property**: The square root of JS divergence, $\sqrt{D_{\text{JS}}(p||q)}$, satisfies the triangle inequality, making it a proper metric.
4. **Relationship to KL divergence**: JS divergence is the average of two KL divergences from p and q to their mixture.

Jensen-Shannon divergence is a **symmetric** alternative to KL divergence, making it more suitable for certain applications.

In **Generative Adversarial Networks (GANs)**, the original GAN objective can be interpreted as minimizing JS divergence between the real data distribution and the generated distribution. However, in practice, GANs often use other objectives (e.g., Wasserstein distance, hinge loss) due to training stability issues. JS divergence is also used in **domain adaptation** where symmetry is desirable, and in **clustering** algorithms where symmetric distance measures are preferred. The boundedness of JS divergence (unlike KL divergence which can be unbounded) makes it more numerically stable in some applications.

14.8 f-Divergence

14.8.1 Definition: f-Divergence

For a convex function $f : (0,\infty) \to \mathbb{R}$ with $f(1) = 0$, and two probability distributions $p(x)$ and $q(x)$, the **f-divergence** is defined as

$$D_f(p||q) = \sum_x q(x) f\left(\frac{p(x)}{q(x)}\right) \tag{14.15}$$

for discrete distributions, or

$$D_f(p||q) = \int q(x) f\left(\frac{p(x)}{q(x)}\right) dx \tag{14.16}$$

for continuous distributions.

14.8.2 Special Cases

1. **KL divergence:** $f(t) = t \log t$ gives $D_f(p||q) = D_{\text{KL}}(p||q)$.
2. **Reverse KL divergence:** $f(t) = -\log t$ gives $D_f(p||q) = D_{\text{KL}}(q||p)$.
3. **Jensen-Shannon divergence:** $f(t) = t \log t - (t+1) \log \frac{t+1}{2}$ gives JS divergence.
4. **Total Variation distance:** $f(t) = \frac{1}{2}|t-1|$ gives $D_f(p||q) = \frac{1}{2}\sum_x |p(x) - q(x)|$.
5. **Chi-squared divergence:** $f(t) = (t-1)^2$ gives $\chi^2(p||q) = \sum_x \frac{(p(x)-q(x))^2}{q(x)}$.

f-divergence provides a **unified framework** for many distribution distance measures. Different choices of f lead to different divergence measures with different properties. In **generative modeling**, different f-divergences have different characteristics: KL divergence is mode-covering, reverse KL is mode-seeking, and JS divergence is symmetric. **f-GANs** explicitly use f-divergences as objectives, allowing practitioners to choose the divergence that best suits their application. The convexity of f ensures that f-divergences satisfy important properties like non-negativity and the data processing inequality. Understanding f-divergences helps unify seemingly different generative model objectives under a single mathematical framework.

14.9 Wasserstein Distance

14.9.1 Definition: Wasserstein Distance (Earth Mover's Distance)

For two probability distributions $p(x)$ and $q(x)$ on a metric space $(\mathcal{X}, d)$ where $d(x, y)$ is a distance function, the **Wasserstein distance** (also called **Earth Mover's Distance**) of order $p \geq 1$ is defined as

$$W_p(p, q) = \left(\inf_{\gamma \in \Gamma(p,q)} \int_{\mathcal{X}\times\mathcal{X}} d(x, y)^p \, d\gamma(x, y)\right)^{1/p} \tag{14.17}$$

where $\Gamma(p, q)$ is the set of all **couplings** (joint distributions) with marginals p and q. A coupling $\gamma(x, y)$ satisfies:

- $\int_{\mathcal{X}} \gamma(x, y)\, dy = p(x)$ (marginal over y equals p)
- $\int_{\mathcal{X}} \gamma(x, y)\, dx = q(y)$ (marginal over x equals q)

14.9.2 Special Case: Wasserstein-1 Distance

The most commonly used Wasserstein distance is **Wasserstein-1** (W_1), which uses $p = 1$:

$$W_1(p, q) = \inf_{\gamma \in \Gamma(p,q)} \int_{\mathcal{X} \times \mathcal{X}} d(x, y)\, d\gamma(x, y).$$

For one-dimensional distributions with cumulative distribution functions F_p and F_q, W_1 has a closed form:

$$W_1(p, q) = \int_{-\infty}^{\infty} |F_p(x) - F_q(x)|\, dx.$$

14.9.3 Kantorovich-Rubinstein Duality

The Wasserstein-1 distance has a dual formulation via the **Kantorovich-Rubinstein duality**:

$$W_1(p, q) = \sup_{f:\|f\|_L \leq 1} \left[\mathbb{E}_{x\sim p}[f(x)] - \mathbb{E}_{x\sim q}[f(x)]\right] \tag{14.18}$$

where the supremum is over all **1-Lipschitz functions** f (functions satisfying $|f(x) - f(y)| \leq d(x, y)$ for all x, y). This dual formulation is the basis for **Wasserstein GANs (WGANs)**.

14.9.4 Properties of Wasserstein Distance

1. **Metric property**: W_p is a proper metric: it satisfies non-negativity, symmetry ($W_p(p, q) = W_p(q, p)$), and the triangle inequality.
2. **Weak convergence**: Convergence in Wasserstein distance implies weak convergence of probability measures.
3. **Sensitivity to support**: Unlike f-divergences (which can be infinite when supports don't overlap), Wasserstein distance is well-defined even when p and q have disjoint supports, making it more suitable for distributions with different supports.
4. **Continuity**: Wasserstein distance is continuous with respect to the underlying metric, making it more stable for optimization compared to f-divergences.
5. **Computational complexity**: Computing Wasserstein distance exactly is computationally expensive (requires solving an optimal transport problem), but efficient approximations exist (e.g., Sinkhorn algorithm, sliced Wasserstein distance).

14.9.5 Relationship to Optimal Transport

Wasserstein distance is the solution to the **optimal transport problem**: given two distributions p and q, find the most efficient way to "transport" mass from p to q, where efficiency is measured by the total distance traveled. The optimal coupling γ^* represents the optimal transport plan.

14.9.6 Sliced Wasserstein Distance

For high-dimensional distributions, computing Wasserstein distance is computationally expensive. **Sliced Wasserstein distance** is an efficient approximation that:

1. Projects the distributions onto random one-dimensional lines
2. Computes the 1D Wasserstein distance for each projection
3. Averages over multiple random projections

This reduces computational complexity while preserving many desirable properties of Wasserstein distance.

Wasserstein distance is fundamental to **optimal transport theory** and has become a standard tool in modern machine learning. Unlike f-divergences (which compare probability densities), Wasserstein distance considers the **geometry** of the underlying space, making it more suitable for distributions with different supports. In **Generative Adversarial Networks (GANs)**, **Wasserstein GANs (WGANs)** use Wasserstein-1 distance as the objective, which provides more stable training compared to JS divergence: WGANs have better convergence properties, avoid mode collapse, and provide meaningful loss values that correlate with sample quality. The Kantorovich-Rubinstein duality allows WGANs to be trained by maximizing the difference between expectations over real and generated samples, subject to a Lipschitz constraint (enforced via gradient penalty or weight clipping). In **domain adaptation**, Wasserstein distance is used to align source and target distributions. In **variational inference**, Wasserstein distance appears in Wasserstein variational inference, which provides better approximations than KL divergence for certain distributions. In **normalizing flows** and **diffusion models**, Wasserstein distance is used to measure how well the learned distribution matches the target distribution. The computational efficiency of sliced Wasserstein distance makes it practical for high-dimensional problems in **deep learning**. Wasserstein distance is also used in **reinforcement learning** for distributional RL, where value functions are modeled as distributions rather than point estimates.

14.10 Fisher Information

14.10.1 Definition: Fisher Information (Scalar)

For a parametric probability distribution $p_\theta(x)$ with a single parameter $\theta \in \mathbb{R}$, the **Fisher Information** is a scalar that measures the amount of information that data provides about the parameter:

$$F(\theta) = \mathbb{E}_{p_\theta(x)}\left[\left(\frac{\partial \log p_\theta(x)}{\partial \theta}\right)^2\right] \tag{14.19}$$

Intuitively, the Fisher information measures how sensitive the log-probability of the data is to changes in the parameter θ. Higher values indicate that the parameter can be estimated more precisely.

14.10.2 Generalization: Fisher Information Matrix

For a parametric probability distribution $p_\theta(\mathbf{x})$ with multiple parameters $\theta \in \mathbb{R}^d$, the **Fisher Information Matrix** (FIM) extends the scalar Fisher information to a $d \times d$ matrix:

$$F(\theta)_{ij} = \mathbb{E}_{p_\theta(\mathbf{x})}\left[\frac{\partial \log p_\theta(\mathbf{x})}{\partial \theta_i}\frac{\partial \log p_\theta(\mathbf{x})}{\partial \theta_j}\right] \tag{14.20}$$

14.10.3 Why This Results in a Matrix

While expectation is typically a scalar operation, here we're computing a separate expectation for each pair of parameters (i, j):

1. **For each element** (i, j): We compute the expectation of the product of two score functions
2. **Expectation of a random variable**: Each product $\frac{\partial \log p}{\partial \theta_i}\frac{\partial \log p}{\partial \theta_j}$ is a random variable (function of the data $\mathbf{x}$)
3. **Result is a scalar per element**: Taking the expectation of this random variable gives a scalar value
4. **Assembling the matrix**: We repeat this for all pairs (i, j) to build the $d \times d$ matrix

Each element of the matrix represents the expected covariation between the score functions of different parameter pairs.

14.10.4 Relationship Between Scalar and Matrix Forms

- The Fisher information matrix is a **generalization** of the scalar Fisher information
- When $d = 1$, the Fisher information matrix reduces to a 1×1 matrix (essentially a scalar)
- In practice, "Fisher information" is often used as a shorthand for the matrix in multi-parameter contexts

14.10.5 Alternative Definition

Under regularity conditions (interchangeability of expectation and differentiation), the Fisher Information Matrix can also be defined as the negative expected Hessian of the log-density:

$$F(\theta)_{ij} = -\mathbb{E}_{p_\theta(\mathbf{x})}\left[\frac{\partial^2 \log p_\theta(\mathbf{x})}{\partial\theta_i \partial\theta_j}\right] \tag{14.21}$$

This form is often easier to compute for complex distributions, as it avoids calculating the square of the score function.

14.10.6 Properties of Fisher Information Matrix

1. **Symmetry**: $F(\theta)$ is symmetric: $F_{ij} = F_{ji}$.
2. **Positive semi-definiteness**: $F(\theta)$ is positive semi-definite.
3. **Cramér-Rao bound**: The covariance matrix of any unbiased estimator $\hat{\theta}$ satisfies $\text{Cov}(\hat{\theta}) \succeq F(\theta)^{-1}$ (in the sense of positive semi-definite ordering).
4. **Connection to KL divergence**: For small perturbations δ,

$$D_{\text{KL}}(p_\theta(\mathbf{x}) || p_{\theta+\delta}(\mathbf{x})) \approx \frac{1}{2}\delta^\top F(\theta)\delta. \tag{14.22}$$

This shows that the Fisher Information Matrix defines a **Riemannian metric** on the parameter space.

14.10.7 Example: Fisher Information for Gaussian Distribution

To illustrate the calculation of the Fisher Information Matrix, consider a simple Gaussian distribution with a single parameter μ representing the mean, and fixed variance $\sigma^2 = 1$:

$$p(x|\mu) = \mathcal{N}(x; \mu, 1)$$

Step 1: Compute the score function

The score function (see Chapter 17), which is the derivative of the log-density with respect to the parameter:

$$\frac{\partial \log p(x|\mu)}{\partial\mu} = \frac{\partial}{\partial\mu}\left(-\frac{1}{2}(x-\mu)^2 - \frac{1}{2}\log(2\pi)\right) = (x-\mu)$$

Step 2: Compute the Fisher Information

The Fisher Information is the expected value of the squared score function:

$$F(\mu) = \mathbb{E}_{p(x|\mu)}\left[\left(\frac{\partial \log p(x|\mu)}{\partial \mu}\right)^2\right] = \mathbb{E}_{p(x|\mu)}\left[(x-\mu)^2\right] = \mathrm{Var}[x] = 1$$

Interpretation: For this simple Gaussian case, the Fisher Information is constant and equals 1, regardless of the value of μ. This means the information content about μ is the same for any value of the mean, which makes intuitive sense since the Gaussian distribution is symmetric around its mean.

14.10.7.1 Extension to Multiple Parameters

If we include the variance σ^2 as a second parameter, the Fisher Information Matrix can be derived step-by-step:

Step 1: Define the log-density

For a Gaussian distribution with both mean μ and variance σ^2 as parameters:

$$\log p(x|\mu, \sigma^2) = -\frac{1}{2\sigma^2}(x-\mu)^2 - \frac{1}{2}\log(2\pi\sigma^2)$$

Step 2: Compute the score functions

The score functions are the partial derivatives of the log-density with respect to each parameter:

- With respect to μ:

$$\frac{\partial \log p}{\partial \mu} = \frac{1}{\sigma^2}(x-\mu)$$

- With respect to σ^2:

$$\frac{\partial \log p}{\partial \sigma^2} = \frac{1}{2\sigma^4}(x-\mu)^2 - \frac{1}{2\sigma^2}$$

Step 3: Compute the Fisher Information Matrix elements

The Fisher Information Matrix F has elements $F_{ij} = \mathbb{E}\left[\frac{\partial \log p}{\partial \theta_i}\frac{\partial \log p}{\partial \theta_j}\right]$:

1. **F_{11} (mean-mean element):**

$$F_{11} = \mathbb{E}\left[\left(\frac{1}{\sigma^2}(x-\mu)\right)^2\right] = \frac{1}{\sigma^4}\mathbb{E}[(x-\mu)^2] = \frac{1}{\sigma^4}\cdot\sigma^2 = \frac{1}{\sigma^2}$$

2. **F_{12} and F_{21} (mean-variance elements):**

$$F_{12} = \mathbb{E}\left[\frac{1}{\sigma^2}(x-\mu)\cdot\left(\frac{1}{2\sigma^4}(x-\mu)^2 - \frac{1}{2\sigma^2}\right)\right] = 0$$

   This is zero because the odd moments of the Gaussian distribution about its mean are zero.

3. **F_{22} (variance-variance element):**

$$F_{22} = \mathbb{E}\left[\left(\frac{1}{2\sigma^4}(x-\mu)^2 - \frac{1}{2\sigma^2}\right)^2\right]$$

   Expanding and using the fourth central moment of the Gaussian ($\mathbb{E}[(x-\mu)^4] = 3\sigma^4$):

$$F_{22} = \frac{1}{4\sigma^8}\cdot 3\sigma^4 - \frac{1}{2\sigma^6}\cdot\sigma^2 + \frac{1}{4\sigma^4} = \frac{3}{4\sigma^4} - \frac{1}{2\sigma^4} + \frac{1}{4\sigma^4} = \frac{1}{2\sigma^4}$$

Step 4: Assemble the Fisher Information Matrix

$$F(\mu, \sigma^2) = \begin{bmatrix} \frac{1}{\sigma^2} & 0 \\ 0 & \frac{1}{2\sigma^4} \end{bmatrix}$$

This shows that the Fisher Information Matrix is diagonal for the Gaussian distribution, reflecting the independence of the mean and variance parameters in terms of information content. The diagonal structure implies that uncertainty in the mean parameter does not affect uncertainty in the variance parameter, and vice versa.

14.10.8 Example: Fisher Information in GPTQ Quantization

The Fisher Information Matrix plays a crucial role in **GPTQ (GPT Quantization)**[1], a state-of-the-art method for quantizing large language models to low bit-widths (e.g., 4 bits) while preserving performance.

GPTQ Context: GPTQ quantizes model weights in **blocks** (typically 128 weights per block) and uses block-wise Fisher Information Matrices to determine which weights are most sensitive to quantization error.

Mathematical Setup: Consider a linear layer with weight matrix $W \in \mathbb{R}^{M \times N}$ and input activation vector $x \in \mathbb{R}^N$. The layer output is $y = Wx$.

Block-Wise Fisher Information Calculation

Step 1: Select a block: Choose a contiguous block of weights $w \in \mathbb{R}^B$ (e.g., $B = 128$) from W

Step 2: Compute the Fisher Information: For each weight w_i in the block, the Fisher Information is approximated as:

$$F_i = \mathbb{E}[g_i^2]$$

where $g_i = \frac{\partial \mathcal{L}}{\partial w_i}$ is the gradient of the loss function with respect to weight w_i, estimated using a small batch of calibration data.

Step 3: Quantization with FIM-Aware Ordering: The weights are quantized in order of increasing Fisher Information:

- Weights with lower F_i (less sensitive) are quantized first
- Weights with higher F_i (more sensitive) are quantized later, allowing them to retain more precision

Intuition: The Fisher Information measures how much each weight contributes to the model's output. By quantizing less sensitive weights first, GPTQ minimizes the cumulative quantization error, as more sensitive weights are given priority in retaining their original values.

Why Fisher Information Works Without Probability: While Fisher Information is traditionally defined in the context of probabilistic models, its usage in GPTQ extends this concept to measure **parameter sensitivity** in neural networks, regardless of the probabilistic interpretation.

[1]GPTQ is also discussed elsewhere: quadratic form $\mathbf{e}^\top H\mathbf{e}$ for quantization error (Chapter~7), block-wise Hessian decomposition and Figure 10.3 (Chapter~10), and Hessian-based quantization with Taylor expansion (Chapter~11).

Traditional Probabilistic Definition:

$$F_i = \mathbb{E}\left[\left(\frac{\partial \log p(y|x, w_i)}{\partial w_i}\right)^2\right]$$

GPTQ's Deterministic Interpretation:

$$F_i = \mathbb{E}\left[\left(\frac{\partial \mathcal{L}}{\partial w_i}\right)^2\right]$$

Even without an explicit probability distribution, the Fisher Information in GPTQ captures:

- **Output Sensitivity**: How much each weight affects the model's output
- **Error Propagation**: How quantization error in each weight will affect overall performance
- **Relative Importance**: Which weights are most critical to preserve accurately

This approach works because it identifies weights that carry the most "information" about the model's output, regardless of whether that output is interpreted probabilistically.

Key Advantage: Using block-wise FIM allows GPTQ to quantize very large models efficiently, as it avoids computing the full Fisher Information Matrix for the entire model, which would be computationally infeasible.

Connection to Linear Algebra: The block-wise approach demonstrates how matrix partitioning and local information (Fisher Information) can be leveraged to solve large-scale optimization problems, a recurring theme in both linear algebra and machine learning.

The Fisher Information Matrix is fundamental to **second-order optimization** and **statistical inference**. In **natural gradient descent**, the update direction is $F(\theta)^{-1}\nabla_\theta \mathcal{L}(\theta)$, which accounts for the geometry of the parameter space defined by the Fisher metric. This is more efficient than standard gradient descent, especially for ill-conditioned problems. In **variational inference**, the Fisher Information Matrix appears when using natural gradients for updating variational parameters.

In **GPTQ (GPT Quantization)**, block-wise Fisher Information Matrices are computed to minimize quantization error. The connection to KL divergence shows that the Fisher Information Matrix measures the **local curvature** of the KL divergence surface, making it crucial for understanding the geometry of probabilistic models. In **Bayesian inference**, the Fisher Information Matrix provides a measure of the information content of the data about the parameters.

14.11 Evidence Lower Bound (ELBO)

14.11.1 Definition: Evidence Lower Bound

For a probabilistic model with observed data $\mathbf{x}$, latent variables $\mathbf{z}$, and parameters θ, the **evidence** (marginal likelihood) is

$$p_\theta(\mathbf{x}) = \int p_\theta(\mathbf{x}, \mathbf{z})\, d\mathbf{z}. \tag{14.23}$$

The **Evidence Lower Bound (ELBO)** is defined as

$$\text{ELBO}(q, \theta) = \mathbb{E}_{q(\mathbf{z})}[\log p_\theta(\mathbf{x}, \mathbf{z})] - \mathbb{E}_{q(\mathbf{z})}[\log q(\mathbf{z})] \tag{14.24}$$

where $q(\mathbf{z})$ is a variational distribution approximating the true posterior $p_\theta(\mathbf{z}|\mathbf{x})$.

14.11.2 Relationship to KL Divergence

The ELBO can be rewritten as

$$\text{ELBO}(q, \theta) = \log p_\theta(\mathbf{x}) - D_{\text{KL}}(q(\mathbf{z})||p_\theta(\mathbf{z}|\mathbf{x})). \tag{14.25}$$

This shows that maximizing the ELBO is equivalent to:

1. Maximizing the log-likelihood $\log p_\theta(\mathbf{x})$
2. Minimizing the KL divergence $D_{\text{KL}}(q(\mathbf{z})||p_\theta(\mathbf{z}|\mathbf{x}))$ between the variational distribution and the true posterior

14.11.3 Alternative Form

The ELBO can also be written as

$$\text{ELBO}(q, \theta) = \mathbb{E}_{q(\mathbf{z})}[\log p_\theta(\mathbf{x}|\mathbf{z})] - D_{\text{KL}}(q(\mathbf{z})||p_\theta(\mathbf{z})) \tag{14.26}$$

which separates the **reconstruction term** (expected log-likelihood) and the **regularization term** (KL divergence from prior).

The Evidence Lower Bound is the fundamental objective function for **variational inference** and **variational autoencoders (VAEs)**. The name "lower bound" comes from the fact that $\text{ELBO}(q, \theta) \leq \log p_\theta(\mathbf{x})$, with equality when $q(\mathbf{z}) = p_\theta(\mathbf{z}|\mathbf{x})$ (the true posterior). In **VAEs**, the ELBO is maximized to train both the encoder (which learns $q(\mathbf{z}|\mathbf{x})$) and decoder (which learns $p_\theta(\mathbf{x}|\mathbf{z})$). The decomposition into reconstruction and regularization terms provides interpretability: the reconstruction term encourages the model to accurately reconstruct the data, while the KL regularization term encourages the latent distribution to match a prior (typically a standard Gaussian). In **Bayesian neural networks**, ELBO is used to approximate the posterior distribution over weights. The ELBO framework unifies many probabilistic models, including VAEs, normalizing flows, and diffusion models. Variational inference and the ELBO are developed in depth in Chapter~16.

14.12 Markov Chain

14.12.1 Definition: Markov Chain

A sequence of random variables $X_0, X_1, X_2, \ldots$ forms a **Markov chain** (also called a **Markov process**, denoted as $X_0 \to X_1 \to X_2 \to \cdots$) if for all $n \geq 1$ and all values $x_0, x_1, \ldots, x_n$:

$$P(X_n = x_n | X_{n-1} = x_{n-1}, X_{n-2} = x_{n-2}, \ldots, X_0 = x_0) = P(X_n = x_n | X_{n-1} = x_{n-1}). \tag{14.27}$$

This is the **Markov property**: the future depends only on the present, not on the past.

Markov chains are fundamental to many areas of mathematics and machine learning. In **probability theory**, Markov chains model stochastic processes where the future state depends only on the current state. In **information theory**, Markov chains appear in the data processing inequality, which states that information can only be lost (never gained) through processing. In **machine learning**, Markov chains are used in **hidden Markov models (HMMs)** for sequence modeling, **Markov decision processes (MDPs)** for reinforcement learning, and **Markov chain Monte Carlo (MCMC)** for sampling. The conditional independence property $P(X, Z|Y) = P(X|Y)P(Z|Y)$ is crucial for understanding information flow: if $X \to Y \to Z$ forms a Markov chain, then $I(X; Z) \leq I(X; Y)$, meaning that processing information through Y can only reduce the information about X.

14.12.2 Finite State Space Representation

For a Markov chain with a finite state space, it can be described by a tuple $(\mathcal{S}, P)$, where:

- $\mathcal{S} = \{s_0, s_1, \dots, s_{N-1}\}$ is a finite set of **states** with $|\mathcal{S}| = N$
- $P \in \mathbb{R}^{N\times N}$ is the **state transition matrix** (also called **transition probability matrix**)

The state transition matrix P defines the transition probabilities between all pairs of states:

$$[P]_{ij} = P(X_{n+1} = s_j | X_n = s_i),$$

where $[P]_{ij}$ is the element in the i-th row and j-th column, representing the probability of transitioning from state s_i to state s_j. This is called the **state transition function**.

14.12.3 Properties of State Transition Matrix

1. **Non-negativity**: $[P]_{ij} \geq 0$ for all i, j (probabilities are non-negative).
2. **Row stochastic**: Each row sums to 1:

$$\sum_{j=0}^{N-1} [P]_{ij} = 1 \quad \forall i \in \{0, 1, \dots, N-1\}.$$

   This ensures that from any state s_i, the probabilities of transitioning to all possible states sum to 1.

3. **Stochastic matrix**: A matrix satisfying both properties above is called a **stochastic matrix** (or **Markov matrix**).

14.12.3.1 Example: Markov Chain with Transition Matrix

Consider a Markov chain with 6 states $\mathcal{S} = \{S_1, S_2, S_3, S_4, S_5, S_6\}$ and transition matrix:

$$P = \begin{bmatrix} 0.9 & 0.1 & 0 & 0 & 0 & 0 \\ 0.5 & 0 & 0.5 & 0 & 0 & 0 \\ 0 & 0 & 0 & 0.6 & 0 & 0.4 \\ 0 & 0 & 0 & 0 & 0.3 & 0.7 \\ 0 & 0.2 & 0.3 & 0.5 & 0 & 0 \\ 0 & 0 & 0 & 0 & 0 & 1 \end{bmatrix},$$

where $[P]_{ij} = P(X_{n+1} = S_j | X_n = S_i)$ represents the probability of transitioning from state S_i to state S_j.

Properties:

1. **Stochastic matrix**: Each row sums to 1: $\sum_{j=1}^{6} [P]_{ij} = 1$ for all $i \in \{1, 2, \dots, 6\}$.
2. **Absorbing state**: State S_6 is an **absorbing state** since $P(X_{n+1} = S_6 | X_n = S_6) = 1$. Once the chain reaches state S_6, it stays there forever.
3. **Communicating classes**: The states can be grouped into communicating classes based on reachability:
   - States $\{S_1, S_2, S_3\}$ can communicate with each other
   - States $\{S_4, S_5\}$ can communicate with each other
   - State $\{S_6\}$ is absorbing and isolated
4. **Stationary distribution**: For an **ergodic Markov chain**, the stationary distribution π satisfies $\pi^\top = \pi^\top P$. For this chain, since state S_6 is absorbing, the long-term behavior is that the chain will eventually be absorbed in state S_6 with probability 1.

This example illustrates key concepts in Markov chain theory. The transition matrix encodes all the probabilistic transitions between states. **Absorbing states** are important in many applications: in **reinforcement learning**, terminal states are absorbing; in **epidemiology**, recovered or deceased states are absorbing. **Communicating classes** partition the state space into sets of states that can reach each other, which is crucial for understanding the long-term behavior of the chain. The stationary distribution (if it exists) describes the long-run proportion of time spent in each state, which is fundamental to **Markov chain Monte Carlo (MCMC)** methods where we want to sample from a target distribution.

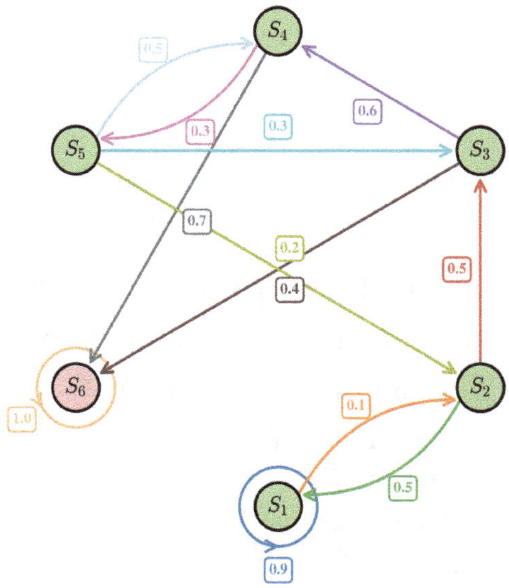

Figure 14.3: Transition Diagram

14.13 Stochastic Matrices and Markov Chains

14.13.1 Definition: Stochastic Matrix

A matrix $P \in \mathbb{R}^{N\times N}$ is called a **stochastic matrix** (or **Markov matrix**) if:

1. **Non-negativity**: $[P]_{ij} \geq 0$ for all $i, j \in \{0, 1, \dots, N-1\}$
2. **Row stochastic**: Each row sums to 1: $\sum_{j=0}^{N-1} [P]_{ij} = 1$ for all i

A stochastic matrix is called **doubly stochastic** if, in addition:

3. **Column stochastic**: Each column also sums to 1: $\sum_{i=0}^{N-1} [P]_{ij} = 1$ for all j

14.13.2 Properties of Stochastic Matrices

1. **Eigenvalue 1**: Every stochastic matrix has eigenvalue $\lambda = 1$ with a corresponding left eigenvector $\pi^\top$ (the stationary distribution) satisfying $\pi^\top P = \pi^\top$.
2. **Spectral radius**: The spectral radius (defined in Chapter~6) $\rho(P) = \max_i |\lambda_i|$ satisfies $\rho(P) = 1$ for stochastic matrices.
3. **Matrix powers**: For a stochastic matrix P, the powers P^k are also stochastic matrices for all $k \geq 1$.
4. **Convex combination**: If P is stochastic, then for any probability vector $\mathbf{p} \in \mathbb{R}^N$ (where $\sum_i p_i = 1$ and $p_i \geq 0$), the product $P\mathbf{p}$ is also a probability vector.

14.13.3 Stationary Distribution via Eigendecomposition

For an **ergodic** (irreducible and aperiodic) Markov chain with transition matrix P, the stationary distribution π can be found by solving:

$$\pi^\top P = \pi^\top,$$

which is equivalent to finding the left eigenvector of P corresponding to eigenvalue $\lambda = 1$. This can be solved using eigendecomposition (Chapter~6) or by solving the linear system:

$$(P^\top - I)\pi = \mathbf{0}, \quad \text{subject to} \quad \sum_{i=0}^{N-1} \pi_i = 1.$$

The stationary distribution π represents the long-run proportion of time the chain spends in each state: $\lim_{k\to\infty} P^k = \mathbf{1}_N \pi^\top$ (where $\mathbf{1}_N$ is a column vector of ones).

14.13.4 Perron-Frobenius Theorem for Stochastic Matrices

For an irreducible stochastic matrix P:

1. **Dominant eigenvalue**: $\lambda = 1$ is a simple eigenvalue (algebraic multiplicity 1).
2. **Positive eigenvector**: The left eigenvector π corresponding to $\lambda = 1$ has all positive entries: $\pi_i > 0$ for all i.
3. **Convergence**: For any initial distribution $\mathbf{p}_0$, the distribution after k steps converges to the stationary distribution:
$$\lim_{k\to\infty} \mathbf{p}_0^\top P^k = \pi^\top.$$

14.13.5 Doubly Stochastic Matrices

A **doubly stochastic matrix** satisfies both row and column stochasticity. Important properties:

1. **Uniform stationary distribution**: For an ergodic doubly stochastic matrix, the stationary distribution is uniform: $\pi = \frac{1}{N}\mathbf{1}_N$.
2. **Birkhoff-von Neumann theorem**: Every doubly stochastic matrix can be expressed as a convex combination of permutation matrices.

   Proof: Let $P \in \mathbb{R}^{N\times N}$ be a doubly stochastic matrix. We prove by induction on the number of positive entries in P.

   Base case: If P has exactly N positive entries (one per row and column), then P is a permutation matrix itself, and the result holds trivially.

   Inductive step: Suppose P has more than N positive entries. We will show that there exists a permutation matrix Π and a scalar $\alpha \in (0, 1]$ such that $P - \alpha\Pi$ is still doubly stochastic with non-negative entries, and has fewer positive entries than P.

   To construct Π and α, consider the bipartite graph G with $2N$ vertices (representing rows and columns) where there is an edge between row i and column j if and only if $[P]_{ij} > 0$. Since P is doubly stochastic, every row and column has at least one positive entry. By **Hall's marriage theorem**, there exists a perfect matching in G, which corresponds to a permutation σ such that $[P]_{i,\sigma(i)} > 0$ for all i.

   Define the permutation matrix Π by $[\Pi]_{ij} = 1$ if $j = \sigma(i)$ and 0 otherwise. Let
$$\alpha = \min_{i=0}^{N-1} [P]_{i,\sigma(i)} > 0.$$

   Then $P - \alpha\Pi$ is doubly stochastic (since subtracting α from matching entries preserves row and column sums) and has at least one fewer positive entry than P (the entry where the minimum was achieved becomes zero).

   By the inductive hypothesis, $P - \alpha\Pi$ can be written as a convex combination of permutation matrices:
$$P - \alpha\Pi = \sum_{k=1}^{m} \beta_k \Pi_k,$$

where $\sum_{k=1}^{m} \beta_k = 1 - \alpha$ and $\beta_k \geq 0$ for all k. Therefore,

$$P = \alpha\Pi + \sum_{k=1}^{m} \beta_k \Pi_k = \alpha\Pi + (1-\alpha) \sum_{k=1}^{m} \frac{\beta_k}{1-\alpha} \Pi_k,$$

which expresses P as a convex combination of permutation matrices. By **Carathéodory's theorem**, this representation can be achieved with at most $N^2 - 2N + 2$ permutation matrices.

3. **Sinkhorn-Knopp algorithm**: Given a non-negative matrix A, the Sinkhorn-Knopp algorithm iteratively normalizes rows and columns to produce a doubly stochastic matrix, which is useful in modern deep learning applications.

> Stochastic matrices are fundamental to **Markov chain theory** and have wide applications in machine learning. In **reinforcement learning**, transition matrices in Markov Decision Processes are stochastic. In **PageRank**, the web graph is represented as a stochastic matrix where entries represent transition probabilities between web pages. In **natural language processing**, stochastic matrices model word-to-word transitions in language models. Recently, **doubly stochastic matrices** have gained attention in deep learning: in **Manifold-Constrained Hyper-Connections (mHC)**[a], doubly stochastic matrices (computed via Sinkhorn-Knopp algorithm) are used to constrain residual connections in neural networks, ensuring stable signal propagation while maintaining the identity mapping property. This application demonstrates how stochastic matrix theory enables stable large-scale training of deep networks. The Perron-Frobenius theorem guarantees convergence to stationary distributions, which is crucial for **Markov chain Monte Carlo (MCMC)** methods used in Bayesian inference and generative modeling.
>
> [a] Xie, Z. etc (2025). mHC: Manifold-Constrained Hyper-Connections. *arXiv preprint arXiv:2512.24880*. Available at: https://arxiv.org/html/2512.24880v1

14.13.6 Example: Computing Stationary Distribution

Consider a Markov chain with transition matrix:

$$P = \begin{bmatrix} 0.7 & 0.3 & 0 \\ 0.2 & 0.5 & 0.3 \\ 0.1 & 0.2 & 0.7 \end{bmatrix}.$$

To find the stationary distribution $\pi = [\pi_0, \pi_1, \pi_2]^\top$, we solve $\pi^\top P = \pi^\top$:

$$\begin{bmatrix} \pi_0 & \pi_1 & \pi_2 \end{bmatrix} \begin{bmatrix} 0.7 & 0.3 & 0 \\ 0.2 & 0.5 & 0.3 \\ 0.1 & 0.2 & 0.7 \end{bmatrix} = \begin{bmatrix} \pi_0 & \pi_1 & \pi_2 \end{bmatrix},$$

which gives the system:

- $\pi_0 = 0.7\pi_0 + 0.2\pi_1 + 0.1\pi_2$
- $\pi_1 = 0.3\pi_0 + 0.5\pi_1 + 0.2\pi_2$
- $\pi_2 = 0.3\pi_1 + 0.7\pi_2$
- $\pi_0 + \pi_1 + \pi_2 = 1$

Solving this system yields $\pi \approx [0.4, 0.35, 0.25]^\top$, which represents the long-run proportion of time spent in each state.

14.14 Information Bottleneck

14.14.1 Definition: Information Bottleneck

For a Markov chain $X \to Z \to Y$ where X is the input, Z is a learned representation, and Y is the target, the **Information Bottleneck principle** seeks to find a representation Z that:

1. **Maximizes** $I(Z;Y)$ (preserves information about the target)
2. **Minimizes** $I(X;Z)$ (compresses the input)

This is formalized as the optimization problem:

$$\min_{p(z|x)} I(X;Z) - \beta I(Z;Y),$$

where $\beta > 0$ is a trade-off parameter controlling the balance between compression and preservation of relevant information.

14.14.2 Interpretation

The Information Bottleneck principle states that optimal representations should:

- **Compress** the input (minimize $I(X;Z)$)
- **Preserve** information relevant to the task (maximize $I(Z;Y)$)

This creates a "bottleneck" that forces the representation to extract only the most relevant information.

The Information Bottleneck principle provides a theoretical foundation for **representation learning** and **feature extraction**.

In **deep learning**, the Information Bottleneck explains why neural networks learn compressed, task-relevant representations: early layers extract general features, while later layers focus on task-specific information. In **CNNs**, the Information Bottleneck explains the hierarchical feature extraction: early convolutional layers compress raw pixel information into local patterns, while deeper layers compress these patterns into high-level semantic features relevant to the task. In **transformers**, the Information Bottleneck perspective explains how attention mechanisms compress sequence information: self-attention layers compress information across tokens, while cross-attention compresses information between different modalities. In **graph neural networks**, the Information Bottleneck explains how message passing compresses node and edge information into task-relevant node representations. In **variational autoencoders**, the KL regularization term can be interpreted through the Information Bottleneck lens: it encourages the latent representation to compress the input while preserving information needed for reconstruction. In **transfer learning**, the Information Bottleneck explains why pre-trained representations transfer well: they compress general features while preserving task-relevant information. The trade-off parameter β controls the compression-preservation balance: large β preserves more information (less compression), while small β prioritizes compression (more information loss).

14.15 Evaluation Metrics for Ranking and Information Retrieval

14.15.1 Definition: Precision at K

For a ranking or information retrieval system, **Precision at K** (denoted as **P@K**) is defined as the proportion of relevant items among the top K retrieved items:

$$\text{P@K} = \frac{\text{Number of relevant items in top } K}{K}.$$

Precision at K measures the quality of the top K results, which is particularly important in search engines and recommendation systems where users typically only examine the first few results.

14.15.2 Definition: Recall at K

For a ranking or information retrieval system, **Recall at K** (denoted as **R@K**) is defined as the proportion of all relevant items that appear in the top K retrieved items:

$$\text{R@K} = \frac{\text{Number of relevant items in top } K}{\text{Total number of relevant items}}.$$

Recall at K measures how many of the relevant items the system successfully retrieved in the top K positions.

14.15.3 Definition: Average Precision (AP)

For a single query or ranking task, **Average Precision (AP)** is defined as the average of precision values computed at each position where a relevant item is found:

$$\text{AP} = \frac{1}{|\text{Relevant}|} \sum_{k=1}^{K} \text{Precision@}k \cdot \mathbf{1}\{\text{item at position } k \text{ is relevant}\},$$

where $|\text{Relevant}|$ is the total number of relevant items, K is the total number of retrieved items, and $\mathbf{1}\{\cdot\}$ is the indicator function.

Equivalently, if we let $r_k \in \{0, 1\}$ indicate whether the item at position k is relevant, and let $P@k = \frac{\sum_{i=1}^{k} r_i}{k}$ be the precision at position k, then:

$$\text{AP} = \frac{1}{|\text{Relevant}|} \sum_{k:r_k=1} P@k = \frac{1}{|\text{Relevant}|} \sum_{k:r_k=1} \frac{\sum_{i=1}^{k} r_i}{k}.$$

Average Precision provides a single metric that considers both the precision and the ranking order of relevant items. It rewards systems that rank relevant items higher.

14.15.4 Definition: Mean Average Precision (MAP)

For a collection of queries or ranking tasks, **Mean Average Precision (MAP)** is defined as the mean of Average Precision scores across all queries:

$$\text{MAP} = \frac{1}{Q} \sum_{q=1}^{Q} \text{AP}_q,$$

where Q is the total number of queries and AP_q is the Average Precision for query q.

MAP is one of the most widely used metrics in information retrieval and ranking systems because it: 1. **Considers ranking order**: Higher-ranked relevant items contribute more to the score. 2. **Balances precision and recall**: It naturally incorporates both metrics through the precision-at-k values. 3. **Provides a single summary metric**: It aggregates performance across multiple queries into one number.

14.15.4.1 Example: Computing MAP

Consider a retrieval system with 2 queries. For Query 1, we retrieve 5 items with relevance labels $[1, 0, 1, 0, 1]$ (1 = relevant, 0 = irrelevant). For Query 2, we retrieve 4 items with relevance labels $[1, 1, 0, 1]$.

For Query 1: - Relevant items: 3 total - Precision@1 = 1/1 = 1.0 (item 1 is relevant) - Precision@3 = 2/3 ≈ 0.667 (items 1 and 3 are relevant) - Precision@5 = 3/5 = 0.6 (items 1, 3, and 5 are relevant) - AP_1 = (1.0 + 0.667 + 0.6) / 3 ≈ 0.756

For Query 2: - Relevant items: 3 total - Precision@1 = 1/1 = 1.0 (item 1 is relevant) - Precision@2 = 2/2 = 1.0 (items 1 and 2 are relevant) - Precision@4 = 3/4 = 0.75 (items 1, 2, and 4 are relevant) - AP_2 = (1.0 + 1.0 + 0.75) / 3 ≈ 0.917

MAP = (AP_1 + AP_2) / 2 = (0.756 + 0.917) / 2 ≈ 0.837

Mean Average Precision (MAP) is fundamental to **information retrieval** and **ranking systems**. In **search engines**, MAP evaluates how well the system ranks relevant documents for user queries. In **recommendation systems**, MAP measures how well the system ranks relevant items (products, movies, etc.) for each user. In **object detection** and **image retrieval**, MAP evaluates how well the system ranks relevant bounding boxes or images. In **neural information retrieval**, MAP is used to evaluate learned ranking functions that map queries and documents to relevance scores. The key advantage of MAP over simple precision or recall is that it **rewards correct ranking**: a system that places all relevant items in the top positions achieves a higher MAP than a system that places them lower, even if both systems retrieve the same number of relevant items. MAP is particularly useful when the number of relevant items varies across queries, as it normalizes by the number of relevant items per query. In practice, MAP is often computed at a cutoff (e.g., MAP@10 considers only the top 10 results), which is more realistic for applications where users only examine a limited number of results.

14.16 Mindmap

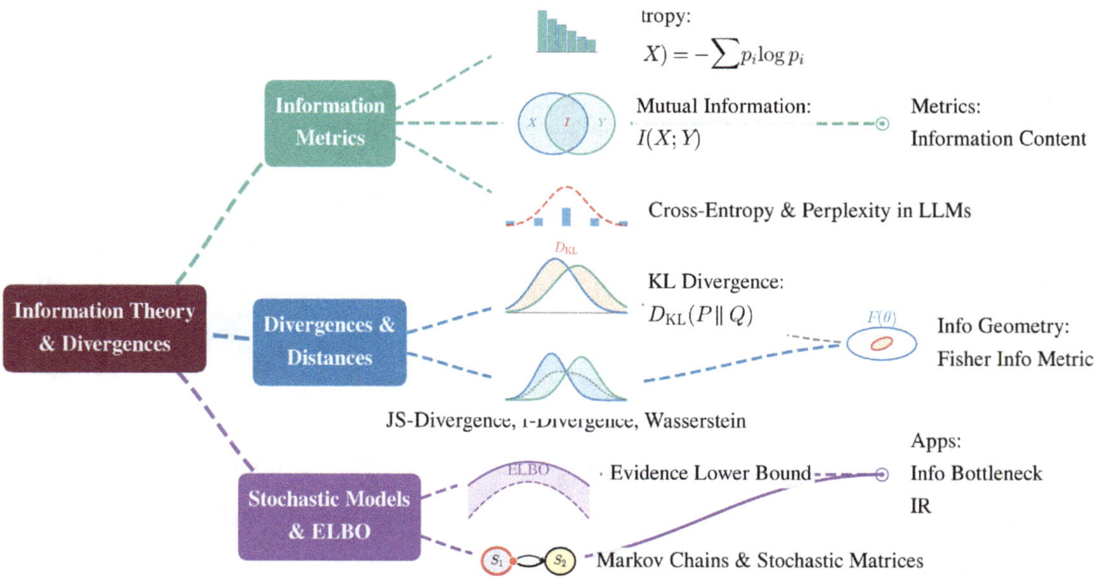

Figure 14.4: Mind map of information theory topics

14.17 Exercises

14.17.1 Information Content and Entropy

1. Compute the information content (in bits) of an outcome with probability $p = 0.25$.
2. For independent events X and Y with probabilities $p(x)$ and $p(y)$, show that $I(x, y) = I(x) + I(y)$.
3. Compute the entropy of a fair coin flip (Bernoulli distribution with $p = 0.5$).
4. For a discrete random variable with N outcomes, prove that entropy is maximized when the distribution is uniform: $p(x) = 1/N$ for all x.
5. For a continuous random variable $X \sim \mathcal{N}(\mu, \sigma^2)$, compute the differential entropy $H(X)$ and show that it depends only on σ^2, not on μ.

14.17.2 Information Gain and Decision Trees

1. For a binary classification problem with target $Y \in \{0, 1\}$ and a binary feature $X \in \{0, 1\}$, show that information gain $\text{IG}(Y; X) = H(Y) - H(Y|X)$ equals mutual information $I(Y; X)$.
2. Consider a dataset with target Y having class distribution $P(Y = 0) = 0.6$, $P(Y = 1) = 0.4$. Compute the entropy $H(Y)$.
3. For the dataset in problem 2, suppose feature X splits the data such that:
   - When $X = 0$: $P(Y = 0|X = 0) = 0.8$, $P(Y = 1|X = 0) = 0.2$
   - When $X = 1$: $P(Y = 0|X = 1) = 0.3$, $P(Y = 1|X = 1) = 0.7$
   - $P(X = 0) = 0.5$, $P(X = 1) = 0.5$

   Compute the conditional entropy $H(Y|X)$ and information gain $\text{IG}(Y; X)$.
4. In a decision tree, explain why maximizing information gain at each split leads to building a tree that minimizes the expected entropy of the target variable at the leaf nodes.
5. Show that information gain is non-negative: $\text{IG}(Y; X) = H(Y) - H(Y|X) \geq 0$, with equality if and only if Y and X are independent.
6. **ID3 Algorithm**: Consider a dataset for binary classification with target $Y \in \{0, 1\}$ and three binary features $X_1, X_2, X_3 \in \{0, 1\}$. The marginal joint distributions are:
   - For X_1: $P(X_1 = 0, Y = 0) = 0.2$, $P(X_1 = 0, Y = 1) = 0.3$, $P(X_1 = 1, Y = 0) = 0.1$, $P(X_1 = 1, Y = 1) = 0.4$
   - For X_2: $P(X_2 = 0, Y = 0) = 0.15$, $P(X_2 = 0, Y = 1) = 0.35$, $P(X_2 = 1, Y = 0) = 0.15$, $P(X_2 = 1, Y = 1) = 0.35$
   - For X_3: $P(X_3 = 0, Y = 0) = 0.25$, $P(X_3 = 0, Y = 1) = 0.25$, $P(X_3 = 1, Y = 0) = 0.05$, $P(X_3 = 1, Y = 1) = 0.45$

   The complete joint distribution $P(X_1, X_2, X_3, Y)$ is given in the table below (only non-zero probabilities are shown; all other combinations have probability 0):

Feature X_1	Feature X_2	Feature X_3	Target Y	Probability
0	0	0	0	0.10
0	1	0	0	0.10
0	1	1	1	0.30
1	0	0	1	0.25

Feature X_1	Feature X_2	Feature X_3	Target Y	Probability
1	0	1	0	0.05
1	0	1	1	0.10
1	1	0	0	0.05
1	1	1	1	0.05

Note: You may use the following approximations: $\log_2 0.1 \approx -3.32$, $\log_2 0.15 \approx -2.74$, $\log_2 0.2 \approx -2.32$, $\log_2 0.25 = -2$, $\log_2 0.3 \approx -1.74$, $\log_2 0.35 \approx -1.51$, $\log_2 0.4 \approx -1.32$, $\log_2 0.45 \approx -1.15$, $\log_2 0.5 = -1$.

(a) Compute the information gain $\text{IG}(Y; X_i)$ for each feature $i \in \{1, 2, 3\}$ at the root node. Show all intermediate steps (marginal distribution of Y, conditional entropies $H(Y|X_i)$, etc.). Round your final answers to 3 decimal places.

(b) According to the ID3 algorithm, which feature should be selected for the root split? Justify your answer.

(c) After selecting the root feature from part (b), consider the branch where the root feature equals 0. In this branch, compute the information gain for the remaining two features. Which feature should be selected for the second split in this branch? Show your work.

(d) In the branch where the root feature equals 1, compute the information gain for the remaining features. Is a further split necessary in this branch? Explain your reasoning.

14.17.3 Cross-Entropy and Perplexity

1. For a binary classification problem with true label $y \in \{0, 1\}$ and predicted probability $\hat{y} \in [0, 1]$, show that the cross-entropy loss is $H(p, q) = -y \log \hat{y} - (1 - y) \log(1 - \hat{y})$.
2. For multi-class classification with C classes, true one-hot label $\mathbf{y} \in \{0, 1\}^C$ (where $\sum_{c=0}^{C-1} y_c = 1$), and predicted probabilities $\hat{\mathbf{y}} \in [0, 1]^C$ (where $\sum_{c=0}^{C-1} \hat{y}_c = 1$), show that the cross-entropy loss is $H(\mathbf{y}, \hat{\mathbf{y}}) = -\sum_{c=0}^{C-1} y_c \log \hat{y}_c$.
3. Prove that minimizing cross-entropy $H(p, q)$ with respect to q (keeping p fixed) is equivalent to minimizing KL divergence $D_{\text{KL}}(p||q)$.
4. For a uniform distribution over N outcomes, compute the perplexity and interpret the result.
5. For a language model with cross-entropy $H = 5$ bits per token, compute the perplexity and explain what it means.
6. Show that perplexity is the geometric mean of the inverse probabilities: Perplexity $= \left(\prod_x p(x)^{-p(x)}\right)^{1/\sum_x p(x)}$.

14.17.4 KL Divergence Properties

1. Prove that $D_{\text{KL}}(p||q) \geq 0$ using Jensen's inequality.
2. For two univariate Gaussian distributions $p(x) = \mathcal{N}(\mu_1, \sigma_1^2)$ and $q(x) = \mathcal{N}(\mu_2, \sigma_2^2)$, compute $D_{\text{KL}}(p||q)$.
3. Show that $D_{\text{KL}}(p||q) \neq D_{\text{KL}}(q||p)$ in general by providing a counterexample.

14.17.5 Fisher Information Matrix

1. For a univariate Gaussian distribution $p(x|\mu, \sigma^2) = \mathcal{N}(\mu, \sigma^2)$ with parameters $\theta = [\mu, \sigma^2]^\top$, compute the Fisher Information Matrix $F(\theta)$.
2. Prove that the Fisher Information Matrix is symmetric and positive semi-definite.
3. Show that for small perturbations δ, $D_{\text{KL}}(p(\mathbf{x}|\theta)||p(\mathbf{x}|\theta + \delta)) \approx \frac{1}{2}\delta^\top F(\theta)\delta$.

14.17.6 ELBO and Variational Inference

1. Prove that $\text{ELBO}(q, \theta) \leq \log p(\mathbf{x}|\theta)$ for any variational distribution $q(\mathbf{z})$.
2. Show that the ELBO can be decomposed as $\text{ELBO}(q, \theta) = \mathbb{E}_{q(\mathbf{z})}[\log p(\mathbf{x}|\mathbf{z}, \theta)] - D_{\text{KL}}(q(\mathbf{z})||p(\mathbf{z}|\theta))$.
3. In a VAE with encoder $q(\mathbf{z}|\mathbf{x})$ and decoder $p(\mathbf{x}|\mathbf{z})$, explain how the ELBO objective balances reconstruction accuracy and latent regularization.

14.17.7 Jensen-Shannon Divergence and f-Divergence

1. Prove that JS divergence is symmetric: $D_{\text{JS}}(p||q) = D_{\text{JS}}(q||p)$.
2. Show that JS divergence is bounded: $0 \leq D_{\text{JS}}(p||q) \leq \ln 2$ nats (equivalently, ≤ 1 bit).
3. For f-divergence with $f(t) = t \log t$, show that it equals KL divergence $D_{\text{KL}}(p||q)$.
4. In GAN training, explain why JS divergence (theoretical objective) is often replaced with Wasserstein distance or other objectives in practice.

14.17.8 Wasserstein Distance

1. For two one-dimensional distributions $p(x)$ and $q(x)$ with cumulative distribution functions F_p and F_q, show that the Wasserstein-1 distance is $W_1(p, q) = \int_{-\infty}^{\infty} |F_p(x) - F_q(x)|\, dx$.
2. Compute the Wasserstein-1 distance between two Dirac delta distributions: $p(x) = \delta(x - a)$ and $q(x) = \delta(x - b)$ on the real line with Euclidean distance $d(x, y) = |x - y|$.
3. For two univariate Gaussian distributions $p(x) = \mathcal{N}(\mu_1, \sigma_1^2)$ and $q(x) = \mathcal{N}(\mu_2, \sigma_2^2)$ with $\sigma_1 = \sigma_2 = \sigma$, show that $W_1(p, q) = |\mu_1 - \mu_2|$.
4. Explain why Wasserstein distance is well-defined even when two distributions have disjoint supports, while KL divergence can be infinite in such cases.
5. In the context of Wasserstein GANs (WGANs), explain the role of the Lipschitz constraint in the Kantorovich-Rubinstein duality formulation. Why is enforcing the Lipschitz constraint important for training stability?
6. Compare and contrast Wasserstein distance with f-divergences (e.g., KL divergence, JS divergence) in terms of:
   - Sensitivity to distribution supports
   - Computational complexity
   - Suitability for gradient-based optimization
   - Interpretation in terms of optimal transport

14.17.9 Stochastic Matrices and Markov Chains

1. Verify that the following matrix is a stochastic matrix:

$$P = \begin{bmatrix} 0.5 & 0.3 & 0.2 \\ 0.2 & 0.6 & 0.2 \\ 0.1 & 0.4 & 0.5 \end{bmatrix}.$$

   Is it doubly stochastic? Explain.

2. For the stochastic matrix in problem 1, find the stationary distribution π by solving $\pi^\top P = \pi^\top$ with the constraint $\sum_i \pi_i = 1$.

3. Prove that if P is a stochastic matrix, then P^k is also a stochastic matrix for any positive integer k.

4. Show that every stochastic matrix P has eigenvalue $\lambda = 1$ with a corresponding left eigenvector $\pi^\top$ (the stationary distribution).

5. For a doubly stochastic matrix P of size $N \times N$, prove that if the Markov chain is ergodic, then the stationary distribution is uniform: $\pi = \frac{1}{N}\mathbf{1}_N$.

6. Consider a Markov chain with transition matrix:

$$P = \begin{bmatrix} 0.8 & 0.2 & 0 \\ 0.3 & 0.5 & 0.2 \\ 0 & 0.4 & 0.6 \end{bmatrix}.$$

   Compute P^2 and P^3, and verify that the rows still sum to 1. What can you say about the convergence of P^k as $k \to \infty$?

7. For an irreducible and aperiodic stochastic matrix P, use the Perron-Frobenius theorem to explain why the stationary distribution π has all positive entries: $\pi_i > 0$ for all i.

14.17.10 Mutual Information and Information Bottleneck

1. Prove that $I(X;Y) = 0$ if and only if X and Y are independent.

2. Show that $I(X;Y) = H(X) - H(X|Y)$.

3. For a Markov chain $X \to Y \to Z$, prove the data processing inequality: $I(X;Z) \leq I(X;Y)$.

4. In the Information Bottleneck framework, explain the trade-off between compression $I(X;Z)$ and information preservation $I(Z;Y)$ as β varies.

14.17.11 MAP and Ranking Metrics

1. A search engine retrieves 8 documents for a query. The relevance labels (1 = relevant, 0 = irrelevant) for the retrieved documents in order are: $[1, 1, 0, 1, 0, 0, 1, 0]$. Compute:

   - Precision@3, Precision@5, Precision@8
   - Recall@3, Recall@5, Recall@8
   - Average Precision (AP)

2. For a recommendation system, three users have the following retrieval results (1 = relevant, 0 = irrelevant):

   - User 1: $[1, 0, 1, 1, 0]$ (5 items retrieved)
   - User 2: $[1, 1, 0, 1]$ (4 items retrieved)
   - User 3: $[0, 1, 1, 1, 0, 1]$ (6 items retrieved)

   Compute the Mean Average Precision (MAP) across all three users.

3. Show that Average Precision (AP) can be written as:

$$\text{AP} = \sum_{k=1}^{K} \frac{r_k}{|\text{Relevant}|} \cdot \frac{\sum_{i=1}^{k} r_i}{k},$$

where $r_k \in \{0, 1\}$ indicates relevance at position k, and $|\text{Relevant}|$ is the total number of relevant items.

4. Compare Precision@K and Recall@K for different values of K. If a system has high Precision@5 but low Recall@5, what does this tell you about the system's performance?

5. In information retrieval, explain why MAP is preferred over simple precision or recall for evaluating ranking systems. Give an example where two systems have the same precision and recall but different MAP values.

14.17.12 Entropy of Common Distributions

1. For a Bernoulli distribution with parameter p, derive the general formula for entropy $H(X) = -p \log p - (1-p)\log(1-p)$. Show that entropy is maximized when $p = 0.5$ (fair coin), and compute the maximum entropy value.

2. For a categorical distribution over K categories with probability vector $\mathbf{p} = (p_0, p_1, ..., p_{K-1})^\top$, show that the entropy is $H(X) = -\sum_{k=0}^{K-1} p_k \log p_k$. What is the maximum entropy, and when is it achieved?

3. For a geometric distribution with parameter p (probability of success), where $P(X = k) = (1-p)^{k-1}p$ for $k = 1, 2, 3, ...$, compute the entropy $H(X)$.

4. For a Poisson distribution with parameter λ, where $P(X = k) = \frac{\lambda^k e^{-\lambda}}{k!}$ for $k = 0, 1, 2, ...$, show that the entropy is approximately $H(X) \approx \frac{1}{2}\log(2\pi e\lambda)$ for large λ. (Hint: Use Stirling's approximation.)

5. Compare the entropy of a uniform distribution over N outcomes with the entropy of a distribution that is highly concentrated (e.g., one outcome has probability 0.9, and the remaining probability is uniformly distributed over the other $N-1$ outcomes). Which has higher entropy, and why?

14.17.13 Conditional Entropy and Chain Rules

1. Prove the chain rule of entropy: $H(X, Y) = H(X) + H(Y|X) = H(Y) + H(X|Y)$.

2. For three random variables X, Y, Z, prove the chain rule: $H(X, Y, Z) = H(X) + H(Y|X) + H(Z|X, Y)$.

3. Show that $H(X|Y) \leq H(X)$ with equality if and only if X and Y are independent. Interpret this result.

4. For random variables X, Y, Z forming a Markov chain $X \to Y \to Z$ (i.e., X and Z are conditionally independent given Y), prove that $H(X|Y, Z) = H(X|Y)$.

5. Consider a dataset with features X_1, X_2, X_3 and target Y. Given:

   - $H(Y) = 1.5$ bits
   - $H(Y|X_1) = 0.8$ bits
   - $H(Y|X_2) = 1.0$ bits
   - $H(Y|X_3) = 1.2$ bits

   Compute the information gain for each feature. Which feature is most informative about Y?

6. For a sequence of random variables $X_1, X_2, \dots, X_n$ where each X_i depends only on X_{i-1} (Markov chain), show that:

$$H(X_1, X_2, \dots, X_n) = H(X_1) + \sum_{i=2}^{n} H(X_i|X_{i-1}).$$

14.17.14 Information Theory Applications in Machine Learning

1. **Feature Selection**: Given a dataset with features X_1, X_2, X_3 and target Y, where:
   - $I(Y; X_1) = 0.5$ bits
   - $I(Y; X_2) = 0.3$ bits
   - $I(Y; X_3) = 0.2$ bits
   - $I(X_1; X_2) = 0.4$ bits
   - $I(X_1; X_3) = 0.1$ bits
   - $I(X_2; X_3) = 0.05$ bits

   If you can only select 2 features, which features should you choose? Justify using mutual information and redundancy considerations.
2. **Language Modeling**: For a language model predicting the next word in a sequence, explain how perplexity relates to the model's uncertainty. If a model has perplexity 100, what does this mean in terms of the effective vocabulary size?
3. **Variational Autoencoders**: In a VAE, the ELBO is ELBO $= \mathbb{E}_{q(\mathbf{z}|\mathbf{x})}[\log p(\mathbf{x}|\mathbf{z})] - D_{\text{KL}}(q(\mathbf{z}|\mathbf{x})||p(\mathbf{z}))$. Explain how the KL divergence term acts as a regularizer that encourages the learned distribution $q(\mathbf{z}|\mathbf{x})$ to be close to the prior $p(\mathbf{z})$. What happens if this term is too large? Too small?
4. **Information Bottleneck in Neural Networks**: Consider a neural network with layers $X \to Z_1 \to Z_2 \to Y$, where X is the input, Z_1 and Z_2 are hidden layers, and Y is the target. Using the data processing inequality, show that $I(X; Y) \leq I(X; Z_1) \leq I(X; Z_2)$. What does this tell you about information flow through the network?
5. **Cross-Entropy Loss**: Explain why minimizing cross-entropy loss is a natural objective for classification tasks. How does it relate to maximum likelihood estimation?
6. **Entropy Regularization**: In some machine learning algorithms, entropy regularization is used to encourage model outputs to be more uniform. For a classification model outputting probabilities $\mathbf{p} = (p_0, p_1, \dots, p_{K-1})^\top$, the entropy is $H(\mathbf{p}) = -\sum_{k=0}^{K-1} p_k \log p_k$. Explain why maximizing $H(\mathbf{p})$ (subject to other constraints) encourages the model to be less overconfident.

14.17.15 Entropy and Coding Theory

1. **Huffman Coding**: For a source with 4 symbols and probabilities $[0.5, 0.25, 0.125, 0.125]$, compute:
   - The entropy of the source
   - The average code length under an optimal Huffman code
   - Verify that the average code length is at least equal to the entropy
2. **Shannon's Source Coding Theorem**: State Shannon's source coding theorem. For a source with entropy $H(X)$ bits per symbol, what is the minimum average number of bits per symbol needed to encode the source losslessly?
3. **Arithmetic Coding**: Explain how arithmetic coding achieves compression rates close to the entropy limit. How does it differ from Huffman coding?

4. **Channel Capacity**: For a binary symmetric channel with error probability p, the channel capacity is $C = 1 - H(p)$, where $H(p) = -p \log p - (1-p)\log(1-p)$ is the binary entropy function. Compute the channel capacity for $p = 0.1$ and $p = 0.5$. Interpret the results.

5. **Data Compression**: A file contains text where each character follows a distribution with entropy $H = 4.5$ bits per character. If the file has 1000 characters, what is the theoretical minimum file size (in bits) after lossless compression? Can you achieve this in practice? Why or why not?

6. **Kolmogorov Complexity**: Explain the relationship between entropy and Kolmogorov complexity. How does entropy provide a lower bound on the compressibility of data?

Chapter 15

Bellman Equations and Operators

Markov decision processes, Bellman operators, and the mathematics of sequential decision making

> *"The best way to predict the future is to create it."*
>
> — Peter Drucker

- ▷ `MDP` $(\mathcal{S}, \mathcal{A}, P, R, \gamma)$ where $P : \mathcal{S}\times\mathcal{A}\times\mathcal{S} \to [0,1]$ with $\sum_{s_1} P(s_1|s,a) = 1\ \forall s, a$, $R : \mathcal{S}\times\mathcal{A} \to \mathbb{R}$, $\gamma \in [0,1)$
- ▷ `Policy` Stochastic: $\pi : \mathcal{S} \times \mathcal{A} \to [0,1]$ with $\sum_a \pi(a|s) = 1$; deterministic: $\pi : \mathcal{S} \to \mathcal{A}$
- ▷ `Discounted return` $G_t = \sum_{k=0}^{\infty} \gamma^k r_{t+1+k}$ for trajectory $(s_0, a_0, r_1, s_1, a_1, r_2, ...)$
- ▷ `Value` $V_\pi(s) = \mathbb{E}_\pi[G_t|S_t = s]$; $Q_\pi(s,a) = \mathbb{E}_\pi[G_t|S_t = s, A_t = a]$; $V_\pi(s) = \sum_a \pi(a|s)Q_\pi(s,a)$; $Q_\pi(s,a) = R(s,a) + \gamma \sum_{s_1} P(s_1|s,a)V_\pi(s_1)$
- ▷ `Bellman(state)` $V_\pi(s) = \sum_a \pi(a|s)[R(s,a) + \gamma \sum_{s_1} P(s_1|s,a)V_\pi(s_1)]$; matrix: $\mathbf{v}_\pi = \mathbf{r}_\pi + \gamma P_\pi \mathbf{v}_\pi = (I - \gamma P_\pi)^{-1}\mathbf{r}_\pi$ (when $\gamma < 1$)
- ▷ `Bellman(action)` $Q_\pi(s,a) = R(s,a) + \gamma \sum_{s_1} P(s_1|s,a) \sum_{a_1} \pi(a_1|s_1)Q_\pi(s_1,a_1)$
- ▷ `Bellman(op)` $T_\pi \mathbf{v} = \mathbf{r}_\pi + \gamma P_\pi \mathbf{v}$; $\mathbf{v}_\pi = T_\pi \mathbf{v}_\pi$; $(T_* \mathbf{v})(s) = \max_a [R(s,a) + \gamma \sum_{s_1} P(s_1|s,a)\mathbf{v}(s_1)]$
- ▷ `Bellman operators are contractions` $\|T_\pi \mathbf{v}_1 - T_\pi \mathbf{v}_2\|_\infty \le \gamma \|\mathbf{v}_1 - \mathbf{v}_2\|_\infty$; same for T_*
- ▷ `Optimal value function` $V_*(s) = \max_\pi V_\pi(s)$; $Q_*(s,a) = \max_\pi Q_\pi(s,a)$; Bellman optimality: $V_*(s) = \max_a [R(s,a) + \gamma \sum_{s_1} P(s_1|s,a)V_*(s_1)]$; $V_* = T_* V_*$
- ▷ `Optimal policy` $\pi_*(s) = \arg\max_a Q_*(s,a)$; deterministic and satisfies $V_{\pi_*}(s) = V_*(s)$
- ▷ `Value iteration` $\mathbf{v}_{k+1} = T_* \mathbf{v}_k$; converges to V_* with $\|\mathbf{v}_k - V_*\|_\infty \le \gamma^k \|\mathbf{v}_0 - V_*\|_\infty$; extract policy: $\pi_*(s) = \arg\max_a [R(s,a) + \gamma \sum_{s_1} P(s_1|s,a)\mathbf{v}_k(s_1)]$
- ▷ `Policy iteration` Alternates (1) Policy evaluation: solve $\mathbf{v}^{\pi_k} = \mathbf{r}^{\pi_k} + \gamma P^{\pi_k} \mathbf{v}^{\pi_k}$; (2) Policy improvement: $\pi_{k+1}(s) = \arg\max_a [R(s,a) + \gamma \sum_{s_1} P(s_1|s,a)\mathbf{v}^{\pi_k}(s_1)]$; converges in finite iterations
- ▷ `State distribution` $\mu_t(s) = P(S_t = s|\mu_0, \pi)$; evolution: $\mu_{t+1}(s_1) = \sum_s \sum_a \mu_t(s)\pi(a|s)P(s_1|s,a)$; matrix: $\mu_{t+1}^\top = \mu_t^\top P_\pi$
- ▷ `Policy gradient,stochastic approx.,func approx.,importance sampling` ...

15.1 Notation and Conventions

This chapter follows the reinforcement learning notation conventions established in the Preface (see "Reinforcement Learning Notation" section). Key conventions include:

- **Random variables**: Uppercase letters (S_t, A_t, R_{t+1}, G_t) for states, actions, rewards, and returns. Following the Sutton & Barto convention[1], we use R_{t+1} to denote the reward received after taking action A_t in state S_t and transitioning to S_{t+1}.
- **Realizations**: Lowercase letters (s, a, r) for specific values
- **Value functions**: $V_\pi(s)$ and $Q_\pi(s,a)$ with policy subscripts
- **Sets**: Calligraphic letters ($\mathcal{S}$, $\mathcal{A}$) for state and action spaces

For detailed notation conventions, see the Preface. The expected reward is $\mathbb{E}[R_{t+1}|S_t = s, A_t = a] = R(s,a)$.

15.2 Markov Decision Processes

In Chapter~14, we studied Markov chains where state transitions follow a fixed probabilistic law with no external control. A Markov Decision Process extends this framework by introducing an **agent** that can influence the process through **actions**. Unlike a Markov chain where transitions are purely stochastic, an MDP allows the agent to choose actions at each step, making the process **controlled** rather than passive. To guide the agent's decision-making, we add a **reward function** that evaluates the desirability of state-action pairs, and a **discount factor** that balances the importance of immediate versus future rewards. This transformation from a passive stochastic process to an active decision-making framework is what makes MDPs the mathematical foundation of reinforcement learning.

15.2.1 Definition: Markov Decision Process (Finite)

A **Markov Decision Process (MDP)** is a 5-element tuple $(\mathcal{S}, \mathcal{A}, P, R, \gamma)$ where:

- $\mathcal{S}$ is a finite set of **states** with $|\mathcal{S}| = N$ (where $|\mathcal{S}|$ denotes the number of elements in $\mathcal{S}$, called the **cardinality**)
- $\mathcal{A}$ is a finite set of **actions** with $|\mathcal{A}| = M$ (where $|\mathcal{A}|$ denotes the number of elements in $\mathcal{A}$)
- $P : \mathcal{S} \times \mathcal{A} \times \mathcal{S} \to [0,1]$ is the **transition probability function** satisfying

$$\sum_{s' \in \mathcal{S}} P(s'|s,a) = 1 \quad \forall s \in \mathcal{S}, a \in \mathcal{A}$$

- $R : \mathcal{S} \times \mathcal{A} \to \mathbb{R}$ is the **reward function**
- $\gamma \in [0,1)$ is the **discount factor**

15.2.2 Intuition

An MDP models sequential decision-making under uncertainty with an **agent** interacting with an **environment**:

---

[1] We follow the **Sutton & Barto convention** in their book *Reinforcement Learning: An Introduction* where R_{t+1} denotes the reward received AFTER transitioning from state S_t to S_{t+1} by taking action A_t. This convention emphasizes that the reward is a consequence of the state transition. Some other textbooks may use R_t to denote the reward associated with the state-action pair (S_t, A_t) at time t. Both conventions are mathematically equivalent; the key difference is whether the reward is indexed at the time of the action (R_t) or at the time after the transition (R_{t+1}). The trajectory representation differs accordingly: $(s_0, a_0, r_1, s_1, a_1, r_2, ...)$ in the Sutton & Barto convention versus $(s_0, a_0, r_0, s_1, a_1, r_1, ...)$ in the alternative convention.

1. At each time step t, the **agent** observes the current state $s_t \in \mathcal{S}$
2. The **agent** selects an action $a_t \in \mathcal{A}$ based on the current state (using a **policy**)
3. The **environment** transitions to the next state $s_{t+1} \sim P(\cdot|s_t, a_t)$ according to the state transition function
4. The **environment** provides a reward r_{t+1} to the agent, where $\mathbb{E}[r_{t+1}|s_t, a_t] = R(s_t, a_t)$
5. The process repeats: the agent observes s_{t+1}, selects a_{t+1}, and so on

The **agent's goal** is to maximize the cumulative reward over time, which we formalize as the **discounted return** $G_t = \sum_{k=0}^{\infty} \gamma^k r_{t+1+k}$ (defined formally in the next section).

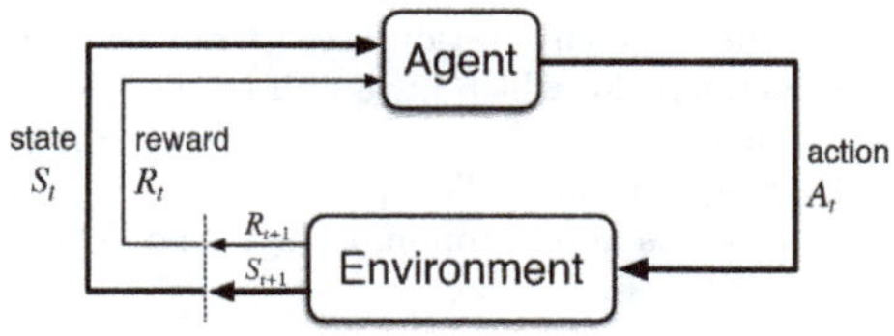

The **Markov property** states that the future depends only on the current state and action, not the history:

$$P(s_{t+1}|s_t, a_t, s_{t-1}, a_{t-1}, \ldots) = P(s_{t+1}|s_t, a_t).$$

This property makes the problem tractable: the agent only needs to consider the current state when making decisions, not the entire history.

> MDPs are the mathematical foundation of **reinforcement learning**. They model environments where an agent learns to make decisions by interacting with the world. The Markov property is crucial: it allows us to reason about the future using only the current state, making the problem tractable. In practice, many real-world problems can be modeled as MDPs: **game playing** (chess, Go), **robotics** (navigation, manipulation), **autonomous systems** (self-driving cars), and **recommendation systems** (sequential recommendations). The discount factor γ ensures that infinite-horizon problems have finite value functions and makes future rewards less important than immediate ones, which is often desirable in practice.

15.2.3 From State Transition Matrix to State Transition Function

In Chapter~14, we introduced the **transition matrix** representation for Markov chains where there is no agent making decisions. The transition probabilities were represented as a matrix $P \in \mathbb{R}^{N\times N}$ where $[P]_{ij} = P(s_j|s_i)$.

In MDPs, where an agent selects actions, we use the **state transition function** $P : \mathcal{S} \times \mathcal{A} \times \mathcal{S} \to [0, 1]$ instead of a matrix. This is because: (1) the state transition now depends on the action as well, making it a three-dimensional structure rather than a two-dimensional matrix; and (2) the functional representation is more general—if the state space is not finite (e.g., continuous state spaces), we cannot use array/matrix representation, but we can still use the functional form $P(s'|s, a)$.

For **finite state spaces**, we can still represent the transition probabilities as a collection of matrices. For each action $a \in \mathcal{A}$, define the **transition matrix** $P^a \in \mathbb{R}^{N\times N}$ where the element in the i-th row and j-th column is

$$P^a_{ij} = P(s_j|s_i, a),$$

where states are indexed as $s_0, s_1, \ldots, s_{N-1}$. Each row of P^a is a probability distribution: $\sum_{j=0}^{N-1} P^a_{ij} = 1$ for all i. However, this matrix representation is a special case that only applies to finite state spaces.

15.2.4 Reward Function Representation

The reward function $R : \mathcal{S} \times \mathcal{A} \to \mathbb{R}$ depends on both state and action. For finite MDPs, it can be represented as a matrix $R \in \mathbb{R}^{N\times M}$ where the element in the i-th row and j-th column is

$$R_{ij} = R(s_i, a_j),$$

where actions are indexed as $a_0, a_1, \dots, a_{M-1}$.

The state transition function $P(s'|s, a)$ is the fundamental representation in MDPs, as it generalizes to continuous state spaces. The matrix representation is a special case for finite state spaces that enables efficient computation using **linear algebra**. Transition matrices are **stochastic matrices** (each row sums to 1), which have well-studied properties. The spectral properties of these matrices determine the long-term behavior of the MDP. In **large-scale RL**, sparse matrix representations are essential for computational efficiency. The functional form $P(s'|s, a)$ is crucial for **continuous control** problems in robotics and autonomous systems, where state spaces are continuous.

15.3 Policies

15.3.1 Definition: Policy

A **policy** is a function that determines which action to take in each state. There are two types:

1. **Stochastic policy**: A function $\pi : \mathcal{S} \times \mathcal{A} \to [0, 1]$ that maps each state-action pair to a probability:
$$\pi(a|s) = P(A_t = a|S_t = s),$$
satisfying $\sum_{a \in \mathcal{A}} \pi(a|s) = 1$ for all $s \in \mathcal{S}$.
2. **Deterministic policy**: A function $\pi : \mathcal{S} \to \mathcal{A}$ that maps each state to a single action:
$$\pi(s) = a.$$
A deterministic policy can be viewed as a special case of a stochastic policy where $\pi(a|s) = 1$ for $a = \pi(s)$ and $\pi(a'|s) = 0$ for $a' \neq \pi(s)$.

15.3.2 Definition: Policy Matrix

For a finite MDP, a policy can be represented as a matrix $\Pi \in \mathbb{R}^{N \times M}$ where the element in the i-th row and j-th column is

$$\Pi_{ij} = \pi(a_j|s_i).$$

Each row of Π is a probability distribution over actions: $\sum_{j=0}^{M-1} \Pi_{ij} = 1$ for all $i \in \{0, 1, \dots, N-1\}$.

15.3.3 Policy Space

The set of all policies forms the **policy space** Π. For finite MDPs, Π is a convex set: if $\pi_1, \pi_2 \in \Pi$ and $\lambda \in [0, 1]$, then $\lambda\pi_1 + (1 - \lambda)\pi_2 \in \Pi$.

Policies are the objects that reinforcement learning algorithms optimize. **Stochastic policies** are more general and enable exploration, while **deterministic policies** are simpler and often optimal in fully observable environments. The policy space is a **probability simplex** for each state, making it a convex set. This convexity is important for optimization: any local optimum in a convex set is also a global optimum. In **policy gradient methods**, we optimize over this policy space using gradient-based methods. The policy matrix representation enables efficient computation and connects policies to **probability distributions**, which are fundamental in statistics and machine learning.

15.4 Value Functions

15.4.1 Definition: Return (Discounted)

For a trajectory $(s_0, a_0, r_1, s_1, a_1, r_2, ...)$ where r_{t+1} is the reward received after taking action a_t in state s_t and transitioning to s_{t+1}, the **discounted return** is

$$G_t = \sum_{k=0}^{\infty} \gamma^k r_{t+1+k} = r_{t+1} + \gamma r_{t+2} + \gamma^2 r_{t+3} + \cdots.$$

15.4.2 Definition: State Value Function

The **state value function** $V_\pi : \mathcal{S} \to \mathbb{R}$ under policy π is the expected return starting from state s:

$$V_\pi(s) = \mathbb{E}_\pi[G_t | S_t = s] = \mathbb{E}_\pi\left[\sum_{k=0}^{\infty} \gamma^k R_{t+1+k} \middle| S_t = s\right],$$

where the expectation is over trajectories generated by following policy π.

15.4.3 Definition: Action Value Function

The **action value function** (Q-function) $Q_\pi : \mathcal{S} \times \mathcal{A} \to \mathbb{R}$ under policy π is the expected return starting from state s, taking action a, and then following policy π:

$$Q_\pi(s,a) = \mathbb{E}_\pi[G_t | S_t = s, A_t = a] = \mathbb{E}_\pi\left[\sum_{k=0}^{\infty} \gamma^k R_{t+1+k} \middle| S_t = s, A_t = a\right].$$

15.4.4 Relationship Between Value Functions

The state and action value functions are related by:

$$V_\pi(s) = \sum_{a \in \mathcal{A}} \pi(a|s) Q_\pi(s,a),$$

and

$$Q_\pi(s,a) = R(s,a) + \gamma \sum_{s' \in \mathcal{S}} P(s'|s,a) V_\pi(s').$$

Value functions are the fundamental objects in reinforcement learning. They measure the "goodness" of states and actions under a given policy. The **state value function** $V_\pi(s)$ answers "How good is it to be in state s?" while the **action value function** $Q_\pi(s,a)$ answers "How good is it to take action a in state s?". These functions enable **value-based methods** (like Q-learning) that learn optimal policies by learning optimal value functions. The relationship between V_π and Q_π shows that we can compute one from the other, providing flexibility in algorithm design. In **deep reinforcement learning**, neural networks approximate these value functions, enabling generalization to large state spaces.

15.4.5 Vector Representation of Value Functions

For finite MDPs, value functions can be represented as vectors. The state value function V_π is a vector $\mathbf{v}_\pi \in \mathbb{R}^N$ where the i-th element is

$$(\mathbf{v}_\pi)_i = V_\pi(s_i).$$

The action value function Q_π is a matrix $\mathbf{Q}_\pi \in \mathbb{R}^{N\times M}$ where the element in the i-th row and j-th column is

$$(\mathbf{Q}_\pi)_{ij} = Q_\pi(s_i, a_j).$$

15.5 Bellman Equations

The Bellman equations are fundamental recursive relationships that characterize value functions in MDPs. There are **two types of Bellman equations**:

1. **Bellman equation for state value function** $V_\pi(s)$: expresses the value of a state in terms of the expected immediate reward and the discounted value of the next state.
2. **Bellman equation for action value function** $Q_\pi(s, a)$: expresses the value of a state-action pair in terms of the immediate reward and the discounted value of the next state-action pairs.

Both equations are fixed-point equations that can be solved iteratively, and they are mathematically equivalent—one can be derived from the other using the relationship between V_π and Q_π.

15.5.1 Theorem: Bellman Equation for State Value Function

For any policy π, the state value function satisfies the **Bellman equation**:

$$\begin{aligned} V_\pi(s) &= \mathbb{E}_\pi[R_{t+1} + \gamma V_\pi(S_{t+1})|S_t = s] \\ &= \sum_{a\in\mathcal{A}} \pi(a|s)\left[R(s,a) + \gamma \sum_{s'\in\mathcal{S}} P(s'|s,a)V_\pi(s')\right]. \end{aligned} \tag{15.1}$$

Proof

Starting from the definition:

$$\begin{aligned} V_\pi(s) &= \mathbb{E}_\pi[G_t|S_t = s] \\ &= \mathbb{E}_\pi[R_{t+1} + \gamma G_{t+1}|S_t = s] \\ &= \sum_{a\in\mathcal{A}} \pi(a|s)\mathbb{E}_\pi[R_{t+1} + \gamma G_{t+1}|S_t = s, A_t = a] \\ &= \sum_{a\in\mathcal{A}} \pi(a|s)\left[R(s,a) + \gamma \sum_{s'\in\mathcal{S}} P(s'|s,a)\mathbb{E}_\pi[G_{t+1}|S_{t+1} = s']\right] \\ &= \sum_{a\in\mathcal{A}} \pi(a|s)\left[R(s,a) + \gamma \sum_{s'\in\mathcal{S}} P(s'|s,a)V_\pi(s')\right]. \end{aligned}$$

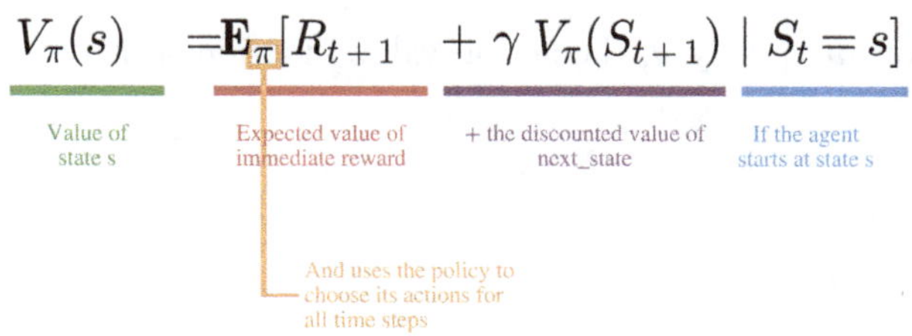

Figure 15.1: Bellman Decomposition

The Bellman equation (*15.1*) expresses a fundamental **recursive relationship** for value functions: the value of a state equals the expected immediate reward plus the discounted expected value of the next state. This decomposition is illustrated in Figure 15.1. The key insight is that we can compute the value of a state by considering: (1) the immediate reward we expect to receive, and (2) the discounted value of where we expect to end up next. This recursive structure makes the Bellman equation a **fixed-point equation**: the value function V_π is the solution to this equation, and once we know V_π, we can verify it satisfies this relationship. This property is crucial for dynamic programming algorithms, which solve for V_π by iteratively applying the Bellman equation until convergence.

15.5.2 Matrix Form of Bellman Equation

For finite MDPs, the Bellman equation can be written in matrix form. Let $\mathbf{v}_\pi \in \mathbb{R}^N$ be the state value vector, $\mathbf{r}_\pi \in \mathbb{R}^N$ be the expected reward vector under policy π where the i-th element is

$$(\mathbf{r}_\pi)_i = \sum_{a \in \mathcal{A}} \pi(a|s_i) R(s_i, a),$$

and $P_\pi \in \mathbb{R}^{N \times N}$ be the state transition matrix under policy π where the element in the i-th row and j-th column is

$$(P_\pi)_{ij} = \sum_{a \in \mathcal{A}} \pi(a|s_i) P(s_j|s_i, a).$$

Then the Bellman equation becomes:

$$\mathbf{v}_\pi = \mathbf{r}_\pi + \gamma P_\pi \mathbf{v}_\pi. \tag{15.2}$$

This is a linear system of equations that can be solved as:

$$\mathbf{v}_\pi = (I - \gamma P_\pi)^{-1} \mathbf{r}_\pi, \tag{15.3}$$

provided that $(I - \gamma P_\pi)$ is invertible (which is guaranteed when $\gamma < 1$).

> The Bellman equation is a **fixed-point equation**: V_π is a fixed point of the Bellman operator. This equation shows that the value of a state equals the expected immediate reward plus the discounted expected value of the next state. The matrix form (equation *15.2*) enables efficient computation using linear algebra: solving the Bellman equation is equivalent to solving a linear system. This connects reinforcement learning to **numerical linear algebra** and **iterative methods**. The invertibility of $(I - \gamma P_\pi)$ is guaranteed by the **Perron-Frobenius theorem** for stochastic matrices, ensuring that the value function is well-defined. The computational complexity of the analytical solution (equation *15.3*) is $O(N^3)$, where N is the number of states, due to matrix inversion. Therefore, this method is only applicable to very small Markov Decision Processes. For solving value functions in larger-scale MDPs, we can use **dynamic programming** algorithms (like value iteration and policy iteration), **Monte Carlo methods**, and **temporal difference** methods, which have better scalability properties.

15.5.3 Theorem: Bellman Equation for Action Value Function

For any policy π, the action value function satisfies:

$$Q_\pi(s,a) = R(s,a) + \gamma \sum_{s' \in \mathcal{S}} P(s'|s,a) \sum_{a' \in \mathcal{A}} \pi(a'|s') Q_\pi(s',a').$$

Proof

Similar to the state value function, using the relationship $Q_\pi(s,a) = R(s,a) + \gamma \sum_{s'} P(s'|s,a) V_\pi(s')$ and $V_\pi(s') = \sum_{a'} \pi(a'|s') Q_\pi(s',a')$.

15.6 Bellman Operators

15.6.1 Definition: Bellman Operator

The **Bellman operator** $T_\pi : \mathbb{R}^N \to \mathbb{R}^N$ for policy π is defined as:

$$(T_\pi \mathbf{v})(s) = \sum_{a \in \mathcal{A}} \pi(a|s) \left[R(s,a) + \gamma \sum_{s' \in \mathcal{S}} P(s'|s,a) \mathbf{v}(s') \right],$$

for any value function vector $\mathbf{v} \in \mathbb{R}^N$.

In matrix form:

$$T_\pi \mathbf{v} = \mathbf{r}_\pi + \gamma P_\pi \mathbf{v}.$$

15.6.2 Fixed Point Property of Bellman Operator

The Bellman equation can be written as:

$$\mathbf{v}_\pi = T_\pi \mathbf{v}_\pi, \tag{15.4}$$

meaning that $\mathbf{v}_\pi$ is a **fixed point** of the operator T_π. This fixed-point property (equation *15.4*) is fundamental to understanding how value iteration and policy iteration algorithms converge.

15.6.3 Definition: Bellman Optimality Operator

The **Bellman optimality operator** $T_* : \mathbb{R}^N \to \mathbb{R}^N$ is defined as:

$$(T_* \mathbf{v})(s) = \max_{a \in \mathcal{A}} \left[R(s,a) + \gamma \sum_{s' \in \mathcal{S}} P(s'|s,a) \mathbf{v}(s') \right].$$

In matrix form, this is:

$$(T_* \mathbf{v})_i = \max_{j \in \{0,\ldots,M-1\}} \left[R_{ij} + \gamma \sum_{k=0}^{N-1} P_{ik}^{a_j} \mathbf{v}_k \right],$$

where $R_{ij} = R(s_i, a_j)$ and $P_{ik}^{a_j} = P(s_k|s_i, a_j)$.

> Bellman operators are **contraction mappings** on the space of value functions, which is crucial for convergence guarantees. The Bellman operator T_π computes the value function for a fixed policy, while the optimality operator T_* computes the optimal value function. These operators transform the problem of finding value functions into a fixed-point problem, enabling iterative solution methods. The contraction property ensures that repeated application of these operators converges to the unique fixed point, which is the basis for **value iteration** and **policy iteration** algorithms. The operator perspective connects reinforcement learning to **functional analysis** and **fixed-point theory**.

15.7 Contraction Mapping Theorem

15.7.1 Definition: Contraction Mapping

Let (X, d) be a metric space. A function $T : X \to X$ is a **contraction mapping** if there exists $\gamma \in [0, 1)$ such that

$$d(T(x), T(y)) \leq \gamma d(x, y)$$

for all $x, y \in X$. The constant γ is called the **contraction factor**.

15.7.2 Theorem: Banach Fixed Point Theorem

If $T : X \to X$ is a contraction mapping on a complete metric space (X, d), then:

1. T has a **unique fixed point** $x^* \in X$ such that $T(x^*) = x^*$
2. For any initial point $x_0 \in X$, the sequence $x_{n+1} = T(x_n)$ converges to x^*
3. The convergence rate is geometric: $d(x_n, x^*) \leq \gamma^n d(x_0, x^*)$

A **Banach space** is a complete normed vector space (as defined in Chapter~1). Every normed vector space is a metric space with the metric $d(x, y) = \|x - y\|$ induced by the norm. The space of value functions $\mathbb{R}^N$ equipped with the supremum norm $\|\cdot\|_\infty$ is a Banach space. The theorem is named after Stefan Banach because it is most commonly applied in Banach spaces, where the metric is induced by a norm. In our context, the space of value functions with the supremum norm forms a Banach space, making the Banach Fixed Point Theorem directly applicable.

15.7.3 Theorem: Bellman Operators are Contractions

The Bellman operator T_π and the Bellman optimality operator T_* are contraction mappings on the space of value functions with the supremum norm:

$$\|T_\pi \mathbf{v}_1 - T_\pi \mathbf{v}_2\|_\infty \leq \gamma \|\mathbf{v}_1 - \mathbf{v}_2\|_\infty,$$

$$\|T_* \mathbf{v}_1 - T_* \mathbf{v}_2\|_\infty \leq \gamma \|\mathbf{v}_1 - \mathbf{v}_2\|_\infty.$$

Proof (for T_π)

For any two value functions $\mathbf{v}_1, \mathbf{v}_2$:

$$\begin{aligned}
|T_\pi \mathbf{v}_1(s) - T_\pi \mathbf{v}_2(s)| &= \left| \sum_a \pi(a|s) \gamma \sum_{s'} P(s'|s,a)(\mathbf{v}_1(s') - \mathbf{v}_2(s')) \right| \\
&\leq \sum_a \pi(a|s) \gamma \sum_{s'} P(s'|s,a) |\mathbf{v}_1(s') - \mathbf{v}_2(s')| \\
&\leq \gamma \max_{s'} |\mathbf{v}_1(s') - \mathbf{v}_2(s')| = \gamma \|\mathbf{v}_1 - \mathbf{v}_2\|_\infty.
\end{aligned}$$

Taking the maximum over s gives the result.

> The contraction property is the mathematical foundation of **value iteration** and **policy iteration**. It guarantees that these algorithms converge to the unique optimal solution, regardless of initialization. The contraction factor γ (the discount factor) determines the convergence rate: smaller γ leads to faster convergence but may not capture long-term dependencies. The Banach fixed point theorem is one of the most important results in **functional analysis** and has applications throughout mathematics, from differential equations to optimization. In reinforcement learning, it ensures that iterative algorithms are well-behaved and predictable.

15.8 Optimal Policies and Value Functions

15.8.1 Definition: Optimal Value Function

The **optimal state value function** $V_* : \mathcal{S} \to \mathbb{R}$ is defined as:

$$V_*(s) = \max_\pi V_\pi(s) \quad \forall s \in \mathcal{S}.$$

The **optimal action value function** $Q_* : \mathcal{S} \times \mathcal{A} \to \mathbb{R}$ is:

$$Q_*(s,a) = \max_\pi Q_\pi(s,a) \quad \forall s \in \mathcal{S}, a \in \mathcal{A}.$$

15.8.2 Theorem: Bellman Optimality Equation

The optimal value function satisfies the **Bellman optimality equation**:

$$V_*(s) = \max_{a \in \mathcal{A}} \left[R(s,a) + \gamma \sum_{s' \in \mathcal{S}} P(s'|s,a) V_*(s') \right], \tag{15.5}$$

or equivalently, V_* is the fixed point of T_*:

$$V_* = T_* V_*. \tag{15.6}$$

15.8.3 Definition: Optimal Policy

A policy π_* is **optimal** if

$$V_{\pi_*}(s) = V_*(s) \quad \forall s \in \mathcal{S}.$$

15.8.4 Theorem: Existence of Optimal Deterministic Policy

For any finite MDP, there exists an optimal deterministic policy $\pi_* : \mathcal{S} \to \mathcal{A}$ such that

$$\pi_*(s) = \arg\max_{a \in \mathcal{A}} Q_*(s,a).$$

Proof

Since $V_*(s) = \max_a Q_*(s,a)$, we can define $\pi_*(s) = \arg\max_a Q_*(s,a)$. This policy is deterministic and satisfies $V_{\pi_*}(s) = V_*(s)$ for all s.

> The existence of optimal deterministic policies simplifies the problem: we only need to search over deterministic policies, which are finite in number for finite MDPs. However, stochastic policies are still important for exploration and in partially observable environments. The optimal policy can be derived from the optimal Q-function by taking the action with the highest Q-value in each state. This is the basis for **Q-learning** and other value-based methods. The optimality equation (equation *15.5*) shows that the optimal value function is the solution to a fixed-point problem (equation *15.6*), which can be solved iteratively using the contraction property of T_*.

15.9 Dynamic Programming

15.9.1 Algorithm: Value Iteration

Algorithm 14 Value Iteration

Input: MDP (S, A, P, R, γ), tolerance $\epsilon > 0$
Output: Optimal value function V_* and optimal policy π_*

```
Initialize v_0 arbitrarily (e.g., v_0 = 0)
k ← 0
repeat
    v_{k+1} ← T_* v_k                          ▷ Apply Bellman optimality operator
    k ← k + 1
until ‖v_{k+1} − v_k‖_∞ < ε
Extract optimal policy: π_*(s) ← argmax_a[R(s,a) + γ Σ_{s'} P(s'|s,a) v_k(s')] for all s ∈ S
return v_k, π_*
```

15.9.2 Convergence of Value Iteration

By the contraction property, value iteration converges to V_*:

$$\lim_{k \to \infty} \mathbf{v}_k = V_*,$$

with geometric convergence rate:

$$\|\mathbf{v}_k - V_*\|_\infty \leq \gamma^k \|\mathbf{v}_0 - V_*\|_\infty.$$

15.9.3 Algorithm: Policy Iteration

Algorithm 15 Policy Iteration

Input: MDP (S, A, P, R, γ)
Output: Optimal policy π_* and optimal value function V_*

```
Initialize policy π_0 arbitrarily
k ← 0
repeat
    Policy Evaluation: Solve v^{π_k} = r^{π_k} + γ P^{π_k} v^{π_k} for v^{π_k}
    Policy Improvement: π_{k+1}(s) ← argmax_a[R(s,a) + γ Σ_{s'} P(s'|s,a) v^{π_k}(s')] for all s ∈ S
    k ← k + 1
until π_{k+1} = π_k
return π_k, v^{π_k}
```

15.9.4 Convergence of Policy Iteration

Policy iteration converges to an optimal policy in a finite number of iterations (at most M^N iterations for finite MDPs), since:

- Each policy improvement step strictly increases the value function (unless the policy is already optimal)
- There are only finitely many policies

Value iteration and **policy iteration** are the fundamental dynamic programming algorithms for solving MDPs. Value iteration is simpler but may require more iterations, while policy iteration typically converges faster but requires solving a linear system at each step. Both algorithms are guaranteed to converge to the optimal solution due to the contraction property. In practice, **modified policy iteration** combines both approaches: it performs a few value iteration steps for policy evaluation instead of solving the system exactly. These algorithms form the foundation for more advanced methods like **approximate dynamic programming** and **reinforcement learning**, which handle large or continuous state spaces.

15.10 State Distribution and Stationarity

15.10.1 Definition: State Distribution

For a policy π and initial state distribution $\mu_0 : \mathcal{S} \to [0, 1]$, the **state distribution** at time t is:

$$\mu_t(s) = P(S_t = s | \mu_0, \pi).$$

15.10.2 State Distribution Evolution

The state distribution evolves according to:

$$\mu_{t+1}(s') = \sum_{s \in \mathcal{S}} \sum_{a \in \mathcal{A}} \mu_t(s) \pi(a|s) P(s'|s, a).$$

In matrix form, with $\mu_t \in \mathbb{R}^N$ as the state distribution vector:

$$\mu_{t+1}^\top = \mu_t^\top P_\pi.$$

15.10.3 Definition: Stationary Distribution

A **stationary distribution** μ^* under policy π satisfies:

$$\mu^{*\top} = \mu^{*\top} P_\pi,$$

or equivalently, μ^* is a left eigenvector of P_π with eigenvalue 1.

15.10.4 Theorem: Existence of Stationary Distribution

For any policy π in a finite MDP, if the Markov chain induced by π is **ergodic** (irreducible and aperiodic), then there exists a unique stationary distribution μ^* such that

$$\lim_{t \to \infty} \mu_t = \mu^*,$$

regardless of the initial distribution μ_0.

> The state distribution is important for understanding the long-term behavior of policies and for **policy gradient methods**, where gradients are computed with respect to the state distribution. The stationary distribution represents the "steady state" of the MDP under a given policy: after many steps, the probability of being in each state converges to the stationary distribution. This connects MDPs to **Markov chain theory** and **ergodic theory**. In **on-policy** reinforcement learning, we often sample trajectories from the stationary distribution, while in **off-policy** methods, we need to account for distribution mismatch. The ergodicity condition ensures that the MDP explores all states, which is important for learning.

15.11 Continuous-Time Markov Chains

The discrete-time Markov chains we have studied so far make transitions at fixed time steps $t = 0, 1, 2, \ldots$. **Continuous-time Markov chains (CTMCs)** extend this framework to allow transitions at any time $t \in [0, \infty)$. This generalization is mathematically natural and connects Markov chains to differential equations.

15.11.1 Definition: Continuous-Time Markov Chain

A **continuous-time Markov chain** on a finite state space $\mathcal{S} = \{s_0, s_1, \ldots, s_{N-1}\}$ is characterized by an **infinitesimal generator** (or **rate matrix**) $Q \in \mathbb{R}^{N \times N}$ satisfying:

1. Off-diagonal entries are non-negative: $Q_{ij} \geq 0$ for $i \neq j$
2. Rows sum to zero: $\sum_{j=0}^{N-1} Q_{ij} = 0$ for all i

Equivalently, the diagonal entries are:

$$Q_{ii} = -\sum_{j \neq i} Q_{ij} \tag{15.7}$$

15.11.2 Interpretation of the Rate Matrix

The entry Q_{ij} for $i \neq j$ represents the **instantaneous transition rate** from state s_i to state s_j:

$$Q_{ij} = \lim_{\Delta \to 0} \frac{P(X_{t+\Delta} = s_j | X_t = s_i)}{\Delta}, \quad i \neq j \tag{15.8}$$

Intuitively, if the system is in state s_i, it transitions to state s_j at rate Q_{ij} per unit time.

The diagonal entry $-Q_{ii} = \sum_{j \neq i} Q_{ij}$ is the **total exit rate** from state s_i.

15.11.3 Relationship to Discrete-Time Chains

For small time step Δ, the transition probability matrix is approximately:

$$P(\Delta) \approx I + \Delta Q \tag{15.9}$$

Compare this to the **Forward Euler** discretization of an ODE (Chapter~20): the rate matrix Q plays the role of the "derivative" matrix A.

15.11.4 Theorem: Kolmogorov Forward Equation

Let $\mathbf{p}(t) \in \mathbb{R}^N$ be the probability distribution over states at time t, with $p_i(t) = P(X_t = s_i)$. The distribution evolves according to the **Kolmogorov forward equation** (also called the **master equation**):

$$\frac{d\mathbf{p}(t)}{dt} = Q^\top \mathbf{p}(t) \tag{15.10}$$

or equivalently, for each state:

$$\frac{dp_j(t)}{dt} = \sum_{i=0}^{N-1} p_i(t) Q_{ij} = \sum_{i \neq j} p_i(t) Q_{ij} - p_j(t) \sum_{i \neq j} Q_{ji}$$

The first term represents **probability flow into** state s_j; the second represents **flow out of** state s_j.

15.11.5 Solution via Matrix Exponential

The solution to the Kolmogorov forward equation with initial distribution $\mathbf{p}(0) = \mathbf{p}_0$ is:

$$\mathbf{p}(t) = \exp(tQ^\top)\mathbf{p}_0 \tag{15.11}$$

The **transition probability matrix** from time 0 to time t is:

$$P(t) = \exp(tQ) \tag{15.12}$$

so that $P_{ij}(t) = P(X_t = s_j | X_0 = s_i)$.

> The matrix exponential $\exp(tQ)$ connects continuous-time Markov chains to the eigendecomposition theory of Chapter~6. If $Q = V\Lambda V^{-1}$, then $P(t) = V\exp(t\Lambda)V^{-1}$, where $\exp(t\Lambda)$ has diagonal entries $e^{t\lambda_k}$. The eigenvalues of Q determine the transient behavior: the rate matrix always has eigenvalue 0 (corresponding to the stationary distribution), and all other eigenvalues have non-positive real parts for a valid CTMC.

15.11.6 Theorem: Stationary Distribution of CTMC

A distribution $\pi \in \mathbb{R}^N$ is **stationary** for a CTMC with generator Q if and only if:

$$Q^\top \pi = \mathbf{0} \tag{15.13}$$

or equivalently, $\pi^\top Q = \mathbf{0}^\top$.

This means π is a **left eigenvector** of Q with eigenvalue 0.

15.11.7 Connection to Discrete-Time Chains

The stationary distribution of a CTMC is also stationary for any discrete-time chain obtained by sampling at fixed intervals. If $\pi^\top Q = \mathbf{0}$, then:

$$\pi^\top P(\Delta) = \pi^\top \exp(\Delta Q) = \pi^\top$$

for any $\Delta > 0$. This follows from the power series: $\pi^\top \exp(\Delta Q) = \pi^\top \sum_{k=0}^{\infty} \frac{(\Delta Q)^k}{k!} = \pi^\top$ since $\pi^\top Q = \mathbf{0}$ implies $\pi^\top Q^k = \mathbf{0}$ for all $k \geq 1$.

15.11.8 Definition: Detailed Balance for CTMCs

A CTMC satisfies **detailed balance** (is **reversible**) if:

$$\pi_i Q_{ij} = \pi_j Q_{ji} \quad \forall i, j \tag{15.14}$$

This says the probability flow from s_i to s_j equals the flow from s_j to s_i in equilibrium.

Continuous-time Markov chains arise naturally in many contexts. In **diffusion models for discrete data** (e.g., text generation), the forward corruption process is often a CTMC where tokens transition according to a rate matrix Q, and the reverse process learns to undo these transitions. The Kolmogorov forward equation is the **discrete-state analog** of the Fokker-Planck equation (Chapter~21), which governs continuous-state diffusion processes. The detailed balance condition connects to **MCMC sampling**: Metropolis-Hastings and Gibbs sampling construct chains satisfying detailed balance to sample from target distributions.

15.11.9 Example: Two-State CTMC

Consider a system with two states $\{0, 1\}$ and rate matrix:

$$Q = \begin{bmatrix} -\alpha & \alpha \\ \beta & -\beta \end{bmatrix}$$

where α is the rate of transitioning from state 0 to 1, and β is the rate from 1 to 0.

Eigenvalues: $\lambda_1 = 0$, $\lambda_2 = -(\alpha + \beta)$

Stationary distribution: Solving $\pi^\top Q = \mathbf{0}$:

$$\pi = \frac{1}{\alpha + \beta} \begin{bmatrix} \beta \\ \alpha \end{bmatrix}$$

Transition matrix:

$$P(t) = \exp(tQ) = \frac{1}{\alpha + \beta} \begin{bmatrix} \beta + \alpha e^{-(\alpha+\beta)t} & \alpha - \alpha e^{-(\alpha+\beta)t} \\ \beta - \beta e^{-(\alpha+\beta)t} & \alpha + \beta e^{-(\alpha+\beta)t} \end{bmatrix}$$

As $t \to \infty$, $P(t) \to \mathbf{1}\pi^\top$, confirming convergence to the stationary distribution.

15.12 Policy Gradient Theorem

15.12.1 Expected Return as Objective

For a policy π_θ parameterized by θ, the expected return starting from initial distribution μ_0 is:

$$J(\theta) = \mathbb{E}_{s \sim \mu_0}[V^{\pi_\theta}(s)] = \sum_{s \in \mathcal{S}} \mu_0(s) V^{\pi_\theta}(s).$$

15.12.2 Theorem: Policy Gradient Theorem

The gradient of the expected return with respect to policy parameters is:

$$\nabla_\theta J(\theta) = \sum_{s\in\mathcal{S}} \mu^{\pi_\theta}(s) \sum_{a\in\mathcal{A}} \nabla_\theta \pi_\theta(a|s) Q_{\pi_\theta}(s,a),$$

where μ^{π_θ} is the stationary state distribution under policy π_θ.

For the infinite-horizon case with discounting, this becomes:

$$\nabla_\theta J(\theta) = \frac{1}{1-\gamma} \sum_{s\in\mathcal{S}} d_{\pi_\theta}(s) \sum_{a\in\mathcal{A}} \nabla_\theta \pi_\theta(a|s) Q_{\pi_\theta}(s,a),$$

where $d_{\pi_\theta}(s) = (1-\gamma)\sum_{t=0}^{\infty} \gamma^t \mu_t(s)$ is the **discounted state visitation distribution**.

> The policy gradient theorem is fundamental to **policy gradient methods** in reinforcement learning. It shows that the gradient of the expected return can be computed using the policy gradient and the action value function. This enables gradient-based optimization of policies, which is the basis for algorithms like **REINFORCE**, **Actor-Critic**, and **Proximal Policy Optimization (PPO)**. The theorem connects policy optimization to **optimization theory** and shows that we can optimize policies directly without first learning value functions (though value functions are often used to reduce variance). The discounted state visitation distribution d_{π_θ} weights states by how often they are visited under the policy, making it a natural distribution for policy gradient estimation.

15.13 Stochastic Approximation

15.13.1 Definition: Stochastic Approximation

A **stochastic approximation** is an iterative method of the form:

$$x_{n+1} = x_n + \alpha_n[h(x_n) + \xi_n],$$

where:

- $x_n \in \mathbb{R}^d$ is the current iterate
- $h : \mathbb{R}^d \to \mathbb{R}^d$ is a deterministic function
- ξ_n is a random noise term with $\mathbb{E}[\xi_n|x_n] = 0$
- $\alpha_n > 0$ is a step size (learning rate)

The goal is to find a fixed point x^* such that $h(x^*) = 0$.

15.13.2 Convergence Conditions

Under appropriate conditions on h and the step size sequence $\{\alpha_n\}$, the stochastic approximation converges to x^* with probability 1 if:

1. **Step size conditions**: $\sum_{n=0}^{\infty} \alpha_n = \infty$ and $\sum_{n=0}^{\infty} \alpha_n^2 < \infty$
2. **Noise conditions**: The noise ξ_n has bounded variance
3. **Stability conditions**: The function h satisfies certain stability properties (e.g., h is a contraction or has a unique zero)

15.13.3 Application to Temporal Difference Learning

Temporal difference (TD) learning can be viewed as stochastic approximation. For Q-learning, the update:

$$Q(s,a) \leftarrow Q(s,a) + \alpha[r + \gamma \max_{a'} Q(s',a') - Q(s,a)]$$

is a stochastic approximation to the Bellman optimality equation $Q_* = T_* Q_*$, where:

- $h(Q) = T_* Q - Q$ (the Bellman error)
- ξ_n is the sampling noise from the MDP

> Stochastic approximation theory provides the mathematical foundation for **temporal difference learning** and **Q-learning**. It explains why these methods converge even when updates are noisy and based on samples rather than exact expectations. The step size conditions ensure that: (1) the algorithm explores enough ($\sum \alpha_n = \infty$) to reach the solution, and (2) the noise is averaged out ($\sum \alpha_n^2 < \infty$) so the algorithm converges. This theory connects reinforcement learning to **stochastic processes** and **martingale theory**. In **deep Q-networks (DQN)**, the stochastic approximation framework explains convergence properties, though function approximation introduces additional challenges.

15.14 Function Approximation

15.14.1 Definition: Function Approximation

Given a target function $f : \mathcal{S} \to \mathbb{R}$ and a function class $\mathcal{F} = \{f_\theta : \theta \in \Theta\}$ parameterized by θ, **function approximation** seeks to find $f_\theta^* \in \mathcal{F}$ that minimizes some approximation error:

$$\theta^* = \arg\min_{\theta \in \Theta} \|f - f_\theta\|,$$

where $\|\cdot\|$ is a norm on the function space.

15.14.2 Definition: Projection Operator

For a function space $\mathcal{V}$ with inner product $\langle \cdot, \cdot \rangle$ and a subspace $\mathcal{F} \subseteq \mathcal{V}$, the **projection operator** $\Pi : \mathcal{V} \to \mathcal{F}$ maps a function $f \in \mathcal{V}$ to its closest function in $\mathcal{F}$:

$$\Pi f = \arg\min_{g \in \mathcal{F}} \|f - g\|.$$

For the L^2 norm with respect to a distribution μ, the projection minimizes:

$$\Pi f = \arg\min_{g \in \mathcal{F}} \mathbb{E}_{s \sim \mu}[(f(s) - g(s))^2].$$

15.14.3 Definition: Approximate Bellman Operator

When value functions are approximated in a function class $\mathcal{F}$, the Bellman operator T_π may map outside $\mathcal{F}$. The **approximate Bellman operator** is defined as:

$$\tilde{T}_\pi = \Pi T_\pi,$$

which first applies T_π and then projects back to $\mathcal{F}$.

15.14.4 Definition: Approximation Error

The **approximation error** is the difference between the true value function and its projection:

$$\epsilon_{\text{approx}} = \|V_\pi - \Pi V_\pi\|.$$

The **estimation error** is the difference between the projected value function and the learned approximation:

$$\epsilon_{\text{est}} = \|\Pi V_\pi - \hat{V}^\pi\|.$$

The total error is bounded by:

$$\|V_\pi - \hat{V}^\pi\| \leq \epsilon_{\text{approx}} + \epsilon_{\text{est}}.$$

Function approximation is essential for scaling reinforcement learning to large state spaces. The projection operator provides a mathematical framework for understanding how approximation affects value functions. The key challenge is that the Bellman operator T_π may not preserve the function class $\mathcal{F}$, leading to **approximation error**. This error can accumulate over iterations, potentially causing divergence. In **linear function approximation**, $\mathcal{F}$ is a linear subspace, and the projection is a linear operator, making analysis tractable. In **neural network approximation** (as in DQN), $\mathcal{F}$ is a nonlinear manifold, and convergence guarantees are more complex. The trade-off between approximation error (bias) and estimation error (variance) is fundamental to understanding the behavior of approximate reinforcement learning methods.

15.15 Importance Sampling

15.15.1 Definition: Importance Sampling Ratio

For two policies π (target) and μ (behavior), the **importance sampling ratio** (or importance weight) at time t is:

$$\rho_t = \frac{\pi(a_t|s_t)}{\mu(a_t|s_t)}.$$

For a trajectory $\tau = (s_0, a_0, r_0, s_1, a_1, r_1, ...)$, the cumulative importance ratio is:

$$\rho_{0:t} = \prod_{k=0}^{t} \frac{\pi(a_k|s_k)}{\mu(a_k|s_k)}.$$

15.15.2 Unbiased Estimation

For an expectation under policy π:

$$\mathbb{E}_\pi[G_t|S_t = s] = \mathbb{E}_\mu\left[\rho_{0:\infty} G_t|S_t = s\right],$$

where G_t is the return and the expectation on the right is over trajectories generated by policy μ.

15.15.3 Variance of Importance Sampling

The variance of the importance sampling estimator is:

$$\mathrm{Var}_\mu[\rho_{0:t}G_t] = \mathbb{E}_\mu[(\rho_{0:t}G_t)^2] - (\mathbb{E}_\mu[\rho_{0:t}G_t])^2.$$

The variance can be large when π and μ are very different, especially for long trajectories where $\rho_{0:t}$ is a product of many ratios.

15.15.4 Definition: Weighted Importance Sampling

To reduce variance, **weighted importance sampling** uses normalized weights:

$$V_\pi(s) \approx \frac{\sum_{i=1}^n \rho_{0:\infty}^{(i)} G_t^{(i)}}{\sum_{i=1}^n \rho_{0:\infty}^{(i)}},$$

where the superscript (i) indexes different trajectories. This estimator is biased but has lower variance than unweighted importance sampling.

> Importance sampling enables **off-policy learning**, where we can evaluate or improve a policy π using data collected under a different policy μ. This is crucial for **efficient exploration** and **data reuse**. The mathematical challenge is that importance sampling ratios can have high variance, especially when policies are very different or trajectories are long. **Weighted importance sampling** trades bias for variance reduction, which is often beneficial in practice. In **off-policy policy gradient methods**, importance sampling corrects the distribution mismatch between the behavior policy (used for data collection) and the target policy (being optimized). The variance analysis is essential for understanding when off-policy methods will work well and when they may fail.

15.16 Trust Region and Natural Policy Gradient

15.16.1 Definition: Trust Region Constraint

A **trust region** constraint limits policy updates to a neighborhood of the current policy. For policies parameterized by θ, a trust region constraint can be expressed as:

$$D_{\mathrm{KL}}(\pi_{\theta_{\mathrm{old}}}||\pi_\theta) \leq \delta,$$

where $\delta > 0$ is the trust region radius and D_{KL} is the KL divergence (see Chapter~15).

15.16.2 Definition: Natural Policy Gradient

The **natural policy gradient** uses the Fisher information matrix $F(\theta)$ to define a metric on the policy space:

$$\tilde{\nabla}_\theta J(\theta) = F(\theta)^{-1}\nabla_\theta J(\theta),$$

where

$$F(\theta)_{ij} = \mathbb{E}_{s\sim d_{\pi_\theta}, a\sim\pi_\theta}\left[\frac{\partial \log \pi_\theta(a|s)}{\partial \theta_i}\frac{\partial \log \pi_\theta(a|s)}{\partial \theta_j}\right].$$

15.16.3 Connection to KL Divergence

For small policy updates δ, the KL divergence between policies is approximately:

$$D_{\text{KL}}(\pi_\theta || \pi_{\theta+\delta}) \approx \frac{1}{2}\delta^\top F(\theta)\delta.$$

This shows that the Fisher information matrix defines a **Riemannian metric** on the policy space, where distances are measured by KL divergence.

15.16.4 Trust Region Optimization

Trust region policy optimization solves:

$$\max_\theta \quad \mathbb{E}_{s\sim d^{\pi_{\theta_{\text{old}}}}, a\sim\pi_{\theta_{\text{old}}}} \left[\frac{\pi_\theta(a|s)}{\pi_{\theta_{\text{old}}}(a|s)} A^{\pi_{\theta_{\text{old}}}}(s,a) \right],$$

subject to:

$$D_{\text{KL}}(\pi_{\theta_{\text{old}}} || \pi_\theta) \leq \delta,$$

where $A^\pi(s,a) = Q^\pi(s,a) - V^\pi(s)$ is the advantage function.

Trust region methods provide a mathematical framework for **safe policy updates** in reinforcement learning. The KL divergence constraint ensures that policy updates are not too large, preventing performance collapse. The natural policy gradient respects the geometry of the policy space (as measured by KL divergence), leading to more stable optimization. In **Proximal Policy Optimization (PPO)**, the trust region constraint is enforced through clipping or adaptive KL penalties, providing a practical approximation to the exact trust region optimization. The Fisher information matrix connects policy gradients to **information geometry**, showing that the policy space has a natural Riemannian structure. This geometric perspective explains why natural gradients often converge faster and more stably than standard gradients in policy optimization.

15.17 Mindmap

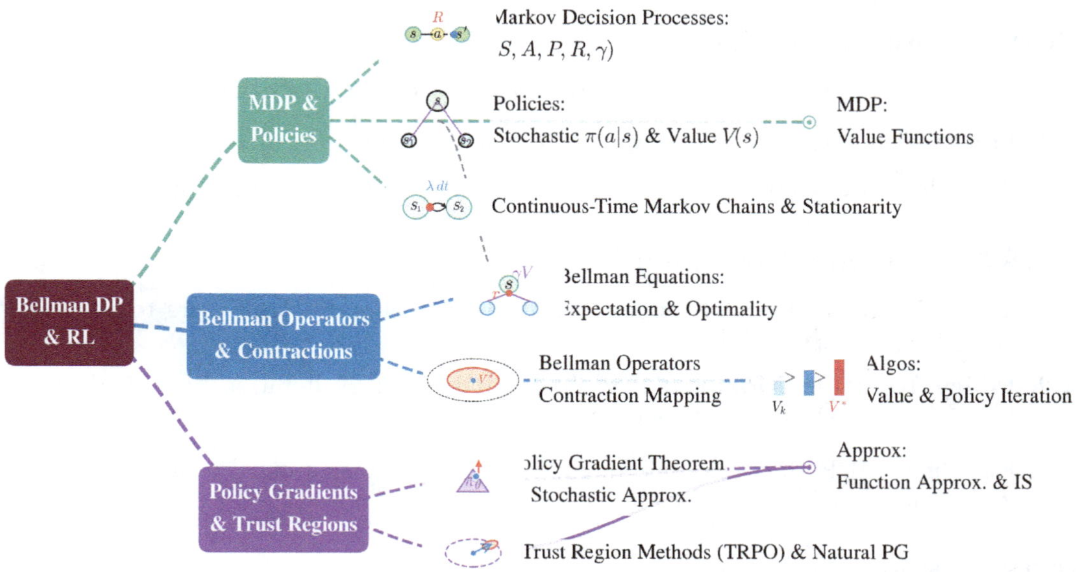

Figure 15.2: Mind map of Bellman dynamic programming topics

15.18 Exercises

15.18.1 MDP Basics

1. For a finite MDP with N states and M actions, how many deterministic policies are there? How many stochastic policies?
2. Show that the policy space Π is convex: if $\pi_1, \pi_2 \in \Pi$ and $\lambda \in [0,1]$, then $\lambda\pi_1 + (1-\lambda)\pi_2 \in \Pi$.
3. For a policy π, show that the state transition matrix P_π is a stochastic matrix (each row sums to 1).

15.18.2 Value Functions

1. Prove the relationship $V_\pi(s) = \sum_{a\in\mathcal{A}} \pi(a|s) Q_\pi(s,a)$.
2. Prove the relationship $Q_\pi(s,a) = R(s,a) + \gamma \sum_{s'\in\mathcal{S}} P(s'|s,a) V_\pi(s')$.
3. For a finite MDP, show that the Bellman equation $\mathbf{v}_\pi = \mathbf{r}_\pi + \gamma P_\pi \mathbf{v}_\pi$ has a unique solution when $\gamma < 1$.

15.18.3 Bellman Operators

1. Prove that the Bellman operator T_π is a contraction mapping with contraction factor γ.
2. Prove that the Bellman optimality operator T_* is a contraction mapping with contraction factor γ.

3. Show that $T_* \mathbf{v} \geq T_\pi \mathbf{v}$ for any policy π and value function $\mathbf{v}$ (component-wise inequality).
4. For two policies π_1 and π_2, if $V_{\pi_1}(s) \geq V_{\pi_2}(s)$ for all s, show that $T_{\pi_1} V_{\pi_1} \geq T_{\pi_2} V_{\pi_2}$.

15.18.4 Optimal Policies

1. Show that if π_* is an optimal policy, then $V_{\pi_*}(s) = \max_a Q_{\pi_*}(s, a)$ for all s.
2. Prove that the policy $\pi_*(s) = \arg\max_a Q_*(s, a)$ is optimal.
3. Show that for any policy π, $V_\pi(s) \leq V_*(s)$ for all s (component-wise inequality).

15.18.5 Dynamic Programming

1. **Value Iteration Convergence**: Show that value iteration converges to V_* with rate $\|\mathbf{v}_k - V_*\|_\infty \leq \gamma^k \|\mathbf{v}_0 - V_*\|_\infty$.
2. **Policy Iteration Monotonicity**: Show that in policy iteration, if $\pi_{k+1} \neq \pi_k$, then $V_{\pi_{k+1}}(s) > V_{\pi_k}(s)$ for at least one state s.
3. For a finite MDP, show that policy iteration converges to an optimal policy in at most M^N iterations.

15.18.6 State Distributions

1. Show that the state distribution evolution $\mu_{t+1}^\top = \mu_t^\top P_\pi$ is a linear dynamical system.
2. For an ergodic MDP, show that the stationary distribution μ_* is unique and satisfies $\mu^{*\top} P_\pi = \mu^{*\top}$.
3. Show that the discounted state visitation distribution $d_{\pi_\theta}(s) = (1-\gamma) \sum_{t=0}^\infty \gamma^t \mu_t(s)$ is a probability distribution (sums to 1).

15.18.7 Policy Gradient

1. Using the policy gradient theorem, derive the gradient for a policy parameterized as $\pi_\theta(a|s) = \frac{\exp(\theta_{sa})}{\sum_{a'} \exp(\theta_{sa'})}$ (softmax parameterization).
2. Show that the policy gradient can be written as:

$$\nabla_\theta J(\theta) = \mathbb{E}_{s \sim d_{\pi_\theta}, a \sim \pi_\theta}[\nabla_\theta \log \pi_\theta(a|s) Q_{\pi_\theta}(s, a)].$$

3. Explain why subtracting a baseline $b(s)$ from $Q_{\pi_\theta}(s, a)$ in the policy gradient (i.e., using $Q_{\pi_\theta}(s, a) - b(s)$) does not change the gradient but can reduce variance.

15.18.8 Stochastic Approximation

1. For a stochastic approximation $x_{n+1} = x_n + \alpha_n[h(x_n) + \xi_n]$ where $h(x) = -x$ and $\mathbb{E}[\xi_n] = 0$, show that if $\sum_{n=0}^\infty \alpha_n = \infty$ and $\sum_{n=0}^\infty \alpha_n^2 < \infty$, then $x_n \to 0$ almost surely.
2. Show that Q-learning update $Q(s, a) \leftarrow Q(s, a) + \alpha[r + \gamma \max_{a'} Q(s', a') - Q(s, a)]$ is a stochastic approximation to the fixed point equation $Q = T_* Q$.
3. For step sizes $\alpha_n = \frac{1}{n}$, verify that $\sum_{n=1}^\infty \alpha_n = \infty$ and $\sum_{n=1}^\infty \alpha_n^2 < \infty$.

15.18.9 Function Approximation

1. For a linear function class $\mathcal{F} = \{f_\theta(s) = \theta^\top \phi(s) : \theta \in \mathbb{R}^d\}$ where $\phi(s)$ are basis functions, show that the projection operator Π is linear.
2. Show that the approximation error $\epsilon_{\text{approx}} = \|V_\pi - \Pi V_\pi\|$ is zero if and only if $V_\pi \in \mathcal{F}$.
3. For the approximate Bellman operator $\tilde{T}_\pi = \Pi T_\pi$, show that $\tilde{T}_\pi$ may not be a contraction even if T_π is a contraction.

15.18.10 Importance Sampling

1. Prove that $\mathbb{E}_\pi[G_t|s_t = s] = \mathbb{E}_\mu[\rho_{0:\infty} G_t|s_t = s]$ where $\rho_{0:\infty} = \prod_{k=0}^{\infty} \frac{\pi(a_k|s_k)}{\mu(a_k|s_k)}$.
2. Show that the variance of the importance sampling estimator $\text{Var}_\mu[\rho_{0:t} G_t]$ increases exponentially with t when π and μ are different.
3. For weighted importance sampling, show that the estimator is biased but has lower variance than unweighted importance sampling.

15.18.11 Trust Region and Natural Policy Gradient

1. For small policy updates δ, show that $D_{\text{KL}}(\pi_\theta||\pi_{\theta+\delta}) \approx \frac{1}{2}\delta^\top F(\theta)\delta$ where $F(\theta)$ is the Fisher information matrix.
2. Show that the natural policy gradient $\tilde{\nabla}_\theta J = F(\theta)^{-1} \nabla_\theta J$ is invariant to reparameterization of the policy.
3. For a trust region constraint $D_{\text{KL}}(\pi_{\theta_{\text{old}}}||\pi_\theta) \leq \delta$, show that this defines a convex constraint set in the policy parameter space.

Chapter 16

Variational Inference and Latent Variables

Latent variable models, variational inference, and the ELBO optimization

> *"All models are wrong, but some are useful."*
>
> — George Box

▷ `Latent model` $p_\theta(\mathbf{x}, \mathbf{z}) = p_\theta(\mathbf{x}|\mathbf{z})p_\theta(\mathbf{z})$

▷ `Marginal Likelihood` $p_\theta(\mathbf{x}) = \int p_\theta(\mathbf{x}|\mathbf{z})p_\theta(\mathbf{z})d\mathbf{z}$ (intractable)

▷ `Posterior` $p_\theta(\mathbf{z}|\mathbf{x}) = \frac{p_\theta(\mathbf{x},\mathbf{z})}{p_\theta(\mathbf{x})}$ (intractable)

▷ `Varational Inference` Approx $p_\theta(\mathbf{z}|\mathbf{x})$ with $q_\phi(\mathbf{z})$; min $D_{\text{KL}}(q_\phi||p_\theta)$

▷ `ELBO` $\text{ELBO} = \mathbb{E}_{q_\phi(\mathbf{z})}[\log p_\theta(\mathbf{x}|\mathbf{z})] - D_{\text{KL}}(q_\phi(\mathbf{z})||p_\theta(\mathbf{z}))$ (rec - reg)

▷ `Mean field` $q(\mathbf{z}) = \prod_{i=0}^{k-1} q_i(z_i)$; $\text{ELBO} = \mathbb{E}_{q(\mathbf{z})}[\log p_\theta(\mathbf{x}|\mathbf{z})] - \sum_i \mathbb{E}_{q_i(z_i)}[\log q_i(z_i)] + \sum_i \mathbb{E}_{q_i(z_i)}[\log p_\theta(z_i)]$

▷ `CAVI` $q_i^*(z_i) \propto \exp(\mathbb{E}_{q_{-i}(\mathbf{z}_{-i})}[\log p_\theta(\mathbf{x}, \mathbf{z})])$ where $q_{-i}(\mathbf{z}_{-i}) = \prod_{j \neq i} q_j(z_j)$

▷ `VAE encoder` $q_\phi(\mathbf{z}|\mathbf{x}) = \mathcal{N}(\mu_\phi(\mathbf{x}), \text{diag}(\sigma_\phi^2(\mathbf{x})))$

▷ `VAE decoder` $p_\theta(\mathbf{x}|\mathbf{z})$; continuous: $\mathcal{N}(\mu_\theta(\mathbf{z}), \Sigma_\theta(\mathbf{z}))$; binary: $\prod_i \text{Bernoulli}(x_i|\sigma(\mu_{\theta,i}(\mathbf{z})))$

▷ `VAE objective` $\mathcal{L} = \mathbb{E}_{q_\phi(\mathbf{z}|\mathbf{x})}[\log p_\theta(\mathbf{x}|\mathbf{z})] - D_{\text{KL}}(q_\phi(\mathbf{z}|\mathbf{x})||\mathcal{N}(\mathbf{0}, \mathbf{I}))$

▷ `Reparam` $\epsilon \sim \mathcal{N}(\mathbf{0}, \mathbf{I})$, $\mathbf{z} = \mu_\phi(\mathbf{x}) + \sigma_\phi(\mathbf{x}) \odot \epsilon$

▷ `KL(N)` $D_{\text{KL}}(\mathcal{N}(\mu, \text{diag}(\sigma^2))||\mathcal{N}(\mathbf{0}, \mathbf{I})) = \frac{1}{2}\sum_i(\mu_i^2 + \sigma_i^2 - 1 - \log \sigma_i^2)$

▷ `Stochastic VI` $\hat{\mathcal{L}} = \frac{1}{|\mathcal{B}|}\sum_n[\mathbb{E}_{q_\phi(\mathbf{z}|\mathbf{x}^{(n)})}[\log p_\theta(\mathbf{x}^{(n)}|\mathbf{z})] - D_{\text{KL}}(q_\phi(\mathbf{z}|\mathbf{x}^{(n)})||p(\mathbf{z}))]$

▷ `MC` $\mathbb{E}_{q_\phi(\mathbf{z}|\mathbf{x})}[\log p_\theta(\mathbf{x}|\mathbf{z})] \approx \frac{1}{S}\sum_s \log p_\theta(\mathbf{x}|\mathbf{z}^{(s)})$, $\mathbf{z}^{(s)} \sim q_\phi$, $S = 1$

▷ `Amortized` Shared $q_\phi(\mathbf{z}|\mathbf{x})$ (VAE) vs per-datum $\phi^{(n)}$ (CAVI)

▷ `Reverse KL` $D_{\text{KL}}(q||p)$ mode-seeking, tractable, underestimates variance

▷ `Forward KL` $D_{\text{KL}}(p||q)$ mode-covering, intractable, overestimates variance

▷ `beta-VAE` $\mathcal{L}_\beta = \mathbb{E}_{q_\phi(\mathbf{z}|\mathbf{x})}[\log p_\theta(\mathbf{x}|\mathbf{z})] - \beta D_{\text{KL}}(q_\phi(\mathbf{z}|\mathbf{x})||p(\mathbf{z}))$, $\beta > 1$

16.1 Latent Variable Models

16.1.1 Definition: Latent Variable Model

A **latent variable model** is a probabilistic model that assumes **observed data x is generated from unobserved (latent) variables z** through a generative process:

$$p_\theta(\mathbf{x}, \mathbf{z}) = p_\theta(\mathbf{x}|\mathbf{z})p_\theta(\mathbf{z}) \tag{16.1}$$

where:

- $\mathbf{x} \in \mathbb{R}^d$ is the observed data (e.g., the specific image(s), document(s) etc)
- $\mathbf{z} \in \mathbb{R}^k$ is the latent variable (typically $k < d$)
- θ are model parameters
- $p_\theta(\mathbf{z})$ is the **prior** distribution over latent variables
- $p_\theta(\mathbf{x}|\mathbf{z})$ is the **likelihood** (generative model)

Latent variable models capture the idea that observed data has an underlying structure that is not directly observable. The latent variables $\mathbf{z}$ represent this hidden structure, and the model describes how the observed data $\mathbf{x}$ is generated from $\mathbf{z}$.

Figure 16.1 visualizes the components of a latent variable model with a focus on the **generative process**. The left panel shows the latent variables z with a light blue background, representing the prior distribution $p(z)$. The middle panel illustrates the generative process ($z \to x$) through the likelihood $p(x|z)$, with a prominent black arrow showing the forward direction. The right panel shows the observed data x, representing the marginal likelihood $p(x)$. This visualization provides a clear overview of how latent variable models structure the relationship between hidden variables and observed data through the forward generative process.

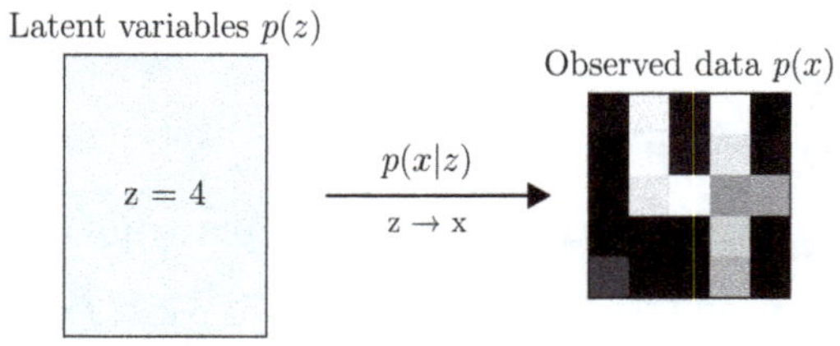

Figure 16.1: Latent Variable Model

While $p_\theta(\mathbf{x}, \mathbf{z}) = p_\theta(\mathbf{z}|\mathbf{x})p_\theta(\mathbf{x})$ is mathematically valid, it's not used in latent variable model definitions because it is not a generative model and also relies on intractable quantities ($p(\mathbf{x})$ and $p(\mathbf{z}|\mathbf{x})$) that are unknown upfront.

Latent variable models are fundamental to many machine learning applications. In **dimensionality reduction**, latent variables represent low-dimensional embeddings of high-dimensional data (e.g., PCA can be viewed as a latent variable model). In **clustering**, latent variables represent cluster assignments. In **topic modeling** (e.g., LDA), latent variables represent topic distributions. In **generative models** like VAEs and GANs, latent variables represent compressed representations that can be decoded to generate new data. In **recommendation systems**, latent variables represent user preferences and item features. The key advantage of latent variable models is that they can capture complex dependencies in data through a simpler latent structure, making them both interpretable and computationally tractable.

16.1.1.1 Example: Gaussian Mixture Model as Latent Variable Model

As defined in ~13.17.2 at Chapter~13, **Gaussian Mixture Model (GMM)** can be viewed as a latent variable model where:

- The latent variable $z \in \{0, 1, \dots, K-1\}$ represents the mixture component (cluster assignment)
- The prior is $p(z = k) = \pi_k$ where $\sum_{k=0}^{K-1} \pi_k = 1$
- The likelihood is $p(\mathbf{x}|z = k) = \mathcal{N}(\mathbf{x}|\mu_k, \Sigma_k)$

The joint distribution is:

$$p(\mathbf{x}, z) = \sum_{k=0}^{K-1} \pi_k \mathcal{N}(\mathbf{x}|\mu_k, \Sigma_k)\delta_{z,k},$$

where $\delta_{z,k}$ is the Kronecker delta (equal to 1 if $z = k$, and 0 otherwise), and the marginal distribution (mixture) is:

$$p(\mathbf{x}) = \sum_{k=0}^{K-1} \pi_k \mathcal{N}(\mathbf{x}|\mu_k, \Sigma_k). \tag{16.2}$$

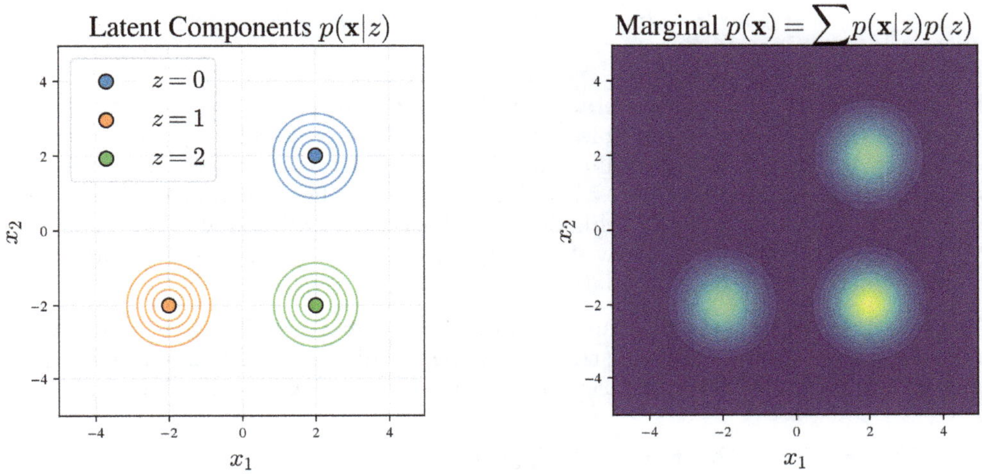

Figure 16.2: GMM Generative Process

Figure 16.2 illustrates the concept of a latent variable model using a Gaussian Mixture Model. The left panel shows the individual components $p(\mathbf{x}|z)$ conditioned on the latent variable $z \in \{0, 1, 2\}$, each represented by a different color and centered at different locations. The right panel shows the resulting marginal distribution $p(\mathbf{x}) = \sum_z p(\mathbf{x}|z)p(z)$, which is a weighted mixture of the three Gaussian components. This visualization makes concrete how a complex marginal distribution (the mixture) can be constructed from simpler conditional distributions (the components) through the latent variable framework. The latent variable z acts as a "selector" that determines which component generates each data point, enabling the model to capture multimodal structure in the data.

16.2 The Inference Problem - p(z|x)

In the context of latent variable models, **inference** refers to the process of reasoning about unobserved (latent) variables $\mathbf{z}$ given observed data $\mathbf{x}$. This is the inverse of the generative process: while the model describes how to generate data from latent variables ($\mathbf{x}$ from $\mathbf{z}$), inference asks what latent variables are most likely given the observed data ($\mathbf{z}$ from $\mathbf{x}$). In **Variational Autoencoders (VAEs)**, the encoder network performs inference by mapping observed data $\mathbf{x}$ to the parameters of the approximate posterior distribution $q_\phi(\mathbf{z}|\mathbf{x})$.

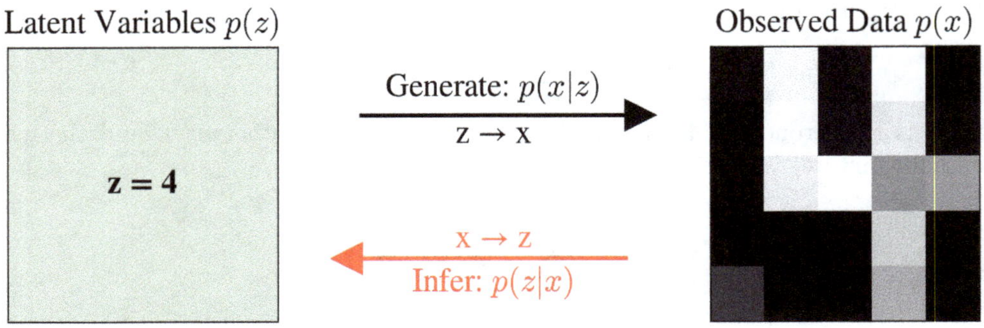

Figure 16.3: Inference Process in Latent Variable Models

Figure 16.3 visualizes both the generative and inference processes in latent variable models. The left panel shows the latent variables z with a light blue background, representing the prior distribution $p(z)$. The middle panel illustrates both the generative process ($z \to x$) with a black arrow and the inference process ($x \to z$) with a red arrow. The right panel shows the observed data x, representing the marginal likelihood $p(x)$. This visualization highlights the bidirectional nature of latent variable models: how data is generated from latent variables, and how latent variables are inferred from observed data.

The inference process involves reasoning backward from observed data x to latent variables z, which is captured by the posterior distribution $p(z|x)$. This is complemented by the generative process that moves from latent variables to observed data through the likelihood $p(x|z)$.

16.2.1 Definition: Posterior Inference

Given observed data $\mathbf{x}$ and a latent variable model $p_\theta(\mathbf{x}, \mathbf{z})$, the **posterior inference** problem is to compute the posterior distribution:

$$p_\theta(\mathbf{z}|\mathbf{x}) = \frac{p_\theta(\mathbf{x}, \mathbf{z})}{p_\theta(\mathbf{x})} = \frac{p_\theta(\mathbf{x}|\mathbf{z})p_\theta(\mathbf{z})}{\int p_\theta(\mathbf{x}|\mathbf{z})p_\theta(\mathbf{z})\, d\mathbf{z}}. \tag{16.3}$$

Computing the posterior is often **intractable** for several reasons, as illustrated by the MNIST handwritten digit dataset:

1. **No analytical solution**: The marginal likelihood (evidence) $p_\theta(\mathbf{x}) = \int p_\theta(\mathbf{x}|\mathbf{z})p_\theta(\mathbf{z})\, d\mathbf{z}$ requires integrating over all possible latent values, which typically has no closed-form solution. For MNIST, even if we model each digit as generated by a latent variable, the integral over all possible digit styles and variations has no analytical expression.

2. **Computational complexity**: For discrete latent variables, the number of possible configurations grows exponentially with the dataset size. For MNIST, if we model each of the 60,000 training images with a 10-dimensional one-hot latent variable (representing digit class), we would need to sum over $10^{60,000}$ configurations—an astronomically large number far beyond any computational possibility.

3. **High-dimensional integration**: For continuous latent variables in high dimensions, numerical integration becomes intractable due to the curse of dimensionality. For MNIST, if we use a 50-dimensional continuous latent space (common in VAEs), we would need to integrate over 50 dimensions, which is infeasible with traditional numerical methods.

4. **Complex models**: For neural network-based generative models or other complex models, exact inference is computationally infeasible even when the mathematical form exists. For MNIST, modern generative models like VAEs use deep neural networks to parameterize the

likelihood $p_\theta(\mathbf{x}|\mathbf{z})$, making exact posterior inference intractable despite the model's explicit mathematical form.

16.2.2 Limitations of Exact Inference Methods

Maximum A Posteriori (MAP) estimation attempts to find the mode of the posterior. For latent variable models, MAP inference seeks the most likely latent variables z^* given the data:

$$\mathbf{z}^* = \arg\max_{\mathbf{z}} p_\theta(\mathbf{z}|\mathbf{x}) = \arg\max_{\mathbf{z}} p_\theta(\mathbf{x}|\mathbf{z})p_\theta(\mathbf{z})$$

16.2.3 Explanation of the MAP Formula

The MAP estimator $\mathbf{z}^*$ is derived using Bayes' rule. Let's break down the formula:

1. **First Equality**: $\mathbf{z}^* = \arg\max_{\mathbf{z}} p_\theta(\mathbf{z}|\mathbf{x})$
   - This defines MAP as the latent variable configuration $\mathbf{z}$ that maximizes the posterior probability $p_\theta(\mathbf{z}|\mathbf{x})$
   - In other words, it finds the most probable explanation for the observed data $\mathbf{x}$
2. **Second Equality**: $\arg\max_{\mathbf{z}} p_\theta(\mathbf{z}|\mathbf{x}) = \arg\max_{\mathbf{z}} p_\theta(\mathbf{x}|\mathbf{z})p_\theta(\mathbf{z})$
   - This follows from Bayes' rule: $p(\mathbf{z}|\mathbf{x}) = \frac{p(\mathbf{x}|\mathbf{z})p(\mathbf{z})}{p(\mathbf{x})}$
   - Since $p(\mathbf{x})$ is constant with respect to $\mathbf{z}$, it doesn't affect the maximization
   - Thus, maximizing the posterior is equivalent to maximizing the product of the likelihood $p_\theta(\mathbf{x}|\mathbf{z})$ and the prior $p_\theta(\mathbf{z})$

16.2.4 Intuition

The MAP estimator balances two competing factors:

- **Likelihood term** $p_\theta(\mathbf{x}|\mathbf{z})$: How well the latent variables explain the observed data
- **Prior term** $p_\theta(\mathbf{z})$: How plausible the latent variables are before seeing any data

This balance makes MAP a compromise between purely data-driven estimates (maximum likelihood) and prior knowledge.

However, MAP has limitations:

- It only provides a **point estimate**, not the full posterior distribution
  - **Example**: Consider a handwritten digit classifier. A traditional Autoencoder (AE) uses MAP-like estimation, outputting a single latent vector for each input image (a point estimate). In contrast, a Variational Autoencoder (VAE) outputs a full posterior distribution over latent variables, including both a mean vector (central tendency) and a variance vector (uncertainty). For a blurry digit image, the VAE's posterior will have higher variance, reflecting greater uncertainty about the latent representation.

  Concrete Numerical Example: Suppose the input is an image of the digit "5":

  - **Traditional Autoencoder output**: A single latent vector $\mathbf{z} = [0.3, -1.2]$ (a fixed point estimate)
  - **VAE Encoder output**:
    * Mean vector $\mu = [0.3, -1.2]$ (central tendency)
    * Log-variance vector $\mathbf{log}\sigma^2 = [-0.5, -0.8]$ (corresponding to variance $\sigma^2 = [0.61, 0.45]$)

  The VAE's output defines a full Gaussian posterior distribution $q_\phi(\mathbf{z}|\mathbf{x}) = \mathcal{N}(\mathbf{z}; \mu, \text{diag}(\sigma^2))$, from which multiple latent vectors can be sampled, reflecting the inherent uncertainty in the latent representation.

- It requires computing the posterior (or its mode), which is often intractable
- It ignores uncertainty quantification, which is crucial for Bayesian inference
- For latent variable models, MAP still requires evaluating the posterior, which depends on the intractable marginal likelihood $p_\theta(\mathbf{x})$

Therefore, we need **approximate inference** methods that can handle intractable posteriors.

16.2.5 Approximate Inference Methods

When exact inference is intractable, we use **approximate inference** methods. There are three main approaches:

16.2.5.1 Sampling-Based Methods

Sampling-based methods generate **samples** from the posterior distribution, which can then be used to estimate expectations and other quantities. The most important class of sampling methods is **Markov Chain Monte Carlo (MCMC)** (see Chapter 18 for the formal definition).

MCMC methods construct a Markov chain whose stationary distribution is the target posterior. A key MCMC algorithm is **Gibbs sampling**, which is a special case of MCMC where proposals are drawn from conditional distributions. Gibbs sampling iteratively samples each dimension from its conditional distribution given all other dimensions, cycling through all dimensions in sequence.

For detailed discussion of MCMC algorithms including Gibbs sampling, Metropolis-Hastings, and Langevin-based methods, see Chapter 18.

Key properties of sampling-based methods (MCMC):

- Provides **samples** from the posterior (asymptotically exact)
- Only requires conditional distributions (often easier to compute than the full posterior)
- Samples are **correlated** (not independent), requiring many samples for accurate estimates
- **Mixing time** can be long for complex distributions
- Computationally expensive (sequential sampling)

16.2.5.2 Optimization-Based Methods

Optimization-based methods optimize a lower bound on the log-likelihood (the ELBO) to find model parameters and approximate posteriors. There are two main approaches:

Expectation-Maximization (EM) EM is an iterative algorithm that maximizes the log-likelihood by alternating between two steps: the **E-step** (Expectation) computes the exact posterior $p_\theta(\mathbf{z}|\mathbf{x})$ when it's tractable, and the **M-step** (Maximization) updates model parameters θ to maximize the expected log-likelihood.

Key properties of EM:

- Uses the **exact posterior** in the E-step (when tractable)
- Alternates between E-step and M-step until convergence
- Guaranteed to increase the log-likelihood at each iteration
- Works well for models with conjugate priors (e.g., GMM, HMM)
- Limited to models where the posterior can be computed exactly

Variational Inference (VI) The term "variational" comes from the **calculus of variations**, a branch of mathematics that studies how to find functions that minimize or maximize functionals (functions of functions). In the context of inference, we're searching for a probability distribution $q(\mathbf{z})$ that minimizes the KL divergence functional $KL(q||p)$. This is in contrast to traditional calculus, which finds scalar values that minimize functions of scalars.

Variational Inference approximates the posterior with a simpler **parametric distribution** $q_\phi(\mathbf{z})$ and optimizes the parameters ϕ to minimize KL divergence from q to the true posterior.

Key properties of VI:

- Provides a **parametric approximation** to the posterior (not exact, but fast)
- Optimizes a lower bound (ELBO) using gradient-based methods
- Computationally efficient (can use stochastic optimization)
- Integrates well with neural networks and deep learning
- Provides a distribution (not just samples), enabling fast inference
- Works for models where the exact posterior is intractable

Key differences: MCMC **samples** from the posterior (asymptotically exact but slow). EM uses the **exact posterior** when tractable (classical statistical models). VI **approximates** the posterior with a simpler distribution (fast but approximate, works for complex models). This chapter focuses on **variational inference** and **expectation-maximization**. The inference problem is central to **Bayesian machine learning**. In **Bayesian neural networks**, we need to infer the posterior distribution over weights given data. In **VAEs**, we need to infer the posterior over latent codes given images. In **topic models** like LDA, we need to infer topic distributions given documents. The intractability of exact inference motivates approximate methods. **Variational inference** is particularly popular in deep learning because it can be combined with gradient-based optimization and neural networks, making it scalable to large datasets and complex models. The trade-off is that variational inference provides a lower bound (ELBO) rather than exact inference, but in practice, this bound is often tight enough for good performance.

16.3 Expectation-Maximization

The **Expectation-Maximization (EM)** algorithm is a classical method for maximum likelihood estimation in latent variable models. Unlike variational inference, which approximates the posterior, EM uses the **exact posterior** when it's tractable. EM alternates between computing the posterior (E-step) and maximizing the expected log-likelihood (M-step).

16.3.1 Definition: Expectation-Maximization Algorithm

The **Expectation-Maximization (EM) algorithm** is an iterative algorithm for maximum likelihood estimation in latent variable models. Given observed data $\mathbf{x}$ and a latent variable model $p_\theta(\mathbf{x}, \mathbf{z})$, EM alternates between:

1. **E-step (Expectation)**: Compute the posterior distribution $p_{\theta^{(t)}}(\mathbf{z}|\mathbf{x})$ using current parameters $\theta^{(t)}$.
2. **M-step (Maximization)**: Update parameters to maximize the expected log-likelihood:

$$\theta^{(t+1)} = \arg\max_\theta \mathbb{E}_{p_{\theta^{(t)}}(\mathbf{z}|\mathbf{x})}[\log p_\theta(\mathbf{x}, \mathbf{z})].$$

The algorithm iterates until convergence (when the log-likelihood $\log p_\theta(\mathbf{x})$ stops increasing).

The EM algorithm was introduced by Dempster, Laird, and Rubin (1977) and has become a fundamental tool in statistics and machine learning. EM is particularly effective for models with **conjugate priors**, where the posterior has a closed-form expression. Common applications include **Gaussian Mixture Models (GMM)**, **Hidden Markov Models (HMM)**, **factor analysis**, and **topic models** (e.g., probabilistic Latent Semantic Analysis). The key advantage of EM is that it uses the exact posterior (when tractable), avoiding approximation errors. However, EM is limited to models where the E-step can be computed exactly, which restricts its applicability to simpler models compared to variational inference.

16.3.2 Derivation: EM as ELBO Maximization

EM can be derived from the ELBO framework. Starting from the log-likelihood:

$$\log p_\theta(\mathbf{x}) = \log \int p_\theta(\mathbf{x}, \mathbf{z}) \, d\mathbf{z}.$$

Introducing a distribution $q(\mathbf{z})$ and applying the same trick as in VI:

$$\log p_\theta(\mathbf{x}) = \log \int q(\mathbf{z}) \frac{p_\theta(\mathbf{x}, \mathbf{z})}{q(\mathbf{z})} \, d\mathbf{z} = \log \mathbb{E}_{q(\mathbf{z})} \left[\frac{p_\theta(\mathbf{x}, \mathbf{z})}{q(\mathbf{z})} \right].$$

By Jensen's inequality (see Chapter 12, Section~12.2.4):

$$\log p_\theta(\mathbf{x}) \geq \mathbb{E}_{q(\mathbf{z})}[\log p_\theta(\mathbf{x}, \mathbf{z})] - \mathbb{E}_{q(\mathbf{z})}[\log q(\mathbf{z})] = \text{ELBO}(q, \theta). \tag{16.4}$$

The key difference from VI is that **EM sets** $q(\mathbf{z}) = p_\theta(\mathbf{z}|\mathbf{x})$ (the exact posterior) in the E-step, rather than approximating it with a restricted family. This is only possible when the posterior is tractable.

E-step: Set $q^{(t+1)}(\mathbf{z}) = p_{\theta^{(t)}}(\mathbf{z}|\mathbf{x})$ to maximize the ELBO with respect to q (keeping θ fixed). This achieves equality in Jensen's inequality, making the bound tight.

M-step: Maximize the ELBO with respect to θ (keeping q fixed):

$$\theta^{(t+1)} = \arg\max_\theta \mathbb{E}_{p_{\theta^{(t)}}(\mathbf{z}|\mathbf{x})}[\log p_\theta(\mathbf{x}, \mathbf{z})].$$

Since the ELBO is a lower bound, and the E-step makes it tight, the M-step increases the log-likelihood: $\log p_{\theta^{(t+1)}}(\mathbf{x}) \geq \log p_{\theta^{(t)}}(\mathbf{x})$.

16.3.3 Algorithm: Expectation-Maximization

Input: Observed data $\{\mathbf{x}^{(i)}\}_{i=1}^N$, initial parameters $\theta^{(0)}$

Output: Estimated parameters θ^*

1. **Initialize** $\theta^{(0)}$ (e.g., randomly or using heuristics)
2. **Iterate** until convergence:
   - **E-step**: For each data point $\mathbf{x}^{(i)}$, compute the posterior:

     $$q^{(t+1)}(\mathbf{z}^{(i)}) = p_{\theta^{(t)}}(\mathbf{z}^{(i)}|\mathbf{x}^{(i)})$$

   - **M-step**: Update parameters to maximize expected log-likelihood:

     $$\theta^{(t+1)} = \arg\max_\theta \sum_{i=1}^N \mathbb{E}_{q^{(t+1)}(\mathbf{z}^{(i)})}[\log p_\theta(\mathbf{x}^{(i)}, \mathbf{z}^{(i)})]$$

3. **Return** $\theta^* = \theta^{(t)}$ (at convergence)

Convergence criterion: Stop when $|\log p_{\theta^{(t+1)}}(\mathbf{x}) - \log p_{\theta^{(t)}}(\mathbf{x})| < \epsilon$ for a small threshold $\epsilon > 0$.

16.3.4 Example: EM for Coin Flipping (Simple Case)

To build intuition, let's start with a simple coin flipping example. Suppose we have two coins (Coin A and Coin B) with unknown biases θ_A and θ_B (probability of heads). We perform the following experiment:

1. **Experiment**: We randomly select one of the two coins (with equal probability), flip it 10 times, and record the number of heads.
2. **Data**: We repeat this experiment 5 times and observe the results: $\mathbf{x} = [6, 7, 3, 2, 8]$ (number of heads in each trial).
3. **Latent Variable**: For each trial, we don't know which coin was used. Let $z_i \in \{A, B\}$ be the coin used in trial i.

Our goal is to estimate θ_A and θ_B using the EM algorithm.

16.3.4.1 E-step

For each trial i and coin $k \in \{A, B\}$, compute the posterior probability that coin k was used, given the data and current parameter estimates:

$$q^{(t+1)}(z_i = k) = p(z_i = k|\mathbf{x}_i, \theta_A^{(t)}, \theta_B^{(t)}) = \frac{p(\mathbf{x}_i|z_i = k, \theta_k^{(t)})p(z_i = k)}{\sum_{j\in\{A,B\}} p(\mathbf{x}_i|z_i = j, \theta_j^{(t)})p(z_i = j)}$$

Since we select coins with equal probability, $p(z_i = A) = p(z_i = B) = 0.5$. The likelihood $p(\mathbf{x}_i|z_i = k, \theta_k)$ is the binomial probability:

$$p(\mathbf{x}_i = h|z_i = k, \theta_k) = \binom{10}{h}\theta_k^h(1-\theta_k)^{10-h}$$

16.3.4.2 M-step

Update the coin biases to maximize the expected log-likelihood:

$$\theta_k^{(t+1)} = \frac{\sum_{i=1}^5 q^{(t+1)}(z_i = k)\cdot h_i}{\sum_{i=1}^5 q^{(t+1)}(z_i = k)\cdot 10}$$

where h_i is the number of heads in trial i.

16.3.4.3 Example Calculation

Suppose we start with initial estimates $\theta_A^{(0)} = 0.6$ and $\theta_B^{(0)} = 0.4$:

E-step (t=0): - For trial 1 (6 heads):

- $p(x_1 = 6|A, 0.6) = \binom{10}{6}0.6^6 0.4^4 \approx 0.251$
- $p(x_1 = 6|B, 0.4) = \binom{10}{6}0.4^6 0.6^4 \approx 0.111$
- $q(z_1 = A) = \frac{0.251\cdot 0.5}{0.251\cdot 0.5+0.111\cdot 0.5} \approx 0.693$
- $q(z_1 = B) = 1 - 0.693 = 0.307$

M-step (t=0→1): - $\theta_A^{(1)} = \frac{0.693\cdot 6+\ldots}{0.693\cdot 10+\ldots}$ (calculated for all trials) - $\theta_B^{(1)} = \frac{0.307\cdot 6+\ldots}{0.307\cdot 10+\ldots}$ (calculated for all trials)

This process repeats until convergence, eventually estimating the true biases of the coins.

16.3.5 Example: EM for Gaussian Mixture Model

Consider a **Gaussian Mixture Model (GMM)** with K components (see Definition~13.17.2 in Chapter~13). The model is:

$$p(\mathbf{x}, z) = \sum_{k=0}^{K-1} \pi_k \mathcal{N}(\mathbf{x}|\mu_k, \Sigma_k)\delta_{z,k}$$

where $z \in \{0, 1, \dots, K-1\}$ is the latent component assignment, and $\theta = \{\pi_k, \mu_k, \Sigma_k\}_{k=0}^{K-1}$ are the parameters.

E-step: Compute the posterior probability (responsibility) that data point $\mathbf{x}^{(i)}$ belongs to component k:

$$q^{(t+1)}(z^{(i)} = k) = p_{\theta^{(t)}}(z^{(i)} = k|\mathbf{x}^{(i)}) = \frac{\pi_k^{(t)} \mathcal{N}(\mathbf{x}^{(i)}|\mu_k^{(t)}, \Sigma_k^{(t)})}{\sum_{j=0}^{K-1} \pi_j^{(t)} \mathcal{N}(\mathbf{x}^{(i)}|\mu_j^{(t)}, \Sigma_j^{(t)})} =: \gamma_{ik}^{(t+1)}. \tag{16.5}$$

M-step: Update parameters to maximize expected log-likelihood:

$$\begin{aligned} \pi_k^{(t+1)} &= \frac{1}{N}\sum_{i=1}^{N} \gamma_{ik}^{(t+1)}, \quad \mu_k^{(t+1)} = \frac{\sum_{i=1}^{N} \gamma_{ik}^{(t+1)} \mathbf{x}^{(i)}}{\sum_{i=1}^{N} \gamma_{ik}^{(t+1)}}, \\ \Sigma_k^{(t+1)} &= \frac{\sum_{i=1}^{N} \gamma_{ik}^{(t+1)} (\mathbf{x}^{(i)} - \mu_k^{(t+1)})(\mathbf{x}^{(i)} - \mu_k^{(t+1)})^\top}{\sum_{i=1}^{N} \gamma_{ik}^{(t+1)}}. \end{aligned} \tag{16.6}$$

The E-step computes the "soft assignment" (posterior probability) of each data point to each component, and the M-step updates the component parameters using these soft assignments as weights.

16.3.6 Comparison: EM vs. Variational Inference

Both EM and VI optimize the ELBO, but they differ in how they handle the E-step:

Aspect	Expectation-Maximization	Variational Inference
E-step	Uses **exact posterior** $p_\theta(\mathbf{z}|\mathbf{x})$	Uses **approximate posterior** $q_\phi(\mathbf{z}) \approx p_\theta(\mathbf{z}|\mathbf{x})$
Posterior	True posterior (when tractable)	Restricted family $\mathcal{Q}$
Tractability	Requires tractable posterior	Works for intractable posteriors
Models	Simple models (conjugate priors)	Complex models (neural networks)
Scalability	Medium	High (stochastic optimization)
Inference	Exact (when possible)	Approximate

When to use EM:

- The posterior $p_\theta(\mathbf{z}|\mathbf{x})$ can be computed exactly (e.g., conjugate priors)
- Models are relatively simple (GMM, HMM, factor analysis)
- Small to medium datasets where exact computation is feasible

When to use VI:

- The posterior is intractable (complex models, neural networks)

- Large-scale data requiring stochastic optimization
- Need amortized inference (fast inference for new data points)
- Deep learning applications

Hybrid approaches: **Variational EM** uses VI for the E-step when the exact posterior is intractable, combining the structure of EM with the flexibility of VI.

16.4 Variational Inference

16.4.1 Definition: Variational Inference

Variational Inference (VI)[1] approximates the intractable posterior $p_\theta(\mathbf{z}|\mathbf{x})$ with a simpler **variational distribution** $q_\phi(\mathbf{z})$ parameterized by **variational parameters** ϕ, chosen from a restricted family of distributions $\mathcal{Q}$.

The variational family $\mathcal{Q}$ is a constrained set of distributions (not all possible distributions) chosen for computational tractability. The choice of $\mathcal{Q}$ balances expressiveness (approximation quality) and computational efficiency (optimization and expectation computation). Common choices include Gaussian distributions, **mean field** distributions (factorized into independent components: $q(\mathbf{z}) = \prod_{i=0}^{k-1} q_i(z_i)$), or distributions parameterized by neural networks.

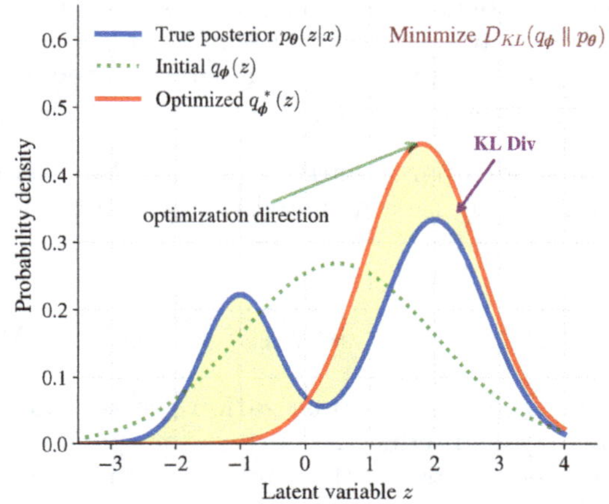

Figure 16.4: Variational Inference

The variational parameters ϕ define the specific distribution within this family. For example, if q is Gaussian, then $\phi = \{\mu, \Sigma\}$; if q is parameterized by a neural network (as in VAEs), then ϕ represents the encoder network's weights and biases.

The goal is to find $q^*_{\phi^*}(\mathbf{z}) \in \mathcal{Q}$ that minimizes the KL divergence from q to the true posterior:

$$q^* = \arg\min_{q \in \mathcal{Q}} D_{\text{KL}}(q_\phi(\mathbf{z})||p_\theta(\mathbf{z}|\mathbf{x})). \tag{16.7}$$

Here, q^* denotes the **optimal variational distribution** (the best approximation within the family $\mathcal{Q}$), and ϕ^* denotes the **optimal variational parameters** that achieve this minimum. The asterisk (*) notation is standard mathematical notation for optimal values.

Figure 16.4 visualizes the core concepts of variational inference. The plot shows the true posterior distribution $p_\theta(z|x)$ (blue solid line), an initial variational approximation $q_\phi(z)$ (green dotted line), and the optimized variational distribution $q^*_\phi(z)$ (red solid line). The yellow shaded area represents the KL divergence $D_{KL}(q_\phi \parallel p_\theta)$ that is minimized during optimization, highlighting the regions where the distributions differ. The green arrow indicates the optimization direction from the initial approximation to the optimized one. The objective function is "minimize $D_{KL}(q_\phi \parallel p_\theta)$". This

[1]The term "variational" comes from the calculus of variations, a branch of mathematics that deals with optimizing over functions rather than just parameters. In variational inference, we optimize over a family of distributions (functions) to find the best approximation to the true posterior, searching through a variational family $\mathcal{Q}$ of tractable distributions.

visualization provides a clear overview of how variational inference approximates an intractable posterior with a tractable variational distribution by minimizing the KL divergence between them.

A common misconception is that variational inference assumes the prior distribution $p(\mathbf{z})$ is Gaussian. In reality, variational inference assumes the **variational distribution** $q(\mathbf{z})$ (our approximation to the posterior $p(\mathbf{z}|\mathbf{x})$) is from a tractable family like Gaussian. The prior $p(\mathbf{z})$ can be **any distribution**; it's the posterior approximation $q(\mathbf{z})$ that we constrain to be tractable.

16.4.2 Posterior Collapse

A critical issue in variational inference is **posterior collapse** (also called **KL vanishing**), where the variational posterior $q_\phi(\mathbf{z}|\mathbf{x})$ collapses to the prior $p(\mathbf{z})$ (typically $\mathcal{N}(\mathbf{0}, \mathbf{I})$), causing the latent variable $\mathbf{z}$ to carry no information about the input $\mathbf{x}$. This occurs when the KL regularization term $D_{\mathrm{KL}}(q_\phi(\mathbf{z}|\mathbf{x})||p(\mathbf{z}))$ dominates the **reconstruction term** (the term that measures how well the model can reconstruct the input data from the latent variable), forcing $q_\phi(\mathbf{z}|\mathbf{x}) \approx p(\mathbf{z})$.

This issue is particularly relevant in **Variational Autoencoders (VAEs)**, where a common solution is to use a weighted KL term. In β**-VAE**, the KL term is weighted by a hyperparameter β:

$$\mathcal{L}_{\beta\text{-VAE}} = \mathbb{E}_{q_\phi(\mathbf{z}|\mathbf{x})}[\log p_\theta(\mathbf{x}|\mathbf{z})] - \beta \cdot D_{\mathrm{KL}}(q_\phi(\mathbf{z}|\mathbf{x})||p(\mathbf{z})). \tag{16.8}$$

When $\beta > 1$, the regularization pressure increases, which can help prevent posterior collapse but may also reduce reconstruction quality. When $\beta < 1$, the model prioritizes reconstruction, potentially allowing the latent code to encode more information. The optimal β balances these trade-offs.

16.4.3 Precision over Recall: The Mode-Seeking of Reverse KL

When we choose to minimize $D_{\mathrm{KL}}(q||p)$ (reverse KL) instead of $D_{\mathrm{KL}}(p||q)$ (forward KL) in Variational Inference and VAEs, we are making a fundamental trade-off: we favor **"precision"** (accuracy within a specific region) over **"recall"** (covering every possible outcome). This choice shapes how our approximating distribution q behaves relative to the true posterior p, and understanding this behavior is crucial for grasping why variational methods work the way they do.

To see why, consider what happens when we use reverse KL. The divergence $D_{\mathrm{KL}}(q||p)$ penalizes q heavily whenever it places probability mass where the true posterior p is nearly zero. To minimize this penalty, q naturally "shrinks" to fit snugly inside a single mode of p. It would rather be narrow and "precise" about one peak than risk spreading into a low-probability gap. This **mode-seeking behavior** leads q to concentrate on the major modes of p rather than spreading out to cover all modes. In contrast, forward KL $D_{\mathrm{KL}}(p||q)$ works differently: it penalizes the model if q is zero anywhere p is high. This forces q to spread out and "recall" every mode of p, often resulting in an over-dispersed approximation that covers the gaps between modes. This **mode-covering behavior** prioritizes recall over precision.

The precision-oriented nature of reverse KL makes it particularly well-suited for VAEs. The Reverse KL objective involves an expectation over q ($\mathbb{E}_q[\cdot]$), and since q is a distribution we define (like a Gaussian), we can easily sample from it and compute gradients. Forward KL, on the other hand, requires expectations over p, which is exactly what we are trying to avoid computing because it is intractable. Beyond tractability, mode-seeking behavior aligns perfectly with what we want from a VAE encoder: we want it to map an input to a specific, compact region in the latent space. This ensures that the latent representation is concentrated and reliable, rather than a blurred average of multiple possible interpretations. Moreover, by seeking a single mode, the optimization process is often more stable and results in tighter lower bounds (ELBO) for that specific region of the data manifold.

Of course, this precision comes at a cost. Reverse KL tends to produce q with smaller variance than p, which means we underestimate uncertainty. For point estimates, this is often acceptable and is outweighed by the computational advantages. However, as we will see in Section~16.4.2, this mode-seeking behavior can lead to posterior collapse in some models, where the variational posterior becomes too simple and fails to capture important structure in the data.

> The choice of KL divergence direction is crucial in variational inference. The precision-recall analogy helps explain why **reverse KL** $D_{\mathrm{KL}}(q||p)$ is preferred: it prioritizes precision (accurate approximation within a specific region) over recall (covering all modes). This mode-seeking behavior is computationally tractable (expectations over q), leads to tighter bounds when q is simpler than p, and produces meaningful, concentrated latent representations in VAEs. However, this mode-seeking behavior can lead to issues in some models (see Section~16.4.2 for details on posterior collapse in VAEs). **Forward KL** $D_{\mathrm{KL}}(p||q)$ exhibits mode-covering behavior (recall-oriented) but requires expectations over the intractable p, making it unsuitable for variational inference.

Figure 16.5 brings this abstract discussion to life by visualizing the crucial difference between forward KL and reverse KL when approximating a bimodal distribution $p(z)$ (blue curve). On the left, we see what happens with **reverse KL** $D_{\mathrm{KL}}(q||p)$ (used in variational inference): the approximation $q(z)$ (red shaded region) concentrates tightly on a single mode, demonstrating the **precision-oriented** (mode-seeking) behavior we just described. On the right, **forward KL** $D_{\mathrm{KL}}(p||q)$ produces a very different result: the approximation $q(z)$ (green shaded region) spreads out to cover both modes, demonstrating **recall-oriented** (mode-covering) behavior. This side-by-side comparison makes concrete why variational inference uses reverse KL: it produces approximations that prioritize precision (accurate fit within a specific region) over recall (covering all modes), which is computationally more tractable and often more useful for point estimates. However, as the figure also suggests, this precision-oriented behavior can miss important structure when the true posterior has multiple significant modes.

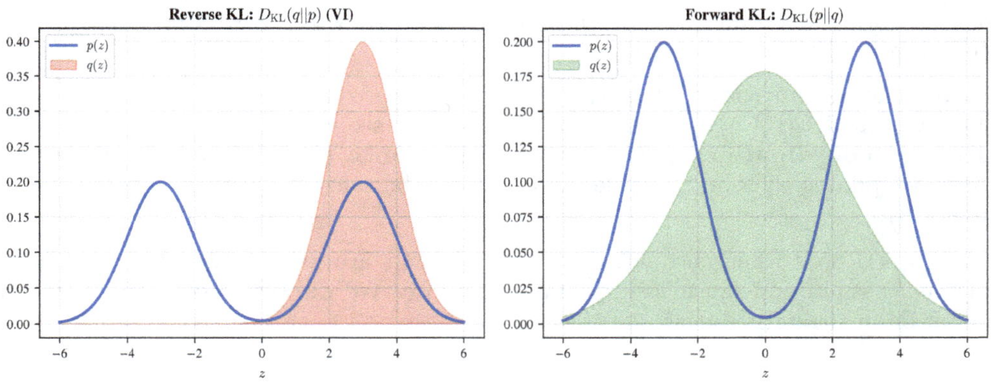

Figure 16.5: Forward vs Reverse KL Divergence

16.4.4 Connection to ELBO

As discussed in Chapter 14, minimizing $D_{\mathrm{KL}}(q(\mathbf{z})||p_\theta(\mathbf{z}|\mathbf{x}))$ is equivalent to maximizing the **Evidence Lower Bound (ELBO)**:

$$\mathrm{ELBO}(q, \theta) = \mathbb{E}_{q(\mathbf{z})}[\log p_\theta(\mathbf{x}, \mathbf{z})] - \mathbb{E}_{q(\mathbf{z})}[\log q(\mathbf{z})]. \tag{16.9}$$

The ELBO can be decomposed as:

$$\mathrm{ELBO}(q, \theta) = \mathbb{E}_{q(\mathbf{z})}[\log p_\theta(\mathbf{x}|\mathbf{z})] - D_{\mathrm{KL}}(q(\mathbf{z})||p_\theta(\mathbf{z})) \tag{16.10}$$

which separates into:

- **Reconstruction term**: $\mathbb{E}_{q(\mathbf{z})}[\log p_\theta(\mathbf{x}|\mathbf{z})]$ encourages accurate data reconstruction
- **Regularization term**: $-D_{\text{KL}}(q(\mathbf{z})||p_\theta(\mathbf{z}))$ encourages q to match the prior

16.4.5 Variational Family

The choice of variational family $\mathcal{Q}$ is crucial:

1. **Mean Field**: $q(\mathbf{z}) = \prod_{i=0}^{k-1} q_i(z_i)$ (factorized, independent components)
2. **Gaussian**: $q(\mathbf{z}) = \mathcal{N}(\mathbf{z}|\mu, \Sigma)$ (full covariance or diagonal)
3. **Neural Networks**: $q_\phi(\mathbf{z}|\mathbf{x}) = \mathcal{N}(\mathbf{z}|\mu_\phi(\mathbf{x}), \Sigma_\phi(\mathbf{x}))$ where μ_ϕ, Σ_ϕ are neural networks (used in VAEs)

> The choice of variational family determines the expressiveness and computational cost of variational inference. **Mean field** assumes independence, making optimization easier but potentially missing dependencies. **Full Gaussian** captures correlations but requires more parameters. **Neural network-based** variational families (as in VAEs) are highly expressive and can learn complex posterior distributions, but require careful design to ensure the KL divergence is computable. In practice, **diagonal Gaussian** distributions parameterized by neural networks are common in VAEs because they balance expressiveness and computational efficiency. The reparameterization trick (discussed later) enables gradient-based optimization of these neural variational families.

16.5 Mean Field Variational Inference

We have seen three choices of variational family; the simplest is **mean field**. Here we assume the latent variables are independent, so $q(\mathbf{z})$ factorizes into a product of univariate marginals. That factorization buys tractability: we can optimize each factor separately, often in closed form. In this section we proceed step by step:

1. **State the assumption**—make the mean field factorization precise.
2. **Derive the ELBO** under this factorization, so we see how the objective simplifies.
3. **Present coordinate-ascent variational inference (CAVI)**—an algorithm that iteratively updates each $q_i(z_i)$ while holding the others fixed, with updates that are often analytic.

Mean field works well when correlations between latents are weak; when they are strong, the independence assumption can miss important structure. We will see a geometric illustration of this limitation in the figure below.

16.5.1 Definition: Mean Field Assumption

The **mean field** assumption assumes the variational distribution $q(\mathbf{z})$ factorizes into **independent** components:

$$q(\mathbf{z}) = \prod_{i=0}^{k-1} q_i(z_i) \tag{16.11}$$

where each $q_i(z_i)$ is a univariate distribution.

The mean field assumption is motivated by the goal of **computational simplicity**. By assuming independence between latent variables, we achieve several key benefits:

1. **Dimensionality reduction**: Instead of optimizing a complex high-dimensional joint distribution, we optimize k separate low-dimensional distributions independently

2. **Closed-form updates**: For many models, the optimal $q_i(z_i)$ can be computed analytically in closed form
3. **Parallel optimization**: Each $q_i(z_i)$ can be updated in parallel, significantly speeding up the process

The term "mean field" originates from statistical physics, where it describes the approximation that each particle experiences the "mean" effect of the entire field created by all other particles, rather than interacting with each individual particle separately. This statistical physics concept directly translates to our setting, where each latent variable is optimized based on the average effect of all other latent variables.

16.5.2 Mean Field ELBO

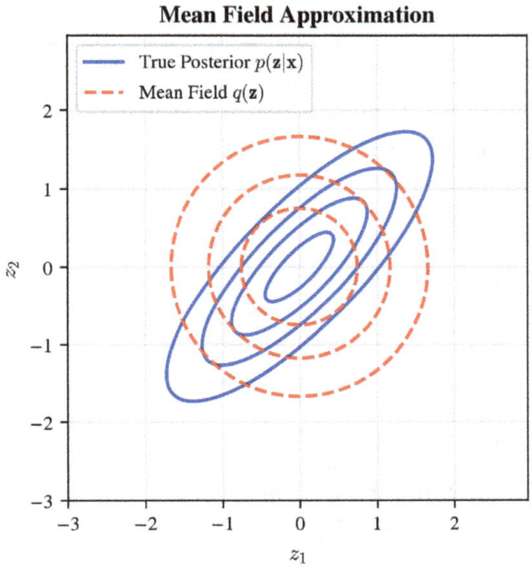

Figure 16.6: Mean Field Approximation Limitation

Under the mean field assumption, the ELBO becomes:

$$\text{ELBO} = \mathbb{E}_{q(\mathbf{z})}[\log p_\theta(\mathbf{x}|\mathbf{z})] - \sum_{i=0}^{k-1} \mathbb{E}_{q_i(z_i)}[\log q_i(z_i)] + \sum_{i=0}^{k-1} \mathbb{E}_{q_i(z_i)}[\log p_\theta(z_i)] \tag{16.12}$$

where the last two terms come from the KL divergence decomposition under independence.

Figure 16.6 demonstrates the limitation of the mean field assumption when the true posterior has correlations. The blue solid contours show a correlated true posterior $p_\theta(\mathbf{z}|\mathbf{x})$ with correlation $\rho = 0.8$ between z_1 and z_2. The red dashed contours show the mean field approximation $q_\phi(\mathbf{z}) = q_1(z_1)q_2(z_2)$, which assumes independence and is axis-aligned. The mean field approximation underestimates variance in correlated directions and cannot capture the elliptical shape of the true posterior. This visualization makes concrete why mean field can be a poor approximation when latent variables are strongly dependent: the independence assumption forces the approximation to miss important correlation structure, leading to larger KL divergence and potentially poor inference quality. In practice, mean field works well when correlations are weak or when the computational simplicity outweighs the approximation error.

16.5.3 Coordinate Ascent Variational Inference (CAVI)

Coordinate Ascent Variational Inference (CAVI) optimizes the mean field ELBO by iteratively updating each $q_i(z_i)$ while keeping others fixed.

For each i, the optimal $q_i^*(z_i)$ is:

$$q_i^*(z_i) \propto \exp\left(\mathbb{E}_{q_{-i}(\mathbf{z}_{-i})}[\log p_\theta(\mathbf{x}, \mathbf{z})]\right) \tag{16.13}$$

where $q_{-i}(\mathbf{z}_{-i}) = \prod_{j \neq i} q_j(z_j)$ is the distribution over all variables except z_i.

The iterative process is summarized in Algorithm~16. The algorithm initializes all variational factors and then iteratively updates each $q_i(z_i)$ according to Equation~16.13 while keeping the others fixed, until the ELBO converges.

Algorithm 16 Coordinate Ascent Variational Inference (CAVI)

Input: Joint distribution $p_\theta(\mathbf{x}, \mathbf{z})$, Variational family $q(\mathbf{z}) = \prod_{i=0}^{k-1} q_i(z_i)$
Output: Optimized variational factors $q_i(z_i)$
  Initialize $q_i(z_i)$ for all i
  while ELBO has not converged **do**
    for $i = 0, \dots, k-1$ **do**
      $q_i(z_i) \propto \exp\left(\mathbb{E}_{q_{-i}(\mathbf{z}_{-i})}[\log p_\theta(\mathbf{x}, \mathbf{z})]\right)$
    end for
    Compute ELBO(q)
  end while

Figure 16.7 illustrates the CAVI optimization process for mean field variational inference. The left panel shows the true correlated posterior $p_\theta(z_1, z_2|x)$ with blue contours. The subsequent panels show how the mean field approximation $q(z_1)q(z_2)$ evolves through each iteration of the CAVI algorithm, gradually approaching the true posterior. The factorized nature of the mean field approximation is evident in the axis-aligned contours, which can only approximate the elliptical shape of the true posterior. This visualization makes concrete how CAVI iteratively optimizes the variational factors to minimize the KL divergence, even though the independence assumption limits the approximation's ability to capture correlations.

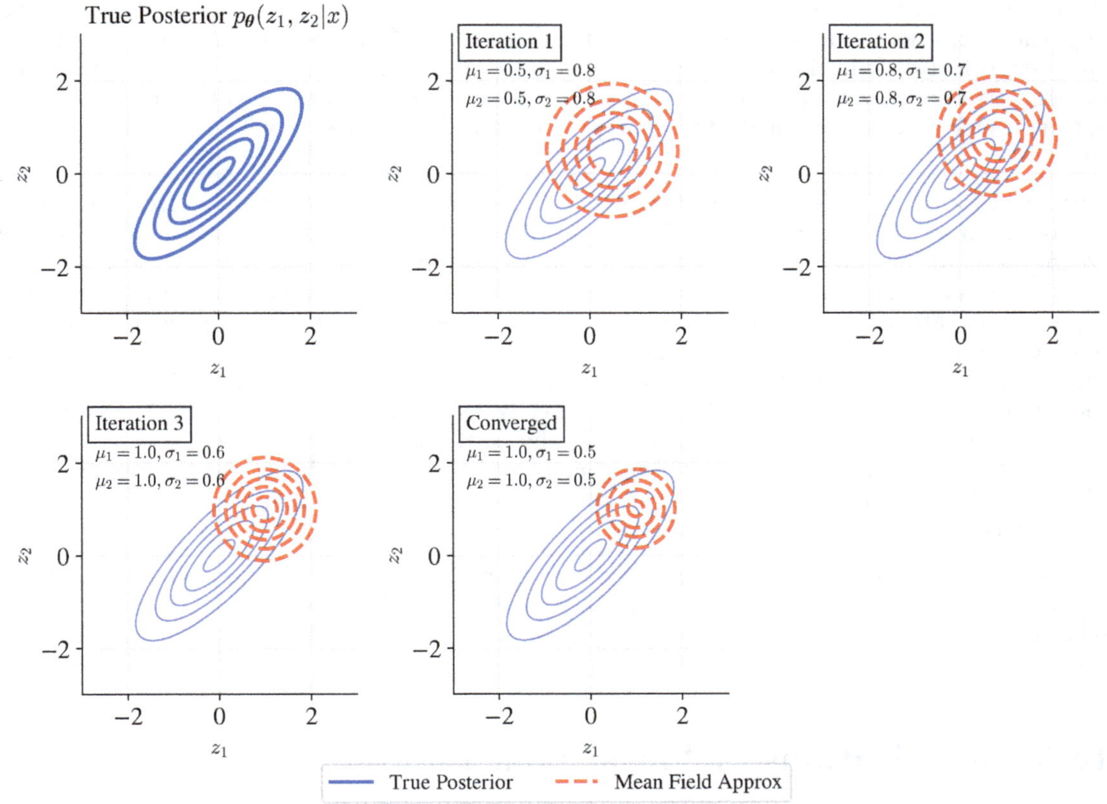

Figure 16.7: CAVI Algorithm: Mean Field Optimization Process

16.5.3.1 Example: Mean Field for Gaussian Mixture Model

Consider a GMM with K components. The latent variable is $z \in \{0, 1, \dots, K-1\}$ (discrete), and we want to approximate the posterior $p(z|\mathbf{x})$.

Using mean field, we set $q(z) = \text{Categorical}(z|\pi_q)$ where $\pi_q = [\pi_{q,0}, \dots, \pi_{q,K-1}]^\top$ with $\sum_{k=0}^{K-1} \pi_{q,k} = 1$.

The CAVI update for $\pi_{q,k}$ is:

$$\pi_{q,k} \propto \pi_k \exp\left(\mathbb{E}_{q(z)}[\log \mathcal{N}(\mathbf{x}|\mu_k, \Sigma_k)]\right),$$

which simplifies to:

$$\pi_{q,k} \propto \pi_k \exp\left(-\frac{1}{2}(\mathbf{x} - \mu_k)^\top \Sigma_k^{-1}(\mathbf{x} - \mu_k) - \frac{1}{2}\log|\Sigma_k|\right).$$

16.5.4 Practical Considerations

Mean field variational inference offers a powerful balance of computational efficiency and approximation quality, with several key advantages: it is computationally efficient with closed-form updates for many models, scalable with linear complexity in the number of latent variables, parallelizable with updates that can be performed independently, and easy to implement compared to structured variational inference. However, it has limitations: it misses correlations between latent variables, provides suboptimal approximation when variables are strongly dependent, and is restricted to factorized distributions with axis-aligned limitations. From a practical perspective, important considerations include distribution choice—using Gaussians for continuous variables and categorical distributions for discrete variables—random initialization (which often works, though better initialization speeds convergence), convergence properties (CAVI guarantees a local optimum, not necessarily global), and monitoring the ELBO to check convergence and diagnose issues.

Mean field variational inference is widely used because it simplifies optimization (each component can be updated independently) and is computationally efficient. However, the independence assumption can be too restrictive, missing important dependencies in the true posterior. In practice, mean field works well when the true posterior is approximately factorized or when dependencies are weak. **Structured variational inference** extends mean field to capture some dependencies (e.g., grouping related variables), trading off between expressiveness and computational cost. CAVI is similar to the EM algorithm (see Section~16.3): both guarantee convergence to a local optimum of the ELBO, but convergence can be slow and both are sensitive to initialization. The key difference is that EM uses the exact posterior in the E-step (when tractable), while CAVI uses an approximate posterior from a restricted family. Modern approaches often use gradient-based optimization instead, especially when the variational family is parameterized by neural networks.

Despite these considerations, mean field remains a foundational technique in variational inference, providing a balance between computational efficiency and approximation quality that makes it suitable for a wide range of applications, from classical statistical models to modern deep learning systems.

16.6 Applications of Variational Inference

The variational inference framework provides a powerful mathematical tool for approximate inference in complex probabilistic models. In this section, we illustrate how VI is applied in practice through an important example: Variational Autoencoders.

Earlier, we saw **Gaussian Mixture Models (GMM)** as an example of a latent variable model with **discrete latent variables** (the mixture component assignment $z \in \{0, 1, \dots, K-1\}$). GMMs represent classical statistical models where the posterior can often be computed exactly, making them suitable for EM algorithms. **Variational Autoencoders (VAEs)** extend this framework to handle **continuous latent variables** in high-dimensional spaces, enabling modern deep learning applications. VAEs demonstrate how the mathematical concepts (ELBO, KL divergence, reparameterization) are used in real-world generative modeling with neural networks.

16.6.1 Definition: Variational Autoencoder

A **Variational Autoencoder (VAE)** is a generative model that combines:

- An **encoder** $q_\phi(\mathbf{z}|\mathbf{x})$ (for approximate posterior inference) that maps data $\mathbf{x}$ to a distribution over latent codes $\mathbf{z}$
- A **decoder** $p_\theta(\mathbf{x}|\mathbf{z})$ (the likelihood/generative model from the latent variable model, see Definition~16.1.1) that maps latent codes $\mathbf{z}$ to a distribution over data $\mathbf{x}$. In practice, the decoder generates reconstructed data $\hat{\mathbf{x}}$ (e.g., by sampling from $p_\theta(\mathbf{x}|\mathbf{z})$ or taking its mean).

The model is trained by **maximizing the ELBO**, which we derive below.

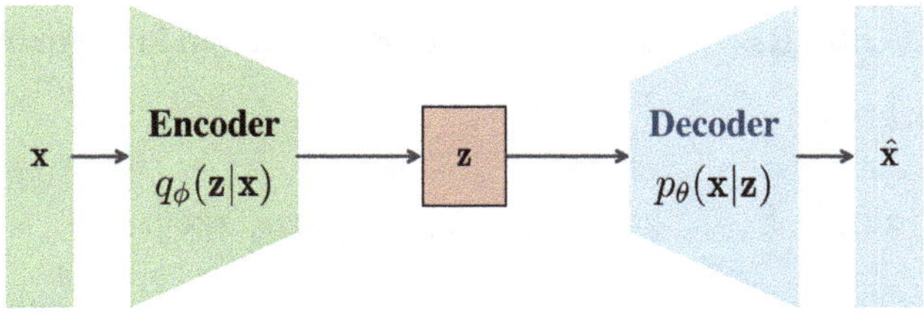

Figure 16.8: VAE Architecture Flow

Figure 16.8 illustrates the VAE architecture flow. The input $\mathbf{x}$ (green rectangle) is encoded by the encoder $q_\phi(\mathbf{z}|\mathbf{x})$ (green trapezoid) into the latent variable $\mathbf{z}$ (red square). The decoder $p_\theta(\mathbf{x}|\mathbf{z})$ (blue trapezoid) then generates the output $\hat{\mathbf{x}}$ (blue rectangle) from the latent code. This diagram shows the forward pass: data flows from input through the encoder to the latent space, then through the decoder to the generated output.

16.6.2 Derivation of VAE Loss Function

16.6.2.1 Step 1: Objective — Maximize Data Log-Likelihood

The generative model (decoder) is defined as:

$$p_\theta(\mathbf{x}, \mathbf{z}) = p(\mathbf{z}) \cdot p_\theta(\mathbf{x}|\mathbf{z})$$

where $p(\mathbf{z})$ is the prior (typically $\mathcal{N}(\mathbf{0}, \mathbf{I})$) and $p_\theta(\mathbf{x}|\mathbf{z})$ is the decoder.

Our goal is to maximize the **marginal log-likelihood** of the observed data:

$$\log p_\theta(\mathbf{x}) = \log \int p_\theta(\mathbf{x}, \mathbf{z})\, d\mathbf{z} = \log \int p(\mathbf{z}) p_\theta(\mathbf{x}|\mathbf{z})\, d\mathbf{z}. \tag{16.14}$$

This integral is typically **intractable** because it requires integrating over all possible latent values $\mathbf{z}$.

16.6.2.2 Step 2: Introduce Variational Distribution (Encoder)

We introduce a **variational distribution** (encoder) $q_\phi(\mathbf{z}|\mathbf{x})$ to approximate the intractable posterior $p_\theta(\mathbf{z}|\mathbf{x})$. The key trick is to multiply and divide by $q_\phi(\mathbf{z}|\mathbf{x})$ inside the integral:

$$\log p_\theta(\mathbf{x}) = \log \int q_\phi(\mathbf{z}|\mathbf{x}) \frac{p_\theta(\mathbf{x}, \mathbf{z})}{q_\phi(\mathbf{z}|\mathbf{x})}\, d\mathbf{z} = \log \mathbb{E}_{q_\phi(\mathbf{z}|\mathbf{x})} \left[\frac{p_\theta(\mathbf{x}, \mathbf{z})}{q_\phi(\mathbf{z}|\mathbf{x})} \right]. \tag{16.15}$$

16.6.2.3 Step 3: Apply Jensen's Inequality to Obtain ELBO

Since log is a **concave function**, by **Jensen's inequality** (see Chapter 12, Section~12.2.4):

$$\log \mathbb{E}_q[f(\mathbf{z})] \geq \mathbb{E}_q[\log f(\mathbf{z})]$$

Applying this to Equation~16.15:

$$\log p_\theta(\mathbf{x}) \geq \mathbb{E}_{q_\phi(\mathbf{z}|\mathbf{x})} \left[\log p_\theta(\mathbf{x}, \mathbf{z}) - \log q_\phi(\mathbf{z}|\mathbf{x}) \right]. \tag{16.16}$$

The right-hand side is the **Evidence Lower Bound (ELBO)**:

$$\mathcal{L}_{\text{ELBO}}(\mathbf{x}; \theta, \phi) = \mathbb{E}_{q_\phi(\mathbf{z}|\mathbf{x})}[\log p_\theta(\mathbf{x}, \mathbf{z})] - \mathbb{E}_{q_\phi(\mathbf{z}|\mathbf{x})}[\log q_\phi(\mathbf{z}|\mathbf{x})]. \tag{16.17}$$

16.6.2.4 Step 4: Decompose ELBO into Reconstruction and Regularization Terms

Expanding $\log p_\theta(\mathbf{x}, \mathbf{z}) = \log p_\theta(\mathbf{x}|\mathbf{z}) + \log p(\mathbf{z})$:

$$\mathcal{L}_{\text{ELBO}}(\mathbf{x}; \theta, \phi) = \mathbb{E}_{q_\phi(\mathbf{z}|\mathbf{x})}[\log p_\theta(\mathbf{x}|\mathbf{z})] + \mathbb{E}_{q_\phi(\mathbf{z}|\mathbf{x})}[\log p(\mathbf{z})] - \mathbb{E}_{q_\phi(\mathbf{z}|\mathbf{x})}[\log q_\phi(\mathbf{z}|\mathbf{x})].$$

The last two terms can be rewritten as:

$$\mathbb{E}_{q_\phi(\mathbf{z}|\mathbf{x})}[\log p(\mathbf{z})] - \mathbb{E}_{q_\phi(\mathbf{z}|\mathbf{x})}[\log q_\phi(\mathbf{z}|\mathbf{x})] = -\mathbb{E}_{q_\phi(\mathbf{z}|\mathbf{x})}\left[\log \frac{q_\phi(\mathbf{z}|\mathbf{x})}{p(\mathbf{z})}\right] = -D_{\text{KL}}(q_\phi(\mathbf{z}|\mathbf{x})||p(\mathbf{z})).$$

Therefore, the ELBO decomposes into:

$$\mathcal{L}_{\text{ELBO}}(\mathbf{x}; \theta, \phi) = \underbrace{\mathbb{E}_{q_\phi(\mathbf{z}|\mathbf{x})}[\log p_\theta(\mathbf{x}|\mathbf{z})]}_{\text{reconstruction term}} - \underbrace{D_{\text{KL}}(q_\phi(\mathbf{z}|\mathbf{x})||p(\mathbf{z}))}_{\text{regularization term}}. \tag{16.18}$$

Interpretation:

- **Reconstruction term**: Encourages the decoder to accurately reconstruct $\mathbf{x}$ from latent codes $\mathbf{z}$ sampled from the encoder.
- **Regularization term**: Encourages the encoder's posterior $q_\phi(\mathbf{z}|\mathbf{x})$ to stay close to the prior $p(\mathbf{z})$, ensuring the latent space is "regular" and can be sampled from for generation.

16.6.2.5 Step 5: VAE Loss — Minimize Negative ELBO

Training a VAE involves **maximizing the ELBO**, which is equivalent to **minimizing the negative ELBO**:

$$\mathcal{J}_{\text{VAE}}(\mathbf{x}; \theta, \phi) = -\mathcal{L}_{\text{ELBO}}(\mathbf{x}; \theta, \phi) = -\mathbb{E}_{q_\phi(\mathbf{z}|\mathbf{x})}[\log p_\theta(\mathbf{x}|\mathbf{z})] + D_{\text{KL}}(q_\phi(\mathbf{z}|\mathbf{x})||p(\mathbf{z})). \tag{16.19}$$

For a dataset $\{\mathbf{x}^{(i)}\}_{i=1}^N$, we minimize:

$$\min_{\theta,\phi} \sum_{i=1}^{N} \mathcal{J}_{\text{VAE}}(\mathbf{x}^{(i)}; \theta, \phi).$$

The first term $-\mathbb{E}_{q_\phi(\mathbf{z}|\mathbf{x})}[\log p_\theta(\mathbf{x}|\mathbf{z})]$ is the **negative log-likelihood (NLL)**. Its specific form depends on the decoder's distributional assumption:

- If $p_\theta(\mathbf{x}|\mathbf{z}) = \mathcal{N}(\hat{\mathbf{x}}_\theta(\mathbf{z}), \sigma^2\mathbf{I})$, then NLL is proportional to **mean squared error (MSE)**.
- If $\mathbf{x}$ is binary (e.g., normalized pixels), using a Bernoulli decoder, then NLL corresponds to **binary cross-entropy (BCE)**.

The Amortization Gap: In a VAE, the encoder $q_\phi(\mathbf{z}|\mathbf{x})$ is **amortized**—it's a single neural network that performs inference for any input $\mathbf{x}$, rather than having separate variational parameters $\phi^{(n)}$ for each data point $\mathbf{x}^{(n)}$. This amortization makes inference computationally efficient, but it introduces an **amortization gap**: the ELBO achieved by the amortized encoder is typically lower than what could be achieved with **per-datum optimization** (where each data point has its own optimized variational parameters, as in CAVI). The gap arises because the encoder must generalize across all data points, constraining its ability to perfectly optimize the ELBO for any individual point. This is a fundamental trade-off: amortization enables scalability and fast inference for new data points, but at the cost of potentially suboptimal ELBO values. The gap can be reduced by using more expressive encoders (e.g., deeper networks, normalizing flows) or by fine-tuning the encoder for specific data points. Understanding this gap is important for interpreting VAE performance and for choosing between amortized and per-datum inference approaches.

16.6.3 The Reparameterization Trick for Gradient-Based Optimization

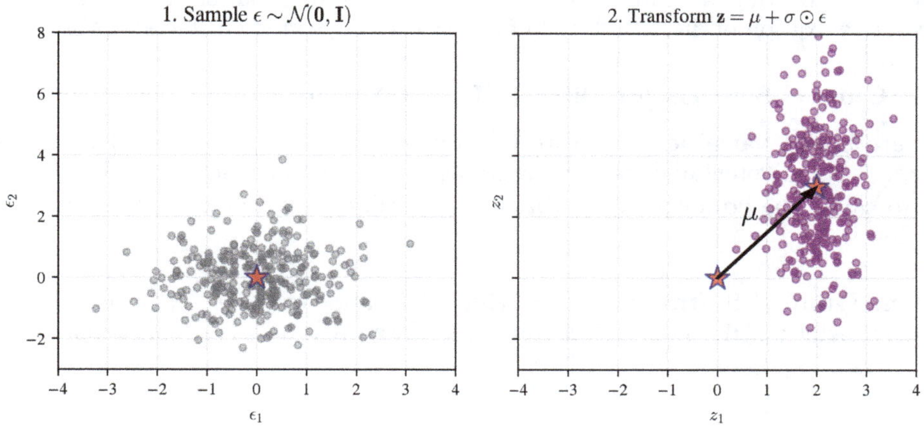

Figure 16.9: Reparameterization

The ELBO contains an expectation over $q_\phi(\mathbf{z}|\mathbf{x})$, which depends on ϕ. To enable gradient-based optimization, we use the **reparameterization trick**:

Instead of sampling $\mathbf{z} \sim q_\phi(\mathbf{z}|\mathbf{x}) = \mathcal{N}(\mu_\phi(\mathbf{x}), \text{diag}(\sigma^2_\phi(\mathbf{x})))$, we:

1. Sample $\epsilon \sim \mathcal{N}(\mathbf{0}, \mathbf{I})$ (independent of ϕ)
2. Compute $\mathbf{z} = \mu_\phi(\mathbf{x}) + \sigma_\phi(\mathbf{x}) \odot \epsilon$ (element-wise multiplication)

This reparameterization makes $\mathbf{z}$ a **differentiable function** of ϕ (through μ_ϕ and σ_ϕ), enabling backpropagation. With reparameterization, the ELBO becomes:

$$\begin{aligned}\mathcal{L}(\theta, \phi) = \mathbb{E}_{\epsilon \sim \mathcal{N}(\mathbf{0},\mathbf{I})} [\log p_\theta(\mathbf{x}|\mathbf{z}(\epsilon, \mathbf{x}, \phi))] \\ - D_{\text{KL}}(q_\phi(\mathbf{z}|\mathbf{x})||p(\mathbf{z}))\end{aligned} \tag{16.20}$$

where

$$\mathbf{z}(\epsilon, \mathbf{x}, \phi) = \mu_\phi(\mathbf{x}) + \sigma_\phi(\mathbf{x}) \odot \epsilon. \tag{16.21}$$

The reparameterization trick enables gradient-based optimization of the ELBO by making the sampling operation differentiable.

Figure 16.9 visualizes the reparameterization trick transformation. The left panel shows samples from the base noise distribution $\epsilon \sim \mathcal{N}(\mathbf{0}, \mathbf{I})$ (gray points), which is independent of the encoder parameters ϕ. The red star marks the mean of this distribution at the origin $(0, 0)$. The right panel shows the transformed latent codes $\mathbf{z} = \mu_\phi(\mathbf{x}) + \sigma_\phi(\mathbf{x}) \odot \epsilon$ (purple points), which are shifted by the mean μ and scaled by the standard deviation σ. The red stars mark the origin $(0, 0)$ and the transformed mean μ, and the black arrow indicates the mean shift from the origin to μ. This visualization makes concrete how the reparameterization trick isolates stochasticity in the noise variable ϵ, making the latent code $\mathbf{z}$ a deterministic, differentiable function of the encoder parameters. This enables gradient-based optimization because gradients can flow through the deterministic transformation, whereas direct sampling from $q_\phi(\mathbf{z}|\mathbf{x})$ would block gradient flow.

Requirements for reparameterization: The reparameterization trick requires the variational distribution $q_\phi(\mathbf{z}|\mathbf{x})$ to belong to a **location-scale family**, where samples can be written as $\mathbf{z} = \mu(\phi) + \sigma(\phi) \odot \epsilon$ for some base distribution $\epsilon \sim p_0(\epsilon)$ that is independent of ϕ. Gaussian distributions naturally satisfy this requirement. For **discrete latent variables**, the reparameterization trick is not directly applicable because discrete sampling is non-differentiable. In such cases, alternative techniques like **Gumbel-Softmax** (also called the concrete distribution) are used, which provide a continuous relaxation of discrete distributions that enables gradient-based optimization.

16.6.3.1 Gumbel-Softmax for Discrete Latent Variables

For **discrete latent variables** (e.g., categorical variables $z \in \{0, 1, \dots, K-1\}$), direct sampling from $q_\phi(z|\mathbf{x})$ is non-differentiable, preventing gradient-based optimization. The **Gumbel-Softmax trick** (also called the **concrete distribution**) provides a continuous relaxation that enables gradient flow.

Definition: Gumbel-Softmax Distribution Given logits $\boldsymbol{\ell} = [\ell_0, \ell_1, \dots, \ell_{K-1}]^\top$ and a temperature parameter $\tau > 0$, the **Gumbel-Softmax distribution** (concrete distribution) is defined as:

$$\mathbf{y} = \text{softmax}\left(\frac{\boldsymbol{\ell} + \mathbf{g}}{\tau}\right) \tag{16.22}$$

where $\mathbf{g} = [g_0, g_1, \dots, g_{K-1}]^\top$ and each $g_k \sim \text{Gumbel}(0, 1)$ is sampled independently from the standard Gumbel distribution. The Gumbel distribution has CDF $F(g) = \exp(-\exp(-g))$ and can be sampled via $g = -\log(-\log(u))$ where $u \sim \text{Uniform}(0, 1)$.

The components of $\mathbf{y}$ are:

$$y_k = \frac{\exp((\ell_k + g_k)/\tau)}{\sum_{j=0}^{K-1} \exp((\ell_j + g_j)/\tau)}, \quad k = 0, 1, \dots, K-1. \tag{16.23}$$

Key properties:

1. **Continuous relaxation:** $\mathbf{y} \in \Delta^{K-1}$ (the $(K-1)$-dimensional probability simplex), making it differentiable with respect to $\boldsymbol{\ell}$.
2. **Temperature parameter**: As $\tau \to 0$, the distribution becomes increasingly concentrated, approaching a one-hot vector (discrete categorical). As $\tau \to \infty$, it approaches a uniform distribution over the simplex.
3. **Differentiable sampling**: The sampling operation $\mathbf{y} = \text{softmax}((\boldsymbol{\ell} + \mathbf{g})/\tau)$ is differentiable with respect to $\boldsymbol{\ell}$ because the randomness comes from $\mathbf{g}$ (independent of $\boldsymbol{\ell}$), and the softmax function is smooth.

Application to Variational Inference For a discrete latent variable $z \in \{0, 1, \dots, K-1\}$ with variational posterior $q_\phi(z|\mathbf{x}) = \text{Categorical}(\pi_\phi(\mathbf{x}))$ where $\pi_\phi(\mathbf{x}) = [\pi_0, \pi_1, \dots, \pi_{K-1}]^\top$ are the category probabilities, we can use Gumbel-Softmax as follows:

1. **Compute logits:** $\boldsymbol{\ell} = \log \pi_\phi(\mathbf{x})$ (element-wise logarithm)
2. **Sample Gumbel noise**: $\mathbf{g} \sim \text{Gumbel}(0, 1)^K$ (sample K independent Gumbel random variables)
3. **Apply Gumbel-Softmax:** $\mathbf{y} = \text{softmax}((\boldsymbol{\ell} + \mathbf{g})/\tau)$
4. **Use $\mathbf{y}$ in place of one-hot z:** The continuous vector $\mathbf{y}$ approximates a one-hot encoding of z and is differentiable with respect to $\pi_\phi(\mathbf{x})$.

The ELBO can then be optimized using gradients that flow through the Gumbel-Softmax transformation. During training, τ is typically annealed from a high value (e.g., $\tau = 1.0$) to a low value (e.g., $\tau = 0.1$) to gradually sharpen the distribution toward discrete values.

Straight-Through Estimator (STE): An alternative approach is to use the **straight-through estimator** (see Chapter 11, Section~11.6.3), where discrete samples are used in the forward pass, but gradients are computed using the continuous Gumbel-Softmax distribution. This provides discrete samples while maintaining differentiability.

> Gumbel-Softmax extends the reparameterization trick to discrete distributions, enabling gradient-based optimization of models with discrete latent variables. The temperature parameter τ controls the trade-off between approximation quality (lower τ gives more discrete-like samples) and gradient quality (higher τ gives smoother gradients). In practice, **temperature annealing** is commonly used: start with $\tau = 1.0$ for stable gradients and gradually decrease to $\tau = 0.1$ or lower to get discrete-like samples. Gumbel-Softmax is widely used in **variational autoencoders with discrete latents**, **attention mechanisms** (e.g., discrete attention), and **reinforcement learning** (e.g., discrete action spaces). The **straight-through estimator** is a simpler alternative that uses discrete samples in the forward pass but continuous gradients in the backward pass, though it introduces bias in the gradient estimates.

16.6.4 Closed-Form KL Divergence for Gaussian

For $q_\phi(\mathbf{z}|\mathbf{x}) = \mathcal{N}(\mu, \text{diag}(\sigma^2))$ and $p(\mathbf{z}) = \mathcal{N}(\mathbf{0}, \mathbf{I})$, the KL divergence has a closed form:

$$D_{\text{KL}}(q_\phi(\mathbf{z}|\mathbf{x})||p(\mathbf{z})) = \frac{1}{2}\sum_{i=0}^{k-1}(\mu_i^2 + \sigma_i^2 - 1 - \log\sigma_i^2)\,. \tag{16.24}$$

This can be computed efficiently without sampling, making the VAE training objective tractable.

Figure 16.10 plots the typical behavior of the two components of the ELBO objective during VAE training. The blue solid line shows the **reconstruction loss** $-\mathbb{E}_{q_\phi(\mathbf{z}|\mathbf{x})}[\log p_\theta(\mathbf{x}|\mathbf{z})]$, which decreases rapidly in early epochs as the model learns to reconstruct data accurately. The orange dashed line shows the **KL divergence** $D_{\text{KL}}(q_\phi(\mathbf{z}|\mathbf{x})||p(\mathbf{z}))$, which increases gradually as the encoder learns to use the latent space. This visualization makes concrete the trade-off between reconstruction accuracy and regularization: the reconstruction term encourages the model to use the latent code effectively, while the KL term acts as a constraint that prevents the posterior from deviating too far from the prior. In practice, this trade-off is often controlled by the β-VAE formulation, where the KL term is weighted by a hyperparameter β to balance reconstruction quality and latent space structure.

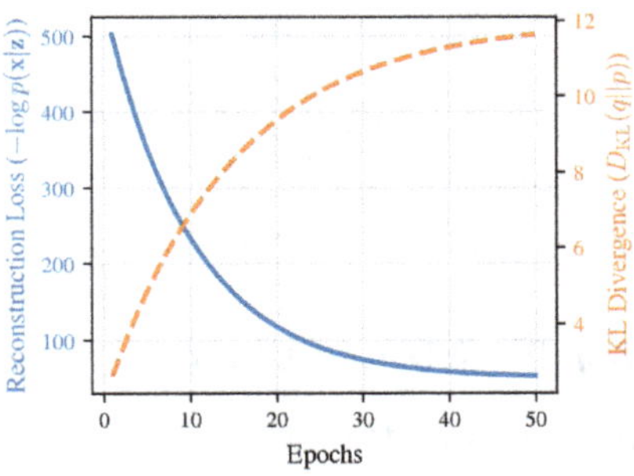

Figure 16.10: VAE ELBO Components during Training

VAEs demonstrate how variational inference enables scalable approximate inference in high-dimensional latent variable models. The key mathematical innovations are: (1) the ELBO derivation using Jensen's inequality, which provides a tractable lower bound on the intractable log-likelihood, and (2) the reparameterization trick, which makes the ELBO differentiable and enables gradient-based optimization. VAEs illustrate the fundamental trade-off between reconstruction accuracy and regularization, which is central to variational inference.

Figure 16.11 shows a 2D scatter plot of simulated latent variables $\mathbf{z}$ colored by class, illustrating how VAEs organize the latent space. The plot shows three distinct clusters corresponding to different classes (A, B, C), demonstrating that the VAE learns a structured, continuous latent representation where similar data points are mapped to nearby regions in latent space. This visualization makes concrete the goal of VAEs: learning a compressed, continuous latent representation that preserves semantic structure. The latent space forms a **manifold** where interpolation between points corresponds to smooth transitions in data space, enabling generative modeling through sampling from the latent space.

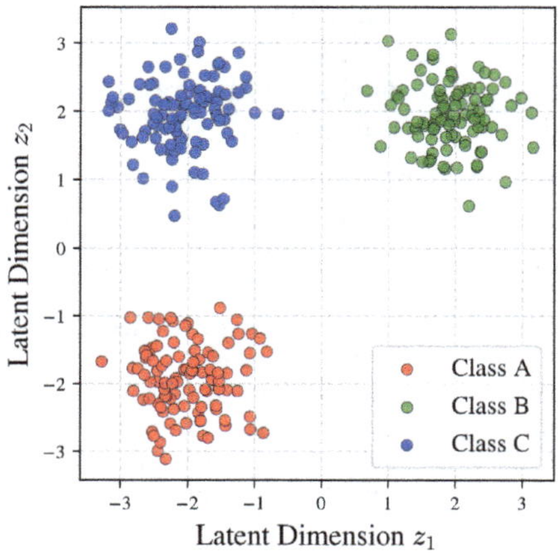

Figure 16.11: Latent Space Manifold

16.7 Stochastic Variational Inference

16.7.1 Mini-Batch Training

For large datasets, we use **stochastic variational inference** where the ELBO is estimated using mini-batches:

$$\hat{\mathcal{L}}(\theta, \phi) = \frac{1}{|\mathcal{B}|} \sum_{\mathbf{x}^{(n)} \in \mathcal{B}} \left[\mathbb{E}_{q(\mathbf{z}|\mathbf{x}^{(n)},\phi)}[\log p_\theta(\mathbf{x}^{(n)}|\mathbf{z}^{(n)})] - D_{\mathrm{KL}}(q(\mathbf{z}^{(n)}|\mathbf{x}^{(n)}, \phi)||p(\mathbf{z}^{(n)}))\right] \tag{16.25}$$

where $\mathcal{B}$ is a mini-batch of data points.

16.7.2 Monte Carlo Estimation

The reconstruction term is estimated using Monte Carlo sampling:

$$\mathbb{E}_{q_\phi(\mathbf{z}|\mathbf{x})}[\log p_\theta(\mathbf{x}|\mathbf{z})] \approx \frac{1}{S} \sum_{s=0}^{S-1} \log p_\theta(\mathbf{x}|\mathbf{z}^{(s)}) \tag{16.26}$$

where $\mathbf{z}^{(s)} \sim q_\phi(\mathbf{z}|\mathbf{x})$ and typically $S = 1$ (single sample) is sufficient due to the reparameterization trick enabling low-variance gradients.

Stochastic variational inference makes VAEs scalable to large datasets. The key insight is that we can estimate the ELBO using mini-batches and Monte Carlo sampling, and the reparameterization trick ensures that gradients have low variance even with a single sample. This is in contrast to score function estimators (REINFORCE) which require many samples or variance reduction techniques. In practice, **single-sample** estimation ($S = 1$) works well for VAEs because the reparameterization trick provides unbiased, low-variance gradients. For more complex models or when higher accuracy is needed, **importance sampling** or **multiple samples** can be used. The mini-batch approach also enables **online learning** where the model is updated as new data arrives, which is important for streaming data or continual learning scenarios.

16.8 Amortized Variational Inference

16.8.1 Definition: Amortized Inference

Amortized variational inference uses a **shared encoder** $q_\phi(\mathbf{z}|\mathbf{x})$ parameterized by neural networks to perform inference for all data points, rather than optimizing separate variational parameters for each data point.

This is in contrast to **per-datum** variational inference (like CAVI) where each data point $\mathbf{x}^{(n)}$ has its own variational parameters $\phi^{(n)}$.

16.8.2 Advantages

1. **Efficiency**: After training, inference for new data points is fast (single forward pass through the encoder).
2. **Generalization**: The encoder learns to perform inference for unseen data points.
3. **Scalability**: Can handle large datasets with millions of examples.

16.8.3 VAE as Amortized Inference

In a VAE, the encoder $q_\phi(\mathbf{z}|\mathbf{x})$ is **amortized**: it's a single neural network that performs inference for any input $\mathbf{x}$, rather than having separate parameters for each data point. Amortized inference is a key innovation that makes variational inference practical for deep learning. Instead of optimizing variational parameters for each data point separately (which would be computationally expensive), we train a neural network to perform inference. This is similar to how **meta-learning** works: the encoder learns to quickly perform inference for new data points. However, amortization can introduce an **amortization gap**: the ELBO achieved by the amortized encoder may be lower than what could be achieved with per-datum optimization. This gap can be reduced by using more expressive encoders or by fine-tuning the encoder for specific data points. In **normalizing flows**, the encoder is invertible, allowing exact likelihood computation and eliminating the amortization gap. In **diffusion models**, the reverse process can be viewed as amortized inference over a sequence of latent variables.

16.9 Other Generative Modeling Approaches

While variational inference is a powerful framework for approximate inference, there are other mathematical approaches to generative modeling with latent variables. Two important alternatives are:

- **Generative Adversarial Networks (GANs)**: Use **minimax optimization** (see Chapter 12, Section~12.11) rather than variational inference. GANs illustrate how game-theoretic optimization can be applied to generative modeling.
- **Normalizing Flows**: Use **invertible transformations** and the change of variables formula (see Chapter 13, Section~13.10) to create expressive variational families. Normalizing flows can be used within variational inference to create more expressive posterior approximations, or as standalone generative models.

These methods demonstrate how different mathematical frameworks (minimax optimization, change of variables) can be applied to the same problem of learning generative models with latent variables.

16.10 Variational Inference vs. Other Methods

The choice between variational inference and MCMC depends on the application. **Variational inference** is preferred in deep learning because it integrates seamlessly with neural networks and gradient-based optimization, enabling end-to-end training. It's also faster and more scalable. However, **MCMC** (see Chapter~18) provides exact inference (asymptotically) and better uncertainty quantification. In practice, many modern methods combine both: use VI for fast approximate inference, then use MCMC to refine the approximation or to get better uncertainty estimates. **Stein Variational Gradient Descent (SVGD)** bridges the gap by using gradient-based optimization to fit a set of particles (samples) rather than a parametric distribution. **Normalizing flows** (see Chapter~13, Section~13.10.3) use invertible neural networks to learn exact posteriors, combining the benefits of both approaches.

The key differences between variational inference and MCMC are summarized in the following comparison:

Aspect	Variational Inference	MCMC
Output	Parametric distribution $q(\mathbf{z})$	Samples from $p(\mathbf{z} \mid \mathbf{x})$
Convergence	Fast (optimization)	Slow (mixing time)
Scalability	Good (stochastic optimization)	Limited (sequential sampling)
Accuracy	Approximate (lower bound)	Exact (asymptotically)
Gradient-based	Yes (reparameterization)	No (sampling-based)

When to use variational inference: Variational inference is the preferred choice when you need fast inference for real-time applications, when working with large datasets that benefit from stochastic optimization, when you want to combine inference with neural networks using gradient-based methods, or when approximate inference is acceptable for your application.

When to use MCMC: MCMC is more suitable when you need exact inference (asymptotically) for small models, when working with small datasets where exact computation is feasible, or when you require uncertainty quantification through samples from the true posterior distribution.

16.11 Variational Inference in Other Domains

We have focused on VAEs as a central application of variational inference, but the same ideas—approximating an intractable posterior with a tractable q, and maximizing the ELBO—recur across many other areas. Below we briefly tour several domains where VI is widely used; in each case, pointers to earlier chapters indicate where the underlying mathematics is developed.

16.11.1 Bayesian Neural Networks (BNN)

Bayesian neural networks replace point estimates of weights with distributions over weights, giving principled uncertainty quantification that is important for safe and reliable systems. The posterior over weights given data is intractable for large networks, so we use a variational distribution $q(\mathbf{w})$ to approximate $p(\mathbf{w}|\mathcal{D})$. This yields uncertainty-aware predictions, regularization through the prior, and better calibration. In practice, methods such as variational dropout, mean field BNNs, and Monte Carlo Dropout implement this idea. The ELBO objective is defined in Chapter~14; for sampling-based alternatives (e.g. SGLD) and posterior sampling over weights, see Chapter~18.

16.11.2 Topic Models

In **Latent Dirichlet Allocation (LDA)**, documents are modeled as mixtures of latent topics; the goal is to infer topic distributions and assignments from text. Exact inference is intractable because of the combinatorial structure of the model. Collapsed variational Bayesian inference approximates the posterior over topics and topic assignments, scaling to large corpora while keeping topic interpretability. Extensions such as the Hierarchical Dirichlet Process (HDP) use

VI for unbounded numbers of topics. The Dirichlet prior used for topic proportions in LDA is introduced in Chapter~13; a related view via non-negative matrix factorization is in Chapter~9.

16.11.3 Recommendation Systems

Recommendation systems must learn latent factors for users and items from sparse interactions. Variational autoencoders are a natural fit: they model user–item data with a latent representation and use the ELBO for training. VAE-based methods (e.g. VAE-CF, hierarchical VAEs for user modeling) capture complex preferences, help with cold start, and can provide uncertainty estimates. The latent-factor and matrix-factorization perspective (e.g. collaborative filtering via low-rank structure) appears in Chapter~8 (randomized SVD) and Chapter~9 (NMF); spherical embeddings for user and item vectors are discussed in Chapter~13.

16.11.4 Reinforcement Learning

In reinforcement learning, variational methods are used to model distributions over policies or value functions, improving exploration and sample efficiency and handling partial observability. VI approximates posteriors over optimal policies or latent states, leading to methods such as Variational Policy Gradients, Stochastic Value Gradients, and VAEs for dynamics modeling. The MDP framework, policies, and value functions that underpin RL are developed in Chapter~15; Markov chains and stochastic matrices (e.g. transition structure) appear in Chapter~14.

16.11.5 Other Applications

Beyond these, variational inference appears in **computer vision** (unsupervised representation learning, image generation, anomaly detection), **natural language processing** (language modeling, machine translation, summarization), **healthcare** (Bayesian models for diagnosis and treatment recommendation), and **robotics** (uncertainty-aware planning and control; the decision-making backbone is again MDPs and policies in Chapter~15).

Taken together, these applications show variational inference as a unifying framework for approximate Bayesian inference: the same trade-off between computational efficiency and expressiveness makes VI useful both in classical statistical models and in modern deep learning.

16.12 Mindmap

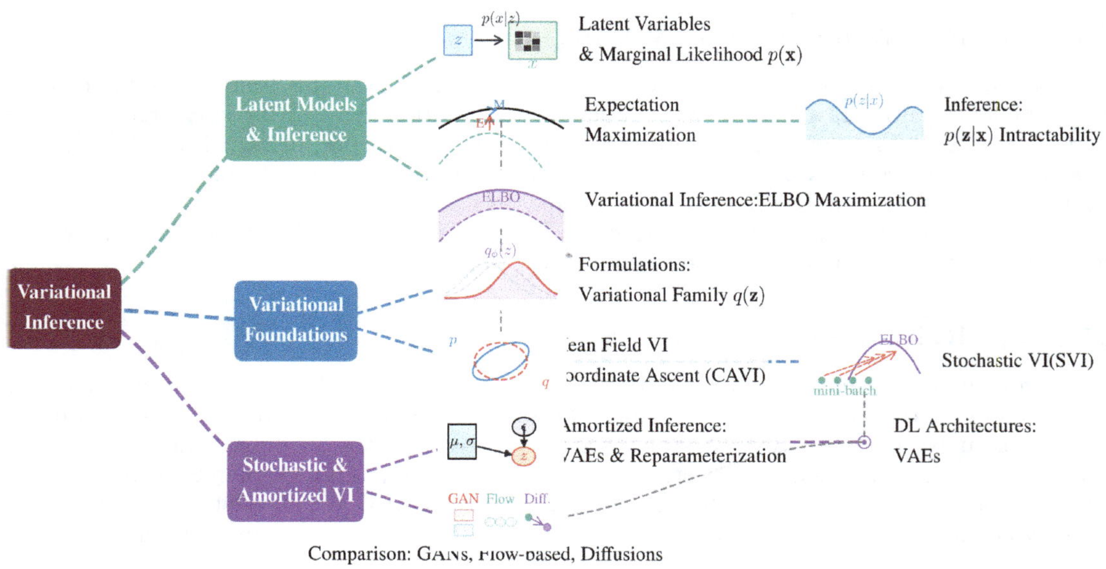

Figure 16.12: Mind map of variational inference topics

16.13 Exercises

16.13.1 Latent Variable Models

1. Show that for a latent variable model $p(\mathbf{x}, \mathbf{z}) = p(\mathbf{x}|\mathbf{z})p(\mathbf{z})$, the marginal likelihood is $p(\mathbf{x}) = \int p(\mathbf{x}|\mathbf{z})p(\mathbf{z})\, d\mathbf{z}$.
2. For a Gaussian Mixture Model with $K = 2$ components, $p(z = 0) = \pi_0$, $p(z = 1) = \pi_1 = 1 - \pi_0$, and $p(\mathbf{x}|z = k) = \mathcal{N}(\mu_k, \Sigma_k)$, derive the marginal distribution $p(\mathbf{x})$.
3. In a topic model (LDA), documents are generated from topic distributions. Explain how this can be viewed as a latent variable model. What are the observed variables, latent variables, prior, and likelihood?

16.13.2 Variational Inference Basics

1. Prove that $D_{\mathrm{KL}}(q(\mathbf{z})||p(\mathbf{z}|\mathbf{x})) = \log p(\mathbf{x}) - \mathrm{ELBO}(q)$.
2. Show that maximizing the ELBO is equivalent to minimizing $D_{\mathrm{KL}}(q(\mathbf{z})||p(\mathbf{z}|\mathbf{x}))$.
3. For a variational distribution $q(\mathbf{z}) = \mathcal{N}(\mu, \Sigma)$ and prior $p(\mathbf{z}) = \mathcal{N}(\mathbf{0}, \mathbf{I})$, compute $D_{\mathrm{KL}}(q||p)$ when Σ is diagonal: $\Sigma = \mathrm{diag}(\sigma_0^2, \dots, \sigma_{k-1}^2)$.
4. Show that for diagonal covariance, the KL divergence simplifies to:

$$D_{\mathrm{KL}}(q||p) = \frac{1}{2}\sum_{i=0}^{k-1}\left(\mu_i^2 + \sigma_i^2 - 1 - \log \sigma_i^2\right).$$

5. **Reverse KL vs. Forward KL**: Consider a bimodal true posterior $p(z|\mathbf{x})$ with two modes at $z = -2$ and $z = 2$, and a unimodal approximation $q(z) = \mathcal{N}(\mu, \sigma^2)$.
   - Explain why minimizing $D_{\text{KL}}(q||p)$ (reverse KL) would lead q to concentrate on one mode (mode-seeking behavior).
   - Explain why minimizing $D_{\text{KL}}(p||q)$ (forward KL) would lead q to spread out to cover both modes (mode-covering behavior).
   - Why is reverse KL computationally tractable in variational inference, while forward KL is not?

16.13.3 Mean Field Variational Inference

1. Under the mean field assumption $q(\mathbf{z}) = \prod_{i=0}^{k-1} q_i(z_i)$, show that the entropy decomposes as $H(q) = \sum_{i=0}^{k-1} H(q_i)$.

2. For a 2D Gaussian posterior $p(z_0, z_1|\mathbf{x}) = \mathcal{N}\left(\begin{bmatrix}\mu_0\\ \mu_1\end{bmatrix}, \begin{bmatrix}\sigma_0^2 & \rho\sigma_0\sigma_1\\ \rho\sigma_0\sigma_1 & \sigma_1^2\end{bmatrix}\right)$ with correlation ρ, explain why mean field $q(z_0, z_1) = q_0(z_0)q_1(z_1)$ might be a poor approximation when $|\rho|$ is large.

3. In CAVI, we iteratively update each $q_i(z_i)$ while keeping others fixed. Explain why this is a coordinate ascent algorithm on the ELBO.

16.13.4 Variational Autoencoders

1. **Reparameterization Trick**: For $q(\mathbf{z}|\mathbf{x}) = \mathcal{N}(\mu_\phi(\mathbf{x}), \sigma_\phi^2(\mathbf{x}))$, show that sampling $\mathbf{z} = \mu_\phi(\mathbf{x}) + \sigma_\phi(\mathbf{x}) \odot \epsilon$ where $\epsilon \sim \mathcal{N}(\mathbf{0}, \mathbf{I})$ is equivalent to sampling from $q(\mathbf{z}|\mathbf{x})$.

2. **Gradient Estimation**: Explain why the reparameterization trick enables gradient-based optimization of the ELBO, while direct sampling $\mathbf{z} \sim q(\mathbf{z}|\mathbf{x}, \phi)$ does not.

3. For a VAE with:
   - Encoder: $q(\mathbf{z}|\mathbf{x}) = \mathcal{N}(\mu_\phi(\mathbf{x}), \text{diag}(\sigma_\phi^2(\mathbf{x})))$ where $\mathbf{z} \in \mathbb{R}^k$
   - Prior: $p(\mathbf{z}) = \mathcal{N}(\mathbf{0}, \mathbf{I})$
   - Decoder: $p(\mathbf{x}|\mathbf{z}) = \mathcal{N}(\mu_\theta(\mathbf{z}), \sigma^2\mathbf{I})$ where σ^2 is fixed

   Derive the ELBO and show how to compute each term.

4. **Posterior Collapse**: In VAEs, the KL term $D_{\text{KL}}(q(\mathbf{z}|\mathbf{x})||p(\mathbf{z}))$ can collapse to zero, meaning $q(\mathbf{z}|\mathbf{x}) \approx p(\mathbf{z})$ and the latent code carries no information about $\mathbf{x}$. Explain why this happens and how β-VAE ($\mathcal{L}_\beta = \mathbb{E}[\log p(\mathbf{x}|\mathbf{z})] - \beta D_{\text{KL}}(q||p)$ with $\beta < 1$) can help prevent it.

5. For binary data $\mathbf{x} \in \{0, 1\}^d$ with decoder $p(\mathbf{x}|\mathbf{z}) = \prod_{i=0}^{d-1} \text{Bernoulli}(x_i|\sigma(\mu_{\theta,i}(\mathbf{z})))$, show that the reconstruction term is:

$$\mathbb{E}_{q(\mathbf{z}|\mathbf{x})}[\log p(\mathbf{x}|\mathbf{z})] = \sum_{i=0}^{d-1} \mathbb{E}_{q(\mathbf{z}|\mathbf{x})}[x_i \log \sigma(\mu_{\theta,i}(\mathbf{z})) + (1 - x_i)\log(1 - \sigma(\mu_{\theta,i}(\mathbf{z})))].$$

16.13.5 Amortized Inference

1. Explain the difference between **per-datum** variational inference (optimizing $\phi^{(n)}$ for each $\mathbf{x}^{(n)}$) and **amortized** variational inference (using a shared encoder $q(\mathbf{z}|\mathbf{x}, \phi)$).

2. What is the **amortization gap**? Why does it occur, and how can it be reduced?

3. In a VAE, after training, how do you perform inference for a new data point $\mathbf{x}_{\text{new}}$? Compare the computational cost with per-datum inference.

16.13.6 Additional Topics in Variational Inference

1. **Expectation-Maximization**: Show that the EM algorithm guarantees that the log-likelihood $\log p_\theta(\mathbf{x})$ increases (or stays the same) at each iteration: $\log p_{\theta^{(t+1)}}(\mathbf{x}) \geq \log p_{\theta^{(t)}}(\mathbf{x})$.
2. **EM for GMM**: For a Gaussian Mixture Model with K components, derive the E-step and M-step updates. Specifically:
    - E-step: Show that the posterior responsibility is $\gamma_{ik} = \frac{\pi_k \mathcal{N}(\mathbf{x}^{(i)}|\mu_k,\Sigma_k)}{\sum_{j=0}^{K-1} \pi_j \mathcal{N}(\mathbf{x}^{(i)}|\mu_j,\Sigma_j)}$.
    - M-step: Derive the parameter updates $\pi_k^{(t+1)}$, $\mu_k^{(t+1)}$, and $\Sigma_k^{(t+1)}$ that maximize the expected log-likelihood.
3. **Gumbel-Softmax**: For a categorical latent variable $z \in \{0, 1, \ldots, K-1\}$ with probabilities $\pi = [\pi_0, \pi_1, \ldots, \pi_{K-1}]^\top$, show that sampling via Gumbel-Softmax: $\mathbf{y} = \text{softmax}((\log \pi + \mathbf{g})/\tau)$ where $\mathbf{g} \sim \text{Gumbel}(0, 1)^K$ is differentiable with respect to π. Explain the role of the temperature parameter τ: what happens as $\tau \to 0$ and as $\tau \to \infty$?
4. **Stochastic Variational Inference**: Show that the mini-batch ELBO estimator $\hat{\mathcal{L}}(\theta, \phi) = \frac{1}{|\mathcal{B}|} \sum_{\mathbf{x}^{(n)} \in \mathcal{B}} \left[\mathbb{E}_{q(\mathbf{z}|\mathbf{x}^{(n)},\phi)}[\log p_\theta(\mathbf{x}^{(n)}|\mathbf{z}^{(n)})] - D_{\text{KL}}(q(\mathbf{z}^{(n)}|\mathbf{x}^{(n)}, \phi)||p(\mathbf{z}^{(n)}))\right]$ is an unbiased estimator of the full-dataset ELBO.
5. **Monte Carlo Estimation**: For the reconstruction term $\mathbb{E}_{q_\phi(\mathbf{z}|\mathbf{x})}[\log p_\theta(\mathbf{x}|\mathbf{z})]$, show that the Monte Carlo estimator $\frac{1}{S} \sum_{s=0}^{S-1} \log p_\theta(\mathbf{x}|\mathbf{z}^{(s)})$ where $\mathbf{z}^{(s)} \sim q_\phi(\mathbf{z}|\mathbf{x})$ is unbiased. Explain why $S = 1$ (single sample) is often sufficient for VAEs when using the reparameterization trick.

16.13.7 Advanced Topics

1. **Importance Weighted Autoencoder (IWAE)**: The IWAE uses importance sampling to get a tighter bound:

$$\mathcal{L}_K = \mathbb{E}_{\mathbf{z}^{(1)},\ldots,\mathbf{z}^{(K)} \sim q(\mathbf{z}|\mathbf{x})} \left[\log \frac{1}{K} \sum_{k=1}^{K} \frac{p(\mathbf{x}, \mathbf{z}^{(k)})}{q(\mathbf{z}^{(k)}|\mathbf{x})}\right].$$

   Show that $\mathcal{L}_K \geq \mathcal{L}_{K-1}$ (tighter bound with more samples) and $\lim_{K\to\infty} \mathcal{L}_K = \log p(\mathbf{x})$.
2. **Normalizing Flows**: Normalizing flows use invertible transformations to learn complex posteriors. If $f : \mathbb{R}^k \to \mathbb{R}^k$ is invertible and $q_0(\mathbf{z}_0)$ is a simple distribution (e.g., $\mathcal{N}(\mathbf{0}, \mathbf{I})$), then $q(\mathbf{z}) = q_0(\mathbf{z}_0) \left|\det \frac{\partial f^{-1}}{\partial \mathbf{z}}\right|$ where $\mathbf{z} = f(\mathbf{z}_0)$. Explain how this enables exact inference (no approximation gap) while maintaining the benefits of amortized inference.
3. **Variational Inference vs. Expectation-Maximization**: Both VI and EM optimize lower bounds on the log-likelihood. Compare and contrast:
    - What lower bound does each optimize?
    - How do they handle the E-step?
    - When is each method preferred?

IV

Score, Dynamics, and Diffusion

Chapters 17–21

Score functions, stochastic dynamics, and continuous-time generative models From optimization to sampling: the mathematics of generation

Generation through evolution: transforming noise into creation

Many modern AI methods can be understood as dynamical systems evolving under randomness. Stochastic differential equations describe the continuous-time limits of iterative algorithms and sampling procedures.

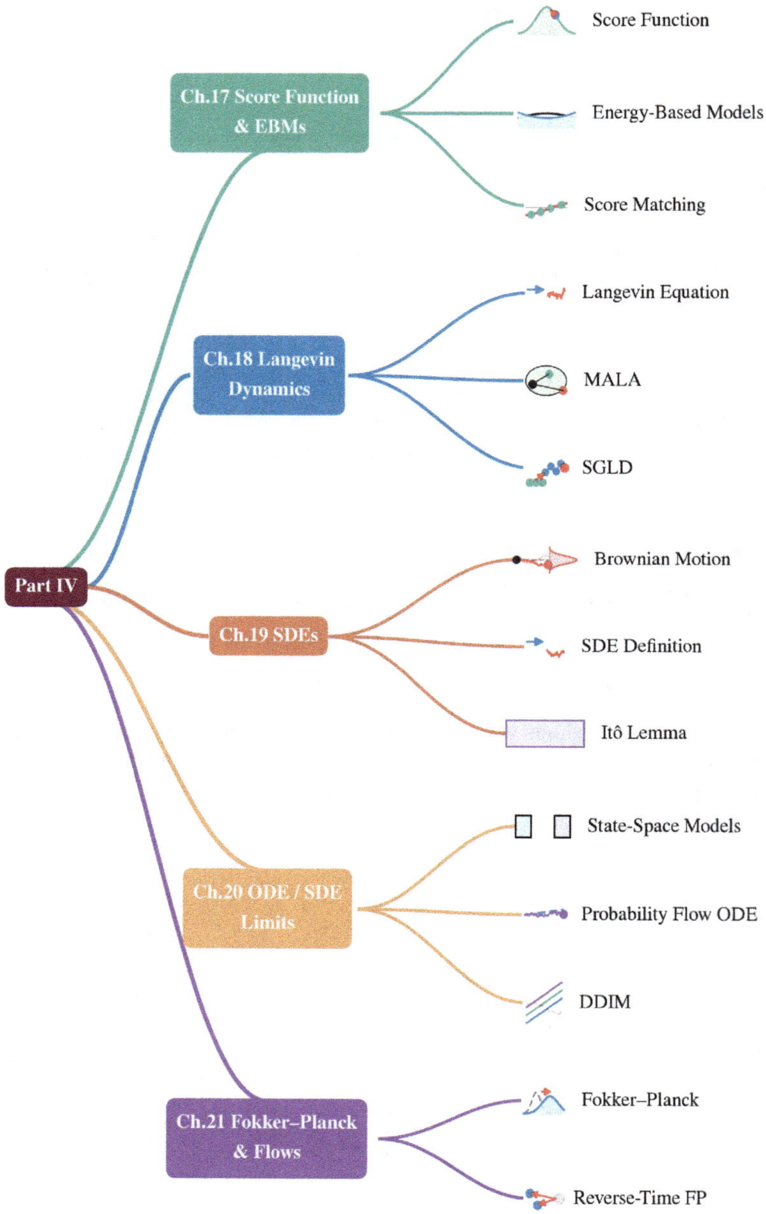
Part IV
Ch.17 Score Function & EBMs
Score Function
Energy-Based Models
Score Matching
Ch.18 Langevin Dynamics
Langevin Equation
MALA
SGLD
Ch.19 SDEs
Brownian Motion
SDE Definition
Itô Lemma
Ch.20 ODE / SDE Limits
State-Space Models
Probability Flow ODE
DDIM
Ch.21 Fokker–Planck & Flows
Fokker–Planck
Reverse-Time FP

Chapter 17

Score Function and Energy-Based Models

Score functions, energy-based models, and score matching for sampling

> *"The theory of probabilities is at bottom nothing but common sense reduced to calculus."*
>
> — Pierre-Simon Laplace

- ▷ `Score function` $\mathbf{s}(\mathbf{x}) \triangleq \nabla_{\mathbf{x}} \log p(\mathbf{x}) \in \mathbb{R}^d$
- ▷ `Normalization invariance` for $p(\mathbf{x}) \propto \tilde{p}(\mathbf{x})$ with $\tilde{p}(\mathbf{x}) = Zp(\mathbf{x})$, $\nabla_{\mathbf{x}} \log \tilde{p}(\mathbf{x}) = \nabla_{\mathbf{x}} \log p(\mathbf{x})$
- ▷ `Gaussian score` if $\mathbf{x} \sim \mathcal{N}(\mu, \Sigma)$, then $\mathbf{s}(\mathbf{x}) = -\Sigma^{-1}(\mathbf{x} - \mu)$
- ▷ `Energy-based model (EBM)` $p_\theta(\mathbf{x}) = \frac{1}{Z_\theta} \exp(-E_\theta(\mathbf{x}))$, with $Z_\theta = \int \exp(-E_\theta(\mathbf{x}))\, d\mathbf{x}$
- ▷ `EBM score` $\nabla_{\mathbf{x}} \log p_\theta(\mathbf{x}) = -\nabla_{\mathbf{x}} E_\theta(\mathbf{x})$ (no Z_θ term)
- ▷ `Score matching (ideal)` $\mathcal{L}_{\mathrm{SM}}(\theta) = \mathbb{E}_{p_{\mathrm{data}}}\left[\|\mathbf{s}_\theta(\mathbf{x}) - \nabla_{\mathbf{x}} \log p_{\mathrm{data}}(\mathbf{x})\|_2^2\right]$
- ▷ `Hyvärinen (IBP)` $\mathcal{L}_{\mathrm{SM}}(\theta) \equiv \mathbb{E}_{p_{\mathrm{data}}}\left[\frac{1}{2}\|\mathbf{s}_\theta(\mathbf{x})\|_2^2 + \nabla_{\mathbf{x}} \cdot \mathbf{s}_\theta(\mathbf{x})\right] + \mathrm{const}$, $\nabla_{\mathbf{x}} \cdot \mathbf{s}_\theta = \sum_{i=1}^d \frac{\partial s_{\theta,i}}{\partial x_i}$
- ▷ `DSM noise model` $\tilde{\mathbf{x}} = \mathbf{x} + \epsilon$, $\epsilon \sim \mathcal{N}(0, \sigma^2 I)$
- ▷ `Conditional score of Gaussian corruption` $\nabla_{\tilde{\mathbf{x}}} \log p(\tilde{\mathbf{x}} \mid \mathbf{x}) = -\frac{1}{\sigma^2}(\tilde{\mathbf{x}} - \mathbf{x})$
- ▷ `Denoising score matching (DSM) objective` $\mathcal{L}_{\mathrm{DSM}}(\theta) = \mathbb{E}_{\mathbf{x} \sim p_{\mathrm{data}}, \epsilon \sim \mathcal{N}(0,\sigma^2 I)}\left[\left\|\mathbf{s}_\theta(\tilde{\mathbf{x}}) + \frac{1}{\sigma^2}(\tilde{\mathbf{x}} - \mathbf{x})\right\|_2^2\right]$
- ▷ `Langevin dynamics (discrete sampler)` $\mathbf{x}_{k+1} = \mathbf{x}_k + \eta\, \mathbf{s}_\theta(\mathbf{x}_k) + \sqrt{2\eta}\, \epsilon_k$, $\epsilon_k \sim \mathcal{N}(0, I)$

17.1 Motivation: From Density to Geometry

The preceding chapters (14–16) developed a framework centered on probability distributions, Kullback-Leibler (KL) divergence, and variational objectives such as the evidence lower bound (ELBO). A fundamental assumption underlying these constructions is the availability of a probability density function $p(\mathbf{x})$ or, at minimum, its log-likelihood $\log p(\mathbf{x})$.

In practice, however, many modern generative models encounter significant limitations and challenges that invalidate this assumption. Specifically, the density $p(\mathbf{x})$ often proves intractable to compute directly, the normalization constant required for proper probability normalization remains unknown, and only local information about the underlying distribution is accessible through practical means.

Here, **local information** refers to properties of the distribution that can be evaluated at specific points (or in small neighborhoods around points) without requiring knowledge of the entire distribution's global structure. Unlike global properties (e.g., the normalization constant, which requires integrating over the entire support), local information describes how the distribution behaves *at a particular location* $\mathbf{x}$ in the input space.

17.1.1 Example: Local Information in a Gaussian Distribution

To illustrate the concept of local information, consider a simple Gaussian distribution $p(\mathbf{x}) = \mathcal{N}(\mathbf{x}; \mu, \Sigma)$ (see Section 13.14 of Chapter 13) with mean μ and covariance Σ.

For this distribution:

1. **Global information** includes the normalization constant $\frac{1}{\sqrt{(2\pi)^d|\Sigma|}}$, which depends on the entire distribution's structure (as it is computed by integrating the unnormalized density over the entire input space).
2. **Local information** at a point $\mathbf{x}$ is captured by the score function:

$$\nabla_{\mathbf{x}} \log p(\mathbf{x}) = -\Sigma^{-1}(\mathbf{x} - \mu)$$

   This gradient vector encodes how the log-probability changes infinitesimally around $\mathbf{x}$—specifically, it points toward the mean μ (the location of highest probability density), with a magnitude determined by the inverse covariance Σ^{-1} (reflecting the distribution's curvature).

Importantly, this local information can be computed directly for any point $\mathbf{x}$ without needing to reference the entire distribution or its normalization constant. This is the key insight that makes score functions valuable for generative modeling.

These limitations (namely, the intractable density and the unknown normalization constant) motivate a crucial shift in perspective. Rather than attempting to model the density function itself—a task complicated by the aforementioned challenges—we instead focus on modeling its gradient field.

This gradient, which encodes local information about the distribution's structure (specifically, how the log-probability changes infinitesimally around $\mathbf{x}$), is known as the **score function**.

Figure 17.1 contrasts what is directly available in practice with what score-based modeling targets. The left panel shows samples from an unknown mixture—we can evaluate properties near observed points but do not have access to the full density $p(\mathbf{x})$ or its normalization constant Z. The right panel shows the score field $\nabla_{\mathbf{x}} \log p(\mathbf{x})$ for the same distribution: at each location, a vector describes how log-probability changes locally, and this gradient does not depend on Z.

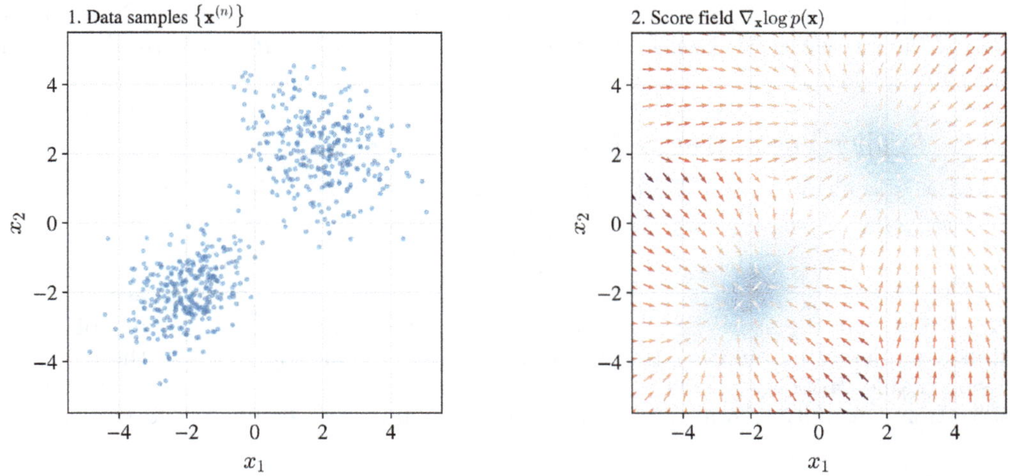

Figure 17.1: Data samples versus the score field

17.1.2 The Challenge of Intractable Densities

Consider a **Variational Autoencoder (VAE)** (see Chapter 16). The VAE learns an approximate posterior $q_\phi(\mathbf{z} \mid \mathbf{x})$ and a decoder $p_\theta(\mathbf{x} \mid \mathbf{z})$, but the true data distribution $p(\mathbf{x})$ remains intractable because it requires marginalizing over the latent space:

$$p(\mathbf{x}) = \int p_\theta(\mathbf{x} \mid \mathbf{z}) p(\mathbf{z}) d\mathbf{z}$$

For high-dimensional latent spaces, this integral is computationally infeasible. Similarly, **Energy-Based Models (EBMs)** (see Section 17.4) define distributions as:

$$p_\theta(\mathbf{x}) = \frac{1}{Z_\theta} \exp(-E_\theta(\mathbf{x}))$$

where $Z_\theta = \int \exp(-E_\theta(\mathbf{x})) d\mathbf{x}$ is the **partition function**. Computing Z_θ requires integrating over the entire data space, which is intractable for high-dimensional data like images.

17.1.3 Why Gradients Are More Accessible

While the density $p(\mathbf{x})$ may be intractable, its **gradient** $\nabla_{\mathbf{x}} \log p(\mathbf{x})$ often has a simpler structure. This can be formalized for any normalized density:

17.1.3.1 General Principle

Consider any probability density that can be written as:

$$p(\mathbf{x}) = \frac{1}{Z} f(\mathbf{x})$$

where $Z = \int f(\mathbf{x})\, d\mathbf{x}$ is the normalization constant (partition function). Taking the gradient of the log-density:

$$\nabla_{\mathbf{x}} \log p(\mathbf{x}) = \nabla_{\mathbf{x}} \left[\log\left(\frac{1}{Z}\right) + \log f(\mathbf{x}) \right] = \nabla_{\mathbf{x}} \log f(\mathbf{x})$$

The key insight is that $\log\left(\frac{1}{Z}\right)$ is a constant with respect to $\mathbf{x}$, so its gradient vanishes. Thus, the gradient of the log-density **does not depend on the normalization constant** Z, even when Z is intractable.

17.1.3.2 Application to Energy-Based Models

For an EBM specifically, where $f(\mathbf{x}) = \exp(-E_\theta(\mathbf{x}))$, this simplification becomes:

$$\nabla_{\mathbf{x}} \log p_\theta(\mathbf{x}) = \nabla_{\mathbf{x}} \log \exp(-E_\theta(\mathbf{x})) = -\nabla_{\mathbf{x}} E_\theta(\mathbf{x})$$

This confirms that the score function of an EBM equals the negative gradient of its energy function, completely eliminating the intractable partition function Z_θ from the computation. This remarkable property allows us to learn the score function without ever evaluating the normalization constant, even for high-dimensional distributions where integration is infeasible.

17.1.4 Geometric Intuition

We can visualize this geometric perspective using the right panel of Figure 17.1. The score function $\mathbf{s}(\mathbf{x}) = \nabla_{\mathbf{x}} \log p(\mathbf{x})$ defines a vector field over the entire space, where each vector encodes the **local geometry** of the probability distribution:

- **Direction**: Points toward regions of higher probability density. In the vector field of Figure 17.1, the arrows converge toward the cluster centers where the data points in the left panel are most concentrated.
- **Magnitude**: Indicates the rate of change of the log-probability (how steep the landscape is). Notice that the arrows are short in very low-density regions far from the modes, grow longer in the steep transition regions, and shrink to zero near the peaks.
- **Zero points**: Correspond to local extrema (modes, minima, or saddle points) where the vector field vanishes.

This geometric perspective connects probability theory to **differential geometry** and provides a natural framework for **sampling algorithms** that navigate the probability landscape by following the score field.

17.2 Score Function

17.2.1 Definition: Score Function

Let $\mathbf{x} \in \mathbb{R}^d$ be a continuous random variable with density $p(\mathbf{x})$. The **score function** is defined as

$$\mathbf{s}(\mathbf{x}) \triangleq \nabla_{\mathbf{x}} \log p(\mathbf{x}) \tag{17.1}$$

This is a vector field over data space.

17.2.2 Interpretation

The score function points in the direction of **increasing probability density**:

- Large magnitude $\Rightarrow$ steep probability landscape
- Zero score $\Rightarrow$ local extrema (modes or saddle points)
- Following the score moves samples toward high-density regions

Geometrically:

- $\log p(\mathbf{x})$ defines a scalar field
- $\mathbf{s}(\mathbf{x})$ is its gradient field

17.2.3 Example: Gaussian Distribution

Let $\mathbf{x} \sim \mathcal{N}(\mu, \Sigma)$.

$$\log p(\mathbf{x}) = -\frac{1}{2}(\mathbf{x}-\mu)^\top \Sigma^{-1}(\mathbf{x}-\mu) + C \tag{17.2}$$

where $C = -\frac{1}{2}\log((2\pi)^d|\Sigma|)$ is a constant independent of $\mathbf{x}$.

Then

$$\mathbf{s}(\mathbf{x}) = \nabla_{\mathbf{x}} \log p(\mathbf{x}) = -\Sigma^{-1}(\mathbf{x}-\mu) \tag{17.3}$$

The score points **toward the mean** μ, and its magnitude is proportional to the distance from the mean, scaled by the inverse covariance matrix.

Geometric interpretation: The score field forms a **radial vector field** centered at μ. At any point $\mathbf{x}$, the score $\mathbf{s}(\mathbf{x})$ points directly toward the mean, creating a "flow" that guides samples toward the high-density region.

Univariate case. For $\mathcal{N}(\mu, \sigma^2)$, the score is $s(x) = -(x-\mu)/\sigma^2$. It is linear in x, zero at the mean, has negative slope $ds/dx = -1/\sigma^2$, and from any x points toward μ (left of μ the score is positive, right of μ it is negative). The slope of the log-density $\log p(x)$ at a point equals the score there. Figure 17.2 shows the PDF with score arrows, the log PDF with tangent lines at sample points, and the score function $s(x)$.

Bivariate case. The formula $\mathbf{s}(\mathbf{x}) = -\Sigma^{-1}(\mathbf{x}-\mu)$ again gives a radial field: every vector points toward μ, with magnitude proportional to distance from the mean and scaled by the inverse covariance. The score at a point is aligned with the direction to the mean (for isotropic Σ they coincide; for non-isotropic Σ the score is the same direction after a linear transform). Figure 17.3 shows the density with the score field, the radial flow with a 1σ ellipse, and one sample point with both the displacement to the mean and the score vector.

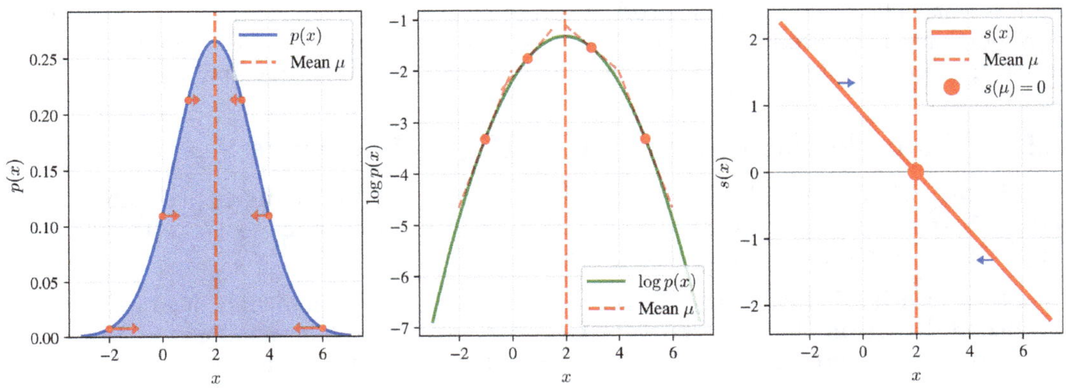

Figure 17.2: Univariate Gaussian score: PDF, log PDF with tangents, and score.

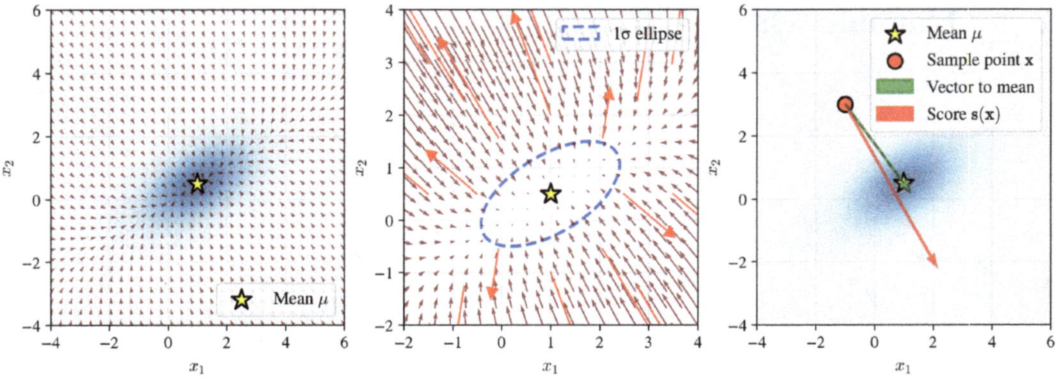

Figure 17.3: Bivariate Gaussian score field and radial flow toward mean.

17.2.4 Example: Mixture of Gaussians

Consider a **mixture of two Gaussians**:

$$p(\mathbf{x}) = \pi_1 \mathcal{N}(\mathbf{x}; \mu_1, \Sigma_1) + \pi_2 \mathcal{N}(\mathbf{x}; \mu_2, \Sigma_2)$$

where $\pi_1 + \pi_2 = 1$ are the mixture weights.

The score function is:

$$\mathbf{s}(\mathbf{x}) = \frac{\pi_1 p_1(\mathbf{x}) \mathbf{s}_1(\mathbf{x}) + \pi_2 p_2(\mathbf{x}) \mathbf{s}_2(\mathbf{x})}{\pi_1 p_1(\mathbf{x}) + \pi_2 p_2(\mathbf{x})}$$

where $p_i(\mathbf{x}) = \mathcal{N}(\mathbf{x}; \mu_i, \Sigma_i)$ and $\mathbf{s}_i(\mathbf{x}) = -\Sigma_i^{-1}(\mathbf{x} - \mu_i)$ are the component scores.

Key insight: The score is a **weighted average** of the component scores, where the weights are proportional to the component densities at $\mathbf{x}$. Near a mode, the score is dominated by that component's score field, creating a multi-modal landscape with multiple attractors.

17.2.5 Example: Score Function in High Dimensions

For high-dimensional data like images ($\mathbf{x} \in \mathbb{R}^{784}$ for 28×28 MNIST digits), the score function $\mathbf{s}(\mathbf{x}) \in \mathbb{R}^{784}$ is a vector field over the 784-dimensional pixel space. While visualizing this directly is impossible, we can understand its structure:

- **Data manifold**: Real images lie on a low-dimensional manifold embedded in $\mathbb{R}^{784}$
- **Score direction**: Points toward the nearest point on the data manifold
- **Score magnitude**: Large near the manifold boundary, small near the manifold interior

This geometric structure explains why **denoising score matching** works: adding noise moves points off the manifold, and the score function learns to "denoise" by pointing back toward the manifold.

17.3 Score vs Density

The score function has several critical advantages:

1. **Normalization-free**

Let $Z > 0$ be an arbitrary positive constant (typically the normalization constant, or partition function, such that $\int Zp(\mathbf{x})d\mathbf{x} = 1$). Then:

$$\nabla_{\mathbf{x}} \log(Zp(\mathbf{x})) = \nabla_{\mathbf{x}} \log p(\mathbf{x}) \tag{17.4}$$

Proof: Using the logarithm product rule and the fact that $\nabla_{\mathbf{x}} \log Z = \mathbf{0}$ (since Z is constant with respect to $\mathbf{x}$):

$$\begin{aligned}\nabla_{\mathbf{x}} \log(Zp(\mathbf{x})) &= \nabla_{\mathbf{x}}(\log Z + \log p(\mathbf{x})) \\ &= \nabla_{\mathbf{x}} \log Z + \nabla_{\mathbf{x}} \log p(\mathbf{x}) \\ &= \mathbf{0} + \nabla_{\mathbf{x}} \log p(\mathbf{x}) \\ &= \nabla_{\mathbf{x}} \log p(\mathbf{x})\end{aligned}$$

This means the score function is **invariant to normalization**: we can compute it from an unnormalized density $\tilde{p}(\mathbf{x}) = Zp(\mathbf{x})$ without knowing Z.

2. **Local** Only local changes in $\mathbf{x}$ matter
3. **Vector-valued** Encodes geometry of the distribution
4. **Directly usable for sampling** (via Langevin dynamics)

This makes score-based modeling attractive for **high-dimensional data** such as images.

> The score function's invariance to normalization is a fundamental property that distinguishes it from the density itself. This property is why score-based models can be trained without computing intractable partition functions, making them practical for high-dimensional generative modeling. The connection between score functions and energy-based models reveals that learning gradients is often easier than learning the full density, a principle that underlies many modern generative models including diffusion models.

17.4 Energy-Based Models (EBMs)

17.4.1 Intuition: Energy as Probability

The core intuition behind energy-based models is elegantly simple: **lower energy corresponds to higher probability**. Imagine the data space as a landscape where each point has an "energy level":

- **Low energy valleys**: Regions where data is likely to occur (high probability)
- **High energy mountains**: Regions where data is unlikely to occur (low probability)

This energy landscape perspective provides a natural way to model complex distributions:

1. **Flexible Shaping**: The energy function $E_\theta(\mathbf{x})$ can be any differentiable function (typically a neural network), allowing it to capture intricate patterns in data
2. **Probability Interpretation**: The exponential function $\exp(-E_\theta(\mathbf{x}))$ transforms energy into an unnormalized probability measure
3. **Automatic Normalization**: The partition function Z_θ ensures the final distribution integrates to 1

This framework unifies many probabilistic models under a single conceptual umbrella, from simple exponential family distributions to complex deep generative models.

17.4.2 Connection to Gaussian Integrals and Normalization

The mathematical structure of energy-based models has deep connections to the Gaussian integral and normalization concepts introduced in Chapter 13.

17.4.2.1 Historical and Mathematical Connection

1. **Gaussian Foundation**: Gauss's derivation of the Gaussian integral $\int_{-\infty}^{\infty} e^{-x^2} dx = \sqrt{\pi}$ provided the mathematical foundation for computing partition functions of exponential-family distributions, including EBMs.
2. **Normal Distribution as EBM**: The normal distribution itself is a simple EBM where:
   - Energy function: $E(x) = \frac{(x-\mu)^2}{2\sigma^2}$
   - Partition function: $Z(\mu, \sigma^2) = \sqrt{2\pi\sigma^2}$
3. **Generalization to Deep Models**: Modern EBMs extend this framework by replacing the fixed quadratic energy function with a flexible neural network, while retaining the same exponential structure and normalization principle.

17.4.2.2 Key Insight

The exponential structure $\exp(-E_\theta(\mathbf{x}))$ ensures that:

- Probabilities are always non-negative
- The energy landscape is smoothly mapped to probabilities
- The mathematical tools developed for Gaussian distributions (like gradient-based optimization) can be extended to more complex energy functions

This connection demonstrates how EBMs represent a natural evolution from classical probability theory to modern deep generative modeling, maintaining mathematical rigor while dramatically increasing expressive power.

17.4.3 Remark: Broader Principles of Mathematical Modeling

The approach underlying energy-based models exemplifies a broader principle in mathematical modeling that extends beyond machine learning to fields like quantitative finance:

17.4.3.1 Transform and Conquer

Many complex modeling problems become tractable through a strategic transformation:

1. **Identify Intractable Quantity**: Direct modeling of the desired quantity (e.g., stock prices, high-dimensional densities) is difficult
2. **Transform to Tractable Domain**: Map the problem to a domain where mathematical tools are available:
   - **Finance**: Use log returns instead of prices
   - **EBMs**: Use energy functions and score functions instead of direct density estimation
   - **Signal Processing**: Use Fourier transforms to simplify convolution
3. **Leverage Domain-Specific Structure**: Exploit mathematical properties of the transformed domain:
   - **Finance**: Additivity of log returns, Gaussian assumptions
   - **EBMs**: Exponential structure, gradient invariance to normalization
   - **Physics**: Symmetry principles, conservation laws

17.4.3.2 Unifying Insight

Across diverse fields, the most successful modeling approaches share a common pattern: **Find the representation where the problem's structure is most simply expressed, and its complexity most naturally managed.**

For energy-based models, this means working with energy landscapes and score functions rather than raw densities. For quantitative finance, it means working with returns rather than prices. This principle of strategic representation choice is a hallmark of effective mathematical modeling.

17.4.4 Definition: Energy-Based Model

An **energy-based model** defines a distribution via an energy function $E_\theta(\mathbf{x})$:

$$p_\theta(\mathbf{x}) = \frac{1}{Z_\theta} \exp\big(- E_\theta(\mathbf{x})\big) \tag{17.5}$$

where:

- $E_\theta(\mathbf{x})$ is an unconstrained function (typically a neural network)
- $Z_\theta = \int \exp(-E_\theta(\mathbf{x}))d\mathbf{x}$ is the **partition function** (see Chapter 13 for the formal definition)

The energy function $E_\theta(\mathbf{x})$ assigns a scalar "energy" to each data point $\mathbf{x}$. Low energy corresponds to high probability, and high energy corresponds to low probability. This is analogous to **statistical physics**, where systems minimize energy to reach equilibrium states.

17.4.5 Historical Context: Boltzmann Machines

The concept of energy-based models originates from **Boltzmann machines** (Hinton & Sejnowski, 1986), which were among the first neural network models for unsupervised learning. A Boltzmann machine defines an energy function over binary variables:

$$E(\mathbf{x}) = -\sum_{i<j} w_{ij}x_i x_j - \sum_i b_i x_i$$

where w_{ij} are connection weights and b_i are biases. The probability distribution is:

$$p(\mathbf{x}) = \frac{1}{Z}\exp(-E(\mathbf{x}))$$

Modern EBMs extend this framework to continuous variables and use deep neural networks as energy functions, enabling modeling of complex high-dimensional distributions like natural images.

17.4.6 Advantages of Energy-Based Models

1. **Flexibility**: Any function $E_\theta(\mathbf{x})$ can be used as an energy function, allowing for highly expressive models
2. **No explicit normalization**: The model structure doesn't require computing Z_θ during training
3. **Multi-modal distributions**: Can naturally represent distributions with multiple modes
4. **Unsupervised learning**: Can learn from unlabeled data by modeling the data distribution directly

17.4.7 Famous Energy-Based Models

Several energy-based models have gained practical importance in real-world applications. **Restricted Boltzmann Machines (RBMs)** emerged as computationally tractable variants of Boltzmann machines, enabling efficient training through Contrastive Divergence. They became foundational building blocks for Deep Belief Networks (DBNs), which were among the first deep learning models to outperform shallow alternatives and helped spark the modern deep learning revolution. **Energy-based GANs (EBGANs)** integrated EBM principles with GAN architecture, improving training stability and generation quality. Most notably, **score-based generative models**—particularly diffusion models—represent the most successful modern EBMs, achieving state-of-the-art results in high-quality image generation, text-to-image synthesis, and other generative tasks. These models leverage the score function's properties to avoid intractable partition functions while producing realistic, diverse samples. Collectively, these EBM variants demonstrate the framework's versatility and practical utility across different generations of machine learning models.

17.4.8 Challenges in Training EBMs

Despite their flexibility, training EBMs faces several challenges:

1. **Intractable partition function**: Computing Z_θ requires integration over the entire data space
2. **Sampling difficulty**: Generating samples requires MCMC methods (see Chapter 18), which can be slow
3. **Mode collapse**: The learned distribution may concentrate on a subset of modes
4. **Training instability**: Energy functions can become very steep, making optimization difficult

These challenges motivated the development of **score-based models**, which avoid computing Z_θ by learning the score function directly.

17.4.9 Score of an EBM

Taking the gradient of the log-probability:

$$\nabla_{\mathbf{x}} \log p_\theta(\mathbf{x}) = \nabla_{\mathbf{x}} \log \left(\frac{1}{Z_\theta} \exp(-E_\theta(\mathbf{x})) \right) = \nabla_{\mathbf{x}} \left(-\log Z_\theta - E_\theta(\mathbf{x})\right) = -\nabla_{\mathbf{x}} E_\theta(\mathbf{x}) \tag{17.6}$$

The partition function Z_θ disappears because it's constant with respect to $\mathbf{x}$.

Key insight:

- Learning the score is equivalent to learning the gradient of the energy.
- No normalization constant appears in the score function.

This is the fundamental advantage of score-based learning: we can train models without ever computing the intractable partition function Z_θ.

17.4.10 Example: Neural Network Energy Function

Consider an energy function parameterized by a neural network:

$$E_\theta(\mathbf{x}) = \text{MLP}_\theta(\mathbf{x})$$

where MLP_θ is a multi-layer perceptron. The score function is:

$$\mathbf{s}_\theta(\mathbf{x}) = -\nabla_{\mathbf{x}} E_\theta(\mathbf{x}) = -\nabla_{\mathbf{x}} \text{MLP}_\theta(\mathbf{x})$$

This can be computed efficiently using **backpropagation** (see Chapter 11). The gradient $\nabla_{\mathbf{x}}\text{MLP}_\theta(\mathbf{x})$ measures how the network output changes with respect to the input, which naturally encodes the local geometry of the learned distribution.

17.4.11 Connection to Discriminative Models

Interestingly, the score function $\mathbf{s}_\theta(\mathbf{x}) = -\nabla_{\mathbf{x}} E_\theta(\mathbf{x})$ has a natural interpretation as a **force field** in physics. The energy $E_\theta(\mathbf{x})$ acts as a potential, and the score (negative gradient) represents the force that would move a particle toward lower energy (higher probability) regions.

This connection to physics provides intuition for why **Langevin dynamics** (see Chapter 18) works: it simulates a particle moving under the influence of the score field, with thermal noise preventing it from getting stuck in local minima.

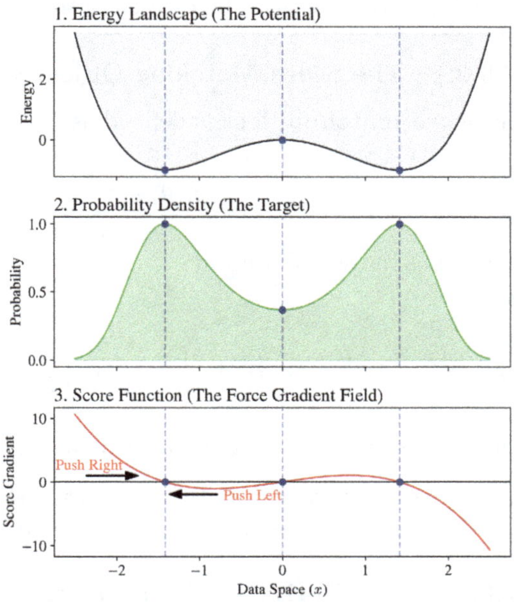

Figure 17.4: Energy, Density, and Score Function

Figure 17.4 illustrates the fundamental relationship between energy, probability density, and the score function using a **double-well potential** $E(x) = \frac{1}{4}x^4 - x^2$. The top panel shows the energy landscape, which has two minima (modes) at $x = \pm\sqrt{2}$ and a local maximum (unstable point) at $x = 0$. The middle panel shows the corresponding probability density $p(x) \propto e^{-E(x)}$, which peaks at the two modes where energy is minimized. The bottom panel shows the score function $s(x) = \nabla_x \log p(x) = -\nabla_x E(x) = 2x - x^3$, which is the negative gradient of the energy. The score function is zero at the critical points (the two modes and the unstable point) and points in the direction of increasing probability. Notice how the score pushes samples away from the unstable point (at $x = 0$) and toward the two stable modes, demonstrating how the score field guides sampling toward high-probability regions.

> The energy-based model framework provides a unified perspective on generative modeling that connects machine learning to statistical physics. The energy function $E_\theta(\mathbf{x})$ acts as a potential, and the score function (negative gradient) represents the force field. This physical analogy helps explain why stochastic sampling algorithms like Langevin dynamics work: they simulate the motion of particles in a potential field, with thermal fluctuations ensuring proper exploration of the energy landscape.

17.5 Score Matching

17.5.1 Motivation

Given data samples $\{\mathbf{x}^{(n)}\}_{n=1}^N$ drawn from an unknown data distribution $p_{\text{data}}(\mathbf{x})$, we want to learn a score function $\mathbf{s}_\theta(\mathbf{x})$ such that:

$$\mathbf{s}_\theta(\mathbf{x}) \approx \nabla_{\mathbf{x}} \log p_{\text{data}}(\mathbf{x}) \tag{17.7}$$

The challenge is that $p_{\text{data}}(\mathbf{x})$ is unknown—we only have samples, not the density itself. We cannot directly compute $\nabla_{\mathbf{x}} \log p_{\text{data}}(\mathbf{x})$ to use as a training target.

Score matching[1] (Hyvärinen, 2005) provides an elegant solution: it derives a training objective that depends only on the data samples, not on the unknown density $p_{\text{data}}(\mathbf{x})$.

17.5.2 The Score Matching Insight

The key insight is to use **integration by parts** to eliminate the unknown density from the objective. Instead of minimizing the distance between score functions directly, we minimize a related quantity that can be estimated from samples.

17.5.2.1 The Score Matching Objective

The **score matching loss** is defined as:

$$\mathcal{L}_{\text{SM}}(\theta) = \mathbb{E}_{p_{\text{data}}}\left[\|\mathbf{s}_\theta(\mathbf{x}) - \nabla_{\mathbf{x}} \log p_{\text{data}}(\mathbf{x})\|_2^2\right] \tag{17.8}$$

Expanding the squared norm:

$$\mathcal{L}_{\text{SM}}(\theta) = \mathbb{E}_{p_{\text{data}}}\left[\sum_{i=1}^{d}\left(s_{\theta,i}(\mathbf{x}) - \frac{\partial \log p_{\text{data}}(\mathbf{x})}{\partial x_i}\right)^2\right]$$

where $s_{\theta,i}(\mathbf{x})$ is the i-th component of $\mathbf{s}_\theta(\mathbf{x})$.

Using **integration by parts**[2] transforms the objective, but introduces a boundary term. To eliminate this term, we assume $p_{\text{data}}(\mathbf{x}) \to 0$ as $\|\mathbf{x}\| \to \infty$. Here's why this assumption works:

17.5.2.2 Full Derivation Using Integration by Parts

Let's derive the score matching objective step-by-step:

Step 1: Start with the objective We begin with the score matching objective expanded earlier (Equation 17.8 with expanded squared norm). For convenience in the derivation, we'll work with the scaled version:

$$\mathcal{L}_{\text{SM}}(\theta) = \frac{1}{2}\mathbb{E}_{p_{\text{data}}}\left[\sum_{i=1}^{d}\left(s_{\theta,i}(\mathbf{x}) - \frac{\partial \log p_{\text{data}}(\mathbf{x})}{\partial x_i}\right)^2\right]$$

The factor of $1/2$ simplifies the derivative calculations and doesn't affect the optimization since it's a constant scalar multiple.

---

[1]Hyvärinen, A. (2005). "Estimation of non-normalized statistical models by score matching." *Journal of Machine Learning Research*, 6, 695-709. This seminal paper introduced the score matching technique for learning unnormalized statistical models without requiring the partition function.

[2]Integration by parts is a fundamental technique from calculus used to transform integrals. For functions $u(x)$ and $v(x)$, it states: $\int u(x)v'(x)dx = u(x)v(x)\Big|_a^b - \int u'(x)v(x)dx$. In the context of score matching, it allows us to eliminate the unknown data density $p_{\text{data}}(\mathbf{x})$ from the objective function by shifting the derivative from the density to the score function.

Step 2: Apply integration by parts to the cross term Consider the cross term involving the true score:

$$\mathbb{E}_{p_{\text{data}}}\left[s_{\theta,i}(\mathbf{x}) \cdot \frac{\partial \log p_{\text{data}}(\mathbf{x})}{\partial x_i}\right] = \int s_{\theta,i}(\mathbf{x}) \cdot \frac{\partial \log p_{\text{data}}(\mathbf{x})}{\partial x_i} p_{\text{data}}(\mathbf{x}) d\mathbf{x}$$

Simplify using the calculus identity for logarithmic derivatives. For any differentiable function $p(x) > 0$:

$$\frac{d}{dx}\log p(x) = \frac{1}{p(x)}\frac{d}{dx}p(x) \implies \frac{d}{dx}\log p(x) \cdot p(x) = \frac{d}{dx}p(x)$$

Applying this identity with $p(x) = p_{\text{data}}(\mathbf{x})$ and differentiating with respect to x_i:

$$\frac{\partial \log p_{\text{data}}(\mathbf{x})}{\partial x_i} p_{\text{data}}(\mathbf{x}) = \frac{\partial p_{\text{data}}(\mathbf{x})}{\partial x_i}$$

$$= \int s_{\theta,i}(\mathbf{x}) \cdot \frac{\partial p_{\text{data}}(\mathbf{x})}{\partial x_i} d\mathbf{x}$$

Step 3: Apply integration by parts formula Using integration by parts with $u = s_{\theta,i}(\mathbf{x})$ and $dv = \frac{\partial p_{\text{data}}(\mathbf{x})}{\partial x_i} d\mathbf{x}$, we have:

$$\int u\, dv = uv\Big|_{-\infty}^{\infty} - \int v\, du$$

Substituting:

$$= s_{\theta,i}(\mathbf{x}) p_{\text{data}}(\mathbf{x})\Big|_{-\infty}^{\infty} - \int p_{\text{data}}(\mathbf{x}) \cdot \frac{\partial s_{\theta,i}(\mathbf{x})}{\partial x_i} d\mathbf{x}$$

Step 4: Apply the boundary condition The first term is the **boundary term**. By our assumption that $p_{\text{data}}(\mathbf{x}) \to 0$ as $\|\mathbf{x}\| \to \infty$, and since $s_{\theta,i}(\mathbf{x})$ is bounded (as a neural network output), this boundary term vanishes:

$$s_{\theta,i}(\mathbf{x}) p_{\text{data}}(\mathbf{x})\Big|_{-\infty}^{\infty} = 0$$

Step 5: Simplify the objective With the boundary term eliminated, the cross term becomes:

$$\mathbb{E}_{p_{\text{data}}}\left[s_{\theta,i}(\mathbf{x}) \cdot \frac{\partial \log p_{\text{data}}(\mathbf{x})}{\partial x_i}\right] = -\mathbb{E}_{p_{\text{data}}}\left[\frac{\partial s_{\theta,i}(\mathbf{x})}{\partial x_i}\right]$$

Substituting back into the original objective and simplifying:

$$\mathcal{L}_{\text{SM}}(\theta) = \mathbb{E}_{p_{\text{data}}}\left[\sum_{i=1}^{d}\left(\frac{1}{2}s_{\theta,i}(\mathbf{x})^2 + \frac{\partial s_{\theta,i}(\mathbf{x})}{\partial x_i}\right)\right] + \text{const}$$

17.5.2.3 Why the Boundary Condition Is Valid

The boundary condition $p_{\text{data}}(\mathbf{x}) \to 0$ as $\|\mathbf{x}\| \to \infty$ is reasonable for most real-world distributions due to three complementary reasons. First, **physical constraints** inherently bound real-world data: images have pixel values limited to 0-255, audio amplitude is constrained by recording equipment limitations, text has finite vocabulary size and sequence length, and tabular data features like age and income have natural bounds. These limitations mean data cannot exist arbitrarily far from the origin in the input space, so the probability density must vanish outside these practical bounds.

Second, from a **probabilistic interpretation**, any valid probability density must integrate to 1 over the entire input space. If a density function did not decay to zero at infinity, its integral would diverge to infinity, violating the fundamental requirement of probability theory. This mathematical necessity ensures all valid probability distributions must satisfy the boundary condition, regardless of the specific data domain.

Third, **empirical observation** confirms that most natural data distributions exhibit rapidly decaying tails. Even for features without strict physical bounds, such as certain economic or biological measurements, empirical distributions typically show exponential or power-law decay. Examples include income distributions (following Pareto's law), word frequencies (obeying Zipf's law), and many biological measurements that cluster around mean values with decreasing probability of extreme values.

Collectively, these considerations establish that the boundary condition is well-justified for practical applications. Examples of distributions satisfying this condition include Gaussian, Laplace, and exponential distributions, as well as the complex distributions learned from real-world data in modern machine learning applications.

With this derivation, the score matching objective becomes:

$$\mathcal{L}_{\text{SM}}(\theta) = \mathbb{E}_{p_{\text{data}}}\left[\sum_{i=1}^{d}\left(\frac{1}{2}s_{\theta,i}(\mathbf{x})^2 + \frac{\partial s_{\theta,i}(\mathbf{x})}{\partial x_i}\right)\right] + \text{const} \tag{17.9}$$

where the constant term is independent of θ and can be ignored during optimization.

Key property: No unknown density $p_{\text{data}}(\mathbf{x})$ appears in the final expression. We can estimate this objective using only data samples:

$$\hat{\mathcal{L}}_{\text{SM}}(\theta) = \frac{1}{N}\sum_{n=1}^{N}\sum_{i=1}^{d}\left(\frac{1}{2}s_{\theta,i}(\mathbf{x}^{(n)})^2 + \frac{\partial s_{\theta,i}(\mathbf{x}^{(n)})}{\partial x_i}\right)$$

17.5.3 Interpretation

Score matching fits a vector field $\mathbf{s}_\theta(\mathbf{x})$ whose:

- **Direction** matches probability flow: $\mathbf{s}_\theta(\mathbf{x})$ points toward higher density regions
- **Divergence** matches data curvature: The term $\sum_i \frac{\partial s_{\theta,i}(\mathbf{x})}{\partial x_i} = \nabla \cdot \mathbf{s}_\theta(\mathbf{x})$ (the divergence of the score field) captures how probability mass flows locally

The divergence term is crucial: it ensures that the learned score field not only points in the right direction but also has the correct local "spreading" behavior, matching the curvature of the true data distribution.

17.5.4 Connection to Fisher Information

The score matching objective is closely related to the **Fisher Information Matrix** (see Chapter 14, Section 14.10). The Fisher Information for a location parameter is:

$$I(\theta) = \mathbb{E}_{p_{\text{data}}} \left[\nabla_{\mathbf{x}} \log p_{\text{data}}(\mathbf{x}) \nabla_{\mathbf{x}} \log p_{\text{data}}(\mathbf{x})^\top \right]$$

Score matching effectively learns to match the score function, which is the gradient of the log-density. When the learned score $\mathbf{s}_\theta(\mathbf{x})$ equals the true score $\nabla_{\mathbf{x}} \log p_{\text{data}}(\mathbf{x})$, the Fisher Information can be estimated from the learned model.

17.5.5 Practical Considerations

While score matching eliminates the need for the unknown density, it requires computing second derivatives:

$$\frac{\partial s_{\theta,i}(\mathbf{x})}{\partial x_i} = \frac{\partial^2 \log p_\theta(\mathbf{x})}{\partial x_i^2}$$

For neural network models, this requires **second-order automatic differentiation**, which can be computationally expensive. However, modern frameworks like PyTorch and JAX support efficient Hessian-vector products, making score matching practical for large models.

17.5.6 Example: Score Matching for a Univariate Gaussian

Suppose the data is generated from a univariate Gaussian distribution:

$$p_{\text{data}}(x) = \mathcal{N}(x; \mu, \sigma^2)$$

For simplicity, set $\mu = 0$ and $\sigma^2 = 1$, so:

$$p_{\text{data}}(x) = \frac{1}{\sqrt{2\pi}} e^{-x^2/2}$$

The **true score function** (gradient of the log-density) is:

$$\nabla_x \log p_{\text{data}}(x) = \frac{d}{dx} \left(-\frac{x^2}{2} - \log \sqrt{2\pi} \right) = -x$$

Define the Score Network

Use a simple linear model for the score function:

$$s_\theta(x) = \theta_1 x + \theta_0$$

This is a minimal model that can perfectly capture the true score (since the true score is linear in x).

Compute the Score Matching Objective

For a univariate distribution, the score matching objective (Equation 17.9) simplifies to:

$$\mathcal{L}_{\text{SM}}(\theta) = \mathbb{E}_{p_{\text{data}}} \left[\frac{1}{2} s_\theta(x)^2 + \frac{ds_\theta(x)}{dx} \right] + \text{const}$$

Substitute the linear score function:

$$s_\theta(x) = \theta_1 x + \theta_0 \quad \text{and} \quad \frac{ds_\theta(x)}{dx} = \theta_1$$

Plugging into the objective:

$$\mathcal{L}_{\text{SM}}(\theta) = \mathbb{E}_{p_{\text{data}}}\left[\frac{1}{2}(\theta_1 x + \theta_0)^2 + \theta_1\right] + \text{const}$$

Expand the squared term:

$$\mathcal{L}_{\text{SM}}(\theta) = \mathbb{E}_{p_{\text{data}}}\left[\frac{1}{2}(\theta_1^2 x^2 + 2\theta_1\theta_0 x + \theta_0^2) + \theta_1\right] + \text{const}$$

Evaluate Expectations

Use the properties of the standard Gaussian $\mathcal{N}(0,1)$:

- $\mathbb{E}[x] = 0$
- $\mathbb{E}[x^2] = 1$

Compute each expectation:

$$\mathbb{E}[x^2] = 1, \quad \mathbb{E}[x] = 0, \quad \mathbb{E}[1] = 1$$

Substitute back:

$$\mathcal{L}_{\text{SM}}(\theta) = \frac{1}{2}(\theta_1^2 \cdot 1 + 2\theta_1\theta_0 \cdot 0 + \theta_0^2 \cdot 1) + \theta_1 + \text{const}$$

Simplify:

$$\mathcal{L}_{\text{SM}}(\theta) = \frac{1}{2}\theta_1^2 + \frac{1}{2}\theta_0^2 + \theta_1 + \text{const}$$

Minimize the Objective

Take partial derivatives with respect to θ_1 and θ_0 , set to zero, and solve:

- Derivative with respect to θ_1 :

$$\frac{\partial \mathcal{L}_{\text{SM}}}{\partial \theta_1} = \theta_1 + 1 = 0 \implies \theta_1 = -1$$

- Derivative with respect to θ_0 :

$$\frac{\partial \mathcal{L}_{\text{SM}}}{\partial \theta_0} = \theta_0 = 0 \implies \theta_0 = 0$$

Verify the Result

The optimal parameters are $\theta_1 = -1$ and $\theta_0 = 0$, so the learned score function is:

$$s_\theta(x) = -x$$

This matches the true score $\nabla_x \log p_{\text{data}}(x) = -x$! The example confirms that score matching correctly recovers the true score function using only the data distribution's moments (via expectations) and the derived objective.

17.5.6.1 Key Takeaways

- **Simplicity**: For a simple linear score function and Gaussian data, score matching exactly recovers the true score.
- **No Density Needed**: The calculation only uses expectations of x and x^2 , which can be estimated from data samples—no knowledge of $p_{\text{data}}(x)$ is required beyond these moments.
- **Divergence Term Importance**: The $\frac{ds_\theta(x)}{dx}$ term (divergence) is crucial; without it, the objective would not correctly identify the true score.

This example demonstrates the core mechanics of score matching in a transparent, calculable setting, making the theoretical derivation more concrete for readers.

17.6 Denoising Score Matching (DSM)

While score matching is theoretically elegant, it faces practical challenges in high dimensions:

- **Manifold structure**: Real data (e.g., natural images) lies on a **low-dimensional manifold** embedded in high-dimensional space. The true data distribution $p_{\text{data}}(\mathbf{x})$ has zero probability mass almost everywhere, making $\nabla \log p_{\text{data}}(\mathbf{x})$ undefined or unstable at most points.
- **Sparse data**: In high dimensions, data samples are sparse. The score function $\nabla \log p_{\text{data}}(\mathbf{x})$ is only well-defined near the data manifold, but we need to learn it everywhere for effective sampling.
- **Numerical instability**: Computing $\frac{\partial s_{\theta,i}(\mathbf{x})}{\partial x_i}$ (second derivatives) can be numerically unstable, especially when the learned distribution is sharp or multi-modal.

17.6.1 Example: Score Undefined on a Low-Dimensional Manifold

To illustrate why vanilla score matching fails on real data, consider a simple 2D circle manifold:

Define the Manifold Suppose our data is sampled from a unit circle in 2D space:

$$\mathcal{M} = \{(x_1, x_2) \in \mathbb{R}^2 \mid x_1^2 + x_2^2 = 1\}$$

This is a **1D manifold** (a circle has only one degree of freedom: angle) embedded in 2D space.

The True Data Density $p_{\text{data}}(\mathbf{x})$ The true density $p_{\text{data}}(\mathbf{x})$ is non-zero only on the circle and zero everywhere else:

$$p_{\text{data}}(\mathbf{x}) = \frac{1}{2\pi}\delta(x_1^2 + x_2^2 - 1)$$

Where:

- $\delta(\cdot)$ is the **Dirac delta function**, a generalized function with two key properties:
  1. It is zero everywhere except at 0: $\delta(x) = 0$ for all $x \neq 0$
  2. Its integral over the entire real line equals 1: $\int_{-\infty}^{\infty} \delta(x)\, dx = 1$

  For our manifold, it enforces that the density is non-zero only when $x_1^2+x_2^2 = 1$ (i.e., on the circle).

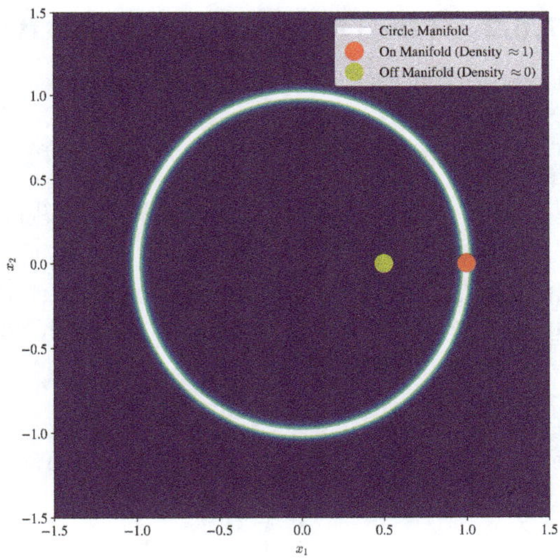

Figure 17.5: Circle Manifold and Its Density

- $\frac{1}{2\pi}$ normalizes the density, since the circle has circumference 2π (the "volume" of the 1D manifold).

Compute the Score Function The score function is the gradient of the log-density:

$$\nabla \log p_{\text{data}}(\mathbf{x}) = \nabla \left(\log \left(\frac{1}{2\pi} \delta(x_1^2 + x_2^2 - 1) \right) \right)$$

Why the Score is Undefined

- **On the manifold** (e.g., $(1, 0)$): The delta function $\delta(0)$ is infinite, so $\log(\delta(0))$ is infinite. Its gradient is undefined.
- **Off the manifold** (e.g., $(0.5, 0)$): The delta function $\delta(-0.75) = 0$, so $\log(0) = -\infty$. Its gradient is undefined.

This demonstrates the problem with real data: the score function is undefined or unstable everywhere except on the low-dimensional manifold, where it's mathematically ill-behaved.

Figure 17.5 visualizes the circle manifold and its density. The white circle represents the 1D manifold embedded in 2D space, while the color gradient shows the density (brighter regions indicate higher density). The red point on the circle shows a point on the manifold where density is high, while the blue point off the circle shows a point where density is nearly zero. This visualization makes concrete why the score function is undefined: it's only well-defined on the manifold itself, where density is non-zero, but undefined everywhere else in the high-dimensional space.

To address the problem with Vanilla Score Matching, the **Solution** is to **add noise** to "smooth out" the distribution, making the score function well-defined everywhere.

17.6.2 The Denoising Perspective

Instead of learning the score of the true data distribution $p_{\text{data}}(\mathbf{x})$, we learn the score of a **noisy version**:

$$p_\sigma(\tilde{\mathbf{x}}) = \int p_{\text{data}}(\mathbf{x}) \mathcal{N}(\tilde{\mathbf{x}}; \mathbf{x}, \sigma^2 I) d\mathbf{x}$$

where $\mathcal{N}(\tilde{\mathbf{x}}; \mathbf{x}, \sigma^2 I)$ is a Gaussian noise distribution. This **convolved distribution** $p_\sigma(\tilde{\mathbf{x}})$ has full support (non-zero probability everywhere), making its score function well-defined and smooth.

17.6.3 Denoising Formulation

Let $\mathbf{x} \sim p_{\text{data}}(\mathbf{x})$ be a clean data sample, and add Gaussian noise:

$$\tilde{\mathbf{x}} = \mathbf{x} + \epsilon, \quad \epsilon \sim \mathcal{N}(\mathbf{0}, \sigma^2 I) \tag{17.10}$$

The conditional distribution of the noisy sample given the clean sample is:

$$p(\tilde{\mathbf{x}} \mid \mathbf{x}) = \mathcal{N}(\tilde{\mathbf{x}}; \mathbf{x}, \sigma^2 I) = \frac{1}{(2\pi\sigma^2)^{d/2}} \exp\left(-\frac{1}{2\sigma^2} \|\tilde{\mathbf{x}} - \mathbf{x}\|_2^2 \right)$$

Taking the log and gradient:

$$\log p(\tilde{\mathbf{x}} \mid \mathbf{x}) = -\frac{1}{2\sigma^2} \|\tilde{\mathbf{x}} - \mathbf{x}\|_2^2 + \text{const}$$

$$\nabla_{\tilde{\mathbf{x}}} \log p(\tilde{\mathbf{x}} \mid \mathbf{x}) = -\frac{1}{\sigma^2}(\tilde{\mathbf{x}} - \mathbf{x}) \tag{17.11}$$

Key insight: The score of the conditional distribution $p(\tilde{\mathbf{x}} \mid \mathbf{x})$ points directly from the noisy sample $\tilde{\mathbf{x}}$ back to the clean sample $\mathbf{x}$. This is exactly the **denoising direction**!

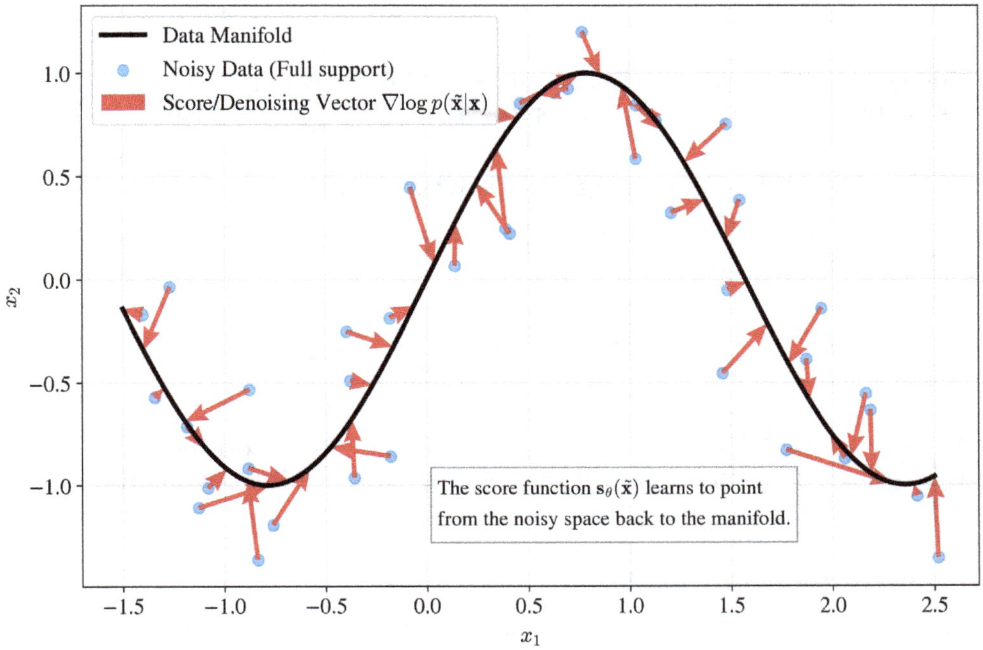

Figure 17.6: Denoising Score Matching and the Manifold Hypothesis

Figure 17.6 visualizes the core idea of denoising score matching using a **1D S-curve manifold** embedded in 2D space. The black curve represents the data manifold—the low-dimensional support where clean data lies. The blue points show noisy samples created by adding Gaussian noise to points on the manifold, which "smears" the data into the full 2D space. The red arrows represent the score vectors (denoising directions) that point from each noisy sample back toward the nearest point on the data manifold. This visualization demonstrates the **manifold hypothesis**: real data lies on a low-dimensional manifold, and adding noise moves points off this manifold. The score function learns to "denoise" by pointing back toward the manifold, making it well-defined everywhere in the high-dimensional space, not just near the sparse data points. This is why denoising score matching works: it learns the score of the noisy distribution $p_\sigma(\tilde{\mathbf{x}})$, which has full support and a well-defined score function everywhere.

17.6.4 Connection to Denoising Autoencoders

This formulation reveals a deep connection to **denoising autoencoders** (Vincent et al., 2008). A denoising autoencoder learns to predict the clean sample $\mathbf{x}$ from a noisy version $\tilde{\mathbf{x}}$. The DSM objective learns to predict the **direction** to the clean sample, which is proportional to the denoising error:

$$\nabla_{\tilde{\mathbf{x}}} \log p(\tilde{\mathbf{x}} \mid \mathbf{x}) = -\frac{1}{\sigma^2}(\tilde{\mathbf{x}} - \mathbf{x}) = -\frac{1}{\sigma^2} \times (\text{denoising error})$$

This connection explains why **denoising is equivalent to learning gradients**: the gradient of the log-conditional-density is the optimal denoising direction.

17.6.5 DSM Objective

The **denoising score matching (DSM)** objective trains $\mathbf{s}_\theta(\tilde{\mathbf{x}})$ to match the conditional score:

$$\mathcal{L}_{\text{DSM}}(\theta) = \mathbb{E}_{\mathbf{x}\sim p_{\text{data}},\epsilon\sim\mathcal{N}(\mathbf{0},\sigma^2 I)}\left[\left\|\mathbf{s}_\theta(\tilde{\mathbf{x}}) + \frac{1}{\sigma^2}(\tilde{\mathbf{x}} - \mathbf{x})\right\|_2^2\right] \tag{17.12}$$

where $\tilde{\mathbf{x}} = \mathbf{x} + \epsilon$.

Why this works: By learning to denoise, the model implicitly learns the score function of the noisy distribution $p_\sigma(\tilde{\mathbf{x}})$. As $\sigma \to 0$, this approaches the true data distribution's score.

17.6.6 Multi-Scale Denoising Score Matching

In practice, we use **multiple noise scales** $\{\sigma_1, \sigma_2, \dots, \sigma_L\}$ to learn a score function that works across different levels of noise:

$$\mathcal{L}_{\text{DSM}}(\theta) = \sum_{l=1}^{L} \lambda_l \mathbb{E}_{\mathbf{x},\epsilon_l}\left[\left\|\mathbf{s}_\theta(\tilde{\mathbf{x}}_l, \sigma_l) + \frac{1}{\sigma_l^2}(\tilde{\mathbf{x}}_l - \mathbf{x})\right\|_2^2\right]$$

where λ_l are weighting factors and $\mathbf{s}_\theta(\mathbf{x}, \sigma)$ is a **time-dependent score function** that depends on the noise scale. This multi-scale approach is the foundation of **diffusion models** (see Chapters 18–21).

17.6.7 Advantages of DSM

1. **Stability**: The noisy distribution has full support, avoiding numerical issues with sparse data
2. **Scalability**: Only requires first derivatives (no second derivatives like vanilla score matching)
3. **Intuitive**: Directly connected to denoising, a well-understood task
4. **Practical**: Easy to implement and train with standard deep learning frameworks

This makes DSM the method of choice for training score-based generative models on high-dimensional data like images.

17.6.8 Example: DSM on the Circle Manifold

To illustrate how DSM solves the manifold problem, let's apply it to our 1D circle manifold example.

Figure 17.7 visualizes the core idea of Denoising Score Matching on the circle manifold. The black circle represents the original data manifold, and the black dots show clean data points on the manifold. The red dots show noisy samples created by adding Gaussian noise to the clean points, which "smears" the data off the manifold into the full 2D space. The blue arrows represent the score vectors (denoising directions) that point from each noisy sample back toward the nearest point on the manifold. This visualization demonstrates how DSM learns a score function that is well-defined everywhere in the high-dimensional space, not just near the sparse data points on the manifold. **The Noisy Manifold Distribution**

Instead of learning the score of the original manifold distribution, we learn the score of a noisy version:

$$p_\sigma(\tilde{\mathbf{x}}) = \int p_{\text{data}}(\mathbf{x})\mathcal{N}(\tilde{\mathbf{x}}; \mathbf{x}, \sigma^2 I)d\mathbf{x}$$

Where $\mathcal{N}(\tilde{\mathbf{x}}; \mathbf{x}, \sigma^2 I)$ is a Gaussian noise distribution with variance σ^2.

Why This Works

1. **Full Support**: The convolved distribution $p_\sigma(\tilde{\mathbf{x}})$ has non-zero probability everywhere in $\mathbb{R}^2$, not just on the circle.
2. **Smooth Score Function**: The score $\nabla_{\tilde{\mathbf{x}}} \log p_\sigma(\tilde{\mathbf{x}})$ is now well-defined and smooth everywhere.
3. **Denoising Interpretation**: The score of $p_\sigma(\tilde{\mathbf{x}})$ points from noisy samples back toward the manifold, effectively denoising them.

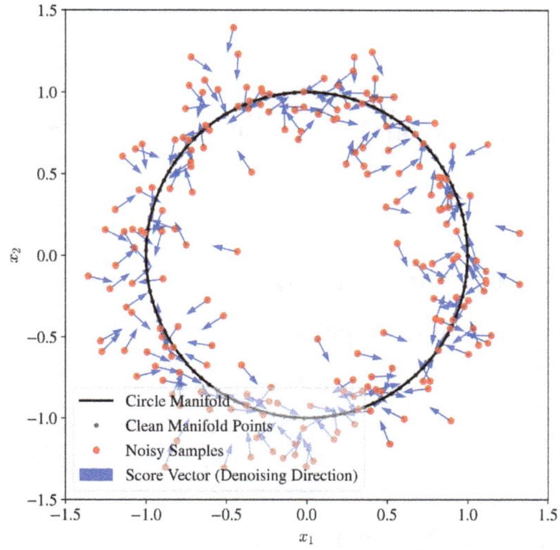

Figure 17.7: DSM on Circle Manifold

Mathematical Formulation

For our circle manifold, the conditional distribution of noisy samples given clean samples is:

$$p(\tilde{\mathbf{x}} \mid \mathbf{x}) = \mathcal{N}(\tilde{\mathbf{x}}; \mathbf{x}, \sigma^2 I)$$

The score of this conditional distribution is:

$$\nabla_{\tilde{\mathbf{x}}} \log p(\tilde{\mathbf{x}} \mid \mathbf{x}) = -\frac{\tilde{\mathbf{x}} - \mathbf{x}}{\sigma^2}$$

This score vector points from the noisy sample $\tilde{\mathbf{x}}$ directly back toward the clean sample $\mathbf{x}$ on the manifold.

The DSM Objective

The DSM objective is to minimize the expected squared error between the model's score and this conditional score:

$$\mathcal{L}_{\text{DSM}}(\theta) = \mathbb{E}_{\mathbf{x},\epsilon}\left[\left\| s_\theta(\tilde{\mathbf{x}}) + \frac{\tilde{\mathbf{x}} - \mathbf{x}}{\sigma^2} \right\|_2^2\right]$$

Where $\tilde{\mathbf{x}} = \mathbf{x} + \epsilon$ and $\epsilon \sim \mathcal{N}(0, \sigma^2 I)$.

Key Insight

By learning to predict the denoising direction (from noisy samples back to the manifold), the model implicitly learns a score function that is well-defined everywhere in the high-dimensional space, not just on the low-dimensional manifold. This is the core advantage of Denoising Score Matching over vanilla score matching.

17.6.9 Example: DSM for a Univariate Gaussian

Define the Clean Data Distribution

Start with the same clean data distribution: a univariate Gaussian with mean 0 and variance 1:

$$x \sim \mathcal{N}(0, 1)$$

Add Gaussian Noise

Create noisy samples by adding zero-mean Gaussian noise with variance σ^2:

$$\tilde{x} = x + \epsilon, \quad \epsilon \sim \mathcal{N}(0, \sigma^2)$$

The conditional distribution of the noisy sample given the clean sample is:

$$p(\tilde{x} \mid x) = \mathcal{N}(\tilde{x}; x, \sigma^2)$$

True Conditional Score

The score of the conditional distribution $p(\tilde{x} \mid x)$ is the gradient of its log-density with respect to $\tilde{x}$:

$$s_{\text{true}}(\tilde{x} \mid x) = \nabla_{\tilde{x}} \log p(\tilde{x} \mid x) = -\frac{\tilde{x} - x}{\sigma^2}$$

This score represents the optimal direction to "denoise" $\tilde{x}$ back toward x.

Define the Score Model

Use a simple linear model for the score function:

$$s_\theta(\tilde{x}) = \theta_1 \tilde{x} + \theta_0$$

This model only depends on the noisy input $\tilde{x}$, not on the clean x (which is unknown during inference).

Compute the DSM Objective

The DSM objective is the expected squared error between the model's prediction and the true conditional score:

$$\mathcal{L}_{\text{DSM}}(\theta) = \mathbb{E}_{x,\epsilon}\left[\left\| s_\theta(\tilde{x}) + \frac{\tilde{x} - x}{\sigma^2} \right\|^2\right]$$

Substitute the linear model and expand the square:

$$\mathcal{L}_{\text{DSM}}(\theta) = \mathbb{E}_{x,\epsilon}\left[\left(\theta_1 \tilde{x} + \theta_0 + \frac{\tilde{x} - x}{\sigma^2}\right)^2\right]$$

Simplify Using Gaussian Moments

Since $x \sim \mathcal{N}(0, 1)$ and $\epsilon \sim \mathcal{N}(0, \sigma^2)$ are independent:

- $\mathbb{E}[x] = 0$, $\mathbb{E}[x^2] = 1$
- $\mathbb{E}[\epsilon] = 0$, $\mathbb{E}[\epsilon^2] = \sigma^2$
- $\mathbb{E}[x\epsilon] = 0$, $\mathbb{E}[\tilde{x}] = 0$, $\mathbb{E}[\tilde{x}^2] = 1 + \sigma^2$
- $\mathbb{E}[x\tilde{x}] = \mathbb{E}[x(x + \epsilon)] = 1$

Substitute these moments and simplify the objective:

$$\mathcal{L}_{\text{DSM}}(\theta) = \theta_1^2(1 + \sigma^2) + 2\theta_1 + \theta_0^2 + \frac{1}{\sigma^2}$$

Minimize the Objective

Take partial derivatives with respect to θ_1 and θ_0, set to zero, and solve:

- Derivative with respect to θ_0:

$$\frac{\partial \mathcal{L}_{\text{DSM}}}{\partial \theta_0} = 2\theta_0 = 0 \implies \theta_0 = 0$$

- Derivative with respect to θ_1:

$$\frac{\partial \mathcal{L}_{\text{DSM}}}{\partial \theta_1} = 2\theta_1(1+\sigma^2) + 2 = 0 \implies \theta_1 = -\frac{1}{1+\sigma^2}$$

Verify the Result

The optimal parameters give the learned score function:

$$s_\theta(\tilde{x}) = -\frac{\tilde{x}}{1+\sigma^2}$$

This is exactly the score of the **marginal distribution** $p(\tilde{x}) = \mathcal{N}(0, 1+\sigma^2)$, which is the convolution of the clean data and noise distributions. The marginal score is well-defined everywhere in the real line, unlike the original data distribution's score which would be undefined on low-dimensional manifolds in high dimensions.

Figure 17.8 visualizes the Denoising Score Matching process for a univariate Gaussian. The top panel shows the clean Gaussian distribution $\mathcal{N}(0,1)$(blue) and the noisy Gaussian distribution $\mathcal{N}(0, 1+\sigma^2)$(red), where $\sigma^2 = 0.25$.

The bottom panel shows the learned score function $s_\theta(x) = -\frac{x}{1+\sigma^2}$(green), which is linear. The red dots represent noisy samples, while the blue dots represent their corresponding clean samples. The red line and arrow show the denoising directions, illustrating how the score function points from noisy sample back toward clean sample.

This example demonstrates how DSM overcomes the practical challenges of vanilla score matching while maintaining the same core principle: learning to match the gradient of the log-density.

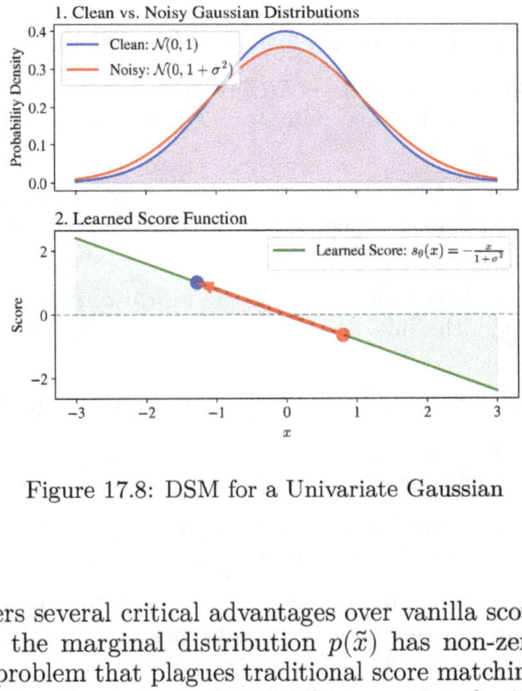

Figure 17.8: DSM for a Univariate Gaussian

Key Takeaways: Denoising Score Matching offers several critical advantages over vanilla score matching. First, it provides **full support**, as the marginal distribution $p(\tilde{x})$ has non-zero probability everywhere, eliminating the manifold problem that plagues traditional score matching. Second, it requires only **first derivatives**, avoiding the numerical instability associated with second derivatives in the original score matching objective. Third, it has a natural **denoising interpretation**: the learned score function implicitly captures the direction from noisy samples back toward clean ones, connecting to well-understood autoencoder concepts. Finally, it exhibits strong **scalability**: while we used a simple linear model here, the framework easily extends to high dimensions where it can be implemented with deep neural networks.

17.7 Applications of Score-Based Models

17.7.1 Image Generation

Score-based models have achieved state-of-the-art results in **image generation**. Models like **NCSN**[3] (Noise Conditional Score Networks, Song & Ermon, 2019) and **DDPM**[4] (Denoising Diffusion Probabilistic Models, Ho et al., 2020) use multi-scale denoising score matching to learn score functions for natural images.

The key insight is to learn a **time-dependent score function** $\mathbf{s}_\theta(\mathbf{x}, t)$ where t indexes the noise scale. At $t = 0$, the score corresponds to the clean data distribution; at $t = T$, it corresponds to pure noise. This creates a continuous path from noise to data, enabling high-quality generation.

17.7.2 Connection to Variational Autoencoders

While VAEs (Chapter 16) learn an approximate posterior $q_\phi(\mathbf{z} \mid \mathbf{x})$ and decoder $p_\theta(\mathbf{x} \mid \mathbf{z})$, score-based models learn the data distribution directly without an explicit latent space. However, both approaches face similar challenges:

- **VAEs**: Require computing the ELBO, which involves an intractable KL divergence
- **Score-based models**: Require sampling via Langevin dynamics, which can be slow

Recent work has combined both approaches, using VAEs to provide a low-dimensional latent space for faster score-based sampling.

17.7.3 Connection to Generative Adversarial Networks

GANs (Generative Adversarial Networks) learn to generate samples by training a generator $G_\theta(\mathbf{z})$ and a discriminator $D_\phi(\mathbf{x})$ in an adversarial game. Score-based models take a different approach:

- **GANs**: Learn an implicit distribution via a generator network
- **Score-based models**: Learn an explicit score function, enabling likelihood estimation and more stable training

Score-based models avoid the **mode collapse** problem common in GANs because they explicitly model the full data distribution rather than learning a mapping from a simple prior.

17.7.4 Uncertainty Quantification

Score functions provide a natural way to quantify **uncertainty** in generative models. The magnitude of the score $\|\mathbf{s}_\theta(\mathbf{x})\|_2$ indicates how confident the model is about the data point $\mathbf{x}$:

- **Large score magnitude**: The point is far from high-density regions (high uncertainty)
- **Small score magnitude**: The point is near a mode (low uncertainty)

This uncertainty information is useful for **out-of-distribution detection** and **anomaly detection** in machine learning systems.

---

[3]Song, Y., & Ermon, S. (2019). "Noise Conditional Score Networks: Learning Conditional Distributions for Noise-Based Generative Modeling." In *Advances in Neural Information Processing Systems* (NeurIPS). This paper introduced the concept of learning time-dependent score functions for noise-based generative modeling.

[4]Ho, J., Jain, A., & Abbeel, P. (2020). "Denoising Diffusion Probabilistic Models." In *Advances in Neural Information Processing Systems* (NeurIPS). This paper proposed the denoising diffusion probabilistic model framework that has become foundational for modern diffusion models.

17.8 From Score to Sampling

Once $\mathbf{s}_\theta(\mathbf{x})$ is learned, we can generate samples by following the score field. The most common approach is **Langevin dynamics**, which we will study in detail in Chapter 18.

17.8.1 Langevin Dynamics Preview

The basic idea is to start from random noise and iteratively update:

$$\mathbf{x}_{k+1} = \mathbf{x}_k + \eta\,\mathbf{s}_\theta(\mathbf{x}_k) + \sqrt{2\eta}\,\epsilon_k \tag{17.13}$$

where $\eta > 0$ is a step size and $\epsilon_k \sim \mathcal{N}(\mathbf{0}, I)$ is Gaussian noise.

Interpretation: This is a **stochastic gradient ascent** on $\log p(\mathbf{x})$:

- The term $\eta\,\mathbf{s}_\theta(\mathbf{x}_k)$ moves toward higher probability regions
- The term $\sqrt{2\eta}\,\epsilon_k$ injects randomness to explore the distribution

Under mild conditions, this process converges to samples from the distribution $p(\mathbf{x})$ as $k \to \infty$ and $\eta \to 0$.

17.8.2 Why Noise Is Necessary

A natural question is: why not just follow the gradient deterministically?

$$\mathbf{x}_{k+1} = \mathbf{x}_k + \eta\,\mathbf{s}_\theta(\mathbf{x}_k)$$

This **deterministic gradient ascent** would converge to a **mode** (local maximum) of the distribution, not sample from it. The noise term is essential for:

1. **Exploration**: Preventing the process from getting stuck in a single mode
2. **Diversity**: Generating diverse samples from multi-modal distributions
3. **Correctness**: Ensuring the stationary distribution matches $p(\mathbf{x})$

The mathematical foundation for why noise is necessary comes from **stochastic differential equations** (Chapter 19) and the **Fokker-Planck equation** (Chapter 21).

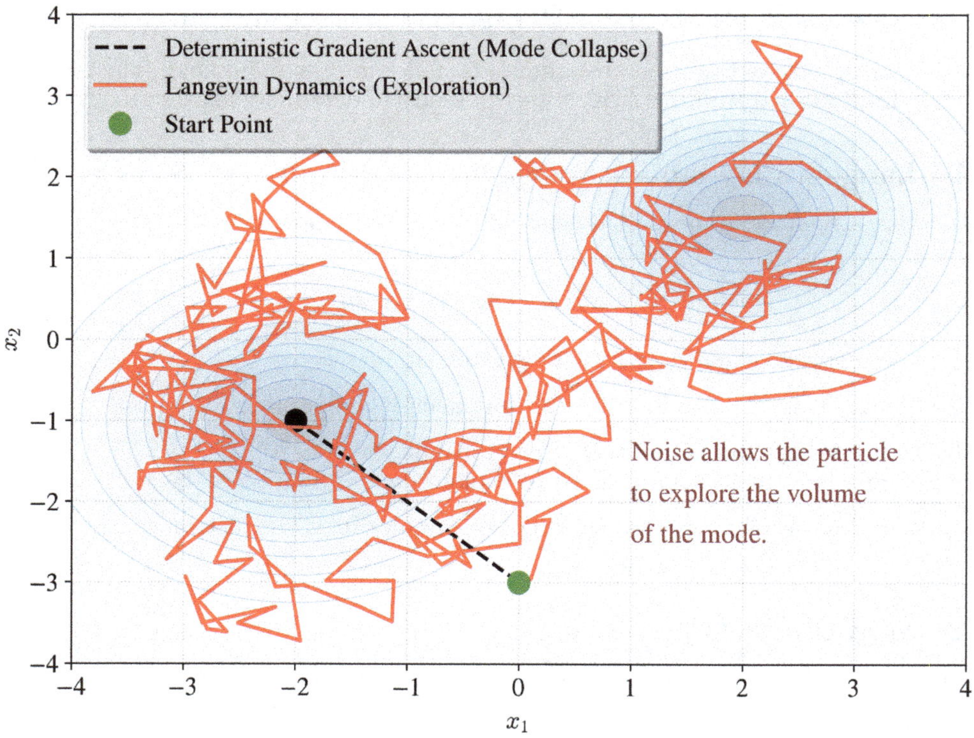

Figure 17.9: Langevin Dynamics vs. Gradient Ascent

Figure 17.9 demonstrates the critical difference between deterministic gradient ascent and stochastic Langevin dynamics for a **mixture of two Gaussians**. The background shows the probability density as contour lines, with two modes (peaks) visible. The black dashed line shows a deterministic gradient ascent trajectory starting from a low-density region. This path follows the score field directly and converges to a single mode, demonstrating **mode collapse**: the deterministic process gets trapped in one mode and cannot explore the other. In contrast, the red solid line shows a Langevin dynamics trajectory with the same starting point. The injected noise allows the process to explore both modes, visiting regions around both peaks. This visualization makes concrete why noise is essential: it enables **exploration** of the entire distribution, **diversity** in generated samples, and **correctness** of the stationary distribution. Without noise, we would only generate samples from a single mode, failing to represent the true multi-modal distribution.

17.8.3 Example: Score Function for a 2D Gaussian Mixture

Consider a 2D mixture of two Gaussians:

$$p(\mathbf{x}) = 0.5 \cdot \mathcal{N}(\mathbf{x}; \mu_1, \Sigma_1) + 0.5 \cdot \mathcal{N}(\mathbf{x}; \mu_2, \Sigma_2)$$

where $\mu_1 = [2, 0]^\top$, $\mu_2 = [-2, 0]^\top$, and both covariance matrices are $\Sigma_1 = \Sigma_2 = 0.5I$.

The score function $\mathbf{s}(\mathbf{x}) = \nabla_{\mathbf{x}} \log p(\mathbf{x})$ can be computed analytically. At a point $\mathbf{x}$:

$$\mathbf{s}(\mathbf{x}) = \frac{w_1(\mathbf{x})\mathbf{s}_1(\mathbf{x}) + w_2(\mathbf{x})\mathbf{s}_2(\mathbf{x})}{w_1(\mathbf{x}) + w_2(\mathbf{x})}$$

where $w_i(\mathbf{x}) = \mathcal{N}(\mathbf{x}; \mu_i, \Sigma_i)$ and $\mathbf{s}_i(\mathbf{x}) = -\Sigma_i^{-1}(\mathbf{x} - \mu_i)$.

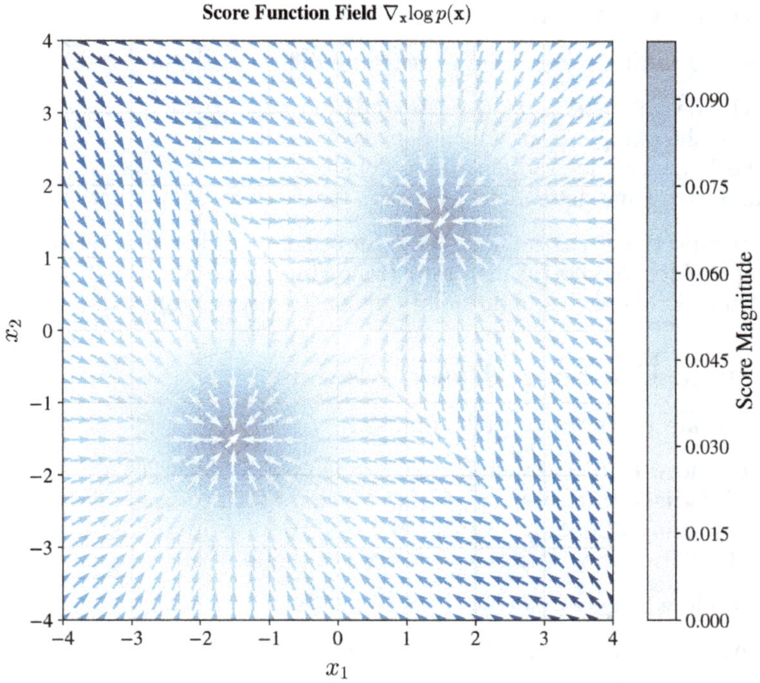

Figure 17.10: Score Function Field for 2D Gaussian Mixture

Figure 17.10 visualizes the score function field $\nabla_{\mathbf{x}} \log p(\mathbf{x})$ for a 2D mixture of two Gaussians. The background shows the probability density as filled contours (darker blue indicates higher density), revealing two distinct modes. The arrows represent the score vectors at each point, colored by magnitude (using the 'autumn' colormap). The score field forms two "attractors" centered at the two means: vectors point inward toward each mode, with larger magnitudes near the boundaries between modes where the probability landscape is steepest. Starting from any point and following the score field (via Langevin dynamics) will guide samples toward one of the two modes, with the noise term ensuring both modes are explored. This visualization makes concrete how the score function encodes the local geometry of the probability distribution, pointing toward regions of higher density and enabling effective sampling through stochastic dynamics.

17.9 Conceptual Unification

The following table summarizes different generative modeling frameworks and what they learn:

Framework	What is learned	Advantages	Challenges
MLE	$\log p(\mathbf{x})$	Direct likelihood optimization	Requires tractable likelihood
VAE	Lower bound on $\log p(\mathbf{x})$	Handles intractable posteriors	Approximate inference, blurry samples
GAN	Implicit distribution	High-quality samples	Mode collapse, training instability
EBM	$E(\mathbf{x})$	Flexible energy functions	Intractable Z_θ, slow sampling
Score-based	$\nabla \log p(\mathbf{x})$	No normalization needed, stable training	Requires MCMC sampling
Diffusion	Time-dependent scores	State-of-the-art quality, stable training	Slow generation

17.9.1 Unified Perspective

All these frameworks can be understood as learning different aspects of the data distribution:

- **MLE/VAE**: Learn the density (or a bound) directly
- **EBM**: Learn the energy, from which density follows
- **Score-based**: Learn the gradient of log-density
- **Diffusion**: Learn time-dependent gradients for a noise-to-data path

The score-based perspective is particularly powerful because: 1. It avoids intractable normalization constants 2. It provides geometric intuition (following gradients) 3. It connects naturally to sampling algorithms 4. It enables high-quality generation on complex data

17.9.2 When to Use Score-Based Models

Score-based models are particularly well-suited for:

- **High-dimensional data**: Images, audio, video where densities are intractable
- **Multi-modal distributions**: Where GANs suffer from mode collapse
- **Uncertainty quantification**: When you need to measure model confidence
- **Likelihood estimation**: When you need to evaluate $p(\mathbf{x})$ (via score integration)

They are less suitable when:

- **Fast generation is critical**: MCMC sampling can be slow
- **Low-dimensional data**: Simpler methods may suffice
- **Real-time applications**: Generation requires multiple iterations

17.10 Mindmap

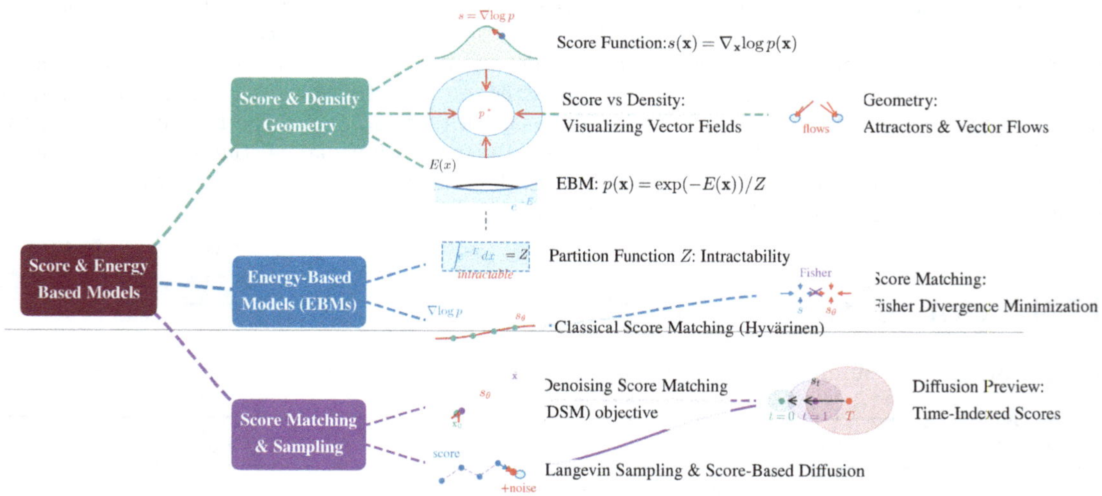

Figure 17.11: Mind map of score and energy-based models topics

17.11 Exercises

17.11.1 Score Function Basics

1. **Normalization invariance**: Let $p(\mathbf{x}) = \frac{1}{Z}\tilde{p}(\mathbf{x})$ for some unknown $Z > 0$. Prove that $\nabla_{\mathbf{x}} \log p(\mathbf{x}) = \nabla_{\mathbf{x}} \log \tilde{p}(\mathbf{x})$.
2. **Gaussian score**: If $\mathbf{x} \sim \mathcal{N}(\mu, \Sigma)$, derive $\mathbf{s}(\mathbf{x}) = \nabla_{\mathbf{x}} \log p(\mathbf{x})$ explicitly and show it points toward μ.
3. **Score transformation**: Let $\mathbf{y} = A\mathbf{x} + \mathbf{b}$ for an invertible matrix A. Express $\nabla_{\mathbf{y}} \log p_{\mathbf{y}}(\mathbf{y})$ in terms of $\nabla_{\mathbf{x}} \log p_{\mathbf{x}}(\mathbf{x})$ and A.

17.11.2 Energy-Based Models

1. **EBM score**: For $p_\theta(\mathbf{x}) = \frac{1}{Z_\theta} \exp(-E_\theta(\mathbf{x}))$, show that

$$\nabla_{\mathbf{x}} \log p_\theta(\mathbf{x}) = -\nabla_{\mathbf{x}} E_\theta(\mathbf{x}),$$

   and explain why Z_θ disappears.
2. **Energy recovery (up to a constant)**: Suppose you are given a smooth vector field $\mathbf{s}(\mathbf{x})$ and you know it equals $-\nabla_{\mathbf{x}} E(\mathbf{x})$. Under what condition(s) on $\mathbf{s}$ does such an E exist globally? (Hint: think "curl-free / conservative field".)

17.11.3 Score Matching

1. **Why naive score matching is impossible**: The "ideal" objective is

$$\mathbb{E}_{p_{\text{data}}} \left[\|\mathbf{s}_\theta(\mathbf{x}) - \nabla_{\mathbf{x}} \log p_{\text{data}}(\mathbf{x})\|_2^2 \right].$$

   Identify precisely what makes this objective intractable from samples alone.
2. **Hyvärinen's integration-by-parts form**: Starting from the ideal objective above, derive (up to an additive constant independent of θ):

$$\mathbb{E}_{p_{\text{data}}} \left[\frac{1}{2} \|\mathbf{s}_\theta(\mathbf{x})\|_2^2 + \nabla_{\mathbf{x}} \cdot \mathbf{s}_\theta(\mathbf{x}) \right].$$

   State the boundary condition(s) needed for the integration by parts step.
3. **1D sanity check**: In one dimension, write $\nabla_{\mathbf{x}} \cdot \mathbf{s}_\theta(\mathbf{x})$ explicitly and verify the Hyvärinen form reduces to

$$\mathbb{E}_{p_{\text{data}}} \left[\frac{1}{2} s_\theta(x)^2 + \frac{d}{dx} s_\theta(x) \right] + \text{const.}$$

17.11.4 Denoising Score Matching (DSM)

1. **Conditional score for Gaussian corruption**: Let $\tilde{\mathbf{x}} = \mathbf{x} + \epsilon$ with $\epsilon \sim \mathcal{N}(0, \sigma^2 I)$. Derive

$$\nabla_{\tilde{\mathbf{x}}} \log p(\tilde{\mathbf{x}} \mid \mathbf{x}) = -\frac{1}{\sigma^2}(\tilde{\mathbf{x}} - \mathbf{x}).$$

2. **DSM objective**: Using the result above, show the DSM objective can be written as

$$\mathbb{E}_{\mathbf{x},\epsilon} \left[\left\| \mathbf{s}_\theta(\tilde{\mathbf{x}}) + \frac{1}{\sigma^2}(\tilde{\mathbf{x}} - \mathbf{x}) \right\|_2^2 \right].$$

3. **Effect of σ**: Explain qualitatively what happens as $\sigma \to 0$ and as $\sigma \to \infty$ (in terms of what information the model is being asked to learn).

17.11.5 From Score to Sampling (Langevin)

1. **Langevin update**: Consider

$$\mathbf{x}_{k+1} = \mathbf{x}_k + \eta\,\mathbf{s}(\mathbf{x}_k) + \sqrt{2\eta}\,\epsilon_k, \qquad \epsilon_k \sim \mathcal{N}(0, I).$$

   Explain why the drift term $\eta\,\mathbf{s}(\mathbf{x}_k)$ can be viewed as a (stochastic) gradient ascent step on $\log p(\mathbf{x})$.

2. **Stationarity intuition**: If $\mathbf{s}(\mathbf{x}) = \nabla_{\mathbf{x}} \log p(\mathbf{x})$ is exact, argue (informally) why Langevin dynamics should have p as its stationary distribution.

Chapter 18

Langevin Dynamics and Sampling

Stochastic gradient descent, Langevin dynamics, and MCMC sampling

> *"One may regard Brownian motion as the result of incessant molecular agitation, manifesting itself through irregular impacts on the particle."*
>
> — Paul Langevin

- ▷ `Score function` $\mathbf{s}(\mathbf{x}) \triangleq \nabla_{\mathbf{x}} \log p(\mathbf{x})$
- ▷ `Gradient ascent (mode-seeking)` $\mathbf{x}_{k+1} = \mathbf{x}_k + \eta\, \mathbf{s}(\mathbf{x}_k)$
- ▷ `Unadjusted Langevin Algorithm (ULA)` $\mathbf{x}_{k+1} = \mathbf{x}_k + \eta\, \mathbf{s}(\mathbf{x}_k) + \sqrt{2\eta}\, \epsilon_k,\ \epsilon_k \sim \mathcal{N}(0, I)$
- ▷ `Overdamped Langevin SDE (continuous limit)` $d\mathbf{X}_t = \mathbf{s}(\mathbf{X}_t)\, dt + \sqrt{2}\, d\mathbf{W}_t$ (stationary density p under mild conditions)
- ▷ `Energy-based model link` if $p(\mathbf{x}) \propto \exp(-E(\mathbf{x}))$, then $\mathbf{s}(\mathbf{x}) = -\nabla_{\mathbf{x}} E(\mathbf{x})$ and ULA becomes $\mathbf{x}_{k+1} = \mathbf{x}_k - \eta\, \nabla E(\mathbf{x}_k) + \sqrt{2\eta}\, \epsilon_k$
- ▷ `MALA proposal` $q(\mathbf{x}' \mid \mathbf{x}) = \mathcal{N}(\mathbf{x} + \eta\, \mathbf{s}(\mathbf{x}),\, 2\eta\, I)$
- ▷ `MALA accept prob.` $\alpha(\mathbf{x}, \mathbf{x}') = \min\Big(1, \frac{p(\mathbf{x}')\, q(\mathbf{x}|\mathbf{x}')}{p(\mathbf{x})\, q(\mathbf{x}'|\mathbf{x})}\Big)$
- ▷ `SGLD update` $\mathbf{x}_{k+1} = \mathbf{x}_k + \eta_k\, \tilde{\mathbf{s}}(\mathbf{x}_k) + \sqrt{2\eta_k}\, \epsilon_k$ (mini-batch / stochastic score)
- ▷ `Robbins-Monro steps (typical)` $\sum_k \eta_k = \infty$, $\sum_k \eta_k^2 < \infty$ (ensures diminishing discretization noise in stochastic approximation settings)

18.1 Motivation: Why Gradients Are Not Enough

In Chapter~17, we learned how to model the **score function**

$$\mathbf{s}(\mathbf{x}) = \nabla_{\mathbf{x}} \log p(\mathbf{x}) \tag{18.1}$$

which encodes the local geometry of a probability distribution. The score function points in the direction of increasing probability density, providing a natural "compass" for navigating the probability landscape.

A natural idea is to **follow the score**:

$$\mathbf{x}_{k+1} = \mathbf{x}_k + \eta \mathbf{s}(\mathbf{x}_k) \tag{18.2}$$

This is simply **gradient ascent** on $\log p(\mathbf{x})$, where $\eta > 0$ is a step size. This deterministic update rule moves iteratively toward regions of higher probability.

However, this approach has fundamental limitations for **sampling**:

- **Mode convergence**: Gradient ascent converges to **local modes** (maxima) of the distribution, not the full distribution. Starting from different initial points, the algorithm may converge to different modes, but it will never explore the entire distribution.
- **No diversity**: Once a mode is reached, the algorithm stays there. For multi-modal distributions, this means we only see samples from one mode, losing the diversity that characterizes the true distribution.
- **Collapse**: In high-dimensional spaces, gradient ascent can get trapped in narrow regions around modes, producing samples that are highly correlated and fail to represent the distribution's true variability.

To **sample** from a distribution, rather than optimize toward its modes, we must inject **randomness**. This randomness serves two critical purposes: (1) it prevents the algorithm from getting stuck in modes, and (2) it ensures that the long-term behavior of the algorithm matches the target distribution, not just its local extrema.

18.1.1 The Sampling vs Optimization Distinction

This distinction between **sampling** and **optimization** is fundamental in machine learning:

- **Optimization** (gradient ascent/descent): Seeks to find a single point (or small set of points) that optimizes an objective function. The goal is convergence to an optimum.
- **Sampling** (Langevin dynamics, MCMC): Seeks to generate a sequence of points whose empirical distribution matches a target distribution. The goal is exploration and representation of the full distribution.

In generative modeling, we need sampling, not optimization. We want to generate diverse samples that collectively represent the data distribution, not just find its modes.

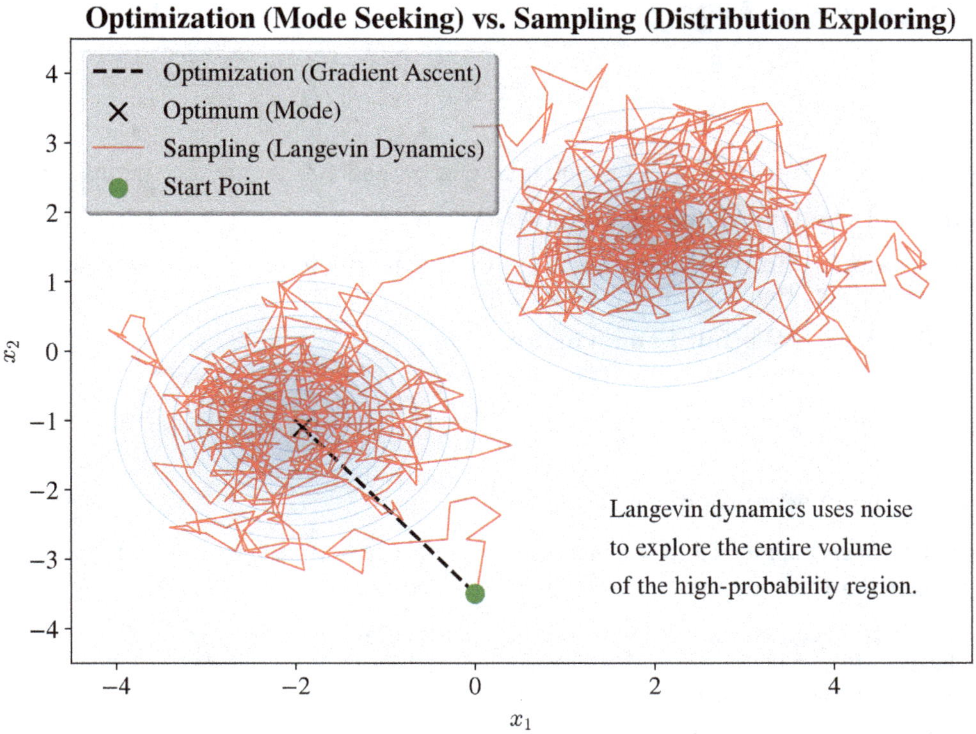

Figure 18.1: Optimization vs. Sampling

Figure 18.1 demonstrates the fundamental difference between optimization and sampling for a **2D mixture of two Gaussians**. The background shows the probability density as contour lines, with two distinct modes. The black dashed line shows a **deterministic gradient ascent** trajectory starting from a low-density region. This path follows the score field directly and converges to a single mode, demonstrating **mode collapse**: the deterministic process gets trapped in one mode and cannot explore the other. In contrast, the red solid line shows a **Langevin dynamics** trajectory with the same starting point. The injected noise allows the process to explore both modes, visiting regions around both peaks. This visualization makes concrete why noise is essential: it enables **exploration** of the entire distribution, **diversity** in generated samples, and **correctness** of the stationary distribution. Without noise, we would only generate samples from a single mode, failing to represent the true multi-modal distribution.

This need for sampling in generative modeling directly motivates the introduction of Langevin dynamics, which provides a principled way to balance gradient guidance with exploratory noise.

18.2 Langevin Dynamics: Optimization + Noise

18.2.1 Historical Origin

Langevin dynamics originates from **statistical physics**, where it was developed by Paul Langevin in 1908 to describe the motion of particles in a fluid. The key insight is that particles are subject to two competing influences:

- **Deterministic forces**: Systematic forces like gravity, electric fields, or energy gradients that push particles in specific directions
- **Random thermal noise**: Random collisions with fluid molecules that cause Brownian motion

In the original physical context, a particle's position $\mathbf{x}(t)$ evolves according to:

$$m\frac{d^2\mathbf{x}}{dt^2} = -\gamma\frac{d\mathbf{x}}{dt} - \nabla V(\mathbf{x}) + \xi(t)$$

where m is mass, γ is friction, $V(\mathbf{x})$ is potential energy, and $\xi(t)$ is random thermal noise.

Derivation of the Overdamped Limit

In the **overdamped limit** (high friction, $m \to 0$), the inertial term becomes negligible compared to the friction term. Here's the derivation:

Step 1: Start with the full Langevin equation

$$m\frac{d^2\mathbf{x}}{dt^2} = -\gamma\frac{d\mathbf{x}}{dt} - \nabla V(\mathbf{x}) + \xi(t)$$

Step 2: In the overdamped regime

When friction is very high ($\gamma \gg 1$) or mass is very small ($m \to 0$), the inertial term $m\frac{d^2\mathbf{x}}{dt^2}$ becomes much smaller than the friction term $-\gamma\frac{d\mathbf{x}}{dt}$. This is because:

- High friction means even small velocities create large opposing forces
- The system can't accelerate significantly before friction dominates
- As $m \to 0$, the inertial term vanishes entirely

Step 3: Set inertial term to zero

$$0 \approx -\gamma\frac{d\mathbf{x}}{dt} - \nabla V(\mathbf{x}) + \xi(t)$$

Step 4: Solve for velocity

Rearranging to solve for $\frac{d\mathbf{x}}{dt}$:

$$\gamma\frac{d\mathbf{x}}{dt} = -\nabla V(\mathbf{x}) + \xi(t)$$

Dividing by γ gives the overdamped Langevin equation:

$$\frac{d\mathbf{x}}{dt} = -\frac{1}{\gamma}\nabla V(\mathbf{x}) + \frac{1}{\gamma}\xi(t)$$

Physical Intuition

In the overdamped regime, the system responds almost instantaneously to changes in potential energy. It doesn't "coast" or exhibit inertial effects because friction dissipates any kinetic energy immediately. This is why the overdamped equation only includes the first derivative (velocity) and not the second derivative (acceleration).

This is the continuous-time Langevin equation, where the deterministic drift $-\nabla V(\mathbf{x})$ competes with random noise.

18.2.2 Example: Langevin Dynamics for a Simple Gaussian Distribution

To illustrate how Langevin dynamics works, consider sampling from a 1D Gaussian distribution $p(x) = \mathcal{N}(x; \mu, \sigma^2)$ with $\mu = 0$ and $\sigma^2 = 1$.

Step 1: Define the energy function

For a Gaussian distribution, the energy function is:

$$E(x) = \frac{1}{2\sigma^2}(x-\mu)^2 = \frac{1}{2}x^2$$

Step 2: Compute the score function

The score function is the negative gradient of the energy:

$$s(x) = -\nabla E(x) = -x$$

Step 3: Apply the Langevin dynamics update

Using the discrete-time update rule with step size $\eta = 0.01$:

$$\begin{aligned} x_{t+1} &= x_t + \eta \cdot s(x_t) + \sqrt{2\eta} \cdot \epsilon_t \\ &= x_t - \eta \cdot x_t + \sqrt{2\eta} \cdot \epsilon_t \end{aligned}$$

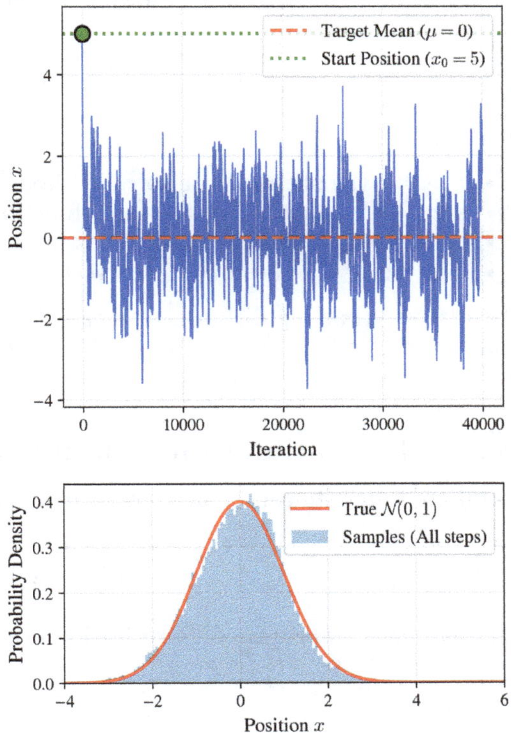

Figure 18.2: Langevin Dynamics Convergence

Step 4: Simulation results

Starting from $x_0 = 5$ (far from the mean), after 40,000 iterations:

- The sample distribution converges to $\mathcal{N}(0, 1)$
- The trajectory shows the particle moving toward the mean while being perturbed by noise
- The algorithm successfully explores the entire distribution

Figure 21.5 demonstrates the convergence of Langevin dynamics to the $\mathcal{N}(0, 1)$ distribution. Starting from $x_0 = 5$ (far from the mean), the top panel shows the trajectory converging toward $\mu = 0$ with noise perturbations, with the starting point marked by a green dot. The bottom panel compares the high-resolution histogram of samples (all steps, 150 bins) with the true $\mathcal{N}(0, 1)$ distribution, confirming successful convergence.

18.2.3 From Physics to Machine Learning

In machine learning, we reinterpret this equation:

- The **potential energy** $V(\mathbf{x})$ becomes the **negative log-probability** $-\log p(\mathbf{x})$
- The **energy gradient** $-\nabla V(\mathbf{x})$ becomes the **score function** $\nabla \log p(\mathbf{x})$
- The **thermal noise** ensures proper exploration of the probability distribution

This provides a principled way to convert gradients (score functions) into **samples** from probability distributions. The mathematical framework from statistical physics guarantees that, under appropriate conditions, the long-term distribution of samples matches the target distribution $p(\mathbf{x})$.

18.2.4 Discrete Langevin Dynamics - ULA

Given a score function $\mathbf{s}(\mathbf{x})$, the **Langevin update** is

$$\mathbf{x}_{k+1} = \mathbf{x}_k + \eta \mathbf{s}(\mathbf{x}_k) + \sqrt{2\eta}\,\epsilon_k, \quad \epsilon_k \sim \mathcal{N}(\mathbf{0}, I) \tag{18.3}$$

where:

- $\eta > 0$ is the **step size** (also called learning rate in optimization contexts)
- $\mathbf{s}(\mathbf{x}_k) = \nabla_{\mathbf{x}} \log p(\mathbf{x}_k)$ is the score function at the current point
- $\epsilon_k \sim \mathcal{N}(\mathbf{0}, I)$ is independent Gaussian noise
- The factor $\sqrt{2\eta}$ ensures the correct variance scaling

This discrete-time algorithm is also called **Unadjusted Langevin Algorithm (ULA)**. The term "unadjusted" refers to the fact that there is no Metropolis-Hastings accept/reject step (see Section on MALA below).

18.2.5 Derivation from Continuous-Time Limit

The discrete update can be derived by discretizing the continuous-time Langevin SDE:

$$d\mathbf{X}_t = \mathbf{s}(\mathbf{X}_t)dt + \sqrt{2}d\mathbf{W}_t$$

Using the Euler-Maruyama discretization (see Chapter 19) with time step η:

$$\mathbf{X}_{t+\eta} \approx \mathbf{X}_t + \mathbf{s}(\mathbf{X}_t)\eta + \sqrt{2}\sqrt{\eta}\epsilon$$

where $\epsilon \sim \mathcal{N}(\mathbf{0}, I)$ and $\sqrt{\eta}d\mathbf{W}_t \approx \sqrt{\eta}\epsilon$. Setting $\sqrt{2\eta} = \sqrt{2}\sqrt{\eta}$ gives the discrete update formula.

The factor $\sqrt{2}$ comes from the requirement that the stationary distribution of the continuous process matches the target distribution $p(\mathbf{x})$. This is a consequence of the **fluctuation-dissipation theorem** from statistical physics.

18.2.6 Interpretation

Each term in the Langevin update plays a distinct role:

Term	Role	Physical Analogy
$\eta \mathbf{s}(\mathbf{x})$	Moves toward high-density regions	Deterministic drift force
$\sqrt{2\eta}\,\epsilon$	Prevents mode collapse	Thermal fluctuations
η	Controls exploration vs stability	Time step / friction parameter

The drift term $\eta \mathbf{s}(\mathbf{x})$: This is the deterministic component that pushes the sample toward regions of higher probability. It acts like a "force" that guides exploration toward the distribution's modes. However, if this were the only term, the algorithm would converge to a single mode and stay there.

The noise term $\sqrt{2\eta}\,\epsilon$: This is the stochastic component that injects randomness. The variance $\sqrt{2\eta}$ is carefully chosen so that, in the continuous-time limit, the stationary distribution matches $p(\mathbf{x})$. The noise prevents the algorithm from getting stuck in modes and ensures proper exploration of the entire distribution.

The step size η: This parameter controls the trade-off between exploration and stability. Large η allows faster movement but can cause instability. Small η is more stable but requires more iterations to explore the distribution. The optimal choice depends on the geometry of the target distribution.

18.2.7 Why the Specific Noise Scaling?

The factor $\sqrt{2\eta}$ in the noise term is not arbitrary. It ensures that the **variance** of the noise matches the variance required for the continuous-time process to have the correct stationary distribution. In the continuous limit, this scaling ensures that the **fluctuation-dissipation relation** is satisfied, which guarantees that the algorithm samples from the correct distribution.

Noise is not a hack — it is **essential** for correct sampling. Without it, we would have deterministic gradient ascent, which converges to modes rather than sampling from the distribution.

18.3 Stationary Distribution

18.3.1 The Fundamental Result

A key theoretical result establishes the connection between Langevin dynamics and the target distribution:

Theorem (Stationary Distribution of Langevin Dynamics): Under mild regularity conditions, the discrete Langevin dynamics algorithm has $p(\mathbf{x})$ as its **stationary distribution**. That is, as $k \to \infty$ and the step size $\eta \to 0$,

$$\mathbf{x}_k \sim p(\mathbf{x}) \tag{18.4}$$

This means that, after a sufficient number of iterations (after the "burn-in" period), the samples $\mathbf{x}_k$ will be distributed according to $p(\mathbf{x})$, regardless of the initial point $\mathbf{x}_0$.

18.3.2 Conditions for Convergence

The convergence result requires several conditions:

1. **Small step size**: The step size η must be sufficiently small relative to the curvature of the distribution. If η is too large, the discretization error can cause the algorithm to diverge or converge to the wrong distribution.
2. **Accurate score field**: The score function $\mathbf{s}(\mathbf{x})$ must accurately approximate $\nabla_{\mathbf{x}} \log p(\mathbf{x})$. Errors in the score function lead to bias in the stationary distribution.
3. **Sufficient mixing**: The Markov chain must be **ergodic**, meaning it can reach any region of the distribution's support. This requires that the distribution is not disconnected into isolated components that the chain cannot traverse.
4. **Regularity conditions**: The target distribution $p(\mathbf{x})$ must satisfy certain smoothness and tail conditions. Typically, we require that $p(\mathbf{x})$ is continuously differentiable and has finite moments.

18.3.3 Intuition: Why This Works

The stationary distribution result can be understood through the lens of **detailed balance**. In the continuous-time limit, the Langevin SDE satisfies:

$$p(\mathbf{x}) \cdot (\text{transition rate from } \mathbf{x} \text{ to } \mathbf{x}') = p(\mathbf{x}') \cdot (\text{transition rate from } \mathbf{x}' \text{ to } \mathbf{x})$$

This detailed balance condition ensures that, at equilibrium, the probability of being at $\mathbf{x}$ and transitioning to $\mathbf{x}'$ equals the probability of being at $\mathbf{x}'$ and transitioning to $\mathbf{x}$. This is exactly what is needed for $p(\mathbf{x})$ to be the stationary distribution.

The discrete algorithm approximates this continuous process, and the approximation becomes exact as $\eta \to 0$.

18.3.4 From Local to Global

This result is remarkable because it transforms **local gradient information** (the score function at each point) into **global sampling** (samples from the entire distribution). We don't need to know the full distribution $p(\mathbf{x})$ or its normalization constant—we only need the score function, which is often much easier to learn or compute.

This is the mathematical foundation that makes score-based generative models practical: we can learn the score function from data, then use Langevin dynamics to generate samples from the learned distribution.

18.4 Connection to Energy-Based Models

Langevin dynamics has a natural connection to energy-based models (EBMs) discussed in Chapter~17. Recall that EBMs define distributions through an energy function:

$$p(\mathbf{x}) \propto \exp(-E(\mathbf{x}))$$

18.4.1 Score Function Connection

As we saw in Chapter~17, the score function of an EBM is the negative gradient of its energy function:

$$\mathbf{s}(\mathbf{x}) = -\nabla E(\mathbf{x}) \tag{18.5}$$

18.4.2 Langevin Dynamics for EBMs

Substituting this score function relationship into the Langevin dynamics update rule gives:

$$\mathbf{x}_{k+1} = \mathbf{x}_k - \eta \nabla E(\mathbf{x}_k) + \sqrt{2\eta}\, \epsilon_k \tag{18.6}$$

This form makes explicit how Langevin dynamics navigates the energy landscape of an EBM:

- The term $-\eta \nabla E(\mathbf{x}_k)$ drives the system toward lower energy states (higher probability regions)
- The noise term $\sqrt{2\eta}\, \epsilon_k$ allows the system to escape local energy minima

18.4.3 Relationship to Stochastic Gradient Descent

This update rule for EBMs bears a striking resemblance to **stochastic gradient descent (SGD)** commonly used in optimization. However, there's a crucial conceptual difference:

- **SGD for EBM training**: Run until convergence to find parameters that minimize the energy of training data
- **Langevin dynamics for EBM sampling**: Run indefinitely to explore the entire energy landscape and generate samples

The key insight is that **sampling from EBMs can be viewed as SGD without convergence**—instead of stopping at a minimum, the algorithm continues to navigate the energy landscape, with the noise term preventing it from getting trapped in local minima.

This connection between optimization and sampling highlights the deep mathematical relationship between these two tasks in the context of energy-based models, where techniques from one domain can often be adapted to the other.

18.5 Markov Chain Monte Carlo (MCMC)

We now place Langevin dynamics within the broader framework of Markov Chain Monte Carlo (MCMC) methods. This provides a more general theoretical foundation and helps us understand how Langevin dynamics relates to other sampling techniques.

18.5.1 Definition: Markov Chain Monte Carlo

Markov Chain Monte Carlo (MCMC) is a class of algorithms for sampling from a target probability distribution $p(\mathbf{x})$ by constructing a **Markov chain** whose stationary distribution is $p(\mathbf{x})$.

The key idea is to generate a sequence of samples $\mathbf{x}_0, \mathbf{x}_1, \mathbf{x}_2, \ldots$ such that:

1. The sequence forms a **Markov chain**: $P(\mathbf{x}_{k+1} \mid \mathbf{x}_k, \mathbf{x}_{k-1}, \ldots, \mathbf{x}_0) = P(\mathbf{x}_{k+1} \mid \mathbf{x}_k)$
2. The **stationary distribution** of the chain is $p(\mathbf{x})$: as $k \to \infty$, the distribution of $\mathbf{x}_k$ converges to $p(\mathbf{x})$

Once the chain has **converged** (reached its stationary distribution), subsequent samples $\mathbf{x}_k$ for large k are approximately distributed according to $p(\mathbf{x})$, even though we cannot sample directly from $p(\mathbf{x})$.

18.5.2 Example: MCMC with Metropolis-Hastings Algorithm

To illustrate how MCMC works, consider using the Metropolis-Hastings algorithm to sample from a 1D bimodal distribution.

Step 1: Define the target distribution

Let's use a bimodal distribution with two peaks:

$$p(x) \propto \exp\left(-\frac{1}{2}(x+2)^2\right) + \exp\left(-\frac{1}{2}(x-2)^2\right)$$

This distribution has peaks at $x = -2$ and $x = 2$.

Step 2: Choose a proposal distribution

Use a Gaussian proposal distribution centered at the current sample:

$q(x' \mid x) = \mathcal{N}(x'; x, \sigma^2)$$ with $\sigma = 0.5$.

Step 3: Metropolis-Hastings algorithm

1. **Initialize**: Start with an initial sample $x_0 = 0$
2. **Iterate**: For each step t:
    a. **Generate proposal**: Sample $x' \sim q(x' \mid x_t)$
    b. **Compute acceptance ratio**:

$$r = \frac{p(x')q(x_t \mid x')}{p(x_t)q(x' \mid x_t)} = \frac{p(x')}{p(x_t)}$$

c. **Accept or reject**: Sample $u \sim \text{Uniform}(0, 1)$
   - If $u \leq r$, accept the proposal: $x_{t+1} = x'$
   - Otherwise, reject the proposal: $x_{t+1} = x_t$

Step 4: Results

After 40,000 iterations:

- **Burn-in period**: Discard the first 1,000 samples to allow the chain to reach stationarity
- **Sample distribution**: The remaining 9,000 samples form a distribution that closely matches the target bimodal distribution
- **Exploration**: The chain successfully explores both modes of the distribution

Comparison with Langevin Dynamics

While Metropolis-Hastings works well for this example, Langevin dynamics would typically converge faster by using gradient information to move toward high-probability regions more efficiently.

18.5.3 Why MCMC?

MCMC is used when:

- **Direct sampling is intractable**: We cannot sample directly from $p(\mathbf{x})$ (e.g., the normalization constant Z in $p(\mathbf{x}) = \frac{1}{Z}\exp(-E(\mathbf{x}))$ is unknown)
- **Only unnormalized density is available**: We can evaluate $p^*(\mathbf{x}) \propto p(\mathbf{x})$ (e.g., $p^*(\mathbf{x}) = \exp(-E(\mathbf{x}))$), but not the normalized $p(\mathbf{x})$
- **High-dimensional spaces**: Direct sampling methods (e.g., rejection sampling) become inefficient in high dimensions

MCMC solves this by constructing a Markov chain that **explores** the distribution through a sequence of dependent samples, rather than requiring independent samples from $p(\mathbf{x})$.

18.5.4 Key Concepts

Transition kernel: The conditional distribution $T(\mathbf{x}' \mid \mathbf{x})$ that defines how to move from state $\mathbf{x}$ to state $\mathbf{x}'$ in one step. The Markov chain is defined by:

$$\mathbf{x}_{k+1} \sim T(\cdot \mid \mathbf{x}_k)$$

Stationary distribution: A distribution $\pi(\mathbf{x})$ is stationary for transition kernel T if:

$$\pi(\mathbf{x}') = \int T(\mathbf{x}' \mid \mathbf{x})\pi(\mathbf{x})\, d\mathbf{x}$$

This means that if $\mathbf{x}_k \sim \pi$, then $\mathbf{x}_{k+1} \sim \pi$ as well.

Detailed balance: A sufficient condition for π to be stationary is **detailed balance** (also called **reversibility**):

$$T(\mathbf{x}' \mid \mathbf{x})\pi(\mathbf{x}) = T(\mathbf{x} \mid \mathbf{x}')\pi(\mathbf{x}')$$

This states that the probability of transitioning from $\mathbf{x}$ to $\mathbf{x}'$ equals the probability of transitioning from $\mathbf{x}'$ to $\mathbf{x}$, when both are weighted by the stationary distribution.

Mixing time: The number of steps required for the chain to converge to its stationary distribution. Fast mixing is desirable for efficient sampling.

18.5.5 Common MCMC Algorithms

1. **Metropolis-Hastings**: General-purpose algorithm that uses an arbitrary proposal distribution and accepts/rejects proposals to ensure detailed balance
2. **Gibbs Sampling**: Special case where proposals are drawn from conditional distributions
3. **Langevin-based methods**: Use gradient information (score function) to guide proposals:
   - **Unadjusted Langevin Algorithm (ULA)**: Direct Langevin updates (biased but fast)
   - **Metropolis-Adjusted Langevin Algorithm (MALA)**: Langevin proposals with Metropolis-Hastings correction (exact but slower)

MCMC is fundamental to **Bayesian inference**, where we need to sample from posterior distributions $p(\theta \mid \mathbf{x})$ that are often intractable. In **generative modeling**, MCMC is used to sample from energy-based models $p(\mathbf{x}) = \frac{1}{Z}\exp(-E(\mathbf{x}))$ where the partition function Z is unknown. The key advantage of MCMC is that it only requires evaluating the **unnormalized** density (or score function), not the normalization constant. However, MCMC has limitations: (1) **mixing time** can be very long for complex distributions, (2) samples are **correlated** (not independent), requiring many samples for accurate estimates, and (3) **convergence** is difficult to diagnose. In practice, MCMC is often combined with **variational inference** (Chapter 16): use VI for fast approximate inference, then use MCMC to refine the approximation or get better uncertainty estimates.

18.6 Metropolis-Adjusted Langevin Algorithm (MALA)

18.6.1 Motivation: Discretization Bias

The discrete Langevin update introduces **discretization bias** because it approximates a continuous process with finite step sizes. Even when the step size η is small, there is a non-zero probability that the proposed update moves the chain to a region that would not be reached by the exact continuous process.

This discretization error can cause the stationary distribution of the discrete chain to differ from the target distribution $p(\mathbf{x})$. The error is typically $O(\eta)$, meaning it decreases linearly with step size, but never completely vanishes for finite η.

To correct this bias and guarantee **exact sampling** from $p(\mathbf{x})$, we add a **Metropolis–Hastings accept/reject step**. This creates a Markov chain that satisfies detailed balance exactly, ensuring the correct stationary distribution regardless of step size (though very large step sizes will lead to low acceptance rates and poor mixing).

18.6.2 MALA Algorithm

The **Metropolis-Adjusted Langevin Algorithm (MALA)** can be formally described as follows:

The MALA algorithm uses the following proposal step:

$$\mathbf{x}' = \mathbf{x} + \eta\,\mathbf{s}(\mathbf{x}) + \sqrt{2\eta}\,\epsilon \tag{18.7}$$

where $\epsilon \sim \mathcal{N}(\mathbf{0}, I)$.

And the acceptance probability is:

$$\alpha = \min\left(1, \frac{p(\mathbf{x}')\,q(\mathbf{x} \mid \mathbf{x}')}{p(\mathbf{x})\,q(\mathbf{x}' \mid \mathbf{x})}\right) \tag{18.8}$$

where $q(\mathbf{x}' \mid \mathbf{x})$ is the proposal density.

The complete algorithm is:

Algorithm 17 Metropolis-Adjusted Langevin Algorithm (MALA)

Input: Target distribution $p(\mathbf{x})$, score function $\mathbf{s}(\mathbf{x})$, step size η, initial state $\mathbf{x}_0$
Output: Sequence of samples $\mathbf{x}_1, \mathbf{x}_2, \ldots$ distributed according to $p(\mathbf{x})$

```
x ← x_0
for k = 0, 1, 2, ... do
    Propose new state:
      Sample ε ~ N(0, I)
      x' ← x + η s(x) + sqrt(2η) ε
    Compute acceptance probability:
      α ← min(1, p(x') q(x|x') / (p(x) q(x'|x)))
    Accept or reject:
      Sample u ~ Uniform(0, 1)
    if u ≤ α then
        x ← x'                         ▷ Accept proposal
    else
        x ← x                          ▷ Reject proposal, stay at current state
    end if
    x_{k+1} ← x
end for
return sequence x_1, x_2, ...
```

18.6.3 Understanding the Acceptance Probability

The acceptance probability ensures **detailed balance**, which guarantees that the Markov chain has $p(\mathbf{x})$ as its stationary distribution. The ratio $\frac{p(\mathbf{x}')}{p(\mathbf{x})}$ compares the target probabilities, while $\frac{q(\mathbf{x}|\mathbf{x}')}{q(\mathbf{x}'|\mathbf{x})}$ accounts for the asymmetry in the proposal distribution.

For the Langevin proposal, the proposal density is:

$$q(\mathbf{x}' \mid \mathbf{x}) = \mathcal{N}(\mathbf{x}'; \mathbf{x} + \eta \mathbf{s}(\mathbf{x}), 2\eta I)$$

This is a Gaussian centered at $\mathbf{x} + \eta \mathbf{s}(\mathbf{x})$ with covariance $2\eta I$. The asymmetry comes from the fact that $q(\mathbf{x}' \mid \mathbf{x}) \neq q(\mathbf{x} \mid \mathbf{x}')$ in general, because the score function $\mathbf{s}(\mathbf{x})$ depends on the current state.

18.6.4 Computational Cost

MALA guarantees **exact sampling** from $p(\mathbf{x})$ (up to Monte Carlo error), but it is computationally expensive because:

1. **Accept/reject step**: Each iteration requires evaluating $p(\mathbf{x}')$ and computing the acceptance probability, which may be expensive if $p(\mathbf{x})$ is complex.
2. **Rejected proposals**: When proposals are rejected, the chain stays at the current state, effectively "wasting" computational effort. The acceptance rate depends on the step size η: too large η leads to many rejections, too small η leads to slow exploration.

3. **Score function evaluation**: The proposal requires evaluating $\mathbf{s}(\mathbf{x})$ at both the current and proposed states, which may be expensive for neural network-based score functions.

In practice, the optimal acceptance rate for MALA is around **50-70%**, which balances exploration (large steps) with acceptance (small steps).

18.6.5 Practical Tradeoff

The choice between ULA and MALA involves a fundamental trade-off:

Method	Accuracy	Speed	Use Case
ULA	Approximate	Fast	When approximate sampling is acceptable
MALA	Exact	Slow	When exact sampling is required

ULA (Unadjusted Langevin Algorithm):

- **Advantages**: Fast, simple, no accept/reject overhead
- **Disadvantages**: Discretization bias, may not converge to exact $p(\mathbf{x})$
- **When to use**: Large-scale applications where speed matters more than exactness, or when the bias is acceptable

MALA (Metropolis-Adjusted Langevin Algorithm):

- **Advantages**: Exact sampling, guaranteed convergence
- **Disadvantages**: Slower due to accept/reject, requires evaluating $p(\mathbf{x})$
- **When to use**: When exact sampling is critical, or when the target distribution is expensive to evaluate (making rejections costly)

18.6.6 Why Diffusion Models Use ULA

Diffusion models (see Chapters 19-21) almost always use **ULA-style updates** rather than MALA for several reasons:

1. **Scale**: Diffusion models operate in very high-dimensional spaces (e.g., 784×784 images = $614,656$ dimensions). The accept/reject step in MALA would be prohibitively expensive.

2. **Approximate score functions**: The score function $\mathbf{s}_\theta(\mathbf{x})$ is learned from data and is already approximate. The discretization bias from ULA is often smaller than the approximation error in the score function itself.

3. **Multiple steps**: Diffusion models use many small steps (hundreds to thousands), so the discretization error accumulates slowly and can be controlled by using small step sizes.

4. **Empirical success**: In practice, ULA works well for diffusion models, producing high-quality samples despite the theoretical bias.

The key insight is that **approximate but fast** sampling (ULA) is often preferable to **exact but slow** sampling (MALA) in large-scale applications, especially when the score function itself is approximate.

18.7 Stochastic Gradient Langevin Dynamics (SGLD)

The standard Langevin dynamics algorithm requires computing the exact gradient of the log-density $\nabla_{\mathbf{x}} \log p(\mathbf{x})$ at each iteration. However, in modern machine learning scenarios with large datasets, this presents significant challenges:

- **Computational Cost**: Computing gradients using the entire dataset (full-batch) is prohibitively expensive for large-scale models and datasets
- **Memory Constraints**: Storing the entire dataset in memory is often impossible
- **Online Learning**: Many applications require learning from streaming data where the full dataset is never available

18.7.1 Stochastic Gradient Estimation

Stochastic Gradient Langevin Dynamics (SGLD) addresses these challenges by replacing full-batch gradients with **mini-batch estimates**:

1. **Mini-batch Selection**: At each iteration, randomly select a small subset (mini-batch) of the dataset
2. **Gradient Estimation**: Compute the gradient using only this mini-batch
3. **Noise Adjustment**: The inherent stochasticity of mini-batch gradients provides additional noise, which interacts with the Langevin noise term

18.7.2 SGLD Update Rule and Implementation

The SGLD update rule modifies the standard Langevin dynamics to use stochastic gradients. Let $\hat{\nabla}_{\mathbf{x}} \log p(\mathbf{x}_k)$ or equivalently $\tilde{\mathbf{s}}(\mathbf{x}_k)$ denote the stochastic gradient estimate using a mini-batch. The update rule is:

$$\mathbf{x}_{k+1} = \mathbf{x}_k + \eta_k \cdot \tilde{\mathbf{s}}(\mathbf{x}_k) + \sqrt{2\eta_k} \cdot \epsilon_k \tag{18.9}$$

Where:

- $\tilde{\mathbf{s}}(\mathbf{x}_k) = \hat{\nabla}_{\mathbf{x}} \log p(\mathbf{x}_k)$ is the stochastic gradient estimate
- η_k is the learning rate, which typically decreases over time and should satisfy the following conditions for proper convergence:

$$\sum_k \eta_k = \infty, \quad \sum_k \eta_k^2 < \infty$$

- $\epsilon_k \sim \mathcal{N}(0, I)$ is Gaussian noise

Several key considerations for implementing SGLD include: carefully choosing a learning rate schedule that decays over time to ensure convergence, balancing batch size (larger batches reduce variance but increase computational cost), understanding how stochastic gradient noise and Langevin noise together help the algorithm explore the parameter space, and recognizing that under appropriate conditions, SGLD still converges to the target distribution despite using stochastic gradients.

SGLD represents a crucial adaptation of Langevin dynamics for modern machine learning, enabling its application to large-scale datasets and models that would be infeasible with standard full-batch methods.

18.7.3 Example: SGLD for Logistic Regression

To illustrate how SGLD works in practice, let's walk through a complete example of training a logistic regression model on a large dataset, step by step.

Step 1: Problem Setup

Imagine we have a dataset with $N = 1,000,000$ samples, and we want to train a logistic regression model to classify this data. Logistic regression is well-suited for classification tasks, and SGLD is particularly effective for handling large datasets like this one.

Step 2: Define the log-posterior

First, we need to define the log-posterior distribution. Assuming a Gaussian prior $p(\mathbf{w}) = \mathcal{N}(\mathbf{w}; 0, \lambda^{-1}I)$ over the model parameters $\mathbf{w}$, the log-posterior is:

$$\log p(\mathbf{w}|\mathcal{D}) \propto \sum_{i=1}^{N} \log p(y_i|\mathbf{x}_i, \mathbf{w}) - \frac{\lambda}{2}\|\mathbf{w}\|^2$$

This expression combines the likelihood of the data given the parameters with our prior beliefs about the parameters.

Step 3: Full-batch vs. Mini-batch

For a dataset of this size:

- **Full-batch training** would be computationally expensive and memory-intensive, as it requires processing all $1,000,000$ samples for each gradient calculation
- **Mini-batch training** offers a more efficient alternative by randomly selecting $B = 128$ samples per iteration, significantly reducing computational cost while maintaining good gradient estimates

Step 4: SGLD Update

Using a mini-batch of size $B = 128$ and initial learning rate $\eta = 0.001$, the SGLD update rule becomes:

$$\mathbf{w}_{t+1} = \mathbf{w}_t + \eta \cdot \left(\frac{N}{B} \sum_{i \in \text{batch}} \nabla_{\mathbf{w}} \log p(y_i|\mathbf{x}_i, \mathbf{w}) - \lambda \mathbf{w}_t \right) + \sqrt{2\eta} \cdot \epsilon_t$$

Notice the $\frac{N}{B}$ scaling factor, which properly adjusts the mini-batch gradient to estimate the full-batch gradient, ensuring we maintain the correct posterior distribution.

Step 5: Learning Rate Schedule

To ensure proper convergence, we implement a decaying learning rate schedule: $\eta_t = \eta_0/(1 + t/\tau)$ where $\tau = 40,000$. This schedule starts with a larger learning rate for faster initial progress and gradually reduces it for finer exploration around the posterior mode.

Step 6: Results and Analysis

After running the SGLD algorithm for $10,000$ iterations (with $2,000$ iterations for warmup), we obtain the following results:

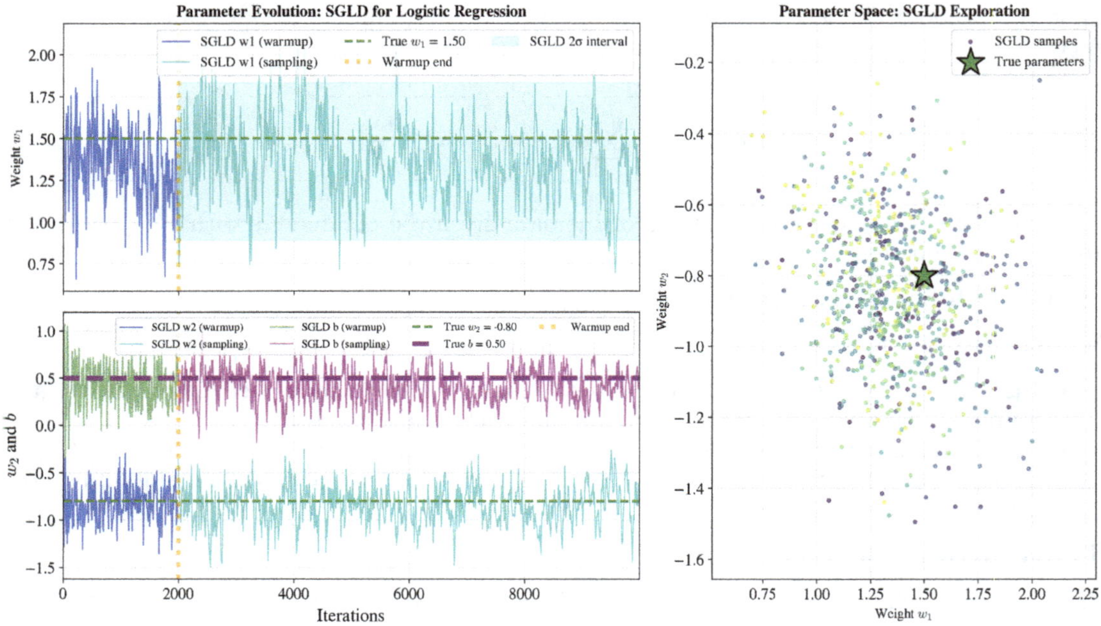

Figure 18.3: SGLD for Logistic Regression

Figure 18.3 visualizes the application of SGLD to logistic regression. The left panels display the evolution of parameters over time: the top panel shows the first weight w_1, while the bottom panel shows the second weight w_2 and bias b. The blue lines represent the warmup phase, where noise strength gradually increases, and the cyan (for weights) and magenta (for bias) lines show the sampling phase with full noise. The orange vertical line marks the end of the warmup phase. The right panel presents the parameter space trajectory, with the color gradient indicating the progression of iterations after warmup and the green star marking the true parameter values $w_1 = 1.5, w_2 = -0.8$. With $N = 500$ data points, SGLD effectively balances convergence to the posterior mode with exploration of the surrounding parameter space, producing posterior mean estimates close to the true values while providing valuable uncertainty estimates. This visualization demonstrates how SGLD can be applied to more complex models like logistic regression, making it a powerful tool for Bayesian inference in practical machine learning tasks.

18.7.4 Interpretation: Optimization Meets Sampling

SGLD blurs the boundary between:

- **Optimization (SGD)**: When $\eta_k \to 0$ and noise is negligible, SGLD becomes SGD, which converges to a local minimum
- **Bayesian inference (sampling)**: When step sizes satisfy the Robbins-Monro conditions, SGLD samples from the posterior distribution

This dual nature explains why **SGD often behaves like implicit posterior sampling**. In Bayesian neural networks, the parameter distribution learned by SGD (with appropriate step size schedules) can approximate the true posterior distribution, even though SGD was designed for optimization, not sampling.

18.7.5 Connection to Bayesian Neural Networks

Figure 18.4 shows the posterior distributions of weight and bias parameters estimated by SGLD. The blue histogram represents the distribution of the weight parameter w, while the cyan histogram represents the bias parameter b. The green dashed lines mark the true parameter values. This visualization demonstrates how SGLD captures uncertainty in the model parameters, which is essential for Bayesian inference.

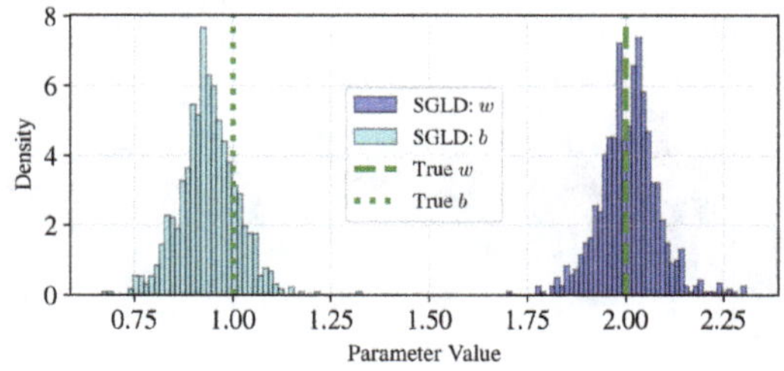

Figure 18.4: SGLD Posterior Distributions

In **Bayesian neural networks**, we want to sample from the posterior distribution over weights:

$$p(\theta \mid \mathcal{D}) \propto p(\mathcal{D} \mid \theta)p(\theta)$$

where $\mathcal{D}$ is the training data and $p(\theta)$ is the prior. The log-posterior is:

$$\log p(\theta \mid \mathcal{D}) = \log p(\mathcal{D} \mid \theta) + \log p(\theta) + \text{const}$$

The gradient is:

$$\nabla_\theta \log p(\theta \mid \mathcal{D}) = \nabla_\theta \log p(\mathcal{D} \mid \theta) + \nabla_\theta \log p(\theta)$$

The first term is the **data likelihood gradient** (computed via backpropagation), and the second term is the **prior gradient** (often a simple regularization term like weight decay).

SGLD can be used to sample from this posterior by using stochastic gradients computed on mini-batches. This provides a practical way to perform Bayesian inference in large neural networks, where exact MCMC methods are intractable.

18.7.6 Example: SGLD for Bayesian Linear Regression

Consider a simple Bayesian linear regression model with:

- Dataset: 100 points generated from $y = 2x + 1 + \mathcal{N}(0, 0.5^2)$
- Prior: $w \sim \mathcal{N}(0, 1)$, $b \sim \mathcal{N}(0, 1)$
- Mini-batch size: $B = 32$
- Learning rate schedule: $\eta_t = 0.01/(1 + t/1000)$

After 10,000 iterations:

- **Convergence**: SGLD converges to the same solution as full-batch methods
- **Computational Efficiency**: Training time reduced by a factor of ~7800 (from 1M to 128 samples per iteration)
- **Sample Quality**: The final parameter distribution captures uncertainty in the model parameters

The resulting parameter distributions center around $w \approx 2$ and $b \approx 1$, with uncertainty quantified by the posterior variance. This demonstrates how SGLD efficiently combines optimization speed with Bayesian uncertainty quantification.

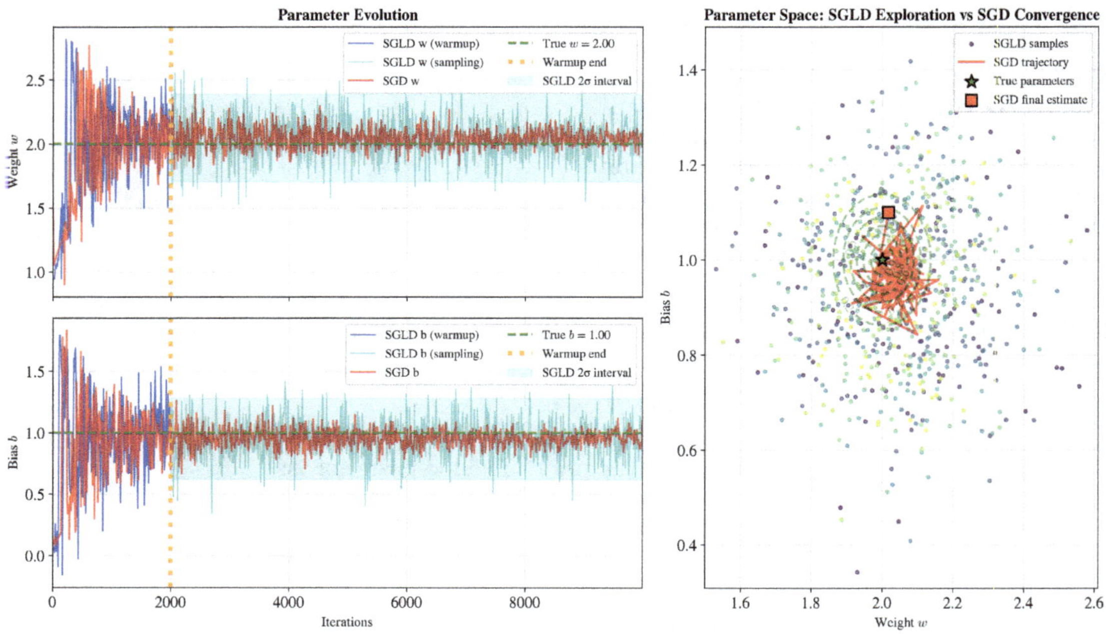

Figure 18.5: SGLD Trajectories with Warmup

Figure 18.5 shows the evolution of parameters over time for SGLD with warmup versus SGD. The left panel displays two time series plots: the top shows the weight parameter w and the bottom shows the bias parameter b. The right panel presents the parameter space trajectory. The blue lines represent the warmup phase, where noise strength gradually increases, and the cyan lines show the sampling phase with full noise. The yellow vertical line marks the end of the warmup phase. With $N = 200$ data points and observation noise $\sigma = 1.0$, the results show SGD converging to $w = 2.018, b = 1.100$, while SGLD explores around the posterior mean $w = 2.025 \pm 0.245, b = 0.953 \pm 0.233$. The green dashed lines indicate the true parameter values, and the right panel's contour lines represent the theoretical posterior distribution with standard deviation ≈ 0.071. This visualization clearly demonstrates how SGLD explores the parameter space (with fluctuations) while SGD converges smoothly to a single point estimate.

18.7.7 Bayesian vs Frequentist Approaches

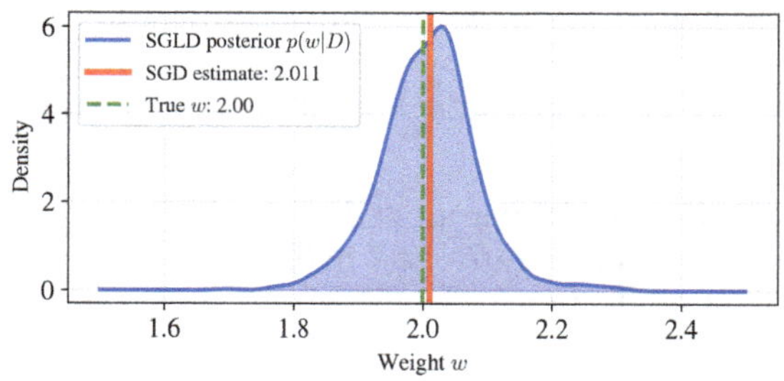

Figure 18.6: Bayesian vs Frequentist: SGLD vs SGD

Figure 18.6 contrasts the Bayesian approach (SGLD) with the frequentist approach (SGD). The blue area represents the posterior distribution $p(w|D)$ estimated by SGLD, which captures the uncertainty in the weight parameter. The red line shows the point estimate provided by SGD.

The green dashed line marks the true parameter value. This visualization highlights the key difference between the two approaches:

- **Bayesian (SGLD)**: Provides a full posterior distribution, quantifying uncertainty
- **Frequentist (SGD)**: Provides a single point estimate, ignoring uncertainty

While SGD is faster and simpler, SGLD's ability to quantify uncertainty makes it more suitable for tasks where uncertainty estimation is important, such as active learning, Bayesian optimization, and safe reinforcement learning.

18.7.8 Step Size Conditions - Robbins-Monro Algorithm

The step size conditions for SGLD are derived from the **Robbins-Monro algorithm**, a stochastic approximation method for solving equations of the form $\mathbb{E}[F(\mathbf{x}, \xi)] = 0$.

Algorithm 18 Robbins-Monro Algorithm

Input: Function $F(\mathbf{x}, \xi)$ where ξ is a random variable
Input: Initial guess $\mathbf{x}_0$
Input: Sequence of positive step sizes $\{\eta_k\}_{k=1}^{\infty}$
Output: Solution to $\mathbb{E}[F(\mathbf{x}, \xi)] = 0$
  $k \leftarrow 0$
  repeat
    Sample $\xi_k \sim p(\xi)$
    Compute $F(\mathbf{x}_k, \xi_k)$
    $\mathbf{x}_{k+1} \leftarrow \mathbf{x}_k - \eta_k F(\mathbf{x}_k, \xi_k)$
    $k \leftarrow k + 1$
  until Convergence

Convergence Conditions (Robbins-Monro Conditions):

$$\sum_{k=1}^{\infty} \eta_k = \infty, \quad \sum_{k=1}^{\infty} \eta_k^2 < \infty \tag{18.10}$$

These conditions ensure that:

- **Infinite exploration**: The first condition ensures the algorithm can explore the entire distribution (step sizes don't decay too fast)
- **Variance control**: The second condition ensures the noise variance decreases fast enough for convergence

Common step size schedules that satisfy these conditions include:

- $\eta_k = \frac{1}{k}$ (harmonic series)
- $\eta_k = \frac{1}{\sqrt{k}}$ (slower decay, more exploration)
- $\eta_k = \frac{a}{(b+k)^\gamma}$ with $\gamma \in (0.5, 1]$ (polynomial decay)

In practice, constant or slowly decreasing step sizes are often used for better mixing, even though they don't satisfy the theoretical conditions for asymptotic convergence.

18.8 Practical Considerations

18.8.1 Step Size Selection

Choosing the step size η is critical for Langevin dynamics. The optimal choice depends on several factors:

1. **Distribution geometry**: Distributions with high curvature (sharp peaks) require smaller step sizes to avoid overshooting. Distributions with low curvature (flat regions) can tolerate larger step sizes.
2. **Score function accuracy**: If the learned score function $\mathbf{s}_\theta(\mathbf{x})$ has errors, smaller step sizes can help reduce the impact of these errors.
3. **Computational budget**: Larger step sizes require fewer iterations but may lead to poor mixing. Smaller step sizes require more iterations but produce better samples.

Heuristic: A common rule of thumb is to choose η such that the acceptance rate (if using MALA) is around 50-70%, or such that the effective sample size (ESS) is maximized.

18.8.2 Mixing and Convergence Diagnostics

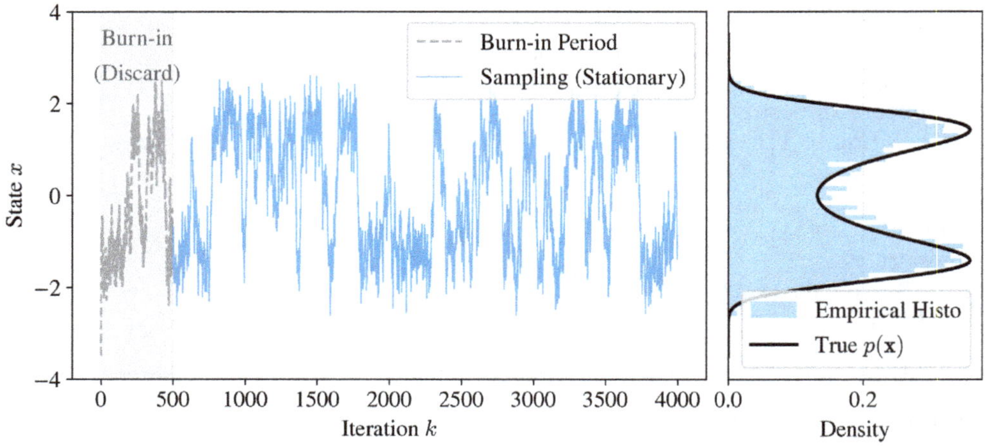

Figure 18.7: Trace Plot and Histogram for Langevin Dynamics

Several diagnostics can help assess whether Langevin dynamics has converged:

1. **Trace plots**: Visualize the trajectory of individual components over time. Good mixing shows random fluctuations around a stable mean.

Figure 18.7 illustrates the practical "health checks" for Langevin dynamics using a **1D double-well potential**. The left panel shows a **trace plot** that visualizes the path of the particle over time. The gray dashed line represents the **burn-in period** (first 500 iterations), where the particle moves from a low-probability initialization toward the high-probability modes. These samples are biased and usually discarded. The blue solid line shows the **sampling phase**, where the particle "hops" between the two modes (at approximately $x = \pm\sqrt{2}$). Good mixing appears as a "fuzzy caterpillar" that explores the full range repeatedly, indicating that the chain has reached its stationary distribution. The right panel shows a **histogram** of the post-burn-in samples (blue bars) overlaid with the theoretical probability density $p(x)$ (black line). The close match confirms that Langevin dynamics correctly samples from the target distribution, validating that the algorithm has converged and is producing representative samples.

2. **Autocorrelation**: Measure the correlation between samples at different lags. Low autocorrelation indicates good mixing.
3. **Effective sample size (ESS)**: Estimate how many independent samples the chain produces. Higher ESS indicates better mixing.
4. **Geweke diagnostic**: Compare means from early and late segments of the chain. Similar means suggest convergence.

5. **R-hat statistic**: Compare variance within and between multiple chains. Values near 1.0 indicate convergence.

18.8.3 Burn-in Period

The initial samples from Langevin dynamics are often not representative of the target distribution, especially if the initialization $\mathbf{x}_0$ is far from high-probability regions. It is common practice to discard the first B samples (the "burn-in" period) and only use samples after iteration B.

The burn-in period should be long enough for the chain to reach the typical set of the distribution, but not so long that it wastes computational resources. In practice, burn-in periods of 10-50% of total iterations are common.

18.8.4 Thinning

To reduce autocorrelation and storage requirements, we can "thin" the chain by keeping only every T-th sample. This is called **thinning** or **subsampling**. However, thinning reduces the effective sample size and is generally not recommended unless storage is a critical constraint.

18.9 Score-Based Sampling Algorithm

Putting Chapters 17–18 together, we can now construct a complete score-based sampling algorithm.

Algorithm 19 Score-Based Sampling

Input: Score function $\mathbf{s}_\theta(\mathbf{x})$, step size η, number of steps K, burn-in B
Output: Sample $\mathbf{x}_K$ from the target distribution

```
Initialize x_0 ~ N(0, I)                          ▷ Random initialization
for k = 0, ..., K − 1 do
    Sample ε_k ~ N(0, I)                          ▷ Independent noise
    x_{k+1} ← x_k + η s_θ(x_k) + √(2η) ε_k        ▷ Langevin update
end for
return x_K          ▷ Return final sample (or samples after burn-in)
```

18.9.1 Complete Pipeline

The full pipeline for score-based generative modeling consists of:

1. **Training**: Learn the score function $\mathbf{s}_\theta(\mathbf{x})$ from data using denoising score matching (Chapter~17)
2. **Sampling**: Use Langevin dynamics to generate samples from the learned distribution
3. **Evaluation**: Assess sample quality using metrics like FID, IS, or human evaluation

This pipeline has achieved state-of-the-art results in image generation, demonstrating that score-based models can produce high-quality samples from complex high-dimensional distributions.

18.9.2 Example: 2D Gaussian Mixture

Consider sampling from a 2D mixture of two Gaussians:

$$p(\mathbf{x}) = 0.5 \cdot \mathcal{N}(\mathbf{x}; \mu_1, \Sigma_1) + 0.5 \cdot \mathcal{N}(\mathbf{x}; \mu_2, \Sigma_2)$$

where $\mu_1 = [2, 0]^\top$, $\mu_2 = [-2, 0]^\top$, and both covariance matrices are $\Sigma_1 = \Sigma_2 = 0.5I$.

The score function is:

$$\mathbf{s}(\mathbf{x}) = \frac{w_1(\mathbf{x})\mathbf{s}_1(\mathbf{x}) + w_2(\mathbf{x})\mathbf{s}_2(\mathbf{x})}{w_1(\mathbf{x}) + w_2(\mathbf{x})}$$

where $w_i(\mathbf{x}) = \mathcal{N}(\mathbf{x}; \mu_i, \Sigma_i)$ and $\mathbf{s}_i(\mathbf{x}) = -\Sigma_i^{-1}(\mathbf{x} - \mu_i)$.

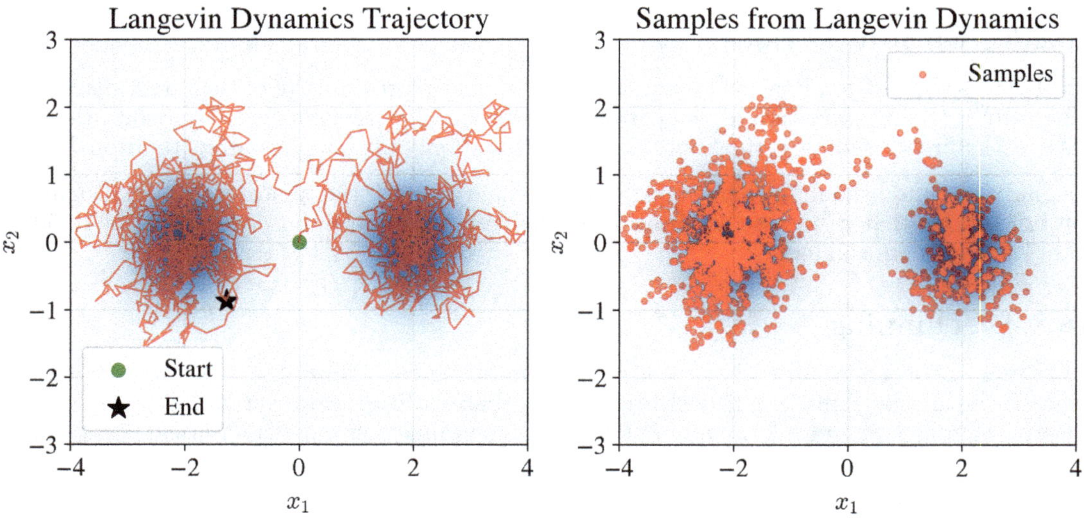

Figure 18.8: Langevin Dynamics for 2D Gaussian Mixture

Figure 18.8 shows the results of running Langevin dynamics with step size $\eta = 0.01$ for 2000 iterations, starting from the origin. The left panel displays the trajectory of the process (red line), beginning at the origin (green dot) and exploring both modes of the distribution. The background contour plot represents the true probability density of the Gaussian mixture. The right panel shows the samples generated after a burn-in period of 500 iterations, which form a distribution that closely matches the true Gaussian mixture.

This clearly demonstrates how the noise term in Langevin dynamics enables the algorithm to explore multiple modes of a distribution, rather than getting trapped in a single mode like deterministic gradient ascent would.

18.10 Why Noise Is Not Optional

18.10.1 A Fundamental Principle

There is a deep mathematical principle that distinguishes deterministic from stochastic dynamics:

- **Deterministic dynamics preserves dimension**: A deterministic flow $\frac{d\mathbf{x}}{dt} = \mathbf{f}(\mathbf{x})$ maps an initial point to a unique trajectory. The dimension of the trajectory (typically 1 for a curve) is preserved. Deterministic systems cannot explore high-dimensional distributions.
- **Stochastic dynamics explores distributions**: A stochastic process $d\mathbf{X}_t = \mathbf{f}(\mathbf{X}_t)dt + \mathbf{G}(\mathbf{X}_t)d\mathbf{W}_t$ generates random trajectories that collectively explore the full distribution. The noise term enables exploration of the entire probability space.

18.10.2 What Happens Without Noise?

If we remove the noise term from Langevin dynamics, we get deterministic gradient ascent:

$$\mathbf{x}_{k+1} = \mathbf{x}_k + \eta \mathbf{s}(\mathbf{x}_k)$$

This deterministic process has several problems:

- **Trajectories collapse to attractors**: All trajectories starting near a mode converge to that mode. The algorithm cannot escape local regions.
- **Entropy decreases**: The entropy of the distribution of samples decreases over time as they concentrate around modes. This is the opposite of what we want for sampling.
- **Sampling fails**: The empirical distribution of samples does not match the target distribution. Instead, it concentrates on a small set of points (the modes).
- **No diversity**: All samples from the same mode are nearly identical, losing the variability that characterizes the true distribution.

18.10.3 What Happens With Noise?

When we add the noise term $\sqrt{2\eta}\,\epsilon_k$:

- **Entropy is injected**: The noise continuously injects entropy into the system, preventing collapse to a single point.
- **Exploration is guaranteed**: The noise ensures that the process can reach any region of the distribution's support, given enough time. This is the **ergodicity** property.
- **Equilibrium emerges**: The competition between the deterministic drift (toward modes) and stochastic noise (exploration) leads to an equilibrium where the stationary distribution matches $p(\mathbf{x})$.
- **Correct sampling**: The long-term behavior of the stochastic process produces samples from the target distribution.

18.10.4 The Mathematical Necessity

The noise is not just helpful—it is **mathematically necessary** for correct sampling. The fluctuation-dissipation theorem from statistical physics guarantees that, for a process to have $p(\mathbf{x})$ as its stationary distribution, the noise variance must be exactly $\sqrt{2\eta}$ (in the appropriate units). This specific scaling ensures that the detailed balance condition is satisfied.

Noise is **what makes probability real**—it is the mechanism that transforms a deterministic optimization algorithm into a stochastic sampling algorithm. Without noise, we have optimization. With noise, we have sampling.

The connection between noise and sampling is fundamental to statistical mechanics and stochastic processes. The fluctuation-dissipation theorem establishes a precise relationship between the deterministic drift and stochastic noise that ensures correct sampling. This principle appears throughout physics, from Brownian motion to quantum mechanics.

> The same principle provides the mathematical foundation for MCMC methods in machine learning.

18.11 Discrete-Time vs Continuous-Time View

Langevin dynamics is a **discretization** of a continuous process:

$$d\mathbf{X}_t = \mathbf{s}(\mathbf{X}_t)\,dt + \sqrt{2}\,d\mathbf{W}_t \tag{18.11}$$

This is a **stochastic differential equation (SDE)**.

Chapters 19–21 will formalize this continuous-time limit (see Figure 21.3 in Chapter 21 for a visual demonstration of particle cloud evolution under continuous dynamics).

18.12 Conceptual Summary

The following table summarizes the core mathematical components of Langevin dynamics and their structural roles in transforming score geometry into exact samples from an unnormalized distribution:

Concept	Role
Score function	Geometry of probability
Langevin dynamics	Turns geometry into samples
Noise	Preserves entropy
Step size	Controls bias vs variance
Stationarity	Guarantees correct distribution

In essence, Langevin dynamics bridges gradient-based optimization with Markov Chain Monte Carlo sampling. While deterministic gradient ascent concentrates probability mass at local modes, the injected Gaussian noise maintains entropy, enabling the sampler to explore multi-modal landscapes and converge to the true target distribution as the step size approaches zero.

18.13 Mindmap

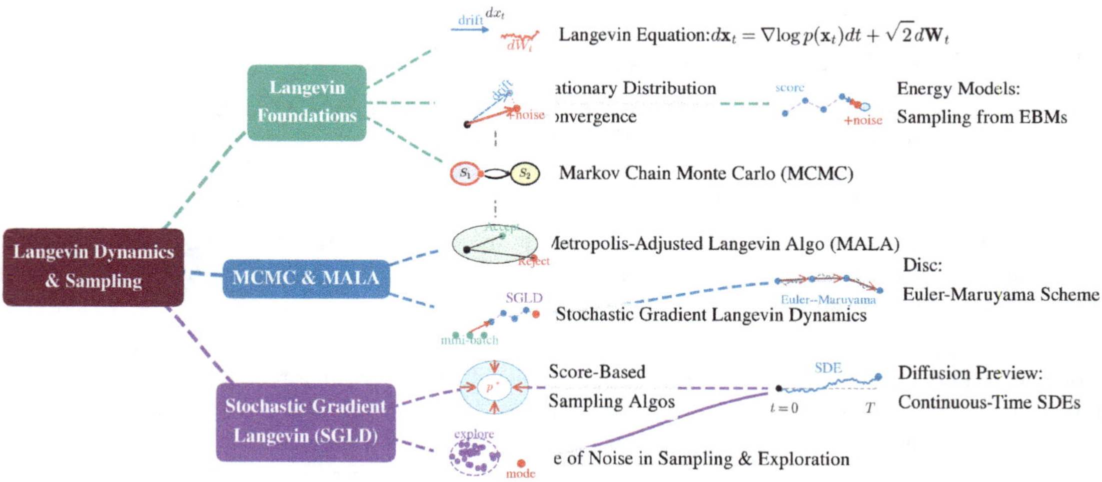

Figure 18.9: Mind map of Langevin dynamics and sampling topics

18.14 Exercises

18.14.1 ULA (Discrete Langevin) Basics

1. **From score to sampler**: Starting from the update

$$\mathbf{x}_{k+1} = \mathbf{x}_k + \eta\,\mathbf{s}(\mathbf{x}_k) + \sqrt{2\eta}\,\epsilon_k, \qquad \epsilon_k \sim \mathcal{N}(0, I),$$

explain (informally) why the drift term moves samples toward high-density regions, while the noise term prevents mode collapse.

2. **1D Gaussian example and discretization bias**: Let the target be $p(x) = \mathcal{N}(0,1)$.
    1. Compute the score $s(x) = \frac{d}{dx}\log p(x)$.
    2. Write the ULA update explicitly as an AR(1) process $x_{k+1} = ax_k + b\epsilon_k$.
    3. Assuming $|a| < 1$, compute the stationary variance $\mathrm{Var}(x_\infty)$ and show it is not exactly 1 for finite η.
3. **Stability condition**: In the Gaussian example above, determine for which step sizes η the Markov chain is stable (i.e., converges in distribution).

18.14.2 Connection to Energy-Based Models

1. **Sign check**: If $p(\mathbf{x}) \propto \exp(-E(\mathbf{x}))$, show that $\mathbf{s}(\mathbf{x}) = -\nabla_{\mathbf{x}}E(\mathbf{x})$. Substitute into ULA and write the update in terms of ∇E.
2. **Quadratic energy**: Let $E(\mathbf{x}) = \frac{1}{2}\mathbf{x}^\top A\mathbf{x}$ with $A \succ 0$.
    1. Compute $\nabla E(\mathbf{x})$.
    2. Write the ULA update.
    3. What Gaussian distribution is the stationary target $p(\mathbf{x})$ proportional to?

18.14.3 MALA (Metropolis-Adjusted Langevin Algorithm)

1. **Proposal density**: Show that the ULA proposal can be written as

$$q(\mathbf{x}' \mid \mathbf{x}) = \mathcal{N}(\mathbf{x} + \eta\,\mathbf{s}(\mathbf{x}),\, 2\eta I)\,.$$

2. **Acceptance ratio (log form)**: Write $\log\alpha(\mathbf{x},\mathbf{x}')$ (before applying the $\min(1,\cdot)$) in terms of $\log p(\cdot)$ and the Gaussian log-density of q.
3. **Why MALA is "exact"**: Explain how the Metropolis–Hastings correction removes discretization bias (what property does it enforce?).

18.14.4 SGLD (Stochastic Gradient Langevin Dynamics)

1. **Step-size conditions**: The classic diminishing-step condition is

$$\sum_{k=1}^{\infty}\eta_k = \infty, \qquad \sum_{k=1}^{\infty}\eta_k^2 < \infty.$$

Verify that $\eta_k = \frac{1}{k}$ satisfies it, and that $\eta_k = \frac{1}{\sqrt{k}}$ fails the second condition.

2. **Why constant step sizes are used in practice**: Give two practical reasons diffusion-style samplers often use fixed η for a finite number of steps, even though theory often assumes decreasing η_k.

18.14.5 Continuous-Time View

1. **Langevin SDE**: Starting from the overdamped Langevin SDE

$$d\mathbf{X}_t = \mathbf{s}(\mathbf{X}_t)\,dt + \sqrt{2}\,d\mathbf{W}_t,$$

explain the relationship between this SDE and the discrete ULA update (what is being discretized?).

Chapter 19

Stochastic Differential Equations

Itô calculus, SDEs, and continuous-time stochastic processes

> *"The theory of probability is the general theory of random phenomena."*
>
> — Andrey Kolmogorov

▷ `Brownian motion` $\{\mathbf{W}_t\}_{t\geq 0}$ with $\mathbf{W}_0 = \mathbf{0}$, indep increments, and $\mathbf{W}_t - \mathbf{W}_s \sim \mathcal{N}(0,(t-s)I)$ for $t > s$

▷ `Itô SDE (forward)` $d\mathbf{X}_t = \mathbf{f}(\mathbf{X}_t, t)\,dt + \mathbf{G}(\mathbf{X}_t, t)\,d\mathbf{W}_t$

▷ `Langevin SDE` $d\mathbf{X}_t = \nabla_{\mathbf{x}} \log p(\mathbf{X}_t)\,dt + \sqrt{2}\,d\mathbf{W}_t$ (stationary density p under mild conditions)

▷ `Itô increment rule` $(dW_t)^2 = dt$, $dW_t\,dt = 0$ (heuristic leading to the extra second-order term)

▷ `Itô discretization (interpretation)` $\mathbf{X}_{t+\Delta t} \approx \mathbf{X}_t + \mathbf{f}(\mathbf{X}_t, t)\Delta t + \mathbf{G}(\mathbf{X}_t, t)(\mathbf{W}_{t+\Delta t} - \mathbf{W}_t)$

▷ `Itô's lemma` for $f(\mathbf{x}, t)$ twice differentiable and $d\mathbf{X}_t = \mathbf{f}\,dt + \mathbf{G}\,d\mathbf{W}_t$,$df(\mathbf{X}_t, t) = \frac{\partial f}{\partial t}\,dt + (\nabla f)^\top d\mathbf{X}_t + \frac{1}{2}\mathrm{Tr}(\mathbf{G}\mathbf{G}^\top \nabla^2 f)\,dt$

▷ `Euler-Maruyama` $\mathbf{X}_{k+1} = \mathbf{X}_k + \mathbf{f}(\mathbf{X}_k, t_k)\Delta t + \mathbf{G}(\mathbf{X}_k, t_k)\sqrt{\Delta t}\,\epsilon_k$, $\epsilon_k \sim \mathcal{N}(0, I)$

▷ `Reverse-time SDE (state-independent diffusion)` if $\mathbf{G} = \mathbf{G}(t)$ and $\mathbf{a}(t) = \mathbf{G}(t)\mathbf{G}(t)^\top$, then (running backward in time) $d\mathbf{X}_t = [\mathbf{f}(\mathbf{X}_t, t) - \mathbf{a}(t)\nabla_{\mathbf{x}} \log p_t(\mathbf{X}_t)]\,dt + \mathbf{G}(t)\,d\bar{\mathbf{W}}_t$

19.1 Motivation: From Algorithms to Processes

In Chapter~18, we studied **Langevin dynamics** as an iterative algorithm:

$$\mathbf{x}_{k+1} = \mathbf{x}_k + \eta \mathbf{s}(\mathbf{x}_k) + \sqrt{2\eta}\epsilon_k, \quad \epsilon_k \sim \mathcal{N}(\mathbf{0}, I) \tag{19.1}$$

This discrete update rule raises a fundamental question: **What happens as the step size $\eta \to 0$?**

The answer is a **continuous-time stochastic process**, described by a **stochastic differential equation (SDE)**. This continuous limit provides:

1. **Theoretical clarity**: The continuous formulation reveals the fundamental mathematical structure underlying discrete algorithms
2. **Unified framework**: Many different discrete algorithms converge to the same continuous process
3. **Analytical tools**: Continuous-time analysis enables the use of powerful mathematical techniques from stochastic calculus
4. **Practical algorithms**: Understanding the continuous limit helps design better discrete approximations

19.1.1 The Continuous-Time Perspective

Viewing discrete algorithms as discretizations of continuous processes is a powerful paradigm in modern machine learning. This perspective:

Each discrete algorithm has a corresponding continuous process that captures its essential behavior in the limit of infinitesimal step sizes. Understanding this connection provides deep insights into algorithm behavior, convergence properties, and the design of new methods.

19.2 Brownian Motion (Wiener Process)

19.2.1 Definition: Brownian Motion

A **Brownian motion** (also called a **Wiener process**) $\{\mathbf{W}_t\}_{t\geq 0}$ in $\mathbb{R}^d$ is a stochastic process satisfying:

1. **Initial condition**: $\mathbf{W}_0 = \mathbf{0}$ almost surely
2. **Independent increments**: For $0 \leq t_0 < t_1 < \cdots < t_n$, the increments $\mathbf{W}_{t_1} - \mathbf{W}_{t_0}, \mathbf{W}_{t_2} - \mathbf{W}_{t_1}, \ldots, \mathbf{W}_{t_n} - \mathbf{W}_{t_{n-1}}$ are independent
3. **Gaussian increments**: For $t > s \geq 0$, $\mathbf{W}_t - \mathbf{W}_s \sim \mathcal{N}(\mathbf{0}, (t-s)I)$
4. **Continuous paths**: The sample paths $t \mapsto \mathbf{W}_t(\omega)$ are almost surely continuous functions

These properties uniquely characterize Brownian motion up to a scaling factor.

19.2.2 Historical Context

Brownian motion is named after Robert Brown, who observed in 1827 that pollen particles in water exhibit random, jittery motion. The mathematical model was developed by Norbert Wiener in the 1920s, and it has since become fundamental to probability theory, physics, and finance.

In physics, Brownian motion models the motion of particles subject to random collisions. In finance, it models stock prices and other random processes. In machine learning, it provides the mathematical foundation for stochastic optimization and generative modeling.

19.2.3 Key Properties

Scaling property: For any $c > 0$, the process $\{\frac{1}{\sqrt{c}}\mathbf{W}_{ct}\}_{t\geq 0}$ is also a Brownian motion. This means Brownian motion is **self-similar**—it looks the same at different time scales.

Markov property: Brownian motion is a **Markov process**—the future depends only on the present, not the past. For any $s < t$:

$$\mathbb{E}[\mathbf{W}_t \mid \{\mathbf{W}_u : u \leq s\}] = \mathbb{E}[\mathbf{W}_t \mid \mathbf{W}_s] = \mathbf{W}_s$$

Martingale property: Brownian motion is a **martingale**—its expected future value equals its current value:

$$\mathbb{E}[\mathbf{W}_t \mid \mathbf{W}_s] = \mathbf{W}_s, \quad \text{for } s < t$$

Quadratic variation: The quadratic variation of Brownian motion is deterministic:

$$[\mathbf{W}, \mathbf{W}]_t = tI$$

This property is crucial for Itô calculus, as it implies $(dW_t)^2 = dt$.

19.2.4 Path Properties

Brownian motion has remarkable and counterintuitive path properties:

- **Nowhere differentiable**: Almost every sample path is nowhere differentiable. This means Brownian motion is extremely "rough" and cannot be analyzed using ordinary calculus.

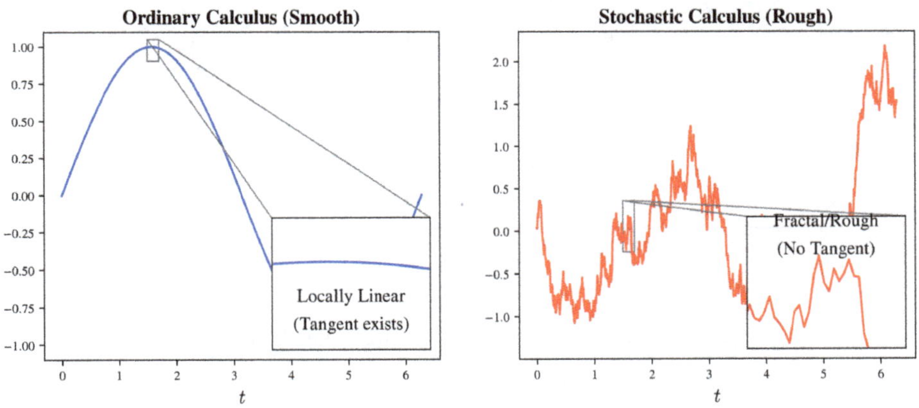

Figure 19.1: Smooth vs. Rough Functions

Figure 19.1 illustrates the fundamental difference between smooth functions (ODEs) and rough functions (SDEs). The left panel shows a smooth, differentiable function (e.g., $\sin(t)$) with a zoomed inset revealing that at any point, the function becomes locally linear—a tangent line exists. This is the foundation of ordinary calculus. The right panel shows a Brownian path W_t, which is a sample path of an SDE. No matter how much we zoom in (as shown in the inset), the path remains jagged and fractal-like—no tangent line exists. This "infinite roughness" is why $(dW_t)^2$ is non-zero and contributes at order dt, breaking the rules of ordinary calculus and necessitating stochastic calculus. This visualization makes concrete why Brownian motion cannot be analyzed using ordinary derivatives and integrals.

- **Infinite variation**: Over any interval, Brownian motion has infinite total variation, meaning it oscillates infinitely often. This is why ordinary Riemann-Stieltjes integrals cannot be used.
- **Hölder continuity**: Sample paths are Hölder continuous with exponent $\alpha < 1/2$, but not with $\alpha \geq 1/2$. This means paths are continuous but very rough.
- **Law of the iterated logarithm**:

$$\limsup_{t \to 0^+} \frac{|\mathbf{W}_t|}{\sqrt{2t \log\log(1/t)}} = 1 \quad \text{a.s.}$$

This describes the extreme behavior of Brownian motion near $t = 0$.

19.2.5 Intuition

Brownian motion can be understood as:

- **The limit of random walks**: A symmetric random walk with step size Δx and time step Δt converges to Brownian motion as $\Delta x, \Delta t \to 0$ with $(\Delta x)^2/\Delta t$ held constant.

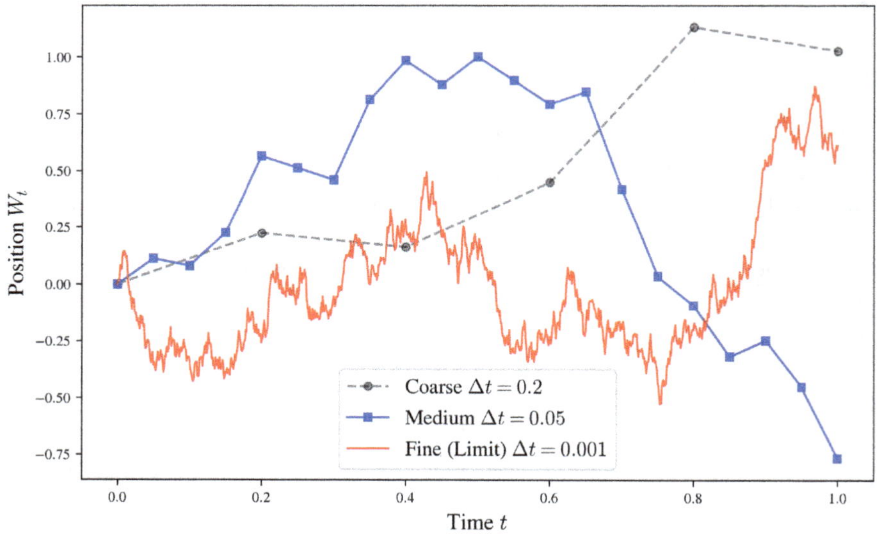

Figure 19.2: Convergence to Brownian Motion

Figure 19.2 demonstrates how discrete random walks converge to continuous Brownian motion as the step size decreases. Three trajectories are shown, all starting from the same random seed but with different time discretizations: coarse ($\Delta t = 0.2$), medium ($\Delta t = 0.05$), and fine ($\Delta t = 0.001$). As $\Delta t \to 0$, the discrete random walk becomes increasingly continuous and exhibits the characteristic "roughness" of Brownian motion. Notice the self-similarity: the rough structure persists at all scales, which is a defining property of Brownian motion. This visualization makes concrete the limit process: as step size goes to zero, the discrete random walk converges to the continuous Wiener process, providing the mathematical foundation for SDEs.

- **Pure noise**: Brownian motion has no drift—it is purely random. It represents the fundamental source of randomness in stochastic processes.
- **Mathematical idealization**: While real physical processes have some smoothness, Brownian motion is the mathematical idealization of pure randomness, making it a fundamental building block for more complex processes.

It is the **source of randomness** in all SDEs—every stochastic differential equation includes a term proportional to $d\mathbf{W}_t$, which represents the infinitesimal increment of Brownian motion.

19.3 Stochastic Differential Equations

19.3.1 Definition: Stochastic Differential Equation

A **stochastic differential equation (SDE)** has the form

$$d\mathbf{X}_t = \mathbf{f}(\mathbf{X}_t, t)\, dt + \mathbf{G}(\mathbf{X}_t, t)\, d\mathbf{W}_t \tag{19.2}$$

where:

- $\mathbf{X}_t \in \mathbb{R}^d$ is the state vector at time t
- $\mathbf{f} : \mathbb{R}^d \times \mathbb{R} \to \mathbb{R}^d$ is the **drift function** (deterministic component)
- $\mathbf{G} : \mathbb{R}^d \times \mathbb{R} \to \mathbb{R}^{d\times m}$ is the **diffusion matrix** (stochastic component)
- $\mathbf{W}_t \in \mathbb{R}^m$ is an m-dimensional Brownian motion
- $d\mathbf{W}_t$ is the infinitesimal increment of Brownian motion

19.3.2 Interpretation

The SDE describes how the state $\mathbf{X}_t$ evolves over time:

- **Drift term $\mathbf{f}(\mathbf{X}_t, t)dt$**: This is the deterministic component that pushes the state in a specific direction. It represents the "mean" or "expected" change.
- **Diffusion term $\mathbf{G}(\mathbf{X}_t, t)d\mathbf{W}_t$**: This is the stochastic component that injects randomness. The matrix $\mathbf{G}$ controls how the Brownian noise affects different components of the state.

The SDE is a **differential equation** because it describes the infinitesimal change $d\mathbf{X}_t$ in terms of infinitesimal time dt and noise $d\mathbf{W}_t$. However, because $d\mathbf{W}_t$ is not a true differential (Brownian motion is not differentiable), the SDE must be interpreted carefully using stochastic calculus.

19.3.3 Integral Form

The SDE is more precisely written in integral form:

$$\mathbf{X}_t = \mathbf{X}_0 + \int_0^t \mathbf{f}(\mathbf{X}_s, s)ds + \int_0^t \mathbf{G}(\mathbf{X}_s, s)d\mathbf{W}_s$$

The first integral is an ordinary Riemann integral. The second integral is a **stochastic integral** (Itô integral), which requires special interpretation because $d\mathbf{W}_s$ is not a true differential.

19.3.4 Existence and Uniqueness

Under appropriate conditions on $\mathbf{f}$ and $\mathbf{G}$ (Lipschitz continuity and linear growth), the SDE has a **unique strong solution**. This means:

1. **Existence**: There exists a process $\mathbf{X}_t$ satisfying the SDE
2. **Uniqueness**: Any two solutions with the same initial condition are almost surely identical
3. **Strong solution**: The solution is adapted to the filtration generated by the Brownian motion

These conditions ensure that the SDE is well-posed and can be solved numerically.

19.3.5 Strong vs Weak Solutions

There are two notions of solution for SDEs:

- **Strong solution**: The solution $\mathbf{X}_t$ is a function of the initial condition and the Brownian motion $\mathbf{W}_t$. The solution is adapted (doesn't use future information).
- **Weak solution**: The solution exists on some probability space with some Brownian motion, but the relationship between $\mathbf{X}_t$ and $\mathbf{W}_t$ may not be explicit.

For diffusion models, we typically work with strong solutions, which are more natural for generative modeling where we explicitly construct the process from noise.

19.3.6 Lipschitz and Linear Growth Conditions

The standard existence and uniqueness theorem (Itô's theorem) requires:

1. **Lipschitz condition**: For all $\mathbf{x}, \mathbf{y} \in \mathbb{R}^d$ and $t \in [0, T]$:

$$\|\mathbf{f}(\mathbf{x}, t) - \mathbf{f}(\mathbf{y}, t)\| + \|\mathbf{G}(\mathbf{x}, t) - \mathbf{G}(\mathbf{y}, t)\| \leq L\|\mathbf{x} - \mathbf{y}\|$$

   for some constant $L > 0$.

2. **Linear growth condition**: For all $\mathbf{x} \in \mathbb{R}^d$ and $t \in [0, T]$:

$$\|\mathbf{f}(\mathbf{x}, t)\|^2 + \|\mathbf{G}(\mathbf{x}, t)\|^2 \leq K(1 + \|\mathbf{x}\|^2)$$

   for some constant $K > 0$.

These conditions ensure that the SDE doesn't explode (solution remains finite) and that solutions are unique. In practice, neural network-based score functions $\mathbf{s}_\theta(\mathbf{x}, t)$ may not satisfy these conditions globally, but they typically do on compact sets where the process evolves.

19.3.7 Example: Langevin SDE

The continuous-time limit of Langevin dynamics (Chapter 18) is:

$$d\mathbf{X}_t = \nabla_{\mathbf{x}} \log p(\mathbf{X}_t)\, dt + \sqrt{2}\, d\mathbf{W}_t \tag{19.3}$$

Here:

- **Drift**: $\mathbf{f}(\mathbf{X}_t, t) = \nabla_{\mathbf{x}} \log p(\mathbf{X}_t)$ is the score function
- **Diffusion**: $\mathbf{G}(\mathbf{X}_t, t) = \sqrt{2}I$ is constant (state-independent)

This SDE has $p(\mathbf{x})$ as its **stationary distribution**, meaning that as $t \to \infty$, the distribution of $\mathbf{X}_t$ converges to $p(\mathbf{x})$. This is the continuous-time version of the result from Chapter~18.

19.3.8 Example: Ornstein-Uhlenbeck Process

The **Ornstein-Uhlenbeck (OU) process** is a fundamental SDE:

$$d\mathbf{X}_t = -\theta(\mathbf{X}_t - \mu)dt + \sigma d\mathbf{W}_t$$

where $\theta > 0$ is the mean reversion rate, μ is the mean, and $\sigma > 0$ is the volatility.

Interpretation: The drift term $-\theta(\mathbf{X}_t - \mu)$ pulls the process toward μ (mean reversion), while the diffusion term $\sigma d\mathbf{W}_t$ adds random fluctuations.

Stationary distribution: The OU process has a Gaussian stationary distribution:

$$\mathbf{X}_t \xrightarrow{t\to\infty} \mathcal{N}\left(\mu, \frac{\sigma^2}{2\theta} I\right)$$

The OU process is widely used in finance (modeling interest rates, volatility) and physics (modeling particle motion in a potential well).

19.3.9 Example: Geometric Brownian Motion

Geometric Brownian motion is used to model stock prices:

$$dX_t = \mu X_t dt + \sigma X_t dW_t$$

where μ is the drift (expected return) and σ is the volatility.

Key feature: The drift and diffusion both scale with X_t, making the process **multiplicative** rather than additive. This ensures $X_t > 0$ (stock prices are positive).

Solution: Using Itô's lemma, the solution is:

$$X_t = X_0 \exp\left((\mu - \frac{\sigma^2}{2})t + \sigma W_t\right)$$

This is the famous **Black-Scholes model** for option pricing in finance.

Figure 19.3 visualizes the Itô correction effect for geometric Brownian motion with drift $\mu = 0.3$ and volatility $\sigma = 0.5$. The gray cloud shows 50 sample paths of the SDE. The red dashed line shows what ordinary calculus would predict: $e^{\mu t} = e^{0.3t}$, which overestimates the growth. The blue solid line shows the correct Itô trend: $e^{(\mu-\sigma^2/2)t} = e^{(0.3-0.125)t} = e^{0.175t}$, which matches the center of mass of the sample paths. The difference between these two lines is the **Itô correction term** $-\frac{\sigma^2}{2}$, which represents the "tax" paid to volatility. This correction arises from the quadratic variation of Brownian motion: $(dW_t)^2 = dt$, which contributes a second-order term that cannot be ignored. The visualization makes concrete why stochastic calculus is necessary: noise drags the drift down, and this effect is captured by Itô's lemma.

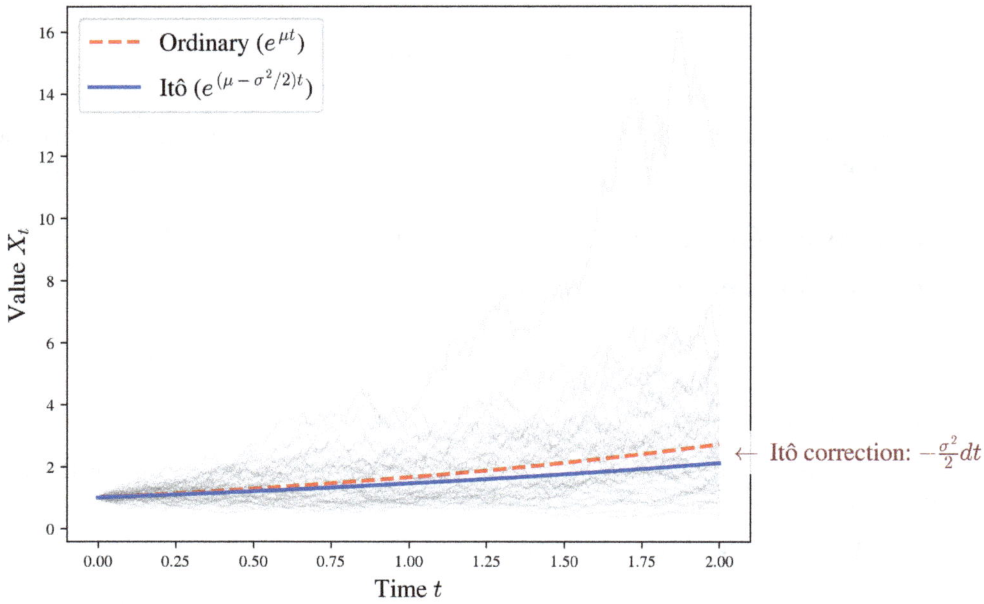

Figure 19.3: Geometric Brownian Motion and the Itô Correction

19.4 Itô vs Stratonovich

19.4.1 Why Ordinary Calculus Fails

The fundamental issue with SDEs is that Brownian motion has **infinite quadratic variation**. In ordinary calculus, we have $(dx)^2 = 0$ for any differential dx. However, for Brownian motion:

$$(dW_t)^2 = dt, \quad dW_t \cdot dt = 0, \quad (dt)^2 = 0 \tag{19.4}$$

This means that terms of order $(dW_t)^2$ cannot be ignored—they contribute at order dt. This breaks the rules of classical calculus and requires a new calculus: **stochastic calculus**.

19.4.2 The Chain Rule Problem

In ordinary calculus, the chain rule for $f(x(t))$ is:

$$df = f'(x)dx$$

For an SDE $dX_t = f(X_t, t)dt + G(X_t, t)dW_t$, if we naively apply the chain rule to $f(X_t)$, we get:

$$df(X_t) \stackrel{?}{=} f'(X_t)dX_t = f'(X_t)[f(X_t, t)dt + G(X_t, t)dW_t]$$

However, this is **incorrect** because we ignored the $(dW_t)^2 = dt$ term. The correct formula (Itô's lemma) includes an extra second-order term:

$$df(X_t) = f'(X_t)dX_t + \frac{1}{2}f''(X_t)(dX_t)^2$$

where $(dX_t)^2 = G(X_t, t)^2 dt$ (using $(dW_t)^2 = dt$ and $dW_t \cdot dt = 0$).

This extra term is the fundamental difference between ordinary and stochastic calculus.

19.4.3 Itô SDE

An **Itô SDE** is interpreted using the **Itô integral**, where the integrand is evaluated at the **left endpoint** of each interval:

$$\mathbf{X}_{t+\Delta t} = \mathbf{X}_t + \mathbf{f}(\mathbf{X}_t, t)\Delta t + \mathbf{G}(\mathbf{X}_t, t)\left(\mathbf{W}_{t+\Delta t} - \mathbf{W}_t\right) \tag{19.5}$$

Key property: The drift and diffusion are evaluated at the **current time** t, not at a future time. This makes Itô integrals **non-anticipating** (adapted to the filtration), which is crucial for modeling physical and financial processes where future information is not available.

19.4.4 Stratonovich SDE

A **Stratonovich SDE** uses the **Stratonovich integral**, where the integrand is evaluated at the **midpoint**:

$$\mathbf{X}_{t+\Delta t} \approx \mathbf{X}_t + \mathbf{f}(\mathbf{X}_t, t)\Delta t + \mathbf{G}\left(\frac{\mathbf{X}_t + \mathbf{X}_{t+\Delta t}}{2}, t\right)\left(\mathbf{W}_{t+\Delta t} - \mathbf{W}_t\right)$$

Key property: Stratonovich integrals satisfy the **ordinary chain rule**, making them more intuitive for some applications. However, they are not adapted (they "look into the future"), which makes them less suitable for modeling.

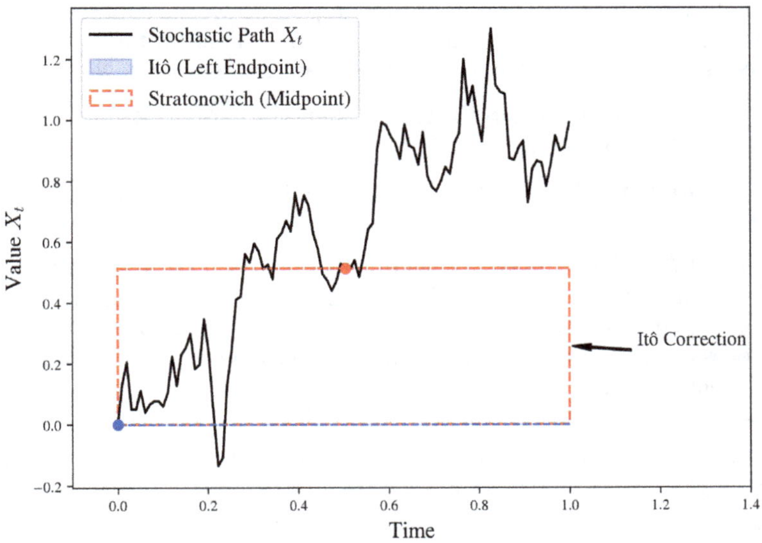

Figure 19.4: Itô vs. Stratonovich Integration

Figure 19.4 illustrates the difference between Itô and Stratonovich integration for a single time step. The black curve shows a highly volatile stochastic path X_t over the interval $[t, t + \Delta t]$. The blue rectangle represents the **Itô integral**, which evaluates the integrand at the **left endpoint** X_t (shown as a blue dot). The red dashed rectangle represents the **Stratonovich integral**, which evaluates the integrand at the **midpoint** $\frac{X_t + X_{t+\Delta t}}{2}$ (shown as a red dot). Because the path is so volatile, the choice of evaluation point significantly changes the area under the curve. This difference is exactly the **Itô correction term** that appears when converting between Itô and

Stratonovich SDEs. The visualization makes concrete why stochastic integrals require careful definition: unlike smooth functions where the choice of evaluation point doesn't matter, the infinite variation of Brownian motion makes this choice crucial.

19.4.5 Conversion Between Itô and Stratonovich

An Itô SDE:

$$d\mathbf{X}_t = \mathbf{f}(\mathbf{X}_t, t)dt + \mathbf{G}(\mathbf{X}_t, t) \circ d\mathbf{W}_t$$

(where $\circ$ denotes Stratonovich) is equivalent to the Itô SDE:

$$d\mathbf{X}_t = \left[\mathbf{f}(\mathbf{X}_t, t) + \frac{1}{2}\sum_{j=1}^{m} \frac{\partial \mathbf{G}}{\partial x_j}(\mathbf{X}_t, t)\mathbf{G}_{:,j}(\mathbf{X}_t, t)\right] dt + \mathbf{G}(\mathbf{X}_t, t)d\mathbf{W}_t$$

The extra drift term $\frac{1}{2}\sum_{j=1}^{m} \frac{\partial \mathbf{G}}{\partial x_j}\mathbf{G}_{:,j}$ is the **Wong-Zakai correction term.**

19.4.6 Why Itô Calculus for Diffusion Models?

All modern diffusion models use **Itô calculus** because:

1. **Non-anticipating**: Itô integrals are adapted, meaning they don't use future information—essential for generative modeling where we generate samples forward in time.
2. **Martingale property**: Itô integrals are martingales, which provides powerful theoretical tools for analysis. A martingale is a stochastic process whose expected future value equals its current value, conditional on past information. This property is crucial for proving convergence results and establishing theoretical guarantees for SDE-based algorithms.
3. **Numerical stability**: Itô discretization (Euler-Maruyama) is more stable and easier to implement than Stratonovich.
4. **Standard framework**: Most of the literature and software use Itô calculus, making it the de facto standard.

The choice between Itô and Stratonovich calculus is not merely a mathematical preference—it reflects different physical interpretations. Itô calculus is natural for processes where the noise is "white" (uncorrelated) and the system cannot anticipate future noise. Stratonovich calculus is natural for processes where the noise has some correlation structure or where the system can "sense" the noise continuously.

> In machine learning applications, Itô calculus is almost universally used because it matches the discrete-time algorithms we implement on computers.

19.5 Itô's Lemma

19.5.1 Theorem: Itô's Lemma

Let $\mathbf{X}_t$ satisfy an Itô SDE, and let $f(\mathbf{x}, t)$ be twice differentiable. Then:

$$df(\mathbf{X}_t, t) = \frac{\partial f}{\partial t}dt + \nabla f^\top d\mathbf{X}_t + \frac{1}{2}\text{Tr}\left(\mathbf{G}\mathbf{G}^\top \nabla^2 f\right) dt \tag{19.6}$$

19.5.2 Interpretation

Itô's lemma is the **chain rule for stochastic calculus**. The key difference from ordinary calculus is the extra second-order term:

$$\frac{1}{2}\text{Tr}\left(\mathbf{G}\mathbf{G}^\top\nabla^2 f\right)dt$$

This term arises from the **quadratic variation** of Brownian motion: $(dW_t)^2 = dt$. In ordinary calculus, second-order terms like $(dx)^2$ are negligible, but in stochastic calculus, they contribute at order dt.

Physical interpretation: The second-order term represents **diffusion**—the spreading of probability mass due to random fluctuations. This is why stochastic processes generate diffusion: the noise causes the distribution to spread out over time, even in the absence of drift.

19.5.3 Example: Applying Itô's Lemma

Consider the geometric Brownian motion $dX_t = \mu X_t dt + \sigma X_t dW_t$ and the function $f(x) = \log x$.

Using Itô's lemma:

$$d\log X_t = \frac{1}{X_t}dX_t + \frac{1}{2}\left(-\frac{1}{X_t^2}\right)(dX_t)^2$$

Substituting $dX_t = \mu X_t dt + \sigma X_t dW_t$ and $(dX_t)^2 = \sigma^2 X_t^2 dt$:

$$d\log X_t = \mu dt + \sigma dW_t - \frac{\sigma^2}{2}dt = \left(\mu - \frac{\sigma^2}{2}\right)dt + \sigma dW_t$$

Integrating gives:

$$\log X_t = \log X_0 + \left(\mu - \frac{\sigma^2}{2}\right)t + \sigma W_t$$

Exponentiating:

$$X_t = X_0 \exp\left(\left(\mu - \frac{\sigma^2}{2}\right)t + \sigma W_t\right)$$

Key observation: The drift in $\log X_t$ is $\mu - \frac{\sigma^2}{2}$, not μ. This is the **Itô correction**—the noise reduces the effective drift due to the convexity of the logarithm function.

19.6 SDEs as Limits of Algorithms

19.6.1 Euler–Maruyama Scheme

The **Euler–Maruyama scheme** is the simplest numerical method for solving SDEs. It discretizes the SDE:

$$d\mathbf{X}_t = \mathbf{f}(\mathbf{X}_t, t)dt + \mathbf{G}(\mathbf{X}_t, t)d\mathbf{W}_t$$

as:

$$\mathbf{X}_{k+1} = \mathbf{X}_k + \mathbf{f}(\mathbf{X}_k, t_k)\Delta t + \mathbf{G}(\mathbf{X}_k, t_k)\sqrt{\Delta t}\,\epsilon_k, \quad \epsilon_k \sim \mathcal{N}(\mathbf{0}, I) \tag{19.7}$$

where Δt is the time step and ϵ_k are independent Gaussian random variables.

Key features:

- The drift term $\mathbf{f}(\mathbf{X}_k, t_k)\Delta t$ is evaluated at the current state
- The diffusion term uses $\sqrt{\Delta t}$ because $dW_t \sim \mathcal{N}(0, dt)$
- This is exactly what we implemented in Chapter~18 for Langevin dynamics

19.6.2 Convergence Properties

The Euler–Maruyama scheme has **strong order 0.5** and **weak order 1.0**:

- **Strong convergence**: $\mathbb{E}[\|\mathbf{X}_T - \mathbf{X}_T^N\|] = O(\sqrt{\Delta t})$, where $\mathbf{X}_T$ is the exact solution and $\mathbf{X}_T^N$ is the numerical approximation with N steps.
- **Weak convergence**: $|\mathbb{E}[g(\mathbf{X}_T)] - \mathbb{E}[g(\mathbf{X}_T^N)]| = O(\Delta t)$ for smooth functions g.

This means that while individual paths converge slowly, **statistical properties** (expectations) converge faster.

19.6.3 Higher-Order Methods

More sophisticated methods exist for solving SDEs:

1. **Milstein scheme**: Strong order 1.0, includes correction terms from Itô's lemma
2. **Runge–Kutta methods**: Extensions of ODE methods to SDEs
3. **Adaptive step sizes**: Automatically adjust Δt based on local error estimates

However, for diffusion models, Euler–Maruyama is often sufficient because:

- The score function $\mathbf{s}_\theta(\mathbf{x}, t)$ is learned and approximate, so higher-order accuracy may not help
- Many small steps are used, so the global error accumulates slowly
- Computational efficiency is more important than high-order accuracy

19.7 Stochastic Optimization and Continuous Limits

19.7.1 Stochastic Gradient Descent as an SDE

Stochastic Gradient Descent (SGD) for minimizing a loss function $\mathcal{L}(\theta)$ can be written as:

$$\theta_{k+1} = \theta_k - \eta_k \nabla_\theta \mathcal{L}(\theta_k) + \eta_k \epsilon_k$$

where ϵ_k is gradient noise from mini-batch sampling.

In the continuous limit with decreasing step sizes $\eta_k \to 0$, this converges to an SDE:

$$d\Theta_t = -\nabla_\theta \mathcal{L}(\Theta_t)dt + \sqrt{2\eta(t)}d\mathbf{W}_t$$

This is a **Langevin-type SDE** where the drift is the negative gradient and the noise comes from stochastic gradient estimates.

19.7.2 Connection to Bayesian Inference

The continuous-time limit of SGD with appropriate noise scaling can be interpreted as sampling from a **posterior distribution** over parameters. The stationary distribution of the SDE is approximately:

$$p(\theta \mid \mathcal{D}) \propto \exp\left(-\frac{1}{\eta}\mathcal{L}(\theta)\right)$$

This connects optimization (SGD) to Bayesian inference (sampling), explaining why SGD often finds solutions that generalize well—they correspond to high-probability regions of the posterior.

19.7.3 Stochastic Approximation

Stochastic approximation is the theory of algorithms that use noisy observations to optimize or estimate quantities. Many machine learning algorithms are stochastic approximation procedures:

- **SGD**: Approximates the true gradient with mini-batch estimates
- **Online learning**: Updates parameters using streaming data
- **Reinforcement learning**: Uses Monte Carlo estimates of value functions

The continuous-time SDE perspective provides a unified framework for analyzing these algorithms, studying their convergence properties, and understanding their long-term behavior.

19.7.4 Ergodicity and Mixing

An important property for SDEs used in sampling is **ergodicity**—the property that time averages equal ensemble averages:

$$\lim_{T\to\infty} \frac{1}{T}\int_0^T f(\mathbf{X}_t)dt = \int f(\mathbf{x})p(\mathbf{x})d\mathbf{x}$$

This ensures that running the SDE for a long time produces samples from the stationary distribution. For Langevin dynamics, ergodicity guarantees that the algorithm eventually explores the entire distribution, not just a single mode.

Mixing time measures how quickly the process reaches its stationary distribution. Fast mixing is desirable for efficient sampling, but can be challenging in high-dimensional spaces with complex geometries.

19.8 Probability Evolution Viewpoint

An SDE defines:

- A **time-indexed density** $p_t(\mathbf{x})$

Key question:

How does $p_t(\mathbf{x})$ evolve over time?

This leads to the **Fokker–Planck equation** (Chapter~21).

19.9 Reversibility and Time Reversal

19.9.1 The Time Reversal Problem

A fundamental question in generative modeling is: **Can we reverse a stochastic process?** Given a forward SDE that transforms data into noise, can we construct a reverse SDE that transforms noise back into data?

The answer is yes, but the reverse process is **not simply the negative of the forward process.** The reverse SDE has a different drift term that depends on the **score function** $\nabla_{\mathbf{x}} \log p_t(\mathbf{x})$.

19.9.2 Reverse-Time SDE

Under mild conditions (smoothness of drift and diffusion, non-degeneracy of diffusion), the reverse process satisfies (in the common **state-independent diffusion** case where $\mathbf{G}$ depends only on t, so $\mathbf{a}(t) = \mathbf{G}(t)\mathbf{G}(t)^\top$):

$$d\mathbf{X}_t = [\mathbf{f}(\mathbf{X}_t, t) - \mathbf{G}(t)\mathbf{G}(t)^\top \nabla_{\mathbf{x}} \log p_t(\mathbf{X}_t)]\, dt + \mathbf{G}(t)\, d\bar{\mathbf{W}}_t \tag{19.8}$$

where $d\bar{\mathbf{W}}_t$ is a Brownian motion running backward in time.

Key observation: The reverse drift has an **extra term** $-\mathbf{G}(t)\mathbf{G}(t)^\top \nabla_{\mathbf{x}} \log p_t(\mathbf{X}_t)$ that depends on the score function. This term corrects for the asymmetry introduced by the forward diffusion.

19.9.3 Why the Score Function Appears

The score function appears in the reverse SDE because reversing a stochastic process requires knowing the **local geometry** of the probability distribution. The forward process spreads probability mass, and reversing it requires "pulling" the mass back together, which requires knowing where the high-probability regions are—exactly what the score function encodes.

19.9.4 Connection to Generative Modeling

This reverse-time SDE is the mathematical foundation of **reverse diffusion** and **score-based generative models**:

1. **Forward process**: Transform data $\mathbf{x}_0 \sim p_{\text{data}}$ into noise $\mathbf{x}_T \sim \mathcal{N}(\mathbf{0}, I)$ using a forward SDE
2. **Learn score function**: Train a neural network $\mathbf{s}_\theta(\mathbf{x}, t)$ to approximate $\nabla_{\mathbf{x}} \log p_t(\mathbf{x})$
3. **Reverse process**: Generate samples by solving the reverse SDE backward in time from noise to data

This framework unifies many generative modeling approaches, including diffusion models, score-based models, and flow-based models.

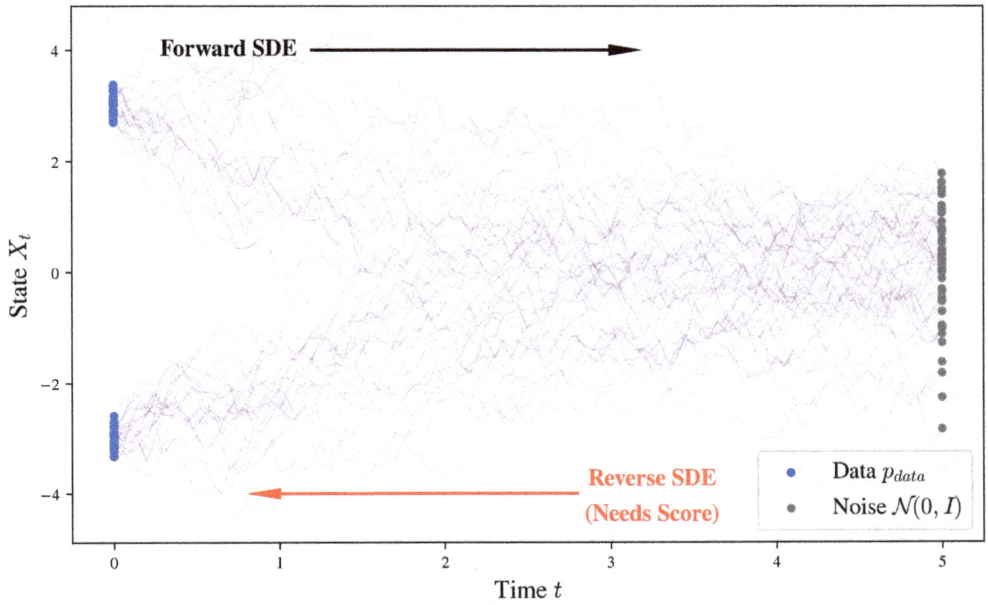

Figure 19.5: Forward and Reverse SDE Processes

Figure 19.5 visualizes the forward and reverse SDE processes using an **Ornstein-Uhlenbeck process** as an example. The plot shows 50 particle trajectories evolving over time. At $t = 0$ (left side), particles start in a structured data distribution (two clusters shown as blue points). As time progresses forward (black arrow), the **forward SDE** adds noise, causing the particles to spread out and mix. By $t = T$ (right side), the particles form a generic Gaussian noise distribution (gray points). The **reverse SDE** (red arrow) would transform noise back into data, but this requires the score function $\nabla_{\mathbf{x}} \log p_t(\mathbf{x})$ to "pull" the particles back together. This visualization demonstrates the core idea of diffusion models: the forward process destroys structure using noise, while the reverse process reconstructs structure using the score function. Without the score function, the reverse process cannot be constructed, which is why score matching (Chapter~17) is essential for generative modeling via time reversal.

19.9.5 The Score Function Requirement

The reverse SDE requires the score function $\nabla_{\mathbf{x}} \log p_t(\mathbf{x})$, which is typically unknown. This is why **score matching** (Chapter~17) is essential—we must learn the score function from data to enable reverse diffusion.

Without the score function, we cannot construct the reverse SDE, and generative modeling via time reversal becomes impossible. This explains why score-based learning is fundamental to modern diffusion models.

19.9.6 Girsanov Theorem and Change of Measure

The reverse-time SDE can be understood through **Girsanov's theorem**, which describes how to change the probability measure (change the drift) of a stochastic process. In the context of diffusion models:

- **Forward measure**: The probability measure under which the forward SDE evolves
- **Reverse measure**: A different probability measure under which the reverse SDE evolves

Girsanov's theorem provides the mathematical framework for constructing the reverse measure, showing that the change of measure introduces the score function term in the reverse drift. This is a deep theoretical result that justifies why the reverse SDE has the specific form it does.

However, for practical purposes, we don't need to explicitly use Girsanov's theorem—we can derive the reverse SDE directly from the forward SDE and the score function. The theorem provides the theoretical foundation but is not needed for implementation.

> Girsanov's theorem is a fundamental result in stochastic calculus that describes how changing the drift of an SDE corresponds to changing the underlying probability measure. In diffusion models, this theorem underlies the theoretical justification for reverse-time processes, but practitioners typically work directly with the reverse SDE formula without explicitly invoking measure changes. The theorem is more important for theoretical analysis and proofs than for practical implementation.

19.10 Conceptual Summary

The continuous-time limit of discrete stochastic algorithms is governed by stochastic differential equations (SDEs). The key components and mathematical objects are summarized below:

Object	Meaning
Brownian motion	Fundamental noise
Drift	Deterministic flow
Diffusion	Random spreading
SDE	Law of random motion
Itô calculus	Correct stochastic analysis

By transitioning from discrete updates to continuous time, SDEs decouple the deterministic trajectory (drift) from stochastic fluctuations (diffusion). Itô calculus provides the necessary analytical machinery to handle non-differentiable Brownian paths, establishing the mathematical foundation for reverse-time generative processes.

19.11 Mindmap

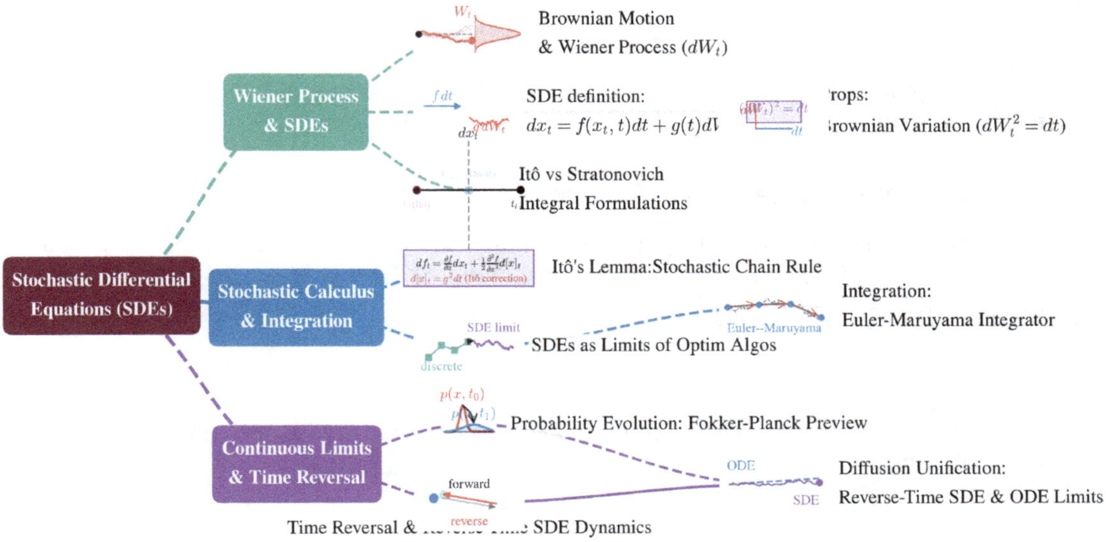

Figure 19.6: Mind map of stochastic differential equations topics

19.12 Exercises

19.12.1 Brownian Motion

1. **Increment distribution**: Using $\mathbf{W}_t - \mathbf{W}_s \sim \mathcal{N}(0, (t-s)I)$, compute $\mathbb{E}[\mathbf{W}_t]$ and $\mathrm{Cov}(\mathbf{W}_t)$.
2. **Scaling property (1D)**: Let W_t be 1D Brownian motion. Show that for any $c > 0$, the process $\tilde{W}_t = \frac{1}{\sqrt{c}} W_{ct}$ is again a Brownian motion.

19.12.2 Itô SDE and Discretization

1. **Euler–Maruyama for Langevin**: For the Langevin SDE

$$d\mathbf{X}_t = \nabla_{\mathbf{x}} \log p(\mathbf{X}_t)\, dt + \sqrt{2}\, d\mathbf{W}_t,$$

write the Euler–Maruyama update with step size Δt.

2. **Noise scaling**: Explain why the diffusion term uses $\sqrt{\Delta t}$ in Euler–Maruyama (i.e., why $\mathbf{W}_{t+\Delta t} - \mathbf{W}_t$ scales like $\sqrt{\Delta t}$).

19.12.3 Itô's Lemma

1. **1D Itô's lemma**: Let X_t satisfy $dX_t = \mu(X_t, t)\, dt + \sigma(X_t, t)\, dW_t$. Write the 1D version of Itô's lemma for $f(X_t, t)$.
2. **Quadratic variation check**: Take $f(x) = x^2$ and apply Itô's lemma to compute $d(X_t^2)$. Identify explicitly where the extra dt term comes from.

19.12.4 From Algorithms to Processes

1. **ULA as a discretization**: Compare the ULA update

$$\mathbf{x}_{k+1} = \mathbf{x}_k + \eta \nabla_{\mathbf{x}} \log p(\mathbf{x}_k) + \sqrt{2\eta}\,\epsilon_k$$

with Euler–Maruyama for the Langevin SDE. What correspondence between η and Δt makes them match?

19.12.5 Reverse-Time SDE (Conceptual)

1. **Why the score appears**: In your own words, explain why reversing an SDE requires knowledge of $\nabla_{\mathbf{x}} \log p_t(\mathbf{x})$ (the score of the marginal at time t).
2. **Sign sanity check**: Consider the reverse-time drift

$$\mathbf{f}(\mathbf{x}, t) - \mathbf{a}(t)\nabla_{\mathbf{x}} \log p_t(\mathbf{x}).$$

Give an intuition for why the score term should point "toward" higher probability regions when running backward.

Chapter 20

ODE, SDE, and Continuous Limits of Algorithms

From discrete algorithms to continuous flows: the mathematical bridge

> *"Quantities, and the ratios of quantities, which in any finite time converge continually to equality, are ultimately equal."*
>
> — Isaac Newton

- ▷ `Gradient descent → ODE` $\mathbf{x}_{k+1} = \mathbf{x}_k - \eta \nabla f(\mathbf{x}_k)$ becomes $\frac{d\mathbf{x}_t}{dt} = -\nabla f(\mathbf{x}_t)$ as $\eta \to 0$
- ▷ `Langevin → SDE` $\mathbf{x}_{k+1} = \mathbf{x}_k + \eta \mathbf{s}(\mathbf{x}_k) + \sqrt{2\eta}\,\epsilon_k$ becomes $d\mathbf{X}_t = \mathbf{s}(\mathbf{X}_t)\,dt + \sqrt{2}\,d\mathbf{W}_t$
- ▷ `Forward SDE (state-independent diffusion)` $d\mathbf{X}_t = \mathbf{f}(\mathbf{X}_t, t)\,dt + \mathbf{G}(t)\,d\mathbf{W}_t$, with $\mathbf{a}(t) = \mathbf{G}(t)\mathbf{G}(t)^\top$
- ▷ `Score at time t` $\nabla_\mathbf{x} \log p_t(\mathbf{x})$ where p_t is the marginal density of $\mathbf{X}_t$
- ▷ `Reverse-time SDE (common form)` $d\mathbf{X}_t = [\mathbf{f}(\mathbf{X}_t, t) - \mathbf{a}(t)\nabla_\mathbf{x} \log p_t(\mathbf{X}_t)]\,dt + \mathbf{G}(t)\,d\bar{\mathbf{W}}_t$
- ▷ `Probability flow ODE` $\frac{d\mathbf{X}_t}{dt} = \mathbf{f}(\mathbf{X}_t, t) - \frac{1}{2}\mathbf{a}(t)\nabla_\mathbf{x} \log p_t(\mathbf{X}_t)$ (shares the same marginals p_t as the SDE)
- ▷ `Key equivalence (marginals)` SDE and probability flow ODE induce the same $p_t(\mathbf{x})$ (noise effect is absorbed into the score term)
- ▷ `Sampling viewpoint` reverse SDE gives stochastic sampling; probability flow ODE gives deterministic sampling (e.g. ODE solvers / DDIM-like)
- ▷ `Continuity (Liouville) equation for ODE` if $\dot{\mathbf{x}} = \mathbf{v}(\mathbf{x}, t)$, then $\partial_t p_t(\mathbf{x}) + \nabla_\mathbf{x} \cdot (p_t(\mathbf{x})\,\mathbf{v}(\mathbf{x}, t)) = 0$
- ▷ `Flow matching objective (generic)` learn $\mathbf{v}_\theta$ by $\min_\theta \mathbb{E}_{t,\mathbf{x}_t}[\|\mathbf{v}_\theta(\mathbf{x}_t, t) - \mathbf{u}(\mathbf{x}_t, t)\|_2^2]$ for a chosen target velocity $\mathbf{u}$
- ▷ `SDE vs ODE at distribution level` SDE $\to$ Fokker–Planck (drift + diffusion), ODE $\to$ continuity equation (transport only)
- ▷ `Linear state-space ODE` $\dot{\mathbf{x}}(t) = A\mathbf{x}(t) + B\mathbf{u}(t)$, $\mathbf{y}(t) = C\mathbf{x}(t)$ (a deterministic flow used by modern state-space sequence models)

20.1 Motivation: Why Remove Noise?

In Chapter~19, we described generative modeling using **stochastic differential equations**:

$$d\mathbf{X}_t = \mathbf{f}(\mathbf{X}_t, t)\, dt + \mathbf{G}(\mathbf{X}_t, t)\, d\mathbf{W}_t \tag{20.1}$$

This formulation is natural, expressive, and theoretically sound. The stochastic noise $d\mathbf{W}_t$ is essential for defining the forward diffusion process and ensuring proper exploration during sampling.

However, this raises a surprising empirical observation:

- If diffusion models rely on randomness, why can they often be sampled **deterministically**?

Empirically, modern diffusion models like **DDIM** (Denoising Diffusion Implicit Models) and **probability flow ODEs** can generate high-quality samples using **deterministic ODE solvers**, without injecting any noise during the reverse process.

20.1.1 The Key Insight

The answer lies in understanding the difference between:

1. **Path-level equivalence**: Individual sample paths from SDEs and ODEs are different (SDE paths are random, ODE paths are deterministic)
2. **Distribution-level equivalence**: The **marginal distributions** $p_t(\mathbf{x})$ can be identical for both processes

This means that while individual trajectories differ, the **ensemble of trajectories** (the probability distribution) can be the same. For generative modeling, we care about the distribution, not individual paths.

20.1.2 Practical Benefits of Deterministic Sampling

Deterministic sampling via ODEs offers several advantages:

1. **Faster generation**: ODE solvers can use larger step sizes and adaptive methods, reducing the number of function evaluations
2. **Reproducibility**: Deterministic sampling produces the same sample for the same initial condition, enabling controlled generation
3. **Latent space interpolation**: Deterministic paths enable smooth interpolation in the latent space
4. **Computational efficiency**: ODE solvers are often more efficient than SDE solvers for the same accuracy

This chapter explains the mathematical foundation that makes deterministic sampling possible and how it connects to the broader framework of continuous-time generative models.

20.2 Discrete Algorithms and Their Continuous Limits

Many algorithms in machine learning are inherently **discrete**—they operate on a sequence of iterations $k = 0, 1, 2, ...$:

- **Gradient descent**: $\mathbf{x}_{k+1} = \mathbf{x}_k - \eta \nabla f(\mathbf{x}_k)$
- **Langevin dynamics**: $\mathbf{x}_{k+1} = \mathbf{x}_k + \eta \mathbf{s}(\mathbf{x}_k) + \sqrt{2\eta}\epsilon_k$
- **SGD**: $\theta_{k+1} = \theta_k - \eta_k \tilde{\nabla}\mathcal{L}(\theta_k)$
- **Diffusion samplers**: Multi-step denoising processes

Yet their behavior becomes clearer when viewed as **time-discretized dynamical systems**. By introducing a continuous time variable t and taking the limit as the step size $\eta \to 0$, we obtain continuous-time processes that reveal the fundamental mathematical structure.

20.2.1 The Continuous-Time Paradigm

The continuous-time perspective provides several advantages:

1. **Unified framework**: Many different discrete algorithms converge to the same continuous process, revealing their underlying connections
2. **Analytical tools**: Continuous-time analysis enables the use of powerful mathematical techniques (differential equations, calculus of variations, etc.)
3. **Algorithm design**: Understanding the continuous limit helps design better discrete approximations
4. **Theoretical insights**: Continuous-time analysis provides insights into convergence, stability, and long-term behavior

This paradigm shift—from discrete iterations to continuous flows—is central to modern machine learning theory and has led to significant advances in understanding optimization, sampling, and generative modeling.

20.2.2 Gradient Descent → ODE

Discrete update:

$$\mathbf{x}_{k+1} = \mathbf{x}_k - \eta \nabla f(\mathbf{x}_k) \tag{20.2}$$

where $\eta > 0$ is the learning rate and $f : \mathbb{R}^d \to \mathbb{R}$ is the objective function.

Derivation of continuous limit:

Introduce continuous time t by setting $t_k = k\eta$ and define $\mathbf{x}(t_k) = \mathbf{x}_k$. The discrete update becomes:

$$\mathbf{x}(t_k + \eta) = \mathbf{x}(t_k) - \eta \nabla f(\mathbf{x}(t_k))$$

Rewriting as a difference quotient:

$$\frac{\mathbf{x}(t_k + \eta) - \mathbf{x}(t_k)}{\eta} = -\nabla f(\mathbf{x}(t_k))$$

Taking the limit as $\eta \to 0$ (with $t_k \to t$):

$$\frac{d\mathbf{x}_t}{dt} = -\nabla f(\mathbf{x}_t) \tag{20.3}$$

This is a **deterministic ODE** called the **gradient flow**. It describes the continuous-time evolution of the optimization trajectory.

Properties:

- The ODE follows the steepest descent direction
- Energy decreases: $\frac{d}{dt} f(\mathbf{x}_t) = \nabla f(\mathbf{x}_t)^\top \frac{d\mathbf{x}_t}{dt} = -\|\nabla f(\mathbf{x}_t)\|^2 \leq 0$
- Under appropriate conditions, $\mathbf{x}_t$ converges to a local minimum as $t \to \infty$

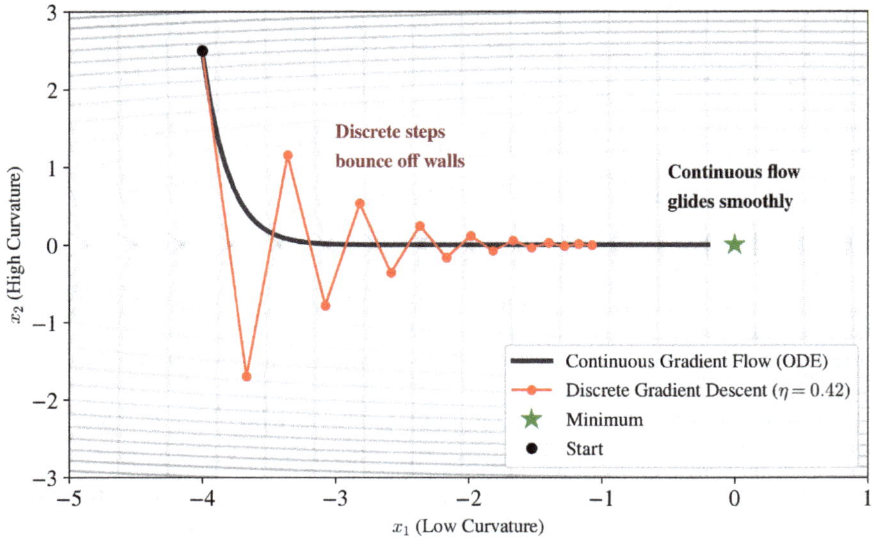

Figure 20.1: Discrete Gradient Descent vs. Continuous Gradient Flow

Figure 20.1 visualizes the convergence of discrete gradient descent to the continuous gradient flow ODE. The plot shows an elongated quadratic valley ($f = 0.1x_1^2 + 2x_2^2$) with contour lines. The red zigzagging path shows discrete gradient descent with a large learning rate ($\eta = 0.42$), which bounces off the valley walls due to the high curvature in the x_2 direction. The black smooth curve shows the continuous gradient flow ODE, which follows the exact path of steepest descent without oscillation. As the step size $\eta \to 0$, the discrete algorithm converges to the deterministic ODE trajectory. This visualization makes concrete the continuous-time limit: discrete iterations become a smooth flow, revealing the fundamental mathematical structure underlying optimization algorithms.

20.2.3 Momentum Methods → Second-Order ODE

Many optimization algorithms can be understood as discretizations of second-order ODEs. For example, **Nesterov's accelerated gradient descent** can be viewed as a discretization of:

$$\ddot{\mathbf{x}}_t + \frac{3}{t}\dot{\mathbf{x}}_t + \nabla f(\mathbf{x}_t) = 0$$

This is a **second-order ODE** with a friction term $\frac{3}{t}\dot{\mathbf{x}}_t$ that provides acceleration. The continuous-time perspective reveals why momentum methods can converge faster than plain gradient descent.

20.2.4 Langevin Dynamics → SDE

Discrete update:

$$\mathbf{x}_{k+1} = \mathbf{x}_k + \eta\,\mathbf{s}(\mathbf{x}_k) + \sqrt{2\eta}\,\epsilon_k, \quad \epsilon_k \sim \mathcal{N}(0, I) \tag{20.4}$$

Continuous limit:

$$d\mathbf{X}_t = \mathbf{s}(\mathbf{X}_t)\,dt + \sqrt{2}\,d\mathbf{W}_t \tag{20.5}$$

This is a **stochastic ODE**, i.e., an SDE.

20.3 Two Views of the Same Evolution

An SDE defines **two equivalent descriptions** of the same stochastic process:

1. **Sample-path view (microscopic)** Individual random trajectories $\mathbf{X}_t(\omega)$ for each realization ω. Each path is a random function of time, representing a single sample from the process.
2. **Distributional view (macroscopic)** Deterministic evolution of the probability density $p_t(\mathbf{x})$. While individual paths are random, the distribution evolves deterministically according to the Fokker-Planck equation (Chapter~21).

20.3.1 The Microscopic-Macroscopic Duality

This duality is fundamental to stochastic processes:

- **Microscopic (paths)**: Random, unpredictable individual trajectories
- **Macroscopic (distributions)**: Deterministic, predictable evolution of probability mass

This is analogous to statistical mechanics, where:

- Individual particles move randomly (microscopic)
- But the gas as a whole follows deterministic laws (macroscopic, e.g., ideal gas law)

20.3.2 Why the Distributional View Matters

For generative modeling, we care about the **distribution** $p_t(\mathbf{x})$, not individual paths. This means:

- We can use **either** SDEs (stochastic paths) **or** ODEs (deterministic paths) to generate the same distribution
- The choice between SDE and ODE is a matter of **implementation preference**, not fundamental necessity
- Deterministic ODEs can be faster and more efficient while producing the same samples (in distribution)

The second view (distributional) allows a deterministic reformulation because the evolution of $p_t(\mathbf{x})$ is deterministic, even though individual paths are random.

Figure 20.2 illustrates the core insight of this chapter: **two ways to generate the same distribution**. The left panel shows stochastic SDE trajectories (500 paths) starting from a tight cluster and evolving toward a standard normal distribution. The paths are rough, jagged, and cross each other frequently—this is the stochastic nature of SDEs. The side histogram shows the final particle positions matching the theoretical Gaussian (black dashed line). The right panel shows probability flow ODE trajectories (500 paths) starting from the same initial conditions. The paths are smooth, deterministic, and **never cross** (due to uniqueness of ODE solutions). The side histogram shows the **exact same** final distribution. Despite these dramatic path-level differences, both processes produce the **same marginal distribution** $p_t(\mathbf{x})$ at each time t. This visualization makes concrete why deterministic sampling works: individual paths differ (stochastic vs. deterministic), but they transport the same probability mass, making the distributional evolution identical.

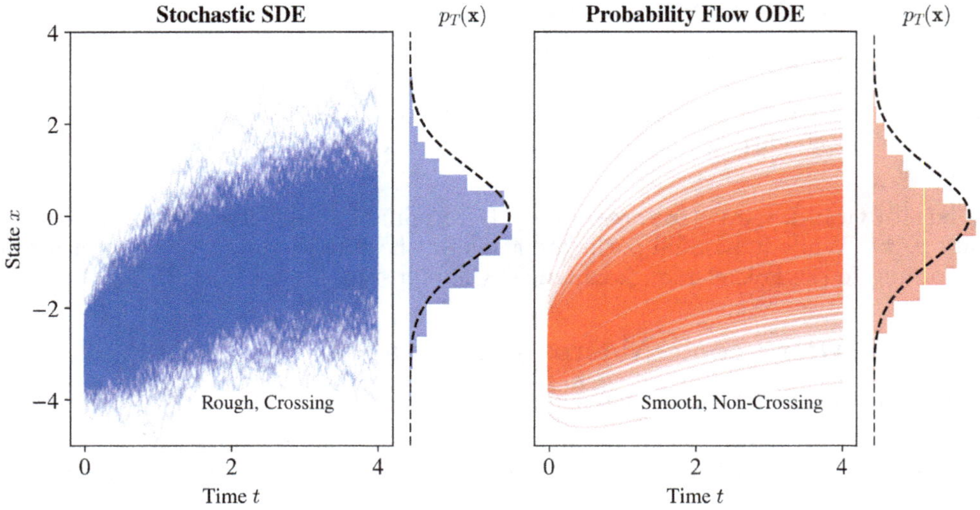

Figure 20.2: SDE vs. Probability Flow ODE: Two Ways to Generate the Same Distribution

20.4 Probability Flow Perspective

20.4.1 The Forward SDE

Let the forward SDE be (with state-independent diffusion for simplicity):

$$d\mathbf{X}_t = \mathbf{f}(\mathbf{X}_t, t)\, dt + \mathbf{G}(t)\, d\mathbf{W}_t \tag{20.6}$$

where $\mathbf{G}(t)$ depends only on time, not on the state $\mathbf{X}_t$. This SDE induces a time-dependent density $p_t(\mathbf{x})$ that evolves according to the Fokker-Planck equation.

20.4.2 The Key Insight

The fundamental insight is that:

- The same family of densities $p_t(\mathbf{x})$ can be generated
- by a **deterministic ODE** (the probability flow ODE)
- even though individual sample paths differ

This means we have two ways to generate samples from the same distribution: 1. **Stochastic method**: Solve the SDE, generating random paths 2. **Deterministic method**: Solve the ODE, generating deterministic paths

Both methods produce samples from the same distribution $p_t(\mathbf{x})$ at each time t.

20.4.3 Why This Is Possible

The reason this works is that the **marginal distribution** $p_t(\mathbf{x})$ depends only on the **statistics** of the process, not on the individual paths. The SDE and ODE can have different path-level behavior but identical distribution-level behavior.

This is analogous to how different physical processes can produce the same equilibrium distribution: the final state depends on the distribution, not the specific path taken to reach it.

20.5 Probability Flow ODE

20.5.1 Definition: Probability Flow ODE

The **probability flow ODE** associated with an SDE is:

$$\frac{d\mathbf{X}_t}{dt} = \mathbf{f}(\mathbf{X}_t, t) - \frac{1}{2}\mathbf{G}(t)\mathbf{G}(t)^\top \nabla_{\mathbf{x}} \log p_t(\mathbf{X}_t) \tag{20.7}$$

where $\mathbf{a}(t) = \mathbf{G}(t)\mathbf{G}(t)^\top$ is the diffusion matrix.

This ODE satisfies:

- **No stochastic term**: The ODE is completely deterministic—no $d\mathbf{W}_t$ term
- **Same marginal distributions**: The marginal density $p_t(\mathbf{x})$ is identical to that of the SDE
- **Deterministic trajectories**: For a given initial condition, the solution is unique and deterministic

20.5.2 Derivation Sketch

The probability flow ODE can be derived by requiring that the ODE and SDE produce the same Fokker-Planck equation. The key is that the Fokker-Planck equation for an SDE:

$$\frac{\partial p_t}{\partial t} = -\nabla \cdot (\mathbf{f} p_t) + \frac{1}{2}\nabla \cdot \nabla(\mathbf{a} p_t)$$

can be rewritten in continuity equation form:

$$\frac{\partial p_t}{\partial t} + \nabla \cdot \mathbf{J}_t = 0$$

where $\mathbf{J}_t$ is the probability current. For the probability flow ODE $\frac{d\mathbf{X}_t}{dt} = \mathbf{v}(\mathbf{X}_t, t)$, the continuity equation is:

$$\frac{\partial p_t}{\partial t} + \nabla \cdot (p_t \mathbf{v}) = 0$$

Matching the two forms and solving for $\mathbf{v}$ yields the probability flow ODE formula.

20.5.3 The Score Function's Role

The score function $\nabla_{\mathbf{x}} \log p_t(\mathbf{X}_t)$ appears in the probability flow ODE because it encodes the **local geometry** of the probability distribution. The term $-\frac{1}{2}\mathbf{a}(t)\nabla_{\mathbf{x}} \log p_t(\mathbf{X}_t)$ corrects for the diffusion that would occur in the SDE, ensuring that the ODE produces the same distribution.

Key insight: The score function **absorbs the effect of noise**. In the SDE, noise spreads probability mass. In the ODE, the score function term pulls probability mass back together, achieving the same net effect without explicit noise.

20.5.4 Interpretation

Each term in the probability flow ODE has a specific role:

Term	Role
$\mathbf{f}(\mathbf{X}_t, t)$	Drift of original SDE—deterministic component that would exist even without noise
$\nabla_{\mathbf{x}} \log p_t(\mathbf{X}_t)$	Score function—encodes local geometry of the probability distribution
$\mathbf{G}(t)\mathbf{G}(t)^\top$	Diffusion strength—measures how much the noise spreads probability mass

The correction term $-\frac{1}{2}\mathbf{a}(t)\nabla_{\mathbf{x}} \log p_t(\mathbf{X}_t)$ compensates for the diffusion that occurs in the SDE. In the SDE, the noise term $\mathbf{G}(t)d\mathbf{W}_t$ spreads probability mass outward. In the ODE, the score function term pulls probability mass inward, achieving the same net distributional effect.

20.5.5 Physical Analogy

This is analogous to **fluid dynamics**:

- **SDE**: Like a fluid with both advection (drift $\mathbf{f}$) and diffusion (noise $\mathbf{G}d\mathbf{W}_t$)
- **Probability flow ODE**: Like a fluid with advection and a **negative diffusion** term (the score correction) that exactly cancels the spreading effect

The score function acts as a "probability current" that transports mass in a way that mimics the combined effect of drift and diffusion in the SDE.

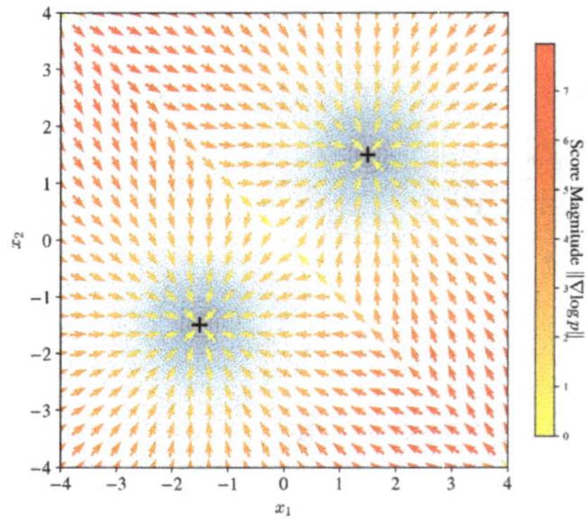

Figure 20.3: The Score Field: Negative Diffusion Vector Field

Figure 20.3 visualizes the mathematical mechanism of the probability flow ODE. The background shows contour lines of a **mixture of two Gaussians** (multimodal distribution), with darker regions indicating higher probability density. The vector field shows the **score function** $\nabla \log p(\mathbf{x})$, which points **uphill** toward the nearest high-density peak (mode). The black crosses mark the two modes. This vector field represents the "negative diffusion" correction term $-\frac{1}{2}\mathbf{a}(t)\nabla_{\mathbf{x}} \log p_t(\mathbf{x})$ in the probability flow ODE. The SDE naturally spreads mass outward (diffusion), while the probability flow ODE applies this velocity field (pointing inward toward modes) that exactly counteracts the spreading, keeping the distribution shape consistent without noise. The visualization makes concrete how the score function "pulls" probability mass back together, achieving the same net effect as the SDE's diffusion but through deterministic transport. Notice the implicit boundary between the two modes where arrows diverge, effectively partitioning the space.

20.6 Why Deterministic Sampling Works

20.6.1 The Training-Sampling Distinction

In diffusion models, there is a crucial distinction between **training** and **sampling**:

During training:

- We learn the score function $\nabla_{\mathbf{x}} \log p_t(\mathbf{x})$ using denoising score matching
- The forward SDE (with noise) is used to create training data
- Noise is essential for defining the forward process and learning the score

During sampling:

- We have a learned score function $\mathbf{s}_\theta(\mathbf{x}, t) \approx \nabla_{\mathbf{x}} \log p_t(\mathbf{x})$

- Noise is **not needed** for sampling—we can use the probability flow ODE
- The score function contains all the information needed to generate samples

20.6.2 Why This Works

The key insight is that:

- **Noise defines the model**: The forward SDE with noise determines what distribution we're learning
- **Scores define the sampler**: Once we know the score function, we can sample using either SDEs or ODEs

The score function encodes the **complete information** about how to reverse the diffusion process. The noise in the forward process is necessary to define the problem, but once we've learned the score, we don't need noise for sampling.

20.6.3 Practical Implications

This explains why several deterministic sampling methods work:

1. **DDIM (Denoising Diffusion Implicit Models)**: A deterministic variant of DDPM that uses the learned denoising function without explicit noise injection
2. **ODE samplers**: Directly solve the probability flow ODE using numerical ODE solvers (Runge-Kutta, etc.)
3. **Probability flow solvers**: Specialized ODE solvers designed for probability flow equations

All of these methods work because they're solving the same underlying mathematical problem: following the probability flow that generates the target distribution.

20.6.4 Trade-offs: Stochastic vs Deterministic Sampling

While both methods produce the same distribution, they have different characteristics:

Aspect	Stochastic (SDE)	Deterministic (ODE)
Sample diversity	Higher (noise adds variability)	Lower (deterministic paths)
Speed	Slower (requires many small steps)	Faster (can use larger steps)
Reproducibility	Non-reproducible (random)	Reproducible (deterministic)
Interpolation	Difficult (random paths)	Easy (smooth deterministic paths)
Robustness	More robust to score errors	Less robust (errors accumulate)

In practice, deterministic sampling is often preferred for speed and reproducibility, while stochastic sampling may be preferred when maximum diversity is needed.

Figure 20.4 demonstrates the practical benefit of deterministic sampling for latent space interpolation. The plot shows 12 interpolation paths between two points in latent space, with colors indicating different interpolation points. The left panel shows **deterministic ODE paths**: smooth, curved trajectories that follow the non-linear geometry of the data manifold (swirl pattern), enabling clean morphing between samples. The paths stay within high-density regions and maintain smooth continuity. The right panel shows **stochastic SDE paths**: noisy, wandering trajectories that may drift off the data manifold, causing jittery transitions. Even though the mean drift follows the manifold, the individual paths wander due to stochastic noise. This visualization makes concrete why deterministic sampling is preferred for tasks like image interpolation, style transfer, and latent space manipulation: the smooth, predictable paths enable controlled generation, while stochastic paths introduce unwanted randomness that breaks smooth transitions.

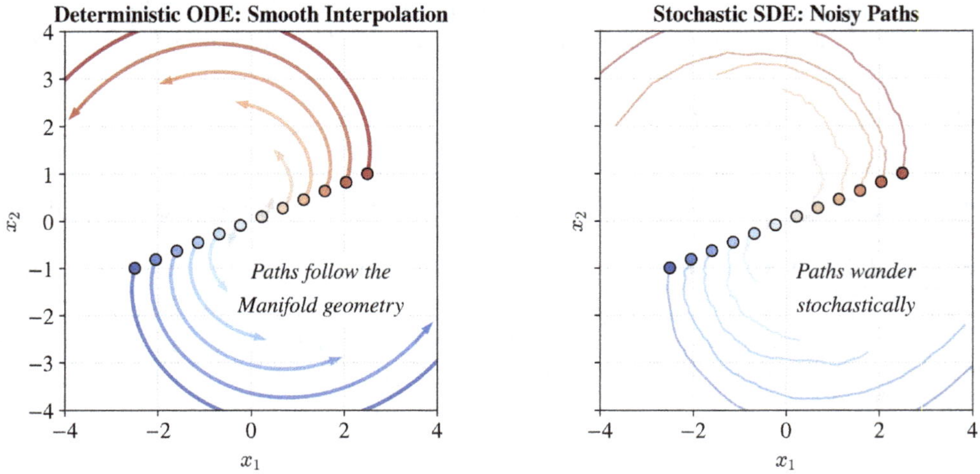

Figure 20.4: Latent Space Interpolation: ODE vs. SDE

20.7 Flow Matching and Deterministic Probability Flows

Diffusion models describe **stochastic transport** of probability mass. Flow matching describes **deterministic transport** of probability mass. The mathematical bridge is the continuity equation.

20.7.1 ODE View: Velocity Field

Consider a deterministic dynamics described by a velocity field:

$$\frac{d\mathbf{x}_t}{dt} = \mathbf{v}(\mathbf{x}_t, t) \tag{20.8}$$

where $\mathbf{v} : \mathbb{R}^d \times [0, 1] \to \mathbb{R}^d$ is a time-dependent vector field that specifies the velocity at each point in space and time.

Let $p_t(\mathbf{x})$ be the density of $\mathbf{x}_t$ (the probability density of the state at time t). Then p_t evolves according to the **continuity equation** (also called the **Liouville equation**):

$$\frac{\partial p_t(\mathbf{x})}{\partial t} + \nabla_{\mathbf{x}} \cdot (p_t(\mathbf{x})\,\mathbf{v}(\mathbf{x}, t)) = 0 \tag{20.9}$$

20.7.2 Derivation of the Continuity Equation

The continuity equation can be derived from the principle of **conservation of probability mass.** Consider a small volume V in state space. The rate of change of probability mass in V equals the flux of probability through the boundary:

$$\frac{d}{dt} \int_V p_t(\mathbf{x}) d\mathbf{x} = -\oint_{\partial V} p_t(\mathbf{x})\mathbf{v}(\mathbf{x}, t) \cdot \mathbf{n} dS$$

where $\mathbf{n}$ is the outward normal. Using the divergence theorem:

$$\int_V \frac{\partial p_t}{\partial t} d\mathbf{x} = -\int_V \nabla \cdot (p_t \mathbf{v}) d\mathbf{x}$$

Since this holds for any volume V, we obtain the continuity equation.

Physical interpretation: The continuity equation states that probability mass is conserved—it flows but is neither created nor destroyed. The divergence term $\nabla \cdot (p_t \mathbf{v})$ represents the net flux of probability mass out of a point.

This is the deterministic analogue of the Fokker–Planck equation (Chapter~21), which includes an additional diffusion term for stochastic processes.

20.7.3 Flow Matching (High-Level)

Flow matching is a framework for learning generative models by directly learning a deterministic vector field that transports one distribution into another. Unlike diffusion models that learn score functions, flow matching learns velocity fields.

The key idea is to learn a time-dependent vector field $\mathbf{v}_\theta(\mathbf{x}, t)$ such that the induced ODE:

$$\frac{d\mathbf{x}_t}{dt} = \mathbf{v}_\theta(\mathbf{x}_t, t)$$

transports an initial distribution $p_0(\mathbf{x})$ (e.g., noise) into a target distribution $p_1(\mathbf{x})$ (e.g., data).

Training objective: Flow matching reduces to supervised regression on velocities along a chosen interpolation path $(\mathbf{x}_t)_{t\in[0,1]}$:

$$\min_\theta \mathbb{E}_{t,\mathbf{x}_t}\left[\left\|\mathbf{v}_\theta(\mathbf{x}_t, t) - \mathbf{u}(\mathbf{x}_t, t)\right\|_2^2\right] \tag{20.10}$$

where $\mathbf{u}(\mathbf{x}, t)$ is a target velocity field determined by the chosen coupling/interpolation strategy.

Key advantages of flow matching: 1. **Direct learning**: Learns the velocity field directly, not the score function 2. **Flexible paths**: Can choose different interpolation paths (optimal transport, simple linear interpolation, etc.) 3. **Fast sampling**: Deterministic ODE can be solved quickly with adaptive solvers 4. **No noise schedule**: Doesn't require designing a noise schedule like diffusion models

20.7.4 Relation to Probability Flow

The probability flow ODE is a **special case** of flow matching where:

- The target velocity field is derived from an SDE
- The marginals match those of the corresponding SDE
- The interpolation path is determined by the forward diffusion process

In that sense:

- **Diffusion / SDE**: Stochastic distribution dynamics (paths are random)
- **Probability flow / ODE**: Deterministic distribution dynamics (paths are deterministic, marginals match SDE)
- **Flow matching**: Learning deterministic dynamics directly (vector field matching, more general than probability flow)

Flow matching generalizes the probability flow concept to arbitrary interpolation paths, not just those derived from SDEs.

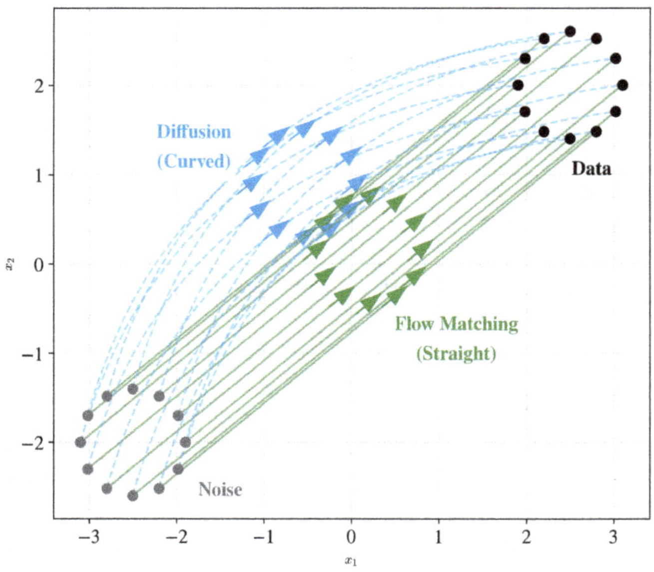

Figure 20.5: Flow Matching (Straight) vs. Diffusion (Curved)

Figure 20.5 differentiates **probability flow** from general **flow matching**. The plot shows trajectories connecting a noise point (gray) to a data point (blue). The green solid line shows **flow matching** with optimal transport: a perfectly straight line connecting the two points, minimizing the transport cost. The blue dashed line shows **diffusion/probability flow**: a curved trajectory that follows the score function, often creating a sigmoid-like shape. Flow matching generalizes probability flow by allowing us to learn **straighter, more efficient paths** for faster sampling. While probability flow paths are determined by the forward diffusion process, flow matching can learn arbitrary interpolation paths, including optimal transport paths that minimize the total "cost" of transporting probability mass. This visualization makes concrete why flow matching can be faster than diffusion models: straighter paths require fewer integration steps to traverse.

20.7.5 Optimal Transport Connection

Flow matching is closely related to **optimal transport** theory. The optimal transport problem seeks the most efficient way to transport one distribution to another. Flow matching can be viewed as learning an approximation to the optimal transport map via a continuous flow.

When the target velocity field $\mathbf{u}(\mathbf{x}, t)$ is chosen according to optimal transport principles, flow matching learns a flow that minimizes the total "cost" of transporting probability mass from p_0 to p_1.

20.8 Linear State-Space Models

Many dynamical systems—from physical systems to modern sequence models—can be described as **linear state-space models**. These provide a bridge between continuous-time differential equations and discrete-time algorithms.

20.8.1 Definition: Continuous-Time State-Space Model

A **continuous-time linear state-space model** is defined by:

$$\dot{\mathbf{h}}(t) = A\mathbf{h}(t) + B\mathbf{u}(t), \quad \mathbf{y}(t) = C\mathbf{h}(t) \tag{20.11}$$

where:

- $\mathbf{h}(t) \in \mathbb{R}^N$ is the **hidden state**
- $\mathbf{u}(t) \in \mathbb{R}^D$ is the **input**
- $\mathbf{y}(t) \in \mathbb{R}^M$ is the **output**
- $A \in \mathbb{R}^{N \times N}$ is the **state matrix**
- $B \in \mathbb{R}^{N \times D}$ is the **input matrix**
- $C \in \mathbb{R}^{M \times N}$ is the **output matrix**

20.8.2 Solution via Matrix Exponential

The solution to the state equation with initial condition $\mathbf{h}(0) = \mathbf{h}_0$ is (see Chapter~6, equation *6.19*):

$$\mathbf{h}(t) = \exp(tA)\mathbf{h}_0 + \int_0^t \exp((t-\tau)A)B\mathbf{u}(\tau)\, d\tau \tag{20.12}$$

The first term is the **homogeneous solution** (response to initial conditions), and the second is the **particular solution** (response to input).

20.8.3 Stability Analysis

The behavior of the state-space model is determined by the eigenvalues of A:

- **Stable**: All eigenvalues have negative real parts ($\text{Re}(\lambda_i) < 0$) — states decay to zero exponentially
- **Unstable**: Any eigenvalue has positive real part — states grow without bound exponentially
- **Marginally stable**: Eigenvalues on imaginary axis — oscillatory behavior without growth or decay

This follows from the matrix exponential: $\exp(tA) = V\exp(t\Lambda)V^{-1}$, where $\exp(t\Lambda)$ has diagonal entries $e^{t\lambda_i}$. The real part of λ_i determines whether $e^{t\lambda_i}$ grows, decays, or oscillates.

20.8.4 Connection to Sequence Modeling

In sequence models, stability is crucial for:

- **Long-range dependencies**: Stable systems can remember information over long distances
- **Gradient flow**: Unstable systems cause exploding gradients during backpropagation
- **Training stability**: Marginally stable or unstable systems are difficult to train

Modern sequence models like S4 and Mamba use **diagonal state matrices with negative real eigenvalues** to ensure stability while maintaining expressiveness. The continuous-time perspective makes it clear why this works: stable continuous systems produce stable discrete systems when properly discretized.

20.8.5 Controllability and Observability

For state-space models, two important properties are:

- **Controllability**: Can we drive the system to any desired state using appropriate inputs?
- **Observability**: Can we determine the initial state from the output history?

These properties are determined by the matrices (A, B, C) and are crucial for understanding what the model can learn and represent. In sequence models, controllability ensures the model can generate diverse outputs, while observability ensures the model can extract information from inputs.

20.9 Discretization of Linear ODEs

To implement continuous-time dynamics on digital computers, we must **discretize** the equations. Given a time step $\Delta > 0$, we seek a discrete-time system:

$$\mathbf{h}_k = \bar{A}\mathbf{h}_{k-1} + \bar{B}\mathbf{u}_k, \quad \mathbf{y}_k = C\mathbf{h}_k \tag{20.13}$$

that approximates the continuous dynamics. Different discretization methods yield different $\bar{A}$ and $\bar{B}$.

20.9.1 Method 1: Forward Euler

The simplest discretization approximates the derivative:

$$\dot{\mathbf{h}}(t) \approx \frac{\mathbf{h}_{k+1} - \mathbf{h}_k}{\Delta}$$

Substituting into the ODE:

$$\frac{\mathbf{h}_{k+1} - \mathbf{h}_k}{\Delta} = A\mathbf{h}_k + B\mathbf{u}_k$$

yields:

$$\bar{A}_{\text{Euler}} = I + \Delta A, \quad \bar{B}_{\text{Euler}} = \Delta B \tag{20.14}$$

Properties:

- Simple to compute
- First-order accurate: error $O(\Delta^2)$ per step
- May be unstable for stiff systems (large eigenvalues of A)

20.9.2 Method 2: Zero-Order Hold (Exact)

Assuming the input $\mathbf{u}(t)$ is constant over each interval $[k\Delta, (k+1)\Delta)$, the **exact** discrete-time equivalent is:

$$\bar{A}_{\text{ZOH}} = \exp(\Delta A), \quad \bar{B}_{\text{ZOH}} = \left(\int_0^\Delta \exp(\tau A)\, d\tau\right) B \tag{20.15}$$

For invertible A:

$$\bar{B}_{\text{ZOH}} = A^{-1}(\exp(\Delta A) - I)B \tag{20.16}$$

Properties:

- Exact for piecewise-constant inputs
- Preserves stability: if $\text{Re}(\lambda_i) < 0$ for all eigenvalues of A, then $|\mu_i| < 1$ for all eigenvalues of $\bar{A}$
- Requires computing matrix exponential (can be expensive)

20.9.3 Method 3: Bilinear Transform (Tustin)

The **bilinear transform** approximates:

$$\frac{d}{dt} \approx \frac{2}{\Delta}\frac{1-z^{-1}}{1+z^{-1}}$$

This yields the implicit scheme:

$$\bar{A}_{\text{bilinear}} = \left(I - \frac{\Delta}{2}A\right)^{-1}\left(I + \frac{\Delta}{2}A\right) \tag{20.17}$$

$$\bar{B}_{\text{bilinear}} = \left(I - \frac{\Delta}{2}A\right)^{-1}\sqrt{\Delta}\,B \tag{20.18}$$

Properties:

- Second-order accurate
- Maps stable continuous systems to stable discrete systems (stability-preserving)
- Maps imaginary axis to unit circle exactly
- Requires matrix inverse (one-time computation)

20.9.4 Comparison of Discretization Methods

Method	$\bar{A}$	Accuracy	Stability	Complexity
Forward Euler	$I + \Delta A$	$O(\Delta)$	Conditional	Low
Zero-Order Hold	$\exp(\Delta A)$	Exact	Preserving	High
Bilinear	$(I - \frac{\Delta}{2}A)^{-1}(I + \frac{\Delta}{2}A)$	$O(\Delta^2)$	Preserving	Medium

The choice of discretization method has significant implications. **Forward Euler** is simple but can be unstable for systems with large eigenvalues (stiff systems). **Zero-order hold** is exact for piecewise-constant inputs but requires computing a matrix exponential. **Bilinear transform** offers a good balance: it is stability-preserving and second-order accurate, requiring only matrix operations. The connection between discretization and the continuous model explains why discrete algorithms can approximate continuous dynamics arbitrarily well as $\Delta \to 0$.

In **state-space sequence models**, efficient discretization is crucial: using diagonal A matrices reduces $\exp(\Delta A)$ to element-wise exponentials, making zero-order hold practical.

20.9.5 Diagonal State Matrices

When A is diagonal (or can be diagonalized), discretization simplifies dramatically. For $A = \text{diag}(\lambda_1, \dots, \lambda_N)$:

$$\exp(\Delta A) = \text{diag}(e^{\Delta\lambda_1}, \dots, e^{\Delta\lambda_N}) \tag{20.19}$$

This reduces the matrix exponential to N scalar exponentials—an $O(N)$ operation instead of $O(N^3)$.

Structured parameterization: In practice, A is often parameterized as diagonal with learnable entries, enabling efficient computation while maintaining expressiveness. This connects to the eigendecomposition perspective: choosing A diagonal is equivalent to working in the eigenbasis.

20.9.6 Connection to Modern Sequence Models

Modern sequence models like **S4** (Structured State Space), **Mamba**, and **RWKV** use this diagonal structure for efficiency:

- **S4**: Uses diagonal-plus-low-rank structure for A, enabling $O(N \log N)$ computation
- **Mamba**: Uses data-dependent diagonal matrices, adapting to input content
- **RWKV**: Uses diagonal state matrices with special initialization for long-range dependencies

The continuous-time perspective provides the theoretical foundation: these models are discretizations of continuous state-space ODEs, and the diagonal structure makes the discretization efficient.

20.9.7 Stability-Preserving Discretization

An important property of zero-order hold and bilinear transform is that they **preserve stability**: if the continuous system is stable (all eigenvalues of A have negative real parts), then the discrete system is also stable (all eigenvalues of $\bar{A}$ have magnitude less than 1).

This is crucial for training stability: unstable discretizations can cause gradients to explode during backpropagation through time. Stability-preserving methods ensure that the learned continuous dynamics remain stable when discretized.

Figure 20.6 illustrates why modern sequence models (S4, Mamba) prefer specific discretizations over simple Euler steps. The plot shows the complex plane (Re vs. Im eigenvalues). The green shaded region represents the **continuous stability region** (left half plane, Re < 0), where stable continuous systems live. The red dashed circle shows the **Forward Euler stable region**: Euler discretization maps stable continuous points to $z = 1 + \lambda\Delta t$, which can map stable points **outside** the unit circle (unstable in discrete time). This is why Euler can be unstable for stiff systems. In contrast, **Bilinear/ZOH** methods map the entire left half plane exactly inside the unit circle (always stable). This visualization explains why modern SSMs use stability-preserving discretizations: they ensure that stable continuous dynamics remain stable when discretized, preventing gradient explosions during training.

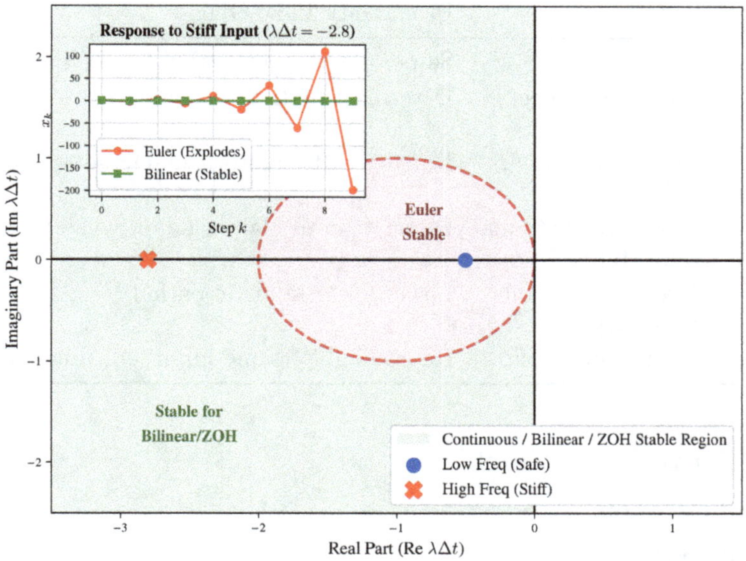

Figure 20.6: Stability of Discretization Methods

The connection between continuous-time state-space models and discrete sequence models is a powerful paradigm in modern deep learning. By viewing RNNs, Transformers, and state-space models as discretizations of continuous ODEs, we gain insights into their behavior, design better architectures, and understand their theoretical properties. The diagonal structure is particularly important because it makes the continuous-to-discrete transition computationally efficient while maintaining expressiveness.

20.10 Reverse-Time Dynamics Revisited

Recall from Chapter~19 the reverse-time SDE:

$$d\mathbf{X}_t = \left[\mathbf{f}(\mathbf{X}_t, t) - \mathbf{G}(t)\mathbf{G}(t)^\top \nabla_{\mathbf{x}} \log p_t(\mathbf{X}_t)\right] dt + \mathbf{G}(t)\, d\bar{\mathbf{W}}_t \tag{20.20}$$

Removing the noise term yields **exactly** the probability flow ODE.

Thus:

- Reverse SDE $\to$ stochastic sampling
- Probability flow ODE $\to$ deterministic sampling

Both are valid.

20.11 Relationship Between ODE and SDE

The following table summarizes the key differences and similarities between SDEs and probability flow ODEs:

Aspect	SDE	Probability Flow ODE
Noise	Explicit ($d\mathbf{W}_t$)	Implicit via score function
Trajectories	Random (different for each realization)	Deterministic (unique for each initial condition)

Aspect	SDE	Probability Flow ODE
Marginals	$p_t(\mathbf{x})$	Same $p_t(\mathbf{x})$ (by construction)
Sampling	Stochastic (injects randomness)	Deterministic (no randomness)
Solvers	Euler–Maruyama, Milstein	Runge–Kutta, adaptive ODE solvers
Speed	Slower (requires many small steps)	Faster (can use larger adaptive steps)
Diversity	Higher (noise adds variability)	Lower (deterministic paths)
Reproducibility	Non-reproducible	Reproducible (same input $\to$ same output)

20.11.1 When to Use Which?

Use SDE sampling when:

- Maximum diversity is needed
- The score function has errors (noise can help average out errors)
- You want to explore the full distribution

Use ODE sampling when:

- Speed is critical
- Reproducibility is needed
- You want smooth interpolation in latent space
- The score function is accurate

In practice, many modern diffusion models use **hybrid approaches**: start with a few stochastic steps for diversity, then switch to deterministic ODE steps for speed.

20.12 Continuous Limits of Diffusion Models

Diffusion models consist of three key components, each of which has both discrete and continuous formulations:

1. **Forward noising process (SDE)**
    - Discrete: Multi-step noise addition $\mathbf{x}_t = \sqrt{\alpha_t}\mathbf{x}_0 + \sqrt{1-\alpha_t}\epsilon$
    - Continuous: Forward SDE $d\mathbf{X}_t = \mathbf{f}(\mathbf{X}_t, t)dt + \mathbf{G}(t)d\mathbf{W}_t$
2. **Learned score function**
    - Discrete: Neural network trained on noisy data $\mathbf{s}_\theta(\mathbf{x}, t)$
    - Continuous: Approximation to $\nabla_{\mathbf{x}} \log p_t(\mathbf{x})$
3. **Reverse-time dynamics**
    - Discrete: Iterative denoising steps
    - Continuous: Reverse SDE or probability flow ODE

20.12.1 The Duality Principle

This duality—discrete algorithms and continuous processes—is central to modern generative modeling:

- **Discrete view**: Practical algorithms that can be implemented on computers
- **Continuous view**: Mathematical framework that reveals fundamental structure

The continuous view provides:

- **Theoretical understanding**: Reveals why algorithms work
- **Algorithm design**: Guides the development of better discrete methods
- **Unified framework**: Connects different methods (DDPM, DDIM, flow matching, etc.)

20.12.2 From Discrete to Continuous and Back

The typical workflow is:

1. **Start with continuous theory**: Define the forward SDE and understand its properties
2. **Discretize for training**: Use discrete steps to generate training data and learn the score function
3. **Use continuous insight for sampling**: Apply probability flow ODE or reverse SDE for generation
4. **Discretize again for implementation**: Use numerical solvers (Euler, Runge-Kutta, etc.) to solve the continuous equations

This back-and-forth between discrete and continuous is a powerful paradigm that has led to many advances in generative modeling.

20.13 Applications in Modern AI/ML

20.13.1 DDIM: Deterministic Diffusion Sampling

DDIM (Denoising Diffusion Implicit Models) was one of the first methods to demonstrate that diffusion models can be sampled deterministically. DDIM uses the learned denoising function without explicit noise injection, effectively solving a discretized version of the probability flow ODE.

The key insight is that the denoising function learned during training already encodes the score function, so we can use it deterministically during sampling.

20.13.2 Fast Diffusion Sampling

Deterministic ODE sampling enables **fast diffusion** methods that can generate high-quality samples in just 10-50 steps (compared to 1000+ steps for standard diffusion). Methods like:

- **DPM-Solver**: Uses high-order ODE solvers for fast sampling
- **DEIS**: Deterministic sampling with improved accuracy
- **PNDM**: Pseudo numerical methods for diffusion models

All rely on the probability flow ODE perspective to achieve speedups.

20.13.3 Latent Space Manipulation

Deterministic paths enable **latent space manipulation**:

- **Interpolation**: Smoothly interpolate between two samples by following deterministic paths
- **Style transfer**: Modify samples by controlling the ODE trajectory
- **Controllable generation**: Guide generation by modifying the velocity field

These applications are much easier with deterministic paths than with random SDE paths.

20.13.4 State-Space Sequence Models

Modern sequence models like **S4**, **Mamba**, and **RWKV** use continuous-time state-space models that are discretized for implementation. The continuous-time perspective provides:

- **Long-range dependencies**: Continuous models can capture dependencies over arbitrary distances
- **Efficient computation**: Diagonal state matrices enable $O(N)$ computation instead of $O(N^2)$
- **Theoretical understanding**: Continuous limits reveal the fundamental structure

These models demonstrate how the continuous-time perspective leads to practical improvements in discrete implementations.

20.14 Conceptual Summary

This chapter unifies stochastic and deterministic continuous-time models through the lens of marginal probability distributions. The key equivalences are summarized below:

Concept	Meaning
ODE	Deterministic flow
SDE	Randomized flow
Score	Geometry of density
Probability flow	Noise-free equivalent
Continuous limit	True mathematical model

Although an SDE and its corresponding Probability Flow ODE generate individual sample paths with fundamentally different stochastic behaviors, both processes produce identical marginal probability densities $p_t(\mathbf{x})$ at every time step t. This duality enables fast, deterministic ODE integration for generative sampling without sacrificing sample quality.

20.15 Mindmap

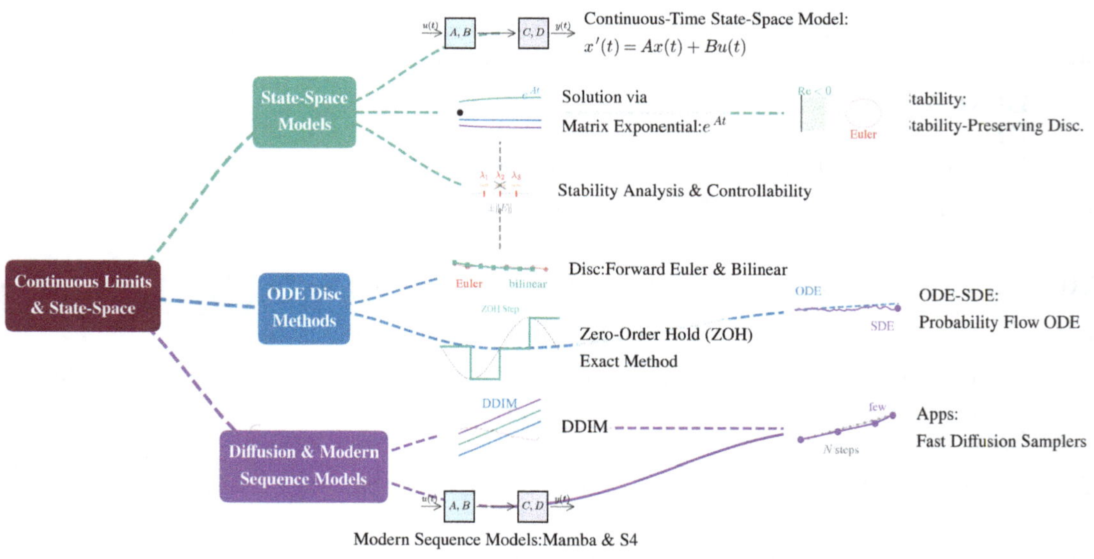

Figure 20.7: Mind map of continuous limits and state-space models topics

20.16 Exercises

20.16.1 Discrete to Continuous Limits

1. **Gradient descent limit**: Starting from $\mathbf{x}_{k+1} = \mathbf{x}_k - \eta \nabla f(\mathbf{x}_k)$, derive the limiting ODE $\frac{d\mathbf{x}}{dt} = -\nabla f(\mathbf{x})$ by setting $t = k\eta$ and letting $\eta \to 0$.
2. **Langevin limit**: Starting from

$$\mathbf{x}_{k+1} = \mathbf{x}_k + \eta \mathbf{s}(\mathbf{x}_k) + \sqrt{2\eta}\, \epsilon_k,$$

explain how it corresponds to Euler–Maruyama applied to the SDE $d\mathbf{X}_t = \mathbf{s}(\mathbf{X}_t)\, dt + \sqrt{2}\, d\mathbf{W}_t$.

20.16.2 Probability Flow ODE

1. **Write the ODE drift**: For the forward SDE $d\mathbf{X}_t = \mathbf{f}(\mathbf{X}_t, t)\, dt + \mathbf{G}(t)\, d\mathbf{W}_t$ with $\mathbf{a}(t) = \mathbf{G}(t)\mathbf{G}(t)^\top$, write the probability flow ODE explicitly.
2. **Factor 1/2 sanity check**: The probability flow drift contains a $\frac{1}{2}\mathbf{a}(t)\nabla \log p_t$ term. Give an intuition for why the factor 1/2 appears (hint: compare reverse SDE drift vs "remove the noise").
3. **Deterministic sampling**: Suppose you have an oracle for the score $\nabla_{\mathbf{x}} \log p_t(\mathbf{x})$. Describe (at a high level) how you would generate samples using an ODE solver.

20.16.3 Continuity Equation (ODE Distribution Dynamics)

1. **Continuity equation derivation (conceptual)**: For the ODE $\dot{\mathbf{x}} = \mathbf{v}(\mathbf{x}, t)$, state the continuity equation for $p_t(\mathbf{x})$ and explain in one paragraph why it expresses conservation of probability mass.
2. **1D transport PDE**: In one dimension, $\dot{x} = v(x, t)$ implies $\partial_t p + \partial_x (pv) = 0$. Expand this PDE and interpret the two terms $\partial_t p$ and $\partial_x (pv)$.
3. **Mass conservation**: Assume $p_t(\mathbf{x})$ decays sufficiently fast at infinity. Show that $\frac{d}{dt}\int_{\mathbb{R}^d} p_t(\mathbf{x})\, d\mathbf{x} = 0$ under the continuity equation.

20.16.4 Flow Matching (Vector Field Matching)

1. **What is being matched?**: In flow matching, we learn $\mathbf{v}_\theta(\mathbf{x}, t)$. Explain what data pairs $(t, \mathbf{x}_t)$ and target velocity $\mathbf{u}(\mathbf{x}_t, t)$ mean conceptually.
2. **Loss form**: Write a generic squared-error objective for flow matching and explain what choice(s) in the method determine $\mathbf{u}$.

20.16.5 State-Space Models as Flows

1. **Linear state-space dynamics**: Consider $\dot{\mathbf{x}}(t) = A\mathbf{x}(t) + B\mathbf{u}(t)$. If $\mathbf{u}(t) \equiv 0$, give a condition on the eigenvalues of A that ensures stability (solutions decay to 0).
2. **Discrete-time connection**: Using Euler discretization with step size Δt, derive a discrete-time update $\mathbf{x}_{k+1} = \bar{A}\mathbf{x}_k + \bar{B}\mathbf{u}_k$ in terms of $A, B, \Delta t$.

20.16.6 ODE vs SDE

1. **Trajectory vs marginal**: Explain the difference between (i) matching sample paths and (ii) matching marginal densities p_t. Why can an SDE and an ODE share the same p_t while having very different trajectories?
2. **When randomness matters**: Give one situation where stochastic sampling may be preferable (or necessary) compared to deterministic ODE sampling, and justify your answer.

Chapter 21

Fokker-Planck and Distribution Dynamics

The evolution of probability distributions: Fokker-Planck equations and diffusion processes

"Entropy is a measure of the probability of a given state."

— Ludwig Boltzmann

- ▷ `Diffusion matrix` $\mathbf{D}(\mathbf{x},t) = \mathbf{G}(\mathbf{x},t)\mathbf{G}(\mathbf{x},t)^\top$
- ▷ `Fokker-Planck (Itô SDE)` for $d\mathbf{X}_t = \mathbf{f}(\mathbf{X}_t,t)\,dt + \mathbf{G}(\mathbf{X}_t,t)\,d\mathbf{W}_t$, $\partial_t p_t(\mathbf{x}) = -\nabla_\mathbf{x}\cdot(\mathbf{f}(\mathbf{x},t)p_t(\mathbf{x})) + \frac{1}{2}\nabla_\mathbf{x}\cdot\nabla_\mathbf{x}(\mathbf{D}(\mathbf{x},t)p_t(\mathbf{x}))$
- ▷ `Continuity form` $\partial_t p_t + \nabla\cdot\mathbf{J}_t = 0$
- ▷ `Probability current` $\mathbf{J}_t(\mathbf{x}) = \mathbf{f}(\mathbf{x},t)p_t(\mathbf{x}) - \frac{1}{2}\nabla\cdot(\mathbf{D}(\mathbf{x},t)p_t(\mathbf{x}))$
- ▷ `Langevin Fokker-Planck` for $d\mathbf{X}_t = \nabla\log p(\mathbf{X}_t)\,dt + \sqrt{2}\,d\mathbf{W}_t$, $\partial_t p_t = -\nabla\cdot(p_t\nabla\log p) + \nabla^2 p_t$
- ▷ `Stationary distribution` p_* is stationary if $\partial_t p_t = 0$ (equivalently $\nabla\cdot\mathbf{J}_* = 0$; detailed balance often strengthens to $\mathbf{J}_* \equiv 0$)
- ▷ `Heat equation (pure diffusion)` if $\mathbf{f} = 0$ and $\mathbf{D}(\mathbf{x},t) = \beta(t)I$, then $\partial_t p_t = \frac{1}{2}\beta(t)\nabla^2 p_t$
- ▷ `Probability flow ODE (state-independent diffusion)` for $d\mathbf{X}_t = \mathbf{f}(\mathbf{X}_t,t)\,dt + \mathbf{G}(t)\,d\mathbf{W}_t$ with $\mathbf{a}(t) = \mathbf{G}(t)\mathbf{G}(t)^\top$, $\frac{d\mathbf{X}_t}{dt} = \mathbf{f}(\mathbf{X}_t,t) - \frac{1}{2}\mathbf{a}(t)\nabla_\mathbf{x}\log p_t(\mathbf{X}_t)$ (induces the same marginals p_t)
- ▷ `Reverse-time density evolution (score-dependent)` reversing diffusion is ill-posed without the score $\nabla\log p_t$

21.1 Motivation: From Paths to Densities

In Chapters 19–20, we studied **stochastic differential equations (SDEs)** and their deterministic ODE counterparts:

$$d\mathbf{X}_t = \mathbf{f}(\mathbf{X}_t, t)\, dt + \mathbf{G}(\mathbf{X}_t, t)\, d\mathbf{W}_t \tag{21.1}$$

This describes **individual sample paths**—random trajectories $\mathbf{X}_t(\omega)$ for each realization ω of the stochastic process.

However, generative modeling ultimately concerns **distributions**, not individual paths. We care about:

- The probability distribution $p_t(\mathbf{x})$ of samples at time t
- How this distribution evolves over time
- Whether it converges to the target distribution

The fundamental question is:

- Given an SDE, how does the probability density $p_t(\mathbf{x})$ evolve over time?

The answer is the **Fokker–Planck equation**—a partial differential equation (PDE) that describes the deterministic evolution of probability distributions under stochastic dynamics.

21.1.1 The Three Levels of Description

We now have a complete three-level description of stochastic processes:

1. **Path level (SDE)**: Individual random trajectories $\mathbf{X}_t(\omega)$
2. **Path level (ODE)**: Deterministic trajectories that produce the same marginals
3. **Distribution level (PDE)**: Evolution of the probability density $p_t(\mathbf{x})$

The Fokker-Planck equation provides the highest-level description, directly describing how probability mass flows and spreads, without reference to individual paths. This perspective is particularly powerful for understanding the long-term behavior and convergence properties of diffusion processes.

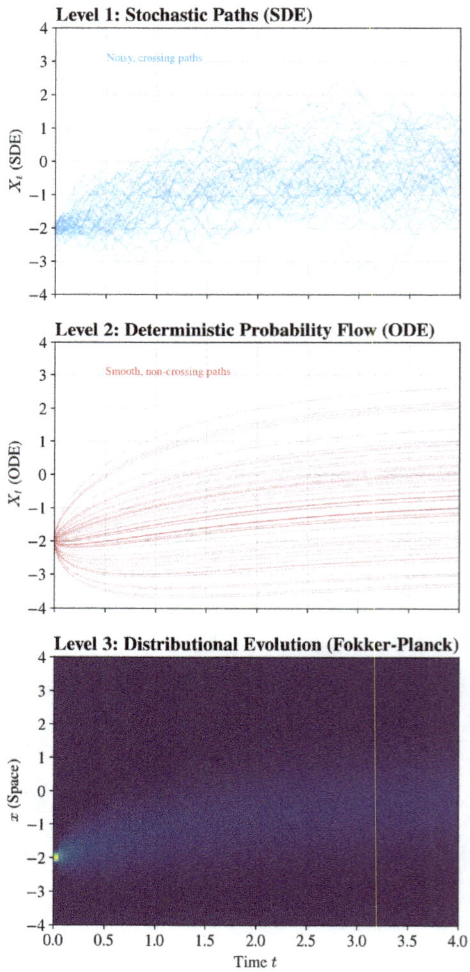

Figure 21.1: Path and Distribution Levels

Figure 21.1 visualizes the complete three-level hierarchy of diffusion process descriptions. The top panel shows **Level 1: Stochastic Paths (SDE)**—individual random trajectories that are noisy and cross each other frequently, representing the physical reality of diffusion. The middle panel shows **Level 2: Deterministic Probability Flow (ODE)**—smooth, non-crossing paths that define the "skeleton" of the distribution, producing the same marginals as the SDE. The bottom panel shows **Level 3: Distributional Evolution (Fokker-Planck PDE)**—a heatmap of the probability density $p_t(x)$ evolving over time. Both the SDE and ODE paths above generate samples that follow this exact density evolution. This visualization makes concrete the hierarchy: from random individual paths (microscopic) to deterministic distribution evolution (macroscopic), with the Fokker-Planck equation providing the highest-level mathematical description that unifies all three perspectives.

21.2 Distributional Viewpoint

An SDE induces two related but distinct objects:

- **A stochastic process $\mathbf{X}_t$**: A collection of random variables indexed by time, describing individual sample paths
- **A deterministic family of densities $p_t(\mathbf{x})$**: The probability density function of $\mathbf{X}_t$ at each time t

21.2.1 The Determinism of Distributions

While individual trajectories $\mathbf{X}_t(\omega)$ are random and unpredictable, the density evolution $p_t(\mathbf{x})$ is **fully deterministic**. This is a remarkable fact: randomness at the microscopic (path) level produces determinism at the macroscopic (distribution) level.

This mirrors a deep principle from statistical mechanics and physics:

- **Random micro-dynamics produce deterministic macro-laws.**

Examples:

- **Brownian motion**: Individual particles move randomly, but the concentration evolves deterministically (diffusion equation)
- **Gas dynamics**: Individual molecules move randomly, but pressure and temperature follow deterministic laws (ideal gas law)
- **Diffusion models**: Individual sample paths are random, but the distribution evolves deterministically (Fokker-Planck equation)

21.2.2 Why This Matters for Generative Modeling

For generative modeling, the distributional viewpoint is often more useful than the path viewpoint because:

1. **We care about distributions**: The goal is to generate samples from a target distribution, not to follow specific paths
2. **Deterministic evolution**: The distribution evolves deterministically, making it easier to analyze and control
3. **Convergence analysis**: We can study convergence to the target distribution without tracking individual paths
4. **Design principles**: Understanding distributional evolution helps design better diffusion processes

The Fokker-Planck equation provides the mathematical tool to work directly with distributions, bypassing the need to analyze individual stochastic paths.

21.3 The Fokker–Planck Equation

21.3.1 Definition: Fokker–Planck Equation

Consider the Itô SDE:

$$d\mathbf{X}_t = \mathbf{f}(\mathbf{X}_t, t)dt + \mathbf{G}(\mathbf{X}_t, t)d\mathbf{W}_t \tag{21.2}$$

where $\mathbf{f}: \mathbb{R}^d \times \mathbb{R} \to \mathbb{R}^d$ is the drift and $\mathbf{G}: \mathbb{R}^d \times \mathbb{R} \to \mathbb{R}^{d \times m}$ is the diffusion matrix.

Let $p_t(\mathbf{x})$ be the probability density function of $\mathbf{X}_t$ (the density of the random variable at time t).

Then p_t satisfies the **Fokker–Planck equation** (also called the **Kolmogorov forward equation**):

$$\frac{\partial p_t(\mathbf{x})}{\partial t} = -\nabla_{\mathbf{x}} \cdot (\mathbf{f}(\mathbf{x}, t) p_t(\mathbf{x})) + \frac{1}{2} \nabla_{\mathbf{x}} \cdot \nabla_{\mathbf{x}} (\mathbf{D}(\mathbf{x}, t) p_t(\mathbf{x})) \tag{21.3}$$

where:

$$\mathbf{D}(\mathbf{x}, t) = \mathbf{G}(\mathbf{x}, t) \mathbf{G}(\mathbf{x}, t)^\top \tag{21.4}$$

is the **diffusion matrix** (also called the **diffusion tensor**).

21.3.2 Historical Context

The Fokker-Planck equation was independently derived by:

- **Adriaan Fokker** (1914): In the context of Brownian motion in a force field
- **Max Planck** (1917): In the context of statistical mechanics

It is a fundamental equation in statistical physics, describing how probability distributions evolve under the combined influence of deterministic drift and random diffusion.

21.3.3 Derivation Sketch

The Fokker-Planck equation can be derived using Itô's lemma and the definition of the probability density. The key steps are:

1. Consider an arbitrary smooth test function $\phi(\mathbf{x})$
2. Compute $\frac{d}{dt}\mathbb{E}[\phi(\mathbf{X}_t)]$ using Itô's lemma
3. Express this in terms of $p_t(\mathbf{x})$ using integration by parts
4. Since ϕ is arbitrary, equate coefficients to obtain the Fokker-Planck equation

The derivation reveals that the Fokker-Planck equation is the **adjoint** of the generator of the Markov process, providing a deep connection between stochastic processes and partial differential equations.

21.3.4 Interpretation of Terms

Each term in the Fokker-Planck equation has a clear physical and probabilistic interpretation:

Term	Meaning	Physical Analogy
$\partial_t p$	Temporal change in probability density	Rate of change of concentration
$-\nabla \cdot (\mathbf{f}p)$	Transport (drift) of probability mass	Advection in fluid flow
$\frac{1}{2}\nabla\nabla(\mathbf{D}p)$	Diffusion (spreading) of probability mass	Molecular diffusion in a fluid

The drift term $-\nabla \cdot (\mathbf{f}p)$: This represents the **advection** or **transport** of probability mass by the deterministic drift field $\mathbf{f}(\mathbf{x}, t)$. Probability mass flows along the drift field, like particles being carried by a current.

The diffusion term $\frac{1}{2}\nabla \cdot \nabla(\mathbf{D}p)$: This represents the **spreading** or **diffusion** of probability mass due to the stochastic noise. The diffusion matrix $\mathbf{D}$ controls how much and in which directions the probability spreads. This is analogous to how heat or particles diffuse in a medium.

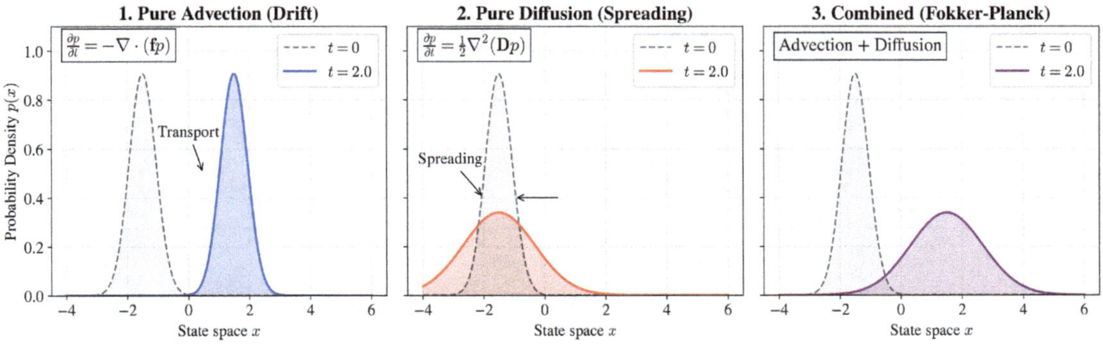

Figure 21.2: Advection, Diffusion, and Combined Effects

Figure 21.2 demonstrates the separate and combined effects of advection and diffusion in the Fokker-Planck equation. The left panel shows **pure advection** (drift only): the initial distribution (dashed line) is transported to the right without changing shape, like particles being carried by a current. The middle panel shows **pure diffusion** (spreading only): the distribution stays centered but spreads outward, with variance increasing linearly with time. The right panel shows **combined Fokker-Planck evolution**: both translation and spreading occur simultaneously. This visualization makes concrete how the two terms in the Fokker-Planck equation work together: the drift term transports probability mass, while the diffusion term spreads it, and their combination determines the net evolution of the distribution.

21.3.5 Particle Cloud: Microscopic Paths to Macroscopic Density

While Figure 21.2 illustrates continuous 1D density evolution theoretically, Figure 21.3 provides a particle-level numerical verification by tracking the evolution of a **cloud of 2000 particles** under Langevin dynamics for a 2D mixture of Gaussians. Instead of following a single trajectory, this visualization shows how the **empirical density of particles** evolves over time:

- $t = 0$ **(Initialization)**: Particles start in a tight cluster in a low-probability region far from the modes.
- $t = 30$ **(Advection-Dominant Phase)**: The **deterministic drift** dominates—the score field acts like a strong force pushing the cloud rapidly toward high-density regions. The stretching of the cloud directly visualizes the **advection term** in the Fokker-Planck equation.
- $t = 1000$ **(Stationary Equilibrium)**: The system reaches **equilibrium**. The **diffusion term** (stochastic noise) prevents particles from collapsing to single points, keeping them spread out to match the theoretical distribution.

This visualization demonstrates that Langevin dynamics on individual samples (microscopic SDE level) is exact equivalent to Fokker-Planck evolution on probability densities (macroscopic PDE level).

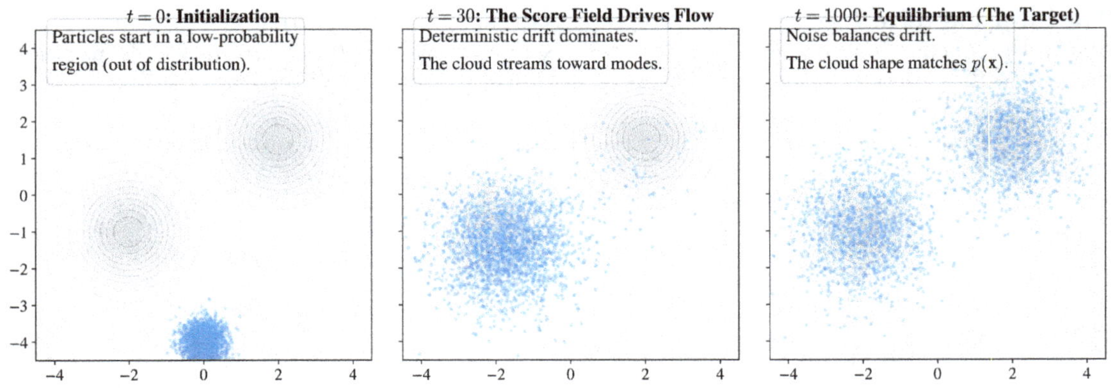

Figure 21.3: Evolution of a Particle Cloud

21.3.6 Conservation of Probability

The Fokker-Planck equation is a **conservation law** for probability mass. It ensures that:

$$\int p_t(\mathbf{x})d\mathbf{x} = 1 \quad \text{for all } t$$

(assuming appropriate boundary conditions). Probability mass is neither created nor destroyed—it only flows and spreads. This is the probabilistic analogue of mass conservation in fluid dynamics.

21.3.7 Connection to Fluid Dynamics

The Fokker-Planck equation is mathematically similar to the **advection-diffusion equation** in fluid dynamics:

$$\frac{\partial c}{\partial t} + \nabla \cdot (c\mathbf{v}) = \nabla \cdot (D\nabla c)$$

where $c(\mathbf{x}, t)$ is concentration, $\mathbf{v}$ is velocity, and D is diffusivity. The Fokker-Planck equation is the probabilistic version, where:

- Concentration $c \to$ Probability density p
- Velocity $\mathbf{v} \to$ Drift $\mathbf{f}$
- Diffusivity $D \to$ Diffusion matrix $\mathbf{D}$

This connection provides intuition: probability behaves like a fluid that is both advected (by drift) and diffused (by noise).

21.4 Continuity Equation Perspective

The Fokker–Planck equation can be written in **continuity equation form**:

$$\frac{\partial p_t}{\partial t} + \nabla \cdot \mathbf{J}_t = 0 \tag{21.5}$$

where $\mathbf{J}_t(\mathbf{x})$ is the **probability current** (also called the **probability flux**):

$$\mathbf{J}_t = \mathbf{f}(\mathbf{x}, t)p_t(\mathbf{x}) - \frac{1}{2}\nabla \cdot (\mathbf{D}(\mathbf{x}, t)p_t(\mathbf{x})) \tag{21.6}$$

21.4.1 Physical Interpretation

This form makes the **conservation of probability** explicit:

- $\frac{\partial p_t}{\partial t}$: Rate of change of probability density at a point
- $\nabla \cdot \mathbf{J}_t$: Net flux of probability into/out of that point
- **Equation**: The change in density equals the negative divergence of the current (inflow minus outflow)

This mirrors **fluid dynamics** and **electromagnetism**, where similar continuity equations describe conservation of mass or charge:

- **Probability behaves like a fluid**: Probability mass flows through space
- **Drift transports mass**: The term $\mathbf{f}(\mathbf{x}, t)p_t(\mathbf{x})$ represents advective transport
- **Diffusion spreads mass**: The term $-\frac{1}{2}\nabla \cdot (\mathbf{D}p_t)$ represents diffusive spreading

21.4.2 Probability Current in Detail

The probability current $\mathbf{J}_t(\mathbf{x})$ has two components:

1. **Advective current** $\mathbf{f}(\mathbf{x}, t)p_t(\mathbf{x})$: Probability mass flowing due to the drift field
2. **Diffusive current** $-\frac{1}{2}\nabla \cdot (\mathbf{D}(\mathbf{x}, t)p_t(\mathbf{x}))$: Probability mass flowing due to diffusion (Fick's law)

The total current $\mathbf{J}_t$ describes the net flow of probability mass at each point in space and time.

Figure 21.4 visualizes the probability current $\mathbf{J}_t$ for a 2D spiraling Ornstein-Uhlenbeck process. The background shows contour lines of the probability density $p_t(\mathbf{x})$ (a Gaussian distribution), with darker regions indicating higher density. The vector field (arrows) shows the total probability current $\mathbf{J}_t = \mathbf{f}p - \frac{1}{2}\nabla(\mathbf{D}p)$, which represents the net flow of probability mass. The current has two competing components: the **advective current** ($\mathbf{f}p$) from the drift field, which spirals inward toward the origin, and the **diffusive current** ($-\frac{1}{2}\nabla(\mathbf{D}p)$), which pushes outward, spreading mass. The total current shows a solenoidal (circulating) flow pattern. This visualization makes concrete the continuity equation perspective: even if the density $p_t(\mathbf{x})$ is stationary (constant in time), there can still be a non-zero solenoidal current if detailed balance does not hold, as seen in this spiraling example. The probability current represents the net flow of probability mass, connecting the microscopic (path-level) dynamics to the macroscopic (distribution-level) evolution.

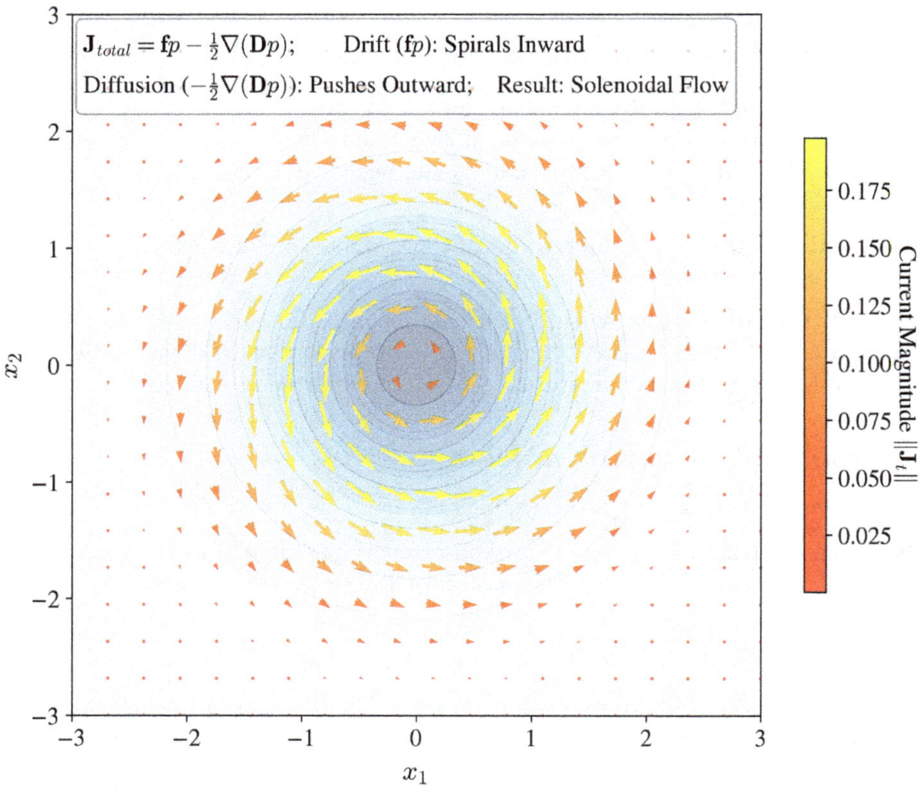

Figure 21.4: The Probability Current

21.4.3 Stationary States

At equilibrium (stationary distribution), $\frac{\partial p_t}{\partial t} = 0$, so:

$$\nabla \cdot \mathbf{J}_t = 0$$

This means the probability current is **divergence-free** (solenoidal). In many cases, we have the stronger condition:

$$\mathbf{J}_t = \mathbf{0}$$

which is called **detailed balance**. This means there is no net flow of probability at equilibrium—the system is in perfect balance.

21.5 Example: Langevin Dynamics

Consider the Langevin SDE (Chapter 18):

$$d\mathbf{X}_t = \nabla_{\mathbf{x}} \log p(\mathbf{X}_t)\, dt + \sqrt{2}\, d\mathbf{W}_t \tag{21.7}$$

where $p(\mathbf{x})$ is the target distribution we want to sample from.

Here:

- **Drift:** $\mathbf{f}(\mathbf{X}_t, t) = \nabla_{\mathbf{x}} \log p(\mathbf{X}_t)$ (the score function)
- **Diffusion:** $\mathbf{G}(\mathbf{X}_t, t) = \sqrt{2}I$ (constant, state-independent)
- **Diffusion matrix:** $\mathbf{D} = \mathbf{G}\mathbf{G}^\top = 2I$

The corresponding Fokker–Planck equation is:

$$\frac{\partial p_t}{\partial t} = -\nabla \cdot (p_t \nabla \log p) + \nabla^2 p_t \tag{21.8}$$

where ∇^2 is the Laplacian operator (sum of second derivatives).

21.5.1 Simplifying the Langevin Fokker-Planck Equation

Using the identity $\nabla \cdot (p_t \nabla \log p) = \nabla \cdot \left(p_t \frac{\nabla p}{p}\right) = \nabla \cdot \nabla p = \nabla^2 p$, the equation simplifies to:

$$\frac{\partial p_t}{\partial t} = -\nabla^2 p + \nabla^2 p_t = \nabla^2 (p_t - p)$$

This shows that the distribution p_t evolves toward the target distribution p. When $p_t = p$, we have $\frac{\partial p_t}{\partial t} = 0$, confirming that p is a stationary solution.

21.5.2 Convergence to Stationary Distribution

The Langevin Fokker-Planck equation shows that p_t converges to p as $t \to \infty$ (under appropriate conditions). This is the distributional justification for why Langevin dynamics works: the Fokker-Planck equation has p as its unique stationary solution.

Figure 21.5 demonstrates why Langevin dynamics works using the distributional viewpoint. The top panel shows the **KL divergence** $D_{KL}(p_t || p_{\text{target}})$ monotonically decreasing toward zero over time, indicating convergence to the target distribution. Three key moments are highlighted: initialization ($t = 0$) with high KL divergence, transient phase ($t \approx 2.0$) with medium KL divergence, and equilibrium ($t = 10.0$) with near-zero KL divergence. The bottom panels show snapshots of the actual probability density $p_t(x)$ (histograms) compared to the theoretical target $p(x)$ (black dashed line) at these three moments. At $t = 0$, the distribution is a Gaussian initialization (wrong shape, high KL). At $t \approx 2.0$, the distribution splits and moves toward the modes of the double-well target (medium KL). At $t = 10.0$, the distribution perfectly matches the target (near-zero KL). This visualization makes concrete the fundamental insight: **Langevin dynamics is an optimization process in the space of probability distributions.** It minimizes the KL divergence to the target distribution, and the Fokker-Planck equation guarantees that this convergence happens deterministically, even though individual sample paths are random.

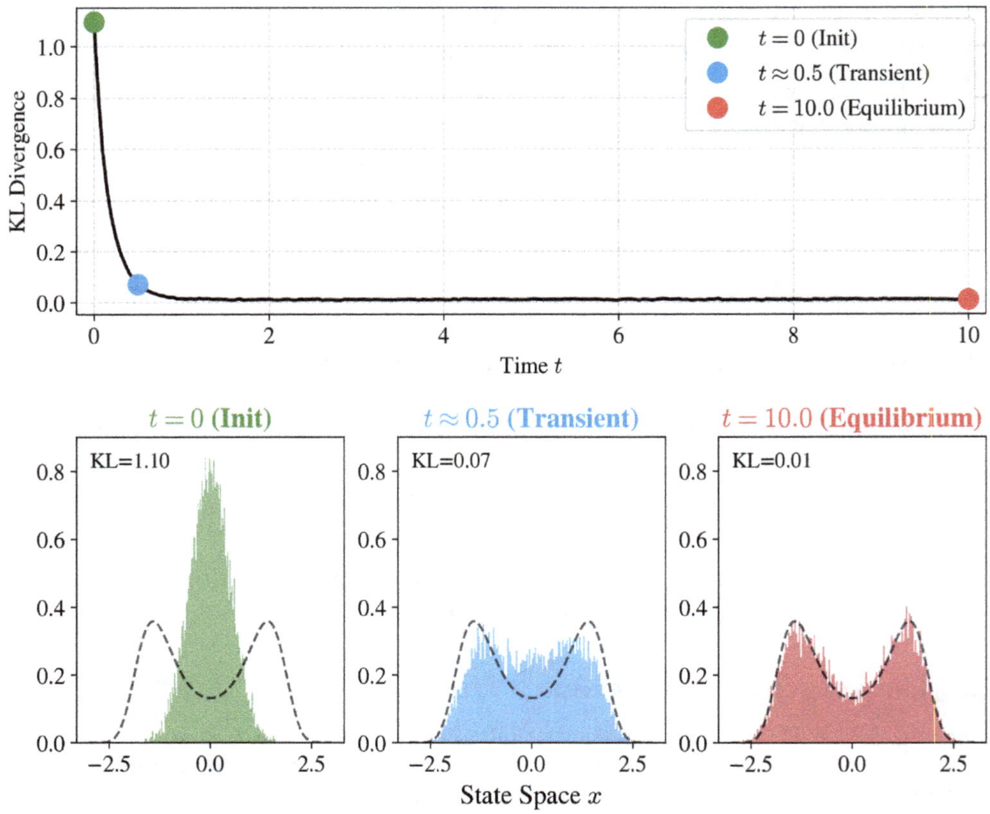

Figure 21.5: Langevin Dynamics Convergence

21.5.3 Stationary Distribution

At equilibrium (stationary state), $\partial_t p_t = 0$, yielding:

$$\nabla^2 p = \nabla \cdot (p \nabla \log p) \tag{21.9}$$

whose solution is exactly $p(\mathbf{x})$, the target distribution.

Verification: For $p_t = p$, we have:

- Left side: $\nabla^2 p$
- Right side: $\nabla \cdot (p\nabla \log p) = \nabla \cdot \nabla p = \nabla^2 p$

So the equation is satisfied, confirming that p is indeed a stationary solution.

21.5.4 Uniqueness and Convergence

Under appropriate conditions (ergodicity, irreducibility), the stationary solution p is **unique**, and any initial distribution p_0 converges to p as $t \to \infty$. This is the theoretical foundation for Langevin dynamics as a sampling algorithm.

Thus:

- **Langevin dynamics converges** because its Fokker–Planck equation has the correct fixed point $p(\mathbf{x})$
- **The convergence is guaranteed** by the mathematical structure of the Fokker-Planck equation

- **The rate of convergence** depends on the geometry of p and can be analyzed using spectral methods

This provides a complete distributional justification for why Langevin dynamics works: it's not just an algorithm, but a process whose distributional evolution is guaranteed to converge to the target.

21.6 Forward Diffusion in Generative Models

In diffusion models, the **forward process** is designed to gradually add noise to data. The forward SDE is typically:

$$d\mathbf{X}_t = -\frac{1}{2}\beta(t)\mathbf{X}_t dt + \sqrt{\beta(t)}d\mathbf{W}_t$$

where $\beta(t) > 0$ is a noise schedule that increases over time.

The corresponding Fokker–Planck equation is:

$$\frac{\partial p_t}{\partial t} = \frac{1}{2}\beta(t)\nabla^2 p_t\} \tag{21.10}$$

This is the **heat equation** (also called the **diffusion equation**), which describes pure diffusion without drift.

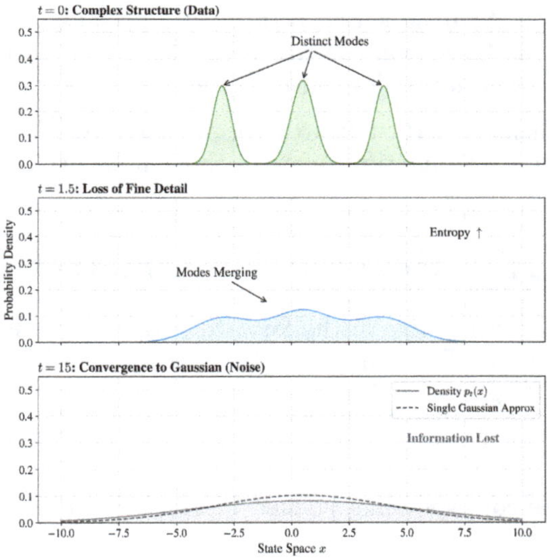

Figure 21.6: The Heat Equation: Forward Diffusion

Figure 21.6 visualizes the forward diffusion process in diffusion models, demonstrating how the heat equation destroys information over time. The top panel shows the initial distribution at $t = 0$: a complex, multimodal structure with three distinct peaks representing structured data with high information content. The middle panel shows the distribution at $t = 1.5$: the peaks begin to merge and blur as fine details are lost, entropy increases, and information is destroyed. The bottom panel shows the distribution at $t = 15$: convergence to a single, broad, featureless Gaussian (pure noise) with zero information. The dashed line shows the theoretical single Gaussian approximation. This visualization makes concrete the fundamental property of diffusion: it acts as a smoothing operator, and complex data structures are inevitably smoothed into trivial noise—this is the Central Limit Theorem in action. For diffusion models, this forward process transforms any data distribution into pure noise, which can then be reversed using the learned score function.

21.6.1 Solution of the Heat Equation

The solution of the heat equation with initial condition $p_0(\mathbf{x})$ is:

$$p_t(\mathbf{x}) = \mathcal{N}(\mathbf{x}; \mathbf{0}, \sigma_t^2 I) * p_0(\mathbf{x}) \tag{21.11}$$

where $*$ denotes convolution and $\sigma_t^2 = \int_0^t \beta(s)ds$ is the accumulated variance.

Interpretation: The distribution at time t is the convolution of the initial distribution p_0 with a Gaussian kernel. This means:

- Each point in p_0 spreads out into a Gaussian
- The variance σ_t^2 increases with time
- The distribution becomes smoother and more spread out

21.6.2 Convergence to Gaussian

As $t \to \infty$ (and $\sigma_t \to \infty$), p_t converges to a Gaussian:

$$p_t(\mathbf{x}) \xrightarrow{t\to\infty} \mathcal{N}(\mathbf{x}; \mathbf{0}, \sigma_\infty^2 I)$$

This is the **central limit theorem** in action: repeated convolution with Gaussians leads to a Gaussian distribution, regardless of the initial distribution p_0.

For diffusion models: This means the forward process transforms any data distribution into pure noise (Gaussian), which can then be reversed using the learned score function.

21.6.3 Why This Design Works

The heat equation design is particularly elegant because:

1. **Analytical solution**: The solution is known in closed form (Gaussian convolution)
2. **Simple score function**: For Gaussian noise, the score function has a simple form
3. **Controllable noise level**: The noise schedule $\beta(t)$ controls how quickly data is destroyed
4. **Theoretical guarantees**: The convergence to Gaussian is mathematically guaranteed

This makes the forward process easy to implement and analyze, while the reverse process (learning the score function) handles the complexity of the data distribution.

21.7 Reverse-Time Fokker–Planck Equation

Reversing time in the forward SDE yields a **reverse-time Fokker–Planck equation**. For the reverse SDE (Chapter~19):

$$d\mathbf{X}_t = [\mathbf{f}(\mathbf{X}_t, t) - \mathbf{D}(t)\nabla_\mathbf{x} \log p_t(\mathbf{X}_t)]\, dt + \mathbf{G}(t)d\bar{\mathbf{W}}_t$$

the corresponding Fokker–Planck equation is:

$$\frac{\partial p_t}{\partial t} = -\nabla \cdot [(\mathbf{f} - \mathbf{D}\nabla \log p_t)\, p_t] \tag{21.12}$$

21.7.1 The Ill-Posed Problem

This equation is **ill-posed** unless the score $\nabla_\mathbf{x} \log p_t$ is known. This means:

- **Forward process**: Well-posed—we know $\mathbf{f}$ and $\mathbf{G}$, so we can solve for p_t forward in time
- **Reverse process**: Ill-posed—we need $\nabla \log p_t$ to solve backward, but p_t is what we're trying to find

This creates a **chicken-and-egg problem**: to reverse the diffusion, we need the score function, but the score function depends on the distribution we're trying to recover.

21.7.2 Why Score Learning Is Essential

This explains why **score learning is essential** for diffusion models:

1. **The forward process** destroys information (data → noise)
2. **The reverse process** requires the score function to reconstruct information (noise → data)
3. **Score matching** (Chapter~17) learns the score function from data, making the reverse process well-posed

Without the score function, we cannot construct the reverse Fokker-Planck equation, and generative modeling via time reversal becomes impossible. This is why score-based learning is fundamental to diffusion models.

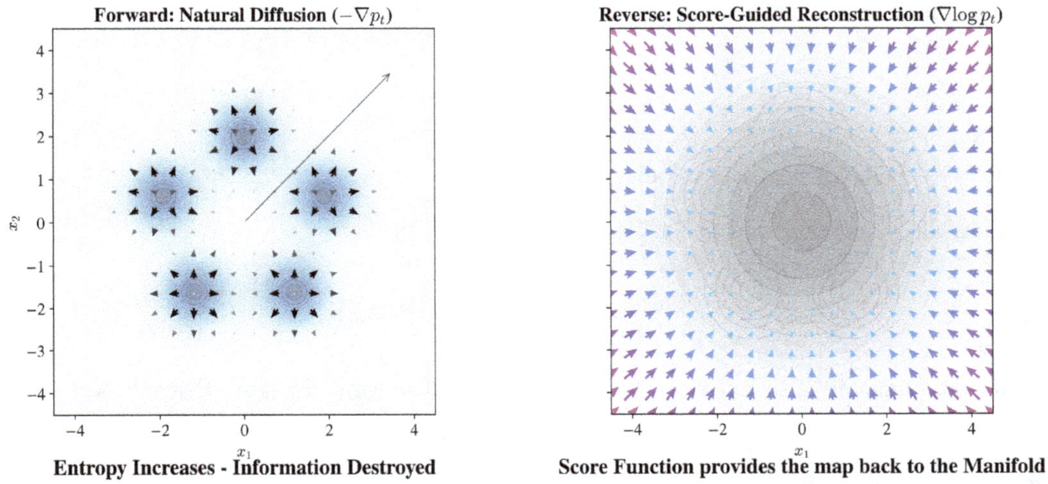

Figure 21.7: The Reverse-Time "Miracle": Score Reconstruction

Figure 21.7 visualizes the fundamental challenge and solution of reverse-time diffusion. The left panel shows the **forward process**: a "star" distribution (five modes arranged in a pentagon) diffusing into a "cloud" (unstructured Gaussian). The arrows represent the **diffusive flux** $-\nabla p_t$, showing how probability mass naturally spreads outward from high-density regions to low-density regions. Nature tends toward disorder—entropy increases and information is destroyed. The right panel shows the **reverse process**: the "cloud" trying to reassemble into a "star". The arrows represent the **score field** $\nabla \log p_t$, which points **uphill** toward high-density regions, against the gradient of entropy. Unlike the diffusive flux, the score function guides probability mass from the unstructured cloud back toward the specific modes of the data distribution. This visualization makes concrete why reversing entropy is ill-posed: you cannot simply "undo" diffusion because information has been destroyed. The score function provides the necessary "map" (vector field) to guide the probability mass back to the data manifold. Without the score function, the reverse process cannot reconstruct the lost information, which is why score learning is mathematically necessary for generative modeling via time reversal.

21.7.3 The Learning-Sampling Distinction

This highlights the distinction between:

- **Learning phase**: Use the forward process to learn the score function via denoising score matching
- **Sampling phase**: Use the learned score function to solve the reverse Fokker-Planck equation (or equivalently, the reverse SDE or probability flow ODE)

The Fokker-Planck perspective makes it clear that learning the score function is not just a convenience—it's mathematically necessary for the reverse process to be well-posed.

21.8 Connection to Probability Flow ODE

From Chapter~20, the probability flow ODE:

$$\frac{d\mathbf{X}_t}{dt} = \mathbf{f} - \frac{1}{2}\mathbf{D}\nabla \log p_t \tag{21.13}$$

induces a continuity equation (Chapter~20, equation *20.9*):

$$\frac{\partial p_t}{\partial t} + \nabla \cdot (p_t \mathbf{v}) = 0$$

where $\mathbf{v} = \mathbf{f} - \frac{1}{2}\mathbf{D}\nabla \log p_t$ is the velocity field.

Expanding the divergence and using $\nabla \log p_t = \frac{\nabla p_t}{p_t}$, this becomes:

$$\frac{\partial p_t}{\partial t} = -\nabla \cdot (\mathbf{f} p_t) + \frac{1}{2}\nabla \cdot (\mathbf{D}\nabla p_t)$$

For state-independent diffusion $\mathbf{D}(t)$, this simplifies to the **same Fokker–Planck equation** as the SDE.

21.8.1 The Unification

This demonstrates a remarkable unification:

- **SDEs and ODEs are different microscopes** (different path-level descriptions)
- **for the same macroscopic density evolution** (same Fokker-Planck equation)

The Fokker-Planck equation is the **common language** that unifies:

- Stochastic paths (SDE)
- Deterministic paths (ODE)
- Distributional evolution (PDE)

All three perspectives describe the same underlying process, just at different levels of detail.

21.8.2 Why This Unification Matters

This unification is crucial for practical applications:

1. **Algorithm design**: We can use either SDE or ODE solvers, knowing they produce the same distribution
2. **Theoretical analysis**: We can analyze distributional properties using PDEs, then implement using SDEs or ODEs
3. **Flexibility**: Choose the most convenient representation for each task (SDE for diversity, ODE for speed)

The Fokker-Planck equation provides the mathematical proof that these different approaches are equivalent at the distributional level.

21.9 Why Fokker–Planck Is the Right Abstraction

The Fokker-Planck equation provides the highest-level mathematical description of diffusion processes. The following hierarchy shows the different levels of description:

Level	Description	Mathematical Object
Path	Random trajectory	$\mathbf{X}_t(\omega)$
SDE	Law of motion (path level)	$d\mathbf{X}_t = \mathbf{f}dt + \mathbf{G}d\mathbf{W}_t$
ODE	Deterministic path (same marginals)	$\frac{d\mathbf{X}_t}{dt} = \mathbf{f} - \frac{1}{2}\mathbf{D}\nabla \log p_t$
Fokker–Planck	Law of distributions	$\frac{\partial p_t}{\partial t} = -\nabla \cdot (\mathbf{f}p_t) + \frac{1}{2}\nabla \cdot \nabla(\mathbf{D}p_t)$
Score	Local geometry	$\nabla_{\mathbf{x}} \log p_t(\mathbf{x})$
Diffusion model	Controlled PDE	Learned score function + Fokker-Planck evolution

21.9.1 The Abstraction Hierarchy

This hierarchy shows increasing levels of abstraction:

1. **Path level (SDE/ODE)**: Describes individual trajectories
2. **Distribution level (Fokker-Planck)**: Describes how probability mass evolves
3. **Geometry level (Score)**: Describes local structure of the distribution
4. **Model level (Diffusion model)**: Combines learned geometry with distributional evolution

Each level provides different insights:

- **Path level**: Understanding individual samples and trajectories
- **Distribution level**: Understanding convergence and long-term behavior
- **Geometry level**: Understanding local structure and learning objectives
- **Model level**: Understanding the complete generative modeling framework

21.9.2 Why Distribution Level Is Fundamental

The Fokker-Planck equation (distribution level) is fundamental because:

1. **Generative modeling cares about distributions**: We want to generate samples from a distribution, not follow specific paths
2. **Convergence analysis**: Distributional evolution directly shows convergence properties
3. **Unification**: It unifies SDE and ODE perspectives
4. **Design principles**: Understanding distributional evolution helps design better diffusion processes

This is the highest-level description of diffusion models, providing the mathematical foundation that connects all the pieces together.

21.10 Analytical Solutions and Special Cases

While the Fokker-Planck equation is difficult to solve in general, there are several important cases where analytical solutions exist.

21.10.1 Linear SDEs

For **linear SDEs** with constant coefficients:

$$d\mathbf{X}_t = A\mathbf{X}_t dt + Bd\mathbf{W}_t$$

where A and B are constant matrices, the solution is a **Gaussian process**:

$$p_t(\mathbf{x}) = \mathcal{N}(\mathbf{x}; \mu_t, \Sigma_t)$$

where μ_t and Σ_t satisfy ordinary differential equations:

$$\frac{d\mu_t}{dt} = A\mu_t, \quad \frac{d\Sigma_t}{dt} = A\Sigma_t + \Sigma_t A^\top + BB^\top$$

This is the basis for **Gaussian diffusion models**, where the forward process is linear and the distribution remains Gaussian throughout.

21.10.2 Ornstein-Uhlenbeck Process

For the OU process $d\mathbf{X}_t = -\theta(\mathbf{X}_t - \mu)dt + \sigma d\mathbf{W}_t$, the Fokker-Planck equation has a Gaussian stationary solution:

$$p_*(\mathbf{x}) = \mathcal{N}\left(\mathbf{x}; \mu, \frac{\sigma^2}{2\theta} I\right)$$

This can be verified by direct substitution into the Fokker-Planck equation.

21.10.3 Heat Equation

For pure diffusion (no drift, $\mathbf{f} = \mathbf{0}$), the Fokker-Planck equation becomes the **heat equation**:

$$\frac{\partial p_t}{\partial t} = \frac{1}{2}\beta(t)\nabla^2 p_t$$

The solution is a **convolution with a Gaussian kernel**:

$$p_t(\mathbf{x}) = \int p_0(\mathbf{y})\mathcal{N}(\mathbf{x}; \mathbf{y}, \sigma_t^2 I)d\mathbf{y}$$

This is exactly what happens in the forward diffusion process of diffusion models.

21.11 Numerical Methods for Fokker-Planck Equations

While the Fokker-Planck equation provides the theoretical foundation, solving it numerically is often necessary for analysis and design of diffusion processes.

21.11.1 Finite Difference Methods

The most common approach is **finite difference methods**, which discretize space and time:

$$\frac{p_t^{n+1} - p_t^n}{\Delta t} = -\nabla \cdot (\mathbf{f} p_t^n) + \frac{1}{2}\nabla \cdot \nabla(\mathbf{D} p_t^n)$$

where p_t^n is the discretized density at time step n.

Challenges in high dimensions:

- The density $p_t(\mathbf{x})$ lives in d-dimensional space

- Discretizing requires $O(N^d)$ grid points, where N is points per dimension
- This becomes intractable for high-dimensional data (e.g., images with $d = 784$ or higher)

21.11.2 Spectral Methods

Spectral methods represent the density as a sum of basis functions:

$$p_t(\mathbf{x}) \approx \sum_{k=1}^{K} a_k(t)\phi_k(\mathbf{x})$$

where $\{\phi_k\}$ are basis functions (e.g., Fourier modes, Hermite polynomials).

Advantages:

- Can be more efficient than finite differences
- Naturally handles periodic or unbounded domains
- Can achieve high accuracy with fewer terms

Disadvantages:

- Still faces curse of dimensionality
- Basis selection is non-trivial

21.11.3 Why We Don't Solve Fokker-Planck Directly

In practice, we **don't solve the Fokker-Planck equation directly** for high-dimensional diffusion models because:

1. **Curse of dimensionality**: The density $p_t(\mathbf{x})$ is too high-dimensional to represent explicitly
2. **We only need samples**: For generative modeling, we need samples from p_t, not the full density
3. **Score-based approach is more efficient**: Learning the score function and using SDE/ODE solvers is more practical

Instead, the Fokker-Planck equation serves as:

- **Theoretical tool**: For understanding and proving properties
- **Design guide**: For designing diffusion processes
- **Analysis framework**: For studying convergence and behavior

The practical implementation uses path-level methods (SDE/ODE solvers) that implicitly solve the Fokker-Planck equation without explicitly representing the density.

The Fokker-Planck equation provides the theoretical foundation for understanding diffusion processes, but practical implementations work at the path level (SDE/ODE solvers) rather than solving the PDE directly. This is a common pattern in high-dimensional problems: the theoretical framework (PDE) guides understanding and design, while practical algorithms (path-level methods) handle the computational challenges. The Fokker-Planck equation is to diffusion models what the Schrödinger equation is to quantum mechanics: the fundamental law that governs the system, even if we don't solve it directly.

21.12 Applications in Modern AI/ML

21.12.1 Diffusion Model Design

The Fokker-Planck equation provides a framework for **designing diffusion processes**:

- **Noise schedules**: The diffusion coefficient $\beta(t)$ in the heat equation controls how quickly data is destroyed. Understanding the Fokker-Planck evolution helps design optimal schedules.

- **Forward processes**: Different forward SDEs lead to different Fokker-Planck equations. Choosing the right forward process is equivalent to choosing the right distributional evolution.
- **Convergence analysis**: The Fokker-Planck equation directly shows convergence properties, helping design processes that converge quickly to the target distribution.

21.12.2 Understanding Model Behavior

The Fokker-Planck perspective helps understand why diffusion models work:

- **Why score learning is sufficient**: The reverse Fokker-Planck equation requires the score function, explaining why learning scores enables generation.
- **Why deterministic sampling works**: The probability flow ODE produces the same Fokker-Planck evolution, explaining why ODE solvers work.
- **Why noise schedules matter**: Different noise schedules lead to different Fokker-Planck evolutions, affecting convergence speed and sample quality.

21.12.3 Connection to Other Fields

The Fokker-Planck equation connects diffusion models to:

- **Statistical physics**: Same equation describes particle diffusion, heat flow, and other physical processes. The connection to statistical mechanics provides insights into equilibrium, entropy, and phase transitions.
- **Fluid dynamics**: Probability behaves like a fluid, with advection and diffusion. The continuity equation form makes this connection explicit, enabling the use of fluid dynamics tools and intuition.
- **Optimal transport**: The continuity equation connects to optimal transport theory, which seeks the most efficient way to transport one distribution to another. Flow matching (Chapter~20) uses this connection.
- **PDE theory**: Provides tools from partial differential equations for analysis, including existence and uniqueness theorems, regularity theory, and numerical methods.

This cross-disciplinary connection enriches our understanding and provides tools from other fields. The Fokker-Planck equation is a universal language that connects probability theory, physics, and mathematics.

21.12.4 Entropy and Information Flow

The Fokker-Planck equation also connects to information theory (Chapter~14). The **entropy** of the distribution:

$$H[p_t] = -\int p_t(\mathbf{x}) \log p_t(\mathbf{x}) d\mathbf{x}$$

evolves according to:

$$\frac{dH[p_t]}{dt} = \int \mathbf{J}_t \cdot \nabla \log p_t d\mathbf{x} + \int \frac{\mathbf{D}}{p_t} \|\nabla p_t\|^2 d\mathbf{x}$$

The first term represents entropy change due to drift (can be positive or negative), while the second term is always non-negative, representing entropy production due to diffusion. This connects the Fokker-Planck equation to the second law of thermodynamics and information-theoretic principles.

21.13 Conceptual Summary

The Fokker–Planck equation provides the highest-level macroscopic description of stochastic systems by modeling how entire probability distributions evolve over continuous time. The structural relationships across path-level and distribution-level descriptions are summarized below:

Concept	Meaning
SDE	Random dynamics (path level)
ODE	Deterministic dynamics (path level, same marginals)
Fokker–Planck	Density evolution (distribution level)
Drift	Transport of probability mass
Diffusion	Spreading of probability mass
Score	Local geometry of p_t
Probability current	Flow of probability mass
Reverse diffusion	Inverting entropy (noise $\rightarrow$ data)
Continuity equation	Conservation of probability mass

Through the Fokker–Planck framework, microscopic stochastic sample trajectories (SDEs) and deterministic trajectories (Probability Flow ODEs) are synthesized into a single Partial Differential Equation (PDE). The advection term governs the directional transport of probability mass along the score field, while the diffusion term models spatial spreading, establishing a direct connection between stochastic dynamics, optimal transport, and generative diffusion models.

21.14 Detailed Balance and Equilibrium

21.14.1 Detailed Balance Condition

A stronger condition than stationarity is **detailed balance**:

$$\mathbf{J}_t(\mathbf{x}) = \mathbf{0} \quad \text{for all } \mathbf{x}$$

This means there is **no net flow of probability** at any point in space. Detailed balance is stronger than $\nabla \cdot \mathbf{J}_t = 0$ (which only requires zero net flux through closed surfaces).

For Langevin dynamics with stationary distribution $p(\mathbf{x})$, detailed balance holds when:

$$\mathbf{f}(\mathbf{x})p(\mathbf{x}) = \frac{1}{2}\nabla \cdot (\mathbf{D}p(\mathbf{x}))$$

This is satisfied when $\mathbf{f}(\mathbf{x}) = \nabla \log p(\mathbf{x})$ (the score function), which is exactly the Langevin SDE.

21.14.2 Reversibility

Detailed balance implies that the process is **reversible**: the forward and reverse processes have the same stationary distribution and the same probability of transitions. This is a fundamental property of equilibrium systems in statistical mechanics.

For diffusion models, reversibility means that the forward process (data $\rightarrow$ noise) and reverse process (noise $\rightarrow$ data) are mathematically symmetric, which is why learning the score function enables perfect reversal.

21.15 Closing Perspective

We have now completed a full conceptual arc that connects multiple areas of mathematics and machine learning:

1. **Information and entropy** (Chapter~14): Measuring uncertainty and information content
2. **Optimization and Bellman equations** (Chapter~15): Finding optimal policies and value functions
3. **Variational inference** (Chapter~16): Approximating intractable posterior distributions
4. **Score functions** (Chapter~17): Learning the geometry of probability distributions
5. **Langevin sampling** (Chapter~18): Converting gradients into samples
6. **Stochastic differential equations** (Chapter~19): Continuous-time stochastic processes
7. **Deterministic probability flows** (Chapter~20): ODEs that generate the same distributions
8. **Fokker–Planck equations** (Chapter~21): PDEs governing distributional evolution

Together, these form a **mathematically complete foundation** for modern diffusion models and score-based generative modeling.

21.15.1 The Three-Level Description

We now have a complete three-level description of diffusion processes:

1. **Path level (SDE)**: Individual random trajectories $\mathbf{X}_t(\omega)$
2. **Path level (ODE)**: Deterministic trajectories with the same marginals
3. **Distribution level (PDE)**: Evolution of probability density $p_t(\mathbf{x})$

Each level provides different insights and enables different applications. The Fokker-Planck equation unifies all three levels by describing the common distributional evolution.

21.15.2 The Fundamental Principle

The fundamental principle that emerges is:

- **Generative modeling is not about samples** (individual paths)
- **It is about governing how probability moves** (distributional evolution)

The Fokker-Planck equation is the mathematical tool that directly describes this probability movement, making it the highest-level and most fundamental description of diffusion processes. Understanding this equation provides deep insights into why diffusion models work, how to design them, and how to analyze their behavior.

This completes the mathematical journey from information theory through stochastic calculus to partial differential equations, providing a unified framework for understanding modern generative AI.

21.16 Mindmap

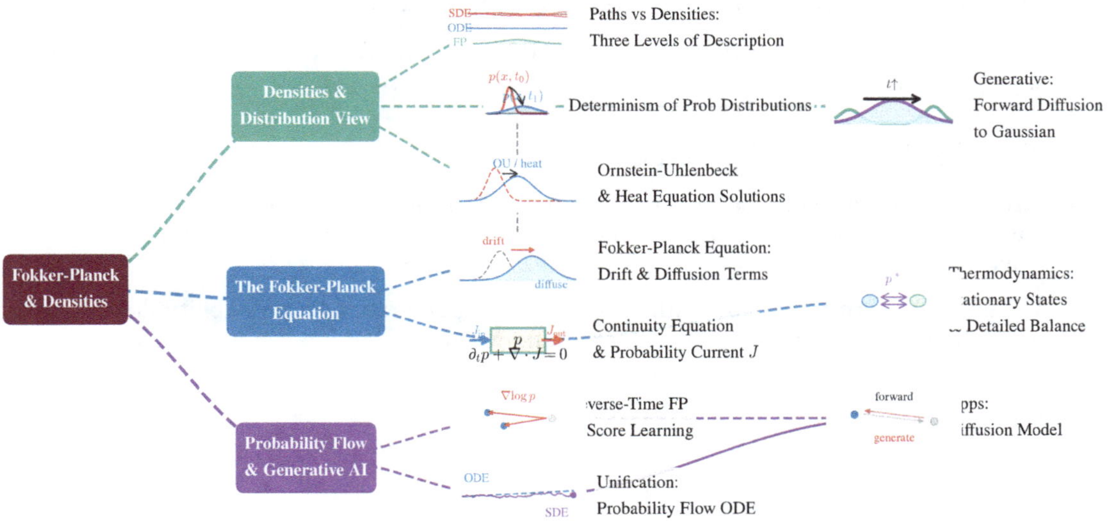

Figure 21.8: Mind map of Fokker-Planck and distribution dynamics topics

21.17 Exercises

21.17.1 Deriving Fokker–Planck

1. **1D Fokker–Planck**: For the 1D Itô SDE $dX_t = f(X_t, t)\,dt + g(X_t, t)\,dW_t$, write the 1D Fokker–Planck equation for $p_t(x)$ in terms of f and $D = g^2$.

2. **Constant diffusion**: Specialize the multidimensional Fokker–Planck equation to the case $\mathbf{D}(\mathbf{x}, t) = \sigma^2 I$ and simplify the diffusion term.

21.17.2 Continuity Equation and Probability Current

1. **Current form**: Starting from the Fokker–Planck equation, verify it can be written as $\partial_t p_t + \nabla \cdot \mathbf{J}_t = 0$ and derive the expression for $\mathbf{J}_t$.

2. **Mass conservation**: Assume p_t decays sufficiently fast at infinity. Show that $\frac{d}{dt} \int_{\mathbb{R}^d} p_t(\mathbf{x})\,d\mathbf{x} = 0$.

21.17.3 Langevin Dynamics and Stationarity

1. **Stationary solution check**: For the Langevin SDE $d\mathbf{X}_t = \nabla \log p(\mathbf{X}_t)\,dt + \sqrt{2}\,d\mathbf{W}_t$, plug $p_t = p$ into the Fokker–Planck equation and verify $\partial_t p_t = 0$.

2. **Detailed balance intuition**: What does $\mathbf{J}_*(\mathbf{x}) \equiv 0$ mean physically? Explain how it relates to reversibility.

21.17.4 Heat Equation (Forward Diffusion)

1. **Gaussian smoothing**: For $\partial_t p_t = \frac{1}{2}\beta \nabla^2 p_t$ with constant $\beta > 0$, show that p_t is the convolution of p_0 with a Gaussian. (You may use the fact that the heat kernel is Gaussian.)

2. **Variance growth (1D)**: If $X_t = X_0 + \sqrt{\beta} W_t$, compute $\mathrm{Var}(X_t)$ in terms of $\mathrm{Var}(X_0)$ and t.

21.17.5 Probability Flow ODE

1. **Same marginals**: State precisely what it means that an SDE and the probability flow ODE have the "same marginals" p_t. Why does that imply they satisfy the same Fokker–Planck equation?

2. **When ODE sampling can fail**: Give one reason deterministic probability flow sampling might be less robust than stochastic sampling when the score model is imperfect.

Appendix

1 Latex Notation

1.1 Sets, Probability, and Numbers

Notation	LaTeX Command	Meaning	Example
$\mathbb{R}$	\mathbb{R}	Real numbers	$\mathbb{R}^d$
$\mathbb{R}^{M\times N}$	\mathbb{R}^{M \times N}	Set of $M \times N$ real matrices	$A \in \mathbb{R}^{M\times N}$
$\mathbb{C}$	\mathbb{C}	Complex numbers	$\lambda \in \mathbb{C}$
$\mathbb{N}$	\mathbb{N}	Natural numbers	$k \in \mathbb{N}$
$\mathbb{Z}$	\mathbb{Z}	Integers	$n \in \mathbb{Z}$
$\subset$	\subset	Subset of	$W \subset \mathbb{R}^d$
$\cup$	\cup	Union	$A \cup B$
$\cap$	\cap	Intersection	$A \cap B$
$\forall$	\forall	For all	$\forall \mathbf{x} \neq 0$
$\exists$	\exists	There exists	$\exists Q$ such that...
$\{\}$	\{ \}	Set	$\{\mathbf{v}_0, \cdots, \mathbf{v}_{N-1}\}$
$[a, b]$	[a, b]	Closed interval	$[\lambda_{\min}, \lambda_{\max}]$
(a, b)	(a, b)	Open interval	$(0, 1)$
$\in$	\in	In set	$1 \in \mathbb{R}$
$\notin$	\notin	Not in set	$i \notin \mathbb{R}$
$\emptyset$	\emptyset	Empty set	$0 \notin \emptyset$
$\mid$	\mid	Such that (conditional)	$P(X \mid Y)$
$\land$	\land	Logical AND	$P(A \land B)$
$\lor$	\lor	Logical OR	$P(A \lor B)$

1.2 Vectors and Matrices

Notation	LaTeX Command	Meaning	Example
$\mathbf{v}$	\mathbf{v}	Bold vector (column vector)	$\mathbf{v} \in \mathbb{R}^d$
$\vec{v}$	\vec{v}	Vector with arrow	$\vec{v}_0, \cdots, \vec{v}_{N-1}$
$[\mathbf{v}]_j$	[\mathbf{v}]_j	j-th component/element of vector	$[\nabla_x f(x)]_j = \frac{\partial f}{\partial x_j}$
$[A]_{i,j}$	[A]_{i,j}	Element in row i, column j of matrix	$[A]_{m,n} = A_{mn}$
$\mathbf{v}^\top$	\mathbf{v}^\top	Transpose of vector	$\mathbf{v}^\top \mathbf{u}$
$\mathbf{v}^*$	\mathbf{v}^*	Conjugate transpose	$\mathbf{x}^* A \mathbf{x}$
A	A	Matrix (capital letter, no \mathbf)	$A \in \mathbb{R}^{M\times N}$
$A^\top$	A^\top	Transpose of matrix	$A = A^\top$
A^{-1}	A^{-1}	Inverse of matrix	$A^{-1} = Q\Lambda^{-1}Q^\top$
A^k	A^k	Matrix power	$A^k \mathbf{x}$
I	I	Identity matrix	$Q^\top Q = I$
Λ	\Lambda	Diagonal matrix (eigenvalues)	$A = Q\Lambda Q^\top$
Σ	\Sigma	Covariance matrix	$\mathbf{x} \sim \mathcal{N}(\mu, \Sigma)$
Θ	\Theta	Parameter matrix	$\Theta \in \mathbb{R}^{d\times d}$

Convention: In this book, matrices are denoted by capital letters (e.g., A, W, X, H, J) without \mathbf, while vectors may use \mathbf{v} for emphasis or lowercase letters (e.g., x, a, b) in regular math mode. The choice depends on context and clarity.

1.3 Operators and Functions

<table>
<tr><th>Notation</th><th>LaTeX Command</th><th>Meaning</th><th>Example</th></tr>
<tr><td>$\langle \mathbf{u}, \mathbf{v} \rangle$</td><td>\langle \mathbf{u}, \mathbf{v}
\rangle</td><td>Inner product</td><td>$\langle \mathbf{u}, \mathbf{v} \rangle = \mathbf{u}^\top \mathbf{v}$</td></tr>
<tr><td>$\|\mathbf{v}\|_2$</td><td>\|\mathbf{v}\|_2</td><td>Euclidean(ℓ_2) norm</td><td>$\|\mathbf{v}\|_2 = \sqrt{\mathbf{v}^\top \mathbf{v}}$</td></tr>
<tr><td>$\|\mathbf{v}\|_p$</td><td>\|\mathbf{v}\|_p</td><td>ℓ_p norm</td><td>$\|\mathbf{v}\|_1, \|\mathbf{v}\|_\infty$</td></tr>
<tr><td>$\|A\|_F$</td><td>\|A\|_F</td><td>Frobenius norm</td><td>$\|A\|_F$</td></tr>
<tr><td>$|\cdot|$</td><td>|\cdot|</td><td>Abs value; set cardinality</td><td>$|x|$</td></tr>
<tr><td>$\det(A)$</td><td>\det(A)</td><td>Determinant</td><td>$\det(A)$</td></tr>
<tr><td>$\mathrm{tr}(A)$</td><td>\mathrm{tr}(A)</td><td>Trace</td><td>$\mathrm{tr}(AA^\top)$</td></tr>
<tr><td>$\mathrm{rank}(A)$</td><td>\mathrm{rank}(A)</td><td>Rank</td><td>$\mathrm{rank}(A) = r$</td></tr>
<tr><td>$\dim(W)$</td><td>\dim(W)</td><td>Dimension</td><td>$\dim(W) = d$</td></tr>
<tr><td>$\ker(A)$</td><td>\ker(A)</td><td>Kernel (null space)</td><td>$\ker(A)$</td></tr>
<tr><td>$\mathrm{col}(A)$</td><td>\mathrm{col}(A)</td><td>Column space</td><td>$\mathrm{col}(A)$</td></tr>
<tr><td>$\mathrm{row}(A)$</td><td>\mathrm{row}(A)</td><td>Row space</td><td>$\mathrm{row}(A)$</td></tr>
<tr><td>$\mathrm{Proj}_{\mathbf{v}}(\mathbf{u})$</td><td>\mathrm{Proj}_{\mathbf{v}}(\mathbf{u})</td><td>Projection</td><td>$\mathrm{Proj}_{\hat{\mathbf{v}}}(\mathbf{u})$</td></tr>
<tr><td>∇f</td><td>\nabla f</td><td>Gradient</td><td>$\nabla Q_A(\mathbf{x}) = 2A\mathbf{x}$</td></tr>
<tr><td>$\nabla^2 f$</td><td>\nabla^2 f</td><td>Hessian</td><td>$\nabla^2 Q_A(\mathbf{x}) = 2A$</td></tr>
<tr><td>$\frac{\partial f}{\partial x}$</td><td>\frac{\partial f}{\partial x}</td><td>Partial derivative</td><td>$\frac{\partial Q_A}{\partial x_k}$</td></tr>
</table>

1.4 Sums and Products

Notation	LaTeX Command	Meaning	Example
$\sum_{i=0}^{N-1}$	\sum_{i=0}^{N-1}	Summation	$\sum_{i=0}^{N-1} c_i \mathbf{v}_i$
$\prod_{i=0}^{N-1}$	\prod_{i=0}^{N-1}	Product	$\prod_{i=0}^{N-1} \lambda_i$
$\int$	\int	Integral	$\int f(x)dx$

1.5 Relations and Comparisons

Notation	LaTeX Command	Meaning	Example
$=$	=	Equals	$A = A^\top$
$\neq$	\neq	Not equals	$\mathbf{x} \neq 0$
$<$	<	Less than	$\lambda_{\min} < \lambda_{\max}$
$\leq$	\le	Less than or equal	$\lambda_0 \leq \lambda_1$
$>$	>	Greater than	$\mathbf{x}^\top A\mathbf{x} > 0$
$\geq$	\ge	Greater than or equal	$\mathbf{x}^\top A\mathbf{x} \geq 0$
$A \succ 0$	A \succ 0	Positive definite	$A \succ 0$ if $\mathbf{x}^\top A\mathbf{x} > 0$
$A \prec 0$	A \prec 0	Negative definite	$A \prec 0$ if $\mathbf{x}^\top A\mathbf{x} < 0$
$A \succeq 0$	A \succeq 0	Positive semi-definite	$A \succeq 0$ if $\mathbf{x}^\top A\mathbf{x} \geq 0$

Notation	LaTeX Command	Meaning	Example
$A \preceq 0$	`A \preceq 0`	Negative semi-definite	$A \preceq 0$ if $\mathbf{x}^\top A\mathbf{x} \leq 0$
$\perp$	`\perp`	Perpendicular/orthogonal	$\mathbf{u} \perp \mathbf{v}$
$\oplus$	`\oplus`	Direct sum	$C(A) \oplus N(A^\top) = \mathbb{R}^M$
$\otimes$	`\otimes`	Kronecker product (tensor product)	$A \otimes B$
$\propto$	`\propto`	Proportional to	$H \propto I$

1.6 Greek Letters (Common)

Notation	LaTeX Command	Meaning	Example
α	`\alpha`	Alpha	$\alpha \in \mathbb{R}$
β	`\beta`	Beta	β
γ	`\gamma`	Gamma	γ
δ	`\delta`	Delta	δ
ϵ	`\epsilon`	Epsilon	ϵ
ε	`\varepsilon`	Variant epsilon	ε
λ	`\lambda`	Lambda (eigenvalue)	$A\mathbf{x} = \lambda\mathbf{x}$
Λ	`\Lambda`	Capital lambda (diagonal)	$A = Q\Lambda Q^\top$
μ	`\mu`	Mu (mean)	$\mu \in \mathbb{R}^d$
σ	`\sigma`	Sigma (singular value)	$\sigma_0, \sigma_1, \ldots$
Σ	`\Sigma`	Capital sigma (covariance)	$\mathbf{x} \sim \mathcal{N}(\mu, \Sigma)$
θ	`\theta`	Theta (angle/parameter)	$\theta \in [0, 2\pi)$
Θ	`\Theta`	Capital theta	Θ
ϕ	`\phi`	Phi	$\phi(\mathbf{x})$
φ	`\varphi`	Variant phi	φ
π	`\pi`	Pi	π
ℓ	`\ell`	Script l (for ℓ_p norm)	ℓ_2 norm

1.7 Special Symbols

Notation	LaTeX Command	Meaning	Example
$\cdot$	`\cdot`	Dot (placeholder)	$\|\cdot\|_2$
$\cdots$	`\cdots`	Horizontal ellipsis	$\mathbf{v}_0, \cdots, \mathbf{v}_{N-1}$
$\ldots$	`\ldots`	Lower ellipsis	$c_0, \cdots, c_{N-1}$
$\vdots$	`\vdots`	Vertical ellipsis	$\vdots$
$\ddots$	`\ddots`	Diagonal ellipsis	$\ddots$
$\hat{\mathbf{v}}$	`\hat{\mathbf{v}}`	Unit vector	$\hat{\mathbf{v}} = \frac{\mathbf{v}}{\|\mathbf{v}\|_2}$
$\bar{x}$	`\bar{x}`	Bar (mean/complement)	$\bar{\mathbf{x}}$
$\overline{\lambda}$	`\overline{\lambda}`	Complex conjugate	$\overline{\lambda}$
$\triangleq$	`\triangleq`	Defined as	$\kappa(\mathbf{x}, \mathbf{y}) \triangleq \phi(\mathbf{x})^\top \phi(\mathbf{y})$
$\iff$	`\iff`	If and only if	$A \succ 0 \iff$ all eigenvalues > 0
$\implies$	`\implies`	Implies	$A = A^\top \implies$ symmetric
	`\quad`	Horizontal space	$\forall\mathbf{x} \neq 0$

1.8 Distributions

Notation	LaTeX Command	Meaning	Example
$\mathcal{N}(\mu, \sigma^2)$	`\mathcal{N}(\mu, \sigma^2)`	Univariate Gaussian	$X \sim \mathcal{N}(\mu, \sigma^2)$
$\mathcal{N}(\mu, \Sigma)$	`\mathcal{N}(\boldsymbol{\mu}, \Sigma)`	Multivariate Gaussian	$X \sim \mathcal{N}(\mu, \Sigma)$
Bernoulli	`\text{Bernoulli}(p)`	Bernoulli	$X \sim \text{Bernoulli}(p)$
Poisson(λ	`\text{Poisson}(\lambda)`	Poisson	$X \sim \text{Poisson}(\lambda)$
Exp(λ)	`\text{Exp}(\lambda)`	Exponential	$X \sim \text{Exp}(\lambda)$
Uniform(	`\text{Uniform}(a, b)`	Uniform	$X \sim \text{Uniform}(a, b)$
Beta(α, β	`\text{Beta}(\alpha, \beta)`	Beta	$X \sim \text{Beta}(\alpha, \beta)$
Gamma(	`\text{Gamma}(\alpha, \beta)`	Gamma	$X \sim \text{Gamma}(\alpha, \beta)$
$\sim$	`\sim`	Distributed as	$X \sim \mathcal{N}(0, I)$

2 Numpy (v2.2+)

Cat	Syntax	Example Usage	Notes
IO	`import numpy as np`	`import numpy as np`	Standard NumPy import
	`np.loadtxt`	`np.loadtxt("file.txt")`	Load data from `.txt`
	`np.genfromtxt`	`np.genfromtxt("file.csv", delimiter=",")`	Load CSV with missing values
	`np.savetxt`	`np.savetxt("file.txt", arr)`	Save array to text
	`np.savetxt`	`np.savetxt("file.csv", arr, delimiter=",")`	Save array to CSV
Init	`np.array`	`np.array([1,2,3])`	Create 1D array
	`np.array`	`np.array([[1,2],[3,4]])`	Create 2D array
	`np.zeros`	`np.zeros(3)`	Zeros array
	`np.ones`	`np.ones((3,4))`	Ones array
	`np.eye`	`np.eye(5)`	Identity matrix
	`np.linspace`	`np.linspace(0,100,6)`	Evenly spaced values
	`np.arange`	`np.arange(0,10,3)`	Range with step
	`np.full`	`np.full((2,3), 8)`	Constant-filled array
	`np.random.rand`	`np.random.rand(4,5)`	Uniform random floats
	`np.random.randint`	`np.random.randint(5, size=(2,3))`	Random integers
Prop	`arr.shape`	`arr.shape`	Dimensions
	`arr.size`	`arr.size`	Number of elements
	`arr.dtype`	`arr.dtype`	Element type
	`arr.astype`	`arr.astype(float)`	Type casting

Cat	Syntax	Example Usage	Notes
	arr.tolist	arr.tolist()	Convert to Python list
	np.info	np.info(np.eye)	Function documentation
Ops	np.copy	np.copy(arr)	Deep copy
	arr.view	arr.view(np.float32)	Shallow view
	arr.sort	arr.sort()	In-place sort
	arr.sort(axis=0)	arr.sort(axis=0)	Axis-wise sort
	flatten	arr.flatten()	2D → 1D
	arr.T	arr.T	Transpose
	reshape	arr.reshape(3,4)	Reshape without data change
	resize	arr.resize((5,6))	Resize with padding
Elem	np.append	np.append(arr, vals)	Append values
	np.insert	np.insert(arr, 2, vals)	Insert before index
	np.delete	np.delete(arr, 3, axis=0)	Delete row
	np.delete	np.delete(arr, 4, axis=1)	Delete column
Comb	np.concatenate	np.concatenate((a,b), axis=0)	Stack rows
	np.concatenate	np.concatenate((a,b), axis=1)	Stack columns
	np.split	np.split(arr, 3)	Split array
	np.hsplit	np.hsplit(arr, 5)	Horizontal split
Idx	arr[i]	arr[5]	Element access
	arr[i,j]	arr[2,5]	2D indexing
	arr[i]=v	arr[1]=4	Assignment
	arr[i:j]	arr[0:3]	Slice
	arr[:,j]	arr[:,1]	Column selection
	arr[i:j,k]	arr[0:3,4]	Submatrix
Bool	arr < c	arr < 5	Boolean mask
	~mask	~arr	Negation
	&	(a<3)&(b>5)	AND
	|	(a<3)|(b>5)	OR
	arr[mask]	arr[arr<5]	Boolean filtering
Scal	np.add	np.add(arr,1)	Add scalar
	np.subtract	np.subtract(arr,2)	Subtract scalar
	np.multiply	np.multiply(arr,3)	Multiply scalar
	np.divide	np.divide(arr,4)	Divide scalar
	np.power	np.power(arr,5)	Power
Vec	np.add	np.add(a,b)	Elementwise add
	np.subtract	np.subtract(a,b)	Elementwise subtract
	np.multiply	np.multiply(a,b)	Elementwise multiply
	np.divide	np.divide(a,b)	Elementwise divide
	np.power	np.power(a,b)	Elementwise power
	np.array_equal	np.array_equal(a,b)	Equality check

Cat	Syntax	Example Usage	Notes
	`np.sqrt`	`np.sqrt(arr)`	Square root
	`np.sin`	`np.sin(arr)`	Sine
	`np.log`	`np.log(arr)`	Natural log
	`np.abs`	`np.abs(arr)`	Absolute value
	`np.ceil`	`np.ceil(arr)`	Ceiling
	`np.floor`	`np.floor(arr)`	Floor
	`np.round`	`np.round(arr)`	Round
Stat	`np.mean`	`np.mean(arr,axis=0)`	Mean
	`sum`	`arr.sum()`	Sum
	`min`	`arr.min()`	Minimum
	`max`	`arr.max(axis=0)`	Maximum
	`np.var`	`np.var(arr)`	Variance
	`np.std`	`np.std(arr,axis=1)`	Standard deviation
	`corrcoef`	`arr.corrcoef()`	Correlation matrix
Data	CSV $\rightarrow$ ndarray	`np.genfromtxt("nyc.csv", delimiter=",", skip_header=1)`	Load dataset
	Column select	`arr[:,1]`	Select column
	Row select	`arr[1]`	Select row
	Multi-column	`arr[:,[1,3,5]]`	Column subset
	Boolean rows	`arr[arr[:,12]>50,:]`	Conditional rows
	Assign slice	`arr[:,7]=arr[:,7].mean()`	Column assignment
	Boolean assign	`arr[arr[:,5]==2,15]=1`	Conditional assignment

2.1 Numpy Einsum

`numpy.einsum()`	Equivalent NumPy Operation
`np.einsum('ij->', a)` `np.einsum('i->', a_1D)`	`np.sum(a)`
`np.einsum('ij,ij->ij', a, b)` `np.einsum('i,i->i', a_1D, b_1D)`	`a * b`
`np.einsum('ij,ij->', a, b)` `np.einsum('i,i->', a_1D, b_1D)`	`np.dot(a, b)` `a @ b`
`np.einsum('ij,ji->ij', a, a)` `np.einsum('i,j->ij', a_1D, a_1D)`	`np.outer(a, a)`
`np.einsum('ij,ji->ij', a, b)` `np.einsum('i,j->ij', a_1D, b_1D)`	`np.outer(a, b)`
`np.einsum('ji', A)` `np.einsum('ij->ji', A)`	`A.T`
`np.einsum('ij->', A)`	`np.sum(A)`
`np.einsum('ij->j', A)`	`np.sum(A, axis=0)`
`np.einsum('ij->i', A)`	`np.sum(A, axis=1)`
`np.einsum('ii->i', A)`	`np.diag(A)`
`np.einsum('ii->', A)`	`np.trace(A)`

numpy.einsum()	Equivalent NumPy Operation
`np.einsum('ij,jk->ik', A, B)`	`A @ B` `np.matmul(A, B)`
`np.einsum('ij,jk->', A, B)`	`np.sum(A @ B)`
`np.einsum('ij,jk->ki', A, B)`	`(A @ B).T`
`np.einsum('ij,ij->ij', A, B)`	`A * B`

PyTorch also provides similar einsum functionality through `torch.einsum()`, which uses the same notation and syntax as NumPy's `einsum()`. The operations shown in the table above can be performed in PyTorch by replacing `np.einsum()` with `torch.einsum()` and using PyTorch tensors instead of NumPy arrays.

3 Pandas (v2.3+)

Cat	API / Syntax	Example Usage	Purpose / Notes
Core	`import pandas as pd`	`import pandas as pd`	Load pandas
	`pd.Series`	`pd.Series([1,2,3])`	1D labeled array
	`pd.DataFrame`	`pd.DataFrame(data)`	2D table
IO	`pd.read_csv`	`pd.read_csv("f.csv")`	Load CSV
	`pd.read_excel`	`pd.read_excel("f.xlsx")`	Load Excel
	`pd.read_json`	`pd.read_json("f.json")`	Load JSON
	`to_csv`	`df.to_csv("f.csv")`	Save CSV
	`to_excel`	`df.to_excel("f.xlsx")`	Save Excel
View	`head / tail`	`df.head(5)`	Preview rows
	`info`	`df.info()`	Schema & nulls
	`describe`	`df.describe()`	Stats summary
	`shape`	`df.shape`	(rows, cols)
	`columns`	`df.columns`	Column names
Sel	`[]`	`df["col"]`	Select column
	`.loc`	`df.loc[r,c]`	Label-based
	`.iloc`	`df.iloc[i,j]`	Index-based
	`.at / .iat`	`df.at[i,"c"]`	Fast scalar
	`filter`	`df.filter(cols)`	Subset
Idx	`set_index`	`df.set_index("id")`	Set index
	`reset_index`	`df.reset_index()`	Reset index
	`rename`	`df.rename(cols={})`	Rename
	`reindex`	`df.reindex(idx)`	Align
Sort	`sort_values`	`df.sort_values("c")`	Sort rows
	`sort_index`	`df.sort_index()`	Sort index
Miss	`isna / notna`	`df.isna()`	Missing mask
	`fillna`	`df.fillna(0)`	Fill missing
	`dropna`	`df.dropna()`	Remove missing
	`interpolate`	`df.interpolate()`	Interpolate
Type	`astype`	`df["c"].astype(int)`	Cast type
	`to_datetime`	`pd.to_datetime(s)`	Parse dates

Cat	API / Syntax	Example Usage	Purpose / Notes
	dt	df.t.dt.year	Date access
Math	sum / mean	df.mean()	Reduction
	min / max	df.max()	Extremes
	std / var	df.std()	Dispersion
	corr	df.corr()	Correlation
	rank	df.rank()	Ranking
Bool	query	df.query("a>0")	Filter rows
	where	df.where(c>0)	Mask
	mask	df.mask(c<0)	Inverse mask
Grp	groupby	df.groupby("k")	Group
	agg	g.agg("mean")	Aggregate
	transform	g.transform("sum")	Broadcast
	apply	g.apply(fn)	Custom
Join	merge	pd.merge(a,b,on="k")	SQL join
	join	a.join(b)	Index join
	concat	pd.concat([a,b])	Stack
Win	rolling	df.rolling(3).mean()	Window
	expanding	df.expanding().sum()	Growing
	ewm	df.ewm(alpha=.1)	EMA
Map	map	s.map(fn)	Element map
	apply	df.apply(fn)	Row/col
	applymap	df.applymap(fn)	Cell-wise
Str	str.lower	s.str.lower()	String ops
	str.contains	s.str.contains("x")	Regex
	str.split	s.str.split(",")	Split
Cat	astype("category")	s.astype("category")	Categorical
	cat.codes	s.cat.codes	Encode
	cat.rename	s.cat.rename()	Rename cats
Time	resample	df.resample("D").mean()	Time agg
	shift	df.shift(1)	Lag
	diff	df.diff()	Difference
Plot	plot	df.plot()	Quick plot
	hist	df.hist()	Histogram
Util	copy	df.copy()	Deep copy
	pipe	df.pipe(fn)	Chaining
	memory_usage	df.memory_usage()	RAM

4 PyTorch (v2.9+)

Cat	API / Syntax	Example Usage	Purpose / Notes
Core	import torch	import torch	Load PyTorch
	torch.device	device=torch.device("cuda")	Device abstraction
	torch.cuda.is_available()	torch.cuda.is_available()	CUDA check
	tensor.to(dev)	x=x.to(dev)	Move tensor

Cat	API / Syntax	Example Usage	Purpose / Notes
	`tensor.cuda() / cpu()`	`x.cuda()`	Explicit move
	`cudnn.deterministic`	`True`	Reproducibility
IO	`torch.save`	`torch.save(model,"m.pt")`	Save model
	`torch.load`	`model=torch.load("m.pt")`	Load model
	`state_dict`	`model.state_dict()`	Weights only
	`load_state_dict`	`model.load_state_dict(sd)`	Restore
	`model.eval()`	`model.eval()`	Inference mode
JIT	`torch.jit.trace`	`trace(model,x)`	Static graph
	`@torch.jit.script`	`@script`	Scripted model
	`torch.onnx.export`	`export(model,x,"m.onnx")`	ONNX export
Grad	`torch.autograd`	`import torch.autograd`	Auto-diff
Init	`torch.tensor`	`torch.tensor([1,2])`	Create tensor
	`zeros / ones`	`torch.zeros(3,4)`	Init tensors
	`eye`	`torch.eye(5)`	Identity
	`full`	`torch.full((2,3),8)`	Constant
	`empty`	`torch.empty(3,3)`	Uninit
	`arange`	`torch.arange(0,10)`	Range
	`linspace`	`torch.linspace(0,1,5)`	Even spacing
Shape	`reshape`	`torch.reshape(x,(3,4))`	Reshape
	`squeeze`	`torch.squeeze(x,1)`	Remove dim
	`unsqueeze`	`torch.unsqueeze(x,0)`	Add dim
	`transpose`	`torch.transpose(x,0,1)`	Swap dims
	`cat`	`torch.cat((a,b),0)`	Concat
	`split`	`torch.split(x,2)`	Split
Rand	`manual_seed`	`torch.manual_seed(42)`	Reproducible
	`rand`	`torch.rand(3,4)`	Uniform
	`randn`	`torch.randn(3,4)`	Normal
	`randint`	`torch.randint(0,10,(2,3))`	Integers
	`randperm`	`torch.randperm(n)`	Permutation
Elem	`add / sub`	`torch.add(a,b)`	Elementwise
	`mul / div`	`torch.mul(a,b)`	Elementwise
	`sqrt`	`torch.sqrt(x)`	$\surd$
	`abs`	`torch.abs(x)`	Absolute
	`ceil / floor`	`torch.ceil(x)`	Rounding
Trig	`sin / cos`	`torch.sin(x)`	Trig
	`exp`	`torch.exp(x)`	Exponential
	`log`	`torch.log(x)`	Log
Redu	`sum`	`torch.sum(x)`	Reduction
	`mean`	`torch.mean(x)`	Average
	`std / var`	`torch.std(x)`	Stats
	`norm`	`torch.norm(x)`	Norm
LA	`dot`	`torch.dot(a,b)`	Dot
	`matmul`	`torch.matmul(A,B)`	Matrix mult
	`svd`	`torch.svd(A)`	SVD
	`inv`	`torch.inverse(A)`	Inverse

Cat	API / Syntax	Example Usage	Purpose / Notes
NN	`nn.Module`	`class M(nn.Module)`	Base class
	`nn.Sequential`	`nn.Sequential(...)`	Stack
	`nn.Parameter`	`nn.Parameter(w)`	Trainable
Lin	`nn.Linear`	`nn.Linear(in,out)`	FC layer
Conv	`nn.Conv{1,2,3}d`	`nn.Conv2d(...)`	CNN
	`ConvTranspose`	`nn.ConvTranspose2d`	Deconv
Pool	`MaxPool`	`nn.MaxPool2d(2)`	Pooling
	`AdaptiveAvgPool`	`nn.AdaptiveAvgPool2d(1)`	Size-free
RNN	`LSTM / GRU`	`nn.LSTM(...)`	Sequence
Attn	`nn.Transformer`	`nn.Transformer(...)`	Attention
	`MultiheadAttention`	`nn.MultiheadAttention`	Core
Norm	`BatchNorm`	`nn.BatchNorm2d(c)`	BN
	`InstanceNorm`	`nn.InstanceNorm2d(c)`	IN
Drop	`nn.Dropout`	`nn.Dropout(0.5)`	Regularize
Act	`ReLU / GELU`	`nn.GELU()`	Activation
	`Sigmoid`	`nn.Sigmoid()`	Logistic
	`Softmax`	`nn.Softmax(1)`	Prob
Loss	`MSELoss`	`nn.MSELoss()`	Regression
	`CrossEntropyLoss`	`nn.CrossEntropyLoss()`	Classif
	`BCELoss`	`nn.BCELoss()`	Binary
Opt	`SGD`	`optim.SGD(p,lr)`	SGD
	`Adam`	`optim.Adam(p,lr)`	Adam
	`AdamW`	`optim.AdamW(p,lr)`	AdamW
	`RMSprop`	`optim.RMSprop(p)`	RMS
LR	`StepLR`	`StepLR(opt,step)`	Step
	`CosineLR`	`CosineAnnealingLR`	Cosine
	`ReduceLR`	`ReduceLROnPlateau`	Metric
Data	`Dataset`	`class D(Dataset)`	Dataset
	`DataLoader`	`DataLoader(ds,batch)`	Batching
Vis	`datasets.*`	`MNIST, CIFAR`	Vision data
	`models.*`	`resnet50(pretrained)`	CNNs
	`transforms.*`	`Compose([...])`	Augment
	`ops.nms`	`nms(boxes,scores)`	Detection
Txt	`datasets.*`	`IMDB, AG_NEWS`	NLP
	`vocab.GloVe`	`GloVe(name,dim)`	Embeddings
Aud	`load / save`	`torchaudio.load(f)`	Audio IO
	`transforms.*`	`MelSpectrogram`	Features
	`kaldi.*`	`mfcc, fbank`	ASR
Misc	`einsum`	`einsum("ij,jk->ik",A,B)`	Tensor alg
	`fft`	`torch.fft(x)`	FFT
	`unique`	`torch.unique(x)`	Unique

About the author

Fuheng Wu is a Principal Machine Learning Tech Lead at Oracle's Generative AI organization. He studied High Performance Computation in Singapore–MIT Alliance(https://smart.mit.edu), including Gilbert Strang's linear algebra course at MIT, which shaped how this book presents the subject for machine learning readers. Earlier in his career he worked in quantitative finance, including convertible-bond pricing at Bloomberg and algorithmic trading at WorldQuant, and later built marketplace pricing models at Uber and deep learning infrastructure for self-driving cars at Uber ATG. His work spans large language models, computer vision, and the mathematical foundations that make modern AI systems tractable—from linear algebra and optimization to probabilistic modeling and diffusion.

Fuheng has contributed to open-source projects including Deepseek, SGLang, pyLLaMA, and chatLLaMA, and is the author of *Distributed AI Systems* (best seller book in Amazon - #1 in memory management, #2 in parallel computing and #33 in all AI books). He teaches Generative AI foundations in programs at the McCombs School of Business, University of Texas at Austin, and mentors students preparing for AI olympiads, including CAIO, USAAIO and IAIO. This book grew out of years of implementing and explaining the core mathematical ideas that recur across deep learning, reinforcement learning, and generative modeling—aimed at practitioners who need rigorous tools they can actually use when reading papers, deriving algorithms, and building systems.

Index

www.ingramcontent.com/pod-product-compliance
Lightning Source LLC
LaVergne TN
LVHW062211200926
848484LV00006B/1370